◆ LIFE SCIENCES

◆ SOCIAL SCIENCES

SECOND EDITION

Essentials of College Mathematics

FOR BUSINESS, ECONOMICS, LIFE SCIENCES, AND SOCIAL SCIENCES

College Mathematics Series

This book is part of a comprehensive series designed to accommodate a wide variety of courses. Difficulty level and topic coverage are determined by the choice of a particular book or a combination of books from the series. Many topics in individual books are independent and may be selected in any order or omitted. All books include reviews of relevant algebraic topics.

Essentials of College Mathematics: One or two semesters
A brief introduction to finite mathematics and calculus that is suitable for a one- or two-semester course or a one- or two-quarter course

College Mathematics: Two semesters
A more comprehensive introduction to finite mathematics and calculus that is suitable for a two-semester course or a two- or three-quarter course

Applied Mathematics: Two or three semesters
An even more comprehensive introduction to finite mathematics and calculus for schools requiring this level and degree of coverage; additional topics are included in both finite mathematics and calculus, and the definite integral is given a more formal treatment

Finite Mathematics: One semester
The finite mathematics portion of *College Mathematics*, with an added chapter on games and decisions; may be used in combination with either *Calculus* or *Applied Calculus* to create a two- or three-semester course

Calculus: One semester
The calculus portion of *College Mathematics*, with an added chapter on trigonometric functions

Applied Calculus: One or two semesters
A more extensive treatment of calculus for schools requiring this level and degree of coverage; the book can be completed in two semesters, however, with appropriate topic selection — since many of the topics are independent — the book also can be used for a strong one-semester course

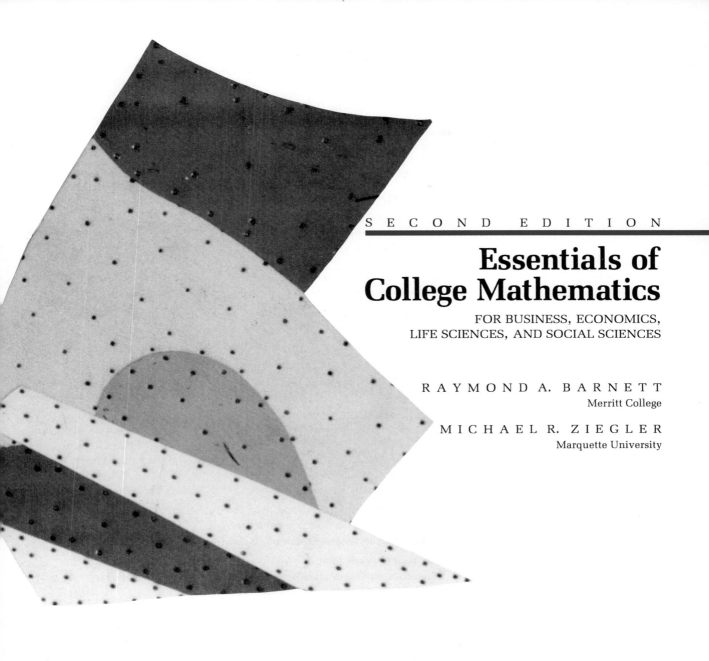

SECOND EDITION

Essentials of College Mathematics

FOR BUSINESS, ECONOMICS, LIFE SCIENCES, AND SOCIAL SCIENCES

RAYMOND A. BARNETT
Merritt College

MICHAEL R. ZIEGLER
Marquette University

DELLEN PUBLISHING COMPANY
an imprint of
MACMILLAN PUBLISHING COMPANY, New York

MAXWELL MACMILLAN CANADA, Toronto

On the cover: The detail on the cover is from a metal collage by Los Angeles artist Tony Berlant. His work can be described as part painting, part sculpture. It is resplendent with color and texture. The collage is formed by nailing fragments of printed and colored tin onto a wood backing. Berlant's work is represented by the L. A. Louver Gallery in Venice, California and the Louver Gallery in New York City. His work may also be seen in the permanent collections at the Whitney Museum of American Art in New York City, the Academy of Fine Art in Philadelphia, the Hirshorn Museum in Washington, D.C., the Art Institute of Chicago, and the Los Angeles County Museum.

© Copyright 1992 by Dellen Publishing Company, an imprint of Macmillan Publishing Co.

Printed in the United States of America

Macmillan Publishing Company
866 Third Avenue, New York, New York 10022

Macmillan Publishing Company is
part of the Maxwell Communication
Group of Companies.

Maxwell Macmillan Canada, Inc.
1200 Eglinton Avenue East
Suite 200
Don Mills, Ontario M3C 3N1

Permissions: Dellen Publishing Company
            400 Pacific Avenue
            San Francisco, California 94133

Orders: Dellen Publishing Company   Maxwell Macmillan Canada, Inc.
        c/o Macmillan Publishing Company   1200 Eglinton Avenue East
        Front and Brown Streets   Suite 200
        Riverside, New Jersey 08075   Don Mills, Ontario M3C 3N1

Library of Congress Cataloging-in-Publication Data

Barnett, Raymond A.
    Essentials of college mathematics for buisness, economics, life
sciences, and social sciences / Raymond A. Barnett, Michael R.
Ziegler. — 2nd ed.
        p.   cm. — (College mathematics series)
    Includes index.
    ISBN 0-02-305921-4
        1. Mathematics.   I. Ziegler, Michael R.   II. Title.   III. Series:
College mathematics series (San Francisco, Calif.)
    QA37.2.B365   1992
    510 — dc20   91-3477
                                                                          CIP

Printing: 1 2 3 4 5 6 7 8 9   Year: 1 2 3 4 5

C O N T E N T S   Preface

Chapter Dependencies

P A R T   O N E :   P R E L I M I N A R I E S

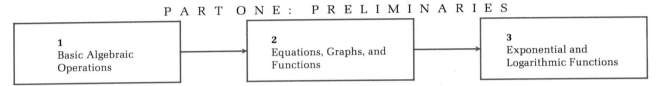

1 Basic Algebraic Operations	→	2 Equations, Graphs, and Functions	→	3 Exponential and Logarithmic Functions

Selected topics from Part One may be referred to as needed in Parts Two and Three or reviewed systematically before starting either part.

P A R T   T W O :   F I N I T E   M A T H E M A T I C S

4 Mathematics of Finance	5 Systems of Linear Equations; Matrices	7 Probability

6 Linear Inequalities and Linear Programming	8 Additional Topics in Probability

P A R T   T H R E E :   C A L C U L U S

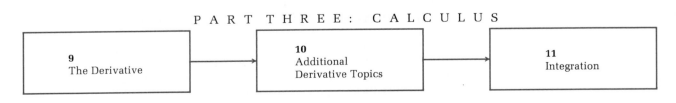

9 The Derivative	→	10 Additional Derivative Topics	→	11 Integration

The second edition of *Essentials of College Mathematics for Business, Economics, Life Sciences, and Social Sciences* is designed for mathematics courses that include topics from algebra, finite mathematics, and calculus, and for students who have had 1 or 2 years of high school algebra, or the equivalent. It is one of six books in the authors' College Mathematics series (see page ii for a brief comparison of all six books). The choice and organization of topics make the book readily adaptable to a variety of courses, including a combined course in finite mathematics and calculus, or a separate course in either area with a strong algebra component (see the chapter dependency chart on page viii).

◆ PRINCIPAL CHANGES FROM THE SECOND EDITION

This edition has been divided into three parts, "Part One: Preliminaries," "Part Two: Finite Mathematics," and "Part Three: Calculus." Part One (Chapters 1–3) deals with algebraic concepts, including a **review of basic algebraic operations,** a **thorough treatment of functions and their graphs,** and a brief but **comprehensive discussion of exponential and logarithmic functions.** Depending on the background of a given class, this material can be totally omitted, referred to as needed, or selected topics can be covered in detail at the beginning or during the course. A few additional review topics are also included in Appendix A for convenient reference. Part Two (Chapters 4–8) covers the topics that have become standard in a finite mathematics course: **mathematics of finance, linear systems, matrices, linear programming, and probability.** And Part Three (Chapters 9–11) provides basic coverage of **differential and integral calculus for functions of one variable,** including the exponential and logarithmic functions.

The following general improvements are found throughout the text: an increased emphasis on and use of **calculators; boxed definitions, results, and step-by-step processes;** and **schematic illustrations. Examples** have been improved, and many new examples have been added. **Exercise sets,** including applications, have been improved and expanded. All **exposition** has been carefully reviewed and fine-tuned or rewritten.

Specific improvements are as follows:

1. The development of the **simplex method** (Sections 6-3 and 6-4) has been completely rewritten. Together, these two revised sections provide a well-motivated, clear, shorter, and less wordy introduction to the simplex process. The important terms "basic variable," "nonbasic variable," "basic solution," and "basic feasible solution," and the main idea of the **simplex algorithm** are introduced through examples in a geometric setting in Section 6-3. In the development of the simplex method in Section 6-4, the **selection of basic and nonbasic variables** in the **simplex tableau** has been made more precise, and the **pivot operation** now emphasizes **entering basic variables** and **exiting basic variables.** Even though the treatment of the simplex method in the last edition was very well-received, we believe that the new treatment is even better.

2. Other topics in **linear algebra** also received quite a bit of attention, and Sections 5-1, 5-2, 5-6, 6-1, and 6-2 were substantially rewritten. There is now a more uniform use of the technical terms "independent," "dependent," "consistent," and "inconsistent." The discussion of the use of **parameters** in representing **infinite solution sets** is clearer, as is the discussion of **finding the inverse of a matrix** and **solving matrix equations.** The discussion on solving **linear inequalities** and **systems of linear inequalities** has been simplified.

3. Important improvements have been made in the chapters on **probability**. Sections 7-1, 7-2, and 8-2 were substantially rewritten. **Tree diagrams** are used effectively to illustrate the difference between **permutations** and **combinations** in Section 7-1. The important section on **experiments** and **events** (Section 7-2) has been simplified, and is now less formal and more concise. The development of **independent** and **dependent events** in Section 8-2 is improved.

4. Chapter 4, on **mathematics of finance,** is an independent chapter and can be covered at any time. It is placed directly after Chapter 3, on exponential and logarithmic functions, for the convenience of those who wish to tie together the material in these two chapters. Exposition has been improved, applications made current, and some new applications added.

5. The first four sections in the introductory **derivative** chapter of the last edition have been completely rewritten and condensed into three sections. The treatment has been both simplified and expanded. The new first section provides an **intuitive geometric introduction to limits and continuity.** Because of numerous reviewer requests, **left- and right-hand limits** have been added. Because of its importance in solving optimization problems, **continuity on a closed interval** is now defined. **Increment notation** has been removed from this introductory chapter to keep the notation as simple as possible. **Computation of difficult limits** has been reduced, and more emphasis has been placed on the **0/0 indeterminate form** because of its relevance to the definition of the derivative and its various interpretations (slope, instantaneous rate, marginal analysis, and so on). **Infinite limits and**

limits at infinity receive more attention. **Vertical and horizontal asymptotes** are defined using these concepts.

6. **Solving inequalities using continuity** has become a subsection of the section on the first derivative and graphs (Section 10-1). This process of solving inequalities is fundamental to the production of **sign charts** throughout the chapter. A new section on **curve sketching** (Section 10-3) presents a simple and straightforward curve sketching strategy that is easy for students to remember and use. [Step 1: Use $f(x)$; Step 2: Use $f'(x)$; Step 3: Use $f''(x)$; Step 4: Graph f.]

7. In Chapter 11, on integration, more attention is given to **substitution techniques** and **area between the graphs of two functions.**

◆ IMPORTANT FEATURES

Emphasis and Style

The text is **written for student comprehension.** Great care has been taken to write a book that is mathematically correct and accessible to students. Emphasis is on computational skills, ideas, and problem solving rather than mathematical theory. Most derivations and proofs are omitted except where their inclusion adds significant insight into a particular concept. General concepts and results are usually presented only after particular cases have been discussed.

Examples and Matched Problems

Over 300 completely worked examples are included. Each example is followed by a similar problem for the student to work while reading the material. This actively involves the student in the learning process. The answers to these matched problems are included at the end of each section for easy reference.

Exercise Sets

The book contains over 3,500 problems. Each exercise set is designed so that an average or below-average student will experience success and a very capable student will be challenged. Exercise sets are mostly divided into A (routine, easy mechanics), B (more difficult mechanics), and C (difficult mechanics and some theory) levels.

Applications

Enough applications are included to convince even the most skeptical student that mathematics is really useful. The majority of the applications are included at the end of exercise sets and are generally divided into business and economics, life science, and social science groupings. An instructor with students from all three disciplines can let them choose applications from their own field of interest; if most students are from one of the three areas, then special emphasis can be placed there. Most of the applications are simplified versions of actual real-world problems taken from professional journals and books. No specialized experience is required to solve any of the applications.

◆ STUDENT AIDS

1. **Think boxes** (dashed boxes) are used to enclose steps that are usually performed mentally (see Sections 1-2 and 1-3).

2. **Annotation** of examples and developments, in color type, is found throughout the text to help students through critical stages (see Sections 1-2 and 1-3).

3. **Functional use of color** improves the clarity of many illustrations, graphs, and developments, and guides students through certain critical steps (see Sections 1-2 and 1-3).

4. **Boldface type** is used to introduce new terms and highlight important comments.

5. **Screened boxes** are used to highlight important definitions, theorems, results, and step-by-step processes.

6. **Answers** to odd-numbered problems are included in the back of the book.

7. **Chapter review** sections include a review of all important terms and symbols and a comprehensive review exercise. Answers to all review exercises are included in the back of the book.

8. A **student's solution manual** is available at a nominal cost through a book store. The manual includes detailed solutions to all odd-numbered problems and all review exercises.

9. A manual for an **Interactive Computer Applications Package (ICAP)** by Carolyn L. Meitler is available at a nominal cost through a book store. The manual contains instructions, examples, and exercises that demonstrate the use of the programs on the *ICAP for Essentials of College Mathematics* disks. The disks containing the programs, for use on APPLE II® and IBM-PC® computers, are distributed free of charge to institutions using this book. No previous computer experience is necessary to use this package.

10. A **Supplemental Applications and Topics** manual by Jon E. Baum is available at a nominal cost through a book store. Part I of the manual expands the application exercises in the text and reinforces the important role of the mathematics presented in the text. These exercises provide the student with a richer and more varied experience in solving real-world problems. Part II of the manual presents some applications that are not covered in the text, including transportation problems, assignment problems, sensitivity analysis, and a variety of finance topics. After completing the prerequisite material in the text, students interested in these more specialized topics will realize substantial benefits by studying this portion of the manual.

◆ INSTRUCTOR AIDS

See page xvi for detailed information regarding examination copy requests and orders for the instructor aids described below.

1. A unique **computer-generated random test system** is available to instructors without cost. The test system utilizes an IBM-PC, XT, or AT Personal Computer® and will produce high-quality output on an IBM-compatible dot-matrix printer or on a Hewlett-Packard Laserjet II®-compatible laser printer. The test system has been greatly expanded and now contains over 300 different problem algorithms directly related to material in the text.

These carefully constructed algorithms use random number generators to produce different, yet equivalent, versions of each of these problems. The test system is available now in both **free-response and multiple-choice editions.** An almost unlimited number of quizzes, review exercises, chapter tests, mid-terms, and final examinations, each different from the other, can be generated quickly and easily. At the same time, the system will produce answer keys and student work sheets, if desired. Upon request, the publisher will supply institutions using this textbook with **DellenTest III (IBM Free-Response Edition or Multiple-Choice Edition)** on 5.25 inch floppy disks. **IBM Edition User Notes** and **Annotated Problem Printouts** are included with the disks. The notes provide step-by-step instructions for using the testing system and a complete description of the options in this menu-driven program. The annotated printouts identify by chapter and number each question the system is capable of generating, and also correlate each question with the prerequisite section from the text. When used in conjunction with the user notes, the annotated printouts enable instructors to select any combination of questions for an examination.

2. An **instructor's test battery** is also available to instructors without cost. The battery, organized by chapter, contains three equivalent versions (with answers) of over 300 different problems.

3. An **instructor's resource manual** provides over 130 transparency and handout masters, a detailed discussion of chapter and topic dependencies, a comparison of this edition with the previous edition, and a detailed topic chart for comparing this book with other books in the authors' College Mathematics series.

4. An **instructor's answer manual** containing all the answers not included in the text is available to instructors without charge.

5. A **student's solution manual** (see Student Aids) is available to instructors without charge from the publisher.

6. An **Interactive Computer Applications Package (ICAP)** by Carolyn L. Meitler (see Student Aids) is available to instructors without charge from the publisher. The programs in this package are available on diskettes for APPLE II® and IBM-PC® computers. Included on these disks are programs related to the mathematics of finance, row operations, matrix arithmetic, the simplex method, limit estimation, function graphing, and numerical integration. The publisher will supply these disks without charge to institutions using this book.

7. A **Supplemental Applications and Topics** manual by Jon E. Baum (see Student Aids) is available to instructors without charge from the publisher. Instructors can use Part I of this manual to supplement the exercise sets in the text, providing students with additional experience in solving applications utilizing the mathematics presented in the text. Part II of the manual can be used to provide coverage of applications not covered in the text, such as transportation problems, assignment problems, sensitivity analysis, and a variety of finance topics, either as part of the syllabus for a course or as subjects for independent study.

8. **Z-graph,** a HyperCard© graphing stack for the APPLE Macintosh® computer, allows a user to graph most of the mathematical functions likely to be encountered, quickly, accurately, and with considerable control over axes, scales, graph size, and labeling. In addition to graphing functions, this program will perform a variety of mathematical operations related to numerical integration, root approximation, interpolating polynomials, least-square polynomials, and approximate solutions of differential equations. Instructors will find this program useful for preparing examination material, transparency masters, and handouts. The publisher will supply this program free of charge to instructors using this book, and the program may be freely distributed to students.

◆ ERROR CHECK

Because of the careful checking and proofing by a number of mathematics instructors (acting independently), the authors and publisher believe this book to be substantially error-free. For any errors remaining, the authors would be grateful if they were sent to: Dellen Publishing Company, 400 Pacific Avenue, San Francisco, CA 94133.

◆ ACKNOWLEDGMENTS

In addition to the authors, many others are involved in the successful publication of a book. We wish to thank personally:

Chris Boldt, Eastfield College
Bob Bradshaw, Ohlone College
Charles E. Cleaver, The Citadel
Barbara Cohen, West Los Angeles College
Richard L. Conlon, University of Wisconsin—Stevens Point
Kenneth A. Dodaro, Florida State University
Martha M. Harvey, Midwestern State University
Louis F. Hoelzle, Bucks County Community College
Robert H. Johnston, Virginia Commonwealth University
Robert Krystock, Mississippi State University
Roy H. Luke, Los Angeles Pierce College
Mel Mitchell, Clarion University of Pennsylvania
Kenneth A. Peters, Jr., University of Louisville
Stephen Rodi, Austin Community College
Daniel E. Scanlon, Orange Coast College
Joan Smith, Vincennes University
Delores A. Williams, Pepperdine University
Caroline Woods, Marquette University

We also wish to thank:

John Williams for a strong and effective cover design.

John Drooyan and Mark McKenna for the many sensitive and beautiful photographs throughout the book.

Stephen Merrill and Susan Pustejovsky for carefully checking all examples and problems (a tedious but extremely important job).

Jeanne Wallace for accurately and efficiently producing most of the manuals that supplement the text.

Jon E. Baum for developing the *Supplemental Applications and Topics* manual.

All the people at IPS Publishing who contributed their efforts to the production of the computerized testing system.

Carolyn L. Meitler for developing the *ICAP for Essentials of College Mathematics*.

Janet Bollow for another outstanding book design.

Phyllis Niklas for guiding the book smoothly through all publication details.

Don Dellen, the publisher, who continues to provide all the support services and encouragement an author could hope for.

Producing this new edition with the help of all these extremely competent people has been a most satisfying experience.

R. A. Barnett
M. R. Ziegler

Ordering Information

When requesting examination copies or placing orders for this text or any of the related supplementary materials listed below, please refer to the corresponding ISBN numbers.

Title	ISBN Number
Essentials of College Mathematics for Business, Economics, Life Sciences, and Social Sciences, Second Edition	0-02-305921-4
Computer-generated random test system for Essentials of College Mathematics, Second Edition:	
DellenTest III Disks (IBM Free-Response Edition)	0-02-306343-2
DellenTest III Disks (IBM Multiple-Choice Edition)	0-02-306342-4
(IBM Edition User Notes and Annotated Problem Printouts are included with each edition of the disks.)	
Instructor's Answer Manual to accompany Essentials of College Mathematics, Second Edition	0-02-306344-0
Instructor's Resource Manual to accompany Essentials of College Mathematics, Second Edition	0-02-305922-2
Instructor's Test Battery to accompany Essentials of College Mathematics, Second Edition	0-02-305925-7
Interactive Computer Applications Package to accompany Essentials of College Mathematics, Second Edition:	
Manual	0-02-380185-9
IBM-PC Disks	0-02-305924-9
APPLE II Disks	0-02-305923-0
Student's Solution Manual to accompany Essentials of College Mathematics, Second Edition	0-02-334385-0
Supplemental Applications and Topics to accompany the Barnett and Ziegler College Mathematics Series	0-02-306770-5
Z-graph Macintosh Disk	0-02-306255-X

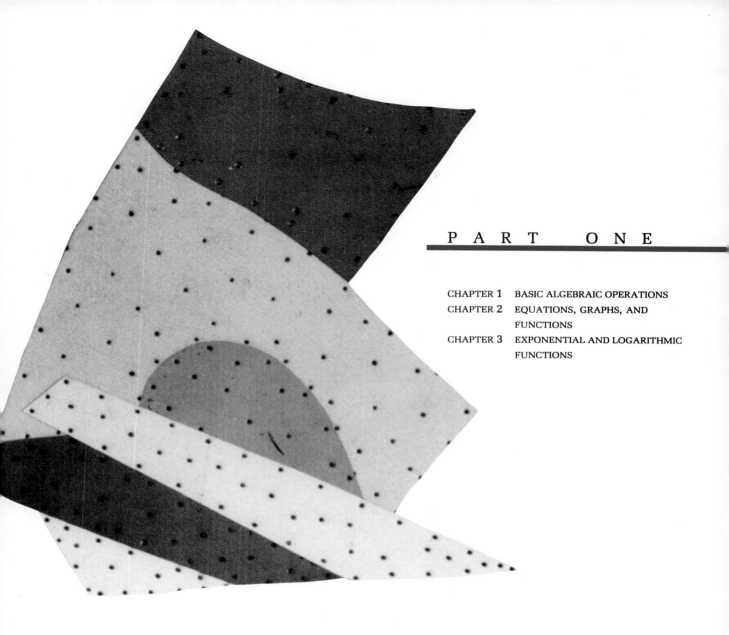

P A R T   O N E

Preliminaries

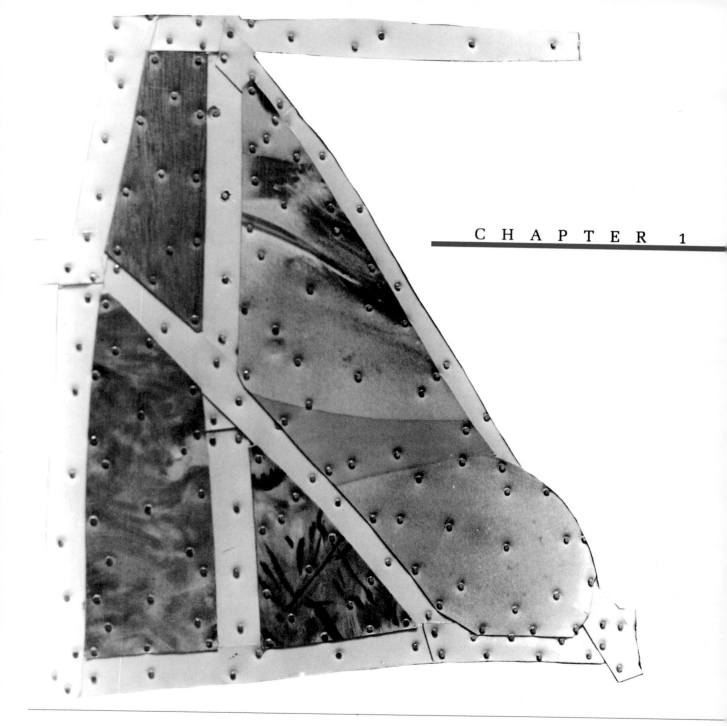

Basic Algebraic Operations

Contents

In this chapter we will review some important basic algebraic operations usually studied in earlier courses. The material may be studied systematically before commencing with the rest of the book or reviewed as needed.

SECTION 1-1

Algebra and Real Numbers

◆ THE SET OF REAL NUMBERS
◆ THE REAL NUMBER LINE
◆ BASIC REAL NUMBER PROPERTIES
◆ FURTHER PROPERTIES
◆ FRACTION PROPERTIES

The rules for manipulating and reasoning with symbols in algebra depend, in large measure, on properties of the real numbers. In this section we will look at some of the important properties of this number system. To make our discussions here and elsewhere in the text clearer and more precise, we will occasionally make use of simple *set* concepts and notation. Refer to Appendix A if you are not yet familiar with the basic ideas concerning sets.

◆ THE SET OF REAL NUMBERS

What number system have you been using most of your life? The *real number system*. Informally, a **real number** is any number that has a decimal representation. Table 1 describes the set of real numbers and some of its important subsets. Figure 1 illustrates how these sets of numbers are related.

The set of integers contains all the natural numbers and something else — their negatives and 0. The set of rational numbers contains all the integers and something else — noninteger ratios of integers. And the set of real numbers contains all the rational numbers and something else — irrational numbers.

The Set of Real Numbers

SYMBOL	NAME	DESCRIPTION	EXAMPLES
N	Natural numbers	Counting numbers (also called positive integers)	1, 2, 3, . . .
Z	Integers	Natural numbers, their negatives, and 0	. . . , $-2, -1, 0, 1, 2,$ . . .
Q	Rational numbers	Numbers that can be represented as a/b, where a and b are integers and $b \neq 0$; decimal representations are repeating or terminating	$-4, 0, 1, 25, \frac{-3}{5}, \frac{2}{3}, 3.67, -0.33\overline{3},$* $5.272\ 7\overline{27}$
I	Irrational numbers	Numbers that can be represented as nonrepeating and nonterminating decimal numbers	$\sqrt{2}, \pi, \sqrt[3]{7}, 1.414\ 213...,$ $2.718\ 281\ 82...$
R	Real numbers	Rational and irrational numbers	

* The overbar indicates that the number (or block of numbers) repeats indefinitely.

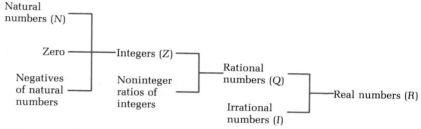

F I G U R E   1
Real numbers and important subsets

◆ THE REAL NUMBER LINE

A one-to-one correspondence exists between the set of real numbers and the set of points on a line. That is, each real number corresponds to exactly one point, and each point corresponds to exactly one real number. A line with a real number associated with each point, and vice versa, as shown in Figure 2, is called a **real number line,** or simply a **real line.** Each number associated with a point is called the **coordinate** of the point.

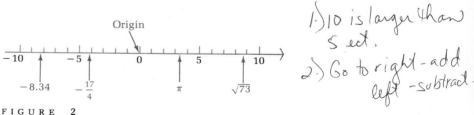

F I G U R E   2
The real number line

1.) 10 is larger than 5 ect.

2.) Go to right - add
    left - subtract.

The point with coordinate 0 is called the **origin.** The arrow on the right end of the line indicates a positive direction. The coordinates of all points to the right of the origin are called **positive real numbers,** and those to the left of the origin are called **negative real numbers.** The real number 0 is neither positive nor negative.

◆ BASIC REAL NUMBER PROPERTIES

We now take a look at some of the basic properties of the real number system that enable us to convert algebraic expressions into *equivalent forms.* These assumed basic properties, called **axioms,** become operational rules in the algebra of real numbers.

✳ A X I O M S

Always in twos

■ **Basic Properties of the Set of Real Numbers**

Let a, b, and c be arbitrary elements in the set of real numbers R.

ADDITION PROPERTIES

CLOSURE:	$a + b$ is a unique element in R.
ASSOCIATIVE:	$(a + b) + c = a + (b + c)$
COMMUTATIVE:	$a + b = b + a$
IDENTITY:	0 is the additive identity; that is, $0 + a = a + 0 = a$ for all a in R, and 0 is the only element in R with this property.
INVERSE:	For each a in R, $-a$ is its unique additive inverse; that is, $a + (-a) = (-a) + a = 0$, and $-a$ is the only element in R relative to a with this property.

MULTIPLICATION PROPERTIES

CLOSURE:	ab is a unique element in R. — #s are infinite
ASSOCIATIVE:	$(ab)c = a(bc)$
COMMUTATIVE:	$ab = ba$
IDENTITY:	1 is the multiplicative identity; that is, $(1)a = a(1) = a$ for all a in R, and 1 is the only element in R with this property.
INVERSE:	For each a in R, $a \neq 0$, $1/a$ is its unique multiplicative inverse; that is, $a(1/a) = (1/a)a = 1$, and $1/a$ is the only element in R relative to a with this property.

DISTRIBUTIVE PROPERTIES

$$a(b + c) = ab + ac \qquad (a + b)c = ac + bc$$

Do not be intimidated by the names of these properties. Most of the ideas presented here are quite simple. In fact, you have been using many of these properties in arithmetic for a long time.

You are already familiar with the **commutative properties** for addition and multiplication. They indicate that the order in which the addition or multiplication of two numbers is performed does not matter. For example,

$$7 + 2 = 2 + 7 \qquad \text{and} \qquad 3 \cdot 5 = 5 \cdot 3$$

Is there a commutative property relative to subtraction or division? That is, does $a - b = b - a$ or does $a \div b = b \div a$ for all real numbers a and b (division by 0 excluded)? The answer is no, since, for example,

$$8 - 6 \neq 6 - 8 \qquad \text{and} \qquad 10 \div 5 \neq 5 \div 10$$

When computing

$$3 + 2 + 6 \qquad \text{or} \qquad 3 \cdot 2 \cdot 6$$

why do we not need parentheses to indicate which two numbers are to be added or multiplied first? The answer is to be found in the **associative properties.** These properties allow us to write

$$(3 + 2) + 6 = 3 + (2 + 6) \qquad \text{and} \qquad (3 \cdot 2) \cdot 6 = 3 \cdot (2 \cdot 6)$$

so it does not matter how we group numbers relative to either operation. Is there an associative property for subtraction or division? The answer is no, since, for example,

$$(12 - 6) - 2 \neq 12 - (6 - 2) \qquad \text{and} \qquad (12 \div 6) \div 2 \neq 12 \div (6 \div 2)$$

Evaluate each side of each equation to see why.

Conclusion

Relative to addition, commutativity and associativity permit us to change the order of addition at will and insert or remove parentheses as we please. The same is true for multiplication, but not for subtraction and division.

What number added to a given number will give that number back again? What number times a given number will give that number back again? The answers are 0 and 1, respectively. Because of this, 0 and 1 are called the **identity elements** for the real numbers. Hence, for any real numbers a and b,

$$0 + 5 = 5 \qquad \text{and} \qquad (a + b) + 0 = a + b$$
$$1 \cdot 4 = 4 \qquad \text{and} \qquad (a + b) \cdot 1 = a + b$$

We now consider **inverses.** For each real number a, there is a unique real number $-a$ such that $a + (-a) = 0$. The number $-a$ is called the **additive inverse** of a, or the **negative** of a. For example, the additive inverse (or negative) of 7 is -7, since $7 + (-7) = 0$. The additive inverse (or negative) of -7 is $-(-7) = 7$, since $-7 + [-(-7)] = 0$. It is important to remember that

$-a$ **is not necessarily a negative number; it is positive if a is negative and negative if a is positive.**

For each number $a \neq 0$, there is a unique real number $1/a$ such that $a(1/a) = 1$. The number $1/a$ is called the **multiplicative inverse** of a, or the **reciprocal** of a. For example, the multiplicative inverse (or reciprocal) of 4 is $\frac{1}{4}$, since $4(\frac{1}{4}) = 1$. (Also note that 4 is the multiplicative inverse of $\frac{1}{4}$.)

We now turn to the **distributive properties,** which involve both multiplication and addition. Consider the following two computations:

$$5(3 + 4) = 5 \cdot 7 = 35 \qquad 5 \cdot 3 + 5 \cdot 4 = 15 + 20 = 35$$

Thus,

$$5(3 + 4) = 5 \cdot 3 + 5 \cdot 4$$

and we say that multiplication by 5 *distributes* over the sum $(3 + 4)$. In general, **multiplication distributes over addition** in the real number system. Two more illustrations are

$$9(m + n) = 9m + 9n \qquad (7 + 2)u = 7u + 2u$$

◆ EXAMPLE 1   State the real number property that justifies the indicated statement.

Statement	Property Illustrated
(A) $x(y + z) = (y + z)x$	Commutative $(\cdot)$
(B) $5(2y) = (5 \cdot 2)y$	Associative $(\cdot)$
(C) $2 + (y + 7) = 2 + (7 + y)$	Commutative $(+)$
(D) $4z + 6z = (4 + 6)z$	Distributive
(E) If $m + n = 0$, then $n = -m$.	Inverse $(+)$ ◆

PROBLEM 1*   State the real number property that justifies the indicated statement.

(A) $8 + (3 + y) = (8 + 3) + y$ (B) $(x + y) + z = z + (x + y)$
(C) $(a + b)(x + y) = a(x + y) + b(x + y)$ (D) $5xy + 0 = 5xy$
(E) If $xy = 1$, $x \neq 0$, then $y = 1/x$.   ◆

◆ FURTHER PROPERTIES

Subtraction and *division* can be defined in terms of addition and multiplication, respectively:

■ **Subtraction and Division**

For all real numbers a and b:

SUBTRACTION: $a - b = a + (-b)$ $\begin{aligned}7 - (-5) &= 7 + [-(-5)] \\ &= 7 + 5 = 12\end{aligned}$

DIVISION: $a \div b = a\left(\dfrac{1}{b}\right), \quad b \neq 0$ $9 \div 4 = 9\left(\dfrac{1}{4}\right) = \dfrac{9}{4}$

---

* Answers to matched problems are found near the end of each section, before the exercise set.

Thus, to subtract b from a, add the negative (the additive inverse) of b to a. To divide a by b, multiply a by the reciprocal (the multiplicative inverse) of b. Note that division by 0 is not defined, since 0 does not have a reciprocal. Thus:

0 can never be used as a divisor!

The following properties of negatives (called **theorems**) can be proved using the preceding axioms and definitions.

THEOREM 1 ▪ **Properties of Negatives**

For all real numbers a and b:

1. $-(-a) = a$
2. $(-a)b = -(ab) = a(-b) = -ab$
3. $(-a)(-b) = ab$
4. $(-1)a = -a$

5. $\dfrac{-a}{b} = -\dfrac{a}{b} = \dfrac{a}{-b}, \quad b \neq 0$

6. $\dfrac{-a}{-b} = -\dfrac{-a}{b} = -\dfrac{a}{-b} = \dfrac{a}{b},$
$\qquad\qquad\qquad\qquad\qquad b \neq 0$

We now state an important theorem involving 0.

THEOREM 2 ▪ **Zero Properties**

For all real numbers a and b:

1. $a \cdot 0 = 0$
2. $ab = 0$ if and only if $a = 0$ or $b = 0$ (or both)

◆ EXAMPLE 2 State the real number property or definition that justifies each statement.

Statement	Property Illustrated
(A) $7 - (-5) = 7 + [-(-5)]$	Definition of subtraction
(B) $-(-5) = 5$	Negatives (Theorem 1)
(C) $\dfrac{-7}{3} = -\dfrac{7}{3}$	Negatives (Theorem 1)
(D) $-\dfrac{2}{-3} = \dfrac{2}{3}$	Negatives (Theorem 1)
(E) If $(3x + 2)(x - 7) = 0$, then either $3x + 2 = 0$ or $x - 7 = 0$.	Zero (Theorem 2)

◆

State the real number property or definition that justifies each statement.

(A) $\dfrac{3}{5} = 3\left(\dfrac{1}{5}\right)$ (B) $(-5)(2) = -(5 \cdot 2)$ (C) $(-1)3 = -3$

(D) $\dfrac{-7}{9} = -\dfrac{7}{9}$ (E) If $x + 5 = 0$, then $(x - 3)(x + 5) = 0$.   ◆

◆ FRACTION PROPERTIES

Recall that the quotient $a \div b \ (b \neq 0)$ written in the form a/b is called a **fraction.** The quantity a is called the **numerator,** and the quantity b is called the **denominator.**

Fraction Properties

For all real numbers a, b, c, d, and k (division by 0 excluded):

1. $\dfrac{a}{b} = \dfrac{c}{d}$ if and only if $ad = bc$

$\dfrac{4}{6} = \dfrac{6}{9}$ since $4 \cdot 9 = 6 \cdot 6$ *Proportions*

2. $\dfrac{ka}{kb} = \dfrac{a}{b}$ 3. $\dfrac{a}{b} \cdot \dfrac{c}{d} = \dfrac{ac}{bd}$ 4. $\dfrac{a}{b} \div \dfrac{c}{d} = \dfrac{a}{b} \cdot \dfrac{d}{c}$

$\dfrac{7 \cdot 3}{7 \cdot 5} = \dfrac{3}{5}$ $\dfrac{3}{5} \cdot \dfrac{7}{8} = \dfrac{3 \cdot 7}{5 \cdot 8}$ $\dfrac{2}{3} \div \dfrac{5}{7} = \dfrac{2}{3} \cdot \dfrac{7}{5}$

5. $\dfrac{a}{b} + \dfrac{c}{b} = \dfrac{a + c}{b}$ 6. $\dfrac{a}{b} - \dfrac{c}{b} = \dfrac{a - c}{b}$ 7. $\dfrac{a}{b} + \dfrac{c}{d} = \dfrac{ad + bc}{bd}$

$\dfrac{3}{6} + \dfrac{5}{6} = \dfrac{3 + 5}{6}$ $\dfrac{7}{8} - \dfrac{3}{8} = \dfrac{7 - 3}{8}$ $\dfrac{2}{3} + \dfrac{3}{5} = \dfrac{2 \cdot 5 + 3 \cdot 3}{3 \cdot 5}$

Answers to Matched Problems

1. (A) Associative (+)   (B) Commutative (+)   (C) Distributive
   (D) Identity (+)   (E) Inverse (·)
2. (A) Definition of division   (B) Negatives (Theorem 1)
   (C) Negatives (Theorem 1)   (D) Negatives (Theorem 1)
   (E) Zero (Theorem 2)

E X E R C I S E   1-1

All variables represent real numbers.

A   In Problems 1–6, replace each question mark with an appropriate expression that will illustrate the use of the indicated real number property.

1. Commutative property (·): $uv = ?$
2. Commutative property (+): $x + 7 = ?$

3. Associative property (+): $3 + (7 + y) = ?$
4. Associative property (·): $x(yz) = ?$
5. Identity property (·): $1(u + v) = ?$
6. Identity property (+): $0 + 9m = ?$

In Problems 7–18, each statement illustrates the use of one of the following properties or definitions; indicate which one:

Commutative (+, ·)   Associative (+, ·)   Distributive
Identity (+, ·)   Inverse (+, ·)   Subtraction
Division   Negatives (Theorem 1)   Zero (Theorem 2)

7. $5(8m) = (5 \cdot 8)m$
8. $a + cb = a + bc$
9. $-\dfrac{m}{-n} = \dfrac{m}{n}$
10. $5x + 7x = (5 + 7)x$
11. $7 - 11 = 7 + (-11)$
12. $(-3)(\frac{1}{-3}) = 1$
13. $9 \div (-4) = 9(\frac{1}{-4})$
14. $q + (-q) = 0$
15. $uv(w + x) = uvw + uvx$
16. $2(3x + y) + 0 = 2(3x + y)$
17. $0(m + 5) = 0$
18. $\dfrac{-u}{-v} = \dfrac{u}{v}$

B   In Problems 19–26, each statement illustrates the use of one of the following properties or definitions; indicate which one:

Commutative (+, ·)   Associative (+, ·)   Distributive
Identity (+, ·)   Inverse (+, ·)   Subtraction
Division   Negatives (Theorem 1)   Zero (Theorem 2)

19. $(8u)(4v) = 8[u(4v)]$
20. $(7z + 4) + 2 = 2 + (7z + 4)$
21. $(x + 8)(x + 6) = (x + 8)x + (x + 8)6$
22. $(4x + 3) + (x + 2) = 4x + [3 + (x + 2)]$
23. $\dfrac{-4}{-(x + y)} = \dfrac{4}{x + y}$
24. $u(u - 2v) + v(u - 2v) = (u + v)(u - 2v)$
25. If $(x - 2)(2x + 3) = 0$, then either $x - 2 = 0$ or $2x + 3 = 0$.
26. $x(4x + 7) = 0$ if and only if $x = 0$ or $4x + 7 = 0$.

27. If $uv = 1$, does either u or v have to be 1?
28. If $uv = 0$, does either u or v have to be 0?
29. Indicate whether the following are true (T) or false (F):

   (A) All integers are natural numbers.
   (B) All rational numbers are real numbers.
   (C) All natural numbers are rational numbers.

30. Indicate whether the following are true (T) or false (F):

   (A) All natural numbers are integers.
   (B) All real numbers are irrational.
   (C) All rational numbers are real numbers.

31. Give an example of a real number that is not a rational number.

32. Give an example of a rational number that is not an integer.

33. Given the sets of numbers N (natural numbers), Z (integers), Q (rational numbers), and R (real numbers), indicate to which set(s) each of the following numbers belongs:

    (A) 8   (B) $\sqrt{2}$ (C) -1.414 (D) $\frac{-5}{2}$

34. Given the sets of numbers $N, Z, Q,$ and R (see Problem 33), indicate to which set(s) each of the following numbers belongs:

    (A) -3 (B) 3.14   (C) π (D) $\frac{2}{3}$

35. Indicate true (T) or false (F), and for each false statement find real number replacements for a, b, and c that will illustrate its falseness. For all real numbers a, b, and c:

    (A) $(a + b) + c = a + (b + c)$ (B) $(a - b) - c = a - (b - c)$
    (C) $a(bc) = (ab)c$ (D) $(a \div b) \div c = a \div (b \div c)$

36. Indicate true (T) or false (F), and for each false statement find real number replacements for a and b that will illustrate its falseness. For all real numbers a and b:

    (A) $a + b = b + a$ (B) $a - b = b - a$ (C) $ab = ba$
    (D) $a \div b = b \div a$

C 37. If $c = 0.151515...$, then $100c = 15.1515...$ and

$$100c - c = 15.1515... - 0.151515...$$
$$99c = 15$$
$$c = \tfrac{15}{99} = \tfrac{5}{33}$$

Proceeding similarly, convert the repeating decimal 0.090909... into a fraction. (All repeating decimals are rational numbers, and all rational numbers have repeating decimal representations.)

38. Repeat Problem 37 for 0.181818... .

39. For a and b real numbers, justify each step below using a property given in this section:

Statement	Reason
1. $(a + b) + (-a) = (-a) + (a + b)$	1.
2. $= [(-a) + a] + b$	2.
3. $= 0 + b$	3.
4. $= b$	4.

40. To see how the distributive property is behind the mechanics of long multiplication, compute each of the following and compare:

    Long Multiplication   Use of the Distributive Property

    $\quad\quad 23 \quad\quad\quad\quad\quad\quad 23 \cdot 12 = 23(2 + 10)$
    $\underline{\times 12} \quad\quad\quad\quad\quad\quad\quad\quad = 23 \cdot 2 + 23 \cdot 10 =$

Use a calculator to express each number as a decimal to the capacity of your calculator. Observe the repeating decimal representation of the rational numbers and the nonrepeating decimal representation of the irrational numbers.

41. (A) $\frac{13}{6}$ (B) $\sqrt{21}$ (C) $\frac{7}{16}$ (D) $\frac{29}{111}$

42. (A) $\frac{8}{9}$ (B) $\frac{3}{11}$ (C) $\sqrt{5}$ (D) $\frac{11}{8}$

SECTION 1-2   Basic Operations on Polynomials

◆ NATURAL NUMBER EXPONENTS
◆ POLYNOMIALS
◆ COMBINING LIKE TERMS
◆ ADDITION AND SUBTRACTION
◆ MULTIPLICATION
◆ COMBINED OPERATIONS

This section covers basic operations on *polynomials*, a mathematical form encountered frequently in mathematics. Our discussion starts with a brief review of natural number exponents. (Integer and rational exponents and their properties will be discussed in detail in subsequent sections.)

◆ NATURAL NUMBER EXPONENTS

We define a **natural number exponent** as follows:

Natural Number Exponent

For n a natural number and b any real number:

$$b^n = b \cdot b \cdot \ \cdots \ \cdot b \qquad n \text{ factors of } b$$
$$3^5 = 3 \cdot 3 \cdot 3 \cdot 3 \cdot 3 \qquad 5 \text{ factors of } 3$$

where n is called the **exponent** and b is called the **base.**

Along with this definition, we state the **first property of exponents:**

THEOREM 3

First Property of Exponents

For any natural numbers m and n, and any real number b:

$$b^m b^n = b^{m+n} \qquad (2t^4)(5t^3) \ = 2 \cdot 5t^{4+3} \ * = 10t^7$$

---

* Dashed boxes are used throughout the book to represent steps that are usually performed mentally.

◆ POLYNOMIALS

Algebraic expressions are formed by using constants and variables and the algebraic operations of addition, subtraction, multiplication, division, raising to powers, and taking roots. Special types of algebraic expressions are called *polynomials*. A **polynomial in one variable** x is constructed by adding or subtracting constants and terms of the form ax^n, where a is a real number and n is a natural number. A **polynomial in two variables** x and y is constructed by adding and subtracting constants and terms of the form ax^my^n, where a is a real number and m and n are natural numbers. Polynomials in three and more variables are defined in a similar manner.

Examples of Polynomials

8	0
$3x^3 - 6x + 7$	$6x + 3$
$2x^2 - 7xy - 8y^2$	$9y^3 + 4y^2 - y + 4$
$2x - 3y + 2$	$u^5 - 3u^3v^2 + 2uv^4 - v^4$

Polynomial forms are encountered frequently in mathematics, and for their more efficient study, it is useful to classify them according to their *degree*. If a term in a polynomial has only one variable as a factor, then the **degree of the term** is the power of the variable. If two or more variables are present in a term as factors, then the **degree of the term** is the sum of the powers of the variables. The **degree of a polynomial** is the degree of the nonzero term with the highest degree in the polynomial. Any nonzero constant is defined to be a **polynomial of degree 0.** The number 0 is also a polynomial but is not assigned a degree.

◆ E X A M P L E 3   (A) The degree of the first term in $5x^3 + \sqrt{3}x - \frac{1}{2}$ is 3, the degree of the second term is 1, the degree of the third term is 0, and the degree of the whole polynomial is 3 (the same as the degree of the term with the highest degree).

(B) The degree of the first term in $8u^3v^2 - \sqrt{7}uv^2$ is 5, the degree of the second term is 3, and the degree of the whole polynomial is 5.   ◆

P R O B L E M 3   (A) Given the polynomial $6x^5 + 7x^3 - 2$, what is the degree of the first term? The second term? The third term? The whole polynomial?

(B) Given the polynomial $2u^4v^2 - 5uv^3$, what is the degree of the first term? The second term? The whole polynomial?   ◆

In addition to classifying polynomials by degree, we also call a single-term polynomial a **monomial,** a two-term polynomial a **binomial,** and a three-term polynomial a **trinomial.**

◆ COMBINING LIKE TERMS

The concept of *coefficient* plays a central role in the process of combining *like terms.* A constant in a term of a polynomial, including the sign that precedes it, is called the **numerical coefficient,** or simply, the **coefficient,** of the term. If a constant does not appear, or only a $+$ sign appears, the coefficient is understood to be 1. If only a $-$ sign appears, the coefficient is understood to be -1. Thus, given the polynomial

$$5x^4 - x^3 - 3x^2 + x - 7 \;\vdots\; = 5x^4 + (-1)x^3 + (-3)x^2 + 1x + (-7) \;\vdots$$

the coefficient of the first term is 5, the coefficient of the second term is -1, the coefficient of the third term is -3, the coefficient of the fourth term is 1, and the coefficient of the fifth term is -7.

It is useful to state two more distributive properties of real numbers that follow from the distributive properties stated in Section 1-1.

Further Distributive Properties

1. $a(b - c) = (b - c)a = ab - ac$
2. $a(b + c + \cdots + f) = ab + ac + \cdots + af$

Two terms in a polynomial are called **like terms** if they have exactly the same variable factors to the same powers. The numerical coefficients may or may not be the same. Since constant terms involve no variables, all constant terms are like terms. If a polynomial contains two or more like terms, these terms can be combined into a single term by making use of distributive properties. The following example illustrates the reasoning behind the process:

$$
\begin{aligned}
3x^2y - 5xy^2 + x^2y - 2x^2y \;&\vdots\; = 3x^2y + x^2y - 2x^2y - 5xy^2 \\
&\vdots\; = (3x^2y + 1x^2y - 2x^2y) - 5xy^2 \\
&\vdots\; = (3 + 1 - 2)x^2y - 5xy^2 \\
&= 2x^2y - 5xy^2
\end{aligned}
$$

It should be clear that free use is made of the real number properties discussed earlier. The steps shown in the dashed box are usually done mentally, and the process is quickly mechanized as follows:

Like terms in a polynomial are combined by adding their numerical coefficients.

How can we simplify expressions such as $4(x - 2y) - 3(2x - 7y)$? We clear the expression of parentheses using distributive properties, and combine like terms:

$$
\begin{aligned}
4(x - 2y) - 3(2x - 7y) &= 4x - 8y - 6x + 21y \\
&= -2x + 13y
\end{aligned}
$$

◆ E X A M P L E 4   Remove parentheses and simplify:

(A) $2(3x^2 - 2x + 5) + (x^2 + 3x - 7)$

$$= 2(3x^2 - 2x + 5) + 1(x^2 + 3x - 7)$$

Think

$$= 6x^2 - 4x + 10 + x^2 + 3x - 7$$
$$= 7x^2 - x + 3$$

(B) $(x^3 - 2x - 6) - (2x^3 - x^2 + 2x - 3)$

$$= 1(x^3 - 2x - 6) + (-1)(2x^3 - x^2 + 2x - 3)$$

Think   Be careful with
                          the sign here.

$$= x^3 - 2x - 6 - 2x^3 + x^2 - 2x + 3$$
$$= -x^3 + x^2 - 4x - 3$$

(C) $[3x^2 - (2x + 1)] - (x^2 - 1) = [3x^2 - 2x - 1] - (x^2 - 1)$ Remove inner
$$= 3x^2 - 2x - 1 - x^2 + 1$$ parentheses first.
$$= 2x^2 - 2x$$ ◆

P R O B L E M 4   Remove parentheses and simplify:

(A) $3(u^2 - 2v^2) + (u^2 + 5v^2)$ (B) $(m^3 - 3m^2 + m - 1) - (2m^3 - m + 3)$
(C) $(x^3 - 2) - [2x^3 - (3x + 4)]$ ◆

◆ ADDITION AND SUBTRACTION

Addition and subtraction of polynomials can be thought of in terms of removing parentheses and combining like terms, as illustrated in Example 4. Horizontal and vertical arrangements are illustrated in the next two examples. You should be able to work either way, letting the situation dictate your choice.

◆ E X A M P L E 5   Add horizontally and vertically: $x^4 - 3x^3 + x^2$, $-x^3 - 2x^2 + 3x$,   and $3x^2 - 4x - 5$

Solution   Add horizontally:

$$(x^4 - 3x^3 + x^2) + (-x^3 - 2x^2 + 3x) + (3x^2 - 4x - 5)$$
$$= x^4 - 3x^3 + x^2 - x^3 - 2x^2 + 3x + 3x^2 - 4x - 5$$
$$= x^4 - 4x^3 + 2x^2 - x - 5$$

Or vertically, by lining up like terms and adding their coefficients:

$$
\begin{array}{l}
x^4 - 3x^3 + \ x^2 \\
\quad - \ x^3 - 2x^2 + 3x \\
\qquad\qquad\ 3x^2 - 4x - 5 \\
\hline
x^4 - 4x^3 + 2x^2 - \ x - 5
\end{array}
$$

◆

P R O B L E M 5   Add horizontally and vertically: $3x^4 - 2x^3 - 4x^2$, $x^3 - 2x^2 - 5x$,   and $x^2 + 7x - 2$ ◆

◆ E X A M P L E   6   Subtract $4x^2 - 3x + 5$ from $x^2 - 8$, both horizontally and vertically.

Solution

$$(x^2 - 8) - (4x^2 - 3x + 5) \qquad \text{or} \qquad \begin{array}{r} x^2 \qquad - 8 \\ -4x^2 + 3x - 5 \\ \hline -3x^2 + 3x - 13 \end{array} \leftarrow \text{Change signs}$$

$$= x^2 - 8 - 4x^2 + 3x - 5$$

$$= -3x^2 + 3x - 13 \qquad\qquad\qquad\qquad\qquad \text{and add.}$$ ◆

P R O B L E M   6   Subtract $2x^2 - 5x + 4$ from $5x^2 - 6$, both horizontally and vertically.   ◆

◆ MULTIPLICATION

Multiplication of algebraic expressions involves the extensive use of distributive properties for real numbers, as well as other real number properties.

◆ E X A M P L E   7   Multiply: $(2x - 3)(3x^2 - 2x + 3)$

Solution

$$(2x - 3)(3x^2 - 2x + 3) \boxed{= 2x(3x^2 - 2x + 3) - 3(3x^2 - 2x + 3)}$$

$$= 6x^3 - 4x^2 + 6x - 9x^2 + 6x - 9$$

$$= 6x^3 - 13x^2 + 12x - 9$$

Or, using a vertical arrangement,

$$\begin{array}{r} 3x^2 - \ 2x + \ 3 \\ 2x - \ 3 \\ \hline 6x^3 - \ 4x^2 + \ 6x \\ - \ 9x^2 + \ 6x - 9 \\ \hline 6x^3 - 13x^2 + 12x - 9 \end{array}$$ ◆

P R O B L E M   7   Multiply: $(2x - 3)(2x^2 + 3x - 2)$ ◆

Thus, to multiply two polynomials, multiply each term of one by each term of the other, and combine like terms.

Products of binomial factors occur frequently, so it is useful to develop procedures that will enable us to write down their products by inspection. To find the product $(2x - 1)(3x + 2)$, we will use the popular **FOIL method.** We multiply each term of one factor by each term of the other factor as follows:

$$\begin{array}{cccc} \text{F} & \text{O} & \text{I} & \text{L} \\ \text{First} & \text{Outer} & \text{Inner} & \text{Last} \\ \text{product} & \text{product} & \text{product} & \text{product} \\ \downarrow & \downarrow & \downarrow & \downarrow \end{array}$$

$$(2x - 1)(3x + 2) = \quad 6x^2 \quad + \quad 4x \quad - \quad 3x \quad - \quad 2$$

The inner and outer products are like terms and hence combine into one term. Thus,

$$(2x - 1)(3x + 2) = 6x^2 + x - 2$$

To speed up the process, we combine the inner and outer product mentally.

Products of certain binomial factors occur so frequently that it is useful to remember formulas for their products. The following formulas are easily verified by multiplying the factors on the left using the FOIL method:

> **Special Products**
>
> 1. $(a - b)(a + b) = a^2 - b^2$ 2. $(a + b)^2 = a^2 + 2ab + b^2$
> 3. $(a - b)^2 = a^2 - 2ab + b^2$

◆ EXAMPLE 8   Multiply mentally:

(A) $(2x - 3y)(5x + 2y) = 10x^2 - 11xy - 6y^2$
(B) $(3a - 2b)(3a + 2b) = 9a^2 - 4b^2$
(C) $(5x - 3)^2 = 25x^2 - 30x + 9$
(D) $(m + 2n)^2 = m^2 + 4mn + 4n^2$ ◆

PROBLEM 8   Multiply mentally:

(A) $(4u - 3v)(2u + v)$ (B) $(2xy + 3)(2xy - 3)$ (C) $(m + 4n)(m - 4n)$
(D) $(2u - 3v)^2$ (E) $(6x + y)^2$ ◆

◆ COMBINED OPERATIONS

We complete this section by considering several examples that use all the operations just discussed. Note that in simplifying, we usually remove grouping symbols starting from the inside. That is, we remove parentheses ( ) first, then brackets [ ], and finally braces { }, if present. Also, multiplication and division precede addition and subtraction.

◆ EXAMPLE 9   Perform the indicated operations and simplify:

(A) $3x - \{5 - 3[x - x(3 - x)]\} = 3x - \{5 - 3[x - 3x + x^2]\}$
$$= 3x - \{5 - 3x + 9x - 3x^2\}$$
$$= 3x - 5 + 3x - 9x + 3x^2$$
$$= 3x^2 - 3x - 5$$

(B) $(x - 2y)(2x + 3y) - (2x + y)^2 = 2x^2 - xy - 6y^2 - (4x^2 + 4xy + y^2)$
$$= 2x^2 - xy - 6y^2 - 4x^2 - 4xy - y^2$$
$$= -2x^2 - 5xy - 7y^2$$ ◆

PROBLEM 9   Perform the indicated operations and simplify:

(A) $2t - \{7 - 2[t - t(4 + t)]\}$ (B) $(u - 3v)^2 - (2u - v)(2u + v)$ ◆

Answers to Matched Problems   3. (A) 5, 3, 0, 5   (B) 6, 4, 6
                                 4. (A) $4u^2 - v^2$ (B) $-m^3 - 3m^2 + 2m - 4$ (C) $-x^3 + 3x + 2$

5. $3x^4 - x^3 - 5x^2 + 2x - 2$ 6. $3x^2 + 5x - 10$ 7. $4x^3 - 13x + 6$
8. (A) $8u^2 - 2uv - 3v^2$ (B) $4x^2y^2 - 9$ (C) $m^2 - 16n^2$
   (D) $4u^2 - 12uv + 9v^2$ (E) $36x^2 + 12xy + y^2$
9. (A) $-2t^2 - 4t - 7$ (B) $-3u^2 - 6uv + 10v^2$

E X E R C I S E   1-2

A Problems 1–8 refer to the following polynomials:

(A) $2x - 3$ (B) $2x^2 - x + 2$ (C) $x^3 + 2x^2 - x + 3$

1. What is the degree of (C)?   2. What is the degree of (A)?
3. Add (B) and (C).   4. Add (A) and (B).
5. Subtract (B) from (C).   6. Subtract (A) from (B).
7. Multiply (B) and (C).   8. Multiply (A) and (C).

In Problems 9–30, perform the indicated operations and simplify.

odd #
and #
47.

9. $2(u - 1) - (3u + 2) - 2(2u - 3)$ 10. $2(x - 1) + 3(2x - 3) - (4x - 5)$
11. $4a - 2a[5 - 3(a + 2)]$ 12. $2y - 3y[4 - 2(y - 1)]$
13. $(a + b)(a - b)$ 14. $(m - n)(m + n)$
15. $(3x - 5)(2x + 1)$ 16. $(4t - 3)(t - 2)$
17. $(2x - 3y)(x + 2y)$ 18. $(3x + 2y)(x - 3y)$
19. $(3y + 2)(3y - 2)$ 20. $(2m - 7)(2m + 7)$
21. $(3m + 7n)(2m - 5n)$ 22. $(6x - 4y)(5x + 3y)$
23. $(4m + 3n)(4m - 3n)$ 24. $(3x - 2y)(3x + 2y)$
25. $(3u + 4v)^2$ 26. $(4x - y)^2$
27. $(a - b)(a^2 + ab + b^2)$ 28. $(a + b)(a^2 - ab + b^2)$
29. $(4x + 3y)^2$ 30. $(3x + 2)^2$

B In Problems 31–38, perform the indicated operations and simplify.

31. $m - \{m - [m - (m - 1)]\}$ 32. $2x - 3\{x + 2[x - (x + 5)] + 1\}$
33. $(x^2 - 2xy + y^2)(x^2 + 2xy + y^2)$ 34. $(2x^2 + x - 2)(x^2 - 3x + 5)$
35. $(3a - b)(3a + b) - (2a - 3b)^2$ 36. $(2x - 1)^2 - (3x + 2)(3x - 2)$
37. $(x - 2y)^3$ 38. $(2m - n)^3$

39. Subtract the sum of the last two polynomials from the sum of the first
    two: $2x^2 - 4xy + y^2$, $3xy - y^2$, $x^2 - 2xy - y^2$, $-x^2 + 3xy - 2y^2$
40. Subtract the sum of the first two polynomials from the sum of the last
    two: $3m^2 - 2m + 5$, $4m^2 - m$, $3m^2 - 3m - 2$, $m^3 + m^2 + 2$

C In Problems 41–44, perform the indicated operations and simplify.

41. $(2x - 1)^3 - 2(2x - 1)^2 + 3(2x - 1) + 7$
42. $2(x - 2)^3 - (x - 2)^2 - 3(x - 2) - 4$
43. $2\{(x - 3)(x^2 - 2x + 1) - x[3 - x(x - 2)]\}$
44. $-3x\{x[x - x(2 - x)] - (x + 2)(x^2 - 3)\}$

45. If you are given two polynomials, one of degree m and the other of degree n, where $m > n$, what is the degree of their product?

46. What is the degree of the sum of the two polynomials in Problem 45?

APPLICATIONS

Business & Economics

47. *Investment.* You have \$10,000 to invest, part at 9% and the rest at 12%. If x is the amount invested at 9%, write an algebraic expression that represents the total annual income from both investments. Simplify the expression.

48. *Gross receipts.* Six thousand tickets are to be sold for a concert, some for \$9 each and the rest for \$15 each. If x is the number of \$9 tickets sold, write an algebraic expression that represents the gross receipts from ticket sales, assuming all tickets are sold. Simplify the expression.

49. *Gross receipts.* Four thousand tickets are to be sold for a musical show. If x tickets are to be sold for \$10 each and three times that number for \$30 each, and if the rest are sold for \$50 each, write an algebraic expression that represents the gross receipts from ticket sales, assuming all tickets are sold. Simplify the expression.

50. *Investment.* A person has \$100,000 to invest. If \$$x$ are invested in a money market account yielding 7% and twice that amount in certificates of deposit yielding 9%, and if the rest is invested in high-grade bonds yielding 11%, write an algebraic expression that represents the total annual income from all three investments. Simplify the expression.

Life Sciences

51. *Nutrition.* Food mix A contains 2% fat, and food mix B contains 6% fat. A 10 kilogram diet mix of foods A and B is formed. If x kilograms of food A are used, write an algebraic expression that represents the total number of kilograms of fat in the final food mix. Simplify the expression.

52. *Nutrition.* Each ounce of food M contains 8 units of calcium, and each ounce of food N contains 5 units of calcium. A 160 ounce diet mix is formed using foods M and N. If x is the number of ounces of food M used, write an algebraic expression that represents the total number of units of calcium in the diet mix. Simplify the expression.

SECTION 1-3 Factoring Polynomials

◆ WHAT DO WE MEAN BY FACTORING?
◆ COMMON FACTORS
◆ FACTORING SECOND-DEGREE POLYNOMIALS
◆ MORE FACTORING

Our objective in this section is to review some of the standard factoring techniques for polynomials with integer coefficients.

◆ WHAT DO WE MEAN BY FACTORING?

If a number is written as the product of other numbers, then each number in the product is called a **factor of the original number.** Similarly, if an algebraic expression is written as the product of other algebraic expressions, then each algebraic expression in the product is called a **factor of the original algebraic expression.** For example,

$42 = 2 \cdot 3 \cdot 7$ 2, 3, and 7 are factors of 42

$x^2 - 9 = (x - 3)(x + 3)$ $(x - 3)$ and $(x + 3)$ are factors of $x^2 - 9$

The process of writing a number or algebraic expression as the product of other numbers or algebraic expressions is called **factoring.** We begin our discussion of factoring with the positive integers.

An integer such as 42 can be represented in a factored form in many ways. The products

$7 \cdot 6$ $21 \cdot 2$ $3 \cdot 14$ $(\frac{1}{2})(12)(7)$ $2 \cdot 3 \cdot 7$

all yield 42. A particularly useful way of factoring integers greater than 1 is in terms of *prime numbers.*

▪ Prime and Composite Numbers

An integer greater than 1 is **prime** if its only positive integer factors are itself and 1. An integer greater than 1 that is not prime is called a **composite number.** The integer 1 is neither prime nor composite.

PRIME NUMBERS: 2, 3, 5, 7, 11, 13, 17
COMPOSITE NUMBERS: 4, 6, 8, 9, 10, 12, 14

[*Note:*   There is no largest prime number.]

A composite number is said to be **factored completely** if it is represented as a product of prime factors. The only factoring of 42 given above that meets this condition is $42 = 2 \cdot 3 \cdot 7$.

◆ E X A M P L E   10   Write 90 in completely factored form.

Solution $90 = 9 \cdot 10 = 3 \cdot 3 \cdot 2 \cdot 5 = 2 \cdot 3^2 \cdot 5$

or

$90 = 6 \cdot 15 = 2 \cdot 3 \cdot 3 \cdot 5 = 2 \cdot 3^2 \cdot 5$

or

$90 = 18 \cdot 5 = 3 \cdot 6 \cdot 5 = 3 \cdot 2 \cdot 3 \cdot 5 = 2 \cdot 3^2 \cdot 5$ ◆

Notice in Example 10 that we end up with the same set of prime factors of 90 no matter how we progress through the factoring process. This illustrates an important property of integers.

THEOREM 4

The Fundamental Theorem of Arithmetic

Each integer greater than 1 is either prime or can be expressed uniquely, except for the order of factors, as a product of prime factors.

PROBLEM 10 Write 180 in completely factored form. ◆

We also can write polynomials in completely factored form. A polynomial such as $3x^2 + 7x - 6$ can be written in factored form in many ways. The products

$$(3x - 2)(x + 3) \qquad 3x\left(x + \frac{7}{3} - \frac{2}{x}\right) \qquad 3\left(x - \frac{2}{3}\right)(x + 3)$$

all yield $3x^2 + 7x - 6$. A particularly useful way of factoring polynomials is in terms of *prime polynomials*.

Prime Polynomials

A polynomial of degree greater than 0 is said to be **prime relative to a given set of numbers:**

1. If all of its coefficients are from that set of numbers
2. If it cannot be written as a product of two polynomials, excluding 1 and itself, having coefficients from that set of numbers

Relative to the set of integers:
$x^2 - 3$ is prime
$x^2 - 4$ is not prime, since $x^2 - 4 = (x - 2)(x + 2)$

[*Note*: The set of integers is the set of numbers used most frequently in factoring polynomials.]

A nonprime polynomial is said to be **factored completely relative to a given set of numbers** if it is written as a product of prime polynomials relative to that set of numbers. The only factoring given above of $3x^2 + 7x - 6$ relative to the integers is $3x^2 + 7x - 6 = (3x - 2)(x + 3)$.

Writing polynomials in completely factored form is often a difficult task. But accomplishing it can lead to the simplification of certain algebraic expressions and to the solution of certain types of equations. The distributive properties for real numbers are central to the factoring process.

◆ COMMON FACTORS

Generally, a first step in any factoring procedure is to factor out all factors common to all terms.

◆ E X A M P L E   11   Factor out, relative to the integers, all factors common to all terms:

(A) $3x^3y - 6x^2y^2 - 3xy^3$ (B) $3y(2y + 5) + 2(2y + 5)$

Solutions   (A) $3x^3y - 6x^2y^2 - 3xy^3 \; \boxed{= (3xy)x^2 - (3xy)2xy - (3xy)y^2}$

$= 3xy(x^2 - 2xy - y^2)$

(B) $3y(2y + 5) + 2(2y + 5) \; \boxed{= 3y(2y + 5) + 2(2y + 5)}$

$= (3y + 2)(2y + 5)$ ◆

◆ P R O B L E M   11   Factor out, relative to the integers, all factors common to all terms:

(A) $2x^3y - 8x^2y^2 - 6xy^3$ (B) $2x(3x - 2) - 7(3x - 2)$ ◆

◆ FACTORING SECOND-DEGREE POLYNOMIALS

We now turn our attention to factoring second-degree polynomials of the form

$$2x^2 - 5x - 3 \quad \text{and} \quad 2x^2 + 3xy - 2y^2$$

into the product of two first-degree polynomials with integer coefficients. The following example will illustrate an approach to the problem.

◆ E X A M P L E   12   Factor each polynomial, if possible, using integer coefficients:

(A) $2x^2 + 3xy - 2y^2$ (B) $x^2 - 3x + 4$ (C) $6x^2 + 5xy - 4y^2$

Solutions   (A) $2x^2 + 3xy - 2y^2 = (2x + \quad y)(x - \quad y)$ Put in what we know. Signs
                                    ↑   ↑   must be opposite. (We can
                                    ?   ?   reverse this choice if we get
                                                           $-3xy$ instead of $+3xy$ for the
                                                           middle term.)

Now, what are the factors of 2 (the coefficient of y^2)?

$$\frac{2}{1 \cdot 2}$$
$$2 \cdot 1$$

The first choice gives us $-3xy$ for the middle term — close, but not there — so we reverse our choice of signs to obtain

$$2x^2 + 3xy - 2y^2 = (2x - y)(x + 2y)$$

(B) $x^2 - 3x + 4 = (x - \quad)(x - \quad)$

$$\begin{array}{c} \uparrow \qquad \uparrow \\ ? \qquad ? \end{array}$$

$$\frac{4}{\begin{array}{l} 2 \cdot 2 \\ 1 \cdot 4 \\ 4 \cdot 1 \end{array}}$$

No choice produces the middle term; hence, $x^2 - 3x + 4$ is not factorable using integer coefficients.

(C) $6x^2 + 5xy - 4y^2 = (\quad x + \quad y)(\quad x - \quad y)$

$$\begin{array}{cccc} \uparrow & \uparrow & \uparrow & \uparrow \\ ? & ? & ? & ? \end{array}$$

The signs must be opposite in the factors, since the third term is negative. (We can reverse our choice of signs later if necessary.) We write all factors of 6 and of 4:

$$\frac{6}{\begin{array}{l} 2 \cdot 3 \\ 3 \cdot 2 \\ 1 \cdot 6 \\ 6 \cdot 1 \end{array}} \qquad \frac{4}{\begin{array}{l} 2 \cdot 2 \\ 1 \cdot 4 \\ 4 \cdot 1 \end{array}}$$

Now try each choice on the left with each choice on the right—a total of 12 combinations that give us the first and last terms in the polynomial $6x^2 + 5xy - 4y^2$. The question is: Does any combination also give us the middle term, $5xy$? After trial and error and, perhaps, some educated guessing among the choices, we find that $3 \cdot 2$ matched with $4 \cdot 1$ gives us the correct middle term. Thus,

$$6x^2 + 5xy - 4y^2 = (3x + 4y)(2x - y)$$

If none of the 24 combinations (including reversing our sign choice) had produced the middle term, then we would conclude that the polynomial is not factorable using integer coefficients. ◆

PROBLEM 12 Factor each polynomial, if possible, using integer coefficients:

(A) $x^2 - 8x + 12$ (B) $x^2 + 2x + 5$ (C) $2x^2 + 7xy - 4y^2$
(D) $4x^2 - 15xy - 4y^2$ ◆

◆ MORE FACTORING

The factoring formulas listed below will enable us to factor certain polynomial forms that occur frequently. These formulas can be established by multiplying the factors on the right.

Special Factoring Formulas

PERFECT SQUARE:	1. $u^2 + 2uv + v^2 = (u + v)^2$
PERFECT SQUARE:	2. $u^2 - 2uv + v^2 = (u - v)^2$
DIFFERENCE OF SQUARES:	3. $u^2 - v^2 = (u - v)(u + v)$
DIFFERENCE OF CUBES:	4. $u^3 - v^3 = (u - v)(u^2 + uv + v^2)$
SUM OF CUBES:	5. $u^3 + v^3 = (u + v)(u^2 - uv + v^2)$

◆ EXAMPLE 13   Factor completely relative to the integers:

(A) $4m^2 - 12mn + 9n^2$ (B) $x^2 - 16y^2$ (C) $z^3 - 1$ (D) $m^3 + n^3$

Solutions   (A) $4m^2 - 12mn + 9n^2 = (2m - 3n)^2$

(B) $x^2 - 16y^2 = x^2 - (4y)^2 = (x - 4y)(x + 4y)$

(C) $z^3 - 1 = (z - 1)(z^2 + z + 1)$

(D) $m^3 + n^3 = (m + n)(m^2 - mn + n^2)$ ◆

PROBLEM 13   Factor completely relative to the integers:

(A) $x^2 + 6xy + 9y^2$ (B) $9x^2 - 4y^2$ (C) $8m^3 - 1$ (D) $x^3 + y^3z^3$ ◆

We complete this section by considering factoring that involves combinations of the preceding techniques.

Factoring Polynomials

Generally speaking, when asked to factor a polynomial, we first take out all factors common to all terms, if they are present, and then proceed as above until all factors are prime.

◆ EXAMPLE 14   Factor completely relative to the integers:

(A) $3x^3 - 48x$ (B) $3u^4 - 3u^3v - 9u^2v^2$ (C) $3m^4 - 24mn^3$
(D) $3x^4 - 5x^2 + 2$

Solutions   (A) $3x^3 - 48x = 3x(x^2 - 16) = 3x(x - 4)(x + 4)$

(B) $3u^4 - 3u^3v - 9u^2v^2 = 3u^2(u^2 - uv - 3v^2)$

(C) $3m^4 - 24mn^3 = 3m(m^3 - 8n^3) = 3m(m - 2n)(m^2 + 2mn + 4n^2)$

(D) $3x^4 - 5x^2 + 2 = (3x^2 - 2)(x^2 - 1) = (3x^2 - 2)(x - 1)(x + 1)$ ◆

PROBLEM 14   Factor completely relative to the integers:

(A) $18x^3 - 8x$ (B) $4m^3n - 2m^2n^2 + 2mn^3$ (C) $2t^4 - 16t$
(D) $2y^4 - 5y^2 - 12$ ◆

10. $2^2 \cdot 3^2 \cdot 5$ **11.** (A) $2xy(x^2 - 4xy - 3y^2)$ (B) $(2x - 7)(3x - 2)$

12. (A) $(x - 2)(x - 6)$ (B) Not factorable using integers
(C) $(2x - y)(x + 4y)$ (D) $(4x + y)(x - 4y)$

13. (A) $(x + 3y)^2$ (B) $(3x - 2y)(3x + 2y)$ (C) $(2m - 1)(4m^2 + 2m + 1)$
(D) $(x + yz)(x^2 - xyz + y^2z^2)$

14. (A) $2x(3x - 2)(3x + 2)$ (B) $2mn(2m^2 - mn + n^2)$
(C) $2t(t - 2)(t^2 + 2t + 4)$ (D) $(2y^2 + 3)(y - 2)(y + 2)$

E X E R C I S E 1-3

A Factor out, relative to the integers, all factors common to all terms.

1. $6m^4 - 9m^3 - 3m^2$ **2.** $6x^4 - 8x^3 - 2x^2$

3. $8u^3v - 6u^2v^2 + 4uv^3$ **4.** $10x^3y + 20x^2y^2 - 15xy^3$

5. $7m(2m - 3) + 5(2m - 3)$ **6.** $5x(x + 1) - 3(x + 1)$

7. $a(3c + d) - 4b(3c + d)$ **8.** $2w(y - 2z) - x(y - 2z)$

Factor completely relative to the integers. If a polynomial is prime relative to the integers, say so.

9. $3y^2 - y - 2$ **10.** $2x^2 + 5x - 3$ **11.** $u^2 - 2uv - 15v^2$

12. $x^2 - 4xy - 12y^2$ **13.** $m^2 - 6m - 3$ **14.** $x^2 + x - 4$

15. $w^2x^2 - y^2$ **16.** $25m^2 - 16n^2$ **17.** $9m^2 - 6mn + n^2$

18. $x^2 + 10xy + 25y^2$ **19.** $y^2 + 16$ **20.** $u^2 + 81$

B **21.** $4z^2 - 28z + 48$ **22.** $6x^2 + 48x + 72$

23. $2x^4 - 24x^3 + 40x^2$ **24.** $2y^3 - 22y^2 + 48y$

25. $4xy^2 - 12xy + 9x$ **26.** $16x^2y - 8xy + y$

27. $6m^2 - mn - 12n^2$ **28.** $6s^2 + 7st - 3t^2$

29. $4u^3v - uv^3$ **30.** $x^3y - 9xy^3$

31. $2x^3 - 2x^2 + 8x$ **32.** $3m^3 - 6m^2 + 15m$

33. $r^3 - t^3$ **34.** $m^3 + n^3$

35. $a^3 + 1$ **36.** $c^3 - 1$

C **37.** $(x + 2)^2 - 9y^2$ **38.** $(a - b)^2 - 4(c - d)^2$

39. $5u^2 + 4uv - 2v^2$ **40.** $3x^2 - 2xy - 4y^2$

41. $6(x - y)^2 + 23(x - y) - 4$ **42.** $4(A + B)^2 - 5(A + B) - 6$

43. $y^4 - 3y^2 - 4$ **44.** $m^4 - n^4$

45. $27a^2 + a^5b^3$ **46.** $s^4t^4 - 8st$

S E C T I O N 1-4   **Basic Operations on Rational Expressions**

◆ REDUCING TO LOWEST TERMS
◆ MULTIPLICATION AND DIVISION
◆ ADDITION AND SUBTRACTION
◆ COMPOUND FRACTIONS

We now turn our attention to fractional forms. A quotient of two algebraic expressions (division by 0 excluded) is called a **fractional expression.** If both the numerator and the denominator are polynomials, the fractional expression is called a **rational expression.** Some examples of rational expressions are

$$\frac{1}{x^3 + 2x} \qquad \frac{5}{x} \qquad \frac{x + 7}{3x^2 - 5x + 1} \qquad \frac{x^2 - 2x + 4}{1}$$

In this section we will discuss basic operations on rational expressions, including multiplication, division, addition, and subtraction.

Since variables represent real numbers in the rational expressions we will consider, the properties of real number fractions summarized in Section 1-1 will play a central role in much of the work that we will do.

Even though not always explicitly stated, we always assume that variables are restricted so that division by 0 is excluded.

◆ REDUCING TO LOWEST TERMS

Central to the process of reducing rational expressions to *lowest terms* is the *fundamental property of fractions,* which we restate here for convenient reference:

Fundamental Property of Fractions

If a, b, and k are real numbers with $b, k \neq 0$, then

$$\frac{ka}{kb} = \frac{a}{b} \qquad \frac{5 \cdot 2}{5 \cdot 7} = \frac{2}{7} \qquad \frac{x(x + 4)}{2(x + 4)} = \frac{x}{2}, \quad x \neq -4$$

Using this property from left to right to eliminate all common factors from the numerator and the denominator of a given fraction is referred to as **reducing a fraction to lowest terms.** We are actually dividing the numerator and denominator by the same nonzero common factor. Unless stated to the contrary, factors will be relative to the integers.

Using the property from right to left—that is, multiplying the numerator and denominator by the same nonzero factor—is referred to as **raising a fraction to higher terms.** We will use the property in both directions in the material that follows.

◆ E X A M P L E 15   Reduce each rational expression to lowest terms.

(A) $\dfrac{6x^2 + x - 1}{2x^2 - x - 1} = \dfrac{(2x + 1)(3x - 1)}{(2x + 1)(x - 1)}$

$\qquad\qquad = \dfrac{3x - 1}{x - 1}$

Factor numerator and denominator completely; divide numerator and denominator by $(2x + 1)$.

(B) $\dfrac{x^4 - 8x}{3x^3 - 2x^2 - 8x} = \dfrac{x(x-2)(x^2+2x+4)}{x(x-2)(3x+4)}$

$$= \dfrac{x^2 + 2x + 4}{3x + 4}$$

◆

PROBLEM 15 Reduce each rational expression to lowest terms.

(A) $\dfrac{x^2 - 6x + 9}{x^2 - 9}$ (B) $\dfrac{x^3 - 1}{x^2 - 1}$

◆

◆ MULTIPLICATION AND DIVISION

Since we are restricting variable replacements to real numbers, multiplication and division of rational expressions follow the rules for multiplying and dividing real number fractions summarized in Section 1-1.

■ Multiplication and Division

If a, b, c, and d are real numbers, then:

1. $\dfrac{a}{b} \cdot \dfrac{c}{d} = \dfrac{ac}{bd}$, $b, d \neq 0$ $\dfrac{3}{5} \cdot \dfrac{x}{x+5} = \dfrac{3x}{5(x+5)}$

2. $\dfrac{a}{b} \div \dfrac{c}{d} = \dfrac{a}{b} \cdot \dfrac{d}{c}$, $b, c, d \neq 0$ $\dfrac{3}{5} \div \dfrac{x}{x+5} = \dfrac{3}{5} \cdot \dfrac{x+5}{x}$

◆ EXAMPLE 16 Perform the indicated operations and reduce to lowest terms.

(A) $\dfrac{10x^3 y}{3xy + 9y} \cdot \dfrac{x^2 - 9}{4x^2 - 12x}$ Factor numerators and denominators; then divide any numerator and any denominator with a like common factor.

$$= \dfrac{\overset{5x^2}{\cancel{10x^3 y}}}{\underset{3 \cdot 1}{\cancel{3y(x+3)}}} \cdot \dfrac{\overset{1 \cdot 1}{\cancel{(x-3)(x+3)}}}{\underset{2 \cdot 1}{\cancel{4x(x-3)}}}$$

$$= \dfrac{5x^2}{6}$$

(B) $\dfrac{4 - 2x}{4} \div (x - 2) = \dfrac{\overset{1}{\cancel{2(2-x)}}}{\underset{2}{\cancel{4}}} \cdot \dfrac{1}{x - 2}$ $x - 2$ is the same as $\dfrac{x-2}{1}$

$$= \dfrac{2 - x}{2(x - 2)} = \dfrac{\overset{-1}{-\cancel{(x-2)}}}{\underset{1}{2\cancel{(x-2)}}}$$ $b - a = -(a - b)$, a useful change in some problems

$$= -\dfrac{1}{2}$$

(C) $\dfrac{2x^3 - 2x^2y + 2xy^2}{x^3y - xy^3} \div \dfrac{x^3 + y^3}{x^2 + 2xy + y^2}$

$$= \dfrac{\overset{2}{\cancel{2x}}\overset{1}{\cancel{(x^2 - xy + y^2)}}}{\underset{y}{\cancel{xy}}\cancel{(x + y)}(x - y)} \cdot \dfrac{\overset{1}{\cancel{(x + y)^2}}}{\underset{1}{\cancel{(x + y)}}\underset{1}{\cancel{(x^2 - xy + y^2)}}}$$

$$= \dfrac{2}{y(x - y)}$$

 ◆

PROBLEM 16 Perform the indicated operations and reduce to lowest terms.

(A) $\dfrac{12x^2y^3}{2xy^2 + 6xy} \cdot \dfrac{y^2 + 6y + 9}{3y^3 + 9y^2}$

(B) $(4 - x) \div \dfrac{x^2 - 16}{5}$

(C) $\dfrac{m^3 + n^3}{2m^2 + mn - n^2} \div \dfrac{m^3n - m^2n^2 + mn^3}{2m^3n^2 - m^2n^3}$

 ◆

◆ ADDITION AND SUBTRACTION

Again, because we are restricting variable replacements to real numbers, addition and subtraction of rational expressions follow the rules for adding and subtracting real number fractions.

▮ **Addition and Subtraction**

For a, b, and c real numbers:

1. $\dfrac{a}{b} + \dfrac{c}{b} = \dfrac{a + c}{b}$, $b \neq 0$ $\dfrac{x}{x + 5} + \dfrac{8}{x + 5} = \dfrac{x + 8}{x + 5}$

2. $\dfrac{a}{b} - \dfrac{c}{b} = \dfrac{a - c}{b}$, $b \neq 0$ $\dfrac{x}{3x^2y^2} - \dfrac{x + 7}{3x^2y^2} = \dfrac{x - (x + 7)}{3x^2y^2}$

Thus, we add rational expressions with the same denominators by adding or subtracting their numerators and placing the result over the common denominator. If the denominators are not the same, we raise the fractions to higher terms, using the fundamental property of fractions to obtain common denominators, and then proceed as described.

Even though any common denominator will do, our work will be simplified if the *least common denominator* (LCD) is used. Often, the LCD is obvious, but if it is not, the steps in the box at the top of the next page describe how to find it.

The LCD of two or more rational expressions is found as follows:

1. Factor each denominator completely.
2. Identify each different prime factor from all the denominators.
3. Form a product using each different factor to the highest power that occurs in any one denominator. This product is the LCD.

◆ E X A M P L E　17　Combine into a single fraction and reduce to lowest terms.

(A) $\dfrac{3}{10} + \dfrac{5}{6} - \dfrac{11}{45}$　(B) $\dfrac{4}{9x} - \dfrac{5x}{6y^2} + 1$　(C) $\dfrac{x+3}{x^2 - 6x + 9} - \dfrac{x+2}{x^2 - 9} - \dfrac{5}{3-x}$

Solutions　(A)　To find the LCD, factor each denominator completely:

$$\left.\begin{array}{l} 10 = 2 \cdot 5 \\ 6 = 2 \cdot 3 \\ 45 = 3^2 \cdot 5 \end{array}\right\} \quad \text{LCD} = 2 \cdot 3^2 \cdot 5 = 90$$

Now use the fundamental property of fractions to make each denominator 90:

$$\frac{3}{10} + \frac{5}{6} - \frac{11}{45} = \frac{9 \cdot 3}{9 \cdot 10} + \frac{15 \cdot 5}{15 \cdot 6} - \frac{2 \cdot 11}{2 \cdot 45}$$

$$= \frac{27}{90} + \frac{75}{90} - \frac{22}{90}$$

$$= \frac{27 + 75 - 22}{90} = \frac{80}{90} = \frac{8}{9}$$

(B)　$\left.\begin{array}{l} 9x = 3^2 x \\ 6y^2 = 2 \cdot 3y^2 \end{array}\right\} \quad \text{LCD} = 2 \cdot 3^2 xy^2 = 18xy^2$

$$\frac{4}{9x} - \frac{5x}{6y^2} + 1 = \frac{2y^2 \cdot 4}{2y^2 \cdot 9x} - \frac{3x \cdot 5x}{3x \cdot 6y^2} + \frac{18xy^2}{18xy^2}$$

$$= \frac{8y^2 - 15x^2 + 18xy^2}{18xy^2}$$

(C)　$\dfrac{x+3}{x^2 - 6x + 9} - \dfrac{x+2}{x^2 - 9} - \dfrac{5}{3-x} = \dfrac{x+3}{(x-3)^2} - \dfrac{x+2}{(x-3)(x+3)} + \dfrac{5}{x-3}$

Note: $-\dfrac{5}{3-x} = \dfrac{5}{-(3-x)} = \dfrac{5}{x-3}$　　We have again used the fact that $a - b = -(b - a)$.

The LCD $= (x - 3)^2(x + 3)$. Thus,

$$\frac{(x+3)^2}{(x-3)^2(x+3)} - \frac{(x-3)(x+2)}{(x-3)^2(x+3)} + \frac{5(x-3)(x+3)}{(x-3)^2(x+3)}$$

$$= \frac{(x^2+6x+9)-(x^2-x-6)+5(x^2-9)}{(x-3)^2(x+3)}$$

$$= \frac{x^2+6x+9-x^2+x+6+5x^2-45}{(x-3)^2(x+3)}$$

$$= \frac{5x^2+7x-30}{(x-3)^2(x+3)}$$
◆

PROBLEM 17 Combine into a single fraction and reduce to lowest terms.

(A) $\dfrac{5}{28} - \dfrac{1}{10} + \dfrac{6}{35}$ (B) $\dfrac{1}{4x^2} - \dfrac{2x+1}{3x^3} + \dfrac{3}{12x}$

(C) $\dfrac{y-3}{y^2-4} - \dfrac{y+2}{y^2-4y+4} - \dfrac{2}{2-y}$
◆

◆ COMPOUND FRACTIONS

A fractional expression with fractions in its numerator, denominator, or both is called a **compound fraction.** It is often necessary to represent a compound fraction as a **simple fraction**—that is (in all cases we will consider), as the quotient of two polynomials. The process does not involve any new concepts. It is a matter of applying old concepts and processes in the right sequence. We will illustrate two approaches to the problem, each with its own merits, depending on the particular problem under consideration. One of the methods makes very effective use of the fundamental property of fractions in the form

$$\frac{a}{b} = \frac{ka}{kb} \qquad b, k \neq 0$$

◆ EXAMPLE 18 Express as a simple fraction: $\dfrac{\dfrac{y}{x^2} - \dfrac{x}{y^2}}{\dfrac{y}{x} - \dfrac{x}{y}}$

Solution *Method 1.* Multiply the numerator and denominator by the LCD of all fractions in the numerator and denominator—in this case, x^2y^2. (We are multiplying by a form of 1.)

$$\frac{x^2y^2\left(\dfrac{y}{x^2} - \dfrac{x}{y^2}\right)}{x^2y^2\left(\dfrac{y}{x} - \dfrac{x}{y}\right)} = \frac{x^2y^2\dfrac{y}{x^2} - x^2y^2\dfrac{x}{y^2}}{x^2y^2\dfrac{y}{x} - x^2y^2\dfrac{x}{y}} = \frac{y^3 - x^3}{xy^3 - x^3y}$$

$$= \frac{\overset{1}{\cancel{(y-x)}}(y^2+xy+x^2)}{xy\underset{1}{\cancel{(y-x)}}(y+x)} = \frac{y^2+xy+x^2}{xy(y+x)} \quad \text{or} \quad \frac{x^2+xy+y^2}{xy(x+y)}$$

Method 2. Write the numerator and denominator as single fractions. Then treat as a quotient.

$$\frac{\dfrac{y}{x^2} - \dfrac{x}{y^2}}{\dfrac{y}{x} - \dfrac{x}{y}} = \frac{\dfrac{y^3 - x^3}{x^2 y^2}}{\dfrac{y^2 - x^2}{xy}} = \frac{y^3 - x^3}{x^2 y^2} \div \frac{y^2 - x^2}{xy}$$

$$= \frac{\overset{1}{\cancel{(y - x)}}(y^2 + xy + x^2)}{\underset{xy}{\cancel{x^2 y^2}}} \cdot \frac{\overset{1}{\cancel{xy}}}{\underset{1}{\cancel{(y - x)}}(y + x)}$$

$$= \frac{x^2 + xy + y^2}{xy(x + y)} \qquad \blacklozenge$$

PROBLEM 18   Express as a simple fraction reduced to lowest terms. Use the two methods described in Example 18.

$$\frac{\dfrac{a}{b} - \dfrac{b}{a}}{\dfrac{a}{b} + 2 + \dfrac{b}{a}} \qquad \blacklozenge$$

Answers to Matched Problems

15. (A) $\dfrac{x - 3}{x + 3}$ (B) $\dfrac{x^2 + x + 1}{x + 1}$ 16. (A) $2x$ (B) $\dfrac{-5}{x + 4}$ (C) mn

17. (A) $\dfrac{1}{4}$ (B) $\dfrac{3x^2 - 5x - 4}{12x^3}$ (C) $\dfrac{2y^2 - 9y - 6}{(y - 2)^2(y + 2)}$ 18. $\dfrac{a - b}{a + b}$

EXERCISE 1-4

A   *Perform the indicated operations and reduce answers to lowest terms.*

1. $\dfrac{d^5}{3a} \div \left(\dfrac{d^2}{6a^2} \cdot \dfrac{a}{4d^3}\right)$ 2. $\left(\dfrac{d^5}{3a} \div \dfrac{d^2}{6a^2}\right) \cdot \dfrac{a}{4d^3}$

3. $\dfrac{x^2}{12} + \dfrac{x}{18} - \dfrac{1}{30}$ 4. $\dfrac{2y}{18} - \dfrac{-1}{28} - \dfrac{y}{42}$

5. $\dfrac{4m - 3}{18m^3} + \dfrac{3}{4m} - \dfrac{2m - 1}{6m^2}$ 6. $\dfrac{3x + 8}{4x^2} - \dfrac{2x - 1}{x^3} - \dfrac{5}{8x}$

7. $\dfrac{x^2 - 9}{x^2 - 3x} \div (x^2 - x - 12)$ 8. $\dfrac{2x^2 + 7x + 3}{4x^2 - 1} \div (x + 3)$

9. $\dfrac{x^2 - 6x + 9}{x^2 - x - 6} \div \dfrac{x^2 + 2x - 15}{x^2 + 2x}$ 10. $\dfrac{m + n}{m^2 - n^2} \div \dfrac{m^2 - mn}{m^2 - 2mn + n^2}$

11. $\dfrac{3}{x^2 - 1} - \dfrac{2}{x^2 - 2x + 1}$ 12. $\dfrac{1}{a^2 - b^2} + \dfrac{1}{a^2 + 2ab + b^2}$

13. $\dfrac{x+1}{x-1} - 1$

14. $m - 3 - \dfrac{m-1}{m-2}$

15. $\dfrac{3}{a-1} - \dfrac{2}{1-a}$

16. $\dfrac{5}{x-3} - \dfrac{2}{3-x}$

17. $\dfrac{2x}{x^2-y^2} + \dfrac{1}{x+y} - \dfrac{1}{x-y}$

18. $\dfrac{2}{y+3} - \dfrac{1}{y-3} + \dfrac{2y}{y^2-9}$

B   *Perform the indicated operations and reduce answers to lowest terms. Represent any compound fractions as simple fractions reduced to lowest terms.*

19. $\dfrac{x^2}{x^2+2x+1} + \dfrac{x-1}{3x+3} - \dfrac{1}{6}$

20. $\dfrac{y}{y^2-y-2} - \dfrac{1}{y^2+5y-14} - \dfrac{2}{y^2+8y+7}$

21. $\dfrac{2-x}{2x+x^2} \cdot \dfrac{x^2+4x+4}{x^2-4}$

22. $\dfrac{9-m^2}{m^2+5m+6} \cdot \dfrac{m+2}{m-3}$

23. $\dfrac{c+2}{5c-5} - \dfrac{c-2}{3c-3} + \dfrac{c}{1-c}$

24. $\dfrac{x+7}{ax-bx} + \dfrac{y+9}{by-ay}$

25. $\left(\dfrac{x^3-y^3}{y^3} \cdot \dfrac{y}{x-y}\right) \div \dfrac{x^2+xy+y^2}{y^2}$

26. $\dfrac{x^2-16}{2x^2+10x+8} \div \dfrac{x^2-13x+36}{x^3+1}$

27. $\left(\dfrac{3}{x-2} - \dfrac{1}{x+1}\right) \div \dfrac{x+4}{x-2}$

28. $\left(\dfrac{x}{x^2-16} - \dfrac{1}{x+4}\right) \div \dfrac{4}{x+4}$

29. $\dfrac{1+\dfrac{3}{x}}{x-\dfrac{9}{x}}$

30. $\dfrac{1-\dfrac{y^2}{x^2}}{1-\dfrac{y}{x}}$

31. $\dfrac{\dfrac{1}{m^2}-1}{\dfrac{1}{m}+1}$

32. $\dfrac{\dfrac{1}{m}+1}{m+1}$

33. $\dfrac{c-d}{\dfrac{1}{c}-\dfrac{1}{d}}$

34. $\dfrac{\dfrac{1}{x}+\dfrac{1}{y}}{x+y}$

35. $\dfrac{\dfrac{x}{y}-2+\dfrac{y}{x}}{\dfrac{x}{y}-\dfrac{y}{x}}$

36. $\dfrac{1+\dfrac{2}{x}-\dfrac{15}{x^2}}{1+\dfrac{4}{x}-\dfrac{5}{x^2}}$

C   *Represent the compound fractions as simple fractions reduced to lowest terms.*

37. $\dfrac{\dfrac{s^2}{s-t}-s}{\dfrac{t^2}{s-t}+t}$

38. $\dfrac{y-\dfrac{y^2}{y-x}}{1+\dfrac{x^2}{y^2-x^2}}$

39. $1 - \dfrac{1}{1-\dfrac{1}{1-\dfrac{1}{x}}}$

40. $2 - \dfrac{1}{1-\dfrac{2}{a+2}}$

Integer Exponents and Square Root Radicals

◆ INTEGER EXPONENTS
◆ SCIENTIFIC NOTATION
◆ SQUARE ROOT RADICALS

We now consider basic operations on integer exponents, working with scientific notation, and basic operations on square root radicals.

◆ INTEGER EXPONENTS

Definitions for **integer exponents** are listed below.

▰ Definition of a^n

For n an integer and a a real number:

1. For n a positive integer,

$$a^n = a \cdot a \cdot \cdots \cdot a \qquad n \text{ factors of } a \qquad 5^4 = 5 \cdot 5 \cdot 5 \cdot 5$$

2. For $n = 0$,

$$a^0 = 1 \qquad a \neq 0 \qquad\qquad 12^0 = 1$$

0^0 is not defined.

3. For n a negative integer,

$$a^n = \frac{1}{a^{-n}} \qquad a \neq 0 \qquad\qquad a^{-3} = \frac{1}{a^{-(-3)}} = \frac{1}{a^3}$$

[If n is negative, then $(-n)$ is positive.]
Note: It can be shown that for *all* integers n,

$$a^{-n} = \frac{1}{a^n} \quad \text{and} \quad a^n = \frac{1}{a^{-n}} \quad a \neq 0 \qquad a^5 = \frac{1}{a^{-5}}, \quad a^{-5} = \frac{1}{a^5}$$

The following integer exponent properties are very useful in manipulating integer exponent forms.

▰ Exponent Properties

For n and m integers and a and b real numbers:

1. $a^m a^n = a^{m+n}$ $\qquad\qquad\qquad a^8 a^{-3} = a^{8+(-3)} = a^5$
2. $(a^n)^m = a^{mn}$ $\qquad\qquad\qquad (a^{-2})^3 = a^{3(-2)} = a^{-6}$
3. $(ab)^m = a^m b^m$ $\qquad\qquad\qquad (ab)^{-2} = a^{-2} b^{-2}$
4. $\left(\dfrac{a}{b}\right)^m = \dfrac{a^m}{b^m} \qquad b \neq 0 \qquad\qquad \left(\dfrac{a}{b}\right)^5 = \dfrac{a^5}{b^5}$
5. $\dfrac{a^m}{a^n} = a^{m-n} = \dfrac{1}{a^{n-m}} \qquad a \neq 0 \qquad \dfrac{a^{-3}}{a^7} = \dfrac{1}{a^{7-(-3)}} = \dfrac{1}{a^{10}}$

Exponent forms are frequently encountered in algebraic applications. You should sharpen your skills in using these forms by reviewing the above basic definitions and properties and the examples that follow.

◆ E X A M P L E   19   Simplify and express the answers using positive exponents only.

(A) $(2x^3)(3x^5) \boxed{= 2 \cdot 3x^{3+5}} = 6x^8$ (B) $x^5 x^{-9} = x^{-4} = \dfrac{1}{x^4}$

(C) $\dfrac{x^5}{x^7} \boxed{= x^{5-7}} = x^{-2} = \dfrac{1}{x^2}$ (D) $\dfrac{x^{-3}}{y^{-4}} = \dfrac{y^4}{x^3}$

or $\dfrac{x^5}{x^7} \boxed{= \dfrac{1}{x^{7-5}}} = \dfrac{1}{x^2}$

(E) $(u^{-3}v^2)^{-2} \boxed{= (u^{-3})^{-2}(v^2)^{-2}} = u^6 v^{-4} = \dfrac{u^6}{v^4}$

(F) $\left(\dfrac{y^{-5}}{y^{-2}}\right)^{-2} \boxed{= \dfrac{(y^{-5})^{-2}}{(y^{-2})^{-2}}} = \dfrac{y^{10}}{y^4} = y^6$

(G) $\dfrac{4m^{-3}n^{-5}}{6m^{-4}n^3} \boxed{= \dfrac{2m^{-3-(-4)}}{3n^{3-(-5)}}} = \dfrac{2m}{3n^8}$

(H) $\left(\dfrac{2x^{-3}x^3}{n^{-2}}\right)^{-3} = \left(\dfrac{2x^0}{n^{-2}}\right)^{-3} = \left(\dfrac{2}{n^{-2}}\right)^{-3} = \dfrac{2^{-3}}{n^6} = \dfrac{1}{2^3 n^6} = \dfrac{1}{8n^6}$ ◆

P R O B L E M   ⓵⑨   Simplify and express the answers using positive exponents only.

(A) $(3y^4)(2y^3)$ (B) $m^2 m^{-6}$ (C) $(u^3 v^{-2})^{-2}$

(D) $\left(\dfrac{y^{-6}}{y^{-2}}\right)^{-1}$ (E) $\dfrac{8x^{-2}y^{-4}}{6x^{-5}y^2}$ (F) $\left(\dfrac{3m^{-3}}{2x^2 x^{-2}}\right)^{-2}$ ◆

◆ SCIENTIFIC NOTATION

Writing and working with very large or very small numbers in standard decimal notation is often awkward, even with calculators. It is often convenient to represent numbers of this type in **scientific notation;** that is, as the product of a number between 1 and 10 and a power of 10.

◆ E X A M P L E   20

Decimal Fractions and Scientific Notation

$7 = 7 \times 10^0$ $0.5 = 5 \times 10^{-1}$
$67 = 6.7 \times 10$ $0.45 = 4.5 \times 10^{-1}$
$580 = 5.8 \times 10^2$ $0.003\ 2 = 3.2 \times 10^{-3}$
$43{,}000 = 4.3 \times 10^4$ $0.000\ 045 = 4.5 \times 10^{-5}$
$73{,}400{,}000 = 7.34 \times 10^7$ $0.000\ 000\ 391 = 3.91 \times 10^{-7}$ ◆

Note that the power of 10 used corresponds to the number of places we move the decimal to form a number between 1 and 10. The power is positive if the decimal is moved to the left and negative if it is moved to the right. Positive exponents are associated with numbers greater than or equal to 10; negative exponents are associated with positive numbers less than 1.

PROBLEM 20   Write each number in scientific notation.

(A) 370   (B) 47,300,000,000   (C) 0.047   (D) 0.000 000 089   ◆

◆ SQUARE ROOT RADICALS

To start, we define a **square root** of a number:

Definition of Square Root

x is a **square root** of y if $x^2 = y$.   2 is a square root of 4 since $2^2 = 4$.
    -2 is a square root of 4 since $(-2)^2 = 4$.

How many square roots of a real number are there? The following theorem, which we state without proof, answers the question.

THEOREM 5   **Square Roots of Real Numbers**

(A) Every positive real number has exactly two real square roots, each the negative of the other.
(B) Negative real numbers have no real number square roots (since no real number squared can be negative — think about this).
(C) The square root of 0 is 0.

Square Root Notation

For a a positive number:

$\sqrt{a}$ is the positive square root of a.

$-\sqrt{a}$ is the negative square root of a.

[Note: $\sqrt{-a}$ is not a real number.]

◆ E X A M P L E   21   (A) $\sqrt{4} = 2$ (B) $-\sqrt{4} = -2$ (C) $\sqrt{-4}$ is not a real number.   (D) $\sqrt{0} = 0$

◆

P R O B L E M   21   Evaluate, if possible.

(A) $\sqrt{9}$ (B) $-\sqrt{9}$ (C) $\sqrt{-9}$ (D) $\sqrt{-0}$ ◆

It can be shown that if a is a positive integer that is not the square of an integer, then

$$-\sqrt{a} \quad \text{and} \quad \sqrt{a}$$

are irrational numbers. Thus,

$$-\sqrt{7} \quad \text{and} \quad \sqrt{7}$$

name irrational numbers that are, respectively, the negative and positive square roots of 7.

■■ **Properties of Radicals**

For a and b nonnegative real numbers:

1. $\sqrt{a^2} = a$ 2. $\sqrt{a}\,\sqrt{b} = \sqrt{ab}$ 3. $\dfrac{\sqrt{a}}{\sqrt{b}} = \sqrt{\dfrac{a}{b}}$

To see that property 2 holds, let $N = \sqrt{a}$ and $M = \sqrt{b}$. Then $N^2 = a$ and $M^2 = b$. Hence,

$$\sqrt{a}\,\sqrt{b} = NM = \sqrt{(NM)^2} = \sqrt{N^2 M^2} = \sqrt{ab}$$

Note how properties of exponents are used. The proof of property 3 is left as an exercise.

◆ E X A M P L E   22   (A) $\sqrt{5}\,\sqrt{10} = \sqrt{5 \cdot 10} = \sqrt{50} = \sqrt{25 \cdot 2} = \sqrt{25}\,\sqrt{2} = 5\sqrt{2}$

(B) $\dfrac{\sqrt{32}}{\sqrt{8}} = \sqrt{\dfrac{32}{8}} = \sqrt{4} = 2$ (C) $\sqrt{\dfrac{7}{4}} = \dfrac{\sqrt{7}}{\sqrt{4}} = \dfrac{\sqrt{7}}{2}$ or $\tfrac{1}{2}\sqrt{7}$ ◆

P R O B L E M   22   Simplify as in Example 22.

(A) $\sqrt{3}\,\sqrt{6}$ (B) $\dfrac{\sqrt{18}}{\sqrt{2}}$ (C) $\sqrt{\dfrac{11}{9}}$ ◆

The foregoing definitions and theorems allow us to change algebraic expressions containing radicals to a variety of equivalent forms. One form that is often useful is called the *simplest radical form*, defined in the box on the next page.

An algebraic expression that contains square root radicals is in **simplest radical form** if all three of the following conditions are satisfied:

1. No radicand (the expression within the radical sign) when expressed in completely factored form contains a factor raised to a power greater than 1.
   $\sqrt{x^3}$ violates this condition.

2. No radical appears in a denominator.
   $3/\sqrt{5}$ violates this condition.

3. No fraction appears within a radical.
   $\sqrt{\frac{2}{3}}$ violates this condition.

It should be understood that forms other than the simplest radical form may be more useful on occasion. The situation dictates what form to choose.

◆ E X A M P L E   23   Change to simplest radical form. All variables represent positive real numbers.

(A) $\sqrt{8x^3}$ (B) $\dfrac{3x}{\sqrt{3}}$ (C) $\dfrac{2\sqrt{x}-1}{\sqrt{x}+2}$ (D) $\sqrt{\dfrac{3x}{8}}$

Solutions   (A) $\sqrt{8x^3}$ violates condition 1. Separate $8x^3$ into a perfect square part (2^2x^2) and what is left over $(2x)$; then use multiplication property 2:

$$\sqrt{8x^3} = \sqrt{(2^2x^2)(2x)}$$
$$= \sqrt{2^2x^2}\,\sqrt{2x}$$
$$= 2x\sqrt{2x}$$

(B) $3x/\sqrt{3}$ has a radical in the denominator; hence, it violates condition 2. To remove the radical from the denominator, we multiply the numerator and denominator by $\sqrt{3}$ to obtain $\sqrt{3^2}$ in the denominator (this is called **rationalizing a denominator**):

$$\frac{3x}{\sqrt{3}} = \frac{3x}{\sqrt{3}} \cdot \frac{\sqrt{3}}{\sqrt{3}}$$
$$= \frac{3x\sqrt{3}}{\sqrt{3^2}}$$
$$= \frac{3x\sqrt{3}}{3} = x\sqrt{3}$$

(C) $(2\sqrt{x}-1)/(\sqrt{x}+2)$ has a radical in the denominator. Multiplying the numerator and denominator by $\sqrt{x}$ does not remove all radicals from the denominator. However, remembering that $(a+b)(a-b)=a^2-b^2$ suggests that if we multiply the numerator and denominator by $\sqrt{x} - 2$, all radicals will disappear from the denominator (and the denominator will be rationalized).

$$\frac{2\sqrt{x} - 1}{\sqrt{x} + 2} = \frac{2\sqrt{x} - 1}{\sqrt{x} + 2} \cdot \frac{\sqrt{x} - 2}{\sqrt{x} - 2}$$

$$= \frac{2x - 5\sqrt{x} + 2}{x - 4}$$

(D) $\sqrt{3x/8}$ has a fraction within the radical; hence, it violates condition 3. To remove the fraction from the radical, we multiply the numerator and denominator of $3x/8$ by 2 to make the denominator a perfect square:

$$\sqrt{\frac{3x}{8}} = \sqrt{\frac{3x \cdot 2}{8 \cdot 2}}$$

$$= \sqrt{\frac{6x}{16}}$$

$$= \frac{\sqrt{6x}}{\sqrt{16}} = \frac{\sqrt{6x}}{4}$$
◆

PROBLEM 23   Change to simplest radical form. All variables represent positive real numbers.

(A) $\sqrt{18y^3}$ (B) $\dfrac{4xy}{\sqrt{2x}}$ (C) $\dfrac{2 + 3\sqrt{y}}{7 - \sqrt{y}}$ (D) $\sqrt{\dfrac{5y}{18x}}$
◆

Answers to Matched Problems

19. (A) $6y^7$ (B) $\dfrac{1}{m^4}$ (C) $\dfrac{v^4}{u^6}$ (D) y^4 (E) $\dfrac{4x^3}{3y^6}$ (F) $\dfrac{4m^6}{9}$

20. (A) 3.7×10^2 (B) 4.73×10^{10} (C) 4.7×10^{-2} (D) 8.9×10^{-8}

21. (A) 3   (B) -3 (C) Not a real number   (D) 0

22. (A) $3\sqrt{2}$ (B) 3   (C) $\sqrt{11}/3$ or $\frac{1}{3}\sqrt{11}$

23. (A) $3y\sqrt{2y}$ (B) $2y\sqrt{2x}$ (C) $\dfrac{14 + 23\sqrt{y} + 3y}{49 - y}$ (D) $\dfrac{\sqrt{10xy}}{6x}$ or $\dfrac{1}{6x}\sqrt{10xy}$

EXERCISE 1-5

A   *Simplify and express answers using positive exponents only. Variables are restricted to avoid division by 0.*

1. $2x^{-9}$ 2. $3y^{-5}$ 3. $\dfrac{3}{2w^{-7}}$ 4. $\dfrac{5}{4x^{-9}}$

5. $2x^{-8}x^5$ 6. $3c^{-9}c^4$ 7. $\dfrac{w^{-8}}{w^{-3}}$ 8. $\dfrac{m^{-11}}{m^{-5}}$

9. $5v^8v^{-8}$ 10. $7d^{-4}d^4$ 11. $(a^{-3})^2$ 12. $(b^4)^{-3}$
13. $(x^6y^{-3})^{-2}$ 14. $(a^{-3}b^4)^{-3}$

Express in simplest radical form. All variables represent positive real numbers.

15. $\sqrt{x^2}$ **16.** $\sqrt{m^2}$ **17.** $\sqrt{a^5}$ **18.** $\sqrt{m^7}$

19. $\sqrt{18x^4}$ **20.** $\sqrt{8x^3}$ **21.** $\dfrac{1}{\sqrt{m}}$ **22.** $\dfrac{1}{\sqrt{A}}$

23. $\sqrt{\dfrac{2}{3}}$ **24.** $\sqrt{\dfrac{3}{5}}$ **25.** $\sqrt{\dfrac{2}{x}}$ **26.** $\sqrt{\dfrac{3}{y}}$

Write in scientific notation.

27. 82,300,000,000   **28.** 5,380,000   **29.** 0.783

30. 0.019   **31.** 0.000 034   **32.** 0.000 000 007 832

B   *Simplify and express answers using positive exponents only. Write compound fractions as simple fractions.*

33. $(22 + 31)^0$ **34.** $(2x^3y^4)^0$ **35.** $\dfrac{10^{-3} \cdot 10^4}{10^{-11} \cdot 10^{-2}}$ **36.** $\dfrac{10^{-17} \cdot 10^{-5}}{10^{-3} \cdot 10^{-14}}$

37. $(5x^2y^{-3})^{-2}$ **38.** $(2m^{-3}n^2)^{-3}$ **39.** $\dfrac{8 \times 10^{-3}}{2 \times 10^{-5}}$ **40.** $\dfrac{18 \times 10^{12}}{6 \times 10^{-4}}$

41. $\dfrac{8x^{-3}y^{-1}}{6x^2y^{-4}}$ **42.** $\dfrac{9m^{-4}n^3}{12m^{-1}n^{-1}}$ **43.** $\left(\dfrac{6xy^{-2}}{3x^{-1}y^2}\right)^{-3}$ **44.** $\left(\dfrac{2x^{-3}y^2}{4xy^{-1}}\right)^{-2}$

45. $\dfrac{1-x}{x^{-1}-1}$ **46.** $\dfrac{1+x^{-1}}{1-x^{-2}}$ **47.** $\dfrac{u+v}{u^{-1}+v^{-1}}$ **48.** $\dfrac{x^{-1}-y^{-1}}{x-y}$

Write each problem in the form $ax^p + bx^q$ or $ax^p + bx^q + cx^r$, where a, b, and c are real numbers and p, q, and r are integers. For example,

$$\frac{2x^4 - 3x^2 + 1}{2x^3} \boxed{= \frac{2x^4}{2x^3} - \frac{3x^2}{2x^3} + \frac{1}{2x^3}} = x - \frac{3}{2}x^{-1} + \frac{1}{2}x^{-3}$$

49. $\dfrac{7x^5 - x^2}{4x^5}$ **50.** $\dfrac{5x^3 - 2}{3x^2}$ **51.** $\dfrac{3x^4 - 4x^2 - 1}{4x^3}$ **52.** $\dfrac{2x^3 - 3x^2 + x}{2x^2}$

Simplify and express answers in simplest radical form. All variables represent positive real numbers.

53. $\sqrt{18x^8y^5z^2}$ **54.** $\sqrt{8p^3q^2r^5}$ **55.** $\dfrac{12}{\sqrt{3x}}$ **56.** $\dfrac{10}{\sqrt{2y}}$

57. $\sqrt{\dfrac{6x}{7y}}$ **58.** $\sqrt{\dfrac{3m}{2n}}$ **59.** $\sqrt{\dfrac{4a^3}{3b}}$ **60.** $\sqrt{\dfrac{9m^5}{2n}}$

61. $\sqrt{18m^3n^4}\,\sqrt{2m^3n^2}$ **62.** $\sqrt{10x^3y}\,\sqrt{5xy}$ **63.** $\dfrac{\sqrt{4a^3}}{\sqrt{3b}}$ **64.** $\dfrac{\sqrt{9m^5}}{\sqrt{2n}}$

65. $\dfrac{5\sqrt{x}}{3 - 2\sqrt{x}}$ **66.** $\dfrac{3\sqrt{y}}{2\sqrt{y} - 3}$ **67.** $\dfrac{3\sqrt{2} - 2\sqrt{3}}{3\sqrt{3} - 2\sqrt{2}}$ **68.** $\dfrac{2\sqrt{5} + 3\sqrt{2}}{5\sqrt{5} + 2\sqrt{2}}$

Convert each numeral to scientific notation and simplify. Express the answer in scientific notation and in standard decimal form.

69. $\dfrac{9,600,000,000}{(1,600,000)(0.000\ 000\ 25)}$

70. $\dfrac{(60,000)(0.000\ 003)}{(0.000\ 4)(1,500,000)}$

71. $\dfrac{(1,250,000)(0.000\ 38)}{0.015\ 2}$

72. $\dfrac{(0.000\ 000\ 82)(230,000)}{(625,000)(0.008\ 2)}$

C Simplify and write answers using positive exponents only. Write compound fractions as simple fractions.

73. $\left[\left(\dfrac{x^{-2}y^3t}{x^{-3}y^{-2}t^2}\right)^2\right]^{-1}$

74. $\left[\left(\dfrac{u^3v^{-1}w^{-2}}{u^{-2}v^{-2}w}\right)^{-2}\right]^2$

75. $\left(\dfrac{2^2x^2y^0}{8x^{-1}}\right)^{-2}\left(\dfrac{x^{-3}}{x^{-5}}\right)^3$

76. $\left(\dfrac{3^3x^0y^{-2}}{2^3x^3y^{-5}}\right)^{-1}\left(\dfrac{3^3x^{-1}y}{2^2x^2y^{-2}}\right)^2$

77. $\dfrac{4(x-3)^{-4}}{8(x-3)^{-2}}$

78. $\dfrac{12(a+2b)^{-3}}{6(a+2b)^{-8}}$

79. $\dfrac{b^{-2}-c^{-2}}{b^{-3}-c^{-3}}$

80. $\dfrac{xy^{-2}-yx^{-2}}{y^{-1}-x^{-1}}$

Express in simplest radical form. All variables are restricted to avoid division by 0 and square roots of negative numbers.

81. $\dfrac{\sqrt{2x}\,\sqrt{5}}{\sqrt{20x}}$

82. $\dfrac{\sqrt{x}\,\sqrt{8y}}{\sqrt{12y}}$

83. $\dfrac{2}{\sqrt{x-2}}$

84. $\sqrt{\dfrac{1}{x-5}}$

Rationalize the numerators; that is, perform operations on the fractions that will eliminate radicals from the numerators.

85. $\dfrac{\sqrt{t}-\sqrt{x}}{t-x}$

86. $\dfrac{\sqrt{x}-\sqrt{y}}{\sqrt{x}+\sqrt{y}}$

87. $\dfrac{\sqrt{x+h}-\sqrt{x}}{h}$

88. $\dfrac{\sqrt{2+h}+\sqrt{2}}{h}$

SECTION 1-6 Rational Exponents and Radicals

◆ nTH ROOTS OF REAL NUMBERS
◆ RATIONAL EXPONENTS AND RADICALS
◆ PROPERTIES OF RADICALS

Square roots may now be generalized to nth roots, and the meaning of exponent may be generalized to include all rational numbers.

◆ nTH ROOTS OF REAL NUMBERS

Recall from Section 1-5 that r is a **square root** of b if $r^2 = b$. There is no reason to stop there. We may also say that r is a **cube root** of b if $r^3 = b$.

In general:

> For any natural number n:
>
> r is an **nth root** of b if $r^n = b$

How many real square roots of 16 exist? Of 7? Of -4? How many real 4th roots of 7 exist? Of -7? How many real cube roots of -8 exist? Of 11? Theorem 6 (which we state without proof) answers these questions.

THEOREM 6

> **Number of Real nth Roots of a Real Number b**
>
> | | n EVEN | n ODD |
> |---|---|---|
> | b POSITIVE: | Two real nth roots | One real nth root |
> | | -2 and 2 are both 4th roots of 16 | 2 is the only real cube root of 8 |
> | b NEGATIVE: | No real nth root | One real nth root |
> | | -4 has no real square roots | -2 is the only real cube root of -8 |
> | b ZERO: | One real nth root | One real nth root |
> | | The nth root of 0 is 0 for any natural number n. | |

On the basis of Theorem 6, we conclude that

7   has two real square roots, two real 4th roots, and so on.

10   has one real cube root, one real 5th root, and so on.

-13 has one real cube root, one real 5th root, and so on.

-8 has no real square roots, no real 4th roots, and so on.

◆ RATIONAL EXPONENTS AND RADICALS

We now turn to the question of what symbols to use to represent the various kinds of real nth roots. For a natural number n greater than 1 we use

$$b^{1/n} \qquad \text{or} \qquad \sqrt[n]{b}$$

to represent one of the **real nth roots of b.** Which one? The symbols represent the real nth root of b if n is odd and the positive real nth root of b if b is positive and n is even. The symbol $\sqrt[n]{b}$ is called an **nth root radical.** The number n is the index of the radical, and the number b is called the **radicand.** Note that we write simply $\sqrt{b}$ to indicate $\sqrt[2]{b}$.

◆ E X A M P L E 24

(A) $4^{1/2} = \sqrt{4} = 2$ $(\sqrt{4} \neq \pm 2)$ (B) $-4^{1/2} = -\sqrt{4} = -2$
(C) $(-4)^{1/2}$ and $\sqrt{-4}$ are not real numbers
(D) $8^{1/3} = \sqrt[3]{8} = 2$ (E) $(-8)^{1/3} = \sqrt[3]{-8} = -2$ ◆

P R O B L E M 24

Evaluate each of the following:

(A) $16^{1/2}$ (B) $-\sqrt{16}$ (C) $\sqrt[3]{-27}$ (D) $(-9)^{1/2}$ (E) $(\sqrt[4]{81})^3$ ◆

We now define b^r for any rational number $r = m/n$.

Rational Exponents

If m and n are natural numbers without common prime factors, b is a real number, and b is nonnegative when n is even, then

$$b^{m/n} = \begin{cases} (b^{1/n})^m = (\sqrt[n]{b})^m \\ (b^m)^{1/n} = \sqrt[n]{b^m} \end{cases}$$

$$8^{2/3} = (8^{1/3})^2 = (\sqrt[3]{8})^2 = 2^2 = 4$$

$$8^{2/3} = (8^2)^{1/3} = \sqrt[3]{8^2} = \sqrt[3]{64} = 4$$

and

$$b^{-m/n} = \frac{1}{b^{m/n}} \qquad b \neq 0 \qquad 8^{-2/3} = \frac{1}{8^{2/3}} = \frac{1}{4}$$

Note that the two definitions of $b^{m/n}$ are equivalent under the indicated restrictions on m, n, and b.

All the properties listed for integer exponents in Section 1-5 also hold for rational exponents, provided b is nonnegative when n is even. Unless stated to the contrary, all variables in the rest of the discussion represent positive real numbers.

◆ E X A M P L E 25

Change rational exponent form to radical form.

(A) $x^{1/7} = \sqrt[7]{x}$
(B) $(3u^2v^3)^{3/5} = \sqrt[5]{(3u^2v^3)^3}$ or $(\sqrt[5]{3u^2v^3})^3$ The first is usually preferred.

(C) $y^{-2/3} = \dfrac{1}{y^{2/3}} = \dfrac{1}{\sqrt[3]{y^2}}$ or $\sqrt[3]{y^{-2}}$ or $\sqrt[3]{\dfrac{1}{y^2}}$

Change radical form to rational exponent form.

(D) $\sqrt[5]{6} = 6^{1/5}$ (E) $-\sqrt[3]{x^2} = -x^{2/3}$
(F) $\sqrt{x^2 + y^2} = (x^2 + y^2)^{1/2}$ Note that $(x^2 + y^2)^{1/2} \neq x + y$. Why?   ◆

P R O B L E M   25   Convert to radical form.

(A) $u^{1/5}$ (B) $(6x^2y^5)^{2/9}$ (C) $(3xy)^{-3/5}$

Convert to rational exponent form.

(D) $\sqrt[4]{9u}$ (E) $-\sqrt[7]{(2x)^4}$ (F) $\sqrt[3]{x^3 + y^3}$ ◆

◆ E X A M P L E   26   Simplify each and express answers using positive exponents only. If rational exponents appear in final answers, convert to radical form.

(A) $(3x^{1/3})(2x^{1/2}) = 6x^{1/3+1/2} = 6x^{5/6} = 6\sqrt[6]{x^5}$

(B) $(-8)^{5/3} = [(-8)^{1/3}]^5 = (-2)^5 = -32$

(C) $(2x^{1/3}y^{-2/3})^3 = 8xy^{-2} = \dfrac{8x}{y^2}$

(D) $\left(\dfrac{4x^{1/3}}{x^{1/2}}\right)^{1/2} = \dfrac{4^{1/2}x^{1/6}}{x^{1/4}} = \dfrac{2}{x^{1/4-1/6}} = \dfrac{2}{x^{1/12}} = \dfrac{2}{\sqrt[12]{x}}$ ◆

P R O B L E M   26   Simplify each and express answers using positive exponents only. If rational exponents appear in final answers, convert to radical form.

(A) $9^{3/2}$ (B) $(-27)^{4/3}$ (C) $(5y^{1/4})(2y^{1/3})$ (D) $(2x^{-3/4}y^{1/4})^4$

(E) $\left(\dfrac{8x^{1/2}}{x^{2/3}}\right)^{1/3}$ ◆

◆ E X A M P L E   27   Multiply, and express answers using positive exponents only.

(A) $3y^{2/3}(2y^{1/3} - y^2)$ (B) $(2u^{1/2} + v^{1/2})(u^{1/2} - 3v^{1/2})$

Solutions   (A) $3y^{2/3}(2y^{1/3} - y^2)$ $\boxed{= 6y^{2/3+1/3} - 3y^{2/3+2}}$

$= 6y - 3y^{8/3}$

(B) $(2u^{1/2} + v^{1/2})(u^{1/2} - 3v^{1/2}) = 2u - 5u^{1/2}v^{1/2} - 3v$ ◆

P R O B L E M   27   Multiply, and express answers using positive exponents only.

(A) $2c^{1/4}(5c^3 - c^{3/4})$ (B) $(7x^{1/2} - y^{1/2})(2x^{1/2} + 3y^{1/2})$ ◆

◆ PROPERTIES OF RADICALS

Changing or simplifying radical expressions is aided by several properties of radicals that follow directly from the properties of exponents considered earlier.

If c, n, and m are natural numbers greater than or equal to 2, and if x and y are positive real numbers, then:

1. $\sqrt[n]{x^n} = x$ $\qquad$ $\sqrt[3]{x^3} = x$
2. $\sqrt[n]{xy} = \sqrt[n]{x}\,\sqrt[n]{y}$ $\qquad$ $\sqrt[5]{xy} = \sqrt[5]{x}\,\sqrt[5]{y}$
3. $\sqrt[n]{\dfrac{x}{y}} = \dfrac{\sqrt[n]{x}}{\sqrt[n]{y}}$ $\qquad$ $\sqrt[4]{\dfrac{x}{y}} = \dfrac{\sqrt[4]{x}}{\sqrt[4]{y}}$
4. $\sqrt[cn]{x^{cm}} = \sqrt[n]{x^m}$ $\qquad$ $\sqrt[12]{x^8} = \sqrt[4\cdot 3]{x^{4\cdot 2}} = \sqrt[3]{x^2}$

The properties of radicals provide us with the means of changing algebraic expressions containing radicals into a variety of equivalent forms. One particularly useful form is the *simplest radical form*. An algebraic expression that contains radicals is said to be in the **simplest radical form** if all four of the following conditions are satisfied:

■ **Simplest Radical Form**

1. A radicand contains no factor to a power greater than or equal to the index of the radical.
   $\sqrt[3]{x^5}$ violates this condition.

2. The power of the radicand and the index of the radical have no common factor other than 1.
   $\sqrt[6]{x^4}$ violates this condition.

3. No radical appears in a denominator.
   $y/\sqrt[3]{x}$ violates this condition.

4. No fraction appears within a radical.
   $\sqrt[4]{\dfrac{3}{5}}$ violates this condition.

◆ E X A M P L E $\quad$ 28 $\qquad$ Write in simplest radical form.

(A) $\sqrt[3]{x^3 y^6} = \sqrt[3]{(xy^2)^3} = xy^2$

$\qquad$ or $\quad \sqrt[3]{x^3 y^6} = (x^3 y^6)^{1/3} \;\boxed{= x^{3/3} y^{6/3}}\; = xy^2$

(B) $\sqrt[3]{32 x^8 y^3} = \sqrt[3]{(2^3 x^6 y^3)(4x^2)} \;\boxed{= \sqrt[3]{2^3 x^6 y^3}\,\sqrt[3]{4x^2}}$

$\qquad\qquad = 2x^2 y \sqrt[3]{4x^2}$

(C) $\dfrac{6x^2}{\sqrt[3]{9x}} = \dfrac{6x^2}{\sqrt[3]{9x}} \cdot \dfrac{\sqrt[3]{3x^2}}{\sqrt[3]{3x^2}} = \dfrac{6x^2\,\sqrt[3]{3x^2}}{\sqrt[3]{3^3x^3}} = \dfrac{6x^2\,\sqrt[3]{3x^2}}{3x} = 2x\,\sqrt[3]{3x^2}$

(D) $6\sqrt[4]{\dfrac{3}{4x^3}} = 6\sqrt[4]{\dfrac{3}{2^2x^3} \cdot \dfrac{2^2x}{2^2x}} = 6\sqrt[4]{\dfrac{12x}{2^4x^4}}$

$\qquad = 6\,\dfrac{\sqrt[4]{12x}}{\sqrt[4]{2^4x^4}} = 6\,\dfrac{\sqrt[4]{12x}}{2x} = \dfrac{3\,\sqrt[4]{12x}}{x}$

(E) $\sqrt[6]{16x^4y^2} = \sqrt[6]{(4x^2y)^2}$

$\qquad = \sqrt[2\cdot3]{(4x^2y)^{2\cdot1}}$

$\qquad = \sqrt[3]{4x^2y}$ ◆

Note that in Examples 28C and 28D, we **rationalized the denominators;** that is, we performed operations to remove radicals from the denominators. This is a useful operation in some problems.

PROBLEM   28   Write in simplest radical form.

(A) $\sqrt{12x^5y^6}$ (B) $\sqrt[3]{-27x^7y^5}$ (C) $\dfrac{8y^3}{\sqrt[4]{2y}}$ (D) $4x^2\sqrt[5]{\dfrac{y^2}{2x^3}}$ (E) $\sqrt[9]{8x^6y^3}$ ◆

Answers to Matched Problems

24. (A) 4   (B) -4 (C) -3 (D) Not a real number   (E) 27
25. (A) $\sqrt[5]{u}$ (B) $\sqrt[9]{(6x^2y^5)^2}$ or $(\sqrt[9]{6x^2y^5})^2$ (C) $1/\sqrt[5]{(3xy)^3}$ (D) $(9u)^{1/4}$
     (E) $-(2x)^{4/7}$ (F) $(x^3+y^3)^{1/3}$ (not $x+y$)
26. (A) 27   (B) 81   (C) $10y^{7/12} = 10\sqrt[12]{y^7}$ (D) $16y/x^3$
     (E) $2/x^{1/18} = 2/\sqrt[18]{x}$
27. (A) $10c^{13/4} - 2c$ (B) $14x + 19x^{1/2}y^{1/2} - 3y$
28. (A) $2x^2y^3\sqrt{3x}$ (B) $-3x^2y\sqrt[3]{xy^2}$ (C) $4y^2\sqrt[4]{8y^3}$ (D) $2x\sqrt[5]{16x^2y^2}$
     (E) $\sqrt[3]{2x^2y}$

EXERCISE   1-6

A   *Change to radical form; do not simplify.*

1. $6x^{3/5}$ 2. $7y^{2/5}$ 3. $(4xy^3)^{2/5}$ 4. $(7x^2y)^{5/7}$
5. $(x^2+y^2)^{1/2}$ 6. $x^{1/2} + y^{1/2}$

Change to rational exponent form; do not simplify.

7. $5\sqrt[4]{x^3}$ 8. $7m\sqrt[5]{n^2}$ 9. $\sqrt[5]{(2x^2y)^3}$ 10. $\sqrt[9]{(3m^4n)^2}$
11. $\sqrt[3]{x} + \sqrt[3]{y}$ 12. $\sqrt[3]{x^2 + y^3}$

Find rational number representations for each, if they exist.

13. $25^{1/2}$ 14. $64^{1/3}$ 15. $16^{3/2}$ 16. $16^{3/4}$
17. $-36^{1/2}$ 18. $-32^{3/5}$ 19. $(-36)^{1/2}$ 20. $(-32)^{3/5}$
21. $\left(\dfrac{4}{25}\right)^{3/2}$ 22. $\left(\dfrac{8}{27}\right)^{2/3}$ 23. $9^{-3/2}$ 24. $8^{-2/3}$

Simplify each expression and write answers using positive exponents only. All variables represent positive real numbers.

25. $x^{4/5}x^{-2/5}$

26. $y^{-3/7}y^{4/7}$

27. $\dfrac{m^{2/3}}{m^{-1/3}}$

28. $\dfrac{x^{1/4}}{x^{3/4}}$

29. $(8x^3y^{-6})^{1/3}$

30. $(4u^{-2}v^4)^{1/2}$

B

31. $\left(\dfrac{4x^{-2}}{y^4}\right)^{-1/2}$

32. $\left(\dfrac{w^4}{9x^{-2}}\right)^{-1/2}$

33. $\dfrac{8x^{-1/3}}{12x^{1/4}}$

34. $\dfrac{6a^{3/4}}{15a^{-1/3}}$

35. $\left(\dfrac{8x^{-4}y^3}{27x^2y^{-3}}\right)^{1/3}$

36. $\left(\dfrac{25x^5y^{-1}}{16x^{-3}y^{-5}}\right)^{1/2}$

Multiply, and express answers using positive exponents only.

37. $3x^{3/4}(4x^{1/4} - 2x^8)$

38. $2m^{1/3}(3m^{2/3} - m^6)$

39. $(3u^{1/2} - v^{1/2})(u^{1/2} - 4v^{1/2})$

40. $(a^{1/2} + 2b^{1/2})(a^{1/2} - 3b^{1/2})$

41. $(5m^{1/2} + n^{1/2})(5m^{1/2} - n^{1/2})$

42. $(2x^{1/2} - 3y^{1/2})(2x^{1/2} + 3y^{1/2})$

43. $(3x^{1/2} - y^{1/2})^2$

44. $(x^{1/2} + 2y^{1/2})^2$

Write each problem in the form $ax^p + bx^q$, where a and b are real numbers and p and q are rational numbers. For example:

$$\dfrac{2x^{1/3} + 4}{4x} \boxed{= \dfrac{2x^{1/3}}{4x} + \dfrac{4}{4x} = \dfrac{1}{2}x^{1/3-1} + x^{-1}} = \dfrac{1}{2}x^{-2/3} + x^{-1}$$

45. $\dfrac{x^{2/3} + 2}{2x^{1/3}}$

46. $\dfrac{12x^{1/2} - 3}{4x^{1/2}}$

47. $\dfrac{2x^{3/4} + 3x^{1/3}}{3x}$

48. $\dfrac{3x^{2/3} + x^{1/2}}{5x}$

49. $\dfrac{2x^{1/3} - x^{1/2}}{4x^{1/2}}$

50. $\dfrac{x^2 - 4x^{1/2}}{2x^{1/3}}$

Write in simplest radical form.

51. $\sqrt[3]{16m^4n^6}$

52. $\sqrt[3]{27x^7y^3}$

53. $\sqrt[4]{32m^9n^7}$

54. $\sqrt[5]{64u^{17}v^9}$

55. $\dfrac{x}{\sqrt[3]{x}}$

56. $\dfrac{u^2}{\sqrt[3]{u^2}}$

57. $\dfrac{4a^3b^2}{\sqrt[3]{2ab^2}}$

58. $\dfrac{8x^3y^5}{\sqrt[3]{4x^2y}}$

59. $\sqrt[4]{\dfrac{3x^3}{4}}$

60. $\sqrt[5]{\dfrac{3x^2}{2}}$

61. $\sqrt[12]{(x-3)^9}$

62. $\sqrt[8]{(t+1)^6}$

63. $\sqrt{x}\,\sqrt[3]{x^2}$

64. $\sqrt[3]{x}\,\sqrt{x}$

65. $\dfrac{\sqrt{x}}{\sqrt[3]{x^2}}$

66. $\dfrac{\sqrt{x}}{\sqrt[3]{x}}$

C Simplify by writing each expression as a simple fraction reduced to lowest terms and without negative exponents.

67. $\dfrac{(x-1)^{1/2} - x(\frac{1}{2})(x-1)^{-1/2}}{x-1}$

68. $\dfrac{(2x-1)^{1/2} - (x+2)(\frac{1}{2})(2x-1)^{-1/2}(2)}{2x-1}$

69. $\dfrac{(x+2)^{2/3} - x(\frac{2}{3})(x+2)^{-1/3}}{(x+2)^{4/3}}$

70. $\dfrac{2(3x-1)^{1/3} - (2x+1)(\frac{1}{3})(3x-1)^{-2/3}(3)}{(3x-1)^{2/3}}$

In Problems 71–76, evaluate using a calculator. (Refer to the instruction book for your calculator to see how exponential forms are evaluated.)

71. $22^{3/2}$ **72.** $15^{5/4}$ **73.** $827^{-3/8}$

74. $103^{-3/4}$ **75.** $37.09^{7/3}$ **76.** $2.876^{8/5}$

SECTION 1-7   Chapter Review

Important Terms and Symbols

1-1 *Algebra and Real Numbers.* Real number system; natural numbers; integers; rational numbers; irrational numbers; real numbers; real number line; coordinate; origin; positive and negative real numbers; associative, commutative, identity, and inverse properties for addition and multiplication; distributive properties; negative and reciprocal; subtraction and division; properties of negatives; zero properties; fractions and fraction properties; numerator; denominator

$$N, Z, Q, I, R; \quad -a; \quad \frac{1}{a}; \quad \frac{a}{b}$$

1-2 *Basic Operations on Polynomials.* Natural number exponent; base; first property of exponents; algebraic expression; polynomial; degree of a term; degree of a polynomial; monomial; binomial; trinomial; coefficient; like terms; addition and subtraction of polynomials; multiplication of polynomials; FOIL method; special products

$$b^n$$

1-3 *Factoring Polynomials.* Factor; prime number; composite number; factored completely; fundamental theorem of arithmetic; prime polynomial; common factors; factoring second-degree polynomials; perfect square; difference of squares; difference of cubes; sum of cubes

1-4 *Basic Operations on Rational Expressions.* Fractional expression; rational expression; fundamental property of fractions; reducing to lowest terms; raising to higher terms; multiplication and division of rational forms; addition and subtraction of rational forms; least common denominator; compound fraction; simple fraction

$$LCD$$

1-5 *Integer Exponents and Square Root Radicals.* Integer exponent; properties of integer exponents; scientific notation; square root; square root radical; properties of radicals; simplest radical form; rationalizing a denominator

$$a^n; \quad a^{-n}; \quad a^0; \quad \sqrt{a}$$

1-6 *Rational Exponents and Radicals.* nth root; nth root radical; index; radicand; rational exponent; properties of radicals; simplest radical form; rationalizing denominators

$$a^{1/n}; \quad \sqrt[n]{a}; \quad a^{m/n}; \quad \sqrt[n]{a^m}; \quad a^{-m/n}$$

E X E R C I S E   1-7 ## Chapter Review

Work through all the problems in this chapter review and check your answers in the back of the book. (Answers to all review problems are there.) Where weaknesses show up, review appropriate sections in the text.

A **1.** Replace each question mark with an appropriate expression that will illustrate the use of the indicated real number property:

(A) Commutative ($\cdot$): $x(y + z) = ?$
(B) Associative (+): $2 + (x + y) = ?$
(C) Distributive: $(2 + 3)x = ?$

Problems 2–6 refer to the following polynomials:

(A) $3x - 4$ (B) $x + 2$ (C) $3x^2 + x - 8$ (D) $x^3 + 8$

2. Add all four.
3. Subtract the sum of (A) and (C) from the sum of (B) and (D).
4. Multiply (C) and (D).
5. What is the degree of (D)?
6. What is the coefficient of the second term in (C)?

Perform the indicated operations and simplify.

7. $5x^2 - 3x[4 - 3(x - 2)]$ **8.** $(3m - 5n)(3m + 5n)$
9. $(2x + y)(3x - 4y)$ **10.** $(2a - 3b)^2$

Write each polynomial in a completely factored form relative to the integers. If the polynomial is prime relative to the integers, say so.

11. $9x^2 - 12x + 4$ **12.** $t^2 - 4t - 6$ **13.** $6n^3 - 9n^2 - 15n$

Perform the indicated operations and reduce to lowest terms. Represent all compound fractions as simple fractions reduced to lowest terms.

14. $\dfrac{2}{5b} - \dfrac{4}{3a^3} - \dfrac{1}{6a^2b^2}$ **15.** $\dfrac{3x}{3x^2 - 12x} + \dfrac{1}{6x}$

16. $\dfrac{y - 2}{y^2 - 4y + 4} \div \dfrac{y^2 + 2y}{y^2 + 4y + 4}$ **17.** $\dfrac{u - \dfrac{1}{u}}{1 - \dfrac{1}{u^2}}$

Simplify Problems 18–23 and write answers using positive exponents only. All variables represent positive real numbers.

18. $6(xy^3)^5$

19. $\dfrac{9u^8v^6}{3u^4v^8}$

20. $(2 \times 10^5)(3 \times 10^{-3})$

21. $(x^{-3}y^2)^{-2}$

22. $u^{5/3}u^{2/3}$

23. $(9a^4b^{-2})^{1/2}$

24. Change to radical form: $3x^{2/5}$
25. Change to rational exponent form: $-3\sqrt[3]{(xy)^2}$

Simplify Problems 26–30 and express in simplest radical form. All variables represent positive real numbers.

26. $3x\sqrt[3]{x^5y^4}$

27. $\sqrt{2x^2y^5}\sqrt{18x^3y^2}$

28. $\dfrac{6ab}{\sqrt{3a}}$

29. $\dfrac{\sqrt{5}}{3-\sqrt{5}}$

30. $\sqrt[8]{y^6}$

B In Problems 31–36, each statement illustrates the use of one of the following real number properties or definitions. Indicate which one.

Commutative $(+, \cdot)$ Associative $(+, \cdot)$ Distributive
Identity $(+, \cdot)$ Inverse $(+, \cdot)$ Subtraction
Division Negatives Zero

31. $(-7) - (-5) = (-7) + [-(-5)]$
32. $5u + (3v + 2) = (3v + 2) + 5u$
33. $(5m - 2)(2m + 3) = (5m - 2)2m + (5m - 2)3$
34. $9 \cdot (4y) = (9 \cdot 4)y$

35. $\dfrac{u}{-(v - w)} = -\dfrac{u}{v - w}$

36. $(x - y) + 0 = (x - y)$

37. Indicate true (T) or false (F):

    (A) A natural number is a rational number.
    (B) A number with a repeating decimal representation is an irrational number.

38. Give an example of an integer that is not a natural number.
39. Given the following algebraic expressions:

    (a) $2x^2 - 3x + 5$ (b) $x^2 - \sqrt{x - 3}$ (c) $x^{-3} + x^{-2} - 3x^{-1}$
    (d) $x^2 - 3xy - y^2$

    (A) Identify all second-degree polynomials.
    (B) Identify all third-degree polynomials.

Perform the indicated operations and simplify.

40. $(2x - y)(2x + y) - (2x - y)^2$
41. $(m^2 + 2mn - n^2)(m^2 - 2mn - n^2)$
42. $-2x\{(x^2 + 2)(x - 3) - x[x - x(3 - x)]\}$
43. $(x - 2y)^3$

Write in a completely factored form relative to the integers. If the polynomial is prime relative to the integers, say so.

44. $(4x - y)^2 - 9x^2$

45. $2x^2 + 4xy - 5y^2$

46. $6x^3y + 12x^2y^2 - 15xy^3$

47. $3x^3 + 24y^3$

Perform the indicated operations and reduce to lowest terms. Represent all compound fractions as simple fractions reduced to lowest terms.

48. $\dfrac{m - 1}{m^2 - 4m + 4} + \dfrac{m + 3}{m^2 - 4} + \dfrac{2}{2 - m}$

49. $\dfrac{y}{x^2} \div \left(\dfrac{x^2 + 3x}{2x^2 + 5x - 3} \div \dfrac{x^3y - x^2y}{2x^2 - 3x + 1} \right)$

50. $\dfrac{a^{-1} - b^{-1}}{ab^{-2} - ba^{-2}}$

Perform the indicated operations in Problems 51–56. Simplify, and write answers using positive exponents only. All variables represent positive real numbers.

51. $\left(\dfrac{8u^{-1}}{2^2 u^2 v^0} \right)^{-2} \left(\dfrac{u^{-5}}{u^{-3}} \right)^3$

52. $\dfrac{5^0}{3^2} + \dfrac{3^{-2}}{2^{-2}}$

53. $\left(\dfrac{27x^2 y^{-3}}{8x^{-4} y^3} \right)^{1/3}$

54. $(a^{-1/3} b^{1/4})(9a^{1/3} b^{-1/2})^{3/2}$

55. $(x^{1/2} + y^{1/2})^2$

56. $(3x^{1/2} - y^{1/2})(2x^{1/2} + 3y^{1/2})$

57. Convert to scientific notation and simplify: $\dfrac{0.000\ 000\ 000\ 52}{(1{,}300)(0.000\ 002)}$

Perform the indicated operations in Problems 58–64 and express answers in simplest radical form. All radicands represent positive real numbers.

58. $-2x \sqrt[5]{3^6 x^7 y^{11}}$

59. $\dfrac{2x^2}{\sqrt[3]{4x}}$

60. $\sqrt[5]{\dfrac{3y^2}{8x^2}}$

61. $\sqrt[9]{8x^6 y^{12}}$

62. $(2\sqrt{x} - 5\sqrt{y})(\sqrt{x} + \sqrt{y})$

63. $\dfrac{3\sqrt{x}}{2\sqrt{x} - \sqrt{y}}$

64. $\dfrac{2\sqrt{u} - 3\sqrt{v}}{2\sqrt{u} + 3\sqrt{v}}$

C   65. Rationalize the numerator: $\dfrac{\sqrt{t} - \sqrt{5}}{t - 5}$

66. Write in the form $ax^p + bx^q$, where a and b are real numbers and p and q are rational numbers:

$$\dfrac{4\sqrt{x} - 3}{2\sqrt{x}}$$

67. Evaluate $x^2 - 4x + 1$ for $x = 2 - \sqrt{3}$.

68. Simplify: $x(2x - 1)(x + 3) - (x - 1)^3$

CALCULATOR PROBLEMS

Evaluate each expression using a calculator.

69. $\dfrac{(20,410)(0.000\ 003\ 477)}{0.000\ 000\ 022\ 09}$

70. 0.1347^5

71. $(-60.39)^{-3}$

72. $82.45^{8/3}$

73. $\sqrt[5]{0.006\ 604}$

74. $\sqrt[3]{3 + \sqrt{2}}$

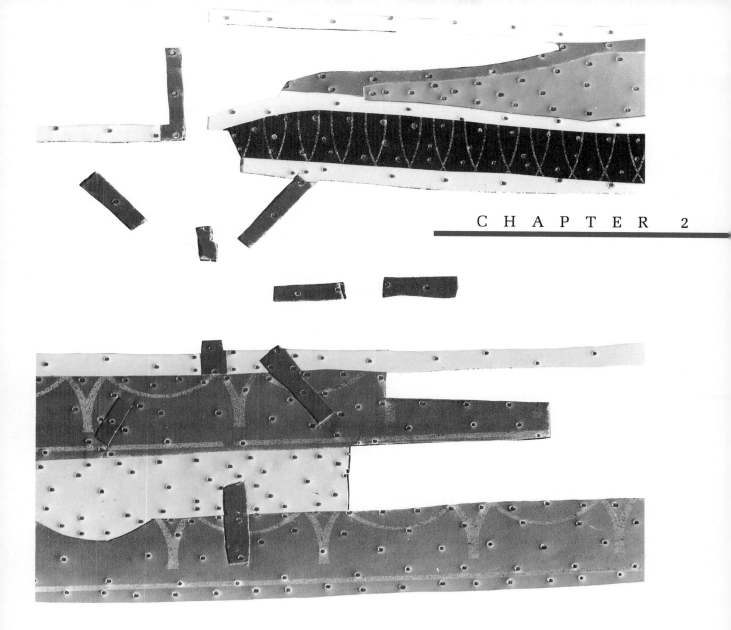

Equations, Graphs, and Functions

Contents

One of the important uses of algebra is the study of relationships between quantities. Many of these relationships can be expressed in terms of equations or inequalities. Additional insight is often obtained by using a graph to illustrate relationships. In this chapter, we will look at techniques for solving linear equations, linear inequalities, and quadratic equations, and we will discuss methods for drawing graphs in one and two dimensions. We will also introduce the important concept of a function and examine the properties of some specific types of functions. Throughout the chapter, we will consider applied problems that can be solved using these techniques.

SECTION 2-1

Linear Equations and Inequalities in One Variable

◆ LINEAR EQUATIONS
◆ LINEAR INEQUALITIES
◆ APPLICATIONS

The equation

$$3 - 2(x + 3) = \frac{x}{3} - 5$$

and the inequality

$$\frac{x}{2} + 2(3x - 1) \geq 5$$

are both first-degree in one variable. In general, a **first-degree, or linear, equation** in one variable is any equation that can be written in the form

STANDARD FORM $\qquad ax + b = 0 \qquad a \neq 0$ $\qquad\qquad$ (1)

If the equality symbol, $=$, in (1) is replaced by $<$, $>$, $\leq$, or $\geq$, then the resulting expression is called a **first-degree, or linear, inequality.**

A **solution** of an equation (or inequality) involving a single variable is a number that when substituted for the variable makes the equation (or inequality)

true. The set of all solutions is called the **solution set.** When we say that we **solve an equation** (or inequality), we mean that we find its solution set.

Knowing what is meant by the solution set is one thing; finding it is another. We start by recalling the idea of equivalent equations and equivalent inequalities. If we perform an operation on an equation (or inequality) that produces another equation (or inequality) with the same solution set, then the two equations (or inequalities) are said to be **equivalent.** The basic idea in solving equations and inequalities is to perform operations on these forms that produce simpler equivalent forms, and to continue the process until we obtain an equation or inequality with an obvious solution.

◆ LINEAR EQUATIONS

The following properties of equality produce equivalent equations when applied:

Equality Properties

For a, b, and c real numbers:

ADDITION PROPERTY:	1. If $a = b$,	then $a + c = b + c$.
SUBTRACTION PROPERTY:	2. If $a = b$,	then $a - c = b - c$.
MULTIPLICATION PROPERTY:	3. If $a = b$,	then $ca = cb$, $c \neq 0$.
DIVISION PROPERTY:	4. If $a = b$,	then $\dfrac{a}{c} = \dfrac{b}{c}$, $c \neq 0$.

Several examples should remind you of the process of solving equations.

◆ E X A M P L E   1   Solve: $8x - 3(x - 4) = 3(x - 4) + 6$

Solution
$$8x - 3(x - 4) = 3(x - 4) + 6$$
$$8x - 3x + 12 = 3x - 12 + 6$$
$$5x + 12 = 3x - 6$$
$$2x = -18$$
$$x = -9 \qquad \qquad ◆$$

P R O B L E M (1)   Solve: $3x - 2(2x - 5) = 2(x + 3) - 8$ ◆

◆ E X A M P L E   2   What operations can we perform on

$$\frac{x + 2}{2} - \frac{x}{3} = 5$$

to eliminate the denominators? If we can find a number that is exactly divisible by each denominator, then we can use the multiplication property of equality to clear the denominators. The LCD (least common denominator) of the fractions,

6, is exactly what we are looking for! Actually, any common denominator will do, but the LCD results in a simpler equivalent equation. Thus, we multiply both sides of the equation by 6:

$$6\left(\frac{x+2}{2} - \frac{x}{3}\right) = 6 \cdot 5$$

$$\overset{3}{6} \cdot \frac{(x+2)}{\underset{1}{2}} - \overset{2}{6} \cdot \frac{x}{\underset{1}{3}} = 30$$

$$3(x+2) - 2x = 30$$

$$3x + 6 - 2x = 30$$

$$x = 24 \qquad \blacklozenge$$

PROBLEM 2   Solve: $\dfrac{x+1}{3} - \dfrac{x}{4} = \dfrac{1}{2}$ $\blacklozenge$

In many applications of algebra, formulas or equations must be changed to alternate equivalent forms. The following examples are typical.

◆ EXAMPLE 3   Solve the amount formula for simple interest, $A = P + Prt$, for:

(A) r in terms of the other variables
(B) P in terms of the other variables

Solutions   (A) $A = P + Prt$ Reverse equation.

         $P + Prt = A$ Now isolate r on the left side.

            $Prt = A - P$ Divide both members by Pt.

$$r = \frac{A - P}{Pt}$$

(B) $A = P + Prt$ Reverse equation.

         $P + Prt = A$ Factor out P (note the use of the distributive property).

       $P(1 + rt) = A$ Divide by $(1 + rt)$.

$$P = \frac{A}{1 + rt} \qquad \blacklozenge$$

PROBLEM 3   Solve $M = Nt + Nr$ for:

(A) t (B) N $\blacklozenge$

◆ LINEAR INEQUALITIES

Before we start solving linear inequalities, let us recall what we mean by $<$ (less than) and $>$ (greater than). If a and b are real numbers, then we write

$a < b$ a is less than b

if there exists a positive number p such that $a + p = b$. Certainly, we would expect that if a positive number was added to any real number, the sum would be larger than the original. That is essentially what the definition states. If $a < b$, we may also write

$b > a$ b is greater than a

◆ **E X A M P L E   4**   (A) $3 < 5$ Since $3 + 2 = 5$

(B) $-6 < -2$ Since $-6 + 4 = -2$

(C) $0 > -10$ Since $-10 < 0$ ◆

P R O B L E M   4   Replace each question mark with either $<$ or $>$.

(A) $2 \,?\, 8$ (B) $-20 \,?\, 0$ (C) $-3 \,?\, -30$ ◆

The inequality symbols have a very clear geometric interpretation on the real number line. If $a < b$, then a is to the left of b on the number line; if $c > d$, then c is to the right of d (Fig. 1).

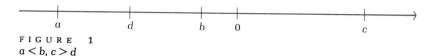

F I G U R E   1
$a < b, c > d$

Now let us turn to the problem of solving linear inequalities in one variable. Recall that a **solution** of an inequality involving one variable is a number that, when substituted for the variable, makes the inequality true. The set of all solutions is called the **solution set**. When we say that we **solve an inequality,** we mean that we find its solution set. The procedures used to solve linear inequalities in one variable are almost the same as those used to solve linear equations in one variable but with two important exceptions (as noted below). The following properties of inequalities produce equivalent inequalities when applied:

Inequality Properties

For a, b, and c real numbers:

1. If $a > b$,   then $a + c > b + c$.
2. If $a > b$,   then $a - c > b - c$.
3. If $a > b$ and c is positive,   then $ca > cb$.$\left.\begin{array}{}\\ \end{array}\right\}$ Note difference
4. If $a > b$ and c is negative,   then $ca < cb$.

5. If $a > b$ and c is positive,   then $\dfrac{a}{c} > \dfrac{b}{c}$.$\left.\begin{array}{}\\ \\ \end{array}\right\}$ Note difference

6. If $a > b$ and c is negative,   then $\dfrac{a}{c} < \dfrac{b}{c}$.

Similar properties hold if each inequality sign is reversed or if $>$ is replaced with $\geq$ **(greater than or equal to)** and $<$ is replaced with $\leq$ **(less than or equal to).** Thus, we can perform essentially the same operations on inequalities that we perform on equations, with the exception that **the sense of the inequality reverses if we multiply or divide both sides by a negative number.** Otherwise, the sense of the inequality does not change. For example, if we start with the true statement

$$-3 > -7$$

and multiply both sides by 2, we obtain

$$-6 > -14$$

and the sense of the inequality stays the same. But if we multiply both sides of $-3 > -7$ by -2, then the left side becomes 6 and the right side becomes 14, so we must write

$$6 < 14$$

to have a true statement. Thus, the sense of the inequality reverses.

If $a < b$, the double inequality $a < x < b$ means that $x > a$ **and** $x < b$; that is, x is between a and b. Other variations, as well as a useful interval notation, are

T A B L E   1

INTERVAL NOTATION	INEQUALITY NOTATION	LINE GRAPH
$[a, b]$	$a \leq x \leq b$	
$[a, b)$	$a \leq x < b$	
$(a, b]$	$a < x \leq b$	
(a, b)	$a < x < b$	
$(-\infty, a]$	$x \leq a$	
$(-\infty, a)$	$x < a$	
$[b, \infty)^*$	$x \geq b$	
(b, ∞)	$x > b$	

* The symbol ∞ (read "infinity") is not a number. When we write $[b, \infty)$, we are simply referring to the interval starting at b and continuing indefinitely to the right. We would never write $[b, \infty]$.

given in Table 1. Note that an end point on a line graph has a square bracket through it if it is included in the inequality and a parenthesis through it if it is not.

◆ E X A M P L E   5

Solve and graph: $2(2x + 3) < 6(x - 2) + 10$

Solution

$$2(2x + 3) < 6(x - 2) + 10$$
$$4x + 6 < 6x - 12 + 10$$
$$4x + 6 < 6x - 2$$
$$-2x + 6 < -2$$
$$-2x < -8$$
$$x > 4 \quad \text{or} \quad (4, \infty)$$

Notice that the sense of the inequality reverses when we divide both sides by -2.

Notice that in the graph of $x > 4$, we use a parenthesis through 4, since the point 4 is not included in the graph.   ◆

P R O B L E M   5

Solve and graph: $3(x - 1) \leq 5(x + 2) - 5$ ◆

◆ E X A M P L E   6

Solve and graph: $-3 < 2x + 3 \leq 9$

Solution

We are looking for all numbers x such that $2x + 3$ is between -3 and 9, including 9 but not -3. We proceed as above except that we try to isolate x in the middle:

$$-3 < 2x + 3 \leq 9$$
$$-3 - 3 < 2x + 3 - 3 \leq 9 - 3$$
$$-6 < 2x \leq 6$$
$$\frac{-6}{2} < \frac{2x}{2} \leq \frac{6}{2}$$
$$-3 < x \leq 3 \quad \text{or} \quad (-3, 3]$$

◆

P R O B L E M   6

Solve and graph: $-8 \leq 3x - 5 < 7$ ◆

Note that a linear equation usually has exactly one solution, while a linear inequality usually has infinitely many solutions.

◆ APPLICATIONS

To realize the full potential of algebra, we must be able to translate real-world problems into mathematical forms. In short, we must be able to do *word problems*.

◆ **EXAMPLE 7**

Break-Even Analysis

It costs a record company $9,000 to prepare a record album—recording costs, album design costs, etc. These costs represent a one-time **fixed cost.** Manufacturing, marketing, and royalty costs (all **variable costs**) are $3.50 per album. If the album is sold to record shops for $5 each, how many albums must be sold for the company to **break even?**

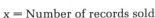

means
divide by

Solution

Let

x = Number of records sold

C = Cost for producing x records

R = Revenue (return) on sales of x records

The company breaks even if $R = C$, with

C = Fixed costs + Variable costs

= \$9,000 + \$3.50x

R = \$5x

Find x such that $R = C$; that is, find x such that

$$5x = 9,000 + 3.5x$$
$$1.5x = 9,000$$
$$x = 6,000$$

Check

For $x = 6,000$,

$$C = 9,000 + 3.5x \qquad \text{and} \qquad R = 5x$$
$$= 9,000 + 3.5(6,000) \qquad\qquad = 5(6,000)$$
$$= \$30,000 \qquad\qquad\qquad = \$30,000$$

Thus, the company must sell 6,000 records to break even; any sales over 6,000 will produce a profit; any sales under 6,000 will result in a loss. ◆

PROBLEM 7

What is the break-even point in Example 7 if fixed costs are $9,900, variable costs are $3.70 per record, and the records are sold for $5.50 each? ◆

Algebra has many different types of applications—so many, in fact, that no single approach applies to all. However, the following suggestions may help you get started:

◆ E X A M P L E 8

Consumer Price Index (CPI)

Solution

Table 2 lists the consumer price index (CPI) for several years. What net monthly salary in 1980 would have the same purchasing power as a net monthly salary of $900 in 1950? Compute the answer to the nearest dollar.

To have the same purchasing power, the ratio of a salary in 1980 to a salary in 1950 would have to be the same as the ratio of the CPI in 1980 to the CPI in 1950. Thus, if x is the net monthly salary in 1980, we solve the equation

$$\frac{x}{900} = \frac{247}{72}$$

$$x = 900 \cdot \frac{247}{72}$$

$$= \$3,088 \text{ per month}$$

◆

T A B L E 2

CPI (1967 = 100)

YEAR	INDEX	YEAR	INDEX
1950	72	1970	116
1955	80	1975	161
1960	89	1980	247
1965	95	1985	322

P R O B L E M 8

Using Table 2, what net monthly salary in 1960 would have the same purchasing power as a net monthly salary of $2,000 in 1975? Compute the answer to the nearest dollar.

◆

Answers to Matched Problems

1. $x = 4$ 2. $x = 2$ 3. (A) $t = \dfrac{M - Nr}{N}$ (B) $N = \dfrac{M}{t + r}$

4. (A) <   (B) <   (C) >   5. $x \geq -4$ or $[-4, \infty)$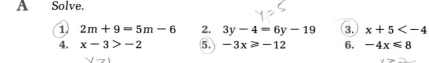

6. $-1 \leq x < 4$ or $[-1, 4)$ 7. $x = 5,500$ 8. $1,106

E X E R C I S E 2-1

A *Solve.*

1. $2m + 9 = 5m - 6$ 2. $3y - 4 = 6y - 19$ 3. $x + 5 < -4$
4. $x - 3 > -2$ 5. $-3x \geq -12$ 6. $-4x \leq 8$

Solve and graph.

7. $-4x - 7 > 5$

8. $-2x + 8 < 4$

   $x > 2$

9. $2 \le x + 3 \le 5$

10. $-3 < y - 5 < 8$

    $2 < y < 13$

Solve.

11. $\dfrac{y}{7} - 1 = \dfrac{1}{7}$

12. $\dfrac{m}{5} - 2 = \dfrac{3}{5}$ $m = 13$

13. $\dfrac{x}{3} > -2$

14. $\dfrac{y}{-2} \le -1$

    $y \ge 2$

15. $\dfrac{y}{3} = 4 - \dfrac{y}{6}$

16. $\dfrac{x}{4} = 9 - \dfrac{x}{2}$ $x = 12$

B

17. $10x + 25(x - 3) = 275$

18. $-3(4 - x) = 5 - (x + 1)$

19. $3 - y \le 4(y - 3)$

20. $x - 2 \ge 2(x - 5)$

21. $\dfrac{x}{5} - \dfrac{x}{6} = \dfrac{6}{5}$

22. $\dfrac{y}{4} - \dfrac{y}{3} = \dfrac{1}{2}$

23. $\dfrac{m}{5} - 3 < \dfrac{3}{5} - m$

24. $u - \dfrac{2}{3} > \dfrac{u}{3} + 2$

25. $0.1(x - 7) + 0.05x = 0.8$

26. $0.4(u + 5) - 0.3u = 17$

Solve and graph.

27. $2 \le 3x - 7 < 14$

28. $-4 \le 5x + 6 < 21$

29. $-4 \le \frac{9}{5}C + 32 \le 68$

30. $-1 \le \frac{2}{3}t + 5 \le 11$

C   *Solve for the indicated variable.*

31. $3x - 4y = 12$,   for y

32. $y = -\frac{2}{3}x + 8$,   for x

33. $Ax + By = C$,   for y $(B \ne 0)$

34. $y = mx + b$,   for m

35. $F = \frac{9}{5}C + 32$,   for C

36. $C = \frac{5}{9}(F - 32)$,   for F

37. $A = Bm - Bn$,   for B

38. $U = 3C - 2CD$,   for C

Solve and graph.

39. $-3 \le 4 - 7x < 18$

40. $-1 < 9 - 2u \le 5$

APPLICATIONS

Business & Economics

41. A jazz concert brought in $60,000 on the sale of 8,000 tickets. If the tickets sold for $6 and $10 each, how many of each type of ticket were sold?

42. An all-day parking meter takes only dimes and quarters. If it contains 100 coins with a total value of $14.50, how many of each type of coin are in the meter?

43. You have $12,000 to invest. If part is invested at 10% and the rest at 15%, how much should be invested at each rate to yield 12% on the total amount?

44. An investor has $20,000 to invest. If part is invested at 8% and the rest at 12%, how much should be invested at each rate to yield 11% on the total amount?

45. *Inflation.* If the price change of cars parallels the change in the CPI (see Table 2 in Example 8), what would a car sell for in 1980 if a comparable model sold for $3,000 in 1965?

46. *Break-even analysis.* For a business to realize a profit, it is clear that revenue R must be greater than costs C; that is, a profit will result only if $R > C$ (the company breaks even when $R = C$). A record manufacturer has a weekly cost equation $C = 300 + 1.5x$ and a revenue equation $R = 2x$, where x is the number of records produced and sold in a week. How many records must be sold for the company to make a profit?

Life Sciences

47. *Wildlife management.* A naturalist for a fish and game department estimated the total number of rainbow trout in a certain lake using the popular capture–mark–recapture technique. He netted, marked, and released 200 rainbow trout. A week later, allowing for thorough mixing, he again netted 200 trout and found 8 marked ones among them. Assuming that the proportion of marked fish in the second sample was the same as the proportion of all marked fish in the total population, estimate the number of rainbow trout in the lake.

48. *Ecology.* If the temperature for a 24 hour period at an Antarctic station ranged between $-49°F$ and $14°F$ (that is, $-49 \leqslant F \leqslant 14$), what was the range in degrees Celsius? [*Note:* $F = \frac{9}{5}C + 32$.]

Social Sciences

49. *Psychology.* The IQ (intelligence quotient) is found by dividing the mental age (MA), as indicated on standard tests, by the chronological age (CA) and multiplying by 100. For example, if a child has a mental age of 12 and a chronological age of 8, the calculated IQ is 150. If a 9-year-old girl has an IQ of 140, compute her mental age.

50. *Anthropology.* In their study of genetic groupings, anthropologists use a ratio called the *cephalic index.* This is the ratio of the width of the head to its length (looking down from above) expressed as a percentage. Symbolically,

$$C = \frac{100W}{L}$$

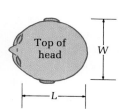

where C is the cephalic index, W is the width, and L is the length. If an Indian tribe in Baja California (Mexico) had an average cephalic index of 66 and the average width of their heads was 6.6 inches, what was the average length of their heads?

Quadratic Equations

- ◆ SOLUTION BY SQUARE ROOT
- ◆ SOLUTION BY FACTORING
- ◆ QUADRATIC FORMULA

A **quadratic equation** in one variable is any equation that can be written in the form

$$ax^2 + bx + c = 0 \qquad a \neq 0$$

where x is a variable and a, b, and c are constants. We will refer to this form as the **standard form.** The equations

$$5x^2 - 3x + 7 = 0 \qquad \text{and} \qquad 18 = 32t^2 - 12t$$

are both quadratic equations, since they are either in the standard form or can be transformed into this form.

We will restrict our review to finding real solutions to quadratic equations.

◆ SOLUTION BY SQUARE ROOT

The easiest type of quadratic equation to solve is the special form where the first-degree term is missing:

$$ax^2 + c = 0 \qquad a \neq 0$$

The method makes use of the definition of square root given in Section 1-5.

◆ E X A M P L E   9   Solve by the square root method.

(A) $x^2 - 7 = 0$ (B) $2x^2 - 10 = 0$ (C) $3x^2 + 27 = 0$

Solutions   (A) $x^2 - 7 = 0$

$\qquad\qquad x^2 = 7 \qquad$ What real number squared is 7?

$\qquad\qquad x = \pm\sqrt{7} \qquad$ Short for $\sqrt{7}$ and $-\sqrt{7}$

(B) $2x^2 - 10 = 0$

$\qquad\qquad 2x^2 = 10$

$\qquad\qquad x^2 = 5 \qquad$ What real number squared is 5?

$\qquad\qquad x = \pm\sqrt{5}$

(C) $3x^2 + 27 = 0$

$\qquad\qquad 3x^2 = -27$

$\qquad\qquad x^2 = -9 \qquad$ What real number squared is -9?

No real solution, since no real number squared is negative.   ◆

P R O B L E M   9   Solve by the square root method.

(A) $x^2 - 6 = 0$ (B) $3x^2 - 12 = 0$ (C) $x^2 + 4 = 0$ ◆

◆ SOLUTION BY FACTORING

If the left side of a quadratic equation when written in standard form can be factored, then the equation can be solved very quickly. The method of solution by factoring rests on the following important property of real numbers (Theorem 2 in Section 1-1):

If a and b are real numbers, then $ab = 0$ if and only if $a = 0$ or $b = 0$ (or both).

◆ **E X A M P L E   10**   Solve by factoring using integer coefficients, if possible.

(A) $3x^2 - 6x - 24 = 0$ (B) $3y^2 = 2y$ (C) $x^2 - 2x - 1 = 0$

Solutions   (A) $3x^2 - 6x - 24 = 0$ Divide both sides by 3, since 3 is a factor of each coefficient.

$\qquad x^2 - 2x - 8 = 0$ Factor the left side, if possible.

$\qquad (x - 4)(x + 2) = 0$

$\qquad x - 4 = 0 \quad \text{or} \quad x + 2 = 0$

$\qquad\qquad x = 4 \quad \text{or} \qquad x = -2$

(B) $\qquad 3y^2 = 2y$

$\qquad 3y^2 - 2y = 0$ We lose the solution $y = 0$ if both sides are divided by y

$\qquad y(3y - 2) = 0$ ($3y^2 = 2y$ and $3y = 2$ are not equivalent).

$\qquad y = 0 \quad \text{or} \quad 3y - 2 = 0$

$\qquad\qquad\qquad\qquad 3y = 2$

$\qquad\qquad\qquad\qquad y = \tfrac{2}{3}$

(C) $x^2 - 2x - 1 = 0$

This equation cannot be factored using integer coefficients. We will solve this type of equation by another method, considered below.   ◆

P R O B L E M   10   Solve by factoring using integer coefficients, if possible.

(A) $2x^2 + 4x - 30 = 0$ (B) $2x^2 = 3x$ (C) $2x^2 - 8x + 3 = 0$ ◆

The factoring and square root methods are fast and easy to use when they apply. However, there are quadratic equations that look simple but cannot be solved by either method. For example, as was noted in Example 10C, the polynomial in

$$x^2 - 2x - 1 = 0$$

cannot be factored using integer coefficients. This brings us to the well-known and widely used *quadratic formula*.

◆ QUADRATIC FORMULA

There is a method called *completing the square* that will work for all quadratic equations. After briefly reviewing this method, we will then use it to develop the famous quadratic formula—a formula that will enable us to solve any quadratic equation quite mechanically.

The method of **completing the square** is based on the process of transforming a quadratic equation in standard form,

$$ax^2 + bx + c = 0$$

into the form

$$(x + A)^2 = B$$

where A and B are constants. Then, this last equation can be solved easily (if it has a real solution) by the square root method discussed above.

Consider the equation

$$x^2 - 2x - 1 = 0 \tag{1}$$

Since the left side does not factor using integer coefficients, we add 1 to each side to remove the constant term from the left side:

$$x^2 - 2x = 1 \tag{2}$$

Now we try to find a number that we can add to each side to make the left side a square of a first-degree polynomial. Note the following two squares:

$$(x + m)^2 = x^2 + 2mx + m^2 \qquad (x - m)^2 = x^2 - 2mx + m^2$$

In each equation, we see that the third term on the right is the square of one-half the coefficient of x in the second term on the right. To complete the square in equation (2), we add the square of one-half the coefficient of x, $(-\frac{2}{2})^2 = 1$, to each side. (This rule works only when the coefficient of x^2 is 1, that is, $a = 1$.) Thus,

$$x^2 - 2x + 1 = 1 + 1$$

The left side is the square of $x - 1$, and we write

$$(x - 1)^2 = 2$$

What number squared is 2?

$$x - 1 = \pm\sqrt{2}$$
$$x = 1 \pm \sqrt{2}$$

And equation (1) is solved!

Let us try the method on the general quadratic equation

$$ax^2 + bx + c = 0 \qquad a \neq 0 \tag{3}$$

and solve it once and for all for x in terms of the coefficients a, b, and c. We start by multiplying both sides of (3) by $1/a$ to obtain

$$x^2 + \frac{b}{a}x + \frac{c}{a} = 0$$

Add $-c/a$ to both sides:

$$x^2 + \frac{b}{a}x = -\frac{c}{a}$$

Now we complete the square on the left side by adding the square of one-half the coefficient of x, that is, $(b/2a)^2 = b^2/4a^2$, to each side:

$$x^2 + \frac{b}{a}x + \frac{b^2}{4a^2} = \frac{b^2}{4a^2} - \frac{c}{a}$$

Writing the left side as a square and combining the right side into a single fraction, we obtain

$$\left(x + \frac{b}{2a}\right)^2 = \frac{b^2 - 4ac}{4a^2}$$

Now we solve by the square root method:

$$x + \frac{b}{2a} = \pm\sqrt{\frac{b^2 - 4ac}{4a^2}}$$

$$x = -\frac{b}{2a} \pm \frac{\sqrt{b^2 - 4ac}}{2a} \qquad \text{Since } \pm\sqrt{4a^2} = \pm 2a \text{ for any real number } a$$

When this is written as a single fraction, it becomes the **quadratic formula:**

Quadratic Formula

If $ax^2 + bx + c = 0$, $a \neq 0$, then

$$x = \frac{-b \pm \sqrt{b^2 - 4ac}}{2a}$$

This formula is generally used to solve quadratic equations when the square root or factoring methods do not work. The quantity $b^2 - 4ac$ under the radical is called the **discriminant**, and it gives us the useful information about solutions listed in Table 3.

TABLE 3

$b^2 - 4ac$	$ax^2 + bx + c = 0$
Positive	Two real solutions
Zero	One real solution
Negative	No real solutions

◆ E X A M P L E   11

Solve $x^2 - 2x - 1 = 0$ using the quadratic formula.

Solution

$$x^2 - 2x - 1 = 0$$

$$x = \frac{-b \pm \sqrt{b^2 - 4ac}}{2a} \qquad a = 1, b = -2, c = -1$$

$$= \frac{-(-2) \pm \sqrt{(-2)^2 - 4(1)(-1)}}{2(1)}$$

$$= \frac{2 \pm \sqrt{8}}{2} = \frac{2 \pm 2\sqrt{2}}{2} = 1 \pm \sqrt{2}$$

Check $\qquad x^2 - 2x - 1 = 0$

When $x = 1 + \sqrt{2}$,

$$(1 + \sqrt{2})^2 - 2(1 + \sqrt{2}) - 1 = 1 + 2\sqrt{2} + 2 - 2 - 2\sqrt{2} - 1 = 0$$

When $x = 1 - \sqrt{2}$,

$$(1 - \sqrt{2})^2 - 2(1 - \sqrt{2}) - 1 = 1 - 2\sqrt{2} + 2 - 2 + 2\sqrt{2} - 1 = 0 \qquad\blacklozenge$$

PROBLEM 11   Solve $2x^2 - 4x - 3 = 0$ using the quadratic formula. $\qquad\blacklozenge$

If we try to solve $x^2 - 6x + 11 = 0$ using the quadratic formula, we obtain

$$x = \frac{6 \pm \sqrt{-8}}{2}$$

which is not a real number. (Why?)

Answers to Matched Problems
9. (A) $\pm\sqrt{6}$ (B) ± 2 (C) No real solution
10. (A) $-5, 3$ (B) $0, \frac{3}{2}$ (C) Cannot be factored using integer coefficients
11. $(2 \pm \sqrt{10})/2$

E X E R C I S E   2-2

Find only real solutions in the problems below. If there are no real solutions, say so.

A   *Solve by the square root method.*

1. $x^2 - 4 = 0$ 2. $x^2 - 9 = 0$ 3. $2x^2 - 22 = 0$
4. $3m^2 - 21 = 0$

Solve by factoring.

5. $2u^2 - 8u - 24 = 0$ 6. $3x^2 - 18x + 15 = 0$ 7. $x^2 = 2x$
8. $n^2 = 3n$

Solve by using the quadratic formula.

9. $x^2 - 6x - 3 = 0$ 10. $m^2 + 8m + 3 = 0$ 11. $3u^2 + 12u + 6 = 0$
12. $2x^2 - 20x - 6 = 0$

B   *Solve, using any method.*

13. $2x^2 = 4x$ 14. $2x^2 = -3x$ 15. $4u^2 - 9 = 0$
16. $9y^2 - 25 = 0$ 17. $8x^2 + 20x = 12$ 18. $9x^2 - 6 = 15x$
19. $x^2 = 1 - x$ 20. $m^2 = 1 - 3m$ 21. $2x^2 = 6x - 3$
22. $2x^2 = 4x - 1$ 23. $y^2 - 4y = -8$ 24. $x^2 - 2x = -3$
25. $(x + 4)^2 = 11$ 26. $(y - 5)^2 = 7$

C   **27.** Solve $A = P(1 + r)^2$ for r in terms of A and P; that is, isolate r on the left side of the equation (with coefficient 1) and end up with an algebraic expression on the right side involving A and P but not r. Write the answer using positive square roots only.

    28. Solve $x^2 + mx + n = 0$ for x in terms of m and n.

APPLICATIONS

Business & Economics   **29.** *Supply and demand.* The demand equation for a certain brand of popular records is $d = 3{,}000/p$. Notice that as the price (p) goes up, the number of records people are willing to buy (d) goes down, and vice versa. The supply equation is given by $s = 1{,}000p - 500$. Notice again, as the price (p) goes up, the number of records a supplier is willing to sell (s) goes up. At what price will supply equal demand; that is, at what price will $d = s$? In economic theory the price at which supply equals demand is called the **equilibrium point**—the point where the price ceases to change.

    30. If P dollars is invested at $100r$ percent compounded annually, at the end of 2 years it will grow to $A = P(1 + r)^2$. At what interest rate will $100 grow to $144 in 2 years? [*Note:* If $A = 144$ and $P = 100$, find r.]

Life Sciences   **31.** *Ecology.* An important element in the erosive force of moving water is its velocity. To measure the velocity v (in feet per second) of a stream, we position a hollow L-shaped tube with one end under the water pointing upstream and the other end pointing straight up a couple of feet out of the water. The water will then be pushed up the tube a certain distance h (in feet) above the surface of the stream. Physicists have shown that $v^2 = 64h$. Approximately how fast is a stream flowing if $h = 1$ foot? If $h = 0.5$ foot?

Social Sciences   **32.** *Safety research.* It is of considerable importance to know the least number of feet d in which a car can be stopped, including reaction time of the driver, at various speeds v (in miles per hour). Safety research has produced the formula $d = 0.044v^2 + 1.1v$. If it took a car 550 feet to stop, estimate the car's speed at the moment the stopping process was started. You might find a calculator of help in this problem.

SECTION 2-3   Cartesian Coordinate System and Straight Lines

    ◆ CARTESIAN COORDINATE SYSTEM
    ◆ GRAPHING LINEAR EQUATIONS IN TWO VARIABLES
    ◆ SLOPE
    ◆ EQUATIONS OF LINES—SPECIAL FORMS
    ◆ APPLICATION

◆ CARTESIAN COORDINATE SYSTEM

Recall that a **Cartesian (rectangular) coordinate system** in a plane is formed by taking two mutually perpendicular real number lines (**coordinate axes**)—one horizontal and one vertical—intersecting at their origins, and then assigning unique **ordered pairs** of numbers **(coordinates)** to each point P in the plane (Fig. 2). The first coordinate **(abscissa)** is the distance of P from the vertical axis, and the second coordinate **(ordinate)** is the distance of P from the horizontal axis. In Figure 2, the coordinates of point P are (a, b). By reversing the process, each ordered pair of real numbers can be associated with a unique point in the plane. The coordinate axes divide the plane into four parts **(quadrants),** numbered I to IV in a counterclockwise direction.

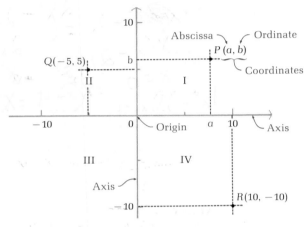

FIGURE 2
The Cartesian coordinate system

◆ GRAPHING LINEAR EQUATIONS IN TWO VARIABLES

A linear equation in two variables is an equation that can be written in the form

STANDARD FORM $Ax + By = C$

with A and B not both 0. For example,

$$2x - 3y = 5 \qquad x = 7 \qquad y = \tfrac{1}{2}x - 3 \qquad y = -3$$

all can be considered linear equations in two variables. The first is in standard form, while the other three can be written in standard form as follows:

	Standard Form
$x = 7$	$x + 0y = 7$
$y = \tfrac{1}{2}x - 3$	$-\tfrac{1}{2}x + y = -3$ or $x - 2y = 6$
$y = -3$	$0x + y = -3$

A **solution** of an equation in two variables is an ordered pair of real numbers that satisfy the equation. For example, $(0, -3)$ is a solution of $3x - 4y = 12$. The **solution set** of an equation in two variables is the set of all solutions of the equation. When we say that we **graph an equation** in two variables, we mean that we graph its solution set on a rectangular coordinate system.

We state the following important theorem without proof:

THEOREM 1

■ **Graph of a Linear Equation in Two Variables**

The graph of any equation of the form

$$\text{STANDARD FORM} \qquad Ax + By = C \tag{1}$$

where A, B, and C are constants (A and B not both 0), is a straight line. Every straight line in a Cartesian coordinate system is the graph of an equation of this type.

Also, the graph of any equation of the form

$$y = mx + b \tag{2}$$

where m and b are constants, is a straight line. Form (2) is simply a special case of (1) for $B \neq 0$. To graph either (1) or (2), we plot any two points of their solution set and use a straightedge to draw the line through these two points. The points where the line crosses the axes — called the **intercepts** — are often the easiest to find when dealing with form (1). To find the **y intercept,** we let $x = 0$ and solve for y; to find the **x intercept,** we let $y = 0$ and solve for x. It is sometimes wise to find a third point as a check.

◆ **EXAMPLE 12**   (A) The graph of $3x - 4y = 12$ is

x	y
0	−3
4	0
8	3

(B) The graph of $y = 2x - 1$ is

x	y
0	-1
4	7
-2	-5

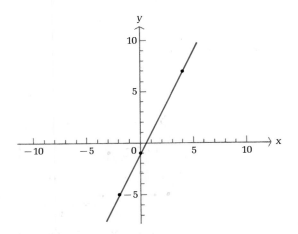

PROBLEM 12 Graph: (A) $4x - 3y = 12$ (B) $y = \dfrac{x}{2} + 2$

◆ SLOPE

It is very useful to have a numerical measure of the "steepness" of a line. The concept of *slope* is widely used for this purpose. The **slope** of a line through the two points (x_1, y_1) and (x_2, y_2) is given by the following formula:

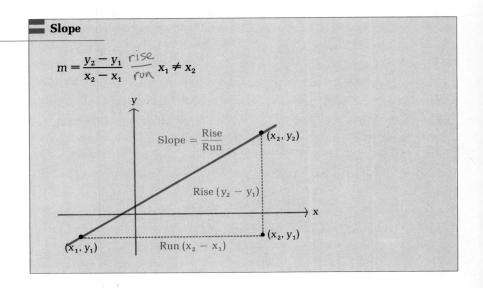

The slope of a vertical line is not defined. (Why? See Example 13B.)

◆ E X A M P L E   13   Find the slope of the line through each pair of points:

(A) $(-2, 5), (4, -7)$ (B) $(-3, -1), (-3, 5)$

Solutions   (A) Let $(x_1, y_1) = (-2, 5)$ and $(x_2, y_2) = (4, -7)$. Then

$$m = \frac{y_2 - y_1}{x_2 - x_1} = \frac{-7 - 5}{4 - (-2)} = \frac{-12}{6} = -2$$

Note that we also could have let $(x_1, y_1) = (4, -7)$ and $(x_2, y_2) = (-2, 5)$, since this simply reverses the sign in both the numerator and denominator and the slope does not change:

$$m = \frac{5 - (-7)}{-2 - 4} = \frac{12}{-6} = -2$$

(B) Let $(x_1, y_1) = (-3, -1)$ and $(x_2, y_2) = (-3, 5)$. Then

$$m = \frac{y_2 - y_1}{x_2 - x_1} = \frac{5 - (-1)}{-3 - (-3)} = \frac{6}{0} \qquad \text{Not defined!}$$

Notice that $x_1 = x_2$. This is always true for a vertical line, since the abscissa (first coordinate) of every point on a vertical line is the same. Thus, the slope of a vertical line is not defined (that is, the slope does not exist).   ◆

P R O B L E M   13   Find the slope of the line through each pair of points:

(A) $(3, -6), (-2, 4)$ (B) $(-7, 5), (3, 5)$ ◆

In general, the slope of a line may be positive, negative, zero, or not defined. Each of these cases is interpreted geometrically in Table 4.

T A B L E   4

Going from Left to Right

LINE	SLOPE	EXAMPLE
Rising	Positive	
Falling	Negative	
Horizontal	Zero	
Vertical	Not defined	

◆ EQUATIONS OF LINES — SPECIAL FORMS

The constants m and b in the equation

$$y = mx + b \qquad\qquad (3)$$

have special geometric significance.

If we let $x = 0$, then $y = b$, and we observe that the graph of (3) crosses the y axis at $(0, b)$. The constant b is the y *intercept*. For example, the y intercept of the graph of $y = -4x - 1$ is -1.

To determine the geometric significance of m, we proceed as follows: If $y = mx + b$, then by setting $x = 0$ and $x = 1$, we conclude that $(0, b)$ and $(1, m + b)$ lie on its graph (a line). Hence, the slope of this graph (line) is given by:

$$\text{Slope} = \frac{y_2 - y_1}{x_2 - x_1} = \frac{(m + b) - b}{1 - 0} = m$$

Thus, m is the slope of the line given by $y = mx + b$.

Slope – Intercept Form

The equation

$$y = mx + b \qquad m = \text{Slope}, \ b = y \text{ intercept} \qquad (4)$$

is called the **slope – intercept form** of an equation of a line.

◆ E X A M P L E   14   (A) Find the slope and y intercept, and graph $y = -\frac{2}{3}x - 3$.
(B) Write the equation of the line with slope $\frac{2}{3}$ and y intercept -2.

Solutions   (A) Slope $= m = -\frac{2}{3}$ (B) $m = \frac{2}{3}$ and $b = -2$; thus, $y = \frac{2}{3}x - 2$
  y intercept $= b = -3$

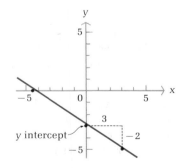

P R O B L E M   14   Write the equation of the line with slope $\frac{1}{2}$ and y intercept -1. Graph.   ◆

Suppose a line has slope m and passes through a fixed point (x_1, y_1). If the point (x, y) is any other point on the line (Fig. 3), then

$$\frac{y - y_1}{x - x_1} = m$$

that is,

$$y - y_1 = m(x - x_1) \tag{5}$$

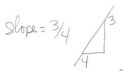

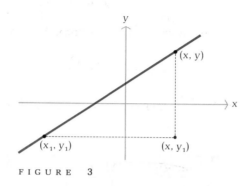

F I G U R E   3

We now observe that (x_1, y_1) also satisfies equation (5) and conclude that equation (5) is an equation of a line with slope m that passes through (x_1, y_1).

Point–Slope Form

An equation of a line with slope m that passes through (x_1, y_1) is

$$y - y_1 = m(x - x_1) \tag{5}$$

which is called the **point–slope form** of an equation of a line.

The point–slope form is extremely useful, since it enables us to find an equation for a line if we know its slope and the coordinates of a point on the line or if we know the coordinates of two points on the line.

◆ **E X A M P L E   15**   (A) Find an equation for the line that has slope $\frac{1}{2}$ and passes through $(-4, 3)$. Write the final answer in the form $Ax + By = C$.

(B) Find an equation for the line that passes through the two points $(-3, 2)$ and $(-4, 5)$. Write the resulting equation in the form $y = mx + b$.

Solutions   (A) Use $y - y_1 = m(x - x_1)$. Let $m = \frac{1}{2}$ and $(x_1, y_1) = (-4, 3)$. Then

$$y - 3 = \frac{1}{2}[x - (-4)]$$
$$y - 3 = \frac{1}{2}(x + 4) \qquad \text{Multiply by 2.}$$
$$2y - 6 = x + 4$$
$$-x + 2y = 10 \quad \text{or} \quad x - 2y = -10$$

(B) First, find the slope of the line by using the slope formula:

$$m = \frac{y_2 - y_1}{x_2 - x_1} = \frac{5 - 2}{-4 - (-3)} = \frac{3}{-1} = -3$$

Now use $y - y_1 = m(x - x_1)$ with $m = -3$ and $(x_1, y_1) = (-3, 2)$:

$$y - 2 = -3[x - (-3)]$$
$$y - 2 = -3(x + 3)$$
$$y - 2 = -3x - 9$$
$$y = -3x - 7 \qquad \blacklozenge$$

PROBLEM   15   (A) Find an equation for the line that has slope $\frac{2}{3}$ and passes through $(6, -2)$. Write the resulting equation in the form $Ax + By = C, A > 0$.
(B) Find an equation for the line that passes through $(2, -3)$ and $(4, 3)$. Write the resulting equation in the form $y = mx + b$. $\qquad \blacklozenge$

The simplest equations of a line are those for horizontal and vertical lines. A **horizontal line** has slope 0; thus, its equation is of the form

$$y = 0x + c \qquad \text{Slope} = 0, y \text{ intercept} = c$$

or, simply,

$$y = c$$

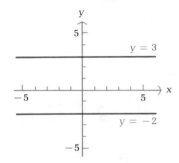

FIGURE   4

Figure 4 illustrates the graphs of $y = 3$ and $y = -2$.

If a line is vertical, then its slope is not defined. All x values (abscissas) of points on a vertical line are equal, while y can take on any value (Fig. 5). Thus, a **vertical line** has an equation of the form

$$x + 0y = c \qquad x \text{ intercept} = c$$

or, simply,

$$x = c$$

Figure 5 illustrates the graphs of $x = -3$ and $x = 4$.

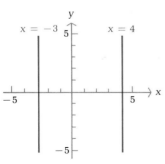

FIGURE   5

HORIZONTAL LINE WITH y INTERCEPT c: $y = c$
VERTICAL LINE WITH x INTERCEPT c: $x = c$

◆ E X A M P L E   16   The equation of a horizontal line through $(-2, 3)$ is $y = 3$, and the equation of a vertical line through the same point is $x = -2$.   ◆

P R O B L E M   16   Find the equations of the horizontal and vertical lines through $(4, -5)$.   ◆

It can be shown that if two nonvertical lines are parallel, then they have the same slope. And if two lines have the same slope, they are parallel. It also can be shown that if two nonvertical lines are perpendicular, then their slopes are the negative reciprocals of each other (that is, $m_2 = -1/m_1$, or, equivalently, $m_1 m_2 = -1$). And if the slopes of two lines are the negative reciprocals of each other, the lines are perpendicular. Symbolically:

Parallel and Perpendicular Lines

Given nonvertical lines L_1 and L_2 with slopes m_1 and m_2, respectively, then:

$L_1 \| L_2$ if and only if $m_1 = m_2$

$L_1 \perp L_2$ if and only if $m_1 m_2 = -1$ or $m_2 = -\dfrac{1}{m_1}$

[Note: $\|$ means "is parallel to" and $\perp$ means "is perpendicular to."]

◆ E X A M P L E   17   Given the line $x - 2y = 4$, find the equation of a line that passes through $(2, -3)$ and is:

(A) Parallel to the given line   (B) Perpendicular to the given line

Write final equations in the form $y = mx + b$.

Solutions   First find the slope of the given line by writing $x - 2y = 4$ in the form $y = mx + b$:

$x - 2y = 4$
$y = \tfrac{1}{2}x - 2$

The slope of the given line is $\tfrac{1}{2}$.

(A) The slope of a line parallel to the given line is also $\frac{1}{2}$. We have only to find the equation of a line through $(2, -3)$ with slope $\frac{1}{2}$ to solve part A:

$$y - y_1 = m(x - x_1) \qquad m = \tfrac{1}{2} \text{ and } (x_1, y_1) = (2, -3)$$
$$y - (-3) = \tfrac{1}{2}(x - 2)$$
$$y + 3 = \tfrac{1}{2}x - 1$$
$$y = \tfrac{1}{2}x - 4$$

(B) The slope of the line perpendicular to the given line is the negative reciprocal of $\frac{1}{2}$; that is, -2. We have only to find the equation of a line through $(2, -3)$ with slope -2 to solve part B:

$$y - y_1 = m(x - x_1) \qquad m = -2 \text{ and } (x_1, y_1) = (2, -3)$$
$$y - (-3) = -2(x - 2)$$
$$y + 3 = -2x + 4$$
$$y = -2x + 1$$

◆

PROBLEM 17   Given the line $2x = 6 - 3y$, find the equation of a line that passes through $(-3, 9)$ and is:

(A) Parallel to the given line   (B) Perpendicular to the given line

Write final equations in the form $y = mx + b$.   ◆

◆ APPLICATION

We will now see how equations of lines occur in certain applications.

◆ EXAMPLE 18

Cost Equation

The management of a company that manufactures roller skates has fixed costs (costs at zero output) of $300 per day and total costs of $4,300 per day at an output of 100 pairs of skates per day. Assume that cost C is linearly related to output x.

(A) Find the slope of the line joining the points associated with outputs of 0 and 100; that is, the line passing through $(0, 300)$ and $(100, 4{,}300)$.

(B) Find an equation of the line relating output to cost. Write the final answer in the form $C = mx + b$.

(C) Graph the cost equation from part B for $0 \leqslant x \leqslant 200$.

Solutions   (A) $m = \dfrac{y_2 - y_1}{x_2 - x_1} = \dfrac{4{,}300 - 300}{100 - 0} = \dfrac{4{,}000}{100} = 40$

(B) We must find an equation of the line that passes through $(0, 300)$ with slope 40. We use the slope–intercept form:

$$C = mx + b$$
$$C = 40x + 300$$

x	C
0	300
100	4,300
200	8,300

(C)

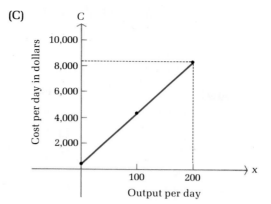

Cost per day in dollars

Output per day

PROBLEM 18 Answer parts A and B in Example 18 for fixed costs of $250 per day and total costs of $3,450 per day at an output of 80 pairs of skates per day.

Answers to Matched Problems

12. (A) 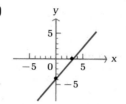 (B)

13. (A) −2
(B) 0 (Zero is a number—it exists! It is the slope of a horizontal line.)

14. $y = \frac{1}{2}x - 1$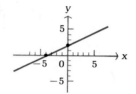

15. (A) $2x - 3y = 18$
(B) $y = 3x - 9$

16. $y = -5$; $x = 4$

17. (A) $y = -\frac{2}{3}x + 7$
(B) $y = \frac{3}{2}x + \frac{27}{2}$

18. (A) $m = 40$
(B) $C = 40x + 250$

$Ax + By = C$

EXERCISE 2-3

X apis crossing is $\frac{C}{A}$

y axis crossing is $\frac{C}{B}$

Join $\frac{C}{A}$, $\frac{C}{B}$ w/ line

Because when

y=0 then $Ax + B(0) = C$
$AX = C$
$X = \frac{C}{A}$

A *Graph in a rectangular coordinate system.*

1. $y = 2x - 3$ 2. $y = \frac{x}{2} + 1$ 3. $2x + 3y = 12$ 4. $8x - 3y = 24$

Find the slope and y intercept of the graph of each equation.

5. $y = 2x - 3$ 6. $y = \frac{x}{2} + 1$ 7. $y = -\frac{2}{3}x + 2$ 8. $y = \frac{3}{4}x - 2$

when X=0 the $A(0) + By = C$
$\frac{By}{B} = \frac{C}{B}$ $y = C/B$

Write an equation of the line with the indicated slope and y intercept.

9. Slope $= -2$
   y intercept $= 4$

10. Slope $= -\frac{2}{3}$
    y intercept $= -2$

11. Slope $= -\frac{3}{5}$
    y intercept $= 3$

12. Slope $= 1$
    y intercept $= -2$

B Graph in a rectangular coordinate system.

13. $y = -\frac{2}{3}x - 2$

14. $y = -\frac{3}{2}x + 1$

15. $3x - 2y = 10$

16. $5x - 6y = 15$

17. $x = 3$ and $y = -2$

18. $x = -3$ and $y = 2$

Find the slope of the graph of each equation. (First write the equation in the form $y = mx + b$.)

19. $3x + y = 5$

20. $2x - y = -3$

21. $2x + 3y = 12$

22. $3x - 2y = 10$

Write an equation of the line through each indicated point with the indicated slope. Transform the equation into the form $y = mx + b$.

23. $m = -3$; $(4, -1)$

24. $m = -2$; $(-3, 2)$

25. $m = \frac{2}{3}$; $(-6, -5)$

26. $m = \frac{1}{2}$; $(-4, 3)$

Find the slope of the line that passes through the given points.

27. $(1, 3)$ and $(7, 5)$

28. $(2, 1)$ and $(10, 5)$

29. $(-5, -2)$ and $(5, -4)$

30. $(3, 7)$ and $(-6, 4)$

Write an equation of the line through each indicated pair of points. Write the final answer in the form $Ax + By = C$, $A > 0$.

31. $(1, 3)$ and $(7, 5)$

32. $(2, 1)$ and $(10, 5)$

33. $(-5, -2)$ and $(5, -4)$

34. $(3, 7)$ and $(-6, 4)$

Write equations of the vertical and horizontal lines through each point.

35. $(3, -5)$ 36. $(-2, 7)$ 37. $(-1, -3)$ 38. $(6, -4)$

Find an equation of the line, given the information in each problem. Write the final answer in the form $y = mx + b$.

39. Line passes through $(-2, 5)$ with slope $-\frac{1}{2}$.

40. Line passes through $(3, -1)$ with slope $-\frac{2}{3}$.

41. Line passes through $(-2, 2)$ and is:

    (A) Parallel   (B) Perpendicular to $y = -\frac{1}{2}x + 5$

42. Line passes through $(-4, -3)$ and is:

    (A) Parallel   (B) Perpendicular to $y = 2x - 3$

43. Line passes through $(-2, -1)$ and is:

    (A) Parallel   (B) Perpendicular to $x - 2y = 4$

44. Line passes through $(-3, 2)$ and is:

(A) Parallel   (B) Perpendicular to $2x + 3y = -6$

C   **45.** Graph $y = mx - 2$ for $m = 2$, $m = \frac{1}{2}$, $m = 0$, $m = -\frac{1}{2}$, and $m = -2$, all on the same coordinate system.

46. Graph $y = -\frac{1}{2}x + b$ for $b = -4$, $b = 0$, and $b = 4$, all on the same coordinate system.

Write an equation of the line through the indicated points. Be careful!

47. $(2, 7)$ and $(2, -3)$ **48.** $(-2, 3)$ and $(-2, -1)$

49. $(2, 3)$ and $(-5, 3)$ **50.** $(-3, -3)$ and $(0, -3)$

APPLICATIONS

Business & Economics   **51.** *Simple interest.* If $P (the principal) is invested at an interest rate of r, then the amount A that is due after t years is given by

$$A = Prt + P$$

If $100 is invested at 6% ($r = 0.06$), then $A = 6t + 100$, $t \geqslant 0$.

(A) What will $100 amount to after 5 years? After 20 years?

(B) Graph the equation for $0 \leqslant t \leqslant 20$.

(C) What is the slope of the graph? (The slope indicates the increase in the amount A for each additional year of investment.)

52. *Cost equation.* The management of a company manufacturing surfboards has fixed costs (zero output) of $200 per day and total costs of $1,400 per day at a daily output of 20 boards.

(A) Assuming the total cost per day (C) is linearly related to the total output per day (x), write an equation relating these two quantities. [*Hint:* Find an equation of the line that passes through $(0, 200)$ and $(20, 1,400)$.]

(B) What are the total costs for an output of 12 boards per day?

(C) Graph the equation for $0 \leqslant x \leqslant 20$.

[*Note:* The slope of the line found in part A is the increase in total cost for each additional unit produced and is called the *marginal cost*. More will be said about this concept later.]

53. *Demand equation.* A manufacturing company is interested in introducing a new power mower. Its market research department gave the management the demand–price forecast listed in the table.

PRICE	ESTIMATED DEMAND
$ 70	7,800
$120	4,800
$160	2,400
$200	0

(A) Plot these points, letting d represent the number of mowers people are willing to buy (demand) at a price of $\$p$ each.

(B) Note that the points in part A lie along a straight line. Find an equation of that line.

[Note: The slope of the line found in part B indicates the decrease in demand for each $1 increase in price.]

54. *Depreciation*. Office equipment was purchased for $20,000 and is assumed to have a scrap value of $2,000 after 10 years. If its value is depreciated linearly (for tax purposes) from $20,000 to $2,000:

(A) Find the linear equation that relates value (V) in dollars to time (t) in years.

(B) What would be the value of the equipment after 6 years?

(C) Graph the equation for $0 \leq t \leq 10$.

[Note: The slope found in part A indicates the decrease in value per year.]

Life Sciences 55. *Nutrition*. In a nutrition experiment, a biologist wants to prepare a special diet for the experimental animals. Two food mixes, A and B, are available. If mix A contains 20% protein and mix B contains 10% protein, what combination of each mix will provide exactly 20 grams of protein? Let x be the amount of A used and let y be the amount of B used. Then write a linear equation relating x, y, and 20. Graph this equation for $x \geq 0$ and $y \geq 0$.

56. *Ecology*. As one descends into the ocean, pressure increases linearly. The pressure is 15 pounds per square inch on the surface and 30 pounds per square inch 33 feet below the surface.

(A) If p is the pressure in pounds and d is the depth below the surface in feet, write an equation that expresses p in terms of d. [Hint: Find an equation of the line that passes through (0, 15) and (33, 30).]

(B) What is the pressure at 12,540 feet (the average depth of the ocean)?

(C) Graph the equation for $0 \leq d \leq 12{,}540$.

[Note: The slope found in part A indicates the change in pressure for each additional foot of depth.]

Social Sciences 57. *Psychology*. In an experiment on motivation, J. S. Brown trained a group of rats to run down a narrow passage in a cage to obtain food in a goal box. Using a harness, he then connected the rats to an overhead wire that was attached to a spring scale. A rat was placed at different distances d (in centimeters) from the goal box, and the pull p (in grams) of the rat toward the food was measured. Brown found that the relationship between these two variables was very close to being linear and could be approximated by the equation

$$p = -\tfrac{1}{5}d + 70 \qquad 30 \leq d \leq 175$$

(See J. S. Brown, *Journal of Comparative and Physiological Psychology*, 1948, 41:450–465.)

(A) What was the pull when $d = 30$? When $d = 175$?

(B) Graph the equation.

(C) What is the slope of the line?

SECTION   2-4   **Functions**

◆ DEFINITION OF A FUNCTION
◆ FUNCTIONS SPECIFIED BY EQUATIONS
◆ FUNCTION NOTATION
◆ APPLICATION

The function concept is one of the most important concepts in mathematics. The idea of correspondence plays a central role in its formulation. You have already had experiences with correspondences in everyday life. For example:

To each person there corresponds an annual income.

To each item in a supermarket there corresponds a price.

To each day there corresponds a maximum temperature.

For the manufacture of x items there corresponds a cost.

For the sale of x items there corresponds a revenue.

To each square there corresponds an area.

To each number there corresponds its cube.

One of the most important aspects of any science (managerial, life, social, physical, etc.) is the establishment of correspondences among various types of phenomena. Once a correspondence is known, predictions can be made. A cost analyst would like to predict costs for various levels of output in a manufacturing process; a medical researcher would like to know the correspondence between heart disease and obesity; a psychologist would like to predict the level of performance after a subject has repeated a task a given number of times; and so on.

◆ DEFINITION OF A FUNCTION

What do all the above examples have in common? Each describes the matching of elements from one set with the elements in a second set. Consider the tables of the cube, square, and square root given in Tables 5–7 at the top of the next page.

TABLE 5	
DOMAIN *Number*	RANGE *Cube*
−2	→ −8
−1	→ −1
0	→ 0
1	→ 1
2	→ 8

TABLE 6	
DOMAIN *Number*	RANGE *Square*
−2	→ 4
−1	→ 1
0	→ 0
1	→ 0
2	

TABLE 7	
DOMAIN *Number*	RANGE *Square Root*
0	→ 0
1	→ 1
	→ −1
4	→ 2
	→ −2
9	→ 3
	→ −3

Tables 5 and 6 specify functions, but Table 7 does not. Why not? The definition of the very important term *function* will explain.

▮ Definition of a Function (Rule Form)

A **function** is a rule (process or method) that produces a correspondence between one set of elements, called the **domain,** and a second set of elements, called the **range,** such that to each element in the domain there corresponds *one and only one* element in the range.

Tables 5 and 6 specify functions, since to each domain value there corresponds exactly one range value (for example, the cube of −2 is −8 and no other number). On the other hand, Table 7 does not specify a function, since to at least one domain value there corresponds more than one range value (for example, to the domain value 9 there corresponds −3 and 3, both square roots of 9).

Since in a function elements in the range are paired with elements in the domain by some rule or process, this correspondence (pairing) can be illustrated by using ordered pairs of elements, where the first component represents a domain element and the second component represents a corresponding range element. Thus, we can write functions 1 and 2 (Tables 5 and 6) as follows:

Function 1 = {(−2, −8), (−1, −1), (0, 0), (1, 1), (2, 8)}

Function 2 = {(−2, 4), (−1, 1), (0, 0), (1, 1), (2, 4)}

In either case, notice that no two ordered pairs have the same first component and different second components. On the other hand, if we list the ordered pairs determined by Table 7, we have

A = {(0, 0), (1, 1), (1, −1), (4, 2), (4, −2), (9, 3), (9, −3)}

In this case, there are ordered pairs with the same first component and different second components — for example, (1, 1) and (1, −1) both belong to the set A — indicating once again that Table 7 does not specify a function.

This suggests an alternative but equivalent way of defining functions that produces additional insight into this concept.

Definition of a Function (Set Form)

A **function** is a set of ordered pairs with the property that no two ordered pairs have the same first element and different second elements. The set of all first components in a function is called the **domain** of the function, and the set of all second components is called the **range**.

◆ E X A M P L E   19   (A) The set $S = \{(-2, 4), (-1, 3), (0, 1), (1, 3), (2, 4)\}$ defines a function since no two ordered pairs have the same first component and different second components. The domain and range are

$$\text{Domain} = \{-2, -1, 0, 1, 2\} \qquad \text{Set of first components}$$
$$\text{Range} = \{1, 3, 4\} \qquad \text{Set of second components}$$

(B) The set $T = \{(2, 1), (1, 3), (0, 2), (2, 3), (1, 4)\}$ does not define a function since there are ordered pairs with the same first component and different second components — for example, $(1, 3)$ and $(1, 4)$.   ◆

P R O B L E M   19   Determine whether each set defines a function. If it does, then state the domain and range.

$$S = \{(1, 2), (2, 4), (3, 6), (2, 8), (1, 10)\}$$
$$T = \{(-3, 1), (-1, 2), (1, 0), (3, 2), (5, 1)\} \qquad ◆$$

◆ FUNCTIONS SPECIFIED BY EQUATIONS

Most of the domains and ranges included in this text will be (infinite) sets of real numbers, and the rules associating range values with domain values will be equations in two variables. Consider the equation

$$y = x^2 - x \qquad x \in R \qquad \text{Recall that } R \text{ is the set of real numbers.}$$

For each **input** x we obtain one **output** y. For example:

If $x = 3$, then $y = 3^2 - 3 = 6$.

If $x = -\frac{1}{2}$, then $y = (-\frac{1}{2})^2 - (-\frac{1}{2}) = \frac{1}{4} + \frac{1}{2} = \frac{3}{4}$.

The input values are domain values, and the output values are range values. The equation (a rule) assigns each domain value x a range value y. The variable x is called an *independent variable* (since values can be "independently" assigned to x from the domain), and y is called a *dependent variable* (since the value of y "depends" on the value assigned to x). In general, any variable used as a placeholder for domain values is called an **independent variable;** any variable that is used as a placeholder for range values is called a **dependent variable.**

When does an equation specify a function?

Equations and Functions

In an equation in two variables, if there corresponds exactly one value of the dependent variable (output) to each value of the independent variable (input), then the equation specifies a function. If there is any value of the independent variable to which there corresponds more than one value of the dependent variable, then the equation does not specify a function.

◆ E X A M P L E 20

Determine which of the following equations specify functions with independent variable x.

(A) $4y - 3x = 8$, $x \in R$ (B) $y^2 - x^2 = 9$, $x \in R$

Solutions

(A) Solving for the dependent variable y, we have

$$4y - 3x = 8 \tag{1}$$
$$4y = 8 + 3x$$
$$y = 2 + \tfrac{3}{4}x$$

Since each input value x corresponds to exactly one output value ($y = 2 + \tfrac{3}{4}x$), we see that equation (1) specifies a function.

(B) Solving for the dependent variable y, we have

$$y^2 - x^2 = 9 \tag{2}$$
$$y^2 = 9 + x^2$$
$$y = \pm\sqrt{9 + x^2}$$

Since $9 + x^2$ is always a positive real number for any real number x and since each positive real number has two square roots, to each input value x there corresponds two output values ($y = -\sqrt{9 + x^2}$ and $y = \sqrt{9 + x^2}$). For example, if $x = 4$, then equation (2) is satisfied for $y = 5$ and for $y = -5$. Thus, equation (2) does not specify a function.   ◆

P R O B L E M 20

Determine which of the following equations specify functions with independent variable x.

(A) $y^2 - x^4 = 9$, $x \in R$ (B) $3y - 2x = 3$, $x \in R$ ◆

Since the graph of an equation is the graph of all the ordered pairs that satisfy the equation, it is very easy to determine whether an equation specifies a function by examining its graph. The graphs of the two equations we considered in Example 20 are shown in Figure 6. (The graph in Figure 6B was obtained using point-by-point plotting, a technique we will discuss later in this chapter.)

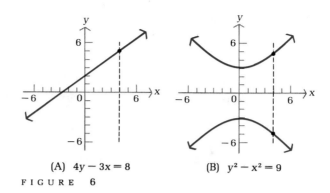

(A) $4y - 3x = 8$ (B) $y^2 - x^2 = 9$

FIGURE 6

In Figure 6A notice that any vertical line will intersect the graph of the equation $4y - 3x = 8$ in exactly one point. This shows that to each x value there corresponds exactly one y value and confirms our conclusion that this equation specifies a function. On the other hand, Figure 6B shows that there exist vertical lines that intersect the graph of $y^2 - x^2 = 9$ in two points. This indicates that there exist x values to which there correspond two different y values and verifies our conclusion that this equation does not specify a function. These observations are generalized in Theorem 2.

THEOREM 2

Vertical Line Test for a Function

An equation specifies a function if each vertical line in the coordinate system passes through at most one point on the graph of the equation. If any vertical line passes through two or more points on the graph of an equation, then the equation does not specify a function.

In Example 20, the domains were explicitly stated along with the given equations. In many cases, this will not be done. Unless stated to the contrary, we shall adhere to the following convention regarding domains and ranges for functions specified by equations.

Agreement on Domains and Ranges

If a function is specified by an equation and the domain is not indicated, then we assume that the domain is the set of all real number replacements of the independent variable (inputs) that produce *real values* for the dependent variable (outputs). The range is the set of all outputs corresponding to input values.

◆ EXAMPLE 21

Find the domain of the function specified by the equation $y = \sqrt{x + 4}$, assuming x is the independent variable.

Solution   For y to be real, $x + 4$ must be greater than or equal to 0; that is,

$$x + 4 \geqslant 0$$
$$x \geqslant -4$$

Thus,

Domain: $x \geqslant -4$ or $[-4, \infty)$ ◆

PROBLEM   21   Find the domain of the function specified by the equation $y = \sqrt{x - 2}$, assuming x is the independent variable.   ◆

◆ FUNCTION NOTATION

We have just seen that a function involves two sets, a domain and a range, and a rule of correspondence that enables us to assign to each element in the domain exactly one element in the range. We use different letters to denote names for numbers; in essentially the same way, we will now use different letters to denote names for functions. For example, f and g may be used to name the functions specified by the equations $y = 2x + 1$ and $y = x^2 + 2x - 3$:

$f:$ $y = 2x + 1$

$g:$ $y = x^2 + 2x - 3$

If x represents an element in the domain of a function f, then we frequently use the symbol

$f(x)$

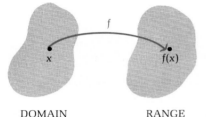

DOMAIN   RANGE

FIGURE   7

in place of y to designate the number in the range of the function f to which x is paired (Fig. 7).

It is important not to think of $f(x)$ as the product of f and x. The symbol $f(x)$ is read "f of x," "f at x," or "the value of f at x." Whenever we write $y = f(x)$, we assume that the variable x is an independent variable and that both y and $f(x)$ are dependent variables. The symbol $f: x \to f(x)$, read "f maps x into $f(x)$," is also used to indicate the relationship between the independent variable x and the dependent variable $f(x)$.

This function notation is extremely important, and its correct use should be mastered as early as possible. For example, in place of the more formal representation of the functions f and g,

$f:$ $y = 2x + 1$ and $g:$ $y = x^2 + 2x - 3$

we can now write

$f(x) = 2x + 1$ and $g(x) = x^2 + 2x - 3$

The function symbols $f(x)$ and $g(x)$ have certain advantages over the variable y in many situations. For example, if we write $f(3)$ and $g(-5)$, then each symbol indicates in a concise way that these are range values of particular functions associated with particular domain values. Let us find $f(3)$ and $g(-5)$.

To find $f(3)$, we replace x by 3 wherever x occurs in $f(x) = 2x + 1$ and evaluate the right side:

$$f(x) = 2x + 1$$
$$f(3) = 2 \cdot 3 + 1$$
$$= 6 + 1 = 7$$

Thus,

$f(3) = 7$ The function f assigns the range value 7 to the domain value 3; the ordered pair (3, 7) belongs to f.

To find $g(-5)$, we replace x by -5 wherever x occurs in $g(x) = x^2 + 2x - 3$ and evaluate the right side:

$$g(x) = x^2 + 2x - 3$$
$$g(-5) = (-5)^2 + 2(-5) - 3$$
$$= 25 - 10 - 3 = 12$$

Thus,

$g(-5) = 12$ The function g assigns the range value 12 to the domain value -5; the ordered pair $(-5, 12)$ belongs to g.

It is very important to understand and remember the definition of $f(x)$:

The Symbol $f(x)$

For any element x in the domain of the function f, the symbol $f(x)$ represents the element in the range of f corresponding to x in the domain of f. If x is an input value, then $f(x)$ is the corresponding output value; or, symbolically, $f: x \rightarrow f(x)$. The ordered pair $(x, f(x))$ belongs to the function f. If x is an element that is not in the domain of f, then f is *not defined at* x and $f(x)$ *does not exist.*

Figure 8 (on the next page), which illustrates a "function machine," may give you additional insight into the nature of functions and the symbol $f(x)$. We can think of a function machine as a device that produces exactly one output (range) value for each input (domain) value on the basis of a set of instructions. (If more than one output value was produced for an input value, then the machine would not be a function machine.)

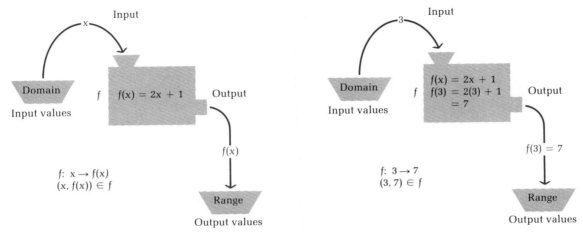

FIGURE 8
Function machine—exactly one output for each input

For the function $f(x) = 2x + 1$, the machine takes each domain value (input), multiplies it by 2, and then adds 1 to the result, to produce the range value (output). Different rules inside the function machine result in different functions.

◆ EXAMPLE 22   If

$$f(x) = \frac{12}{x - 2} \qquad g(x) = 1 - x^2 \qquad h(x) = \sqrt{x - 1}$$

then:

(A) $f(6)$ $= \dfrac{12}{6 - 2} = \dfrac{12}{4} = 3$

(B) $g(-2)$ $= 1 - (-2)^2 = 1 - 4 = -3$

(C) $h(-2)$ $= \sqrt{-2 - 1} = \sqrt{-3}$

Since $\sqrt{-3}$ is not a real number, -2 is not in the domain of h and $h(-2)$ is not defined.

(D) $f(0) + g(1) - h(10)$ $= \dfrac{12}{0 - 2} + (1 - 1^2) - \sqrt{10 - 1}$

$$= \frac{12}{-2} + 0 - \sqrt{9}$$

$$= -6 - 3 = -9$$

◆

PROBLEM 22   Use the functions in Example 22 to find:

(A) $f(-2)$ (B) $g(-1)$ (C) $h(-8)$ (D) $\dfrac{f(3)}{h(5)}$ ◆

◆ **EXAMPLE 23**   Find the domains of functions f, g, and h:

$$f(x) = \frac{12}{x-2} \qquad g(x) = 1 - x^2 \qquad h(x) = \sqrt{x-1}$$

Domain of f $12/(x-2)$ represents a real number for all replacements of x by real numbers except for $x = 2$ (division by 0 is not defined). Thus, $f(2)$ does not exist, and the domain of f is the set of all real numbers except 2. We often indicate this by writing

$$f(x) = \frac{12}{x-2} \qquad x \neq 2$$

Domain of g   The domain is R, the set of all real numbers, since $1 - x^2$ represents a real number for all replacements of x by real numbers.

Domain of h   The domain is the set of all real numbers x such that $\sqrt{x-1}$ is a real number— that is, such that

$$x - 1 \geqslant 0$$
$$x \geqslant 1 \quad \text{or} \quad [1, \infty)$$
◆

PROBLEM 23   Find the domains of functions F, G, and H:

$$F(x) = x^2 - 3x + 1 \qquad G(x) = \frac{5}{x+3} \qquad H(x) = \sqrt{2-x}$$ ◆

◆ **EXAMPLE 24**   For $f(x) = x^2 - 2x + 7$, find:

(A) $f(a)$ (B) $f(a+h)$ (C) $\dfrac{f(a+h) - f(a)}{h}$

Solutions   (A) $f(a) = a^2 - 2a + 7$
(B) $f(a+h) = (a+h)^2 - 2(a+h) + 7$
$\qquad\qquad = a^2 + 2ah + h^2 - 2a - 2h + 7$

(C) $\dfrac{f(a+h) - f(a)}{h} = \dfrac{(a^2 + 2ah + h^2 - 2a - 2h + 7) - (a^2 - 2a + 7)}{h}$

$$= \frac{2ah + h^2 - 2h}{h} \quad \boxed{= \frac{h(2a + h - 2)}{h}} = 2a + h - 2$$ ◆

PROBLEM 24   Repeat Example 24 for $f(x) = x^2 - 4x + 9$.   ◆

♦ APPLICATION

The market research department of a company that manufactures memory chips for microcomputers has determined that the demand equation for 256k chips is

$$x = 10{,}000 - 50p$$

where x is the number of chips that can be sold at a price of $\$p$ per chip.

(A) Express the revenue $R(x)$ in terms of the demand x.
(B) What is the domain of the function R?

Solutions

(A) The revenue from the sale of x units at $\$p$ per unit is $R = xp$. To express R in terms of x, we first solve the demand equation for p in terms of x:

$$x = 10{,}000 - 50p$$
$$50p = 10{,}000 - x$$
$$p = 200 - \tfrac{1}{50}x$$

Thus,

$$R(x) = xp = x(200 - \tfrac{1}{50}x) = 200x - \tfrac{1}{50}x^2$$

(B) Since neither price nor demand can be negative, x must satisfy

$$x \geqslant 0 \qquad \text{and} \qquad 200 - \tfrac{1}{50}x \geqslant 0$$

Solving the second inequality and combining it with the first, the domain of R is

$$0 \leqslant x \leqslant 10{,}000 \quad \text{or} \quad [0, 10{,}000] \qquad ♦$$

PROBLEM 25   Repeat Example 25 for the demand equation $x = 15{,}000 - 30p$.   ♦

Answers to Matched Problems

19. S does not define a function; T defines a function with domain $\{-3, -1, 1, 3, 5\}$ and range $\{0, 1, 2\}$
20. (A) Does not specify a function   (B) Specifies a function
21. $x \geqslant 2$ (inequality notation) or $[2, \infty)$ (interval notation)
22. (A) -3 (B) 0 (C) Does not exist   (D) 6
23. Domain of F: R; Domain of G: All real numbers except -3; Domain of H: $x \leqslant 2$ (inequality notation) or $(-\infty, 2]$ (interval notation)
24. (A) $a^2 - 4a + 9$ (B) $a^2 + 2ah + h^2 - 4a - 4h + 9$ (C) $2a + h - 4$
25. (A) $R(x) = x(500 - \tfrac{1}{30}x) = 500x - \tfrac{1}{30}x^2$
     (B) Domain: $0 \leqslant x \leqslant 15{,}000$ (inequality notation) or $[0, 15{,}000]$ (interval notation)

A   *Indicate whether each table specifies a function.*

1.
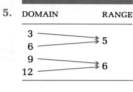

2.

3.
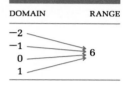

4.
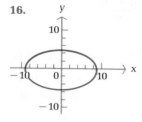

5.
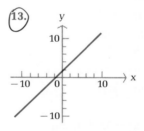

6.
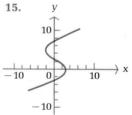

Indicate whether each set in Problems 7 – 12 specifies a function. Find the domain and range of each function.

7.   {(1, 3), (2, 5), (4, 7), (5, 9)}
8.   {(−3, 5), (−2, 4), (−1, 3), (0, 2)}
9.   {(6, −6), (3, −3), (0, 0), (3, 3), (6, 6)}
10.   {(−6, 6), (−3, 3), (0, 0), (3, 3), (6, 6)}
11.   {(1, 3), (2, 3), (3, 3), (4, 4), (5, 4), (6, 4)}
12.   {(0, 2), (1, 2), (2, 2), (0, 1), (1, 1), (2, 1)}

Indicate whether each graph specifies a function.

13.
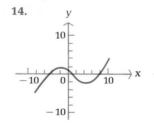

14.

15.

16.
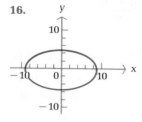

17.
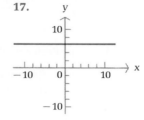

18.
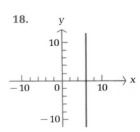

If $f(x) = 3x - 2$ and $g(x) = x - x^2$, find each of the following:

19. $f(2)$ 20. $f(1)$ 21. $f(-1)$ 22. $f(-2)$
23. $g(3)$ 24. $g(1)$ 25. $f(0)$ 26. $f(\frac{1}{3})$
27. $g(-3)$ 28. $g(-2)$ 29. $f(1) + g(2)$ 30. $g(1) + f(2)$
31. $g(2) - f(2)$ 32. $f(3) - g(3)$ 33. $g(3) \cdot f(0)$ 34. $g(0) \cdot f(-2)$

35. $\dfrac{g(-2)}{f(-2)}$ 36. $\dfrac{g(-3)}{f(2)}$

B   Find the domain of each function in Problems 37–48.

37. $f(x) = \sqrt{x}$ 38. $f(x) = \dfrac{1}{\sqrt{x}}$ 39. $f(x) = \dfrac{x - 3}{(x - 5)(x + 3)}$

40. $f(x) = \dfrac{x + 1}{x - 2}$ 41. $f(x) = \sqrt{x + 5}$ 42. $f(x) = \sqrt{7 - x}$

43. $f(x) = \dfrac{x^2 + 1}{x^2 - 1}$ 44. $f(x) = \dfrac{x^2 + 5}{x^2 - 9}$ 45. $f(x) = \dfrac{x}{x^2 + 3x - 4}$

46. $f(x) = \dfrac{x}{x^2 + x - 6}$ 47. $f(x) = \dfrac{x + 4}{x^2 - 4x + 5}$ 48. $f(x) = \dfrac{x - 7}{x^2 + 6x + 10}$

Determine which of the equations in Problems 49–58 specify functions with independent variable x. For those that do, find the domain. For those that do not, find a value of x to which there corresponds more than one value of y.

49. $4x - 5y = 20$ 50. $3y - 7x = 15$ 51. $x^2 - y = 1$
52. $x - y^2 = 1$ 53. $x + y^2 = 10$ 54. $x^2 + y = 10$
55. $xy - 4y = 1$ 56. $xy + y - x = 5$ 57. $x^2 + y^2 = 25$
58. $x^2 - y^2 = 16$

59. If $F(t) = 4t + 7$, find: 60. If $G(r) = 3 - 5r$, find:

   $\dfrac{F(3 + h) - F(3)}{h}$ $\dfrac{G(2 + h) - G(2)}{h}$

61. If $g(w) = w^2 - 4$, find: 62. If $f(m) = 2m^2 + 5$, find:

   $\dfrac{g(1 + h) - g(1)}{h}$ $\dfrac{f(4 + h) - f(4)}{h}$

63. If $Q(x) = x^2 - 5x + 1$, find: 64. If $P(x) = 2x^2 - 3x - 7$, find:

   $\dfrac{Q(2 + h) - Q(2)}{h}$ $\dfrac{P(3 + h) - P(3)}{h}$

C   In Problems 65–72, find and simplify: $\dfrac{f(a + h) - f(a)}{h}$

65. $f(x) = 4x - 3$ 66. $f(x) = -3x + 9$ 67. $f(x) = 4x^2 - 7x + 6$
68. $f(x) = 3x^2 + 5x - 8$ 69. $f(x) = x^3$ 70. $f(x) = x^3 - x$

71. $f(x) = \sqrt{x}$ 72. $f(x) = \dfrac{1}{x}$

$$A = \ell w$$
$$P = 2\ell + 2w$$

w

ℓ

Problems 73–76 refer to the area A and perimeter P of a rectangle with length ℓ and width w (see the figure).

73. The area of a rectangle is 25 square inches. Express the perimeter $P(w)$ as a function of the width w, and state the domain of this function.

74. The area of a rectangle is 81 square inches. Express the perimeter $P(\ell)$ as a function of the length ℓ, and state the domain of this function.

75. The perimeter of a rectangle is 100 meters. Express the area $A(\ell)$ as a function of the length ℓ, and state the domain of this function.

76. The perimeter of a rectangle is 160 meters. Express the area $A(w)$ as a function of the width w, and state the domain of this function.

APPLICATIONS

Business & Economics

77. *Cost function.* The fixed costs (tooling and overhead) for manufacturing a particular stereo system are $96,000, and the variable costs per unit (labor, material, etc.) are $80. If x units are manufactured, express the cost $C(x)$ as a function of x. Find the cost of producing 500 stereos.

78. *Cost function.* A company that specializes in manufacturing reproductions of classic automobiles has fixed costs of $100,000 and variable costs of $15,000 per automobile produced. If x cars are manufactured, express the cost $C(x)$ as a function of x. Find the cost of producing 48 automobiles.

79. *Revenue function.* After extensive surveys, the research department of a stereo manufacturing company produced the demand equation

$$x = 8{,}000 - 40p$$

where x is the number of units that retailers are likely to purchase at a price of $p per unit. Express the revenue $R(x)$ in terms of the demand x. Find the domain of R.

80. *Revenue function.* Repeat Problem 79 for the demand equation

$$x = 9{,}000 - 60p$$

81. *Packaging.* A candy box is to be made out of a piece of cardboard that measures 8 by 12 inches. Equal-sized squares x inches on a side will be cut out of each corner, and then the ends and sides will be folded up to form a rectangular box.

(A) Express the volume of the box $V(x)$ in terms of x.

(B) What is the domain of the function V (determined by the physical restrictions)?

(C) Complete the table:

x	V(x)
1	
2	
3	

Notice how the volume changes with different choices of x.

82. *Packaging.* A parcel delivery service will only deliver packages with length plus girth (distance around) not exceeding 108 inches. A rectangular shipping box with square ends x inches on a side is to be used.

(A) If the full 108 inches is to be used, express the volume of the box V(x) in terms of x.

(B) What is the domain of the function V (determined by the physical restrictions)?

(C) Complete the table:

x	V(x)
5	
10	
15	
20	
25	

Notice how the volume changes with different choices of x.

83. *Construction.* A veterinarian wants to construct a kennel with five individual pens, as indicated in the figure. Local ordinances require that each pen have a gate that is 3 feet wide and an area of 45 square feet. If x is the width of one pen, express the total amount of fencing P(x) (excluding the gates) required for construction of the kennel as a function of x, and complete the table in the margin.

x	P(x)
3	
4	
5	
6	

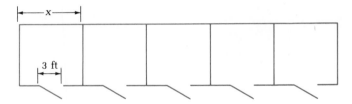

84. *Construction.* A horse breeder wants to construct three adjacent corrals, as indicated in the figure. If each corral must have an area of 600 square feet and x is the width of one corral, express the total amount of fencing P(x) as a function of x, and complete the following table:

x	P(x)
10	
15	
20	
25	

Life Sciences

85. *Muscle contraction.* In a study of the speed of muscle contraction in frogs under various loads, noted British biophysicist and Nobel prize winner A. W. Hill determined that the weight w (in grams) placed on the muscle

and the speed of contraction v (in centimeters per second) are approximately related by an equation of the form

$$(w + a)(v + b) = c$$

where a, b, and c are constants. Suppose that for a certain muscle, $a = 15$, $b = 1$, and $c = 90$. Express v as a function of w. Find the speed of contraction if a weight of 16 grams is placed on the muscle.

Social Sciences

86. *Politics.* The percentage s of seats in the House of Representatives won by Democrats and the percentage v of votes cast for Democrats (when expressed as decimal fractions) are related by the equation

$$5v - 2s = 1.4 \qquad 0 < s < 1, \quad 0.28 < v < 0.68$$

(A) Express v as a function of s, and find the percentage of votes required for the Democrats to win 51% of the seats.

(B) Express s as a function of v, and find the percentage of seats won if Democrats receive 51% of the votes.

SECTION 2-5 Linear and Quadratic Functions

◆ GRAPHS OF FUNCTIONS
◆ LINEAR FUNCTIONS AND THEIR GRAPHS
◆ QUADRATIC FUNCTIONS AND THEIR GRAPHS
◆ PIECEWISE-DEFINED FUNCTIONS
◆ APPLICATION: MARKET RESEARCH

◆ GRAPHS OF FUNCTIONS

Each function that has a real number domain and range has a graph — the graph of the ordered pairs of real numbers that constitute the function. When functions are graphed, domain values are usually associated with the horizontal axis and range values with the vertical axis. Thus, the **graph of a function f** is the graph of the set

$$\{(x, y) | y = f(x), \quad x \in \text{Domain of } f\}$$

where x is the independent variable and the abscissa of a point on the graph of f, and y and $f(x)$ are dependent variables and either is the ordinate of a point on the graph of f (see Fig. 9 at the top of the next page).

The first coordinate (abscissa) of a point where the graph of a function crosses the x axis is called an **x intercept** of the function. The x intercepts are determined by finding the real solutions of the equation $f(x) = 0$, if any exist. The second coordinate (ordinate) of a point where the graph of a function crosses the y axis is called the **y intercept** of the function. The y intercept is given by $f(0)$,

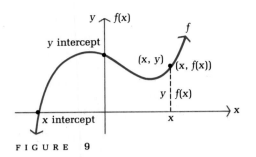

FIGURE 9

provided 0 is in the domain of *f*. Note that a function can have more than one *x* intercept, but can never have more than one *y* intercept (see the vertical line test in the preceding section).

◆ LINEAR FUNCTIONS AND THEIR GRAPHS

A function specified by an equation of the form

$$f(x) = mx + b$$

is called a **linear function.** In the special case *m* = 0, the function has the form

$$f(x) = b$$

and is also called a **constant function.** Graphing linear functions is equivalent to graphing the equation

$$y = mx + b \qquad \text{Slope} = m, \; y \text{ intercept} = b$$

which we discussed in detail in Section 2-3. Since the expression *mx* + *b* represents a real number for all real number replacements of *x*, the domain of any linear function is *R*, the set of all real numbers. The range of a nonconstant linear function is also *R*, while the range of a constant function is the single real number *b*. See the graphs in the box.

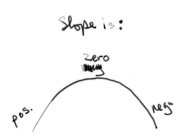

■ **Graph of** $f(x) = mx + b$

The graph of a linear function *f* is a nonvertical straight line with slope *m* and *y* intercept *b*.

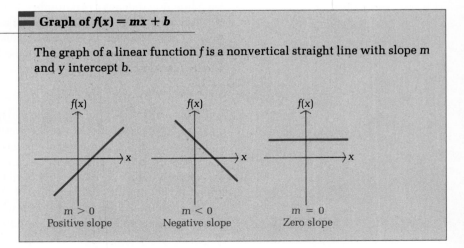

◆ E X A M P L E   26   Graph the linear function defined by

$$f(x) = -\frac{x}{2} + 3$$

and indicate its slope and intercepts.

Solution

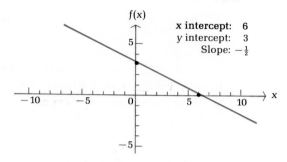

x intercept: 6
y intercept: 3
Slope: $-\frac{1}{2}$

P R O B L E M   26   Graph the linear function defined by

$$f(x) = \frac{x}{3} + 1$$

and indicate its slope and intercepts.

◆ QUADRATIC FUNCTIONS AND THEIR GRAPHS

Any function defined by an equation of the form

$$f(x) = ax^2 + bx + c \qquad a \neq 0$$

where a, b, and c are constants and x is a variable, is called a **quadratic function.**
Let us start by graphing the simple quadratic function

$$f(x) = x^2$$

We evaluate this function for integer values from its domain, find corresponding range values, then plot the resulting ordered pairs listed in Table 8, and join these points with a smooth curve, as shown in Figure 10. The first two steps are usually done mentally or on scratch paper.

FIGURE   10

TABLE   8

Graphing $f(x) = x^2$

DOMAIN VALUES x	RANGE VALUES $y = f(x)$	ELEMENTS OF f $(x, f(x))$
-2	$y = f(-2) = (-2)^2 = 4$	$(-2, 4)$
-1	$y = f(-1) = (-1)^2 = 1$	$(-1, 1)$
0	$y = f(0) = 0^2 = 0$	$(0, 0)$
1	$y = f(1) = 1^2 = 1$	$(1, 1)$
2	$y = f(2) = 2^2 = 4$	$(2, 4)$

The curve shown in Figure 10 is called a **parabola.** It can be shown (in a course in analytic geometry) that the graph of any quadratic function is also a parabola. In general:

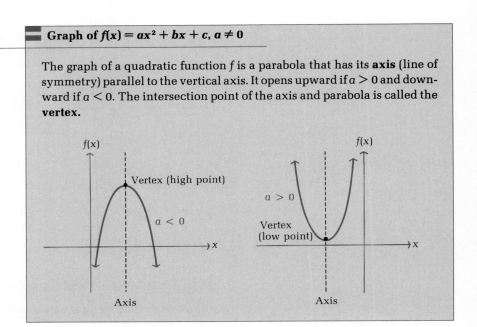

Graph of $f(x) = ax^2 + bx + c, a \neq 0$

The graph of a quadratic function f is a parabola that has its **axis** (line of symmetry) parallel to the vertical axis. It opens upward if $a > 0$ and downward if $a < 0$. The intersection point of the axis and parabola is called the **vertex.**

In addition to the point-by-point method of graphing quadratic functions described above, let us consider another approach that will give us added insight into these functions. (A brief review of completing the square, which is discussed in Section 2-2, may prove useful first.) We illustrate the method through an example, and then generalize the results.

Consider the quadratic function given by

$$f(x) = 2x^2 - 8x + 5$$

If we can find the vertex of the graph, then the rest of the graph can be sketched with relatively few points. In addition, we will then have found the maximum or minimum value of the function. We start by transforming the equation into the form

$$f(x) = a(x - h)^2 + k \qquad a, h, k \text{ constants}$$

by completing the square:

$$f(x) = 2x^2 - 8x + 5 \qquad \text{Factor the coefficient of } x^2 \text{ out of the first}$$
$$= 2(x^2 - 4x) + 5 \qquad \text{two terms.}$$

$$f(x) = 2(x^2 - 4x + ?) + 5$$

Complete the square within parentheses.

$$= 2(x^2 - 4x + 4) + 5 - 8$$

We added 4 to complete the square inside the parentheses; but because of the 2 on the outside, we have actually added 8, so we must subtract 8.

$$= 2(x - 2)^2 - 3$$

The transformation is complete.

Thus,

$$f(x) = \underbrace{2(x - 2)^2}_{} - 3$$

Never negative
(Why?)

When $x = 2$, the first term on the right vanishes, and we add 0 to -3. For *any* other value of x we will add a positive number to -3, thus making $f(x)$ larger. Therefore, $f(2) = -3$ is the minimum value of $f(x)$ for all x. A very important result!

The point $(2, -3)$ is the lowest point on the parabola and is also the vertex. The vertical line $x = 2$ is the axis of the parabola. We plot the vertex and the axis and a couple of points on either side of the axis to complete the graph (Fig. 11).

x	f(x)
2	−3
1	−1
3	−1
0	5
4	5

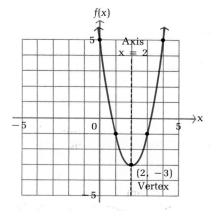

FIGURE 11

Examining the graph, we see that $f(x)$ can assume any value greater than or equal to -3, but no values less than -3. Thus,

Range of $f = \{y | y \geq -3\} = [-3, \infty)$

We also note that f has two x intercepts, which are solutions of the equation $f(x) = 0$. Since we have already completed the square, we use this method to find the x intercepts:

$$f(x) = 0$$
$$2(x - 2)^2 - 3 = 0$$

$$2(x - 2)^2 = 3$$
$$(x - 2)^2 = \tfrac{3}{2}$$
$$x - 2 = \pm\sqrt{\tfrac{3}{2}}$$
$$= \pm\tfrac{1}{2}\sqrt{6}$$
$$x = 2 \pm \tfrac{1}{2}\sqrt{6}$$

Thus, the x intercepts are

$$x = 2 + \tfrac{1}{2}\sqrt{6} \approx 3.22 \qquad \text{and} \qquad x = 2 - \tfrac{1}{2}\sqrt{6} \approx 0.78$$

For completeness, we observe that the y intercept is $f(0) = 5$, obtained as part of the table of values of $f(x)$.

Note all the important results we have with this approach:

1. Axis of the parabola
2. Vertex of the parabola
3. Minimum value of $f(x)$
4. Graph of $y = f(x)$
5. Range of f
6. x intercepts

If we start with the general quadratic function defined by

$$f(x) = ax^2 + bx + c \qquad a \neq 0$$

and complete the square (see Section 2-2), we obtain

$$f(x) = a\left(x + \frac{b}{2a}\right)^2 + \frac{4ac - b^2}{4a}$$

Using the same reasoning as above, we obtain the following general results:

Properties of $f(x) = ax^2 + bx + c$, $a \neq 0$

1. Axis of symmetry: $\quad x = -\dfrac{b}{2a}$

2. Vertex: $\quad \left(-\dfrac{b}{2a}, f\left(-\dfrac{b}{2a}\right)\right)$

3. Maximum or minimum value of $f(x)$:

$$f\left(-\frac{b}{2a}\right) = \begin{cases} \text{Minimum} & \text{if } a > 0 \\ \text{Maximum} & \text{if } a < 0 \end{cases}$$

4. Domain: All real numbers
   Range: Determine from graph

5. y intercept: $f(0) = c$
   x intercepts: Solutions of $f(x) = 0$, if any exist

To graph a quadratic function using this method, we can actually complete the square as in the preceding example or use the properties in the box. Some of you can probably more readily remember a formula, others a process. We use the properties in the box in the next example.

◆ **E X A M P L E   27**

Graph, finding the axis, vertex, maximum or minimum of $f(x)$, range, and intercepts:

$$f(x) = -x^2 - 4x - 5$$ *when* $x = 0$, $y = -5$ *(the y intercept).*

Solution

Note that $a = -1$, $b = -4$, and $c = -5$.

Axis of symmetry: $x = -\dfrac{b}{2a} = -\dfrac{-4}{2(-1)} = -2$

Vertex: $\left(-\dfrac{b}{2a}, f\left(-\dfrac{b}{2a}\right)\right) = (-2, f(-2)) = (-2, -1)$

Maximum value of $f(x)$ (since $a = -1 < 0$):   Max $f(x) = f(-2) = -1$

To graph f, locate the axis and vertex; then plot several points on either side of the axis:

x	f(x)
-4	-5
-3	-2
-2	-1
-1	-2
0	-5

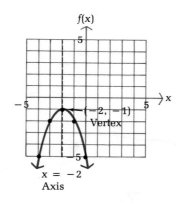

when $A > 0$ minimum
when $A < 0$ maximum

From the graph, we see that the range of f is

$$(-\infty, f(-2)] = (-\infty, -1]$$

The y intercept is $f(0) = -5$. Since the graph of f does not cross the x axis, there are no x intercepts. The equation $f(x) = 0$ has no real solutions (verify this). ◆

P R O B L E M

Graph, finding the axis, vertex, maximum or minimum of $f(x)$, range, and intercepts:

$$f(x) = x^2 - 4x + 4$$ ◆

◆ PIECEWISE-DEFINED FUNCTIONS

The **absolute value** of a real number a, denoted $|a|$, is the (positive) distance on a number line from a to the origin. Thus, $|4| = 4$ and $|-5| = 5$. More formally, the **absolute value function** is defined by

$$f(x) = |x| = \begin{cases} -x & \text{if } x < 0 \\ x & \text{if } x \geq 0 \end{cases}$$

Notice that this function is defined by different formulas for different parts of its domain. A function whose definition involves more than one formula is called a **piecewise-defined function.** As the next example illustrates, piecewise-defined functions occur naturally in many applications.

◆ EXAMPLE 28

Service Charges

The time charges for a service call by a telephone company are $16.20 for the first 6 minute period and $5.40 for each additional 6 minute period (or fraction thereof). Let $C(x)$ be the total time charges for a service call that lasts x minutes. Graph C for $0 < x \leq 30$.

Solution

The total time charges are given by the following piecewise definition:

$$C(x) = \begin{cases} \$16.20 & \text{if } 0 < x \leq 6 \\ \$21.60 & \text{if } 6 < x \leq 12 \\ \$27.00 & \text{if } 12 < x \leq 18 \\ \text{And so on} \end{cases}$$

To graph C, we graph each rule in this definition for the indicated values of x:

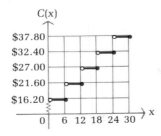

Note: A solid dot on the graph indicates that the point is part of the graph; an open dot indicates that the point is not part of the graph. ◆

PROBLEM 28

The time charges for a company that services home appliances are $24 for the first 15 minute period and $12 for each additional 15 minute period (or fraction thereof). Let $C(x)$ be the total time charges for a service call that lasts x minutes. Graph C for $0 < x \leq 60$. ◆

◆ APPLICATION: MARKET RESEARCH

The market research department of a company recommended to management that the company manufacture and market a promising new product. After extensive surveys, the research department backed up the recommendation with the **demand equation**

$$x = f(p) = 6{,}000 - 30p \qquad (1)$$

where x is the number of units that retailers are likely to buy per month at \$$p$ per unit. Notice that as the price goes up, the number of units goes down. From the financial department, the following **cost equation** was obtained:

$$C = g(x) = 72{,}000 + 60x \qquad (2)$$

where \$72,000 is the fixed cost (tooling and overhead) and \$60 is the variable cost per unit (materials, labor, marketing, transportation, storage, etc.). The **revenue equation** (the amount of money, R, received by the company for selling x units at \$$p$ per unit) is

$$R = xp \qquad (3)$$

And, finally, the **profit equation** is

$$P = R - C \qquad (4)$$

where P is profit, R is revenue, and C is cost.

Notice that the cost equation (2) expresses C as a function of x and the demand equation (1) expresses x as a function of p. Substituting (1) into (2), we obtain cost C as a linear function of price p:

$$\begin{aligned} C &= 72{,}000 + 60(6{,}000 - 30p) \\ &= 432{,}000 - 1{,}800p \end{aligned} \qquad \text{Linear function} \qquad (5)$$

Similarly, substituting (1) into (3), we obtain revenue R as a quadratic function of price p:

$$\begin{aligned} R &= (6{,}000 - 30p)p \\ &= 6{,}000p - 30p^2 \end{aligned} \qquad \text{Quadratic function} \qquad (6)$$

Now let us graph equations (5) and (6) in the same coordinate system. We obtain Figure 12 (page 106). Notice how much information is contained in this graph.

Let us compute the **break-even points;** that is, the prices at which cost equals revenue (the points of intersection of the two graphs in Figure 12). Find p so that

$$C = R$$
$$432{,}000 - 1{,}800p = 6{,}000p - 30p^2$$
$$30p^2 - 7{,}800p + 432{,}000 = 0$$
$$p^2 - 260p + 14{,}400 = 0 \qquad \text{Solve using the}$$
$$\qquad\qquad\qquad\qquad\qquad\qquad \text{quadratic formula}$$
$$p = \frac{260 \pm \sqrt{260^2 - 4(14{,}400)}}{2} \qquad \text{(Section 2-2).}$$
$$= \frac{260 \pm 100}{2}$$
$$= \$80, \quad \$180$$

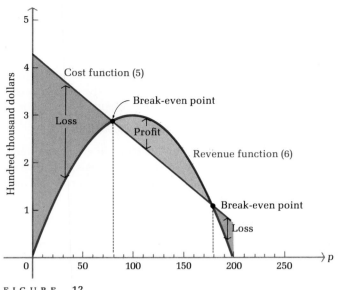

FIGURE 12

Thus, at a price of $80 or $180 per unit the company will break even. Between these two prices it is predicted that the company will make a profit.

At what price will a maximum profit occur? To find out, we write

$$P = R - C$$
$$= (6{,}000p - 30p^2) - (432{,}000 - 1{,}800p)$$
$$= -30p^2 + 7{,}800p - 432{,}000$$

Since this is a quadratic function, the maximum profit occurs at

$$p = -\frac{b}{2a} = -\frac{7{,}800}{2(-30)} = \$130$$

Note that this is not the price at which the maximum revenue occurs. The latter occurs at $p = \$100$, as shown in Figure 12.

Answers to Matched Problems   26.   Slope: $\frac{1}{3}$
y intercept: 1
x intercept: -3

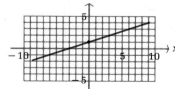

27. Minimum: $f(2) = 0$
Range: $[0, \infty)$
y intercept: 4
x intercept: 2

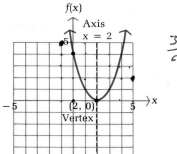

28.

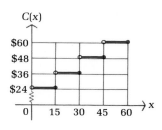

$\dfrac{3}{6}$

A   *Graph each linear function, and indicate its slope and intercepts.*

① $f(x) = 2x - 4$ 2. $g(x) = \dfrac{x}{2}$ ③ $h(x) = 4 - 2x$

4. $f(x) = -\dfrac{x}{2} + 3$ 5. $g(x) = -\dfrac{2}{3}x + 4$ 6. $f(x) = 3$

Find a linear function with the indicated slope and y intercept.

⑦ Slope -2, y intercept 6   8. Slope 3, y intercept -5

Find a linear function whose graph passes through the indicated points.

9. $(-1, 5)$, $(5, 2)$ 10. $(1, -2)$, $(7, 6)$

B   *Graph, finding the axis, vertex, maximum or minimum, intercepts, and range.*

11. $f(x) = (x - 3)^2 - 1$ 12. $g(x) = -(x + 2)^2 + 4$
13. $h(x) = -(x + 1)^2 + 9$ 14. $k(x) = (x - 2)^2 - 16$
15. $f(x) = x^2 + 8x + 16$ 16. $h(x) = x^2 - 2x - 3$
17. $f(u) = u^2 - 2u + 4$ 18. $f(x) = x^2 - 10x + 25$
19. $h(x) = 2 + 4x - x^2$ 20. $g(x) = -x^2 - 6x - 4$
21. $f(x) = 6x - x^2$ 22. $G(x) = 16x - 2x^2$
23. $F(s) = s^2 - 4$ 24. $g(t) = t^2 + 4$
25. $F(x) = 4 - x^2$ 26. $G(x) = 9 - x^2$

Graph each function, and state its domain and range.

27. $f(x) = \begin{cases} 1 & \text{if } 0 \leqslant x \leqslant 2 \\ 3 & \text{if } 2 < x \leqslant 3 \\ 5 & \text{if } 3 < x \leqslant 5 \end{cases}$ 28. $g(x) = \begin{cases} -2 & \text{if } -4 \leqslant x \leqslant -2 \\ 1 & \text{if } -2 < x < 2 \\ 2 & \text{if } 2 \leqslant x \leqslant 4 \end{cases}$

29. $f(x) = \begin{cases} x & \text{if } -2 \leqslant x < 1 \\ -x + 2 & \text{if } 1 \leqslant x \leqslant 2 \end{cases}$

30. $f(x) = \begin{cases} x + 1 & \text{if } -1 \leqslant x < 0 \\ -x + 1 & \text{if } 0 \leqslant x \leqslant 1 \end{cases}$

31. $h(x) = \begin{cases} -x^2 - 2 & \text{if } x < 0 \\ x^2 + 2 & \text{if } x > 0 \end{cases}$

32. $g(x) = \begin{cases} x^2 + 1 & \text{if } x < 0 \\ -x^2 - 1 & \text{if } x > 0 \end{cases}$

33. $G(x) = \begin{cases} -3 & \text{if } x < -3 \\ x & \text{if } -3 \leqslant x \leqslant 3 \\ 3 & \text{if } x > 3 \end{cases}$

34. $F(x) = \begin{cases} 1 & \text{if } x < -2 \\ 5 - x^2 & \text{if } -2 \leqslant x \leqslant 2 \\ 1 & \text{if } x > 2 \end{cases}$

C   *Graph, finding the axis, vertex, maximum or minimum, intercepts, and range.*

35. $f(x) = x^2 - 7x + 10$

36. $g(t) = t^2 - 5t + 2$

37. $h(x) = 2 - 5x - x^2$

38. $g(t) = 4 + 3t - t^2$

Graph f and g on the same set of axes and find any points of intersection.

39. $f(x) = 2 + \frac{1}{2}x;\quad g(x) = 8 - x$

40. $f(x) = -2x - 1;\quad g(x) = x - 4$

41. $f(x) = x^2 - 6x + 8;\quad g(x) = x - 2$

42. $f(x) = 8 + 2x - x^2;\quad g(x) = x + 6$

43. $f(x) = 11 + 2x - x^2;\quad g(x) = x^2 - 1$

44. $f(x) = x^2 - 4x - 10;\quad g(x) = 14 - 2x - x^2$

APPLICATIONS

Business & Economics

45. *Service charges.* On weekends and holidays, an emergency plumbing repair service charges $60 for the first 30 minute period (or fraction thereof) of a service call and $20 for each additional 15 minute period (or fraction thereof). Let $C(x)$ be the cost of a service call that lasts x minutes. Graph $C(x)$ for $0 < x \leqslant 90$.

46. *Delivery charges.* A nationwide package delivery service charges $20.00 for overnight delivery of packages weighing 1 pound or less. Each additional pound (or fraction thereof) costs an additional $2.50. Let $D(x)$ be the charge for overnight delivery of a package weighing x pounds. Graph $D(x)$ for $0 < x \leqslant 10$.

47. *Market research.* Suppose that in the market research example in this section the demand equation (1) is changed to $x = 9{,}000 - 30p$ and the cost equation (2) is changed to $C = 90{,}000 + 30x$.

(A) Express cost C as a linear function of price p.

(B) Express revenue R as a quadratic function of price p.

(C) Graph the cost and revenue functions found in parts A and B in the same coordinate system, and identify the regions of profit and loss on your graph.

(D) Find the break-even points; that is, find the prices to the nearest dollar at which $R = C$. (A calculator might prove useful here.)

(E) Find the price that produces the maximum revenue.

48. *Market research.* Repeat Problem 47 if the demand equation (1) is changed to $x = 5{,}000 - 50p$ and the cost equation (2) is changed to $C = 40{,}000 + 12x$.

49. *Break-even analysis.* A publisher is planning to produce a new textbook. The fixed costs (reviewing, editing, typesetting, etc.) are $240,000, and the variable costs (printing, sales commissions, etc.) are $20 per book. The wholesale price (the amount received by the publisher) will be $35 per book. Let x be the number of books.

(A) Express the cost C as a linear function of x.
(B) Express the revenue R as a linear function of x.
(C) Graph the cost and revenue functions found in parts A and B on the same set of axes.
(D) Find the number of books the publisher has to sell in order to break even.

50. *Break-even analysis.* A computer software company is planning to market a new word processor for a microcomputer. The fixed costs (programming, debugging, etc.) are $300,000, and the variable costs (disk duplication, manual production, etc.) are $25 per unit. The wholesale price of the product will be $100 per unit. Let x be the number of units.

(A) Express the cost C as a linear function of x.
(B) Express the revenue R as a linear function of x.
(C) Graph the cost and revenue functions found in parts A and B on the same set of axes.
(D) Find the number of units the company has to sell in order to break even.

Life Sciences

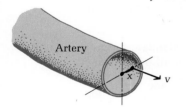

51. *Medicine.* The French physician Poiseuille was the first to discover that blood flows faster near the center of an artery than near the edge. Experimental evidence has shown that the rate of flow v (in centimeters per second) at a point x centimeters from the center of an artery (see the figure) is given by

$$v = f(x) = 1{,}000(0.04 - x^2) \qquad 0 \leqslant x \leqslant 0.2$$

Graph this quadratic function for the indicated values of x.

52. *Air pollution.* On an average summer day in a large city, the pollution index at 8:00 AM is 20 parts per million, and it increases linearly by 15 parts per million each hour until 3:00 PM. Let $P(x)$ be the amount of pollutants in the air x hours after 8:00 AM.

(A) Express $P(x)$ as a linear function of x.
(B) What is the air pollution index at 1:00 PM?
(C) Graph the function P for $0 \leqslant x \leqslant 7$.
(D) What is the slope of the graph? (The slope is the amount of increase in pollution for each additional hour of time.)

53. *Psychology — sensory perception.* One of the oldest studies in psychology concerns the following question: Given a certain level of stimulation (light, sound, weight lifting, electric shock, and so on), how much should the stimulation be increased for a person to notice the difference? In the middle of the nineteenth century, E. H. Weber (a German physiologist) formulated a law that still carries his name: If Δs is the change in stimulus that will just be noticeable at a stimulus level s, then the ratio of Δs to s is a constant:

$$\frac{\Delta s}{s} = k$$

Hence, the amount of change that will be noticed is a linear function of the stimulus level, and we note that the greater the stimulus, the more it takes to notice a difference. In an experiment on weight lifting, the constant k for a given individual was found to be $\frac{1}{30}$.

(A) Find Δs (the difference that is just noticeable) at the 30 pound level; at the 90 pound level.

(B) Graph $\Delta s = s/30$ for $0 \leqslant s \leqslant 120$.

(C) What is the slope of the graph?

SECTION 2-6   Chapter Review

Important Terms and Symbols

2-1 *Linear Equations and Inequalities in One Variable.* Linear equation; linear inequality; solution; solution set; solution of an equation; equivalent equations; equality properties; inequality properties; interval notation; fixed cost; variable cost; break-even analysis

$$a < b; \quad a > b; \quad a \leqslant x \leqslant b; \quad [a, b]; \quad [a, b); \quad (a, b]; \quad (a, b);$$
$$(-\infty; a]; \quad (-\infty, a); \quad [b, \infty); \quad (b, \infty)$$

2-2 *Quadratic Equations.* Quadratic equation; standard form; completing the square; quadratic formula; discriminant; equilibrium point

$$ax^2 + bx + c = 0, a \neq 0; \quad x = \frac{-b \pm \sqrt{b^2 - 4ac}}{2a}; \quad b^2 - 4ac$$

2-3 *Cartesian Coordinate System and Straight Lines.* Cartesian coordinate system; rectangular coordinate system; coordinate axes; ordered pair; coordinates; abscissa; ordinate; quadrants; solution of an equation in two variables; solution set; graph of an equation; standard form; x intercept; y intercept; slope; slope-intercept form; point-slope form; horizontal line; vertical line; parallel lines; perpendicular lines

$$Ax + By = C; \quad m = \frac{y_2 - y_1}{x_2 - x_1}; \quad y = mx + b;$$
$$y - y_1 = m(x - x_1); \quad y = c; \quad x = c$$

2-4 *Functions.* Function; domain; range; input; output; independent variable; dependent variable; vertical line test; function notation; revenue function

$$f(x); \quad f: x \to f(x); \quad (x, f(x))$$

2-5 *Linear and Quadratic Functions.* Graph of a function; x intercept; y intercept; linear function; constant function; quadratic function; parabola; axis of a parabola; vertex of a parabola; maximum; minimum; absolute value; absolute value function; piecewise-defined function; demand equation; cost equation; revenue equation; profit equation; break-even point

$$f(x) = ax + b; \quad f(x) = ax^2 + bx + c, \ a \neq 0; \quad |x|$$

Chapter Review

Work through all the problems in this chapter review and check your answers in the back of the book. (Answers to all review problems are there.) Where weaknesses show up, review appropriate sections in the text.

A

1. Solve: $\dfrac{u}{5} = \dfrac{u}{6} + \dfrac{6}{5}$

2. Solve and graph on a real number line: $2(x + 4) > 5x - 4$
3. Solve: $x^2 = 5x$
4. Graph the equation below in a rectangular coordinate system. Indicate the slope and the y intercept.

$$y = \frac{x}{2} - 2$$

5. Write the equation of a line that passes through (4, 3) with slope $\frac{1}{2}$. Write the final answer in the form $y = mx + b$.
6. Graph $x - y = 2$ in a rectangular coordinate system. Indicate the slope.
7. For $f(x) = 2x - 1$ and $g(x) = x^2 - 2x$, find $f(-2) + g(-1)$.
8. Graph the linear function f given by the equation

$$f(x) = \tfrac{2}{3}x - 1$$

Indicate the slope and intercepts.
9. Find the maximum or minimum value of $f(x) = x^2 - 8x + 7$ without graphing. What are the coordinates of the vertex of the parabola?

B

10. Solve: $\dfrac{x}{12} - \dfrac{x - 3}{3} = \dfrac{1}{2}$

Solve and graph on a real number line.

11. $1 - \dfrac{x - 3}{3} \leq \dfrac{1}{2}$

12. $-2 \leq \dfrac{x}{2} - 3 < 3$

Solve for y in terms of x.

13. $2x - 3y = 6$

14. $xy - y = 3$

Solve Problems 15–17.

15. $3x^2 - 21 = 0$

16. $x^2 - x - 20 = 0$

17. $2x^2 = 3x + 1$

18. Graph $3x + 6y = 18$ in a rectangular coordinate system. Indicate the slope and intercepts.

19. Find an equation of the line that passes through $(-2, 3)$ and $(6, -1)$. Write the answer in the form $Ax + By = C$, $A > 0$. What is the slope of the line?

20. Write the equations of the vertical line and the horizontal line that pass through $(-5, 2)$. Graph both equations on the same coordinate system.

21. Find an equation of the line that passes through $(-2, 5)$ and $(2, -1)$. Write the answer in the form $y = mx + b$.

22. For $f(x) = 10x - 7$, $g(t) = 6 - 2t$, $F(u) = 3u^2$, and $G(v) = v - v^2$, find:

(A) $2g(-1) - 3G(-1)$

(B) $4G(-2) - g(-3)$

(C) $\dfrac{f(2) \cdot g(-4)}{G(-1)}$

(D) $\dfrac{F(-1) \cdot G(2)}{g(-1)}$

23. Find the domains of the functions f and g if

$$f(x) = 2x - x^2 \qquad g(x) = \frac{1}{x - 2}$$

24. For $f(x) = 2x - 1$, find: $\dfrac{f(3 + h) - f(3)}{h}$

25. Determine which of the following equations specify functions with independent variable x. For those that do, find the domain. For those that do not, find a value of x to which there corresponds more than one value of y.

(A) $4x - 3y = 11$

(B) $y^2 - 4x = 1$

(C) $xy + 3y + 5x = 4$

26. Graph $g(x) = 8x - 2x^2$, finding the axis, vertex, maximum or minimum, range, and intercepts.

27. Sketch the graph and find the domain and range of

$$f(x) = \begin{cases} 1 - x^2 & \text{if } -1 \leqslant x < 0 \\ 1 + x^2 & \text{if } 0 \leqslant x \leqslant 1 \end{cases}$$

C **28.** Solve $x^2 + jx + k = 0$ for x in terms of j and k.

29. Write an equation of the line that passes through the points $(4, -3)$ and $(4, 5)$.

30. Write an equation of the line that passes through $(2, -3)$ and is:

(A) Parallel to $2x - 4y = 5$

(B) Perpendicular to $2x - 4y = 5$

Write the final answers in the form $Ax + By = C$, $A > 0$.

31. Find the domain of the function f specified by each equation.

(A) $f(x) = \dfrac{5}{x - 3}$ (B) $f(x) = \sqrt{x - 1}$

32. For $f(x) = x^2 + 7x - 9$, find and simplify: $\dfrac{f(a + h) - f(a)}{h}$

33. Graph $f(x) = 2x - 7$ and $g(x) = x^2 - 6x + 5$ on the same set of axes, and find any points of intersection.

APPLICATIONS

Business & Economics

34. *Investment.* An investor has $60,000 to invest. If part is invested at 8% and the rest at 14%, how much should be invested at each rate to yield 12% on the total amount?

35. *Inflation.* If the CPI was 89 in 1960 and 247 in 1980, how much would a net salary of $800 in 1960 have to be in 1980 in order to keep up with inflation? Set up an equation and solve.

36. *Finance.* If P dollars is invested at $100r\%$ compounded annually, at the end of 2 years it will grow to $A = P(1 + r)^2$. At what interest rate will $1,000 grow to $1,210 in 2 years?

37. *Linear depreciation.* A word-processing system was purchased by a company for $12,000 and is assumed to have a salvage value of $2,000 after 8 years (for tax purposes). If its value is depreciated linearly from $12,000 to $2,000:

(A) Find the linear equation that relates value V in dollars to time t in years.

(B) What would be the value of the system after 5 years?

38. *Pricing.* A sporting goods store sells a tennis racket that cost $30 for $48 and a pair of jogging shoes that cost $20 for $32.

(A) If the markup policy of the store for items that cost over $10 is assumed to be linear and is reflected in the pricing of these two items, write an equation that relates retail price R to cost C.

(B) What should be the retail price of a pair of skis that cost $105?

39. *Market equilibrium.* After extensive research, a retail firm has determined that the supply and demand equations for a certain product are

$$s = 10{,}000p - 25{,}000 \quad \text{and} \quad d = \frac{90{,}000}{p}$$

where p is the price in dollars. Find the equilibrium point (the price where supply equals demand).

40. *Break-even analysis.* A video production company is planning to produce an instructional videotape. The producer estimates that it will cost $84,000

to shoot the video and $15 per unit to copy and distribute the tape. The wholesale price of the tape is $50 per unit. How many tapes must be sold for the company to break even?

41. *Market research.* The market research department of an electronics company has determined that the demand and cost equations for the production of an AM/FM clock radio are, respectively,

$$x = 500 - 10p \quad \text{and} \quad C = 3{,}000 + 10x$$

(A) Express the revenue as a function of the price p.
(B) Express the cost as a function of the price p.
(C) Graph R and C on the same set of axes and identify the regions of profit and loss.
(D) Find the break-even points.
(E) Find the price that will produce a maximum profit.

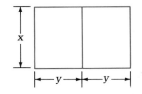

42. *Construction.* A farmer has 180 feet of fencing to be used in the construction of two identical rectangular pens sharing a common side (see the figure in the margin). Assuming all the fencing is used:

(A) Express the total area $A(x)$ enclosed by both pens as a function of the width x.
(B) From physical considerations, what is the domain of the function A?
(C) Find the dimensions of the pens that will make the total enclosed area maximum.

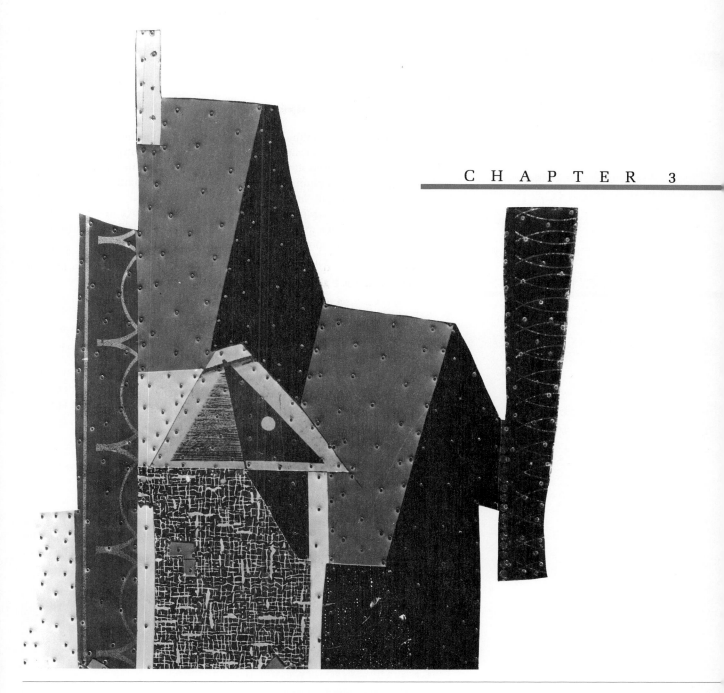

C H A P T E R   3

Exponential and Logarithmic Functions

Contents

CHAPTER 3

In this chapter we will define and investigate the properties of two new and important classes of functions called *exponential functions* and *logarithmic functions*. These functions are used in describing and solving a wide variety of real-world problems, including growth of money at compound interest; growth of populations of people, animals, and bacteria; radioactive decay (negative growth); and learning curves associated with the mastery of a new computer or an assembly process in a manufacturing plant. We will consider applications in these areas plus many more in sections that follow.

SECTION   3-1

Exponential Functions

◆   EXPONENTIAL FUNCTIONS
◆   BASIC EXPONENTIAL GRAPHS
◆   ADDITIONAL PROPERTIES
◆   APPLICATIONS

In this section we will define exponential functions, look at some of their important properties — including graphs — and consider several significant applications.

◆ EXPONENTIAL FUNCTIONS

Let us start by noting that the functions f and g given by

$$f(x) = 2^x \qquad \text{and} \qquad g(x) = x^2$$

are not the same function. Whether a variable appears as an exponent with a constant base or as a base with a constant exponent, makes a big difference. The function g is a quadratic function, which we have already discussed. The function f is a new type of function called an *exponential function*.

Exponential Function

The equation

$$f(x) = b^x \qquad b > 0, b \neq 1$$

defines an **exponential function** for each different constant b, called the **base.** The independent variable x may assume any real value.

Thus, the **domain of f** is the set of all real numbers, and it can be shown that the **range of f** is the set of all positive real numbers. We require the base b to be positive to avoid nonreal numbers such as $(-2)^{1/2}$. We exclude the case $b = 1$ since $f(x) = 1^x = 1$ is a constant function, not an exponential function.

◆ BASIC EXPONENTIAL GRAPHS

Many students, if asked to graph equations such as $y = 2^x$ or $y = 2^{-x}$, would not hesitate at all. [*Note:* $2^{-x} = 1/2^x = (\frac{1}{2})^x$.] They would likely make up tables by assigning integers to x, plot the resulting points, and then join these points with a smooth curve as in Figure 1. The only catch is that 2^x has not been defined at this point for all real numbers. From Section 1-6 we know what 2^5, 2^{-3}, $2^{2/3}$, $2^{-3/5}$, $2^{1.4}$, and $2^{-3.15}$ mean (that is, 2^p, where p is a rational number), but what does

$$2^{\sqrt{2}}$$

mean? The question is not easy to answer at this time. In fact, a precise definition of $2^{\sqrt{2}}$ must wait for more advanced courses, where we can show that

$$2^x$$

names a real number for x any real number, and that the graph of $y = 2^x$ is as indicated in Figure 1A. We also can show that for x irrational, 2^x can be approximated as closely as we like by using rational number approximations for x. Since $\sqrt{2} = 1.414\ 213...$, for example, the sequence

$$2^{1.4},\ 2^{1.41},\ 2^{1.414},\ \ldots$$

approximates $2^{\sqrt{2}}$, and as we move to the right, the approximation improves.

It is useful to note that the graph of

$$f(x) = b^x \qquad b > 1$$

will look very much like Figure 1A, and the graph of

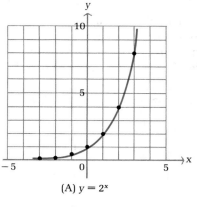

(A) $y = 2^x$

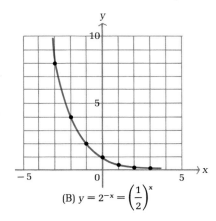

(B) $y = 2^{-x} = \left(\dfrac{1}{2}\right)^x$

FIGURE 1

$$f(x) = b^x \qquad 0 < b < 1$$

will look very much like Figure 1B. Observe that in both cases the graphs approach the x axis, but never touch it. The x axis is called a *horizontal asymptote* for each graph. In general, a line $y = c$ is a **horizontal asymptote** for the graph of the equation $y = f(x)$ if $f(x)$ approaches c as x increases without bound or as x decreases without bound. Asymptotes, if they exist, provide a useful aid in graphing some functions.

The graphs in Figure 1 suggest the following important general properties of exponential functions, which we state without proof:

▬ Basic Properties of the Graph of $f(x) = b^x$, $b > 0$, $b \neq 1$

1. All graphs will pass through the point (0, 1). $b^0 = 1$ for any permissible base b.
2. All graphs are continuous curves, with no holes or jumps.
3. The x axis is a horizontal asymptote.
4. If $b > 1$, then b^x increases as x increases.
5. If $0 < b < 1$, then b^x decreases as x increases.

The use of the $\boxed{y^x}$ key on a scientific calculator makes the plotting of accurate graphs of exponential functions almost routine. Example 1 illustrates the process.

◆ E X A M P L E   1   Graph $y = (\frac{1}{2})4^x$ for $-3 \leqslant x \leqslant 3$.

Solution

x	y
−3	0.01
−2	0.03
−1	0.13
0	0.50
1	2.00
2	8.00
3	32.00

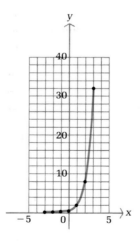

P R O B L E M   1   Graph $y = (\frac{1}{2})4^{-x}$ for $-3 \leqslant x \leqslant 3$.

◆ ADDITIONAL PROPERTIES

Exponential functions, which include irrational exponents, obey the familiar laws of exponents we discussed earlier for rational exponents. We summarize these exponent laws and add two other important and useful properties in the following box:

Exponential Function Properties

For a and b positive, $a \neq 1$, $b \neq 1$, and x and y real:

1. Exponent laws:

$$a^x a^y = a^{x+y} \qquad \frac{a^x}{a^y} = a^{x-y} \qquad (a^x)^y = a^{xy}$$

$$(ab)^x = a^x b^x \qquad \left(\frac{a}{b}\right)^x = \frac{a^x}{b^x} \qquad \frac{4^{2y}}{4^{5y}} \boxed{= 4^{2y-5y}} = 4^{-3y}$$

2. $a^x = a^y$ if and only if $x = y$ If $7^{5t+1} = 7^{3t-3}$, then $5t + 1 = 3t - 3$, and $t = -2$.

3. For $x \neq 0$,

$$a^x = b^x \qquad \text{if and only if} \qquad a = b \qquad \text{If } a^5 = 2^5, \text{ then } a = 2.$$

◆ APPLICATIONS

We will now consider three applications that utilize exponential functions in their analysis: population growth (an example of exponential growth), radioactive decay (an example of negative exponential growth), and compound interest (another example of exponential growth).

Our first example involves the growth of populations, such as people, animals, insects, and bacteria. Populations tend to grow exponentially and at different rates. A convenient and easily understood measure of growth rate is the **doubling time**—that is, the time it takes for a population to double. Over short periods of time the **doubling-time growth model** is often used to model population growth:

$$P = P_0 2^{t/d}$$

where

    P = Population at time t

    P_0 = Population at time $t = 0$

    d = Doubling time

Note that when $t = d$,

$$P = P_0 2^{d/d} = P_0 2$$

and the population is double the original, as it should be. We use this model to solve a population growth problem in Example 2.

◆ EXAMPLE 2

Population Growth

Ethiopia has a population of around 42 million people, and it is estimated that the population will double in 22 years. If population growth continues at the same rate, what will be the population (to the nearest million):

(A) 10 years from now?   (B) 35 years from now?

Solutions

We use the doubling-time growth model:

$$P = P_0 2^{t/d}$$

Substituting $P_0 = 42$ and $d = 22$, we obtain

$$P = 42(2^{t/22})$$ See the graph in the margin.

(A) Find P when $t = 10$ years:

$$P = 42(2^{10/22})$$ Use a calculator.

$$P \approx 58 \text{ million people}$$

(B) Find P when $t = 35$ years:

$$P = 42(2^{36/22})$$ Use a calculator.

$$P \approx 127 \text{ million people}$$ ◆

P

Million people

200
150
100
50

10 20 30 40 50   t
Years

PROBLEM 2

The bacterium *Escherichia coli* (*E. coli*) is found naturally in the intestines of many mammals. In a particular laboratory experiment, the doubling time for *E. coli* is found to be 25 minutes. If the experiment starts with a population of 1,000 *E. coli* and there is no change in growth rate, how many bacteria will be present:

(A) In 10 minutes?   (B) In 5 hours?   ◆

Our second application involves radioactive decay (negative growth). Radioactive materials are used extensively in medical diagnosis and therapy, as power sources in satellites, and as power sources (although controversial) in many countries. If we start with an amount A_0 of a particular radioactive isotope, the amount declines exponentially in time; the rate of decay varies from isotope to isotope. A convenient and easily understood measure of the rate of decay is the **half-life** of the isotope—that is, the time it takes for half of a particular material to decay. In this section we use the following **half-life decay model:**

$$A = A_0 (\tfrac{1}{2})^{t/h} = A_0 2^{-t/h}$$

where

$$A = \text{Amount at time } t$$
$$A_0 = \text{Amount at time } t = 0$$
$$h = \text{Half-life}$$

Note that when $t = h$,

$$A = A_0 2^{-h/h} = A_0 2^{-1} = \frac{A_0}{2}$$

and the amount of isotope is half the original amount, as it should be.

◆ E X A M P L E   3

Radioactive Decay

Radioactive gold-198 (^{198}Au), used in imaging the structure of the liver, has a half-life of 2.67 days. If we start with 50 milligrams of the isotope, how many milligrams will be left after:

(A) $\frac{1}{2}$ day?   (B) 1 week?

Compute answers to two decimal places.

Solutions

We use the half-life decay model:

$$A = A_0\left(\tfrac{1}{2}\right)^{t/h} = A_0 2^{-t/h}$$

Using $A_0 = 50$ and $h = 2.67$, we obtain

$$A = 50(2^{-t/2.67}) \qquad \text{See the graph in the margin.}$$

(A) Find A when $t = 0.5$ day:

$$A = 50(2^{-0.5/2.67}) \qquad \text{Use a calculator.}$$
$$\approx 43.91 \text{ milligrams}$$

(B) Find A when $t = 7$ days:

$$A = 50(2^{-7/2.67}) \qquad \text{Use a calculator.}$$
$$\approx 8.12 \text{ milligrams}$$

◆

P R O B L E M   3

The radioactive isotope gallium-67 (^{67}Ga), used in the diagnosis of malignant tumors, has a biological half-life of 46.5 hours. If we start with 100 milligrams of the isotope, how many milligrams will be left after:

(A) 24 hours?   (B) 1 week?

Compute answers to two decimal places.   ◆

Our third application deals with the growth of money at compound interest. This is just a brief introduction to a topic that is treated in detail in Chapter 4.

The fee paid to use another's money is called **interest.** It is usually computed as a percent (called **interest rate**) of the principal over a given period of time. If, at the end of a payment period, the interest due is reinvested at the same rate, then the interest earned as well as the principal will earn interest during the next payment period. Interest paid on interest reinvested is called **compound interest.** In Chapter 4 we develop another variation of the following compound interest formula:

> ### Compound Interest
>
> If a **principal** P is invested at an annual **rate** r (expressed as a decimal) compounded m times a year, then the **amount** A **(future value)** in the account at the end of t years is given by
>
> $$A = P\left(1 + \frac{r}{m}\right)^{mt}$$

◆ **E X A M P L E 4**

Compound Growth

If $1,000 is invested in an account paying 10% compounded monthly, how much will be in the account at the end of 10 years? Compute the answer to the nearest cent.

Solution

We use the compound interest formula as follows:

$$A = P\left(1 + \frac{r}{m}\right)^{mt}$$

$$= 1,000\left(1 + \frac{0.10}{12}\right)^{(12)(10)} \qquad \text{Use a calculator.}$$

$$= \$2,707.04$$

The graph of

$$A = 1,000\left(1 + \frac{0.10}{12}\right)^{12t}$$

for $0 \leqslant t \leqslant 20$ is shown in the margin. ◆

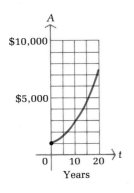

A

$10,000

$5,000

0   10   20 t

Years

P R O B L E M 4

If you deposit $5,000 in an account paying 9% compounded daily, how much will you have in the account in 5 years? Compute the answer to the nearest cent. ◆

Answers to Matched Problems

1. $y = (\frac{1}{2})4^{-x}$

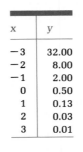

x	y
-3	32.00
-2	8.00
-1	2.00
0	0.50
1	0.13
2	0.03
3	0.01

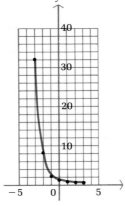

2. (A) $P \approx 1,320$ bacteria
    (B) $P \approx 4,096,000$ bacteria
3. (A) $A \approx 69.92$ mg
    (B) $A \approx 8.17$ mg
4. $7,841.13

A Graph each function over the indicated interval.

1. $y = 5^x$; $[-2, 2]$
2. $y = 3^x$; $[-3, 3]$
3. $y = (\frac{1}{5})^x = 5^{-x}$; $[-2, 2]$
4. $y = (\frac{1}{3})^x = 3^{-x}$; $[-3, 3]$
5. $f(x) = -5^x$; $[-2, 2]$
6. $g(x) = -3^{-x}$; $[-3, 3]$
7. $f(x) = 4(5^x)$; $[-2, 2]$
8. $h(x) = 5(3^x)$; $[-3, 3]$
9. $y = 5^{x+2} + 4$; $[-4, 0]$
10. $y = 3^{x+3} - 5$; $[-6, 0]$

Simplify.

11. $(4^{3x})^{2y}$
12. $10^{3x-1}10^{4-x}$
13. $\dfrac{5^{x-3}}{5^{x-4}}$
14. $\dfrac{3^x}{3^{1-x}}$

15. $(2^x 3^y)^z$
16. $\left(\dfrac{4^x}{5^y}\right)^{3z}$

B Solve for x.

17. $10^{2-3x} = 10^{5x-6}$
18. $5^{3x} = 5^{4x-2}$
19. $4^{5x-x^2} = 4^{-6}$
20. $7^{x^2} = 7^{2x+3}$
21. $5^3 = (x+2)^3$
22. $(1-x)^5 = (2x-1)^5$
23. $9^{x-1} = 3^x$
24. $2^x = 4^{x+1}$

Graph each function over the indicated interval.

25. $f(t) = 2^{t/10}$; $[-30, 30]$
26. $G(t) = 3^{t/100}$; $[-200, 200]$
27. $y = 7(2^{-2x})$; $[-2, 2]$
28. $y = 11(3^{-x/2})$; $[-9, 9]$
29. $f(x) = 2^{|x|}$; $[-3, 3]$
30. $g(x) = 2^{-|x|}$; $[-3, 3]$
31. $y = 100(1.03)^x$; $[0, 20]$
32. $y = 1,000(1.08)^x$; $[0, 10]$
33. $y = 3^{-x^2}$; $[-2, 2]$
34. $y = 2^{-x^2}$; $[-2, 2]$

C Simplify.

35. $(3^x - 3^{-x})(3^x + 3^{-x})$
36. $(6^x + 6^{-x})(6^x - 6^{-x})$
37. $(3^x - 3^{-x})^2 + (3^x + 3^{-x})^2$
38. $(6^x + 6^{-x})^2 - (6^x - 6^{-x})^2$

Graph each function over the indicated interval.

39. $h(x) = x(2^x)$; $[-5, 0]$
40. $m(x) = x(3^{-x})$; $[0, 3]$
41. $g(x) = \dfrac{3^x + 3^{-x}}{2}$; $[-3, 3]$
42. $f(x) = \dfrac{2^x + 2^{-x}}{2}$; $[-3, 3]$

CALCULATOR PROBLEMS

Calculate each using a scientific calculator. Compute answers to four decimal places.

43. $3^{-\sqrt{2}}$
44. $5^{\sqrt{3}}$
45. $\pi^{-\sqrt{3}}$

46. $\pi^{\sqrt{2}}$
47. $\dfrac{3^\pi - 3^{-\pi}}{2}$
48. $\dfrac{2^\pi + 2^{-\pi}}{2}$

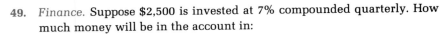

Business & Economics

49. *Finance.* Suppose $2,500 is invested at 7% compounded quarterly. How much money will be in the account in:

(A) $\frac{3}{4}$ year?   (B) 15 years?

Compute answers to the nearest cent.

50. *Finance.* Suppose $4,000 is invested at 11% compounded weekly. How much money will be in the account in:

(A) $\frac{1}{2}$ year?   (B) 10 years?

Compute answers to the nearest cent.

51. *Finance.* A person wishes to have $15,000 cash for a new car 5 years from now. How much should be placed in an account now, if the account pays 9.75% compounded weekly? Compute the answer to the nearest dollar.

52. *Finance.* A couple just had a new child. How much should they invest now at 8.25% compounded daily in order to have $40,000 for the child's education 17 years from now? Compute the answer to the nearest dollar.

Life Sciences

53. *Population growth.* If Kenya has a population of 23,000,000 people and a doubling time of 19 years, and if the growth continues at the same rate, find the population in:

(A) 10 years   (B) 30 years

Compute answers to the nearest million.

54. *Bacterial growth.* If bacteria in a certain culture double every $\frac{1}{2}$ hour, write an equation that gives the number of bacteria N in the culture after t hours, assuming the culture has 100 bacteria at the start. Graph the equation for $0 \leq t \leq 5$.

55. *Radioactive tracers.* The radioactive isotope technetium-99m (^{99m}Tc) is used in imaging the brain. This isotope has a half-life of 6 hours. If 12 milligrams are used, how much will be present after:

(A) 3 hours?   (B) 24 hours?

Compute answers to two decimal places.

56. *Insecticides.* The use of the insecticide DDT is no longer allowed in many countries because of its long-term adverse effects. If a farmer uses 25 pounds of active DDT, assuming its half-life is 12 years, how much will still be active after:

(A) 5 years?   (B) 20 years?

Compute answers to the nearest pound.

The Exponential Function with Base *e*

◆ BASE *e* EXPONENTIAL FUNCTION
◆ GROWTH AND DECAY APPLICATIONS REVISITED
◆ CONTINUOUS COMPOUND INTEREST

The number π is probably the most important irrational number you have encountered until now. In this section we will introduce another irrational number, *e*, that is just as important in mathematics and its applications.

◆ BASE *e* EXPONENTIAL FUNCTION

Consider the following expression:

$$\left(1 + \frac{1}{m}\right)^m$$

What happens to the value of the expression as *m* increases without bound? (Think about this for a moment before proceeding.) Maybe you guessed that the value approaches 1 using the following reasoning: As *m* gets large, $1 + (1/m)$ approaches 1 (since $1/m$ approaches 0), and 1 raised to any power is 1. Let us see if this reasoning is correct by actually calculating the value of the expression for larger and larger values of *m*. Table 1 summarizes the results.

Interestingly, the value of $[1 + (1/m)]^m$ is never close to 1 but seems to be approaching a number close to 2.7183. As m increases without bound, the value of $[1 + (1/m)]^m$ approaches an irrational number that we call **e**. The irrational number *e* to twelve decimal places is

$$e = 2.718\ 281\ 828\ 459$$

Exactly who discovered the constant *e* is still being debated. It is named after the great Swiss mathematician Leonhard Euler (1707–1783). This constant turns out to be an ideal base for an exponential function, because in calculus and higher mathematics many operations take on their simplest form using this base. This is why you will see *e* used extensively in expressions and formulas that model real-world phenomena.

TABLE 1

m	$\left(1 + \dfrac{1}{m}\right)^m$
1	2
10	2.593 74...
100	2.704 81...
1,000	2.716 92...
10,000	2.718 14...
100,000	2.718 27...
1,000,000	2.718 28...
.	.
.	.
.	.

Exponential Function with Base *e*

For x a real number, the equation

$$f(x) = e^x$$

defines the **exponential function with base e.**

The exponential function with base e is used so frequently that it is often referred to as *the* exponential function. Because of its importance, all scientific calculators have an $\boxed{e^x}$ key or its equivalent—consult your user's manual.

The important constant e, along with the two other important constants, $\sqrt{2}$ and π, are shown on the number line in Figure 2A. Using the properties of the graph of $f(x) = b^x$ discussed in Section 3-1, we obtain the graphs of $y = e^x$ and $y = (1/e)^x = e^{-x}$ shown in Figure 2B. Notice that neither graph crosses the x axis. Thus, we conclude that $e^x > 0$ and $e^{-x} > 0$ for all real numbers x, an important observation when solving certain equations involving the exponential function.

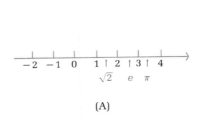

(A)

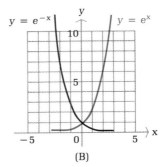

(B)

FIGURE 2

◆ GROWTH AND DECAY APPLICATIONS REVISITED

Most exponential growth and decay problems are modeled using base e exponential functions. We present two applications here and many more in Exercise 3-2.

◆ EXAMPLE 5

Exponential Growth

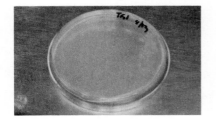

Cholera, an intestinal disease, is caused by a cholera bacterium that multiplies exponentially by cell division as given approximately by

$$N = N_0 e^{1.386t}$$

where N is the number of bacteria present after t hours and N_0 is the number of bacteria present at $t = 0$. If we start with 10 bacteria, how many bacteria will be present:

(A) In 0.6 hour?   (B) In 3.5 hours?

Solutions

Substituting $N_0 = 10$ into the above equation, we obtain

$N = 10e^{1.386t}$ The graph is shown in the margin.

(A) Solve for N when $t = 0.6$:

$N = 10e^{1.386(0.6)}$ Use a calculator.

$= 23$ bacteria

(B) Solve for N when $t = 3.5$:

$N = 10e^{1.386(3.5)}$ Use a calculator.

$= 1,279$ bacteria   ◆

PROBLEM 5

Refer to the exponential growth model for cholera in Example 5. If we start with 50 bacteria, how many bacteria will be present:

(A) In 0.85 hour?   (B) In 7.25 hours?   ◆

◆ **EXAMPLE 6**

Exponential Decay

Cosmic-ray bombardment of the atmosphere produces neutrons, which in turn react with nitrogen to produce radioactive carbon-14 (^{14}C). Radioactive ^{14}C enters all living tissues through carbon dioxide, which is first absorbed by plants. As long as a plant or animal is alive, ^{14}C is maintained in the living organism at a constant level. Once the organism dies, however, ^{14}C decays according to the equation

$A = A_0e^{-0.000124t}$

where A is the amount present after t years and A_0 is the amount present at time $t = 0$. If 500 milligrams of ^{14}C are present in a sample from a skull at the time of death, how many milligrams will be present in the sample in:

(A) 15,000 years?   (B) 45,000 years?

Compute answers to two decimal places.

Solutions

Substituting $A_0 = 500$ in the decay equation, we have

$A = 500e^{-0.000124t}$ See the graph in the margin.

(A) Solve for A when $t = 15,000$:

$A = 500e^{-0.000124(15,000)}$ Use a calculator.

$= 77.84$ milligrams

(B) Solve for A when $t = 45,000$:

$A = 500e^{-0.000124(45,000)}$ Use a calculator.

$= 1.89$ milligrams   ◆

PROBLEM 6　　Refer to the exponential decay model in Example 6. How many milligrams of ^{14}C would have to be present at the beginning in order to have 25 milligrams present after 18,000 years? Compute the answer to the nearest milligram. ◆

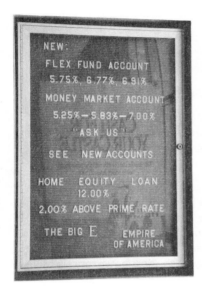

◆ CONTINUOUS COMPOUND INTEREST

The constant e occurs naturally in the study of compound interest. Returning to the compound interest formula discussed in Section 3-1,

COMPOUND INTEREST　　$A = P\left(1 + \dfrac{r}{m}\right)^{mt}$

recall that P is the principal invested at an annual rate r compounded m times a year and A is the amount in the account after t years. Suppose P, r, and t are held fixed and m is increased without bound. Will the amount A increase without bound or will it tend to some limiting value?

Starting with $P = \$100$, $r = 0.08$, and $t = 2$ years, we construct Table 2 for several values of m with the aid of a calculator. Notice that the largest gain appears in going from annual to semiannual compounding. Then, the gains slow down as m increases. It appears that A gets closer and closer to $\$117.35$ as m gets larger and larger.

TABLE 2

COMPOUNDING FREQUENCY	m	$A = 100\left(1 + \dfrac{0.08}{m}\right)^{2m}$
Annually	1	\$116.6400
Semiannually	2	116.9859
Quarterly	4	117.1659
Weekly	52	117.3367
Daily	365	117.3490
Hourly	8,760	117.3510

In Section 10-5, we show that

$$P\left(1 + \frac{r}{m}\right)^{mt}$$

gets closer and closer to Pe^{rt} as the number of compounding periods m gets larger and larger. The latter is referred to as the **continuous compound interest formula,** a formula that is widely used in business, banking, and economics.

> ▬ **Continuous Compound Interest Formula**
>
> If a principal P is invested at an annual rate r (expressed as a decimal) compounded continuously, then the amount A in the account at the end of t years is given by
>
> $$A = Pe^{rt}$$

◆ E X A M P L E   7   What amount will an account have after 2 years if \$5,000 is invested at an annual rate of 8%:

(A) Compounded daily?   (B) Compounded continuously?

Compute answers to the nearest cent.

Solutions   (A) Use the compound interest formula

$$A = P\left(1 + \frac{r}{m}\right)^{mt}$$

with $P = 5{,}000$, $r = 0.08$, $m = 365$, and $t = 2$:

$$A = 5{,}000\left(1 + \frac{0.08}{365}\right)^{(365)(2)} \qquad \text{Use a calculator.}$$

$$= \$5{,}867.45$$

(B) Use the continuous compound interest formula

$$A = Pe^{rt}$$

with $P = 5{,}000$, $r = 0.08$, and $t = 2$:

$$A = 5{,}000e^{(0.08)(2)} \qquad \text{Use a calculator.}$$

$$= \$5{,}867.55 \qquad\qquad\qquad\qquad\qquad ◆$$

P R O B L E M   7   What amount will an account have after 1.5 years if \$8,000 is invested at an annual rate of 9%:

(A) Compounded weekly?   (B) Compounded continuously?

Compute answers to the nearest cent.   ◆

Answers to Matched Problems   5. (A) 162 bacteria   (B) 1,156,054 bacteria
6. 233 mg   7. (A) \$9,155.23   (B) \$9,156.29

A   *Graph over the indicated interval.*

1. $y = -e^{-x}$; $[-3, 3]$
2. $y = -e^x$; $[-3, 3]$
3. $y = 100e^{0.1x}$; $[-5, 5]$
4. $y = 10e^{0.2x}$; $[-10, 10]$
5. $g(t) = 10e^{-0.2t}$; $[-5, 5]$
6. $f(t) = 100e^{-0.1t}$; $[-5, 5]$

B   7. $y = -3 + e^{1+x}$; $[-4, 2]$
8. $y = 2 + e^{x-2}$; $[-1, 5]$
9. $y = e^{|x|}$; $[-3, 3]$
10. $y = e^{-|x|}$; $[-3, 3]$

11. $C(x) = \dfrac{e^x + e^{-x}}{2}$; $[-5, 5]$
12. $M(x) = e^{x/2} + e^{-x/2}$; $[-5, 5]$

Simplify.

13. $e^x(e^{-x} + 1) - e^{-x}(e^x + 1)$
14. $(e^x + e^{-x})^2 + (e^x - e^{-x})^2$

15. $\dfrac{e^x(e^x + e^{-x}) - (e^x - e^{-x})e^x}{e^{2x}}$
16. $\dfrac{e^{-x}(e^x - e^{-x}) + e^{-x}(e^x + e^{-x})}{e^{-2x}}$

C   *Solve each equation (remember, $e^x \neq 0$ and $e^{-x} \neq 0$).*

17. $(x - 3)e^x = 0$
18. $2xe^{-x} = 0$
19. $3xe^{-x} + x^2e^{-x} = 0$
20. $x^2e^x - 5xe^x = 0$

Graph over the indicated interval.

21. $N = \dfrac{100}{1 + e^{-t}}$; $[0, 5]$
22. $N = \dfrac{200}{1 + 3e^{-t}}$; $[0, 5]$

APPLICATIONS

Business & Economics

23. *Money growth.* If you invest $7,500 in an account paying 8.35% compounded continuously, how much money will be in the account at the end of:

(A) 5.5 years?   (B) 12 years?

24. *Money growth.* If you invest $5,250 in an account paying 11.38% compounded continuously, how much money will be in the account at the end of:

(A) 6.25 years?   (B) 17 years?

25. *Money growth.* Barron's (a national business and financial weekly) published the following "Top Savings Deposit Yields" for 1 year certificate of deposit accounts:

(A) Alamo Savings, 8.25% compounded quarterly
(B) Lamar Savings, 8.05% compounded continuously

Compute the value of $10,000 invested in each account at the end of 1 year.

26. *Money growth.* Refer to Problem 25. In another issue of *Barron's*, $2\frac{1}{2}$ year certificate of deposit accounts included the following:

(A) Gill Saving, 8.30% compounded continuously
(B) Richardson Savings and Loan, 8.40% compounded quarterly
(C) USA Savings, 8.25% compounded daily

Compute the value of $1,000 invested in each account at the end of $2\frac{1}{2}$ years.

27. *Present value.* A promissory note will pay $50,000 at maturity $5\frac{1}{2}$ years from now. How much should you be willing to pay for the note now if money is worth 10% compounded continuously?

28. *Present value.* A promissory note will pay $30,000 at maturity 10 years from now. How much should you be willing to pay for the note now if money is worth 9% compounded continuously?

29. *Advertising.* A company is trying to introduce a new product to as many people as possible through television advertising in a large metropolitan area with 2 million possible viewers. A model for the number of people N (in millions) who are aware of the product after t days of advertising was found to be

$$N = 2(1 - e^{-0.037t})$$

Graph this function for $0 \leqslant t \leqslant 50$. What value does N tend to as t increases without bound?

30. *Learning curve.* People assigned to assemble circuit boards for a computer manufacturing company undergo on-the-job training. From past experience it was found that the learning curve for the average employee is given by

$$N = 40(1 - e^{-0.12t})$$

where N is the number of boards assembled per day after t days of training. Graph this function for $0 \leqslant t \leqslant 30$. What is the maximum number of boards an average employee can be expected to produce in 1 day?

Life Sciences

31. *Marine biology.* Marine life is dependent upon the microscopic plant life that exists in the *photic* zone, a zone that goes to a depth where about 1% of the surface light still remains. In some waters with a great deal of sediment, the photic zone may go down only 15–20 feet. In some murky harbors, the intensity of light d feet below the surface is given approximately by

$$I = I_0 e^{-0.23d}$$

What percent of the surface light will reach a depth of:

(A) 10 feet?   (B) 20 feet?

32. *Marine biology.* Refer to Problem 31. Light intensity I relative to depth d (in feet) for one of the clearest bodies of water in the world, the Sargasso Sea in the West Indies, can be approximated by

$$I = I_0 e^{-0.00942d}$$

where I_0 is the intensity of light at the surface. What percent of the surface light will reach a depth of:

(A) 50 feet? (B) 100 feet?

33. *AIDS epidemic.* As of this writing, AIDS among intravenous drug users in the United States is spreading at the rate of about 46% compounded continuously. Assuming this rate does not change and there were 12,000 cases of AIDS among intravenous drug users at the end of 1987, how many cases should we expect by the end of:

(A) 1990? (B) 1995?

34. *AIDS epidemic.* At the end of 1987 there were approximately 45,000 diagnosed cases of AIDS among the general population in the United States. As of this writing, the disease is estimated to be spreading at the rate of 38% compounded continuously. Assuming this rate does not change, how many cases of AIDS should be expected by the end of:

(A) 1992? (B) 2000?

Social Sciences

35. *Population growth.* If the population in Mexico was around 100 million people in 1987 and if the population continues to grow at 2.3% compounded continuously, what will the population be in 1995? Compute the answer to the nearest million.

36. *Population growth.* If the world population was 5 billion people in 1988 and if the population continues to grow at 1.7% compounded continuously, what will the population be in 1999? Compute the answer to the nearest billion.

S E C T I O N   3-3   **Logarithmic Functions**

◆ DEFINITION OF LOGARITHMIC FUNCTIONS
◆ FROM LOGARITHMIC TO EXPONENTIAL FORM AND VICE VERSA
◆ PROPERTIES OF LOGARITHMIC FUNCTIONS
◆ CALCULATOR EVALUATION OF COMMON AND NATURAL LOGARITHMS
◆ APPLICATION

Now we are ready to consider logarithmic functions, which are closely related to exponential functions.

◆ DEFINITION OF LOGARITHMIC FUNCTIONS

If we start with an exponential function f defined by

$$y = 2^x \tag{1}$$

and interchange the variables, we obtain an equation that defines a new relation g defined by

$$x = 2^y \tag{2}$$

Any ordered pair of numbers that belongs to f will belong to g if we interchange the order of the components. For example, (3, 8) satisfies equation (1) and (8, 3) satisfies equation (2). Thus, the domain of f becomes the range of g and the range of f becomes the domain of g. Graphing f and g on the same coordinate system (Fig. 3), we see that g is also a function. We call this new function the **logarithmic function with base 2,** and write

$$y = \log_2 x \quad \text{if and only if} \quad x = 2^y$$

Note that if we fold the paper along the dashed line $y = x$ in Figure 3, the two graphs match exactly.

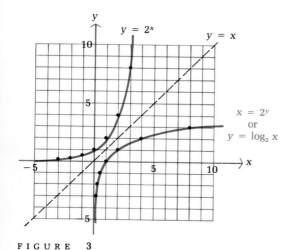

EXPONENTIAL FUNCTION		LOGARITHMIC FUNCTION	
x	$y = 2^x$	$x = 2^y$	y
-3	$\frac{1}{8}$	$\frac{1}{8}$	-3
-2	$\frac{1}{4}$	$\frac{1}{4}$	-2
-1	$\frac{1}{2}$	$\frac{1}{2}$	-1
0	1	1	0
1	2	2	1
2	4	4	2
3	8	8	3

Ordered pairs reversed

FIGURE 3

In general, we define the **logarithmic functions with base b** as follows:

■ Definition of Logarithmic Function

For $b > 0$ and $b \neq 1$,

$\quad y = \log_b x \qquad$ is equivalent to $\qquad x = b^y$

$\quad\quad y = \log_{10} x \qquad$ is equivalent to $\qquad x = 10^y$

$\quad\quad y = \log_e x \qquad$ is equivalent to $\qquad x = e^y$

(The log to the base b of x is the exponent to which b must be raised to obtain x.)

Remember: A logarithm is an exponent.

Since the domain of an exponential function includes all real numbers and its range is the set of positive real numbers, the **domain** of a logarithmic function is the set of all positive real numbers and its **range** is the set of all real numbers. Thus, $\log_{10} 3$ is defined, but $\log_{10} 0$ and $\log_{10}(-5)$ are not defined (3 is a logarithmic domain value, but 0 and -5 are not). Typical logarithmic curves are shown in Figure 4.

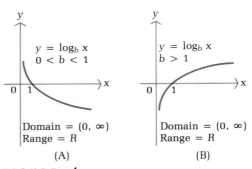

$$\begin{array}{ll}
\text{Domain} = (0, \infty) & \text{Domain} = (0, \infty) \\
\text{Range} = R & \text{Range} = R \\
\quad\quad\text{(A)} & \quad\quad\text{(B)}
\end{array}$$

FIGURE 4
Typical logarithmic graphs

◆ FROM LOGARITHMIC TO EXPONENTIAL FORM AND VICE VERSA

We now consider the matter of converting logarithmic forms to equivalent exponential forms and vice versa.

◆ EXAMPLE 8   Change from logarithmic form to exponential form:

(A) $\log_5 25 = 2 \qquad$ is equivalent to $\quad 25 = 5^2$

(B) $\log_9 3 = \frac{1}{2} \qquad$ is equivalent to $\quad 3 = 9^{1/2}$

(C) $\log_2\left(\frac{1}{4}\right) = -2 \quad$ is equivalent to $\quad \frac{1}{4} = 2^{-2}$ ◆

PROBLEM 8   Change to an equivalent exponential form:

(A) $\log_3 9 = 2$ (B) $\log_4 2 = \frac{1}{2}$ (C) $\log_3(\frac{1}{9}) = -2$ ◆

EXAMPLE 9   Change from exponential form to logarithmic form:

(A) $64 = 4^3$ is equivalent to $\log_4 64 = 3$
(B) $6 = \sqrt{36}$ is equivalent to $\log_{36} 6 = \frac{1}{2}$
(C) $\frac{1}{8} = 2^{-3}$ is equivalent to $\log_2(\frac{1}{8}) = -3$ ◆

PROBLEM 9   Change to an equivalent logarithmic form:

(A) $49 = 7^2$ (B) $3 = \sqrt{9}$ (C) $\frac{1}{3} = 3^{-1}$ ◆

EXAMPLE 10   Find y, b, or x, as indicated.

(A) Find y: $y = \log_4 16$ (B) Find x: $\log_2 x = -3$
(C) Find y: $y = \log_8 4$ (D) Find b: $\log_b 100 = 2$

Solutions   (A) $y = \log_4 16$ is equivalent to $16 = 4^y$. Thus,

$$y = 2$$

(B) $\log_2 x = -3$ is equivalent to $x = 2^{-3}$. Thus,

$$x = \frac{1}{2^3} = \frac{1}{8}$$

(C) $y = \log_8 4$ is equivalent to

$$4 = 8^y \qquad \text{or} \qquad 2^2 = 2^{3y}$$

Thus,

$$3y = 2$$
$$y = \tfrac{2}{3}$$

(D) $\log_b 100 = 2$ is equivalent to $100 = b^2$. Thus,

$$b = 10 \qquad \text{Recall that } b \text{ cannot be negative.}$$ ◆

PROBLEM 10   Find y, b, or x, as indicated.

(A) Find y: $y = \log_9 27$ (B) Find x: $\log_3 x = -1$
(C) Find b: $\log_b 1{,}000 = 3$ ◆

EXAMPLE 11   Graph $y = \log_2(x + 1)$ by converting to an equivalent exponential form first. Do not use a calculator or table.

Solution   Changing $y = \log_2(x + 1)$ to an equivalent exponential form, we have

$$x + 1 = 2^y \qquad \text{or} \qquad x = 2^y - 1$$

Even though x is the independent variable and y is the dependent variable, it is easier to assign y values and solve for x.

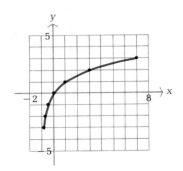

x	y
$-\frac{7}{8}$	-3
$-\frac{3}{4}$	-2
$-\frac{1}{2}$	-1
0	0
1	1
3	2
7	3

◆

PROBLEM 11   Graph $y = \log_3(x - 1)$ by converting to an equivalent exponential form first. ◆

◆ PROPERTIES OF LOGARITHMIC FUNCTIONS

Logarithmic functions have several very useful properties that follow directly from their definitions. These properties will enable us to convert multiplication problems into addition problems, division problems into subtraction problems, and power and root problems into multiplication problems. We will also be able to solve exponential equations such as $2 = 1.06^n$.

Logarithmic Properties ($b > 0$, $b \neq 1$, $M > 0$, $N > 0$)

1. $\log_b b^x = x$
2. $\log_b MN = \log_b M + \log_b N$
3. $\log_b \dfrac{M}{N} = \log_b M - \log_b N$
4. $\log_b M^p = p \log_b M$
5. $\log_b M = \log_b N$ if and only if $M = N$
6. $\log_b 1 = 0$

The first property follows directly from the definition of a logarithmic function. Here, we will sketch a proof for property 2. The other properties are established in a similar way. Let

$$u = \log_b M \quad \text{and} \quad v = \log_b N$$

Or, in equivalent exponential form,

$$M = b^u \quad \text{and} \quad N = b^v$$

Now, see if you can provide reasons for each of the following steps:

$$\log_b MN = \log_b b^u b^v = \log_b b^{u+v} = u + v = \log_b M + \log_b N$$

◆ E X A M P L E   12   (A) $\log_b \dfrac{wx}{yz} = \log_b wx - \log_b yz$

$$= \log_b w + \log_b x - (\log_b y + \log_b z)$$

$$= \log_b w + \log_b x - \log_b y - \log_b z$$

(B) $\log_b(wx)^{3/5} = \frac{3}{5} \log_b wx = \frac{3}{5}(\log_b w + \log_b x)$ ◆

P R O B L E M   12   Write in simpler logarithmic forms, as in Example 12.

(A) $\log_b \dfrac{R}{ST}$ (B) $\log_b \left(\dfrac{R}{S}\right)^{2/3}$ ◆

The following examples and problems, though somewhat artificial, will give you additional practice in using basic logarithmic properties.

◆ E X A M P L E   13   Find x so that: $\frac{3}{2} \log_b 4 - \frac{2}{3} \log_b 8 + \log_b 2 = \log_b x$

Solution

$$\frac{3}{2} \log_b 4 - \frac{2}{3} \log_b 8 + \log_b 2 = \log_b x$$

$$\log_b 4^{3/2} - \log_b 8^{2/3} + \log_b 2 = \log_b x \qquad \text{Property 4}$$

$$\log_b 8 - \log_b 4 + \log_b 2 = \log_b x$$

$$\log_b \frac{8 \cdot 2}{4} = \log_b x \qquad \text{Properties 2 and 3}$$

$$\log_b 4 = \log_b x$$

$$x = 4 \qquad \text{Property 5}$$ ◆

P R O B L E M   13   Find x so that: $3 \log_b 2 + \frac{1}{2} \log_b 25 - \log_b 20 = \log_b x$ ◆

◆ E X A M P L E   14   Solve: $\log_{10} x + \log_{10}(x + 1) = \log_{10} 6$

Solution

$$\log_{10} x + \log_{10}(x + 1) = \log_{10} 6$$

$$\log_{10} x(x + 1) = \log_{10} 6 \qquad \text{Property 2}$$

$$x(x + 1) = 6 \qquad \text{Property 5}$$

$$x^2 + x - 6 = 0 \qquad \text{Solve by factoring.}$$

$$(x + 3)(x - 2) = 0$$

$$x = -3, 2$$

We must exclude $x = -3$, since the domain of the function $\log_{10}(x + 1)$ is $x > -1$ or $(-1, \infty)$; hence, $x = 2$ is the only solution. ◆

P R O B L E M   14   Solve: $\log_3 x + \log_3(x - 3) = \log_3 10$ ◆

◆ CALCULATOR EVALUATION OF COMMON AND NATURAL LOGARITHMS

Of all possible logarithmic bases, the base e and the base 10 are used almost exclusively. Before we can use logarithms in certain practical problems, we need to be able to approximate the logarithm of any positive number either to base 10 or to base e. And conversely, if we are given the logarithm of a number to base 10 or base e, we need to be able to approximate the number. Historically, tables were used for this purpose, but now calculators make computations faster and far more accurate.

Common logarithms (also called **Briggsian logarithms**) are logarithms with base 10. **Natural logarithms** (also called **Napierian logarithms**) are logarithms with base e. Most scientific calculators have a key labeled "log" (or "LOG") and a key labeled "ln" (or "LN"). The former represents a common (base 10) logarithm and the latter a natural (base e) logarithm. In fact, "log" and "ln" are both used extensively in mathematical literature, and whenever you see either used in this book without a base indicated they will be interpreted as follows:

Logarithmic Notation

COMMON LOGARITHM:	$\log x = \log_{10} x$
NATURAL LOGARITHM:	$\ln x = \log_e x$

Finding the common or natural logarithm using a scientific calculator is very easy: You simply enter a number from the domain of the function and push $\boxed{\log}$ or $\boxed{\ln}$.

◆ **EXAMPLE 15**   Use a scientific calculator to evaluate each to six decimal places:

(A) $\log 3{,}184$ (B) $\ln 0.000\ 349$ (C) $\log(-3.24)$

Solutions

Enter	Press	Display
(A) 3184	$\boxed{\log}$	3.502973
(B) 0.000 349	$\boxed{\ln}$	−7.960439
(C) −3.24	$\boxed{\log}$	Error

An error is indicated in part C because -3.24 is not in the domain of the log function. [*Note:* The manner in which error messages are displayed varies from one brand of calculator to the next.]   ◆

PROBLEM 15   Use a scientific calculator to evaluate each to six decimal places:

(A) $\log 0.013\ 529$ (B) $\ln 28.693\ 28$ (C) $\ln(-0.438)$ ◆

We now turn to the second problem mentioned above: Given the logarithm of a number, find the number. We make direct use of the logarithmic–exponential relationships, which follow from the definition of logarithmic function given at the beginning of this section.

> **Logarithmic–Exponential Relationships**
>
> $\log x = y$ is equivalent to $x = 10^y$
> $\ln x = y$ is equivalent to $x = e^y$

◆ E X A M P L E   16   Find x to four decimal places, given the indicated logarithm:

(A) $\log x = -2.315$ (B) $\ln x = 2.386$

Solutions   (A) $\log x = -2.315$ Change to equivalent exponential form.

$\qquad x = 10^{-2.315}$ Evaluate with a calculator.

$\qquad x = 0.0048$

(B) $\ln x = 2.386$ Change to equivalent exponential form.

$\qquad x = e^{2.386}$ Evaluate with a calculator.

$\qquad x = 10.8699$ ◆

P R O B L E M   16   Find x to four decimal places, given the indicated logarithm:

(A) $\ln x = -5.062$ (B) $\log x = 2.0821$ ◆

◆ APPLICATION

A convenient and easily understood way of comparing different investments is to use their **doubling times** — the length of time it takes the value of an investment to double. Logarithm properties, as you will see in Example 17, provide us with just the right tool for solving some doubling-time problems.

◆ E X A M P L E   17   How long (to the next whole year) will it take money to double if it is invested at 20% compounded annually?

Solution   We use the compound interest formula discussed in Section 3-1:

$$A = P\left(1 + \frac{r}{m}\right)^{mt} \qquad \text{Compound interest}$$

The problem is to find t, given $r = 0.20$, $m = 1$, and $A = 2P$; that is,

$\qquad 2P = P(1 + 0.2)^t$

$\qquad 2 = 1.2^t$

$\qquad 1.2^t = 2$ Solve for t by taking the natural or common

$\qquad\qquad$ logarithm of both sides (we choose the natural

$\qquad \ln 1.2^t = \ln 2$ logarithm).

$\qquad t \ln 1.2 = \ln 2$ Property 4

$$t = \frac{\ln 2}{\ln 1.2}$$

Use a calculator. [*Note:*
$(\ln 2)/(\ln 1.2) \neq \ln 2 - \ln 1.2$]

= 3.8 years

≈ 4 years   To the next whole year

When interest is paid at the end of 3 years, the money will not be doubled; when paid at the end of 4 years, the money will be slightly more than doubled.   ◆

PROBLEM   17   How long (to the next whole year) will it take money to double if it is invested at 13% compounded annually?   ◆

It is interesting and instructive to graph the doubling times for various rates compounded annually. We proceed as follows:

$$A = P(1 + r)^t$$
$$2P = P(1 + r)^t$$
$$2 = (1 + r)^t$$
$$(1 + r)^t = 2$$
$$\ln(1 + r)^t = \ln 2$$
$$t \ln(1 + r) = \ln 2$$
$$t = \frac{\ln 2}{\ln(1 + r)}$$

Figure 5 shows the graph of this equation (doubling time in years) for interest rates compounded annually from 1% to 70% (expressed as decimals). Note the dramatic change in doubling time as rates change from 1% to 20% (from 0.01 to 0.20).

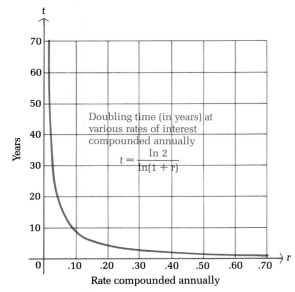

FIGURE   5

Answers to Matched Problems

8. (A) $9 = 3^2$ (B) $2 = 4^{1/2}$ (C) $\frac{1}{9} = 3^{-2}$
9. (A) $\log_7 49 = 2$ (B) $\log_9 3 = \frac{1}{2}$ (C) $\log_3(\frac{1}{3}) = -1$
10. (A) $y = \frac{3}{2}$ (B) $x = \frac{1}{3}$ (C) $b = 10$
11. $y = \log_3(x - 1)$ is equivalent to $x = 3^y + 1$

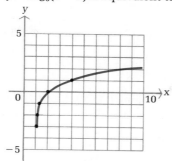

12. (A) $\log_b R - \log_b S - \log_b T$ (B) $\frac{2}{3}(\log_b R - \log_b S)$
13. $x = 2$ 14. $x = 5$
15. (A) $-1.868\ 734$ (B) $3.356\ 663$ (C) Not defined
16. (A) 0.0063 (B) 120.8092 17. 6 yr

E X E R C I S E   3-3

A   *Rewrite in exponential form.*

1. $\log_3 27 = 3$ 2. $\log_2 32 = 5$ 3. $\log_{10} 1 = 0$
4. $\log_e 1 = 0$ 5. $\log_4 8 = \frac{3}{2}$ 6. $\log_9 27 = \frac{3}{2}$

Rewrite in logarithmic form.

7. $49 = 7^2$ 8. $36 = 6^2$ 9. $8 = 4^{3/2}$
10. $9 = 27^{2/3}$ 11. $A = b^u$ 12. $M = b^x$

Evaluate each of the following:

13. $\log_{10} 10^3$ 14. $\log_{10} 10^{-5}$ 15. $\log_2 2^{-3}$
16. $\log_3 3^5$ 17. $\log_{10} 1,000$ 18. $\log_6 36$

Write in terms of simpler logarithmic forms, as in Example 12.

19. $\log_b \dfrac{P}{Q}$ 20. $\log_b FG$ 21. $\log_b L^5$

22. $\log_b w^{15}$ 23. $\log_b \dfrac{p}{qrs}$ 24. $\log_b PQR$

B   *Find x, y, or b.*

25. $\log_3 x = 2$ 26. $\log_2 x = 2$ 27. $\log_7 49 = y$
28. $\log_3 27 = y$ 29. $\log_b 10^{-4} = -4$ 30. $\log_b e^{-2} = -2$

31. $\log_4 x = \frac{1}{2}$ **32.** $\log_{25} x = \frac{1}{2}$ **33.** $\log_{1/3} 9 = y$

34. $\log_{49}(\frac{1}{7}) = y$ **35.** $\log_b 1{,}000 = \frac{3}{2}$ **36.** $\log_b 4 = \frac{2}{3}$

Write in terms of simpler logarithmic forms, going as far as you can with logarithmic properties (see Example 12).

37. $\log_b \dfrac{x^5}{y^3}$ **38.** $\log_b x^2 y^3$ **39.** $\log_b \sqrt[3]{N}$

40. $\log_b \sqrt[5]{Q}$ **41.** $\log_b(x^2 \sqrt[3]{y})$ **42.** $\log_b \sqrt[3]{\dfrac{x^2}{y}}$

43. $\log_b(50 \cdot 2^{-0.2t})$ **44.** $\log_b(100 \cdot 1.06^t)$ **45.** $\log_b P(1 + r)^t$

46. $\log_e A e^{-0.3t}$ **47.** $\log_e 100 e^{-0.01t}$ **48.** $\log_{10}(67 \cdot 10^{-0.12x})$

Find x.

49. $\log_b x = \frac{2}{3} \log_b 8 + \frac{1}{2} \log_b 9 - \log_b 6$

50. $\log_b x = \frac{2}{3} \log_b 27 + 2 \log_b 2 - \log_b 3$

51. $\log_b x = \frac{3}{2} \log_b 4 - \frac{2}{3} \log_b 8 + 2 \log_b 2$

52. $\log_b x = 3 \log_b 2 + \frac{1}{2} \log_b 25 - \log_b 20$

53. $\log_b x + \log_b(x - 4) = \log_b 21$

54. $\log_b(x + 2) + \log_b x = \log_b 24$

55. $\log_{10}(x - 1) - \log_{10}(x + 1) = 1$

56. $\log_{10}(x + 6) - \log_{10}(x - 3) = 1$

Graph by converting to exponential form first.

57. $y = \log_2(x - 2)$ **58.** $y = \log_3(x + 2)$

In Problems 59 and 60, evaluate to five decimal places using a scientific calculator.

59. (A) $\log 3{,}527.2$ (B) $\log 0.006\ 913\ 2$
     (C) $\ln 277.63$ (D) $\ln 0.040\ 883$

60. (A) $\log 72.604$ (B) $\log 0.033\ 041$
     (C) $\ln 40{,}257$ (D) $\ln 0.005\ 926\ 3$

In Problems 61 and 62, find x to four decimal places.

61. (A) $\log x = 1.1285$ (B) $\log x = -2.0497$
     (C) $\ln x = 2.7763$ (D) $\ln x = -1.8879$

62. (A) $\log x = 2.0832$ (B) $\log x = -1.1577$
     (C) $\ln x = 3.1336$ (D) $\ln x = -4.3281$

Evaluate each of the following to three decimal places using a calculator.

63. $n = \dfrac{\log 2}{\log 1.15}$ **64.** $n = \dfrac{\log 2}{\log 1.12}$ **65.** $n = \dfrac{\ln 3}{\ln 1.15}$

66. $n = \dfrac{\ln 4}{\ln 1.2}$ **67.** $x = \dfrac{\ln 0.5}{-0.21}$ **68.** $x = \dfrac{\ln 0.1}{-0.0025}$

Graph, using a calculator.

69. $y = \ln x$
70. $y = -\ln x$
71. $y = |\ln x|$
72. $y = \ln|x|$
73. $y = 2 \ln(x + 2)$
74. $y = 2 \ln x + 2$
75. $y = 4 \ln x - 3$
76. $y = 4 \ln(x - 3)$

C

77. Find the logarithm of 1 for any permissible base.
78. Why is 1 not a suitable logarithmic base? [*Hint:* Try to find $\log_1 8$.]
79. Write $\log_{10} y - \log_{10} c = 0.8x$ in an exponential form that is free of logarithms.
80. Write $\log_e x - \log_e 25 = 0.2t$ in an exponential form that is free of logarithms.

APPLICATIONS

Business & Economics

81. *Doubling time.* How long (to the next whole year) will it take money to double if it is invested at 6% interest compounded annually?

82. *Doubling time.* How long (to the next whole year) will it take money to double if it is invested at 3% interest compounded annually?

83. *Tripling time.* Write a formula similar to the doubling-time formula in Figure 5 for the tripling time of money invested at 100r% interest compounded annually.

84. *Tripling time.* How long (to the next whole year) will it take money to triple if it is invested at 15% interest compounded annually?

Life Sciences

85. *Sound intensity — decibels.* Because of the extraordinary range of sensitivity of the human ear (a range of over 1,000 million millions to 1), it is helpful to use a logarithmic scale, rather than an absolute scale, to measure sound intensity over this range. The unit of measure is called the *decibel*, after the inventor of the telephone, Alexander Graham Bell. If we let N be the number of decibels, I the power of the sound in question (in watts per square centimeter), and I_0 the power of sound just below the threshold of hearing (approximately 10^{-16} watt per square centimeter), then

$$I = I_0 10^{N/10}$$

Show that this formula can be written in the form

$$N = 10 \log \frac{I}{I_0}$$

86. *Sound intensity — decibels.* Use the formula in Problem 85 (with $I_0 = 10^{-16}$ watt/cm²) to find the decibel ratings of the following sounds:

(A) Whisper: 10^{-13} watt/cm²
(B) Normal conversation: 3.16×10^{-10} watt/cm²
(C) Heavy traffic: 10^{-8} watt/cm²
(D) Jet plane with afterburner: 10^{-1} watt/cm²

87. *World population.* If the world population is now 4 billion (4×10^9) people and if it continues to grow at 2% per year compounded annually, how long will it be before there is only 1 square yard of land per person? (The earth contains approximately 1.68×10^{14} square yards of land.)

88. *Archaeology—carbon-14 dating.* The radioactive carbon-14 (^{14}C) in an organism at the time of its death decays according to the equation

$$A = A_0 e^{-0.000124t}$$

where t is time in years and A_0 is the amount of ^{14}C present at time $t = 0$. (See Example 6 in Section 3-2.) Estimate the age of a skull uncovered in an archaeological site if 10% of the original amount of ^{14}C is still present. [*Hint:* Find t such that $A = 0.1A_0$.]

SECTION   3-4   Chapter Review

Important Terms and Symbols

3-1   *Exponential Functions.* Exponential function; base; domain; range; basic graphs; horizontal asymptote; basic properties; exponential growth; exponential decay; doubling-time growth model; half-life decay model; compound interest

$$b^x; \quad P = P_0 2^{t/d}; \quad A = A_0 (\tfrac{1}{2})^{t/h}; \quad A = P\left(1 + \frac{r}{m}\right)^{mt}$$

3-2   *The Exponential Function with Base e.* Irrational number e; exponential function with base e; exponential growth; exponential decay; continuous compound interest

$$N = N_0 e^{kt}; \quad A = A_0 e^{-kt}; \quad A = P e^{rt}$$

3-3   *Logarithmic Functions.* Logarithmic function; base; domain; range; exponential form; properties; common logarithm; natural logarithm; calculator evaluation; doubling time

$$\log_b x; \quad \log x; \quad \ln x$$

EXERCISE   3-4   Chapter Review

Work through all the problems in this chapter review and check your answers in the back of the book. (Answers to all review problems are there.) Where weaknesses show up, review appropriate sections in the text.

A

1. Write in logarithmic form using base e: $u = e^v$
2. Write in logarithmic form using base 10: $x = 10^y$

3. Write in exponential form using base e: $\ln M = N$
4. Write in exponential form using base 10: $\log u = v$

Simplify.

5. $\dfrac{5^{x+4}}{5^{4-x}}$

6. $\left(\dfrac{e^u}{e^{-u}}\right)^u$

Solve for x exactly without the use of a calculator.

7. $\log_3 x = 2$

8. $\log_x 36 = 2$

9. $\log_2 16 = x$

Solve for x to three decimal places.

10. $10^x = 143.7$
11. $e^x = 503{,}000$
12. $\log x = 3.105$
13. $\ln x = -1.147$

B   *Solve for x exactly without the use of a calculator.*

14. $\log(x + 5) = \log(2x - 3)$
15. $2 \ln(x - 1) = \ln(x^2 - 5)$
16. $9^{x-1} = 3^{1+x}$
17. $e^{2x} = e^{x^2-3}$
18. $2x^2 e^x = 3xe^x$
19. $\log_{1/3} 9 = x$
20. $\log_x 8 = -3$
21. $\log_9 x = \frac{3}{2}$

Solve for x to four decimal places.

22. $x = 3(e^{1.49})$
23. $x = 230(10^{-0.161})$
24. $\log x = -2.0144$
25. $\ln x = 0.3618$
26. $35 = 7(3^x)$
27. $0.01 = e^{-0.05x}$
28. $8{,}000 = 4{,}000(1.08^x)$
29. $5^{2x-3} = 7.08$

Simplify.

30. $e^x(e^{-x} + 1) - (e^x + 1)(e^{-x} - 1)$
31. $(e^x - e^{-x})^2 - (e^x + e^{-x})(e^x - e^{-x})$

Graph over the indicated interval.

32. $y = 2^{x-1}$; $[-2, 4]$
33. $f(t) = 10e^{-0.08t}$; $t \geq 0$
34. $y = \ln(x + 1)$; $(-1, 10]$

C   *Solve Problems 35–38 exactly without the use of a calculator.*

35. $\log x - \log 3 = \log 4 - \log(x + 4)$
36. $\ln(2x - 2) - \ln(x - 1) = \ln x$
37. $\ln(x + 3) - \ln x = 2 \ln 2$
38. $\log 3x^2 = 2 + \log 9x$

39. Write $\ln y = -5t + \ln c$ in an exponential form free of logarithms. Then solve for y in terms of the remaining variables.
40. Explain why 1 cannot be used as a logarithmic base.

The two formulas below will be of use in some of the problems that follow:

$$A = P\left(1 + \frac{r}{m}\right)^{mt} \qquad \text{Compound interest}$$

$$A = Pe^{rt} \qquad \text{Continuous compound interest}$$

Business & Economics

41. *Money growth.* If \$5,000 is invested at 12% compounded weekly, how much (to the nearest cent) will be in the account 6 years from now?

42. *Money growth.* If \$5,000 is invested at 12% compounded continuously, how much (to the nearest cent) will be in the account 6 years from now?

43. *Finance.* Find the tripling time (to the next whole year) for money invested at 15% compounded annually.

44. *Finance.* Find the doubling time (to two decimal places) for money invested at 10% compounded continuously.

Life Sciences

45. *Medicine.* One leukemic cell injected into a healthy mouse will divide into 2 cells in about $\frac{1}{2}$ day. At the end of the day these 2 cells will divide into 4. This doubling continues until 1 billion cells are formed; then the animal dies with leukemic cells in every part of the body.

(A) Write an equation that will give the number N of leukemic cells at the end of t days.

(B) When, to the nearest day, will the mouse die?

46. *Marine biology.* The intensity of light entering water is reduced according to the exponential equation

$$I = I_0 e^{-kd}$$

where I is the intensity d feet below the surface, I_0 is the intensity at the surface, and k is the coefficient of extinction. Measurements in the Sargasso Sea in the West Indies have indicated that half of the surface light reaches a depth of 73.6 feet. Find k (to five decimal places), and find the depth (to the nearest foot) at which 1% of the surface light remains.

Social Sciences

47. *Population growth.* Many countries have a population growth rate of 3% (or more) per year. At this rate, how many years (to the nearest tenth of a year) will it take a population to double? Use the annual compounding growth model $P = P_0(1 + r)^t$.

48. *Population growth.* Repeat Problem 47 using the continuous compounding growth model $P = P_0 e^{rt}$.

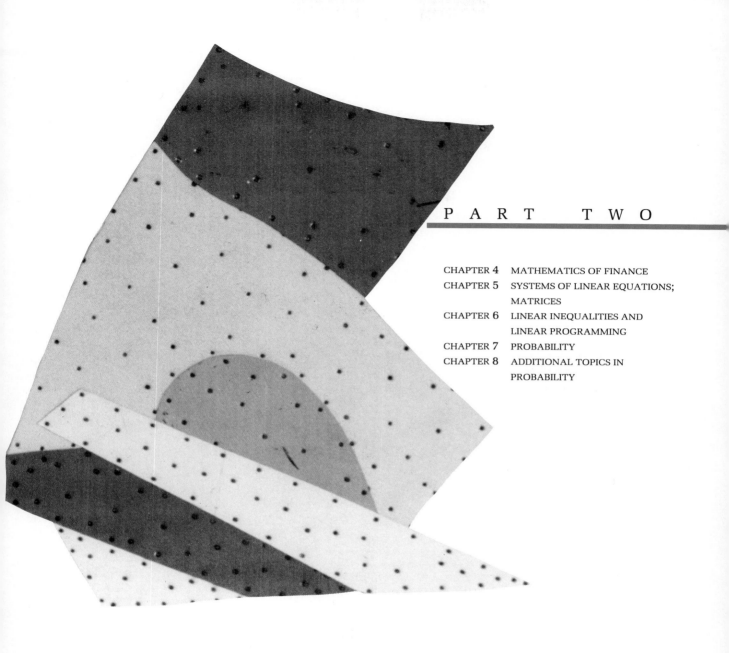

Finite Mathematics

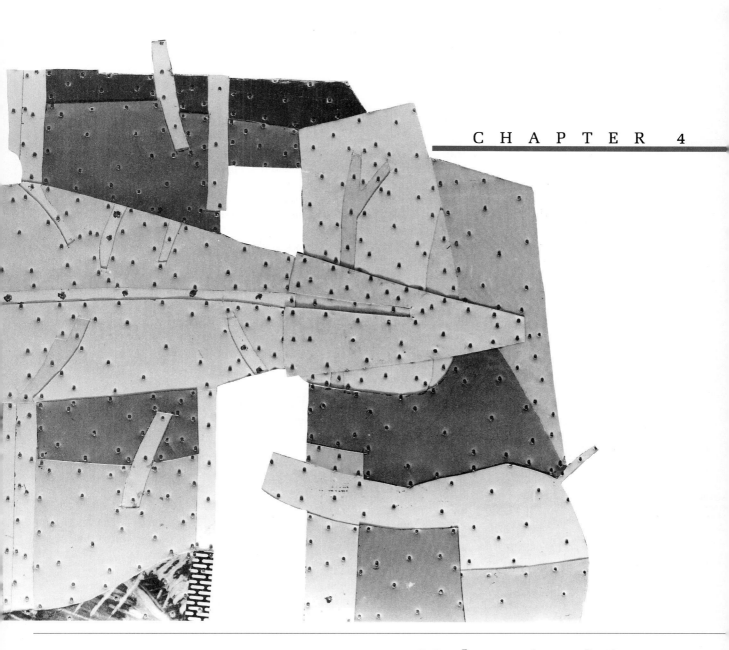

Mathematics of Finance

Contents

This chapter is independent of the others; you can study it at any time. In particular, we do not assume that you have studied Chapter 3, where a few of the topics in this chapter were briefly discussed as applications of exponential and logarithmic functions. Table II in Appendix B can be used to solve most of the problems on compound interest, annuities, amortization, and so on. However, if you have a financial or scientific calculator (both types are now available for under $20), you will be able to work all the problems without tables. Even if you use tables, an inexpensive calculator that has at least $+$, $-$, $\times$, $\div$, and y^x keys will take most of the drudgery out of the calculations. This chapter includes a number of problems that do require the use of a financial or scientific calculator, and these are clearly indicated.

If time permits, you may wish to cover arithmetic and geometric progressions, discussed in Appendix A, before beginning this chapter. Though not necessary, these topics will provide additional insight into some of the topics covered.

To avoid repeating the statement many times, we now point out:

Remark
_____

Throughout the chapter, interest rates are to be converted to decimal form before they are used in a formula.

_____

SECTION 4-1   Simple Interest

Simple interest is generally used only on short-term notes—often of duration less than 1 year. The concept of simple interest, however, forms the basis of much of the rest of the material developed in this chapter, for which time periods may be much longer than a year.

If you deposit a sum of money P in a savings account or if you bottow a sum of money P from a lending agent, then P is referred to as the **principal.** When money is borrowed—whether it is a savings institution borrowing from you when you deposit money in your account or you borrowing from a lending agent—a fee is charged for the money borrowed. This fee is rent paid for the use of another's money, just as rent is paid for the use of another's house. The fee is

called **interest**. It is usually computed as a percentage (called the **interest rate**)*
of the principal over a given period of time. The interest rate, unless otherwise
stated, is an annual rate. **Simple interest** is given by the following formula:

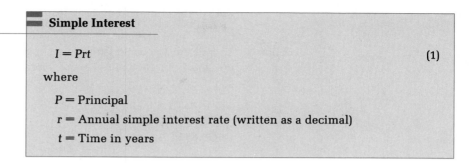

Simple Interest

$$I = Prt \tag{1}$$

where

$P =$ Principal

$r =$ Annual simple interest rate (written as a decimal)

$t =$ Time in years

For example, the interest on a loan of $100 at 12% for 9 months would be

$I = Prt$
$\quad = (100)(0.12)(0.75)$ Convert 12% to a decimal (0.12)
$\quad = \$9$ and 9 months to years $(\frac{9}{12} = 0.75)$.

At the end of 9 months, the borrower would repay the principal ($100) plus the
interest ($9), or a total of $109.

In general, if a principal P is borrowed at a rate r, then after t years the
borrower will owe the lender an amount A that will include the principal P (the
face value of the note) plus the interest I (the rent paid for the use of the money).
Since P is the amount that is borrowed now and A is the amount that must be
paid back in the future, P is often referred to as the **present value** and A as the
future value. The formula relating A and P is as follows:

Amount — Simple Interest

$$A = P + Prt$$
$$\quad = P(1 + rt) \tag{2}$$

where

$P =$ Principal, or present value

$r =$ Annual simple interest rate (written as a decimal)

$t =$ Time in years

$A =$ Amount, or future value

---

* If r is the interest rate written as a decimal, then 100r% would be the rate using %. For
example, if $r = 0.12$, then using the percent symbol, %, we would have 100r% =
100(0.12)% = 12%. The expressions 0.12 and 12% are equivalent.

Given any three of the four variables A, P, r, and t in (2), we can solve for the fourth. The following examples illustrate several types of common problems that can be solved by using formula (2).

◆ EXAMPLE 1   Find the total amount due on a loan of $800 at 18% simple interest at the end of 4 months.

Solution   To find the amount A (future value) due in 4 months, we use formula (2) with $P = 800$, $r = 0.18$, and $t = \frac{4}{12} = \frac{1}{3}$ year. Thus,

$$A = P(1 + rt)$$
$$= 800[1 + 0.18(\tfrac{1}{3})]$$
$$= 800(1.06)$$
$$= \$848$$ ◆

PROBLEM 1   Find the total amount due on a loan of $500 at 12% simple interest at the end of 30 months.   ◆

◆ EXAMPLE 2   If you want to earn an annual rate of 10% on your investments, how much (to the nearest cent) should you pay for a note that will be worth $5,000 in 9 months?

Solution   We again use formula (2), but now we are interested in finding the principal P (present value), given $A = \$5,000$, $r = 0.1$, and $t = \frac{9}{12} = 0.75$ year. Thus,

$$A = P(1 + rt)$$
$$5,000 = P[1 + 0.1(0.75)] \qquad \text{Replace } A, r, \text{ and } t \text{ with the given values,}$$
$$5,000 = (1.075)P \qquad\qquad \text{and solve for } P.$$
$$P = \$4,651.16$$ ◆

PROBLEM 2   Repeat Example 2 with a time period of 6 months.   ◆

◆ EXAMPLE 3   If you must pay $960 for a note that will be worth $1,000 in 6 months, what annual simple interest rate will you earn? (Express the answer as a percentage, correct to two decimal places.)

Solution   Again we use formula (2), but this time we are interested in finding r, given $P = \$960$, $A = \$1,000$, and $t = \frac{6}{12} = 0.5$ year. Thus,

$$A = P(1 + rt) \qquad\qquad \text{Replace } P, A, \text{ and } t \text{ with the given values, and}$$
$$1,000 = 960[1 + r(0.5)] \qquad \text{solve for } r.$$
$$1,000 = 960 + 960r(0.5)$$
$$40 = 480r$$
$$r = \frac{40}{480} \approx 0.0833 \quad \text{or} \quad 8.33\%$$ ◆

PROBLEM 3   Repeat Example 3 assuming you have paid $952 for the note.   ◆

◆ E X A M P L E 4   Suppose after buying a new car you decide to sell your old car to a friend. You accept a 270 day note for $3,500 at 10% simple interest as payment. (Both principal and interest will be paid at the end of 270 days.) Sixty days later you find that you need the money and sell the note to a third party for $3,550. What annual interest rate will the third party receive for the investment? (Express the answer as a percentage, correct to three decimal places.)

Solution   Step 1.   Find the amount that will be paid at the end of 270 days to whomever has the note. (Some financial institutions use a 365 day year and others a 360 day year. In all the problems in this section involving days, we assume a 360 day year. In other sections we will use a 365 day year. The choice will always be clearly stated.)

$$A = P(1 + rt)$$

$$= \$3,500 \left[1 + (0.1)\left(\frac{270}{360}\right) \right]$$

$$= \$3,762.50$$

Step 2.   For the third party we are to find the annual rate of interest r required to make $3,550 grow to $3,762.50 in 210 days (270 − 60); that is, we are to find r (which is to be converted to 100r%), given $A = \$3,762.50$, $P = \$3,550$, and $t = \frac{210}{360}$.

$$A = P + Prt \qquad \text{Solve for } r.$$

$$r = \frac{A - P}{Pt}$$

$$r = \frac{3,762.50 - 3,550}{(3,550)(\frac{210}{360})} = 0.102\ 62 \quad \text{or} \quad 10.262\%$$

◆

P R O B L E M 4   Repeat Example 4 assuming that 90 days after it was initially signed, the note was sold to a third party for $3,500.   ◆

Answers to Matched Problems   1. $650   2. $4,761.90   3. 10.08%   4. 15.0%

E X E R C I S E   4-1

A   Using formula (1) for simple interest, find each of the indicated quantities.

1. $P = \$500$; $r = 8\%$; $t = 6$ months; $I = ?$
2. $P = \$900$; $r = 10\%$; $t = 9$ months; $I = ?$
3. $I = \$80$; $P = \$500$; $t = 2$ years; $r = ?$
4. $I = \$40$; $P = \$400$; $t = 4$ years; $r = ?$

B   Use formula (2) in an appropriate form to find the indicated quantities.

5. $P = \$100$; $r = 8\%$; $t = 18$ months; $A = ?$
6. $P = \$6,000$; $r = 6\%$; $t = 8$ months; $A = ?$
7. $A = \$1,000$; $r = 10\%$; $t = 15$ months; $P = ?$
8. $A = \$8,000$; $r = 12\%$; $t = 7$ months; $P = ?$

C   Solve each formula for the indicated variable.

9. $I = Prt$;   for r
10. $I = Prt$;   for P
11. $A = P + Prt$;   for P
12. $A = P + Prt$;   for r

APPLICATIONS *

Business & Economics

In all problems involving days, a 360 day year is assumed. When annual rates are requested as an answer, express the rate as a percentage, correct to three decimal places.

13.   If $3,000 is loaned for 4 months at a 14% annual rate, how much interest is earned?
14.   If $5,000 is loaned for 10 months at a 10% annual rate, how much interest is earned?
15.   How much interest will you have to pay for a credit card balance of $554 that is 1 month overdue, if a 20% annual rate is charged?
16.   A department store charges an 18% annual rate for overdue accounts. How much interest will be owed on an $835 account that is 2 months overdue?
17.   A loan of $7,250 was repaid at the end of 8 months. What size repayment check (principal and interest) was written, if a 9% annual rate of interest was charged?
18.   A loan of $10,000 was repaid at the end of 14 months. What amount (principal and interest) was repaid if a 12% annual rate of interest was charged?
19.   A loan of $4,000 was repaid at the end of 8 months with a check for $4,270. What annual rate of interest was charged?
20.   A check for $3,262.50 was used to retire a 15 month $3,000 loan. What annual rate of interest was charged?
21.   If you paid $30 to a loan company for the use of $1,000 for 60 days, what annual rate of interest did they charge?
22.   If you paid $120 to a loan company for the use of $2,000 for 90 days, what annual rate of interest did they charge?
23.   A radio commercial for a loan company states: "You only pay 50¢ a day for each $500 borrowed." If you borrow $1,500 for 120 days, what amount will you repay, and what annual interest rate is the company actually charging?

---

* The authors wish to thank Professor Roy Luke of Pierce College for his many useful suggestions of applications in this chapter.

24. George finds a company that charges 70¢ per day for each $1,000 borrowed. If he borrows $3,000 for 60 days, what amount will he repay, and what annual interest rate will he be paying the company?

25. You are interested in buying a 13 week T-bill (treasury bill) from the U.S. Treasury Department. If you buy a T-bill with a maturity value of $10,000 for $9,776.94, what annual interest rate will you earn?

26. If you buy a 26 week T-bill with a maturity value of $10,000 for $9,562.56 from the U.S. Treasury Department, what annual interest rate will you earn?

27. If an investor wants to earn an annual interest rate of 12.63% on a 13 week T-bill with a maturity value of $10,000, how much should the investor pay for the bill?

28. If an investor wants to earn an annual interest rate of 10.58% on a 26 week T-bill with a maturity value of $10,000, how much should the investor pay for the bill?

29. An attorney accepts a 90 day note for $5,500 at 12% simple interest from a client for services rendered. (Both interest and principal will be repaid at the end of 90 days.) Wishing to have use of her money sooner, the attorney sells the note to a third party for $5,500 after 30 days. What annual interest rate will the third party receive for the investment?

30. To complete the sale of a house, the seller accepts a 180 day note for $10,000 at 10% simple interest. (Both interest and principal will be repaid at the end of 180 days.) Wishing to have use of the money sooner for the purchase of another house, the seller sells the note to a third party for $10,100 after 60 days. What annual interest rate will the third party receive for the investment?

In Problems 31 and 32, use the following buying and selling commission schedule (from a well-known discount brokerage house). Example: The commission on 500 shares at $15 per share is $84.50.

DOLLAR RANGE PER TRANSACTION	COMMISSION
$0–2,500	$19 + 1.6% of principal amount
$2,501–6,000	$44 + 0.6% of principal amount
$6,001–22,000	$62 + 0.3% of principal amount
$22,001–50,000	$84 + 0.2% of principal amount
$50,001–500,000	$134 + 0.1% of principal amount
$500,001+	$234 + 0.08% of principal amount

31. An investor purchased 500 shares of a stock at $14.20 a share. After holding the stock for 39 weeks, it was sold for $16.84 a share. Taking into consideration the buying and selling commissions charged by the discount brokerage house (see the table), what annual rate of interest was earned by the investor?

32. An investor purchased 450 shares of a stock at $64.84 a share. After holding the stock for 26 weeks, it was sold for $72.08 a share. Taking into consideration the buying and selling commissions charged by the discount brokerage house (see table), what annual rate of interest was earned by the investor?

Compound Interest

◆ COMPOUND INTEREST
◆ EFFECTIVE RATE
◆ GROWTH AND TIME

◆ COMPOUND INTEREST

If at the end of a payment period the interest due is reinvested at the same rate, then the interest as well as the original principal will earn interest during the next payment period. Interest paid on interest reinvested is called **compound interest.**

For example, suppose you deposit $1,000 in a bank that pays 8% compounded quarterly. How much will the bank owe you at the end of a year? *Compounding quarterly* means that earned interest is paid to your account at the end of each 3 month period and that interest as well as the principal earns interest for the next quarter. Using the simple interest formula (2) from the previous section, we compute the amount in the account at the end of the first quarter after interest has been paid:

$$A = P(1 + rt)$$
$$= 1,000[1 + 0.08(\tfrac{1}{4})]$$
$$= 1,000(1.02) = \$1,020$$

Now, $1,020 is your new principal for the second quarter. At the end of the second quarter, after interest is paid, the account will have

$$A = \$1,020[1 + 0.08(\tfrac{1}{4})]$$
$$= \$1,020(1.02) = \$1,040.40$$

Similarly, at the end of the third quarter, you will have

$$A = \$1,040.40[1 + 0.08(\tfrac{1}{4})]$$
$$= \$1,040.40(1.02) = \$1,061.21$$

Finally, at the end of the fourth quarter, the account will have

$$A = \$1,061.21[1 + 0.08(\tfrac{1}{4})]$$
$$= \$1,061.21(1.02) = \$1,082.43$$

How does this compound amount compare with simple interest? The amount with simple interest would be

$$A = P(1 + rt)$$
$$= \$1,000[1 + 0.08(1)]$$
$$= \$1,000(1.08) = \$1,080$$

We see that compounding quarterly yields $2.43 more than simple interest would provide.

Let us look over the above calculations for compound interest to see if we can uncover a pattern that might lead to a general formula for computing compound interest for arbitrary cases:

$A = 1,000(1.02)$	End of first quarter
$A = [1,000(1.02)](1.02) = 1,000(1.02)^2$	End of second quarter
$A = [1,000(1.02)^2](1.02) = 1,000(1.02)^3$	End of third quarter
$A = [1,000(1.02)^3](1.02) = 1,000(1.02)^4$	End of fourth quarter

It appears that at the end of n quarters, we would have

$$A = 1,000(1.02)^n \qquad \text{End of nth quarter}$$

or

$$A = 1,000[1 + 0.08(\tfrac{1}{4})]^n$$
$$= 1,000[1 + \tfrac{0.08}{4}]^n$$

where $\tfrac{0.08}{4} = 0.02$ is the interest rate per quarter. Since interest rates are generally quoted as *annual nominal rates*, the **rate per compounding period** is found by dividing the annual nominal rate by the number of compounding periods per year.

In general, if P is the principal earning interest compounded m times a year at an annual rate of r, then (by repeated use of the simple interest formula, using $i = r/m$, the rate per period) the amount A at the end of each period is

$A = P(1 + i)$	End of first period
$A = [P(1 + i)](1 + i) = P(1 + i)^2$	End of second period
$A = [P(1 + i)^2](1 + i) = P(1 + i)^3$	End of third period
$\cdot$	
$\cdot$	
$\cdot$	
$A = [P(1 + i)^{n-1}](1 + i) = P(1 + i)^n$	End of nth period

We summarize this important result in the following box:

▬ Amount—Compound Interest

$$A = P(1 + i)^n \qquad\qquad (1)$$

where $i = r/m$ and

$r = $ Annual nominal rate*

$m = $ Number of compounding periods per year

$i = $ Rate per compounding period

$n = $ Total number of compounding periods

$P = $ Principal (present value)

$A = $ Amount (future value) at the end of n periods

---

* This is often shortened to "annual rate" or just "rate."

Several examples will illustrate different uses of formula (1). If any three of the four variables in (1) are given, we can solve for the fourth. The power form $(1 + i)^n$ in formula (1) can be evaluated for various values of i and n by using any calculator with a $\boxed{y^x}$ key or Table II in Appendix B. The exponent n can be found using logarithm properties and a calculator.

◆ E X A M P L E   5   If $1,000 is invested at 8% compounded

(A) annually   (B) semiannually   (C) quarterly

what is the amount after 5 years? Write answers to the nearest cent.

Solutions   (A) Compounding annually means that there is one interest payment period per year. Thus, $n = 5$ and $i = r = 0.08$.

$$\begin{aligned}
A &= P(1 + i)^n \\
&= 1{,}000(1 + 0.08)^5 \qquad \text{Use a calculator (or Table II).} \\
&= 1{,}000(1.469\ 328) \\
&= \$1{,}469.33 \qquad\qquad \text{Interest earned} = A - P = \$469.33
\end{aligned}$$

(B) Compounding semiannually means that there are two interest payment periods per year. Thus, the number of payment periods in 5 years is $n = 2(5) = 10$, and the interest rate per period is

$$i = \frac{r}{m} = \frac{0.08}{2} = 0.04$$

So,

$$A = P(1 + i)^n$$
$$= 1,000(1 + 0.04)^{10} \qquad \text{Use a calculator (or Table II).}$$
$$= 1,000(1.480\ 244)$$
$$= \$1,480.24 \qquad \text{Interest earned} = A - P = \$480.24$$

(C) Compounding quarterly means that there are four interest payments per year. Thus, $n = 4(5) = 20$ and $i = \frac{0.08}{4} = 0.02$. So,

$$A = P(1 + i)^n$$
$$= 1,000(1 + 0.02)^{20} \qquad \text{Use a calculator (or Table II).}$$
$$= 1,000(1.485\ 947)$$
$$= \$1,485.95 \qquad \text{Interest earned} = A - P = \$485.95 \qquad \blacklozenge$$

P R O B L E M 5 Repeat Example 5 with an annual interest rate of 6% over an 8 year period. ◆

Notice the rather significant increase in interest earned in going from annual compounding to quarterly compounding. One might wonder what happens if we compound daily, or every minute, or every second, and so on. Figure 1 shows the relative effect of increasing the number of compounding periods in a year. A limit is reached at compounding *continuously*, which is not a great deal larger than that obtained through monthly compounding. (Continuous compounding is discussed in Section 3-2.) Compare the results in Figure 1 with simple interest earned over the same time period:

$$I = Prt = 1,000(0.08)5 = \$400$$

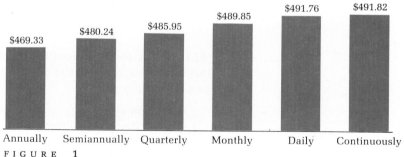

FIGURE 1
Interest on $1,000 for 5 years at 8% compounded at different periods.

Another use of the compound interest formula is in determining how much you should invest now to have a given amount at a future date.

◆ EXAMPLE 6   How much should you invest now at 10% compounded quarterly to have $8,000 toward the purchase of a car in 5 years?

Solution   We are given a future value $A = \$8{,}000$ for a compound interest investment, and we need to find the present value (principal P) given $i = 0.10/4 = 0.025$ and $n = 4(5) = 20$.

$$A = P(1 + i)^n$$

$$8{,}000 = P(1 + 0.025)^{20}$$

$$P = \frac{8{,}000}{(1 + 0.025)^{20}} \qquad \text{Use a calculator (or Table II).}$$

$$= \frac{8{,}000}{1.638\ 616} = \$4{,}882.17 \qquad\qquad ◆$$

PROBLEM 6   How much should new parents invest now at 8% compounded semiannually to have $16,000 toward their child's college education in 17 years?   ◆

◆ EFFECTIVE RATE

Suppose you read in the newspaper that one investment pays 15% compounded monthly and another pays 15.2% compounded semiannually. Which has the better return? A good way to compare investments is to determine their **effective rates**—the simple interest rates that would produce the same returns in *1 year* if the same principal had been invested at simple interest without compounding. (Effective rates are also called **annual yields** or **true interest rates**.)

If principal P is invested at an annual (nominal) rate r compounded m times a year, then in 1 year,

$$A = P\left(1 + \frac{r}{m}\right)^m$$

What simple interest rate will produce the same amount A in 1 year? We call this simple interest rate the effective rate, and denote it by r_e. To find r_e we proceed as follows:

$$\begin{pmatrix} \text{Amount at} \\ \text{simple interest} \\ \text{after 1 year} \end{pmatrix} = \begin{pmatrix} \text{Amount at} \\ \text{compound interest} \\ \text{after 1 year} \end{pmatrix}$$

$$P(1 + r_e) = P\left(1 + \frac{r}{m}\right)^m \qquad \text{Divide both sides by } P.$$

$$1 + r_e = \left(1 + \frac{r}{m}\right)^m \qquad \text{Isolate } r_e \text{ on the left side.}$$

$$r_e = \left(1 + \frac{r}{m}\right)^m - 1$$

Effective Rate

If principal P is invested at the annual (nominal) rate r compounded m times a year, then the effective rate r_e is given by

$$r_e = \left(1 + \frac{r}{m}\right)^m - 1$$

EXAMPLE 7 A savings and loan pays 8% compounded quarterly. What is the effective rate? (Express the answer as a percentage, correct to three decimal places.)

Solution
$$r_e = \left(1 + \frac{r}{m}\right)^m - 1$$
$$= (1 + \tfrac{0.08}{4})^4 - 1$$
$$= (1.02)^4 - 1 \qquad \text{Use a calculator (or Table II).}$$
$$= 1.082\ 432 - 1$$
$$= 0.082\ 432 \quad \text{or} \quad 8.243\%$$

This shows that money invested at 8.243% simple interest earns the same amount of interest *in 1 year* as money invested at 8% compounded quarterly. Thus, the effective rate of 8% compounded quarterly is 8.243%. ◆

PROBLEM 7 What is the effective rate of money invested at 6% compounded quarterly? ◆

EXAMPLE 8 An investor has an opportunity to purchase two different notes: Note *A* pays 15% compounded monthly, and note *B* pays 15.2% compounded semiannually. Which is the better investment, assuming all else is equal?

Solution Nominal rates with different compounding periods cannot be compared directly. We must first find the effective rate of each nominal rate and then compare the effective rates to determine which investment will yield the larger return.

Effective Rate for Note A
$$r_e = \left(1 + \frac{r}{m}\right)^m - 1$$
$$= (1 + \tfrac{0.15}{12})^{12} - 1$$
$$= (1.0125)^{12} - 1 \qquad \text{Use a calculator (or Table II).}$$
$$= 1.160\ 755 - 1$$
$$= 0.160\ 755 \quad \text{or} \quad 16.076\%$$

Effective Rate for Note B
$$r_e = \left(1 + \frac{r}{m}\right)^m - 1$$
$$= (1 + \tfrac{0.152}{2})^2 - 1$$
$$= (1.076)^2 - 1$$
$$= 1.157\ 776 - 1$$
$$= 0.157\ 776 \quad \text{or} \quad 15.778\%$$

Since the effective rate for note A is greater than the effective rate for note B, note A is the preferred investment. ◆

PROBLEM 8 Repeat Example 8 if note A pays 9% compounded monthly and note B pays 9.2% compounded semiannually. ◆

◆ GROWTH AND TIME

Investments are also compared by computing their **growth time**—the time it takes a given principal to grow to a particular value (the shorter the time, the greater the return on the investment). Example 9 illustrates two methods for making this calculation.

◆ EXAMPLE 9 How long will it take $10,000 to grow to $12,000 if it is invested at 9% compounded monthly?

Solution
$$A = P(1 + i)^n$$
$$12,000 = 10,000(1 + \tfrac{0.09}{12})^n$$
$$1.2 = (1.0075)^n$$

Now solve for n:

Method 1. Use Table II in Appendix B. Look down the $(1 + i)^n$ column on the page that gives values for $i = 0.0075$ ($\tfrac{3}{4}$%). Find the value in this column that is closest to and greater than 1.2 and take the n value that corresponds to it. In this case, $n = 25$ months, or 2 years and 1 month.

Method 2. Use logarithms and a calculator.

$$1.2 = (1.0075)^n$$
$$\ln 1.2 = \ln(1.0075)^n \qquad \text{Log to any base can be used; here we choose the}$$
$$\ln 1.2 = n \ln 1.0075 \qquad \text{natural logarithm (base } e \text{) and use the property}$$
$$\log_b M^p = p \log_b M.$$
$$n = \frac{\ln 1.2}{\ln 1.0075} \qquad \text{Use a calculator.}$$
$$= \frac{0.1823}{0.0075} = 24.31 \approx 25 \text{ months, or 2 years and 1 month}$$

[*Note:* 24.31 is rounded up to 25 to guarantee reaching $12,000, since interest is paid at the end of each month.] ◆

PROBLEM 9 How long will it take $10,000 to grow to $25,000 if it is invested at 18% compounded quarterly? ◆

Answers to Matched Problems
5. (A) $1,593.85   (B) $1,604.71   (C) $1,610.32
6. $4,216.83   7. 6.136%
8. Note B (effective rate of note A is 9.381% and of note B is 9.412%)
9. 20.78 ≈ 21 quarters, or 5 years and 3 months

Find all dollar amounts to the nearest cent. When an interest rate is requested as an answer, express the rate as a percentage, correct to two decimal places, unless directed otherwise. Use either Table II or a calculator, unless specifically instructed to use a calculator. (A small discrepancy may result between answers obtained using a calculator and answers obtained using a table because of roundoff errors.)

A In Problems 1–8, use compound interest formula (1) to find each of the indicated values.

1. $P = \$100$; $i = 0.01$; $n = 12$; $A = ?$
2. $P = \$1,000$; $i = 0.015$; $n = 20$; $A = ?$
3. $P = \$800$; $i = 0.06$; $n = 25$; $A = ?$
4. $P = \$10,000$; $i = 0.08$; $n = 30$; $A = ?$
5. $A = \$10,000$; $i = 0.03$; $n = 48$; $P = ?$
6. $A = \$1,000$; $i = 0.015$; $n = 60$; $P = ?$
7. $A = \$18,000$; $i = 0.01$; $n = 90$; $P = ?$
8. $A = \$50,000$; $i = 0.005$; $n = 70$; $P = ?$

Given the annual rate and the compounding period, find i, the interest rate per compounding period.

9. 9% compounded monthly
10. 15% compounded annually
11. 7% compounded quarterly
12. 11% compounded semiannually

Given the rate per compounding period, find r, the annual rate.

13. 0.8% per month
14. 5% per year
15. 4.5% per half-year
16. 2.3% per quarter

B 17. If $100 is invested at 6% compounded

 (A) annually (B) quarterly (C) monthly

 what is the amount after 4 years? How much interest is earned?

18. If $2,000 is invested at 7% compounded

 (A) annually (B) quarterly (C) monthly

 what is the amount after 5 years? How much interest is earned?

19. If $5,000 is invested at 18% compounded monthly, what is the amount after

 (A) 2 years? (B) 4 years?

20. If $20,000 is invested at 6% compounded monthly, what is the amount after

 (A) 5 years? (B) 8 years?

21. If an investment company pays 8% compounded semiannually, how much should you deposit now to have $10,000

 (A) 5 years from now? (B) 10 years from now?

22. If an investment company pays 10% compounded quarterly, how much should you deposit now to have $6,000

(A) 3 years from now?   (B) 6 years from now?

23. What is the effective rate of interest for money invested at

(A) 10% compounded quarterly?   (B) 12% compounded monthly?

24. What is the effective rate of interest for money invested at

(A) 6% compounded monthly?   (B) 14% compounded semiannually?

25. How long will it take $4,000 to grow to $9,000 if it is invested at 15% compounded monthly?

26. How long will it take $5,000 to grow to $7,000 if it is invested at 8% compounded quarterly?

C In Problems 27 and 28, use the compound interest formula (1) to find n to the nearest larger integer value.

27. $A = 2P$; $i = 0.06$; $n = ?$ 28. $A = 2P$; $i = 0.05$; $n = ?$

29. How long will it take money to double if it is invested at

(A) 10% compounded quarterly?   (B) 12% compounded quarterly?

30. How long will it take money to double if it is invested at

(A) 14% compounded semiannually?

(B) 10% compounded semiannually?

APPLICATIONS

Business & Economics

31. A newborn child receives a $5,000 gift toward a college education from her grandparents. How much will the $5,000 be worth in 17 years if it is invested at 9% compounded quarterly?

32. A person with $8,000 is trying to decide whether to purchase a car now, or to invest the money at 12% compounded semiannually and then buy a more expensive car. How much will be available for the purchase of a car at the end of 3 years?

33. What will a $110,000 house cost 10 years from now if the inflation rate over that period averages 6% compounded annually?

34. If the inflation rate averages 8% per year compounded annually for the next 5 years, what will a car costing $10,000 now cost 5 years from now?

35. Rental costs for office space have been going up at 7% per year compounded annually for the past 5 years. If office space rent is now $20 per square foot per month, what were the rental rates 5 years ago?

36. In a suburb of a city, housing costs have been increasing at 8% per year compounded annually for the past 8 years. A house with a $160,000 value now would have had what value 8 years ago?

37. If the population in a particular third-world country is growing at 4% compounded annually, how long will it take the population to double? (Round up to the next higher year if not exact.)

38. If the world population is now about 5 billion people and is growing at 2% compounded annually, how long will it take the population to grow to 8 billion people? (Round up to the next higher year if not exact.)

39. Which is the better investment and why: 9% compounded monthly or 9.3% compounded annually?

40. Which is the better investment and why: 8% compounded quarterly or 8.3% compounded annually?

41. You have saved $7,000 toward the purchase of a car costing $9,000. How long will the $7,000 have to be invested at 9% compounded monthly to grow to $9,000? (Round up to the next higher month if not exact.)

42. A newly married couple has $15,000 toward the purchase of a house. For the type of house they are interested in buying, they estimate that a $20,000 down payment will be necessary. How long will the money have to be invested at 10% compounded quarterly to grow to $20,000? (Round up to the next higher quarter if not exact.)

The following problems require the use of a calculator with a $\boxed{y^x}$ key. Use a 365 day year.

43. An Individual Retirement Account (IRA) has $20,000 in it and the owner decides not to add any more money to the account other than interest earned at 8% compounded daily. How much will be in the account 35 years from now when the owner reaches retirement age?

44. If $1 had been placed in a bank account at the birth of Christ and forgotten until now, how much would be in the account at the end of 1990 if money earned 2% interest compounded annually? 2% simple interest? (Now you can see the power of compounding and see why inactive accounts are closed after a relatively short period of time.)

45. How long will it take money to double if it is invested at 14% compounded daily? 15% compounded annually? (Compute answers in years to three decimal places.)

46. How long will it take money to triple if it is invested at 10% compounded daily? 11% compounded annually? (Compute answers in years to three decimal places.)

*Problems 47–50 refer to zero coupon bonds. A **zero coupon bond** is a bond that is sold now at a discount and will pay its face value at some time in the future when it matures — no interest payments are made.*

47. Parents wishing to have enough money for their child's college education 17 years from now decide to buy a $30,000 face value zero coupon bond. If money is worth 10% compounded annually, what should they pay for the bond?

48. How much should a $20,000 face value zero coupon bond, maturing in 10 years, be sold for now if its rate of return is to be 8% compounded annually?

49. If the parents in Problem 47 pay $6,844.79 for the $30,000 face value zero coupon bond, what annual compound rate of return will they be receiving?

50. If you pay $5,893.24 for a $12,000 face value zero coupon bond that matures in 7 years, what is your annual compound rate of return?

51. *Barron's* (a national business and financial weekly) published the following "Top Savings Deposit Yields" for money market deposit accounts:

   (A) Virginia Beach Federal S&L   8.28% compounded monthly
   (B) Franklin Savings   8.25% compounded daily
   (C) Guaranty Federal Savings   8.25% compounded monthly

   What is the effective yield for each?

52. *Barron's* also published the following "Top Savings Deposit Yields" for 1 year certificate of deposit (CD) accounts:

   (A) Spindletop Savings   9.00% compounded daily
   (B) Alamo Savings of Texas   9.10% compounded quarterly
   (C) Atlas Savings and Loan   9.00% compounded quarterly

   What is the effective yield for each?

53. If you just sold a stock for $32,456.32 (net) that cost you $24,766.81 (net) 2 years ago, what annual compound rate of return did you make on your investment?

54. If you just sold a stock for $27,339.79 (net) that cost you $14,664.76 (net) 4 years ago, what annual compound rate of return did you make on your investment?

SECTION 4-3 Future Value of an Annuity; Sinking Funds

◆ FUTURE VALUE OF AN ANNUITY
◆ SINKING FUNDS

◆ FUTURE VALUE OF AN ANNUITY

An **annuity** is any sequence of equal periodic payments. If payments are made at the end of each time interval, then the annuity is called an **ordinary annuity.** We will consider only ordinary annuities in this book. The amount, or **future value,** of an annuity is the sum of all payments plus all interest earned.

Suppose you decide to deposit $100 every 6 months into an account that pays 6% compounded semiannually. If you make six deposits, one at the end of each interest payment period, over 3 years, how much money will be in the account after the last deposit is made? To solve this problem, let us look at it in terms of a time line. Using the compound amount formula $A = P(1 + i)^n$, we can find the value of each deposit after it has earned compound interest up through the sixth deposit, as shown in Figure 2.

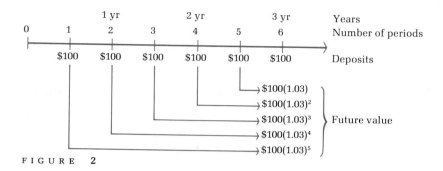

FIGURE 2

We could, of course, evaluate each of the future values in Figure 2 using Table II or a calculator and then add the results to find the amount in the account at the time of the sixth deposit—a tedious project at best. Instead, we take another approach that leads directly to a formula that will produce the same result in a few steps (even when the number of deposits is very large). We start by writing the total amount in the account after the sixth deposit in the form

$$S = 100 + 100(1.03) + 100(1.03)^2 + 100(1.03)^3 + 100(1.03)^4 + 100(1.03)^5 \qquad (1)$$

We would like a simple way to sum these terms. Let us multiply each side of (1) by 1.03 to obtain

$$1.03S = 100(1.03) + 100(1.03)^2 + 100(1.03)^3 + 100(1.03)^4 + 100(1.03)^5 + 100(1.03)^6 \quad (2)$$

Subtracting (1) from (2), left side from left side and right side from right side, we obtain

$$1.03S - S = 100(1.03)^6 - 100 \qquad \text{Notice how many terms drop out.}$$

$$0.03S = 100[(1.03)^6 - 1]$$

$$S = 100\,\frac{(1 + 0.03)^6 - 1}{0.03} \qquad \text{We write } S \text{ in this form to observe a general pattern.} \qquad (3)$$

In general, if R is the periodic deposit, i the rate per period, and n the number of periods, then the future value is given by

$$S = R + R(1 + i) + R(1 + i)^2 + \cdots + R(1 + i)^{n-1} \qquad \text{Note how this compares to (1).}$$

and proceeding as in the above example, we obtain the general formula for the future value of an ordinary annuity:*

$$S = R\,\frac{(1 + i)^n - 1}{i} \qquad \text{Note how this compares to (3).} \qquad (4)$$

---

* This formula also can be obtained by using the formula in Appendix A for the sum of the first n terms in a geometric progression.

Returning to the example above, we use a calculator with a $\boxed{y^x}$ key to complete the problem:

$$S = 100 \frac{(1.03)^6 - 1}{0.03}$$

Compute $(1.03)^6$ first; then complete the problem following the arithmetic operations indicated.

$$= 100 \frac{(1.194\ 052...) - 1}{0.03}$$

$$= \$646.84$$

For improved accuracy, keep all values in the calculator until the end; then round to the required number of decimal places.

Using Table II in place of a calculator, it is common practice to use the symbol

$$s_{\overline{n}|i} = \frac{(1 + i)^n - 1}{i}$$

for the fractional part of (4). The symbol $s_{\overline{n}|i}$, read "s angle n at i," is evaluated in Table II for various values of n and i. (The advantage of a calculator, of course, is that it can handle many more situations than any table, no matter how large the table.)

Returning to the example above, we now use Table II to complete the problem:

$$S = 100 \frac{(1.03)^6 - 1}{0.03}$$

$$= 100 s_{\overline{6}|0.03}$$ Use Table II with $i = 0.03$ and $n = 6$.

$$= 100(6.468\ 410)$$

$$= \$646.84$$

It is common to use FV (future value) for S and PMT (payment) for R in formula (4). Making these changes, we have the following:

Future Value of an Ordinary Annuity

$$FV = PMT \frac{(1 + i)^n - 1}{i} = PMT s_{\overline{n}|i} \tag{5}$$

where

PMT = Periodic payment

$\quad i$ = Rate per period

$\quad n$ = Number of payments (periods)

FV = Future value (amount)

[Note: Payments are made at the end of each period.]

◆ EXAMPLE 10   What is the value of an annuity at the end of 5 years if $100 per month is deposited into an account earning 9% compounded monthly? How much of this value is interest?

Solution   To find the value of the annuity, use formula (5) with $PMT = \$100$, $i = \frac{0.09}{12} = 0.0075$, and $n = 12(5) = 60$.

$$FV = PMT \frac{(1+i)^n - 1}{i} \quad \text{or} \quad PMTs_{\overline{n}|i}$$

$$= 100 \frac{(1.0075)^{60} - 1}{0.0075} \quad \text{or} \quad 100s_{\overline{60}|0.0075} \qquad \text{Use a calculator (or Table II).}$$

$$= 100(75.424\ 137)$$

$$= \$7,542.41$$

To find the interest, subtract the total amount deposited in the annuity (60 payments of $100) from the total value of the annuity after the 60th payment:

$$\text{Deposits} = 60(100)$$
$$= \$6,000$$
$$\text{Interest} = \text{Value} - \text{Deposits}$$
$$= 7,542.41 - 6,000$$
$$= \$1,542.41 \qquad \qquad \qquad \quad ◆$$

PROBLEM 10   What is the value of an annuity at the end of 10 years if $1,000 is deposited every 6 months into an account earning 8% compounded semiannually? How much of this value is interest?   ◆

◆ SINKING FUNDS

The formula for the future value of an ordinary annuity has another important application. Suppose the parents of a newborn child decide that on each of the child's birthdays up to the 17th year, they will deposit $PMT in an account that pays 6% compounded annually. The money is to be used for college expenses. What should the annual deposit PMT be in order for the amount in the account to be $16,000 after the 17th deposit?

We are given FV, i, and n in formula (5), and our problem is to find PMT. Thus,

$$FV = PMT \frac{(1+i)^n - 1}{i} \quad \text{or} \quad PMTs_{\overline{n}|i}$$

$$16,000 = PMT \frac{(1.06)^{17} - 1}{0.06} \quad \text{or} \quad PMTs_{\overline{17}|0.06} \qquad \text{Solve for } PMT.$$

$$PMT = 16,000 \frac{0.06}{(1.06)^{17} - 1} \quad \text{or} \quad 16,000 \frac{1}{s_{\overline{17}|0.06}} \qquad \text{Use a calculator or Table II.}$$

$$= 16,000(0.035\ 445)$$

$$= \$567.12 \text{ per year}$$

An annuity of 17 annual deposits of $567.12 at 6% compounded annually will amount to approximately $16,000 in 17 years.

This is one of many examples of a similar type that are referred to as *sinking fund problems*. In general, any account that is established for accumulating funds to meet future obligations or debts is called a **sinking fund.** If the payments are to be made in the form of an ordinary annuity, then we have only to solve for *PMT* in formula (5) to find the periodic payment into the fund. Doing this, we obtain the following general formula:

■ Sinking Fund Payment — A Variation of Formula (5)

$$PMT = FV \frac{i}{(1+i)^n - 1} = \frac{FV}{s_{\overline{n}|i}} \qquad (6)$$

where

PMT = Sinking fund payment

FV = Value of annuity after n payments (future value)

n = Number of payments (periods)

i = Rate per period

[*Note:* Payments are made at the end of each period.]

◆ E X A M P L E   11

A company estimates that it will have to replace a piece of equipment at a cost of $10,000 in 5 years. To have this money available in 5 years, a sinking fund is established by making fixed monthly payments into an account paying 6% compounded monthly. How much should each payment be?

Solution

To find *PMT*, we can use either formula (5) or (6). We choose formula (6) with $FV = \$10,000$, $i = \frac{0.06}{12} = 0.005$, and $n = 5(12) = 60$:

$$PMT = FV \frac{i}{(1+i)^n - 1} \quad \text{or} \quad \frac{FV}{s_{\overline{n}|i}}$$

$$PMT = (10{,}000) \frac{0.005}{(1.005)^{60} - 1} \quad \text{or} \quad 10{,}000 \frac{1}{s_{\overline{60}|0.005}} \qquad \text{Use a calculator or Table II.}$$

$$PMT = 10{,}000(0.014\ 333)$$

$$= \$143.33 \text{ per month} \qquad \qquad \qquad ◆$$

P R O B L E M   11

A bond issue is approved for building a marina in a city. The city is required to make regular payments every 6 months into a sinking fund paying 6% compounded semiannually. At the end of 10 years, the bond obligation will be retired at a cost of $5,000,000. What should each payment be? ◆

E X E R C I S E 4-3

In Problems 1–20 use formula (5) or (6) and a calculator or Table II (or both) to solve each problem. (Answers may vary slightly depending on whether you use a calculator or Table II.)

A
1. $FV = ?$; $n = 20$; $i = 0.03$; $PMT = \$500$
2. $FV = ?$; $n = 25$; $i = 0.04$; $PMT = \$100$
3. $FV = ?$; $n = 40$; $i = 0.02$; $PMT = \$1,000$
4. $FV = ?$; $n = 30$; $i = 0.01$; $PMT = \$50$

B
5. $FV = \$3,000$; $n = 20$; $i = 0.02$; $PMT = ?$
6. $FV = \$8,000$; $n = 30$; $i = 0.03$; $PMT = ?$
7. $FV = \$5,000$; $n = 15$; $i = 0.01$; $PMT = ?$
8. $FV = \$2,500$; $n = 10$; $i = 0.08$; $PMT = ?$

C
9. $FV = \$4,000$; $i = 0.02$; $PMT = 200$; $n = ?$
10. $FV = \$8,000$; $i = 0.04$; $PMT = 500$; $n = ?$

APPLICATIONS

Business & Economics

11. What is the value of an ordinary annuity at the end of 10 years if $500 per quarter is deposited into an account earning 8% compounded quarterly? How much of this value is interest?

12. What is the value of an ordinary annuity at the end of 20 years if $1,000 per year is deposited into an account earning 7% compounded annually? How much of this value is interest?

13. In order to accumulate enough money for a down payment on a house, a couple deposits $300 per month into an account paying 6% compounded monthly. If payments are made at the end of each period, how much money will be in the account in 5 years?

14. A self-employed person has a Keogh retirement plan. (This type of plan is free of taxes until money is withdrawn.) If deposits of $7,500 are made each year into an account paying 8% compounded annually, how much will be in the account after 20 years?

15. In 5 years a couple would like to have $25,000 for a down payment on a house. What fixed amount should be deposited each month into an account paying 9% compounded monthly?

16. A person wishes to have $200,000 in an account for retirement 15 years from now. How much should be deposited quarterly in an account paying 8% compounded quarterly?

17. A company estimates it will need $100,000 in 8 years to replace a computer. If it establishes a sinking fund by making fixed monthly payments into an account paying 12% compounded monthly, how much should each payment be?

18. Parents have set up a sinking fund in order to have $30,000 in 15 years for their children's college education. How much should be paid semiannually into an account paying 10% compounded semiannually?

19. Beginning in January, a person plans to deposit $100 at the end of each month into an account earning 9% compounded monthly. Each year taxes must be paid on the interest earned during that year. Find the interest earned during each year for the first 3 years.

20. If $500 is deposited each quarter into an account paying 12% compounded quarterly for 3 years, find the interest earned during each of the 3 years.

Use a calculator with a $\boxed{y^x}$ *key (or a financial calculator) to solve Problems 21–26.*

21. Why does it make sense to open an Individual Retirement Account (IRA) early in one's life? (For people in certain income brackets, money deposited into an IRA and earnings from an IRA are tax-deferred until withdrawal.) Compare the following:

(A) Jane deposits $2,000 a year into an IRA earning 9% compounded annually. She makes her first deposit on her 24th birthday and her last deposit on her 31st birthday (8 deposits in all). Making no additional deposits, she leaves the accumulated amount from the 8 deposits in the account, earning interest at 9% compounded annually, until her 65th birthday. How much (to the nearest dollar) will be in her account on her 65th birthday?

(B) John procrastinates and does not make his first $2,000 deposit into an IRA until he is 32, but then he continues to deposit $2,000 on every birthday until he is 65 (34 deposits in all). If his account also earns 9% compounded annually, how much (to the nearest dollar) will he have in his account when he makes his last deposit on his 65th birthday?

(*Surprise*—Jane will have more money than John!)

22. Starting on his 24th birthday, and continuing on every birthday up to and including his 65th, a person deposits $2,000 a year into an IRA. How much (to the nearest dollar) will be in the account on the 65th birthday, if the account earns:

(A) 6% compounded annually?   (B) 8% compounded annually?
(C) 10% compounded annually?   (D) 12% compounded annually?

23. You wish to have $10,000 in 4 years to buy a car. How much should you deposit each month into an account paying 8% compounded monthly? How much interest will the account earn in the 4 years?

24. A company establishes a sinking fund to upgrade a plant in 5 years at an estimated cost of $1,500,000. How much should be invested each quarter into an account paying 9.15% compounded quarterly? How much interest will the account earn in the 5 years?

You can afford monthly deposits of only $150 into an account that pays 8.5% compounded monthly. How long will it be until you have $7,000 to buy a boat? (Round to the next higher month if not exact.)

A company establishes a sinking fund for upgrading office equipment with monthly payments of $1,000 into an account paying 10% compounded monthly. How long will it be before the account has $100,000? (Round up to the next higher month if not exact.)

Present Value of an Annuity; Amortization

◆ PRESENT VALUE OF AN ANNUITY
◆ AMORTIZATION
◆ AMORTIZATION SCHEDULES

◆ PRESENT VALUE OF AN ANNUITY

How much should you deposit in an account paying 6% compounded semi-annually in order to be able to withdraw $1,000 every 6 months for the next 3 years? (After the last payment is made, no money is to be left in the account.)

Actually, we are interested in finding the **present value** of each $1,000 that is paid out during the 3 years. We can do this by solving for P in the compound interest formula:

$$A = P(1 + i)^n$$

$$P = \frac{A}{(1 + i)^n} = A(1 + i)^{-n}$$

The rate per period is $i = \frac{0.06}{2} = 0.03$. The present value P of the first payment is $1,000(1.03)^{-1}$, the second payment is $1,000(1.03)^{-2}$, and so on. Figure 3 shows this in terms of a time line.

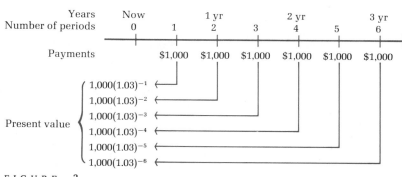

FIGURE 3

We could evaluate each of the present values in Figure 3 using a calculator or Table II and add the results to find the total present values of all the payments (which will be the amount that is needed now to buy the annuity). Since this is generally a tedious process, particularly when the number of payments is large, we will use the same device we used in the last section to produce a formula that will accomplish the same result in a couple of steps. We start by writing the sum of the present values in the form

$$P = 1{,}000(1.03)^{-1} + 1{,}000(1.03)^{-2} + \cdots + 1{,}000(1.03)^{-6} \tag{1}$$

Multiplying both sides of (1) by 1.03, we obtain

$$1.03P = 1{,}000 + 1{,}000(1.03)^{-1} + \cdots + 1{,}000(1.03)^{-5} \tag{2}$$

Now subtract (1) from (2):

$$1.03P - P = 1{,}000 - 1{,}000(1.03)^{-6} \qquad \text{Notice how many terms drop out.}$$

$$0.03P = 1{,}000[1 - (1 + 0.03)^{-6}]$$

$$P = 1{,}000\,\frac{1 - (1 + 0.03)^{-6}}{0.03} \qquad \begin{array}{l}\text{We write } P \text{ in this form to} \\ \text{observe a general pattern.}\end{array} \tag{3}$$

In general, if R is the periodic payment, i the rate per period, and n the number of periods, then the present value of all payments is given by

$$P = R(1 + i)^{-1} + R(1 + i)^{-2} + \cdots + R(1 + i)^{-n} \qquad \begin{array}{l}\text{Note how this compares} \\ \text{to (1).}\end{array}$$

Proceeding as in the above example, we obtain the general formula for the present value of an ordinary annuity:*

$$P = R\,\frac{1 - (1 + i)^{-n}}{i} \qquad \text{Note how this compares to (3).} \tag{4}$$

Returning to the example above, we use a calculator with a $\boxed{y^x}$ key to complete the problem:

$$P = 1{,}000\,\frac{1 - (1.03)^{-6}}{0.03} \qquad \begin{array}{l}\text{Compute } (1.03)^{-6} \text{ first; then complete} \\ \text{the problem following the arithmetic} \\ \text{operations indicated.}\end{array}$$

$$= 1{,}000\,\frac{1 - (0.837\ 484...)}{0.03} \qquad \begin{array}{l}\text{For improved accuracy, keep all values} \\ \text{in the calculator until the end; then} \\ \text{round to the required number of} \\ \text{decimal places.}\end{array}$$

$$= \$5{,}417.19$$

---

* This formula also can be obtained by using the formula in Appendix A for the sum of the first n terms in a geometric progression.

Using Table II in place of a calculator, it is common practice to use the symbol

$$a_{\overline{n}|i} = \frac{1 - (1 + i)^{-n}}{i}$$

for the fractional part of (4). The symbol $a_{\overline{n}|i}$, read "a angle n at i," is evaluated for various values of n and i in Table II. (As we said before, a calculator can handle far more situations than any table, no matter how large the table.)

Returning to the example above, we now use Table II to complete the problem:

$$P = 1{,}000 \, \frac{1 - (1.03)^{-6}}{0.03}$$

$$= 1{,}000 a_{\overline{6}|0.03} \qquad \text{Use Table II with } i = 0.03 \text{ and } n = 6.$$

$$= 1{,}000(5.417 \ 191)$$

$$= \$5{,}417.19$$

It is common to use PV (present value) for P and PMT (payment) for R in formula (4). Making these changes, we have the following:

Present Value of an Ordinary Annuity

$$PV = PMT \, \frac{1 - (1 + i)^{-n}}{i} = PMT a_{\overline{n}|i} \qquad (5)$$

where

PMT = Periodic payment

i = Rate per period

n = Number of periods

PV = Present value of all payments

[Note: Payments are made at the end of each period.]

◆ E X A M P L E   12

What is the present value of an annuity that pays $200 per month for 5 years if money is worth 6% compounded monthly?

Solution

To solve this problem, use formula (5) with $PMT = \$200$, $i = \frac{0.06}{12} = 0.005$, and $n = 12(5) = 60$:

$$PV = PMT \, \frac{1 - (1 + i)^{-n}}{i} \quad \text{or} \quad PMT a_{\overline{n}|i}$$

$$= 200 \, \frac{1 - (1.005)^{-60}}{0.005} \quad \text{or} \quad 200 a_{\overline{60}|0.005} \qquad \text{Use a calculator (or Table II).}$$

$$= 200(51.725 \ 561)$$

$$= \$10{,}345.11$$

◆

How much should you deposit in an account paying 8% compounded quarterly in order to receive quarterly payments of $1,000 for the next 4 years?   ◆

◆ AMORTIZATION

The present value formula for an ordinary annuity (5) has another important use. Suppose you borrow $5,000 from a bank to buy a car and agree to repay the loan in 36 equal monthly payments, including all interest due. If the bank charges 1% per month on the unpaid balance (12% per year compounded monthly), how much should each payment be to retire the total debt including interest in 36 months?

Actually, the bank has bought an annuity from you. The question is: If the bank pays you $5,000 (present value) for an annuity paying them $PMT per month for 36 months at 12% interest compounded monthly, what are the monthly payments (PMT)? (Note that the value of the annuity at the end of 36 months is zero.) To find PMT, we have only to use formula (5) with $PV = \$5,000$, $i = 0.01$, and $n = 36$:

$$PV = PMT \frac{1 - (1 + i)^{-n}}{i} \quad \text{or} \quad PMTa_{\overline{n}|i}$$

$$5,000 = PMT \frac{1 - (1.01)^{-36}}{0.01} \quad \text{or} \quad PMTa_{\overline{36}|0.01} \qquad \text{Solve for } PMT \text{ and use a calculator or Table II.}$$

$$PMT = 5,000(0.033\ 214)$$
$$= \$166.07 \text{ per month}$$

At $166.07 per month, the car will be yours after 36 months. That is, you have *amortized* the debt in 36 equal monthly payments. (*Mort* means "death"; you have "killed" the loan in 36 months.) In general, **amortizing a debt** means that the debt is retired in a given length of time by equal periodic payments that include compound interest. We are usually interested in computing the equal periodic payments. Solving the present value formula (5) for PMT in terms of the other variables, we obtain the following amortization formula:

Amortization Formula — A Variation of Formula (5)

$$PMT = PV \frac{i}{1 - (1 + i)^{-n}} = PV \frac{1}{a_{\overline{n}|i}} \qquad (6)$$

where

   $PV =$ Amount of loan (present value)

   $i =$ Rate per period

   $n =$ Number of payments (periods)

   $PMT =$ Periodic payment

[Note: Payments are made at the end of each period.]

EXAMPLE 13

Assume that you buy a television set for $800 and agree to pay for it in 18 equal monthly payments at $1\frac{1}{2}\%$ interest per month on the unpaid balance.

(A) How much are your payments?   (B) How much interest will you pay?

Solutions

(A) Use formula (6) with $PV = \$800$, $i = 0.015$, and $n = 18$:

$$PMT = PV \frac{i}{1 - (1 + i)^{-n}} \quad \text{or} \quad PV \frac{1}{a_{\overline{n}|i}}$$

$$= 800 \frac{0.015}{1 - (1.015)^{-18}} \quad \text{or} \quad 800 \frac{1}{a_{\overline{18}|0.015}} \qquad \text{Use a calculator or Table II.}$$

$$= 800(0.063\ 806)$$

$$= \$51.04 \text{ per month}$$

(B) Total interest paid = Amount of all payments − Initial loan
$$= 18(\$51.04) - \$800$$
$$= \$118.72$$

PROBLEM 13

If you sell your car to someone for $2,400 and agree to finance it at 1% per month on the unpaid balance, how much should you receive each month to amortize the loan in 24 months? How much interest will you receive?

♦ **AMORTIZATION SCHEDULES**

What happens if you are amortizing a debt with equal periodic payments and at some point decide to pay off the remainder of the debt in one lump-sum payment? This occurs each time a home with an outstanding mortgage is sold. In order to understand what happens in this situation, we must take a closer look at the amortization process. We begin with an example that is simple enough to allow us to examine the effect each payment has on the debt.

EXAMPLE 14

If you borrow $500 that you agree to repay in 6 equal monthly payments at 1% interest per month on the unpaid balance, how much of each monthly payment is used for interest and how much is used to reduce the unpaid balance?

Solution

First, we compute the required monthly payment using formula (5) or (6). We choose formula (6) with $PV = \$500$, $i = 0.01$, and $n = 6$:

$$PMT = PV \frac{i}{1 - (1 + i)^{-n}} \quad \text{or} \quad PV \frac{1}{a_{\overline{n}|i}}$$

$$= 500 \frac{0.01}{1 - (1.01)^{-6}} \quad \text{or} \quad 500 \frac{1}{a_{\overline{6}|0.01}} \qquad \text{Use a calculator or Table II.}$$

$$= 500(0.172\ 548)$$

$$= \$86.27 \text{ per month}$$

At the end of the first month, the interest due is

$$\$500(0.01) = \$5.00$$

The amortization payment is divided into two parts, payment of the interest due and reduction of the unpaid balance (repayment of principal):

Monthly payment	Interest due	Unpaid balance reduction
$86.27 =	$5.00	+ $81.27

The unpaid balance for the next month is

Previous unpaid balance	Unpaid balance reduction	New unpaid balance
$500.00 −	$81.27	= $418.73

At the end of the second month, the interest due on the unpaid balance of $418.73 is

$$\$418.73(0.01) = \$4.19$$

Thus, at the end of the second month, the monthly payment of $86.27 covers interest and unpaid balance reduction as follows:

$$\$86.27 = \$4.19 + \$82.08$$

and the unpaid balance for the third month is

$$\$418.73 - \$82.08 = \$336.65$$

This process continues until all payments have been made and the unpaid balance is reduced to zero. The calculations for each month are listed in Table 1, which is referred to as an **amortization schedule.**

TABLE 1

Amortization Schedule

PAYMENT NUMBER	PAYMENT	INTEREST	UNPAID BALANCE REDUCTION	UNPAID BALANCE
0				$500.00
1	$ 86.27	$ 5.00	$ 81.27	418.73
2	86.27	4.19	82.08	336.65
3	86.27	3.37	82.90	253.75
4	86.27	2.54	83.73	170.02
5	86.27	1.70	84.57	85.45
6	86.30	0.85	85.45	0.00
Totals	$517.65	$17.65	$500.00	

Notice that the last payment had to be increased by $0.03 in order to reduce the unpaid balance to zero. This small discrepancy is due to round-off errors that occur in the computations. In almost all cases, the last payment must be adjusted slightly in order to obtain a final unpaid balance of exactly zero. ◆

PROBLEM 14 Construct the amortization schedule for a $1,000 debt that is to be amortized in 6 equal monthly payments at 1.25% interest per month on the unpaid balance.

♦

◆ **EXAMPLE 15** A family purchased a home 10 years ago for $80,000. The home was financed by paying 20% down and signing a 30 year mortgage at 9% on the unpaid balance. The net market value of the house (amount received after subtracting all costs involved in selling the house) is now $120,000, and the family wishes to sell the house. How much equity (to the nearest dollar) does the family have in the house now after making 120 monthly payments? [**Equity** = (Current net market value) − (Unpaid loan balance).]*

Solution How can we find the unpaid loan balance after 10 years or 120 monthly payments? One way to proceed would be to construct an amortization schedule, but this would require a table with 120 lines. Fortunately, there is an easier way. The unpaid balance after 120 payments is the amount of the loan that can be paid off with the remaining 240 monthly payments (20 remaining years on the loan). Since the lending institution views a loan as an annuity that they bought from the family, **the unpaid balance of a loan with *n* remaining payments is the present value of that annuity and can be computed by using formula (5).** Since formula (5) requires knowledge of the monthly payment, we compute *PMT* first using formula (6).

Step 1. Find the monthly payment:

$$PMT = PV \frac{i}{1 - (1 + i)^{-n}}$$

$$PV = (0.80)(\$80,000) = \$64,000$$
$$i = \frac{0.09}{12} = 0.0075$$
$$n = 12(30) = 360$$

$$= 64,000 \frac{0.0075}{1 - (1.0075)^{-360}}$$ Use a calculator.

$$= \$514.96 \text{ per month}$$

Step 2. Find the present value of a $514.96 per month 20 year annuity:

$$PV = PMT \frac{1 - (1 + i)^{-n}}{i}$$

$$PMT = \$514.96$$
$$n = 12(20) = 240$$
$$i = \frac{0.09}{12} = 0.0075$$

$$= 514.96 \frac{1 - (1.0075)^{-240}}{0.0075}$$ Use a calculator.

$$= \$57,235$$ Unpaid loan balance

---

* We use the word "equity" in keeping with common usage. If a family wants to sell a house and buy another more expensive house, then the price of a new house that they can afford to buy will often depend on their equity in the first house, where equity is defined by the equation given here. In refinancing a house or taking out an "equity loan," the new mortgage (or second mortgage) often will be based on the equity in the house. Other, more technical definitions of equity do not concern us here.

Step 3. Find the equity:

$$\text{Equity} = (\text{Current net market value}) - (\text{Unpaid loan balance})$$
$$= \$120{,}000 - \$57{,}235$$
$$= \$62{,}765$$

Thus, if the family sells the house for $120,000 net, they will have $62,765 after paying off the unpaid loan balance of $57,235. ◆

PROBLEM 15 A couple purchased a home 20 years ago for $65,000. The home was financed by paying 20% down and signing a 30 year mortgage at 8% on the unpaid balance. The net market value of the house is now $130,000, and the couple wishes to sell the house. How much equity (to the nearest dollar) does the couple have in the house now after making 240 monthly payments? ◆

The answer to Example 15 may seem a surprisingly large amount to owe after having made payments for 10 years, but long-term amortizations start out with very small reductions in the unpaid balance. For example, the interest due at the end of the very first period of the loan in Example 15 was

$$\$64{,}000(0.0075) = \$480.00$$

The first monthly payment was divided into

Monthly payment	Interest due	Unpaid balance reduction
$514.96	− $480.00	= $34.96

Thus, only $34.96 was applied to the unpaid balance.

Answers to Matched Problems **12.** $13,577.71   **13.** *PMT* = $112.98 per month; Total interest = $311.52
14.

PAYMENT NUMBER	PAYMENT	INTEREST	UNPAID BALANCE REDUCTION	UNPAID BALANCE
0				$1,000.00
1	$ 174.03	$12.50	$ 161.53	838.47
2	174.03	10.48	163.55	674.92
3	174.03	8.44	165.59	509.33
4	174.03	6.37	167.66	341.67
5	174.03	4.27	169.76	171.91
6	174.06	2.15	171.91	0.00
Totals	$1,044.21	$44.21	$1,000.00	

15. $98,551

Use formula (5) or (6) and a calculator or Table II (or both) to solve each problem. (Answers may vary slightly depending on whether you use a calculator or Table II.)

A

1. $PV = ?;$ $n = 30;$ $i = 0.04;$ $PMT = \$200$
2. $PV = ?;$ $n = 40;$ $i = 0.01;$ $PMT = \$400$
3. $PV = ?;$ $n = 25;$ $i = 0.025;$ $PMT = \$250$
4. $PV = ?;$ $n = 60;$ $i = 0.0075;$ $PMT = \$500$

B

5. $PV = \$6,000;$ $n = 36;$ $i = 0.01;$ $PMT = ?$
6. $PV = \$1,200;$ $n = 40;$ $i = 0.025;$ $PMT = ?$
7. $PV = \$40,000;$ $n = 96;$ $i = 0.0075;$ $PMT = ?$
8. $PV = \$14,000;$ $n = 72;$ $i = 0.005;$ $PMT = ?$
9. $PV = \$5,000;$ $i = 0.01;$ $PMT = \$200;$ $n = ?$
10. $PV = \$20,000;$ $i = 0.0175;$ $PMT = \$500;$ $n = ?$

APPLICATIONS

Business & Economics

11. A relative wills you an annuity paying $4,000 per quarter for the next 10 years. If money is worth 8% compounded quarterly, what is the present value of this annuity?

12. How much should you deposit in an account paying 12% compounded monthly in order to receive $1,000 per month for the next 2 years?

13. Parents of a college student wish to set up an annuity that will pay $350 per month to the student for 4 years. How much should they deposit now at 9% interest compounded monthly to establish this annuity? How much will the student receive in the 4 years?

14. A person pays $120 per month for 48 months for a car, making no down payment. If the loan costs 1.5% interest per month on the unpaid balance, what was the original cost of the car? How much total interest will be paid?

15. (A) If you buy a stereo set for $600 and agree to pay for it in 18 equal installments at 1% interest per month on the unpaid balance, how much are your monthly payments? How much interest will you pay?
    (B) Repeat part A for 1.5% interest per month on the unpaid balance.

16. (A) A company buys a large copy machine for $12,000 and finances it at 12% interest compounded monthly. If the loan is to be amortized in 6 years in equal monthly payments, how much is each payment? How much interest will be paid?
    (B) Repeat part A with 18% interest compounded monthly.

17. A sailboat costs $16,000. You pay 25% down and amortize the rest with equal monthly payments over a 6 year period. If you must pay 1.5% interest per month on the unpaid balance (18% compounded monthly), what is your monthly payment? How much interest will you pay over the 6 years?

18. A law firm buys a computerized word-processing system costing $10,000. If it pays 20% down and amortizes the rest with equal monthly payments over 5 years at 9% compounded monthly, what will be the monthly payment? How much interest will the firm pay?

19. Construct the amortization schedule for a $5,000 debt that is to be amortized in 8 equal quarterly payments at 4.5% interest per quarter on the unpaid balance.

20. Construct the amortization schedule for a $10,000 debt that is to be amortized in 6 equal quarterly payments at 3.5% interest per quarter on the unpaid balance.

21. A woman borrows $6,000 at 12% compounded monthly, which is to be amortized over 3 years in equal monthly payments. For tax purposes, she needs to know the amount of interest paid during each year of the loan. Find the interest paid during the first year, the second year, and the third year of the loan. [Hint: Find the unpaid balance after 12 payments and after 24 payments.]

22. A man establishes an annuity for retirement by depositing $50,000 into an account that pays 9% compounded monthly. Equal monthly withdrawals will be made each month for 5 years, at which time the account will have a zero balance. Each year taxes must be paid on the interest earned by the account during that year. How much interest was earned during the first year? [Hint: The amount in the account at the end of the first year is the present value of a 4 year annuity.]

Use a financial or scientific calculator to solve each of the following problems.

23. Some friends tell you that they paid $25,000 down on a new house and are to pay $525 per month for 30 years. If interest is 9.8% compounded monthly, what was the selling price of the house? How much interest will they pay in 30 years?

24. A family is thinking about buying a new house costing $120,000. They must pay 20% down, and the rest is to be amortized over 30 years in equal monthly payments. If money costs 9.6% compounded monthly, what will their monthly payment be? How much total interest will be paid over the 30 years?

25. A student receives a federally backed student loan of $6,000 at 3.5% interest compounded monthly. After finishing college in 2 years, the student must amortize the loan in the next 4 years by making equal monthly payments. What will the payments be and what total interest will the student pay? [Hint: This is a two-part problem. First find the amount of the debt at the end of the first 2 years; then amortize this amount over the next 4 years.]

26. A person establishes a sinking fund for retirement by contributing $7,500 per year at the end of each year for 20 years. For the next 20 years, equal yearly payments are withdrawn, at the end of which time the account will have a zero balance. If money is worth 9% compounded annually, what yearly payments will the person receive for the last 20 years?

27. A family has a $75,000, 30 year mortgage at 13.2% compounded monthly. Find the monthly payment. Also find the unpaid balance after:

(A) 10 years   (B) 20 years   (C) 25 years

28. A family has a $50,000, 20 year mortgage at 10.8% compounded monthly. Find the monthly payment. Also find the unpaid balance after:

(A) 5 years   (B) 10 years   (C) 15 years

29. A family has a $30,000, 20 year mortgage at 15% compounded monthly.

(A) Find the monthly payment and the total interest paid.
(B) Suppose the family decides to add an extra $100 to its mortgage payment each month starting with the very first payment. How long will it take the family to pay off the mortgage? How much interest will the family save?

30. At the time they retire, a couple has $200,000 in an account that pays 8.4% compounded monthly.

(A) If they decide to withdraw equal monthly payments for 10 years, at the end of which time the account will have a zero balance, how much should they withdraw each month?
(B) If they decide to withdraw $3,000 a month until the balance in the account is zero, how many withdrawals can they make?

31. A couple wishes to borrow money using the equity in their home for collateral. A loan company will loan them up to 70% of their equity. They purchased their home 12 years ago for $79,000. The home was financed by paying 20% down and signing a 30 year mortgage at 12% on the unpaid balance. Equal monthly payments were made to amortize the loan over the 30 year period. The net market value of the house is now $100,000. After making their 144th payment, they applied to the loan company for the maximum loan. How much (to the nearest dollar) will they receive?

32. A person purchased a house 10 years ago for $100,000. The house was financed by paying 20% down and signing a 30 year mortgage at 9.6% on the unpaid balance. Equal monthly payments were made to amortize the loan over a 30 year period. The owner now (after the 120th payment) wishes to refinance the house because of the need of additional cash. If the loan company agrees to a new 30 year mortgage of 80% of the new appraised value of the house, which is $136,000, how much cash (to the nearest dollar) will the owner receive after repaying the balance of the original mortgage?

Chapter Review

4-1   *Simple Interest.* Principal; interest; interest rate; simple interest; face value; present value; future value

$$I = Prt; \quad A = P(1 + rt)$$

4-2 *Compound Interest.* Compound interest; rate per compounding period; principal (present value); amount (future value); annual nominal rate; effective rate (or annual yield); zero coupon bond

$$A = P(1 + i)^n; \quad i = \frac{r}{m}; \quad r_e = \left(1 + \frac{r}{m}\right)^m - 1$$

4-3 *Future Value of an Annuity; Sinking Funds.* Annuity; ordinary annuity; future value; sinking fund

FUTURE VALUE $$FV = PMT \frac{(1 + i)^n - 1}{i} = PMTs_{\overline{n}|i}$$

SINKING FUND $$PMT = FV \frac{i}{(1 + i)^n - 1} = \frac{FV}{s_{\overline{n}|i}}$$

4-4 *Present Value of an Annuity; Amortization.* Present value; amortizing a debt; amortization schedule; equity; current value

PRESENT VALUE $$PV = PMT \frac{1 - (1 + i)^{-n}}{i} = PMTa_{\overline{n}|i}$$

AMORTIZATION $$PMT = PV \frac{i}{1 - (1 + i)^{-n}} = \frac{PV}{a_{\overline{n}|i}}$$

Chapter Review

Work through all the problems in this chapter review and check your answers in the back of the book. (Answers to all review problems are there.) Where weaknesses show up, review appropriate sections in the text.

Solve each problem using a calculator or Table II (or both).

A *Find the indicated quantity, given A = P(1 + rt).*

1. $A = ?;$ $P = \$100;$ $r = 9\%;$ $t = 6$ months
2. $A = \$808;$ $P = ?;$ $r = 12\%;$ $t = 1$ month
3. $A = \$212;$ $P = \$200;$ $r = 8\%;$ $t = ?$
4. $A = \$4,120;$ $P = \$4,000;$ $r = ?;$ $t = 6$ months

B *Find the indicated quantity, given $A = P(1 + i)^n$ and $P = A/(1 + i)^n$.*

5. $A = ?;$ $P = \$1,200;$ $i = 0.005;$ $n = 30$
6. $A = \$5,000;$ $P = ?;$ $i = 0.0075;$ $n = 60$

Find the indicated quantity, given

$$FV = PMT \frac{(1 + i)^n - 1}{i} = PMTs_{\overline{n}|i} \quad and \quad PMT = FV \frac{i}{(1 + i)^n - 1} = \frac{FV}{s_{\overline{n}|i}}$$

7. $FV = ?;$ $PMT = \$1,000;$ $i = 0.005;$ $n = 60$
8. $FV = \$8,000;$ $PMT = ?;$ $i = 0.015;$ $n = 48$

Find the indicated quantity, given

$$PV = PMT\frac{1 - (1 + i)^{-n}}{i} = PMTa_{\overline{n}|i} \quad \text{and} \quad PMT = PV\frac{i}{1 - (1 + i)^{-n}} = \frac{PV}{a_{\overline{n}|i}}$$

9. $PV = ?$; $PMT = \$2,500$; $i = 0.02$; $n = 16$
10. $PV = \$8,000$; $PMT = ?$; $i = 0.0075$; $n = 60$

C   Use a calculator or Table II (or both) to solve for n to the nearest integer.

11. $2,500 = 1,000(1.06)^n$ 12. $5,000 = 100\dfrac{(1.01)^n - 1}{0.01} = 100s_{\overline{n}|0.01}$

APPLICATIONS

Business & Economics
Find all dollar amounts correct to the nearest cent. When an interest rate is requested as an answer, express the rate as a percentage, correct to two decimal places.

Solve Problems 13–36 using a calculator or Table II.

13. If you borrow $3,000 at 14% simple interest for 10 months, how much will you owe in 10 months? How much interest will you pay?
14. A credit card company charges a 22% annual rate for overdue accounts. How much interest will be owed on a $635 account 1 month overdue?
15. A loan of $2,500 was repaid at the end of 10 months with a check for $2,812.50. What annual rate of interest was charged?
16. If you paid $100 to a loan company for the use of $1,500 for 120 days, what annual rate of interest did they charge? (Use a 360 day year.)
17. A loan company advertises in the paper that you will pay only 8¢ a day for each $100 borrowed. What annual rate of interest are they charging? (Use a 360 day year.)
18. If you buy a 13 week T-bill with a maturity value of $5,000 for $4,899.08 from the U.S. Treasury Department, what annual interest rate will you earn?
19. If an investor wants to earn an annual interest rate of 10.76% on a 26 week T-bill with a maturity value of $5,000, how much should the investor pay for the bill?
20. Grandparents deposited $6,000 into a grandchild's account toward a college education. How much money (to the nearest dollar) will be in the account 17 years from now if the account earns 9% compounded monthly?
21. How much should you deposit initially in an account paying 10% compounded semiannually in order to have $25,000 in 10 years?
22. What will an $8,000 car cost (to the nearest dollar) 5 years from now if the inflation rate over that period averages 5% compounded annually?
23. What would the $8,000 car in Problem 22 have cost (to the nearest dollar) 5 years ago if the inflation rate over that period had averaged 5% compounded annually?

24. You have $2,500 toward the purchase of a boat that will cost $3,000. How long will it take the $2,500 to grow to $3,000 if it is invested at 9% compounded quarterly? (Round up to the next higher quarter if not exact.)

25. How long will it take money to double if it is invested at 12% compounded monthly? 18% compounded monthly? (Round up to the next higher month if not exact.)

26. A savings and loan company pays 9% compounded monthly. What is the effective rate?

27. Which is the better investment and why: 9% compounded quarterly or 9.25% compounded annually?

28. What is the value of an ordinary annuity at the end of 8 years if $200 per month is deposited into an account earning 9% compounded monthly? How much of this value is interest?

29. A company decides to establish a sinking fund to replace a piece of equipment in 6 years at an estimated cost of $50,000. To accomplish this, they decide to make fixed monthly payments into an account that pays 9% compounded monthly. How much should each payment be?

30. In order to save enough money for the down payment on a condominium, a young couple deposits $200 each month into an account that pays 9% interest compounded monthly. If they need $10,000 for a down payment, how many deposits will they have to make?

31. A scholarship committee wishes to establish a scholarship that will pay $1,500 per quarter to a student for 2 years. How much should they deposit now at 8% compounded quarterly to establish this scholarship? How much will the student receive in the 2 years?

32. A state-of-the-art compact disk stereo system costs $3,000. You pay one-third down and amortize the rest with equal monthly payments over a 2 year period. If you are charged 1.5% interest per month on the unpaid balance, what is your monthly payment? How much interest will you pay over the 2 years?

33. Construct the amortization schedule for a $1,000 debt that is to be amortized in 4 equal quarterly payments at 2.5% interest per quarter on the unpaid balance.

34. Two years ago you borrowed $10,000 at 12% interest compounded monthly, which was to be amortized over 5 years. Now you have acquired some additional funds and decide that you want to pay off this loan. What is the unpaid balance after making equal monthly payments for 2 years?

35. A business borrows $80,000 at 15% interest compounded monthly for 8 years.

    (A) What is the monthly payment?
    (B) What is the unpaid balance at the end of the first year?
    (C) How much interest was paid during the first year?

36. An individual wants to establish an annuity for retirement purposes. He wants to make quarterly deposits for 20 years so that he can then make

quarterly withdrawals of $5,000 for 10 years. The annuity earns 12% interest compounded quarterly.

(A) How much will have to be in the account at the time he retires?
(B) How much should be deposited each quarter for 20 years in order to accumulate the required amount?
(C) What is the total amount of interest earned during the 30 year period?

Problems 37–50 require the use of a scientific or financial calculator.

37. A $10,000 retirement account is left to earn interest at 7% compounded daily. How much money will be in the account 40 years from now when the owner reaches 65? (Use a 365 day year and round answer to the nearest dollar.)

38. How long will it take money to double if it is invested at 10% compounded daily? 10% compounded annually? (Express answers in years to two decimal places and assume a 365 day year.)

39. Security Savings & Loan pays 9.38% compounded monthly and West Lake Savings & Loan pays 9.35% compounded daily. Which is the better investment? For each investment, express the effective rate as a percentage, correct to three decimal places. Use a 365 day year.

40. How much should a $5,000 face value zero coupon bond, maturing in 5 years, be sold for now, if its rate of return is to be 9.5% compounded annually?

41. If you pay $4,476.20 for a $10,000 face value zero coupon bond that matures in 10 years, what is your annual compound rate of return?

42. If you just sold a stock for $17,388.17 (net) that cost you $12,903.28 (net) 3 years ago, what annual compound rate of return did you make on your investment?

43. Starting on his 21st birthday, and continuing on every birthday up to and including his 65th, John deposits $2,000 a year into an IRA. How much (to the nearest dollar) will be in the account on the 65th birthday, if the account earns:

(A) 7% compounded annually?   (B) 11% compounded annually?

44. A company establishes a sinking fund for plant retooling in 6 years at an estimated cost of $850,000. How much should be invested semiannually into an account paying 8.76% compounded semiannually? How much interest will the account earn in the 6 years?

45. You can afford monthly deposits of only $200 into an account that pays 7.98% compounded monthly. How long will it be until you will have $2,500 to purchase a used car? (Round to the next higher month if not exact.)

46. A car salesperson tells you that you can buy the car you are looking at for $3,000 down and $200 a month for 48 months. If interest is 14% compounded monthly, what is the selling price of the car and how much interest will you pay during the 48 months?

47. A student receives a student loan for $8,000 at 5.5% interest compounded monthly to help her finish the last 1.5 years of college. One year after finishing college, the student must amortize the loan in the next 5 years by making equal monthly payments. What will the payments be and what total interest will the student pay?

48. (A) A man deposits $2,000 in an IRA on his 21st birthday and on each subsequent birthday up to, and including, his 29th (9 deposits in all). The account earns 8% compounded annually. If he then leaves the money in the account without making any more deposits, how much will he have on his 65th birthday, assuming the account continues to earn the same rate of interest?

    (B) How much would be in the account (to the nearest dollar) on his 65th birthday if he had started the deposits on his 30th birthday and continued making deposits on each birthday until (and including) his 65th birthday?

49. In a new housing development, the houses are selling for $100,000 and require a 20% down payment. The buyer is given a choice of 30 year or 15 year financing, both at 10.75% compounded monthly.

    (A) What is the monthly payment for the 30 year choice? For the 15 year choice?

    (B) What is the unpaid balance after 10 years for the 30 year choice? For the 15 year choice?

50. A loan company will loan up to 60% of the equity in a home. A family purchased their home 8 years ago for $83,000. The home was financed by paying 20% down and signing a 30 year mortgage at 11.25% for the balance. Equal monthly payments were made to amortize the loan over the 30 year period. The market value of the house is now $95,000. After making their 96th payment, the family applied to the loan company for the maximum loan. How much (to the nearest dollar) will they receive?

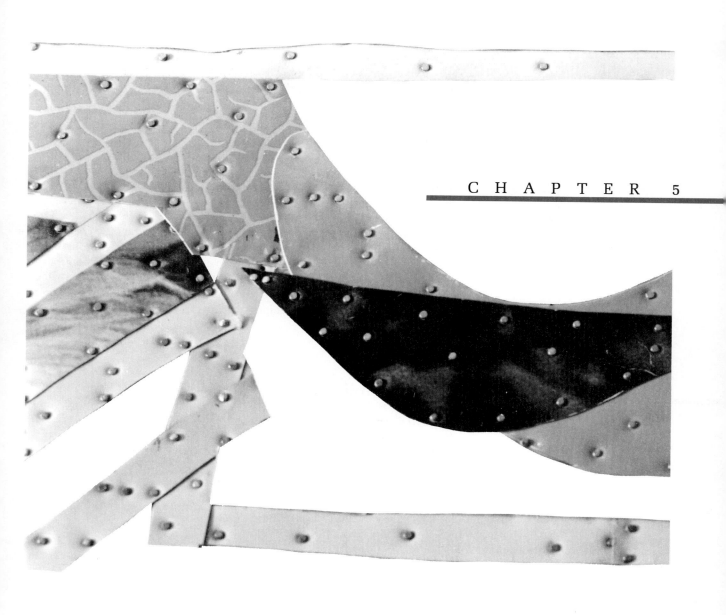

Systems of Linear Equations; Matrices

Contents

In this chapter we first review how systems of linear equations involving two or three variables are solved using techniques learned in elementary algebra. Because these techniques are not suitable for linear systems involving larger numbers of equations and variables, we then turn to a different method of solution involving the concept of an *augmented matrix*, which arises quite naturally when dealing with larger linear systems. We then study *matrices* and *matrix operations* in their own right as a new mathematical form. With this new tool, we return to systems of equations from a fresh point of view. The matrix techniques introduced in this chapter form the basis for computer solutions of large-scale systems.

SECTION 5-1   Review: Systems of Linear Equations

◆   SYSTEMS IN TWO VARIABLES
◆   APPLICATIONS
◆   SYSTEMS IN THREE VARIABLES
◆   APPLICATIONS

◆ SYSTEMS IN TWO VARIABLES

To establish basic concepts, consider the following simple example: If 2 adult tickets and 1 child ticket cost \$8, and if 1 adult ticket and 3 child tickets cost \$9, what is the price of each?

Let

x = Price of adult ticket

y = Price of child ticket

Then

$$2x + \ \ y = 8$$
$$x + 3y = 9$$

We now have a system of two linear equations in two variables. To solve this system, we find all ordered pairs of real numbers that satisfy both equations at the same time. In general, we are interested in solving linear systems of the type

$$ax + by = h$$
$$cx + dy = k$$

where a, b, c, d, h, and k are real constants. A pair of numbers $x = x_0$ and $y = y_0$ [also written as an ordered pair (x_0, y_0)] is a **solution** of this system if each equation is satisfied by the pair. The set of all such ordered pairs of numbers is called the **solution set** for the system. To **solve** a system is to find its solution set. We will consider three methods of solving such systems: *graphing*, *substitution*, and *elimination by addition*. Each method has certain advantages, depending on the situation.

Solution by Graphing

To solve the ticket problem above by graphing, we graph both equations in the same coordinate system. The coordinates of any points that the graphs have in common must be solutions to the system, since they must satisfy both equations.

◆ E X A M P L E   1

Solve the ticket problem by graphing:

$$2x + y = 8$$
$$x + 3y = 9$$

Solution

Graph each equation in the same rectangular coordinate system (both graphs are straight lines). Then estimate the coordinates of any common points on the two lines.

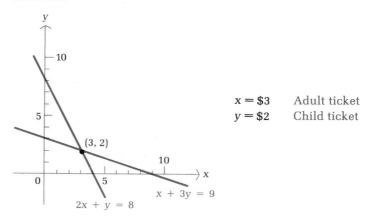

$x = \$3$ Adult ticket
$y = \$2$ Child ticket

Check

$2x + y = 8$	$x + 3y = 9$	(3, 2) must satisfy each original
$2(3) + 2 \stackrel{?}{=} 8$	$3 + 3(2) \stackrel{?}{=} 9$	equation for a complete check.
$8 \stackrel{\checkmark}{=} 8$	$9 \stackrel{\checkmark}{=} 9$	

◆

PROBLEM   1   Solve by graphing and check:

$$2x - y = -3$$
$$x + 2y = -4$$ ◆

It is clear that Example 1 has exactly one solution, since the lines have exactly one point of intersection. In general, lines in a rectangular coordinate system are related to each other in one of the three ways illustrated in the next example.

◆ EXAMPLE   2   Solve each of the following systems by graphing:

(A) $x - 2y = 2$ (B) $x + 2y = -4$ (C) $2x + 4y = 8$
    $x + y = 5$ $2x + 4y = 8$ $x + 2y = 4$

Solutions

(A)

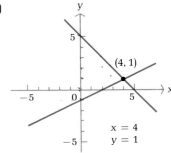

x = 4
y = 1

Intersection at one point
only—exactly one solution

(B)

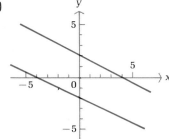

Lines are parallel (each
has slope $-\frac{1}{2}$)—no solutions

(C)

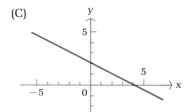

Lines coincide—infinite
number of solutions

◆

PROBLEM   2   Solve each of the following systems by graphing:

(A) $x + y = 4$ (B) $6x - 3y = 9$ (C) $2x - y = 4$
    $2x - y = 2$ $2x - y = 3$ $6x - 3y = -18$ ◆

We now define some terms that we can use to describe the different types of solutions to systems of equations that we will encounter. A system of linear equations is **consistent** if it has one or more solutions and **inconsistent** if no solutions exist. Furthermore, a consistent system is said to be **independent** if it has exactly one solution (often referred to as the **unique solution**) and **dependent** if it has more than one solution. Referring to the three systems in Example 2, the system in part A is a consistent and independent system with the unique solution $x = 4$ and $y = 1$. The system in part B is inconsistent. And the system in part C is consistent and dependent.

By geometrically interpreting a system of two linear equations in two variables, we gain useful information about what to expect in the way of solutions to the system. In general, any two lines in a coordinate plane must intersect in exactly one point, be parallel, or coincide (have identical graphs). Thus, the systems in Example 2 illustrate the only three possible types of solutions for systems of two linear equations in two variables. These ideas are summarized in Theorem 1.

THEOREM 1

Possible Solutions to a Linear System

The linear system

$$ax + by = h$$
$$cx + dy = k$$

must have:

(A) Exactly one solution   CONSISTENT AND INDEPENDENT

Or:

(B) No solution   INCONSISTENT

Or:

(C) Infinitely many solutions   CONSISTENT AND DEPENDENT

There are no other possibilities.

In addition, graphs frequently reveal relationships in problems that might otherwise be hidden. Generally, however, graphic methods give us only rough approximations of solutions. The methods of substitution and elimination by addition, which we consider next, will yield solutions to any desired decimal accuracy — assuming solutions exist.

Solution by Substitution

In this method, we choose one of two equations in a system and solve for one variable in terms of the other. (We make a choice that avoids fractions, if possible.) Then, we substitute the result into the other equation and solve the resulting linear equation in one variable. Finally, we substitute this result back into either of the original equations to find the second variable. An example should make the process clear.

◆ EXAMPLE 3

Solve by substitution:

$$5x + y = 4$$
$$2x - 3y = 5$$

Solution

Solve either equation for one variable in terms of the other; then substitute into the remaining equation. In this problem we can avoid fractions by choosing the first equation and solving for y in terms of x:

$$5x + y = 4 \qquad \text{Solve the first equation for } y \text{ in terms of } x.$$

$$y = \underbrace{4 - 5x} \qquad \text{Substitute into second equation.}$$

$$2x - 3y = 5 \qquad \text{Second equation}$$

$$2x - 3(4 - 5x) = 5 \qquad \text{Solve for } x.$$

$$2x - 12 + 15x = 5$$

$$17x = 17$$

$$x = 1$$

Now, replace x with 1 in $y = 4 - 5x$ to find y:

$$y = 4 - 5x$$

$$y = 4 - 5(1)$$

$$y = -1$$

Check

$$5x + \quad y = 4 \qquad\qquad 2x - \quad 3y = 5$$

$$5(1) + (-1) \overset{?}{=} 4 \qquad\qquad 2(1) - 3(-1) \overset{?}{=} 5$$

$$4 \overset{\checkmark}{=} 4 \qquad\qquad\qquad 5 \overset{\checkmark}{=} 5 \qquad\qquad\qquad\qquad ◆$$

PROBLEM 3   Solve by substitution:

$$3x + 2y = -2$$

$$2x - \quad y = -6 \qquad\qquad\qquad\qquad\qquad\qquad\qquad\qquad ◆$$

Solution Using Elimination by Addition

Now we turn to **elimination by addition.** This is probably the most important method of solution, since it is readily generalized to higher-order systems. The method involves replacing systems of equations with simpler *equivalent systems* by performing appropriate operations, until we obtain a system with an obvious solution. **Equivalent systems** of equations are, as you would expect, systems that have exactly the same solution set. Theorem 2 lists the operations that produce equivalent systems.

THEOREM 2

> ### ▪▪ Operations That Produce Equivalent Systems
>
> A system of linear equations is transformed into an equivalent system if:
>
> (A) Two equations are interchanged.
> (B) An equation is multiplied by a nonzero constant.
> (C) A constant multiple of one equation is added to another equation.

Any one of the three operations in Theorem 2 can be used to produce an equivalent system, but the operations in parts B and C will be of most use to us now. Part A becomes useful when we apply the theorem to larger systems. The use of Theorem 2 is best illustrated by examples.

◆ E X A M P L E   4   Solve the following system using elimination by addition:

$$3x - 2y = 8$$
$$2x + 5y = -1$$

Solution   We use Theorem 2 to eliminate one of the variables, thus obtaining a system with an obvious solution:

$$3x - 2y = 8$$
$$2x + 5y = -1$$

Multiply the top equation by 5 and the bottom equation by 2 (Theorem 2B).

$$5(3x - 2y) = 5(8)$$
$$2(2x + 5y) = 2(-1)$$

$$15x - 10y = 40$$
$$\underline{4x + 10y = -2}$$

Add the top equation to the bottom equation (Theorem 2C).

$$19x = 38$$

Divide both sides by 19, which is the same as multiplying the equation by $\frac{1}{19}$ (Theorem 2B).

$$x = 2$$

This equation paired with either of the two original equations produces a system equivalent to the original system.

Knowing that $x = 2$, we substitute this number back into either of the two original equations (we choose the second) to solve for y:

$$2(2) + 5y = -1$$
$$5y = -5$$
$$y = -1$$

Check

$$3x - 2y = 8 \qquad 2x + 5y = -1$$
$$3(2) - 2(-1) \stackrel{?}{=} 8 \qquad 2(2) + 5(-1) \stackrel{?}{=} -1$$
$$8 \stackrel{\checkmark}{=} 8 \qquad\qquad -1 \stackrel{\checkmark}{=} -1$$

◆

P R O B L E M   4   Solve the following system using elimination by addition:

$$5x - 2y = 12$$
$$2x + 3y = 1$$

◆

Let us see what happens in the elimination process when a system has either no solution or infinitely many solutions. Consider the following system:

$$2x + 6y = -3$$
$$x + 3y = 2$$

Multiplying the second equation by -2 and adding, we obtain

$$\begin{array}{r} 2x + 6y = -3 \\ \underline{-2x - 6y = -4} \\ 0 = -7 \end{array} \qquad \text{Not possible}$$

We have obtained a contradiction. The assumption that the original system has solutions must be false (otherwise, we have proved that $0 = -7$). Thus, the system has no solutions and its solution set is the empty set. The graphs of the equations are parallel and the system is inconsistent.

Now consider the system

$$\begin{array}{r} x - \tfrac{1}{2}y = 4 \\ -2x + y = -8 \end{array}$$

If we multiply the top equation by 2 and add the result to the bottom equation, we obtain

$$\begin{array}{r} 2x - y = 8 \\ \underline{-2x + y = -8} \\ 0 = 0 \end{array}$$

Obtaining $0 = 0$ by addition implies that the equations are equivalent; that is, their graphs coincide and the system is dependent. If we let $x = k$, where k is any real number, and solve either equation for y, we obtain $y = 2k - 8$. Thus, $(k, 2k - 8)$ is a solution for any real number k. The variable k is called a **parameter,** and replacing k with a real number produces a **particular solution** to the system. For example, some particular solutions to this system are

$k = -1$	$k = 2$	$k = 5$	$k = 9.4$
$(-1, -10)$	$(2, -4)$	$(5, 2)$	$(9.4, 10.8)$

◆ APPLICATIONS

Many real-world problems are readily solved by applying two-equation–two-variable methods. We shall discuss two applications in detail.

◆ E X A M P L E   5

Diet

A dietitian in a hospital is to arrange a special diet comprised of two foods, M and N. Each ounce of food M contains 8 units of calcium and 2 units of iron. Each ounce of food N contains 5 units of calcium and 4 units of iron. How many ounces of foods M and N should be used to obtain a food mix that has exactly 74 units of calcium and 35 units of iron?

It is convenient to first summarize the quantities involved in a table, as shown in the margin.

Let

	FOOD M	FOOD N	TOTAL NEEDED
CALCIUM	8	5	74
IRON	2	4	35

x = Number of ounces of food M

y = Number of ounces of food N

$$\begin{pmatrix} \text{Calcium in} \\ x \text{ oz of food } M \end{pmatrix} + \begin{pmatrix} \text{Calcium in} \\ y \text{ oz of food } N \end{pmatrix} = \begin{pmatrix} \text{Total calcium} \\ \text{needed} \end{pmatrix}$$

$$\begin{pmatrix} \text{Iron in } x \text{ oz} \\ \text{of food } M \end{pmatrix} + \begin{pmatrix} \text{Iron in } y \text{ oz} \\ \text{of food } N \end{pmatrix} = \begin{pmatrix} \text{Total iron} \\ \text{needed} \end{pmatrix}$$

$8x$	$+$	$5y$	$=$	74	Units of calcium
$2x$	$+$	$4y$	$=$	35	Units of iron

Solve using elimination by addition:

$$8x + 5y = 74 \qquad\qquad 2x + 4(6) = 35$$
$$\underline{-8x - 16y = -140} \qquad\qquad 2x = 11$$
$$-11y = -66 \qquad\qquad x = 5.5 \text{ oz of food } M$$
$$y = 6 \text{ oz of food } N$$

Check

$$8x + 5y = 74 \qquad\qquad 2x + 4y = 35$$
$$8(5.5) + 5(6) \overset{?}{=} 74 \qquad\qquad 2(5.5) + 4(6) \overset{?}{=} 35$$
$$74 \overset{\checkmark}{=} 74 \qquad\qquad 35 \overset{\checkmark}{=} 35 \qquad\qquad\qquad \blacklozenge$$

P R O B L E M   5

Repeat Example 5 given that each ounce of food M contains 10 units of calcium and 4 units of iron, each ounce of food N contains 6 units of calcium and 4 units of iron, and the mix of M and N must have exactly 92 units of calcium and 44 units of iron. $\qquad\qquad\blacklozenge$

◆ **E X A M P L E   6**

Supply and Demand

The quantity of a product that people are willing to buy during some period of time depends on its price. Generally, the higher the price, the less the demand; the lower the price, the greater the demand. Similarly, the quantity of a product that a supplier is willing to sell during some period of time also depends on the price. Generally, a supplier will be willing to supply more of a product at higher prices and less of a product at lower prices. The simplest supply and demand model is a linear model where the graphs of a demand equation and a supply equation are straight lines.

Suppose that in a city on a particular day, the supply and demand equations for cherries are

$$p = -0.2q + 4 \qquad \text{Demand equation (consumer)}$$
$$p = 0.07q + 0.76 \qquad \text{Supply equation (supplier)}$$

where q represents the quantity in thousands of pounds and p represents the price in dollars. For example, we see that consumers will purchase 10 thousand pounds ($q = 10$) when the price is $p = -0.2(10) + 4 = \$2$ per pound. On the other hand, suppliers will be willing to supply 17.714 thousand pounds of cherries at \$2 per pound (solve $2 = 0.07q + 0.76$). Thus, at \$2 per pound the suppliers are willing to supply more cherries than consumers are willing to purchase. The supply exceeds the demand at that price and the price will come down. At what price will cherries stabilize for the day? That is, at what price will supply equal demand? This price, if it exists, is called the **equilibrium price,** and the quantity sold at that price is called the **equilibrium quantity.** How do we find these quantities? We solve the linear system

$$p = -0.2q + 4 \qquad \text{Demand equation}$$
$$p = 0.07q + 0.76 \qquad \text{Supply equation}$$

We solve this system using substitution (substituting $p = -0.2q + 4$ into the second equation):

$$-0.2q + 4 = 0.07q + 0.76$$
$$-0.27q = -3.24$$
$$q = 12 \text{ thousand pounds} \qquad \text{Equilibrium quantity}$$

Now substitute $q = 12$ back into either of the original equations in the system and solve for p (we choose the first equation):

$$p = -0.2(12) + 4$$
$$p = \$1.60 \text{ per pound} \qquad \text{Equilibrium price}$$

These results are interpreted geometrically in the figure.

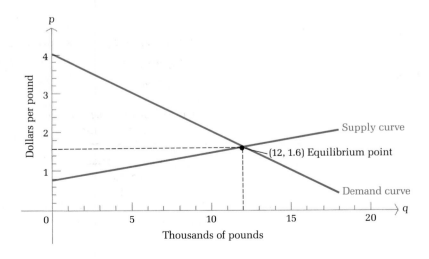

If the price was above the equilibrium price of $1.60 per pound, the supply would exceed the demand and the price would come down. If the price was below the equilibrium price of $1.60 per pound, the demand would exceed the supply and the price would rise. Thus, the price would reach equilibrium at $1.60. At this price, suppliers would supply 12 thousand pounds of cherries and consumers would purchase 12 thousand pounds. ◆

PROBLEM 6 Repeat Example 6 (including drawing the graph) given:

$$p = -0.1q + 3 \qquad \text{Demand equation}$$
$$p = 0.08q + 0.66 \qquad \text{Supply equation}$$

◆

◆ SYSTEMS IN THREE VARIABLES

Any equation that can be written in the form

$$ax + by = c \qquad \text{For example, } 2x - 3y = -1.$$

where a, b, and c are constants (not both a and b equal 0), is called a **linear equation in two variables.** Similarly, any equation that can be written in the form

$$ax + by + cz = k \qquad \text{For example, } 5x + 2y - 3z = 2.$$

where a, b, c, and k are constants (not all a, b, and c equal 0), is called a **linear equation in three variables.** (A similar definition holds for a linear equation in four or more variables.)

Now that we know how to solve systems of linear equations in two variables, there is no reason to stop there. Systems of the form

$$
\begin{aligned}
a_1 x + b_1 y + c_1 z &= k_1 \\
a_2 x + b_2 y + c_2 z &= k_2 \\
a_3 x + b_3 y + c_3 z &= k_3
\end{aligned}
\qquad
\begin{aligned}
&\text{For example,} \\
2x - 3y + z &= -2 \\
x + y - z &= 1 \\
-x + 2y + 6z &= 0
\end{aligned}
\qquad (1)
$$

as well as higher-order systems, are encountered frequently. In fact, systems of equations are so important in solving real-world problems that whole courses are devoted to this one topic. We now restrict our attention to systems having the same number of equations as variables. Later, we will relax this restriction. A triplet of numbers $x = x_0$, $y = y_0$, and $z = z_0$ [also written as an ordered triplet (x_0, y_0, z_0)] is a **solution** of system (1) if each equation is satisfied by this triplet. The set of all such ordered triplets of numbers is called the **solution set** of the system. If operations are performed on a system and the new system has the same solution set as the original, then both systems are said to be **equivalent.**

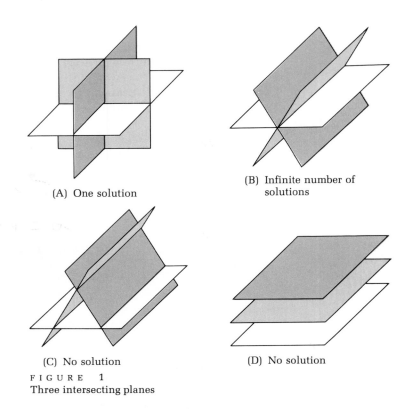

(A) One solution

(B) Infinite number of
solutions

(C) No solution

(D) No solution

FIGURE 1
Three intersecting planes

A linear equation in three variables represents a plane in a three-dimensional space. Trying to visualize how three planes can intersect will give you insight as to what kinds of solution sets are possible for system (1). Figure 1 shows several of the many ways in which three planes can intersect. It turns out that the results in Theorem 1 hold for systems in the form of (1).

Possible Solutions to a Linear System

It can be shown that any linear system must have exactly one solution, no solution, or an infinite number of solutions, regardless of the number of equations or number of variables in the system. The terms *unique solution*, *consistent*, *inconsistent*, *dependent*, and *independent* are used to describe these solutions, just as in the two-variable case.

In this section we will use an extension of the method of elimination discussed above to solve systems in the form of (1). In the next section we will consider techniques for solving linear systems that are more compatible with solving such systems with computers. In practice, most linear systems involving more than three variables are solved with the aid of a computer.

Step 1. Choose two equations from the system and eliminate one of the three variables using elimination by addition. The result is generally one equation in two variables.

Step 2. Now eliminate the same variable from the unused equation and one of those used in step 1. We (generally) obtain another equation in the same two variables.

Step 3. The two equations from steps 1 and 2 form a system of two equations in two variables. Solve as described in the earlier part of this section.

Step 4. Substitute the solution from step 3 into any of the three original equations and solve for the third variable to complete the solution of the original system.

◆ E X A M P L E   7   Solve:

$$3x - 2y + 4z = 6 \tag{2}$$
$$2x + 3y - 5z = -8 \tag{3}$$
$$5x - 4y + 3z = 7 \tag{4}$$

Solution   *Step 1.* We look at the coefficients of the variables and choose to eliminate y from equations (2) and (4) because of the convenient coefficients -2 and -4. Multiply equation (2) by -2 and add to equation (4):

$$
\begin{array}{ll}
-6x + 4y - 8z = -12 & -2[\text{Equation (2)}] \\
\underline{5x - 4y + 3z = 7} & \text{Equation (4)} \\
-x - 5z = -5 & \tag{5}
\end{array}
$$

Step 2. Now we eliminate y (the same variable) from equations (2) and (3):

$$
\begin{array}{ll}
9x - 6y + 12z = 18 & 3[\text{Equation (2)}] \\
\underline{4x + 6y - 10z = -16} & 2[\text{Equation (3)}] \\
13x + 2z = 2 & \tag{6}
\end{array}
$$

Step 3. From steps 1 and 2 we obtain the system

$$-x - 5z = -5 \tag{5}$$
$$13x + 2z = 2 \tag{6}$$

[It follows from Theorem 2 that equations (5) and (6) along with (2), (3), or (4) form a system equivalent to the original system.] We solve system (5) and (6) as in the earlier part of this section:

$$
\begin{array}{rl}
-13x - 65z = -65 & \quad 13[\text{Equation (5)}] \\
\underline{13x + 2z = 2} & \quad \text{Equation (6)} \\
-63z = -63 & \\
z = 1 &
\end{array}
$$

Substitute $z = 1$ back into either equation (5) or (6) [we choose equation (5)] to find x:

$$
\begin{array}{rl}
-x - 5z = -5 & \qquad (5) \\
-x - 5(1) = -5 & \\
-x = 0 & \\
x = 0 &
\end{array}
$$

Step 4. Substitute $x = 0$ and $z = 1$ back into any of the three original equations [we choose equation (2)] to find y:

$$
\begin{array}{rl}
3x - 2y + 4z = 6 & \qquad (2) \\
3(0) - 2y + 4(1) = 6 & \\
-2y + 4 = 6 & \\
-2y = 2 & \\
y = -1 &
\end{array}
$$

Thus, the solution to the original system is $(0, -1, 1)$, or $x = 0$, $y = -1$, $z = 1$.

Check   To check the solution, we must check *each* equation in the original system:

$$
\begin{array}{lll}
3x - 2y + 4z = 6 & 2x + 3y - 5z = -8 & 5x - 4y + 3z = 7 \\
3(0) - 2(-1) + 4(1) \overset{?}{=} 6 & 2(0) + 3(-1) - 5(1) \overset{?}{=} -8 & 5(0) - 4(-1) + 3(1) \overset{?}{=} 7 \\
6 \overset{\checkmark}{=} 6 & -8 \overset{\checkmark}{=} -8 & 7 \overset{\checkmark}{=} 7
\end{array}
$$

$\blacklozenge$

PROBLEM   7   Solve:

$$
\begin{array}{l}
2x + 3y - 5z = -12 \\
3x - 2y + 2z = 1 \\
4x - 5y - 4z = -12
\end{array}
$$

$\blacklozenge$

In the process described above, if we encounter an equation that states a contradiction, such as $0 = -2$, then we must conclude that the system has no solution; that is, the system is inconsistent. (Figures 1C and 1D illustrate inconsistent systems.) On the other hand, if one of the equations turns out to be $0 = 0$, the system has either infinitely many solutions or none. (Figures 1B, 1C, and 1D illustrate these cases.) We must proceed further to determine which. Notice how this last result differs from the two-equation–two-variable case. There, when we obtained $0 = 0$, we *knew* that there were infinitely many solutions. We shall have more to say about this in Section 5-3.

◆ APPLICATIONS

Now let us consider a real-world problem that leads to a system of three equations in three variables.

◆ E X A M P L E   8

Production Scheduling

A garment factory manufactures three shirt styles. Each style shirt requires the services of three departments, as listed in the table. The cutting, sewing, and packaging departments have available a maximum of 1,160, 1,560, and 480 labor-hours per week, respectively. How many of each style shirt must be produced each week for the plant to operate at full capacity?

	STYLE A	STYLE B	STYLE C	TIME AVAILABLE
CUTTING DEPARTMENT	0.2 hr	0.4 hr	0.3 hr	1,160 hr
SEWING DEPARTMENT	0.3 hr	0.5 hr	0.4 hr	1,560 hr
PACKAGING DEPARTMENT	0.1 hr	0.2 hr	0.1 hr	480 hr

Solution

Let

x = Number of style A produced per week

y = Number of style B produced per week

z = Number of style C produced per week

Then

$$0.2x + 0.4y + 0.3z = 1{,}160 \quad \text{Cutting department}$$
$$0.3x + 0.5y + 0.4z = 1{,}560 \quad \text{Sewing department}$$
$$0.1x + 0.2y + 0.1z = \phantom{1{,}}480 \quad \text{Packaging department}$$

We clear the system of decimals by multiplying each side of each equation by 10. Thus,

$$2x + 4y + 3z = 11{,}600 \tag{7}$$
$$3x + 5y + 4z = 15{,}600 \tag{8}$$
$$x + 2y + z = 4{,}800 \tag{9}$$

Let us start by eliminating z from equations (7) and (9):

$$
\begin{array}{ll}
2x + 4y + 3z = 11{,}600 & \text{Equation (7)} \\
\underline{-3x - 6y - 3z = -14{,}400} & -3[\text{Equation (9)}] \\
{-x} - 2y \phantom{{}+ 3z} = -2{,}800 & \tag{10}
\end{array}
$$

We now eliminate z from equations (8) and (9):

$$\begin{array}{rr} 3x + 5y + 4z = & 15{,}600 \qquad \text{Equation (8)} \\ \underline{-4x - 8y - 4z = -19{,}200} \qquad -4[\text{Equation (9)}] \\ -x - 3y \quad\quad = & -3{,}600 \end{array}$$ (11)

Equations (10) and (11) form a system of two equations in two variables:

$$-x - 2y = -2{,}800$$ (10)
$$-x - 3y = -3{,}600$$ (11)

We solve as in the earlier part of this section:

$$\begin{array}{rr} -x - 2y = -2{,}800 \qquad \text{Equation (10)} \\ \underline{x + 3y = \quad 3{,}600} \qquad (-1)[\text{Equation (11)}] \\ y = \quad \mathbf{800} \end{array}$$

Substitute $y = 800$ into either (10) or (11) to find x:

$$-x - \quad 2y = -2{,}800$$ (10)
$$-x - 2(\mathbf{800}) = -2{,}800$$
$$-x - 1{,}600 = -2{,}800$$
$$-x = -1{,}200$$
$$x = \quad \mathbf{1{,}200}$$

Now use either (7), (8), or (9) to find z:

$$2x + \quad 4y + 3z = 11{,}600$$ (7)
$$2(\mathbf{1{,}200}) + 4(\mathbf{800}) + 3z = 11{,}600$$
$$2{,}400 + 3{,}200 + 3z = 11{,}600$$
$$3z = \quad 6{,}000$$
$$z = \quad \mathbf{2{,}000}$$

Thus, each week, the company should produce 1,200 style A shirts, 800 style B shirts, and 2,000 style C shirts to operate at full capacity. The check of the solution is left to the reader. ◆

PROBLEM 8 Repeat Example 8 with the cutting, sewing, and packaging departments having available a maximum of 1,180, 1,560, and 510 labor-hours per week, respectively. ◆

Answers to Matched Problems

1. $x = -2, y = -1$

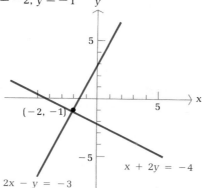

Check:
$$2x - y = -3$$
$$2(-2) - (-1) \stackrel{?}{=} -3$$
$$-3 \stackrel{\checkmark}{=} -3$$
$$x + 2y = -4$$
$$(-2) + 2(-1) \stackrel{?}{=} -4$$
$$-4 \stackrel{\checkmark}{=} -4$$

2. (A) $x = 2, y = 2$ (B) Infinitely many solutions   (C) No solution
3. $x = -2, y = 2$ 4. $x = 2, y = -1$
5. 6.5 oz of food M, 4.5 oz of food N
6. Equilibrium quantity = 13 thousand pounds; Equilibrium price = \$1.70 per pound

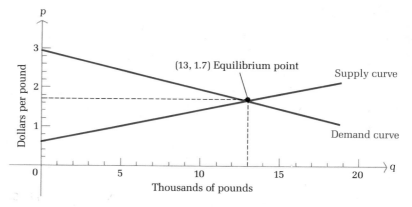

7. $x = -1, y = 0, z = 2$ 8. 900 style A; 1,300 style B; 1,600 style C

A   *Solve by graphing.*

1. $x + y = 5$
   $x - y = 1$

2. $x - y = 2$
   $x + y = 6$

3. $3x - y = 2$
   $x + 2y = 10$

4. $3x - 2y = 12$
   $7x + 2y = 8$

5. $m + 2n = 4$
   $2m + 4n = -8$

6. $3u + 5v = 15$
   $6u + 10v = -30$

Solve using substitution.

7. $y = 2x - 3$
$x + 2y = 14$

8. $y = x - 4$
$x + 3y = 12$

9. $2x + y = 6$
$x - y = -3$

10. $3x - y = 7$
$2x + 3y = 1$

Solve using elimination by addition.

11. $3u - 2v = 12$
$7u + 2v = 8$

12. $2x - 3y = -8$
$5x + 3y = 1$

13. $2m - n = 10$
$m - 2n = -4$

14. $2x + 3y = 1$
$3x - y = 7$

Solve using substitution or elimination by addition.

15. $9x - 3y = 24$
$11x + 2y = 1$

16. $4x + 3y = 26$
$3x - 11y = -7$

17. $2x - 3y = -2$
$-4x + 6y = 7$

18. $3x - 6y = -9$
$-2x + 4y = 12$

19. $3x + 8y = 4$
$15x + 10y = -10$

20. $7m + 12n = -1$
$5m - 3n = 7$

21. $-6x + 10y = -30$
$3x - 5y = 15$

22. $2x + 4y = -8$
$x + 2y = 4$

23. $y = 0.07x$
$y = 80 + 0.05x$

24. $y = 0.08x$
$y = 100 + 0.04x$

B *Solve using elimination by addition.*

25. $0.2x - 0.5y = 0.07$
$0.8x - 0.3y = 0.79$

26. $0.3u - 0.6v = 0.18$
$0.5u + 0.2v = 0.54$

27. $4y - z = -13$
$3y + 2z = 4$
$6x - 5y - 2z = 0$

28. $2x + z = -5$
$x - 3z = -6$
$4x + 2y - z = -9$

29. $2x + y - z = 5$
$x - 2y - 2z = 4$
$3x + 4y + 3z = 3$

30. $x - 3y + z = 4$
$-x + 4y - 4z = 1$
$2x - y + 5z = -3$

31. $2a + 4b + 3c = 6$
$a - 3b + 2c = -7$
$-a + 2b - c = 5$

32. $3u - 2v + 3w = 11$
$2u + 3v - 2w = -5$
$u + 4v - w = -5$

C *Solve using elimination by addition.*

33. $2x - 3y + 3z = -15$
$3x + 2y - 5z = 19$
$5x - 4y - 2z = -2$

34. $3x - 2y - 4z = -8$
$4x + 3y - 5z = -5$
$6x - 5y + 2z = -17$

35. $x - 8y + 2z = -1$
$x - 3y + z = 1$
$2x - 11y + 3z = 2$

36. $-x + 2y - z = -4$
$4x + y - 2z = 1$
$x + y - z = -4$

Business & Economics

37. *Supply and demand.* Suppose the supply and demand equations for printed T-shirts in a resort town for a particular week are

$$p = \ \ \ 0.7q + 3 \qquad \text{Supply equation}$$
$$p = -1.7q + 15 \qquad \text{Demand equation}$$

where p is the price in dollars and q is the quantity in hundreds.

(A) Find the equilibrium price and quantity.

(B) Graph the two equations in the same coordinate system and identify the equilibrium point, supply curve, and demand curve.

38. *Supply and demand.* Repeat Problem 37 with the following supply and demand equations:

$$p = \ \ \ 0.4q + 3.2 \qquad \text{Supply equation}$$
$$p = -1.9q + 17 \qquad \text{Demand equation}$$

39. *Break-even analysis.* A small company manufactures portable home computers. The plant has fixed costs (leases, insurance, and so on) of $48,000 per month and variable costs (labor, materials, and so on) of $1,400 per unit produced. The computers are sold for $1,800 each. Thus, the cost and revenue equations are

$$C = 48,000 + 1,400x$$
$$R = 1,800x$$

where x is the total number of computers produced and sold each month, and C and R are, respectively, monthly costs and revenue in dollars.

(A) How many units must be manufactured and sold each month for the company to break even? (This is actually a three-equation–three-variable problem with the third equation $R = C$. It can be solved by using the substitution method.)

(B) Graph both equations in the same coordinate system and show the break-even point. Interpret the regions between the lines to the left and to the right of the break-even point.

40. *Break-even analysis.* Repeat Problem 39 with the cost and revenue equations

$$C = 65,000 + 1,100x$$
$$R = 1,600x$$

41. *Production scheduling.* A small manufacturing plant makes three types of inflatable boats: one-person, two-person, and four-person models. Each boat requires the services of three departments, as listed in the table on the next page. The cutting, assembly, and packaging departments have available a maximum of 380, 330, and 120 labor-hours per week, respectively.

How many boats of each type must be produced each week for the plant to operate at full capacity?

	ONE-PERSON BOAT	TWO-PERSON BOAT	FOUR-PERSON BOAT
CUTTING DEPARTMENT	0.6 hr	1.0 hr	1.5 hr
ASSEMBLY DEPARTMENT	0.6 hr	0.9 hr	1.2 hr
PACKAGING DEPARTMENT	0.2 hr	0.3 hr	0.5 hr

42. *Production scheduling.* Repeat Problem 41 assuming the cutting, assembly, and packaging departments have available a maximum of 260, 234, and 82 labor-hours per week, respectively.

Life Sciences

43. *Nutrition.* Animals in an experiment are to be kept under a strict diet. Each animal is to receive, among other things, 20 grams of protein and 6 grams of fat. The laboratory technician is able to purchase two food mixes of the following compositions: Mix A has 10% protein and 6% fat; mix B has 20% protein and 2% fat. How many grams of each mix should be used to obtain the right diet for a single animal?

44. *Diet.* In an experiment involving mice, a zoologist needs a food mix that contains, among other things, 23 grams of protein, 6.2 grams of fat, and 16 grams of moisture. She has on hand mixes of the following compositions: Mix A contains 20% protein, 2% fat, and 15% moisture; mix B contains 10% protein, 6% fat, and 10% moisture; and mix C contains 15% protein, 5% fat, and 5% moisture. How many grams of each mix should be used to get the desired diet mix?

Social Sciences

45. *Psychology — approach and avoidance.* People often approach certain situations with "mixed emotions." For example, public speaking often brings forth the positive response of recognition and the negative response of failure. Which dominates? J. S. Brown, in an experiment on approach and avoidance, trained rats by feeding them from a goal box. Then the rats received mild electric shocks from the same goal box. This established an approach–avoidance conflict relative to the goal box. Using appropriate apparatus, Brown arrived at the following relationships:

$$p = -\tfrac{1}{5}d + 70$$
$$a = -\tfrac{4}{3}d + 230$$

$$30 \leqslant d \leqslant 172.5$$

Here, p is the pull (in grams) toward the food goal box when the rat is placed d centimeters from it. The quantity a is the pull (in grams) away from the shock goal box when the rat is placed d centimeters from it.

(A) Graph the two equations above in the same coordinate system.
(B) Find d when $p = a$. (This is actually a three-equation–three-variable problem with the third equation $p = a$. It can be solved by using the substitution method.)

(C) What do you think the rat would do when placed the distance d from the box found in part B?

(For additional discussion of this phenomenon, see J. S. Brown, "Gradients of Approach and Avoidance Responses and Their Relation to Motivation," *Journal of Comparative and Physiological Psychology*, 1948, 41:450–465.)

SECTION 5-2 Systems of Linear Equations and Augmented Matrices

◆ AUGMENTED MATRICES
◆ SOLVING LINEAR SYSTEMS USING AUGMENTED MATRICES
◆ SUMMARY

Most linear systems of any consequence involve large numbers of equations and variables. These systems are solved with computers, since hand methods would be impractical. (Try solving even a five-equation–five-variable problem and you will understand why.) However, even if you have a computer facility to help solve a problem, it is still important for you to know how to formulate the problem so that it can be solved by a computer. In addition, it is helpful to have at least a general idea of how computers solve these problems. Finally, it is important for you to know how to interpret the results.

Even though the procedures and notation introduced in this and the next section are more involved than those used in the preceding section, it is important to keep in mind that our objective is not to find an efficient hand method for solving large-scale systems (there are none), but rather to find a process that generalizes readily for computer use. It turns out that you will receive an added bonus for your efforts, since several of the processes developed in this and the next section will be of considerable value in Sections 5-6 and 6-4.

◆ AUGMENTED MATRICES

In solving systems of equations by elimination, the coefficients of the variables and the constant terms played a central role. The process can be made more efficient for generalization and computer work by the introduction of a mathematical form called a *matrix*. A **matrix** is a rectangular array of numbers written within brackets. Some examples are

$$\begin{bmatrix} 3 & 5 \\ 0 & -2 \end{bmatrix} \quad \begin{bmatrix} 2 \\ -3 \\ 0 \end{bmatrix} \quad \begin{bmatrix} 1 & -1 & 0 & 5 \end{bmatrix}$$

$$\begin{bmatrix} -1 & 2 & -5 & 0 \\ 0 & 3 & 2 & 1 \end{bmatrix} \quad \begin{bmatrix} 1 & 0 & 0 \\ 0 & 1 & 0 \\ 0 & 0 & 1 \end{bmatrix}$$

Each number in a matrix is called an **element** of the matrix.

Associated with the system

$$2x - 3y = 5$$
$$x + 2y = -3$$

$$(1)$$

is the *augmented matrix*

$$\begin{bmatrix} 2 & -3 & | & 5 \\ 1 & 2 & | & -3 \end{bmatrix}$$

which contains the essential parts of the system—namely, the coefficients of the variables and the constant terms. (The vertical bar is included only to separate the coefficients of the variables from the constant terms.)

For ease of generalization to the larger systems in the following sections, we are now going to change the notation for the variables in (1) to a subscript form (we would soon run out of letters, but we will not run out of subscripts). That is, in place of x and y, we will use x_1 and x_2, and (1) will be written as

$$2x_1 - 3x_2 = 5$$
$$x_1 + 2x_2 = -3$$

In general, associated with each linear system of the form

$$a_1x_1 + b_1x_2 = k_1$$
$$a_2x_1 + b_2x_2 = k_2$$

$$(2)$$

where x_1 and x_2 are variables, is the **augmented matrix** of the system:

$$\begin{bmatrix} a_1 & b_1 & | & k_1 \\ a_2 & b_2 & | & k_2 \end{bmatrix}$$

Column 1 (C_1)
Column 2 (C_2)
Column 3 (C_3)
← Row 1 (R_1)
← Row 2 (R_2)

This matrix contains the essential parts of system (2). Our objective is to learn how to manipulate augmented matrices in order to solve system (2), if a solution exists. The manipulative process is a direct outgrowth of the elimination process discussed in Section 5-1.

Recall that two linear systems are said to be **equivalent** if they have exactly the same solution set. How did we transform linear systems into equivalent linear systems? We used Theorem 2, which we restate here.

Operations That Produce Equivalent Systems

A system of linear equations is transformed into an equivalent system if:

(A) Two equations are interchanged.
(B) An equation is multiplied by a nonzero constant.
(C) A constant multiple of one equation is added to another equation.

Paralleling the earlier discussion, we say that two augmented matrices are **row-equivalent**, denoted by the symbol ~ placed between the two matrices, if they are augmented matrices of equivalent systems of equations. (Think about this.) How do we transform augmented matrices into row-equivalent matrices? We use Theorem 3, which is a direct consequence of Theorem 2.

THEOREM 3

An augmented matrix is transformed into a row-equivalent matrix by performing any of the following **row operations:**

(A) Two rows are interchanged $(R_i \leftrightarrow R_j)$.
(B) A row is multiplied by a nonzero constant $(kR_i \rightarrow R_i)$.
(C) A constant multiple of one row is added to another row
$(R_i + kR_j \rightarrow R_i)$.

[*Note:* The arrow $\rightarrow$ means "replaces."]

◆ SOLVING LINEAR SYSTEMS USING AUGMENTED MATRICES

The use of Theorem 3 in solving systems in the form of (2) is best illustrated by examples.

◆ EXAMPLE 9 Solve using augmented matrix methods:

$$3x_1 + 4x_2 = 1$$
$$x_1 - 2x_2 = 7$$

(3)

Solution We start by writing the augmented matrix corresponding to (3)

$$\begin{bmatrix} 3 & 4 & | & 1 \\ 1 & -2 & | & 7 \end{bmatrix}$$

(4)

Our objective is to use row operations from Theorem 3 to try to transform (4) into the form

$$\begin{bmatrix} 1 & 0 & \vline & m \\ 0 & 1 & \vline & n \end{bmatrix} \tag{5}$$

where m and n are real numbers. The solution to system (3) will then be obvious, since matrix (5) will be the augmented matrix of the following system (a row in an augmented matrix always corresponds to an equation in a linear system):

$$x_1 = m \qquad x_1 + 0x_2 = m$$
$$x_2 = n \qquad 0x_1 + x_2 = n$$

We now proceed to use row operations to transform (4) into form (5).

Step 1. To get a 1 in the upper left corner, we interchange Rows 1 and 2 (Theorem 3A):

$$\begin{bmatrix} 3 & 4 & \vline & 1 \\ 1 & -2 & \vline & 7 \end{bmatrix} \quad R_1 \underset{\sim}{\leftrightarrow} R_2 \quad \begin{bmatrix} 1 & -2 & \vline & 7 \\ 3 & 4 & \vline & 1 \end{bmatrix}$$

Now you see why we wanted Theorem 2A!

Step 2. To get a 0 in the lower left corner, we multiply R_1 by (-3) and add to R_2 (Theorem 3C) — this changes R_2 but not R_1. Some people find it useful to write $(-3)R_1$ outside the matrix to help reduce errors in arithmetic, as shown:

$$\begin{bmatrix} 1 & -2 & \vline & 7 \\ 3 & 4 & \vline & 1 \end{bmatrix} \quad R_2 + \underbrace{(-3)R_1}_{} \to R_2 \quad \begin{bmatrix} 1 & -2 & \vline & 7 \\ 0 & 10 & \vline & -20 \end{bmatrix}$$
$$-3 \quad 6 \quad -21 \quad \longleftarrow$$

Step 3. To get a 1 in the second row, second column, we multiply R_2 by $\frac{1}{10}$ (Theorem 3B):

$$\begin{bmatrix} 1 & -2 & \vline & 7 \\ 0 & 10 & \vline & -20 \end{bmatrix} \quad \tfrac{1}{10}R_2 \underset{\sim}{\to} R_2 \quad \begin{bmatrix} 1 & -2 & \vline & 7 \\ 0 & 1 & \vline & -2 \end{bmatrix}$$

Step 4. To get a 0 in the first row, second column, we multiply R_2 by 2 and add the result to R_1 (Theorem 3C) — this changes R_1 but not R_2:

$$0 \quad 2 \quad -4 \longleftarrow$$
$$\begin{bmatrix} 1 & -2 & \vline & 7 \\ 0 & 1 & \vline & -2 \end{bmatrix} \quad R_1 \overset{\frown}{+ 2R_2} \to R_1 \quad \begin{bmatrix} 1 & 0 & \vline & 3 \\ 0 & 1 & \vline & -2 \end{bmatrix}$$

We have accomplished our objective! The last matrix is the augmented matrix for the system

$$x_1 = 3 \qquad x_1 + 0x_2 = 3$$
$$x_2 = -2 \qquad 0x_1 + x_2 = -2 \tag{6}$$

Since system (6) is equivalent to system (3), our starting system, we have solved (3); that is, $x_1 = 3$ and $x_2 = -2$.

Check

$$3x_1 + 4x_2 = 1 \qquad\qquad x_1 - 2x_2 = 7$$
$$3(3) + 4(-2) \overset{?}{=} 1 \qquad\qquad 3 - 2(-2) \overset{?}{=} 7$$
$$1 \overset{\checkmark}{=} 1 \qquad\qquad\qquad 7 \overset{\checkmark}{=} 7$$

The above process may be written more compactly as follows:

Step 1:
Need a 1 here.

$$\begin{bmatrix} 3 & 4 & | & 1 \\ 1 & -2 & | & 7 \end{bmatrix} \quad R_1 \leftrightarrow R_2$$

Step 2:
Need a 0 here.

$$\sim \begin{bmatrix} 1 & -2 & | & 7 \\ 3 & 4 & | & 1 \end{bmatrix} \quad R_2 + (-3)R_1 \rightarrow R_2$$
$$-3 \quad 6 \qquad -21$$

Step 3:
Need a 1 here.

$$\sim \begin{bmatrix} 1 & -2 & | & 7 \\ 0 & 10 & | & -20 \end{bmatrix} \quad \tfrac{1}{10}R_2 \rightarrow R_2$$
$$0 \quad 2 \qquad -4$$

Step 4:
Need a 0 here.

$$\sim \begin{bmatrix} 1 & -2 & | & 7 \\ 0 & 1 & | & -2 \end{bmatrix} \quad R_1 + 2R_2 \rightarrow R_1$$

$$\sim \begin{bmatrix} 1 & 0 & | & 3 \\ 0 & 1 & | & -2 \end{bmatrix}$$

Therefore, $x_1 = 3$ and $x_2 = -2$.

PROBLEM 9   Solve using augmented matrix methods:

$$2x_1 - x_2 = -7$$
$$x_1 + 2x_2 = 4$$

◆ EXAMPLE 10   Solve using augmented matrix methods:

$$2x_1 - 3x_2 = 6$$
$$3x_1 + 4x_2 = \tfrac{1}{2}$$

Solution

Step 1:
Need a 1 here.

$$\begin{bmatrix} 2 & -3 & | & 6 \\ 3 & 4 & | & \frac{1}{2} \end{bmatrix} \quad \frac{1}{2}R_1 \rightarrow R_1$$

Step 2:
Need a 0 here.

$$\sim \begin{bmatrix} 1 & -\frac{3}{2} & | & 3 \\ 3 & 4 & | & \frac{1}{2} \end{bmatrix} \quad R_2 + (-3)R_1 \rightarrow R_2$$

$$\begin{matrix} -3 & \frac{9}{2} & -9 \end{matrix}$$

Step 3:
Need a 1 here.

$$\sim \begin{bmatrix} 1 & -\frac{3}{2} & | & 3 \\ 0 & \frac{17}{2} & | & -\frac{17}{2} \end{bmatrix} \quad \frac{2}{17}R_2 \rightarrow R_2$$

$$\begin{matrix} 0 & \frac{3}{2} & -\frac{3}{2} \end{matrix}$$

Step 4:
Need a 0 here.

$$\sim \begin{bmatrix} 1 & -\frac{3}{2} & | & 3 \\ 0 & 1 & | & -1 \end{bmatrix} \quad R_1 + \frac{3}{2}R_2 \rightarrow R_1$$

$$\sim \begin{bmatrix} 1 & 0 & | & \frac{3}{2} \\ 0 & 1 & | & -1 \end{bmatrix}$$

Thus, $x_1 = \frac{3}{2}$ and $x_2 = -1$. The check is left to the reader. ◆

PROBLEM 10   Solve using augmented matrix methods:

$$5x_1 - 2x_2 = 11$$
$$2x_1 + 3x_2 = \frac{5}{2}$$ ◆

◆ EXAMPLE 11   Solve using augmented matrix methods:

$$2x_1 - x_2 = 4$$
$$-6x_1 + 3x_2 = -12$$ (7)

Solution

$$\begin{bmatrix} 2 & -1 & | & 4 \\ -6 & 3 & | & -12 \end{bmatrix} \quad \begin{matrix} \frac{1}{2}R_1 \rightarrow R_1 \text{ (to get a 1 in the upper left corner)} \\ \frac{1}{3}R_2 \rightarrow R_2 \text{ (this simplifies } R_2) \end{matrix}$$

$$\sim \begin{bmatrix} 1 & -\frac{1}{2} & | & 2 \\ -2 & 1 & | & -4 \end{bmatrix} \quad R_2 + 2R_1 \rightarrow R_2 \text{ (to get a 0 in the lower left corner)}$$

$$\begin{matrix} 2 & -1 & 4 \end{matrix}$$

$$\sim \begin{bmatrix} 1 & -\frac{1}{2} & | & 2 \\ 0 & 0 & | & 0 \end{bmatrix}$$

The last matrix corresponds to the system

$$\begin{matrix} x_1 - \frac{1}{2}x_2 = 2 & \quad x_1 - \frac{1}{2}x_2 = 2 \\ 0 = 0 & \quad 0x_1 + 0x_2 = 0 \end{matrix}$$ (8)

This system is equivalent to the original system. Geometrically, the graphs of the two original equations coincide and there are infinitely many solutions. In general, if we end up with a row of zeros in an augmented matrix for a two-equation–two-variable system, the system is dependent and there are infinitely many solutions.

We represent the infinitely many solutions using the same method that was used in Section 5-1; that is, by introducing a parameter. We start by solving $x_1 - \frac{1}{2}x_2 = 2$, the first equation in (8), for either variable in terms of the other. We choose to solve for x_1 in terms of x_2 because it is easier:

$$x_1 = \tfrac{1}{2}x_2 + 2 \tag{9}$$

Now we introduce a parameter t (we can use other letters, such as k, s, p, q, and so on, to represent a parameter just as well). If we let $x_2 = t$, then for t any real number,

$$\begin{aligned} x_1 &= \tfrac{1}{2}t + 2 \\ x_2 &= t \end{aligned} \tag{10}$$

represents a solution of system (7). Using ordered pair notation, we may also write: For any real number t,

$$(\tfrac{1}{2}t + 2,\ t) \tag{11}$$

is a solution of system (7). And more formally, we may write:

$$\text{Solution set} = \{(\tfrac{1}{2}t + 2,\ t) | t \in R\} \tag{12}$$

We will generally use the less formal forms (10) and (11) to represent the solution set for problems of this type.

Check   The following is a check that (10) provides a solution for system (7) for any real number t:

$$\begin{array}{cc} 2x_1 - x_2 = 4 & -6x_1 + 3x_2 = -12 \\ 2(\tfrac{1}{2}t + 2) - t \overset{?}{=} 4 & -6(\tfrac{1}{2}t + 2) + 3t \overset{?}{=} -12 \\ t + 4 - t \overset{?}{=} 4 & -3t - 12 + 3t \overset{?}{=} -12 \\ 4 \overset{\checkmark}{=} 4 & -12 \overset{\checkmark}{=} -12 \end{array}$$ ◆

PROBLEM   11   Solve using augmented matrix methods:

$$\begin{aligned} -2x_1 + 6x_2 &= 6 \\ 3x_1 - 9x_2 &= -9 \end{aligned}$$ ◆

◆ EXAMPLE   12   Solve using augmented matrix methods:

$$\begin{aligned} 2x_1 + 6x_2 &= -3 \\ x_1 + 3x_2 &= 2 \end{aligned}$$

Solution
$$\begin{bmatrix} 2 & 6 & | & -3 \\ 1 & 3 & | & 2 \end{bmatrix} \quad R_1 \leftrightarrow R_2$$

$$\sim \begin{bmatrix} 1 & 3 & | & 2 \\ 2 & 6 & | & -3 \\ -2 & -6 & & -4 \end{bmatrix} \quad R_2 + (-2)R_1 \rightarrow R_2$$

$$\sim \begin{bmatrix} 1 & 3 & | & 2 \\ 0 & 0 & | & -7 \end{bmatrix} \quad R_2 \text{ implies the contradiction } 0 = -7.$$

This is the augmented matrix of the system

$$\begin{array}{ll} x_1 + 3x_2 = 2 & x_1 + 3x_2 = 2 \\ 0 = -7 & 0x_1 + 0x_2 = -7 \end{array}$$

The second equation is not satisfied by any ordered pair of real numbers. Hence, the original system is inconsistent and has no solution—otherwise, we have proved that $0 = -7$! Thus, if in a row of an augmented matrix we obtain all zeros to the left of the vertical bar and a nonzero number to the right, then the system is inconsistent and there are no solutions. ◆

PROBLEM 12 Solve using augmented matrix methods:

$$\begin{array}{r} 2x_1 - x_2 = 3 \\ 4x_1 - 2x_2 = -1 \end{array}$$
◆

◆ SUMMARY

█ **Summary**

Form 1	Form 2	Form 3
A Unique Solution (Consistent and Independent)	Infinitely Many Solutions (Consistent and Dependent)	No Solution (Inconsistent)
$\begin{bmatrix} 1 & 0 & | & m \\ 0 & 1 & | & n \end{bmatrix}$	$\begin{bmatrix} 1 & m & | & n \\ 0 & 0 & | & 0 \end{bmatrix}$	$\begin{bmatrix} 1 & m & | & n \\ 0 & 0 & | & p \end{bmatrix}$

m, n, p real numbers; $p \neq 0$

The process of solving systems of equations described in this section is referred to as **Gauss–Jordan elimination.** We will formalize this method in the next section so that it will apply to systems of any size, including systems where the number of equations and the number of variables are not the same.

9. $x_1 = -2, x_2 = 3$ 10. $x_1 = 2, x_2 = -\frac{1}{2}$

11. The system is dependent. For t any real number, a solution is

$$x_1 = 3t - 3$$
$$x_2 = t$$

12. Inconsistent — no solution

E X E R C I S E 5-2

A Perform each of the indicated row operations on the following matrix:

$$\begin{bmatrix} 1 & -3 & 2 \\ 4 & -6 & -8 \end{bmatrix}$$

1. $R_1 \leftrightarrow R_2$ 2. $\frac{1}{2}R_2 \rightarrow R_2$ 3. $-4R_1 \rightarrow R_1$
4. $-2R_1 \rightarrow R_1$ 5. $2R_2 \rightarrow R_2$ 6. $-1R_2 \rightarrow R_2$
7. $R_2 + (-4)R_1 \rightarrow R_2$ 8. $R_1 + (-\frac{1}{2})R_2 \rightarrow R_1$ 9. $R_2 + (-2)R_1 \rightarrow R_2$
10. $R_2 + (-3)R_1 \rightarrow R_2$ 11. $R_2 + (-1)R_1 \rightarrow R_2$ 12. $R_2 + (1)R_1 \rightarrow R_2$

Solve using augmented matrix methods.

13. $\begin{aligned} x_1 + x_2 &= 5 \\ x_1 - x_2 &= 1 \end{aligned}$ 14. $\begin{aligned} x_1 - x_2 &= 2 \\ x_1 + x_2 &= 6 \end{aligned}$

B Solve using augmented matrix methods.

15. $\begin{aligned} x_1 - 2x_2 &= 1 \\ 2x_1 - x_2 &= 5 \end{aligned}$ 16. $\begin{aligned} x_1 + 3x_2 &= 1 \\ 3x_1 - 2x_2 &= 14 \end{aligned}$ 17. $\begin{aligned} x_1 - 4x_2 &= -2 \\ -2x_1 + x_2 &= -3 \end{aligned}$

18. $\begin{aligned} x_1 - 3x_2 &= -5 \\ -3x_1 - x_2 &= 5 \end{aligned}$ 19. $\begin{aligned} 3x_1 - x_2 &= 2 \\ x_1 + 2x_2 &= 10 \end{aligned}$ 20. $\begin{aligned} 2x_1 + x_2 &= 0 \\ x_1 - 2x_2 &= -5 \end{aligned}$

21. $\begin{aligned} x_1 + 2x_2 &= 4 \\ 2x_1 + 4x_2 &= -8 \end{aligned}$ 22. $\begin{aligned} 2x_1 - 3x_2 &= -2 \\ -4x_1 + 6x_2 &= 7 \end{aligned}$ 23. $\begin{aligned} 2x_1 + x_2 &= 6 \\ x_1 - x_2 &= -3 \end{aligned}$

24. $\begin{aligned} 3x_1 - x_2 &= -5 \\ x_1 + 3x_2 &= 5 \end{aligned}$ 25. $\begin{aligned} 3x_1 - 6x_2 &= -9 \\ -2x_1 + 4x_2 &= 6 \end{aligned}$ 26. $\begin{aligned} 2x_1 - 4x_2 &= -2 \\ -3x_1 + 6x_2 &= 3 \end{aligned}$

27. $\begin{aligned} 4x_1 - 2x_2 &= 2 \\ -6x_1 + 3x_2 &= -3 \end{aligned}$ 28. $\begin{aligned} -6x_1 + 2x_2 &= 4 \\ 3x_1 - x_2 &= -2 \end{aligned}$ 29. $\begin{aligned} 2x_1 + x_2 &= 1 \\ 4x_1 - x_2 &= -7 \end{aligned}$

30. $\begin{aligned} 2x_1 - x_2 &= -8 \\ 2x_1 + x_2 &= 8 \end{aligned}$ 31. $\begin{aligned} 4x_1 - 6x_2 &= 8 \\ -6x_1 + 9x_2 &= -10 \end{aligned}$ 32. $\begin{aligned} 2x_1 - 4x_2 &= -4 \\ -3x_1 + 6x_2 &= 4 \end{aligned}$

33. $\begin{aligned} -4x_1 + 6x_2 &= -8 \\ 6x_1 - 9x_2 &= 12 \end{aligned}$ 34. $\begin{aligned} -2x_1 + 4x_2 &= 4 \\ 3x_1 - 6x_2 &= -6 \end{aligned}$

C Solve using augmented matrix methods.

35. $\begin{aligned} 3x_1 - x_2 &= 7 \\ 2x_1 + 3x_2 &= 1 \end{aligned}$ 36. $\begin{aligned} 2x_1 - 3x_2 &= -8 \\ 5x_1 + 3x_2 &= 1 \end{aligned}$

37. $\begin{aligned} 3x_1 + 2x_2 &= 4 \\ 2x_1 - x_2 &= 5 \end{aligned}$ 38. $\begin{aligned} 4x_1 + 3x_2 &= 26 \\ 3x_1 - 11x_2 &= -7 \end{aligned}$

39. $\begin{aligned} 0.2x_1 - 0.5x_2 &= 0.07 \\ 0.8x_1 - 0.3x_2 &= 0.79 \end{aligned}$ 40. $\begin{aligned} 0.3x_1 - 0.6x_2 &= 0.18 \\ 0.5x_1 - 0.2x_2 &= 0.54 \end{aligned}$

Gauss–Jordan Elimination

◆ REDUCED MATRICES
◆ SOLVING SYSTEMS BY GAUSS–JORDAN ELIMINATION
◆ APPLICATION

Now that you have had some experience with row operations on simple augmented matrices, we will consider systems involving more than two variables. In addition, we will not require that a system have the same number of equations as variables.

◆ REDUCED MATRICES

Our objective is to start with the augmented matrix of a linear system and transform it by using row operations from Theorem 3 (in the preceding section) into a simple form where the solution can be read by inspection. The simple form so obtained is called the *reduced form*, and we define it as follows:

■ Reduced Matrix

A matrix is in **reduced form** if:

1. Each row consisting entirely of zeros is below any row having at least one nonzero element.
2. The leftmost nonzero element in each row is 1.
3. All other elements in the column containing the leftmost 1 of a given row are zeros.
4. The leftmost 1 in any row is to the right of the leftmost 1 in the row above.

The following matrices are in reduced form. Check each one carefully to convince yourself that the conditions in the definition are met.

$$\left[\begin{array}{cc|c} 1 & 0 & 2 \\ 0 & 1 & -3 \end{array}\right] \qquad \left[\begin{array}{ccc|c} 1 & 0 & 0 & 2 \\ 0 & 1 & 0 & -1 \\ 0 & 0 & 1 & 3 \end{array}\right] \qquad \left[\begin{array}{cc|c} 1 & 0 & 3 \\ 0 & 1 & -1 \\ 0 & 0 & 0 \end{array}\right]$$

$$\left[\begin{array}{cccc|c} 1 & 4 & 0 & 0 & -3 \\ 0 & 0 & 1 & 0 & 2 \\ 0 & 0 & 0 & 1 & 6 \end{array}\right] \qquad \left[\begin{array}{ccc|c} 1 & 0 & 4 & 0 \\ 0 & 1 & 3 & 0 \\ 0 & 0 & 0 & 1 \end{array}\right]$$

◆ E X A M P L E   13   The matrices below are not in reduced form. Indicate which condition in the definition is violated for each matrix.

(A) $\begin{bmatrix} 0 & 1 & \vline & -2 \\ 1 & 0 & \vline & 3 \end{bmatrix}$ (B) $\begin{bmatrix} 1 & 2 & -2 & \vline & 3 \\ 0 & 0 & 1 & \vline & -1 \end{bmatrix}$

(C) $\begin{bmatrix} 0 & 0 & \vline & 0 \\ 1 & 0 & \vline & -3 \\ 0 & 1 & \vline & -2 \end{bmatrix}$ (D) $\begin{bmatrix} 1 & 0 & 0 & \vline & -1 \\ 0 & 2 & 0 & \vline & 3 \\ 0 & 0 & 1 & \vline & -5 \end{bmatrix}$

Solutions   (A) Condition 4 is violated: The leftmost 1 in Row 2 is not to the right of the leftmost 1 in Row 1.
(B) Condition 3 is violated: The column containing the leftmost 1 in Row 2 has a nonzero element above the 1.
(C) Condition 1 is violated: The first row contains all zeros and it is not below any row having at least one nonzero element.
(D) Condition 2 is violated: The leftmost nonzero element in Row 2 is not 1. ◆

P R O B L E M   13   The matrices below are not in reduced form. Indicate which condition in the definition is violated for each matrix.

(A) $\begin{bmatrix} 1 & 0 & \vline & 2 \\ 0 & 3 & \vline & -6 \end{bmatrix}$ (B) $\begin{bmatrix} 1 & 5 & 4 & \vline & 3 \\ 0 & 1 & 2 & \vline & -1 \\ 0 & 0 & 0 & \vline & 0 \end{bmatrix}$

(C) $\begin{bmatrix} 0 & 1 & 0 & \vline & -3 \\ 1 & 0 & 0 & \vline & 0 \\ 0 & 0 & 1 & \vline & 2 \end{bmatrix}$ (D) $\begin{bmatrix} 1 & 2 & 0 & \vline & 3 \\ 0 & 0 & 0 & \vline & 0 \\ 0 & 0 & 1 & \vline & 4 \end{bmatrix}$ ◆

◆ E X A M P L E   14   Write the linear system corresponding to each reduced augmented matrix and solve.

(A) $\begin{bmatrix} 1 & 0 & 0 & \vline & 2 \\ 0 & 1 & 0 & \vline & -1 \\ 0 & 0 & 1 & \vline & 3 \end{bmatrix}$ (B) $\begin{bmatrix} 1 & 0 & 4 & \vline & 0 \\ 0 & 1 & 3 & \vline & 0 \\ 0 & 0 & 0 & \vline & 1 \end{bmatrix}$

(C) $\begin{bmatrix} 1 & 0 & 2 & \vline & -3 \\ 0 & 1 & -1 & \vline & 8 \\ 0 & 0 & 0 & \vline & 0 \end{bmatrix}$ (D) $\begin{bmatrix} 1 & 4 & 0 & 0 & 3 & \vline & -2 \\ 0 & 0 & 1 & 0 & -2 & \vline & 0 \\ 0 & 0 & 0 & 1 & 2 & \vline & 4 \end{bmatrix}$

Solutions   (A) $x_1 \quad = \quad 2$
$\qquad x_2 \quad = -1$
$\qquad \quad x_3 = \quad 3$

The solution is obvious: $x_1 = 2$, $x_2 = -1$, $x_3 = 3$.

(B) $x_1 \qquad + 4x_3 = 0$

$\qquad x_2 + 3x_3 = 0$

$0x_1 + 0x_2 + 0x_3 = 1$

The last equation implies $0 = 1$, which is a contradiction. Hence, the system is inconsistent and has no solution.

(C) $x_1 \quad + 2x_3 = -3$ We disregard the equation corresponding to the third

$\qquad x_2 - \ x_3 = \ \ 8$ row in the matrix, since it is satisfied by all values of x_1, x_2, and x_3.

When a reduced system (a system corresponding to a reduced augmented matrix) has more variables than equations and contains no contradictions, the system is dependent and has infinitely many solutions. To represent these solutions, we note that the first variable in each equation (x_1 and x_2) appears in only one equation in the reduced system. Since these variables correspond to leftmost 1's in the reduced augmented matrix, we call the first variable in each equation of a reduced system a **leftmost variable.** The definition of reduced form ensures that each leftmost variable will appear in exactly one equation of the reduced system and that no two leftmost variables will appear in the same equation. Thus, it is easy to solve for each leftmost variable in terms of the remaining variables. Returning to our original system, we solve for the leftmost variables x_1 and x_2 in terms of the remaining variable x_3:

$x_1 = -2x_3 - 3$

$x_2 = x_3 + 8$

If we let $x_3 = t$, then for any real number t,

$x_1 = -2t - 3$

$x_2 = t + 8$

$x_3 = t$

is a solution. For example:

If $t = 0$, then   If $t = -2$, then

$x_1 = -2(0) - 3 = -3$ $x_1 = -2(-2) - 3 = 1$

$x_2 = 0 + 8 = 8$ $x_2 = -2 + 8 = 6$

$x_3 = 0$ $x_3 = -2$

is a solution.   is a solution.

(D) $x_1 + 4x_2 \qquad + 3x_5 = -2$

$\qquad x_3 \ - 2x_5 = \ \ 0$

$\qquad\qquad x_4 + 2x_5 = \ \ 4$

Solve for x_1, x_3, and x_4 (leftmost variables) in terms of x_2 and x_5 (remaining variables):

$$x_1 = -4x_2 - 3x_5 - 2$$
$$x_3 = 2x_5$$
$$x_4 = -2x_5 + 4$$

If we let $x_2 = s$ and $x_5 = t$, then for any real numbers s and t,

$$x_1 = -4s - 3t - 2$$
$$x_2 = s$$
$$x_3 = 2t$$
$$x_4 = -2t + 4$$
$$x_5 = t$$

is a solution. The system is dependent and has infinitely many solutions.

◆

PROBLEM 14   Write the linear system corresponding to each reduced augmented matrix and solve.

(A) $\begin{bmatrix} 1 & 0 & 0 & | & -5 \\ 0 & 1 & 0 & | & 3 \\ 0 & 0 & 1 & | & 6 \end{bmatrix}$ (B) $\begin{bmatrix} 1 & 2 & -3 & | & 0 \\ 0 & 0 & 0 & | & 1 \\ 0 & 0 & 0 & | & 0 \end{bmatrix}$

(C) $\begin{bmatrix} 1 & 0 & -2 & | & 4 \\ 0 & 1 & 3 & | & -2 \\ 0 & 0 & 0 & | & 0 \end{bmatrix}$ (D) $\begin{bmatrix} 1 & 0 & 3 & 2 & | & 5 \\ 0 & 1 & -2 & -1 & | & 3 \\ 0 & 0 & 0 & 0 & | & 0 \end{bmatrix}$ ◆

◆ SOLVING SYSTEMS BY GAUSS–JORDAN ELIMINATION

We are now ready to outline the Gauss–Jordan elimination method for solving systems of linear equations. The method systematically transforms an augmented matrix into a reduced form from which we can write the solution to the original system by inspection, if a solution exists. The method will also reveal when a solution fails to exist (see Example 17).

The Gauss–Jordan elimination method is named after the German mathematician Carl Friedrich Gauss (1777–1855) and the French mathematician Camille Jordan (1838–1922). Gauss, one of the greatest mathematicians of all time, used a method of solving systems of equations that was later generalized by Jordan to the method we present here.

◆ EXAMPLE 15   Solve by Gauss–Jordan elimination:

$$2x_1 - 2x_2 + x_3 = 3$$
$$3x_1 + x_2 - x_3 = 7$$
$$x_1 - 3x_2 + 2x_3 = 0$$

Solution   Write the augmented matrix and follow the steps indicated at the right.

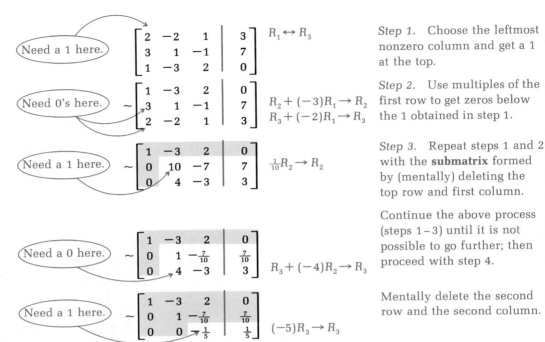

Need a 1 here.

$$\begin{bmatrix} 2 & -2 & 1 & | & 3 \\ 3 & 1 & -1 & | & 7 \\ 1 & -3 & 2 & | & 0 \end{bmatrix} \quad R_1 \leftrightarrow R_3$$

Step 1. Choose the leftmost nonzero column and get a 1 at the top.

Need 0's here.

$$\sim \begin{bmatrix} 1 & -3 & 2 & | & 0 \\ 3 & 1 & -1 & | & 7 \\ 2 & -2 & 1 & | & 3 \end{bmatrix} \quad \begin{matrix} R_2 + (-3)R_1 \rightarrow R_2 \\ R_3 + (-2)R_1 \rightarrow R_3 \end{matrix}$$

Step 2. Use multiples of the first row to get zeros below the 1 obtained in step 1.

Need a 1 here.

$$\sim \begin{bmatrix} 1 & -3 & 2 & | & 0 \\ 0 & 10 & -7 & | & 7 \\ 0 & 4 & -3 & | & 3 \end{bmatrix} \quad \tfrac{1}{10}R_2 \rightarrow R_2$$

Step 3. Repeat steps 1 and 2 with the **submatrix** formed by (mentally) deleting the top row and first column.

Continue the above process (steps 1–3) until it is not possible to go further; then proceed with step 4.

Need a 0 here.

$$\sim \begin{bmatrix} 1 & -3 & 2 & | & 0 \\ 0 & 1 & -\tfrac{7}{10} & | & \tfrac{7}{10} \\ 0 & 4 & -3 & | & 3 \end{bmatrix} \quad R_3 + (-4)R_2 \rightarrow R_3$$

Mentally delete the second row and the second column.

Need a 1 here.

$$\sim \begin{bmatrix} 1 & -3 & 2 & | & 0 \\ 0 & 1 & -\tfrac{7}{10} & | & \tfrac{7}{10} \\ 0 & 0 & \tfrac{1}{5} & | & \tfrac{1}{5} \end{bmatrix} \quad (-5)R_3 \rightarrow R_3$$

Need 0's here.

$$\sim \begin{bmatrix} 1 & -3 & 2 & | & 0 \\ 0 & 1 & -\tfrac{7}{10} & | & \tfrac{7}{10} \\ 0 & 0 & 1 & | & -1 \end{bmatrix} \quad \begin{matrix} R_1 + (-2)R_3 \rightarrow R_1 \\ R_2 + \tfrac{7}{10}R_3 \rightarrow R_2 \end{matrix}$$

Since steps 1–3 cannot be carried further, proceed to step 4.

Step 4. Consider the whole matrix. Begin with the bottom nonzero row and use appropriate multiples of it to get zeros above the leftmost 1. Continue the process, moving up row by row, until the matrix is in reduced form.

Need a 0 here.

$$\sim \begin{bmatrix} 1 & -3 & 0 & | & 2 \\ 0 & 1 & 0 & | & 0 \\ 0 & 0 & 1 & | & -1 \end{bmatrix} \quad R_1 + 3R_2 \rightarrow R_1$$

$$\sim \begin{bmatrix} 1 & 0 & 0 & | & 2 \\ 0 & 1 & 0 & | & 0 \\ 0 & 0 & 1 & | & -1 \end{bmatrix}$$

The matrix is in reduced form, and we can write the solution to the original system by inspection.

Solution: $x_1 = 2$, $x_2 = 0$, $x_3 = -1$. It is left to the reader to check this solution.

◆

Steps 1–4 outlined in the solution of Example 15 are referred to as *Gauss–Jordan elimination*. The steps are summarized below for easy reference:

Gauss–Jordan Elimination

Step 1. Choose the leftmost nonzero column and use appropriate row operations to get a 1 at the top.

Step 2. Use multiples of the first row to get zeros in all places below the 1 obtained in step 1.

Step 3. Delete (mentally) the top row and first column of the matrix. Repeat steps 1 and 2 with the **submatrix** (the matrix that remains after deleting the top row and first column). Continue this process (steps 1–3) until it is not possible to go further.

Step 4. Now consider the whole matrix. Begin with the bottom nonzero row and use appropriate multiples of it to get zeros above the leftmost 1. Continue this process, moving up row by row, until the matrix is finally in reduced form.

[*Note:* If at any point in this process we obtain a row with all zeros to the left of the vertical line and a nonzero number n to the right, we can stop, since we will have a contradiction: $0 = n$, $n \neq 0$. We can then conclude that the system has no solution.]

Remark

The sequence of steps (algorithm) presented here for transforming a matrix into a reduced form is not unique. That is, other sequences of steps (using row operations) can produce a reduced matrix. (For example, it is possible to use row operations in such a way that fractions can be avoided.) But we emphasize again that we are not interested in the most efficient hand methods for transforming a matrix into a reduced form. Our main interest is in giving you a little experience with a method that is suitable for computer use with large-scale systems.

PROBLEM 15   Solve by Gauss–Jordan elimination:

$$3x_1 + x_2 - 2x_3 = 2$$
$$x_1 - 2x_2 + x_3 = 3$$
$$2x_1 - x_2 - 3x_3 = 3$$

◆

◆ E X A M P L E   16   Solve by Gauss–Jordan elimination:

$$2x_1 - x_2 + 4x_3 = -2$$
$$3x_1 + 2x_2 - x_3 = 1$$

Solution

Need a 1 here.
$$\begin{bmatrix} 2 & -1 & 4 & | & -2 \\ 3 & 2 & -1 & | & 1 \end{bmatrix} \quad \tfrac{1}{2}R_1 \to R_1$$

Need a 0 here.
$$\sim \begin{bmatrix} 1 & -\tfrac{1}{2} & 2 & | & -1 \\ 3 & 2 & -1 & | & 1 \end{bmatrix} \quad R_2 + (-3)R_1 \to R_2$$

Need a 1 here.
$$\sim \begin{bmatrix} 1 & -\tfrac{1}{2} & 2 & | & -1 \\ 0 & \tfrac{7}{2} & -7 & | & 4 \end{bmatrix} \quad \tfrac{2}{7}R_2 \to R_2$$

Need a 0 here.
$$\sim \begin{bmatrix} 1 & -\tfrac{1}{2} & 2 & | & -1 \\ 0 & 1 & -2 & | & \tfrac{8}{7} \end{bmatrix} \quad R_1 + \tfrac{1}{2}R_2 \to R_1$$

$$\sim \begin{bmatrix} 1 & 0 & 1 & | & -\tfrac{3}{7} \\ 0 & 1 & -2 & | & \tfrac{8}{7} \end{bmatrix}$$

The matrix is now in reduced form. Write the corresponding system and the solution.

$$x_1 \quad + \quad x_3 = -\tfrac{3}{7}$$
$$x_2 - 2x_3 = \tfrac{8}{7}$$

Solve for the leftmost variables x_1 and x_2 in terms of the remaining variable x_3:

$$x_1 = -x_3 - \tfrac{3}{7}$$
$$x_2 = 2x_3 + \tfrac{8}{7}$$

If $x_3 = t$, then for t any real number,

$$x_1 = -t - \tfrac{3}{7}$$
$$x_2 = 2t + \tfrac{8}{7}$$
$$x_3 = t$$

is a solution. Checking that $(-t - \tfrac{3}{7}, 2t + \tfrac{8}{7}, t)$ is a solution for any real t is left to the reader. ◆

Remark

In general, it can be proved that a system with more variables than equations cannot have a unique solution.

P R O B L E M   16   Solve by Gauss–Jordan elimination:

$$3x_1 + 6x_2 - 3x_3 = 2$$
$$2x_1 - x_2 + 2x_3 = -1$$

◆

◆ E X A M P L E   17   Solve by Gauss–Jordan elimination:

$$2x_1 - x_2 = -4$$
$$2x_1 + 4x_2 = 6$$
$$3x_1 - x_2 = -1$$

Solution

$$\begin{bmatrix} 2 & -1 & -4 \\ 2 & 4 & 6 \\ 3 & -1 & -1 \end{bmatrix}$$

$R_1 \leftrightarrow R_2$ Interchanging R_1 and R_2 and then multiplying by $\frac{1}{2}$ will avoid fractions and simplify calculations.

$$\sim \begin{bmatrix} 2 & 4 & 6 \\ 2 & -1 & -4 \\ 3 & -1 & -1 \end{bmatrix}$$

$\frac{1}{2}R_1 \rightarrow R_1$

$$\sim \begin{bmatrix} 1 & 2 & 3 \\ 2 & -1 & -4 \\ 3 & -1 & -1 \end{bmatrix}$$

$R_2 + (-2)R_1 \rightarrow R_2$
$R_3 + (-3)R_1 \rightarrow R_3$

$$\sim \begin{bmatrix} 1 & 2 & 3 \\ 0 & -5 & -10 \\ 0 & -7 & -10 \end{bmatrix}$$

$-\frac{1}{5}R_2 \rightarrow R_2$

$$\sim \begin{bmatrix} 1 & 2 & 3 \\ 0 & 1 & 2 \\ 0 & -7 & -10 \end{bmatrix}$$

$R_3 + 7R_2 \rightarrow R_3$

$$\sim \begin{bmatrix} 1 & 2 & 3 \\ 0 & 1 & 2 \\ 0 & 0 & 4 \end{bmatrix}$$

We stop the Gauss–Jordan elimination, even though the matrix is not in a reduced form, since the last row produces a contradiction.

The last row implies $0 = 4$, which is a contradiction; therefore, the system has no solution.   ◆

P R O B L E M   17   Solve by Gauss–Jordan elimination:

$$3x_1 + x_2 = 5$$
$$2x_1 + 3x_2 = 1$$
$$2x_1 - 2x_2 = 6$$

◆

◆ APPLICATION

◆ E X A M P L E   18

Production Scheduling

A casting company produces three different bronze sculptures. The casting department has available a maximum of 350 labor-hours per week, and the finishing department has a maximum of 150 labor-hours available per week. Sculpture A requires 30 hours for casting and 10 hours for finishing; sculpture B requires 10 hours for casting and 10 hours for finishing; and sculpture C requires 10 hours for casting and 30 hours for finishing. If the plant is to operate at maximum capacity, how many of each sculpture should be produced each week?

Solution First, we summarize the relevant manufacturing data in a table:

	LABOR-HOURS PER SCULPTURE			MAXIMUM LABOR-HOURS AVAILABLE PER WEEK
	A	B	C	
CASTING DEPARTMENT	30	10	10	350
FINISHING DEPARTMENT	10	10	30	150

Let

x_1 = Number of sculpture A produced per week

x_2 = Number of sculpture B produced per week

x_3 = Number of sculpture C produced per week

Then

$$30x_1 + 10x_2 + 10x_3 = 350 \quad \text{Casting department}$$
$$10x_1 + 10x_2 + 30x_3 = 150 \quad \text{Finishing department}$$

Now we can form the augmented matrix of the system and solve by using Gauss–Jordan elimination:

$$\begin{bmatrix} 30 & 10 & 10 & | & 350 \\ 10 & 10 & 30 & | & 150 \end{bmatrix} \quad \begin{matrix} \frac{1}{10}R_1 \to R_1 \\ \frac{1}{10}R_2 \to R_2 \end{matrix} \quad \text{Simplify each row.}$$

$$\sim \begin{bmatrix} 3 & 1 & 1 & | & 35 \\ 1 & 1 & 3 & | & 15 \end{bmatrix} \quad R_1 \leftrightarrow R_2$$

$$\sim \begin{bmatrix} 1 & 1 & 3 & | & 15 \\ 3 & 1 & 1 & | & 35 \end{bmatrix} \quad R_2 + (-3)R_1 \to R_2$$

$$\sim \begin{bmatrix} 1 & 1 & 3 & | & 15 \\ 0 & -2 & -8 & | & -10 \end{bmatrix} \quad -\frac{1}{2}R_2 \to R_2$$

$$\sim \begin{bmatrix} 1 & 1 & 3 & | & 15 \\ 0 & 1 & 4 & | & 5 \end{bmatrix} \quad R_1 + (-1)R_2 \to R_1$$

$$\sim \begin{bmatrix} 1 & 0 & -1 & | & 10 \\ 0 & 1 & 4 & | & 5 \end{bmatrix} \quad \text{Matrix is in reduced form.}$$

$$\begin{matrix} x_1 & - & x_3 = 10 \\ & x_2 + 4x_3 = & 5 \end{matrix} \quad \text{or} \quad \begin{matrix} x_1 = & x_3 + 10 \\ x_2 = -4x_3 + & 5 \end{matrix}$$

Let $x_3 = t$. Then for t any real number,

$$x_1 = t + 10$$
$$x_2 = -4t + 5$$
$$x_3 = t$$

is a solution—or is it? We cannot produce a negative number of sculptures. If we also assume that we cannot produce a fractional number of sculptures, then t

must be a nonnegative whole number. And because of the middle equation $(x_2 = -4t + 5)$, t can only assume the values 0 and 1. Thus, for $t = 0$, we have $x_1 = 10$, $x_2 = 5$, $x_3 = 0$; and for $t = 1$, we have $x_1 = 11$, $x_2 = 1$, $x_3 = 1$. These are the only possible production schedules that utilize the full capacity of the plant.

◆

PROBLEM 18 Repeat Example 18 given a casting capacity of 400 labor-hours per week and a finishing capacity of 200 labor-hours per week.

◆

Answers to Matched Problems

13. (A) Condition 2 is violated: The 3 in the second row should be a 1.
   (B) Condition 3 is violated: In the second column, the 5 should be a 0.
   (C) Condition 4 is violated: The leftmost 1 in the second row is not to the right of the leftmost 1 in the first row.
   (D) Condition 1 is violated: The all-zero second row should be at the bottom.

14. (A)
$$\begin{aligned} x_1 &= -5 \\ x_2 &= 3 \\ x_3 &= 6 \end{aligned}$$

Solution:
$$x_1 = -5,\ x_2 = 3,\ x_3 = 6$$

(B)
$$\begin{aligned} x_1 + 2x_2 - 3x_3 &= 0 \\ 0x_1 + 0x_2 + 0x_3 &= 1 \\ 0x_1 + 0x_2 + 0x_3 &= 0 \end{aligned}$$

Inconsistent; no solution

(C)
$$\begin{aligned} x_1 - 2x_3 &= 4 \\ x_2 + 3x_3 &= -2 \end{aligned}$$

Dependent: Let $x_3 = t$. Then for any real t,
$$\begin{aligned} x_1 &= 2t + 4 \\ x_2 &= -3t - 2 \\ x_3 &= t \end{aligned}$$

is a solution.

(D)
$$\begin{aligned} x_1 + 3x_3 + 2x_4 &= 5 \\ x_2 - 2x_3 - x_4 &= 3 \end{aligned}$$

Dependent: Let $x_3 = s$ and $x_4 = t$. Then for any real s and t,
$$\begin{aligned} x_1 &= -3s - 2t + 5 \\ x_2 &= 2s + t + 3 \\ x_3 &= s \\ x_4 &= t \end{aligned}$$

is a solution.

15. $x_1 = 1$, $x_2 = -1$, $x_3 = 0$
16. $x_1 = -\frac{3}{5}t - \frac{4}{15}$, $x_2 = \frac{4}{5}t + \frac{7}{15}$, $x_3 = t$, t any real number
17. $x_1 = 2$, $x_2 = -1$
18. $x_1 = t + 10$, $x_2 = -4t + 10$, $x_3 = t$, where $t = 0$, 1, 2; that is, $(x_1, x_2, x_3) = (10, 10, 0)$, (11, 6, 1), or (12, 2, 2)

EXERCISE 5-3

A *Indicate whether each matrix is in reduced form.*

1. $\begin{bmatrix} 1 & 0 & | & 2 \\ 0 & 1 & | & -1 \end{bmatrix}$

2. $\begin{bmatrix} 0 & 1 & | & 2 \\ 1 & 0 & | & -1 \end{bmatrix}$

3. $\begin{bmatrix} 1 & 0 & 2 & | & 3 \\ 0 & 0 & 0 & | & 0 \\ 0 & 1 & -1 & | & 4 \end{bmatrix}$

4. $\begin{bmatrix} 1 & 0 & 0 & | & -2 \\ 0 & 1 & 0 & | & 0 \\ 0 & 0 & 1 & | & 1 \end{bmatrix}$

5. $\begin{bmatrix} 0 & 1 & 0 & | & 2 \\ 0 & 0 & 3 & | & -1 \\ 0 & 0 & 0 & | & 0 \end{bmatrix}$

6. $\begin{bmatrix} 1 & 3 & 0 & | & 0 \\ 0 & 0 & 1 & | & 0 \\ 0 & 0 & 0 & | & 1 \end{bmatrix}$

7. $\begin{bmatrix} 1 & 2 & 0 & 3 & | & 2 \\ 0 & 0 & 1 & -1 & | & 0 \end{bmatrix}$

8. $\begin{bmatrix} 0 & 1 & 2 & | & 1 \\ 1 & 0 & -3 & | & 2 \end{bmatrix}$

Write the linear system corresponding to each reduced augmented matrix and solve.

9. $\begin{bmatrix} 1 & 0 & 0 & | & -2 \\ 0 & 1 & 0 & | & 3 \\ 0 & 0 & 1 & | & 0 \end{bmatrix}$

10. $\begin{bmatrix} 1 & 0 & 0 & 0 & | & -2 \\ 0 & 1 & 0 & 0 & | & 0 \\ 0 & 0 & 1 & 0 & | & 1 \\ 0 & 0 & 0 & 1 & | & 3 \end{bmatrix}$

11. $\begin{bmatrix} 1 & 0 & -2 & | & 3 \\ 0 & 1 & 1 & | & -5 \\ 0 & 0 & 0 & | & 0 \end{bmatrix}$

12. $\begin{bmatrix} 1 & -2 & 0 & | & -3 \\ 0 & 0 & 1 & | & 5 \\ 0 & 0 & 0 & | & 0 \end{bmatrix}$

13. $\begin{bmatrix} 1 & 0 & | & 0 \\ 0 & 1 & | & 0 \\ 0 & 0 & | & 1 \end{bmatrix}$

14. $\begin{bmatrix} 1 & 0 & | & 5 \\ 0 & 1 & | & -3 \\ 0 & 0 & | & 0 \end{bmatrix}$

15. $\begin{bmatrix} 1 & -2 & 0 & -3 & | & -5 \\ 0 & 0 & 1 & 3 & | & 2 \end{bmatrix}$

16. $\begin{bmatrix} 1 & 0 & -2 & 3 & | & 4 \\ 0 & 1 & -1 & 2 & | & -1 \end{bmatrix}$

B Use row operations to change each matrix to reduced form.

17. $\begin{bmatrix} 1 & 2 & | & -1 \\ 0 & 1 & | & 3 \end{bmatrix}$

18. $\begin{bmatrix} 1 & 3 & | & 1 \\ 0 & 2 & | & -4 \end{bmatrix}$

19. $\begin{bmatrix} 1 & 0 & -3 & | & 1 \\ 0 & 1 & 2 & | & 0 \\ 0 & 0 & 3 & | & -6 \end{bmatrix}$

20. $\begin{bmatrix} 1 & 0 & 4 & | & 0 \\ 0 & 1 & -3 & | & -1 \\ 0 & 0 & -2 & | & 2 \end{bmatrix}$

21. $\begin{bmatrix} 1 & 2 & -2 & | & -1 \\ 0 & 3 & -6 & | & 1 \\ 0 & -1 & 2 & | & -\frac{1}{3} \end{bmatrix}$

22. $\begin{bmatrix} 0 & -2 & 8 & | & 1 \\ 2 & -2 & 6 & | & -4 \\ 0 & -1 & 4 & | & \frac{1}{2} \end{bmatrix}$

Solve using Gauss–Jordan elimination.

23. $\begin{aligned} 2x_1 + 4x_2 - 10x_3 &= -2 \\ 3x_1 + 9x_2 - 21x_3 &= 0 \\ x_1 + 5x_2 - 12x_3 &= 1 \end{aligned}$

24. $\begin{aligned} 3x_1 + 5x_2 - x_3 &= -7 \\ x_1 + x_2 + x_3 &= -1 \\ 2x_1 + 11x_3 &= 7 \end{aligned}$

25. $\begin{aligned} 3x_1 + 8x_2 - x_3 &= -18 \\ 2x_1 + x_2 + 5x_3 &= 8 \\ 2x_1 + 4x_2 + 2x_3 &= -4 \end{aligned}$

26. $\begin{aligned} 2x_1 + 7x_2 + 15x_3 &= -12 \\ 4x_1 + 7x_2 + 13x_3 &= -10 \\ 3x_1 + 6x_2 + 12x_3 &= -9 \end{aligned}$

27. $\begin{aligned} 2x_1 - x_2 - 3x_3 &= 8 \\ x_1 - 2x_2 &= 7 \end{aligned}$

28. $\begin{aligned} 2x_1 + 4x_2 - 6x_3 &= 10 \\ 3x_1 + 3x_2 - 3x_3 &= 6 \end{aligned}$

29. $\begin{aligned} 2x_1 + 3x_2 - x_3 &= 1 \\ x_1 - 2x_2 + 2x_3 &= -2 \end{aligned}$

30. $\begin{aligned} x_1 - 3x_2 + 2x_3 &= -1 \\ 3x_1 + 2x_2 - x_3 &= 2 \end{aligned}$

31. $\begin{aligned} 2x_1 + 2x_2 &= 2 \\ x_1 + 2x_2 &= 3 \\ -3x_2 &= -6 \end{aligned}$

32. $\begin{aligned} 2x_1 - x_2 &= 0 \\ 3x_1 + 2x_2 &= 7 \\ x_1 - x_2 &= -1 \end{aligned}$

33. $\begin{aligned} 2x_1 - x_2 &= 0 \\ 3x_1 + 2x_2 &= 7 \\ x_1 - x_2 &= -2 \end{aligned}$

34. $\begin{aligned} x_1 - 3x_2 &= 5 \\ 2x_1 + x_2 &= 3 \\ x_1 - 2x_2 &= 5 \end{aligned}$

35. $\begin{aligned} 3x_1 - 4x_2 - x_3 &= 1 \\ 2x_1 - 3x_2 + x_3 &= 1 \\ x_1 - 2x_2 + 3x_3 &= 2 \end{aligned}$

36. $\begin{aligned} 3x_1 + 7x_2 - x_3 &= 11 \\ x_1 + 2x_2 - x_3 &= 3 \\ 2x_1 + 4x_2 - 2x_3 &= 10 \end{aligned}$

37. $\begin{aligned} 3x_1 - 2x_2 + x_3 &= -7 \\ 2x_1 + x_2 - 4x_3 &= 0 \\ x_1 + x_2 - 3x_3 &= 1 \end{aligned}$

38. $\begin{aligned} 2x_1 + 3x_2 + 5x_3 &= 21 \\ x_1 - x_2 - 5x_3 &= -2 \\ 2x_1 + x_2 - x_3 &= 11 \end{aligned}$

39. $\begin{aligned} 2x_1 + 4x_2 - 2x_3 &= 2 \\ -3x_1 - 6x_2 + 3x_3 &= -3 \end{aligned}$

40. $\begin{aligned} 3x_1 - 9x_2 + 12x_3 &= 6 \\ -2x_1 + 6x_2 - 8x_3 &= -4 \end{aligned}$

C Solve using Gauss–Jordan elimination.

41. $\begin{aligned} 2x_1 - 3x_2 + 3x_3 &= -15 \\ 3x_1 + 2x_2 - 5x_3 &= 19 \\ 5x_1 - 4x_2 - 2x_3 &= -2 \end{aligned}$

42. $\begin{aligned} 3x_1 - 2x_2 - 4x_3 &= -8 \\ 4x_1 + 3x_2 - 5x_3 &= -5 \\ 6x_1 - 5x_2 + 2x_3 &= -17 \end{aligned}$

43. $\begin{aligned} 5x_1 - 3x_2 + 2x_3 &= 13 \\ 2x_1 + 4x_2 - 3x_3 &= -9 \\ 4x_1 - 2x_2 + 5x_3 &= 13 \end{aligned}$

44. $\begin{aligned} 4x_1 - 2x_2 + 3x_3 &= 0 \\ 3x_1 - 5x_2 - 2x_3 &= -12 \\ 2x_1 + 4x_2 - 3x_3 &= -4 \end{aligned}$

45. $\begin{aligned} x_1 + 2x_2 - 4x_3 - x_4 &= 7 \\ 2x_1 + 5x_2 - 9x_3 - 4x_4 &= 16 \\ x_1 + 5x_2 - 7x_3 - 7x_4 &= 13 \end{aligned}$

46. $\begin{aligned} 2x_1 + 4x_2 + 5x_3 + 4x_4 &= 8 \\ x_1 + 2x_2 + 2x_3 + x_4 &= 3 \end{aligned}$

APPLICATIONS

Solve all the following problems using Gauss–Jordan elimination.

Business & Economics

47. *Production scheduling.* A small manufacturing plant makes three types of inflatable boats: one-person, two-person, and four-person models. Each boat requires the services of three departments, as listed in the table. The cutting, assembly, and packaging departments have available a maximum of 380, 330, and 120 labor-hours per week, respectively. How many boats of each type must be produced each week for the plant to operate at full capacity?

	ONE-PERSON BOAT	TWO-PERSON BOAT	FOUR-PERSON BOAT
CUTTING DEPARTMENT	0.5 hr	1.0 hr	1.5 hr
ASSEMBLY DEPARTMENT	0.6 hr	0.9 hr	1.2 hr
PACKAGING DEPARTMENT	0.2 hr	0.3 hr	0.5 hr

48. *Production scheduling.* Repeat Problem 47 assuming the cutting, assembly, and packaging departments have available a maximum of 350, 330, and 115 labor-hours per week, respectively.

49. *Production scheduling.* Work Problem 47 assuming the packaging department is no longer used.

50. *Production scheduling.* Work Problem 48 assuming the packaging department is no longer used.

51. *Production scheduling.* Work Problem 47 assuming the four-person boat is no longer produced.

52. *Production scheduling.* Work Problem 48 assuming the four-person boat is no longer produced.

53. *Income tax.* A company has a taxable income of $1,664,000. The federal income tax is 25% of the portion of the income that remains after the state and local taxes have been deducted. The state income tax is 10% of the portion of the income that remains after the federal and local taxes have been deducted, and the local tax is 5% of the portion of the income that remains after the federal and state taxes have been deducted. Find the company's federal, state, and local income taxes.

54. *Income tax.* Repeat Problem 53 if local taxes are not deducted before computing the federal and state taxes.

| UNITS PER OUNCE | | |
	Food A	Food B	Food C
CALCIUM	30	10	20
IRON	10	10	20
VITAMIN A	10	30	20

55. *Nutrition.* A dietitian in a hospital is to arrange a special diet composed of three basic foods. The diet is to include exactly 340 units of calcium, 180 units of iron, and 220 units of vitamin A. The number of units per ounce of each special ingredient for each of the foods is indicated in the table. How many ounces of each food must be used to meet the diet requirements?

56. *Nutrition.* Repeat Problem 55 if the diet is to include exactly 400 units of calcium, 160 units of iron, and 240 units of vitamin A.

57. *Nutrition.* Solve Problem 55 with the assumption that food C is no longer available.

58. *Nutrition.* Solve Problem 56 with the assumption that food C is no longer available.

59. *Nutrition.* Solve Problem 55 with the assumption that the vitamin A requirement is deleted.

60. *Nutrition.* Solve Problem 56 with the assumption that the vitamin A requirement is deleted.

61. *Nutrition — plants.* A farmer can buy four types of plant food. Each barrel of mix A contains 30 pounds of phosphoric acid, 50 pounds of nitrogen, and 30 pounds of potash; each barrel of mix B contains 30 pounds of phosphoric acid, 75 pounds of nitrogen, and 20 pounds of potash; each barrel of mix C contains 30 pounds of phosphoric acid, 25 pounds of nitrogen, and 20 pounds of potash; and each barrel of mix D contains 60 pounds of phosphoric acid, 25 pounds of nitrogen, and 50 pounds of potash. Soil tests indicate that a particular field needs 900 pounds of phosphoric acid, 750 pounds of nitrogen, and 700 pounds of potash. How many barrels of each type of food should the farmer mix together to supply the necessary nutrients for the field?

62. *Nutrition — animals.* In a laboratory experiment, rats are to be fed 5 packets of food containing a total of 80 units of vitamin E. There are four different brands of food packets that can be used. A packet of brand *A* contains 5 units of vitamin E, a packet of brand *B* contains 10 units of vitamin E, a packet of brand *C* contains 15 units of vitamin E, and a packet of brand *D* contains 20 units of vitamin E. How many packets of each brand should be mixed and fed to the rats?

Social Sciences

63. *Sociology.* Two sociologists have grant money to study school busing in a particular city. They wish to conduct an opinion survey using 600 telephone contacts and 400 house contacts. Survey company *A* has personnel to do 30 telephone and 10 house contacts per hour; survey company *B* can handle 20 telephone and 20 house contacts per hour. How many hours should be scheduled for each firm to produce exactly the number of contacts needed?

64. *Sociology.* Repeat Problem 63 if 650 telephone contacts and 350 house contacts are needed.

SECTION 5-4

Matrices — Addition and Multiplication by a Number

♦ DIMENSIONS OF A MATRIX
♦ ADDITION AND SUBTRACTION
♦ PRODUCT OF A NUMBER k AND A MATRIX M
♦ APPLICATION

In the last two sections we introduced the important new idea of matrices. In this and the following sections, we shall develop this concept further. Matrices are both a very ancient and a very current mathematical concept. References to matrices and systems of equations can be found in Chinese manuscripts dating back to around 200 B.C. More recently, the advent of personal and large-scale computers has made matrices a very useful tool for a wide variety of applications.

♦ DIMENSIONS OF A MATRIX

Recall that we defined a **matrix** as any rectangular array of real numbers enclosed within brackets and that each number in the array is called an **element** of the matrix. The **size** or **dimension of a matrix** is important to operations on matrices. We define an **$m \times n$ matrix** (read "m by n matrix") to be one with m rows and n columns. It is important to note that the number of rows is always given first. If a matrix has the same number of rows and columns, it is called a

square matrix. A matrix with only one column is called a **column matrix,** and a matrix with only one row is called a **row matrix.** These definitions are illustrated by the following:

$$
\begin{array}{c}
3 \times 2 \\
\begin{bmatrix} -2 & 5 \\ 0 & -2 \\ 3 & 6 \end{bmatrix}
\end{array}
\qquad
\begin{array}{c}
3 \times 3 \\
\begin{bmatrix} 0.5 & 0.2 & 1.0 \\ 0.0 & 0.3 & 0.5 \\ 0.7 & 0.0 & 0.2 \end{bmatrix} \\
\text{Square matrix}
\end{array}
\qquad
\begin{array}{c}
4 \times 1 \\
\begin{bmatrix} 3 \\ -2 \\ 1 \\ 0 \end{bmatrix} \\
\text{Column matrix}
\end{array}
\qquad
\begin{array}{c}
1 \times 4 \\
[2 \quad \tfrac{1}{2} \quad 0 \quad -\tfrac{2}{3}] \\
\text{Row matrix}
\end{array}
$$

Two matrices are **equal** if they have the same dimension and their corresponding elements are equal. For example,

$$
\begin{array}{c}
2 \times 3 \\
\begin{bmatrix} a & b & c \\ d & e & f \end{bmatrix}
\end{array}
=
\begin{array}{c}
2 \times 3 \\
\begin{bmatrix} u & v & w \\ x & y & z \end{bmatrix}
\end{array}
\qquad \text{if and only if} \qquad
\begin{array}{l}
a = u \quad b = v \quad c = w \\
d = x \quad e = y \quad f = z
\end{array}
$$

◆ ADDITION AND SUBTRACTION

The **sum of two matrices of the same dimension** is the matrix with elements that are the sum of the corresponding elements of the two given matrices.

Addition is not defined for matrices with different dimensions.

◆ **EXAMPLE   19**

(A) $\begin{bmatrix} a & b \\ c & d \end{bmatrix} + \begin{bmatrix} w & x \\ y & z \end{bmatrix} = \begin{bmatrix} (a+w) & (b+x) \\ (c+y) & (d+z) \end{bmatrix}$

(B) $\begin{bmatrix} 2 & -3 & 0 \\ 1 & 2 & -5 \end{bmatrix} + \begin{bmatrix} 3 & 1 & 2 \\ -3 & 2 & 5 \end{bmatrix} = \begin{bmatrix} 5 & -2 & 2 \\ -2 & 4 & 0 \end{bmatrix}$ ◆

PROBLEM   19   Add: $\begin{bmatrix} 3 & 2 \\ -1 & -1 \\ 0 & 3 \end{bmatrix} + \begin{bmatrix} -2 & 3 \\ 1 & -1 \\ 2 & -2 \end{bmatrix}$ ◆

Because we add two matrices by adding their corresponding elements, it follows from the properties of real numbers that matrices of the same dimension are commutative and associative relative to addition. That is, if A, B, and C are matrices of the same dimension, then

COMMUTATIVE $A + B = B + A$
ASSOCIATIVE $(A + B) + C = A + (B + C)$

A matrix with elements that are all zeros is called a **zero matrix.** For example,

$$
[0 \quad 0 \quad 0] \qquad
\begin{bmatrix} 0 & 0 \\ 0 & 0 \end{bmatrix} \qquad
\begin{bmatrix} 0 \\ 0 \\ 0 \\ 0 \end{bmatrix} \qquad
\begin{bmatrix} 0 & 0 & 0 & 0 \\ 0 & 0 & 0 & 0 \\ 0 & 0 & 0 & 0 \end{bmatrix}
$$

are zero matrices of different dimensions. [*Note:* "0" is often used to denote the zero matrix of an arbitrary dimension.] The **negative of a matrix M,** denoted by $-M$, is a matrix with elements that are the negatives of the elements in M. Thus, if

$$M = \begin{bmatrix} a & b \\ c & d \end{bmatrix} \qquad \text{then} \qquad -M = \begin{bmatrix} -a & -b \\ -c & -d \end{bmatrix}$$

Note that $M + (-M) = 0$ (a zero matrix).

If A and B are matrices of the same dimension, then we define **subtraction** as follows:

$$A - B = A + (-B)$$

Thus, to subtract matrix B from matrix A, we simply add the negative of B to A.

◆ E X A M P L E 20 $\begin{bmatrix} 3 & -2 \\ 5 & 0 \end{bmatrix} - \begin{bmatrix} -2 & 2 \\ 3 & 4 \end{bmatrix} = \begin{bmatrix} 3 & -2 \\ 5 & 0 \end{bmatrix} + \begin{bmatrix} 2 & -2 \\ -3 & -4 \end{bmatrix} = \begin{bmatrix} 5 & -4 \\ 2 & -4 \end{bmatrix}$ ◆

P R O B L E M 20 Subtract: $[2 \quad -3 \quad 5] - [3 \quad -2 \quad 1]$ ◆

◆ PRODUCT OF A NUMBER k AND A MATRIX M

Finally, the **product of a number k and a matrix M,** denoted by kM, is a matrix formed by multiplying each element of M by k. This definition is partly motivated by the fact that if M is a matrix, then we would like $M + M$ to equal $2M$.

◆ E X A M P L E 21 $-2 \begin{bmatrix} 3 & -1 & 0 \\ -2 & 1 & 3 \\ 0 & -1 & -2 \end{bmatrix} = \begin{bmatrix} -6 & 2 & 0 \\ 4 & -2 & -6 \\ 0 & 2 & 4 \end{bmatrix}$ ◆

P R O B L E M 21 Find: $10 \begin{bmatrix} 1.3 \\ 0.2 \\ 3.5 \end{bmatrix}$ ◆

◆ APPLICATION

◆ E X A M P L E 22

Sales Commissions

Ms. Smith and Mr. Jones are salespeople in a new-car agency that sells only two models. August was the last month for this year's models, and next year's models were introduced in September. Gross dollar sales for each month are given in the following matrices:

	August sales		September sales	
	Compact	Luxury	Compact	Luxury
Ms. Smith	$18,000	$36,000	$72,000	$144,000
Mr. Jones	$36,000	0	$90,000	$108,000

$\begin{bmatrix} \$18{,}000 & \$36{,}000 \\ \$36{,}000 & 0 \end{bmatrix} = A$ $\begin{bmatrix} \$72{,}000 & \$144{,}000 \\ \$90{,}000 & \$108{,}000 \end{bmatrix} = B$

(For example, Ms. Smith had $18,000 in compact sales in August, and Mr. Jones had $108,000 in luxury car sales in September.)

(A) What were the combined dollar sales in August and September for each person and each model?
(B) What was the increase in dollar sales from August to September?
(C) If both salespeople receive 5% commissions on gross dollar sales, compute the commission for each person for each model sold in September.

Solutions

$$\text{(A) } A + B = \begin{matrix} & \text{Compact} & \text{Luxury} & \\ & \left[\begin{matrix} \$90{,}000 \\ \$126{,}000 \end{matrix}\right. & \left.\begin{matrix} \$180{,}000 \\ \$108{,}000 \end{matrix}\right] & \begin{matrix} \text{Ms. Smith} \\ \text{Mr. Jones} \end{matrix} \end{matrix}$$

$$\text{(B) } B - A = \begin{matrix} & \text{Compact} & \text{Luxury} & \\ & \left[\begin{matrix} \$54{,}000 \\ \$54{,}000 \end{matrix}\right. & \left.\begin{matrix} \$108{,}000 \\ \$108{,}000 \end{matrix}\right] & \begin{matrix} \text{Ms. Smith} \\ \text{Mr. Jones} \end{matrix} \end{matrix}$$

$$\text{(C) } 0.05B = \begin{bmatrix} (0.05)(\$72{,}000) & (0.05)(\$144{,}000) \\ (0.05)(\$90{,}000) & (0.05)(\$108{,}000) \end{bmatrix}$$

$$= \begin{bmatrix} \$3{,}600 & \$7{,}200 \\ \$4{,}500 & \$5{,}400 \end{bmatrix} \begin{matrix} \text{Ms. Smith} \\ \text{Mr. Jones} \end{matrix}$$

◆

PROBLEM 22 Repeat Example 22 with

$$A = \begin{bmatrix} \$36{,}000 & \$36{,}000 \\ \$18{,}000 & \$36{,}000 \end{bmatrix} \quad \text{and} \quad B = \begin{bmatrix} \$90{,}000 & \$108{,}000 \\ \$72{,}000 & \$108{,}000 \end{bmatrix}$$

◆

Example 22 involved an agency with only 2 salespeople and 2 models. A more realistic problem might involve 20 salespeople and 15 models. Problems of this size are often solved with the aid of an electronic spreadsheet program on a personal computer. Figure 2 illustrates such a computer solution for Example 22.

	1	2	3	4	5	6	7
1		August Sales		September Sales		September Commissions	
2		Compact	Luxury	Compact	Luxury	Compact	Luxury
3	Smith	$18,000	$36,000	$72,000	$144,000	$3,600	$7,200
4	Jones	$36,000	$0	$90,000	$108,000	$4,500	$5,400
5		Combined Sales		Sales Increase			
6	Smith	$90,000	$180,000	$54,000	$108,000		
7	Jones	$126,000	$108,000	$54,000	$108,000		

FIGURE 2

Answers to Matched Problems

19. $\begin{bmatrix} 1 & 5 \\ 0 & -2 \\ 2 & 1 \end{bmatrix}$ 20. $[-1 \quad -1 \quad 4]$ 21. $\begin{bmatrix} 13 \\ 2 \\ 35 \end{bmatrix}$

22. (A) $\begin{bmatrix} \$126{,}000 & \$144{,}000 \\ \$90{,}000 & \$144{,}000 \end{bmatrix}$ (B) $\begin{bmatrix} \$54{,}000 & \$72{,}000 \\ \$54{,}000 & \$72{,}000 \end{bmatrix}$

(C) $\begin{bmatrix} \$4{,}500 & \$5{,}400 \\ \$3{,}600 & \$5{,}400 \end{bmatrix}$

A Problems 1–20 refer to the following matrices:

$$A = \begin{bmatrix} 2 & -1 \\ 3 & 0 \end{bmatrix} \qquad B = \begin{bmatrix} -3 & 1 \\ 2 & -3 \end{bmatrix} \qquad C = \begin{bmatrix} 2 \\ -3 \\ 0 \end{bmatrix}$$

$$D = \begin{bmatrix} 1 \\ 3 \\ 5 \end{bmatrix} \qquad E = \begin{bmatrix} -4 & 1 & 0 & -2 \end{bmatrix} \qquad F = \begin{bmatrix} 2 & -3 \\ -2 & 0 \\ 1 & 2 \\ 3 & 5 \end{bmatrix}$$

1. What are the dimensions of B? Of E?
2. What are the dimensions of F? Of D?
3. What element is in the third row and second column of matrix F?
4. What element is in the second row and first column of matrix F?
5. Write a zero matrix of the same dimension as B.
6. Write a zero matrix of the same dimension as E.
7. Identify all column matrices.
8. Identify all row matrices.
9. Identify all square matrices.
10. How many additional columns would F have to have to be a square matrix?
11. Find $A + B$.   12. Find $C + D$.   13. Find $E + F$.   14. Find $B + C$.
15. Write the negative of matrix C.
16. Write the negative of matrix B.
17. Find $D - C$.   18. Find $A - A$.   19. Find $5B$.   20. Find $-2E$.

B In Problems 21–28, perform the indicated operations.

21. $\begin{bmatrix} 3 & -2 & 0 & 1 \\ 2 & -3 & -1 & 4 \\ 0 & 2 & -1 & 6 \end{bmatrix} + \begin{bmatrix} -2 & 5 & -1 & 0 \\ -3 & -2 & 8 & -2 \\ 4 & 6 & 1 & -8 \end{bmatrix}$

22. $\begin{bmatrix} 4 & -2 & 8 \\ 0 & -1 & -4 \\ -6 & 5 & 2 \\ 1 & 3 & -6 \end{bmatrix} + \begin{bmatrix} -6 & -2 & -3 \\ 5 & 2 & 4 \\ 8 & 3 & -4 \\ 1 & -5 & 0 \end{bmatrix}$

23. $\begin{bmatrix} 1.3 & 2.5 & -6.1 \\ 8.3 & -1.4 & 6.7 \end{bmatrix} - \begin{bmatrix} -4.1 & 1.8 & -4.3 \\ 0.7 & 2.6 & -1.2 \end{bmatrix}$

24. $\begin{bmatrix} 2.6 & 3.8 \\ -1.9 & 7.3 \\ 5.6 & -0.4 \end{bmatrix} - \begin{bmatrix} 4.8 & -2.1 \\ 3.2 & 5.9 \\ -1.5 & 2.2 \end{bmatrix}$

25. $1,000 \begin{bmatrix} 0.25 & 0.36 \\ 0.04 & 0.35 \end{bmatrix}$

26. $100 \begin{bmatrix} 0.32 & 0.05 & 0.17 \\ 0.22 & 0.03 & 0.21 \end{bmatrix}$

27. $0.08 \begin{bmatrix} 24{,}000 & 35{,}000 \\ 12{,}000 & 24{,}000 \end{bmatrix} + 0.03 \begin{bmatrix} 12{,}000 & 22{,}000 \\ 14{,}000 & 13{,}000 \end{bmatrix}$

28. $0.05 \begin{bmatrix} 430 & 212 \\ 210 & 165 \\ 435 & 315 \end{bmatrix} + 0.07 \begin{bmatrix} 234 & 436 \\ 160 & 212 \\ 410 & 136 \end{bmatrix}$

29. Find a, b, c, and d so that

$$\begin{bmatrix} a & b \\ c & d \end{bmatrix} + \begin{bmatrix} 2 & -3 \\ 0 & 1 \end{bmatrix} = \begin{bmatrix} 1 & -2 \\ 3 & -4 \end{bmatrix}$$

30. Find w, x, y, and z so that

$$\begin{bmatrix} 4 & -2 \\ -3 & 0 \end{bmatrix} + \begin{bmatrix} w & x \\ y & z \end{bmatrix} = \begin{bmatrix} 2 & -3 \\ 0 & 5 \end{bmatrix}$$

31. Find x and y so that

$$\begin{bmatrix} 2x & 4 \\ -3 & 5x \end{bmatrix} + \begin{bmatrix} 3y & -2 \\ -2 & -y \end{bmatrix} = \begin{bmatrix} -5 & 2 \\ -5 & 13 \end{bmatrix}$$

32. Find x and y so that

$$\begin{bmatrix} 5 & 3x \\ 2x & -4 \end{bmatrix} + \begin{bmatrix} 1 & -4y \\ 7y & 4 \end{bmatrix} = \begin{bmatrix} 6 & -7 \\ 5 & 0 \end{bmatrix}$$

C *In Problems 33–36, is it possible to find 2×2 matrices A, B, and C and a real number k such that the indicated equation fails to hold? If so, state an example.*

33. $A + B = B + A$

34. $(A + B) + C = A + (B + C)$

35. $k(A + B) = kA + kB$

36. $k(A - B) = kA - kB$

APPLICATIONS

Business & Economics

37. *Cost analysis.* A company with two different plants manufactures guitars and banjos. Its production costs for each instrument are given in the following matrices:

	Plant X			Plant Y	
	Guitar	Banjo		Guitar	Banjo
Materials	$30	$25 $= A$		$36	$27 $= B$
Labor	$60	$80		$54	$74

Find $\frac{1}{2}(A + B)$, the average cost of production for the two plants.

38. *Cost analysis.* If both labor and materials at plant X in Problem 37 are increased by 20%, find $\frac{1}{2}(1.2A + B)$, the new average cost of production for the two plants.

39. *Markup.* An import car dealer sells three models of a car. The retail prices and the current dealer invoice prices (costs) for the basic models and indi-

3,32

cated options are given in the following two matrices (where "Air" means air-conditioning):

Retail price

	Basic car	Air	AM/FM radio	Cruise control	
Model A	$10,900	$683	$253	$195	
Model B	$13,000	$738	$382	$206	= M
Model C	$16,300	$867	$537	$225	

Dealer invoice price

	Basic car	Air	AM/FM radio	Cruise control	
Model A	$9,400	$582	$195	$160	
Model B	$11,500	$621	$295	$171	= N
Model C	$14,100	$737	$420	$184	

We define the markup matrix to be $M - N$ (**markup** is the difference between the retail price and the dealer invoice price). Suppose the value of the dollar has had a sharp decline and the dealer invoice price is to have an across-the-board 15% increase next year. In order to stay competitive with domestic cars, the dealer increases the retail prices only 10%. Calculate a markup matrix for next year's models and the indicated options. (Compute results to the nearest dollar.)

40. *Markup.* Referring to Problem 39, what is the markup matrix resulting from a 20% increase in dealer invoice prices and an increase in retail prices of 15%? (Compute results to the nearest dollar.)

Life Sciences

41. *Heredity.* Gregor Mendel (1822–1884), an Austrian monk and botanist, made discoveries that revolutionized the science of genetics. In one experiment, he crossed dihybrid yellow round peas (yellow and round are dominant characteristics; the peas also contained genes for the recessive characteristics green and wrinkled) and obtained 560 peas of the types indicated in the matrix:

	Round	Wrinkled	
Yellow	319	101	= M
Green	108	32	

Suppose he carried out a second experiment of the same type and obtained 640 peas of the types indicated in this matrix:

	Round	Wrinkled	
Yellow	370	124	= N
Green	110	36	

If the results of the two experiments are combined, write the resulting matrix $M + N$. Compute the percentage of the total number of peas (1,200) in each category of the combined results.

42. *Psychology.* Two psychologists independently carried out studies on the relationship between height and aggressive behavior in women over 18 years of age. The results of the studies are summarized in the following matrices:

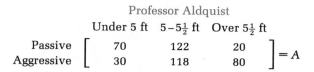

Professor Aldquist

	Under 5 ft	5–5$\frac{1}{2}$ ft	Over 5$\frac{1}{2}$ ft	
Passive	70	122	20	$\Big] = A$
Aggressive	30	118	80	

Professor Kelley

	Under 5 ft	5–5$\frac{1}{2}$ ft	Over 5$\frac{1}{2}$ ft	
Passive	65	160	30	$\Big] = B$
Aggressive	25	140	75	

The two psychologists decided to combine their results and publish a joint paper. Write the matrix $A + B$ illustrating their combined results. Compute the percentage of the total sample in each category of the combined study.

SECTION 5-5 Matrix Multiplication

◆ DOT PRODUCT
◆ MATRIX PRODUCT
◆ MULTIPLICATION PROPERTIES
◆ APPLICATION

In this section, we are going to introduce two types of matrix multiplication that will seem rather strange at first. In spite of this apparent strangeness, these operations are well-founded in the general theory of matrices and, as we will see, are extremely useful in many practical problems.

Historically, matrix multiplication was introduced by the English mathematician Arthur Cayley (1821–1895) in studies of systems of linear equations and linear transformations. In the next section, you will see how matrix multiplication is central to the process of expressing systems of linear equations as matrix equations and to the process of solving matrix equations. Matrix equations and their solutions provide us with an alternate method of solving linear systems with the same number of variables as equations.

◆ DOT PRODUCT

We start by defining the *dot product* of two special matrices.

Dot Product

The **dot product** of a $1 \times n$ row matrix and an $n \times 1$ column matrix is a real number given by

$$\underset{1 \times n}{[a_1 \quad a_2 \quad \cdots \quad a_n]} \cdot \underset{n \times 1}{\begin{bmatrix} b_1 \\ b_2 \\ \cdot \\ \cdot \\ \cdot \\ b_n \end{bmatrix}} = a_1 b_1 + a_2 b_2 + \cdots + a_n b_n \quad \text{A real number}$$

Note that the number of elements in the row matrix and in the column matrix must be the same for the dot product to be defined. Also note that the dot product is a *real number*, not a matrix. The dot between the two matrices is important. If the dot is omitted, the multiplication is of another type, which we will consider below.

◆ E X A M P L E 23

$$[2 \quad -3 \quad 0] \cdot \begin{bmatrix} -5 \\ 2 \\ -2 \end{bmatrix} = (2)(-5) + (-3)(2) + (0)(-2)$$

$$= -10 - 6 + 0 = -16 \qquad ◆$$

P R O B L E M 23

$$[-1 \quad 0 \quad 3 \quad 2] \cdot \begin{bmatrix} 2 \\ 3 \\ 4 \\ -1 \end{bmatrix} = ? \qquad ◆$$

◆ E X A M P L E 24

Labor Costs

A factory produces a slalom water ski that requires 4 labor-hours in the fabricating department and 1 labor-hour in the finishing department. Fabricating personnel receive $8 per hour and finishing personnel receive $6 per hour. Total labor cost per ski is given by the dot product:

$$[4 \quad 1] \cdot \begin{bmatrix} 8 \\ 6 \end{bmatrix} = (4)(8) + (1)(6) = 32 + 6 = \$38 \text{ per ski} \qquad ◆$$

P R O B L E M 24

If the factory in Example 24 also produces a trick water ski that requires 6 labor-hours in the fabricating department and 1.5 labor-hours in the finishing department, write a dot product between appropriate row and column matrices that will give the total labor cost for this ski. Compute the cost. ◆

◆ MATRIX PRODUCT

It is important to remember that the dot product of a row matrix and a column matrix is a real number and not a matrix. We now use the dot product to define a matrix product for certain matrices.

▬ Matrix Product

The **product of two matrices** A and B is defined only on the assumption that the number of columns in A is equal to the number of rows in B. If A is an $m \times p$ matrix and B is a $p \times n$ matrix, then the matrix product of A and B, denoted by AB (with *no* dot), is an $m \times n$ matrix whose element in the ith row and jth column is the dot product of the ith row matrix of A and the jth column matrix of B.

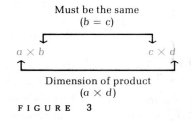

Must be the same
$(b = c)$

$a \times b \qquad c \times d$

Dimension of product
$(a \times d)$

F I G U R E   3

It is important to check dimensions before starting the multiplication process. If matrix A has dimension $a \times b$ and matrix B has dimension $c \times d$, then if $b = c$, the product AB will exist and will have dimension $a \times d$. This is shown schematically in Figure 3. The definition is not as complicated as it might first seem. An example should help to clarify the process. For

$$A = \begin{bmatrix} 2 & 3 & -1 \\ -2 & 1 & 2 \end{bmatrix} \quad \text{and} \quad B = \begin{bmatrix} 1 & 3 \\ 2 & 0 \\ -1 & 2 \end{bmatrix}$$

A is 2×3, B is 3×2, and so AB is 2×2. The four dot products used to produce the four elements in AB (usually calculated mentally or with the aid of a calculator) are shown in the dashed box below. The shaded portions highlight the steps involved in computing the element in the first row and second column of the product matrix.

$$
\underset{2 \times 3}{\begin{bmatrix} 2 & 3 & -1 \\ -2 & 1 & 2 \end{bmatrix}}
\underset{3 \times 2}{\begin{bmatrix} 1 & 3 \\ 2 & 0 \\ -1 & 2 \end{bmatrix}}
=
\begin{bmatrix}
[2 \ 3 \ -1] \cdot \begin{bmatrix} 1 \\ 2 \\ -1 \end{bmatrix} & [2 \ 3 \ -1] \cdot \begin{bmatrix} 3 \\ 0 \\ 2 \end{bmatrix} \\
[-2 \ 1 \ 2] \cdot \begin{bmatrix} 1 \\ 2 \\ -1 \end{bmatrix} & [-2 \ 1 \ 2] \cdot \begin{bmatrix} 3 \\ 0 \\ 2 \end{bmatrix}
\end{bmatrix}
=
\underset{2 \times 2}{\begin{bmatrix} 9 & 4 \\ -2 & -2 \end{bmatrix}}
$$

◆ E X A M P L E   25 (A) $\underset{3 \times 2}{\begin{bmatrix} 2 & 1 \\ 1 & 0 \\ -1 & 2 \end{bmatrix}} \underset{2 \times 4}{\begin{bmatrix} 1 & -1 & 0 & 1 \\ 2 & 1 & 2 & 0 \end{bmatrix}} = \underset{3 \times 4}{\begin{bmatrix} 4 & -1 & 2 & 2 \\ 1 & -1 & 0 & 1 \\ 3 & 3 & 4 & -1 \end{bmatrix}}$

$$2 \times 4 \qquad\qquad 3 \times 2$$

(B) $\begin{bmatrix} 1 & -1 & 0 & 1 \\ 2 & 1 & 2 & 0 \end{bmatrix} \begin{bmatrix} 2 & 1 \\ 1 & 0 \\ -1 & 2 \end{bmatrix}$ (C) $\begin{bmatrix} 2 & 6 \\ -1 & -3 \end{bmatrix} \begin{bmatrix} 1 & 2 \\ 3 & 6 \end{bmatrix} = \begin{bmatrix} 20 & 40 \\ -10 & -20 \end{bmatrix}$

Not defined

(D) $\begin{bmatrix} 1 & 2 \\ 3 & 6 \end{bmatrix} \begin{bmatrix} 2 & 6 \\ -1 & -3 \end{bmatrix} = \begin{bmatrix} 0 & 0 \\ 0 & 0 \end{bmatrix}$ (E) $[2 \;\; -3 \;\; 0] \begin{bmatrix} -5 \\ 2 \\ -2 \end{bmatrix} = [-16]$

(F) $\begin{bmatrix} -5 \\ 2 \\ -2 \end{bmatrix} [2 \;\; -3 \;\; 0] = \begin{bmatrix} -10 & 15 & 0 \\ 4 & -6 & 0 \\ -4 & 6 & 0 \end{bmatrix}$ ◆

PROBLEM   25   Find each product, if it is defined:

(A) $\begin{bmatrix} -1 & 0 & 3 & -2 \\ 1 & 2 & 2 & 0 \end{bmatrix} \begin{bmatrix} -1 & 1 \\ 2 & 3 \\ 1 & 0 \end{bmatrix}$ (B) $\begin{bmatrix} -1 & 1 \\ 2 & 3 \\ 1 & 0 \end{bmatrix} \begin{bmatrix} -1 & 0 & 3 & -2 \\ 1 & 2 & 2 & 0 \end{bmatrix}$

(C) $\begin{bmatrix} 1 & 2 \\ -1 & -2 \end{bmatrix} \begin{bmatrix} -2 & 4 \\ 1 & -2 \end{bmatrix}$ (D) $\begin{bmatrix} -2 & 4 \\ 1 & -2 \end{bmatrix} \begin{bmatrix} 1 & 2 \\ -1 & -2 \end{bmatrix}$

(E) $[3 \;\; -2 \;\; 1] \begin{bmatrix} 4 \\ 2 \\ 3 \end{bmatrix}$ (F) $\begin{bmatrix} 4 \\ 2 \\ 3 \end{bmatrix} [3 \;\; -2 \;\; 1]$ ◆

Notice that the matrix product of a $1 \times n$ row matrix and an $n \times 1$ column matrix is a 1×1 matrix (see Example 25E), whereas their dot product is a real number (see Example 23). (This is a technical distinction, and it is common to see 1×1 matrices written as real numbers.)

◆ MULTIPLICATION PROPERTIES

In the arithmetic of real numbers it does not matter in which order we multiply; for example, $5 \times 7 = 7 \times 5$. In matrix multiplication, however, it does make a difference. That is, AB does not always equal BA, even if both multiplications are defined (see Examples 25C and 25D). Thus, **matrix multiplication is not commutative.** Also, AB may be zero with neither A nor B equal to zero (see Example 25D). Thus, **the zero property does not hold for matrix multiplication** (see Section 1-1 for a discussion of the zero property for real numbers).

Matrix multiplication does have general properties, many of which are similar to the properties of real numbers. Some of these are listed in Theorem 4 (page 242). In the next section we will use these properties to solve matrix equations. Notice that since matrix multiplication is not commutative, there are two distributive properties and two multiplication properties.

Properties of Matrix Multiplication

Assuming all products and sums are defined for the indicated matrices A, B, and C, then for k a real number:

ASSOCIATIVE PROPERTY:	1. $A(BC) = (AB)C$
LEFT DISTRIBUTIVE PROPERTY:	2. $A(B + C) = AB + AC$
RIGHT DISTRIBUTIVE PROPERTY:	3. $(B + C)A = BA + CA$
LEFT MULTIPLICATION PROPERTY:	4. If $A = B$, then $CA = CB$.
RIGHT MULTIPLICATION PROPERTY:	5. If $A = B$, then $AC = BC$.
	6. $k(AB) = (kA)B = A(kB)$

◆ APPLICATION

The next example illustrates the use of the dot and matrix products in a business application.

◆ E X A M P L E 26

Labor Costs

Let us combine the time requirements for slalom and trick water skis discussed in Example 24 and Problem 24 into one matrix:

$$\begin{array}{c} \\ \text{Trick ski} \\ \text{Slalom ski} \end{array} \begin{array}{cc} \text{Fabricating} & \text{Finishing} \\ \text{department} & \text{department} \\ \begin{bmatrix} 6 \text{ hr} & 1.5 \text{ hr} \\ 4 \text{ hr} & 1 \text{ hr} \end{bmatrix} = A \end{array}$$

Now suppose the company has two manufacturing plants, X and Y, in different parts of the country and that their hourly rates for each department are given in the following matrix:

$$\begin{array}{c} \\ \text{Fabricating department} \\ \text{Finishing department} \end{array} \begin{array}{cc} \text{Plant } X & \text{Plant } Y \\ \begin{bmatrix} \$8 & \$7 \\ \$6 & \$4 \end{bmatrix} = B \end{array}$$

To find the total labor costs for each ski at each factory, we multiply A and B:

$$AB = \overset{2 \times 2}{\begin{bmatrix} 6 & 1.5 \\ 4 & 1 \end{bmatrix}} \overset{2 \times 2}{\begin{bmatrix} 8 & 7 \\ 6 & 4 \end{bmatrix}} = \overset{X \quad Y}{\begin{bmatrix} \$57 & \$48 \\ \$38 & \$32 \end{bmatrix}} \begin{array}{l} \text{Trick ski} \\ \text{Slalom ski} \end{array}$$

Notice that the dot product of the first row matrix of A and the first column matrix of B gives us the labor costs, $57, for a trick ski manufactured at plant X. The dot product of the second row matrix of A and the second column matrix of B gives us the labor costs, $32, for manufacturing a slalom ski at plant Y. And so on. ◆

Example 26 is, of course, over simplified. Companies manufacturing many different items in many different plants deal with matrices that have very large numbers of rows and columns.

PROBLEM 26   Repeat Example 26 with

$$A = \begin{bmatrix} 7 \text{ hr} & 2 \text{ hr} \\ 5 \text{ hr} & 1.5 \text{ hr} \end{bmatrix} \quad \text{and} \quad B = \begin{bmatrix} \$10 & \$8 \\ \$6 & \$4 \end{bmatrix}$$ ◆

Answers to Matched Problems

23. 8   24. $[6 \quad 1.5] \cdot \begin{bmatrix} 8 \\ 6 \end{bmatrix} = \$57$

25. (A) Not defined   (B) $\begin{bmatrix} 2 & 2 & -1 & 2 \\ 1 & 6 & 12 & -4 \\ -1 & 0 & 3 & -2 \end{bmatrix}$ (C) $\begin{bmatrix} 0 & 0 \\ 0 & 0 \end{bmatrix}$

(D) $\begin{bmatrix} -6 & -12 \\ 3 & 6 \end{bmatrix}$ (E) $[11]$ (F) $\begin{bmatrix} 12 & -8 & 4 \\ 6 & -4 & 2 \\ 9 & -6 & 3 \end{bmatrix}$

             X   Y

26. $\begin{bmatrix} \$82 & \$64 \\ \$59 & \$46 \end{bmatrix}$ Trick
                        Slalom

EXERCISE 5-5

A   *Find the dot products.*

1. $[2 \quad 4] \cdot \begin{bmatrix} 3 \\ 1 \end{bmatrix}$ 2. $[3 \quad 1] \cdot \begin{bmatrix} 2 \\ 4 \end{bmatrix}$ 3. $[-3 \quad 2] \cdot \begin{bmatrix} -1 \\ -2 \end{bmatrix}$

4. $[3 \quad -2] \cdot \begin{bmatrix} -4 \\ -1 \end{bmatrix}$

Find the matrix products.

5. $[2 \quad 5]\begin{bmatrix} 1 & -1 \\ 2 & 3 \end{bmatrix}$ 6. $[1 \quad 3]\begin{bmatrix} 2 & 3 \\ 1 & -4 \end{bmatrix}$

7. $\begin{bmatrix} 3 & 4 \\ -1 & -2 \end{bmatrix}\begin{bmatrix} -1 \\ 2 \end{bmatrix}$ 8. $\begin{bmatrix} -1 & 1 \\ 2 & -3 \end{bmatrix}\begin{bmatrix} 4 \\ -2 \end{bmatrix}$

9. $\begin{bmatrix} 2 & -3 \\ 1 & 2 \end{bmatrix}\begin{bmatrix} 1 & -1 \\ 0 & -2 \end{bmatrix}$ 10. $\begin{bmatrix} -3 & 2 \\ 4 & -1 \end{bmatrix}\begin{bmatrix} -2 & 5 \\ -1 & 3 \end{bmatrix}$

11. $\begin{bmatrix} 1 & -1 \\ 0 & -2 \end{bmatrix}\begin{bmatrix} 2 & -3 \\ 1 & 2 \end{bmatrix}$ 12. $\begin{bmatrix} -2 & 5 \\ -1 & 3 \end{bmatrix}\begin{bmatrix} -3 & 2 \\ 4 & -1 \end{bmatrix}$

13. $[5 \quad -2]\begin{bmatrix} -3 \\ -4 \end{bmatrix}$ 14. $[-4 \quad 3]\begin{bmatrix} -2 \\ 1 \end{bmatrix}$

15. $\begin{bmatrix} -3 \\ -4 \end{bmatrix}[5 \quad -2]$ 16. $\begin{bmatrix} -2 \\ 1 \end{bmatrix}[-4 \quad 3]$

B *Find the dot products.*

17. $[-1 \quad -2 \quad 2] \cdot \begin{bmatrix} 2 \\ -1 \\ 3 \end{bmatrix}$

18. $[-2 \quad 4 \quad 0] \cdot \begin{bmatrix} -1 \\ -3 \\ 2 \end{bmatrix}$

19. $[-1 \quad -3 \quad 0 \quad 5] \cdot \begin{bmatrix} 4 \\ -3 \\ -1 \\ 2 \end{bmatrix}$

20. $[-1 \quad 2 \quad 3 \quad -2] \cdot \begin{bmatrix} 3 \\ -2 \\ 0 \\ 4 \end{bmatrix}$

Find each matrix product, if it is defined.

21. $\begin{bmatrix} 2 & -1 & 1 \\ 1 & 3 & -2 \end{bmatrix} \begin{bmatrix} 1 & 3 \\ 0 & -1 \\ -2 & 2 \end{bmatrix}$

22. $\begin{bmatrix} -1 & -4 & 3 \\ 2 & 0 & 1 \end{bmatrix} \begin{bmatrix} 2 & -3 \\ 1 & 2 \\ 0 & -1 \end{bmatrix}$

23. $\begin{bmatrix} 1 & 3 \\ 0 & -1 \\ -2 & 2 \end{bmatrix} \begin{bmatrix} 2 & -1 & 1 \\ 1 & 3 & -2 \end{bmatrix}$

24. $\begin{bmatrix} 2 & -3 \\ 1 & 2 \\ 0 & -1 \end{bmatrix} \begin{bmatrix} -1 & -4 & 3 \\ 2 & 0 & 1 \end{bmatrix}$

25. $[3 \quad -2 \quad -4] \begin{bmatrix} 1 \\ 2 \\ -3 \end{bmatrix}$

26. $[1 \quad -2 \quad 2] \begin{bmatrix} 2 \\ -1 \\ 1 \end{bmatrix}$

27. $\begin{bmatrix} 1 \\ 2 \\ -3 \end{bmatrix} [3 \quad -2 \quad -4]$

28. $\begin{bmatrix} 2 \\ -1 \\ 1 \end{bmatrix} [1 \quad -2 \quad 2]$

29. $\begin{bmatrix} 1 & 2 \\ 2 & -1 \\ -3 & 1 \end{bmatrix} [3 \quad -2 \quad -4]$

30. $\begin{bmatrix} 1 \\ 2 \\ -3 \end{bmatrix} \begin{bmatrix} 3 & -2 & 4 \\ 1 & -2 & 2 \end{bmatrix}$

31. $\begin{bmatrix} 1 & 2 & -1 \\ 3 & -1 & 4 \\ 2 & -4 & 5 \end{bmatrix} \begin{bmatrix} 4 \\ 5 \\ 7 \end{bmatrix}$

32. $\begin{bmatrix} 2 & 1 & -1 \\ 1 & 3 & 1 \\ -4 & 2 & 5 \end{bmatrix} \begin{bmatrix} -1 \\ 2 \\ -3 \end{bmatrix}$

33. $\begin{bmatrix} 2 & -1 & 3 & 0 \\ -3 & 4 & 2 & -1 \\ 0 & -2 & 1 & 4 \end{bmatrix} \begin{bmatrix} 2 & -3 & -2 \\ 1 & 0 & 1 \\ -1 & 2 & 0 \\ 2 & -2 & -3 \end{bmatrix}$

34. $\begin{bmatrix} 2 & -3 & -2 \\ 1 & 0 & 1 \\ -1 & 2 & 0 \\ 2 & -2 & -3 \end{bmatrix} \begin{bmatrix} 2 & -1 & 3 & 0 \\ -3 & 4 & 2 & -1 \\ 0 & -2 & 1 & 4 \end{bmatrix}$

C *In Problems 35–38, find $A^2 = AA$.*

35. $\begin{bmatrix} 6 & 9 \\ -4 & -6 \end{bmatrix}$

36. $\begin{bmatrix} -2 & 4 \\ -1 & 2 \end{bmatrix}$

37. $\begin{bmatrix} \frac{1}{3} & \frac{1}{3} \\ \frac{2}{3} & \frac{2}{3} \end{bmatrix}$

38. $\begin{bmatrix} \frac{3}{4} & \frac{3}{4} \\ \frac{1}{4} & \frac{1}{4} \end{bmatrix}$

In Problems 39–44, verify each statement by using the following matrices:

$$A = \begin{bmatrix} 1 & 2 \\ 0 & 1 \end{bmatrix} \qquad B = \begin{bmatrix} 1 & 1 \\ 2 & 3 \end{bmatrix} \qquad C = \begin{bmatrix} -3 & 1 \\ -1 & 2 \end{bmatrix}$$

39. $AB \neq BA$

40. $(AB)C = A(BC)$

41. $A(B + C) = AB + AC$

42. $(B + C)A = BA + CA$

43. $A^2 - B^2 \neq (A - B)(A + B)$

44. $A^2 + 2AB + B^2 \neq (A + B)(A + B)$

APPLICATIONS

Business & Economics

45. *Labor costs.* A company with manufacturing plants located in different parts of the country has labor-hour and wage requirements for the manufacturing of three types of inflatable boats as given in the following two matrices:

Labor-hours per boat

	Cutting department	Assembly department	Packaging department	
$M =$	0.6 hr	0.6 hr	0.2 hr	One-person boat
	1.0 hr	0.9 hr	0.3 hr	Two-person boat
	1.5 hr	1.2 hr	0.4 hr	Four-person boat

Hourly wages

	Plant I	Plant II	
$N =$	$8	$9	Cutting department
	$10	$12	Assembly department
	$5	$6	Packaging department

(A) Find the labor costs for a one-person boat manufactured at plant I. That is, find the dot product

$$[0.6 \quad 0.6 \quad 0.2] \cdot \begin{bmatrix} 8 \\ 10 \\ 5 \end{bmatrix}$$

(B) Find the labor costs for a four-person boat manufactured at plant II. Set up a dot product as in part A and multiply.

(C) What is the dimension of MN?

(D) Find MN and interpret.

46. *Inventory value.* A personal computer retail company sells five different computer models through three stores located in a large metropolitan area. The inventory of each model on hand in each store is summarized in matrix M. Wholesale (W) and retail (R) values of each model computer are summarized in matrix N.

Model

$$M = \begin{array}{c} \\ \\ \\ \end{array} \begin{matrix} A & B & C & D & E \\ \left[\begin{matrix} 4 & 2 & 3 & 7 & 1 \\ 2 & 3 & 5 & 0 & 6 \\ 10 & 4 & 3 & 4 & 3 \end{matrix}\right] & & & & \end{matrix} \begin{matrix} \text{Store 1} \\ \text{Store 2} \\ \text{Store 3} \end{matrix}$$

$$N = \begin{matrix} W & R \\ \left[\begin{matrix} \$700 & \$840 \\ \$1,400 & \$1,800 \\ \$1,800 & \$2,400 \\ \$2,700 & \$3,300 \\ \$3,500 & \$4,900 \end{matrix}\right] \end{matrix} \begin{matrix} A \\ B \\ C \\ D \\ E \end{matrix}$$

(A) What is the retail value of the inventory at store 2?

(B) What is the wholesale value of the inventory at store 3?

(C) Compute MN and interpret.

47. (A) Multiply M in Problem 46 by [1 1 1] and interpret. (The multiplication makes sense in only one direction.)

(B) Multiply MN in Problem 46 by [1 1 1] and interpret. (The multiplication makes sense in only one direction.)

Life Sciences 48. *Nutrition.* A nutritionist for a cereal company blends two cereals in three different mixes. The amounts of protein, carbohydrate, and fat (in grams per ounce) in each cereal are given by matrix M. The amounts of each cereal used in the three mixes are given by matrix N.

$$M = \begin{matrix} \text{Cereal A} & \text{Cereal B} \\ \left[\begin{matrix} 4 \text{ g/oz} & 2 \text{ g/oz} \\ 20 \text{ g/oz} & 16 \text{ g/oz} \\ 3 \text{ g/oz} & 1 \text{ g/oz} \end{matrix}\right] \end{matrix} \begin{matrix} \text{Protein} \\ \text{Carbohydrate} \\ \text{Fat} \end{matrix}$$

$$N = \begin{matrix} \text{Mix } X & \text{Mix } Y & \text{Mix } Z \\ \left[\begin{matrix} 15 \text{ oz} & 10 \text{ oz} & 5 \text{ oz} \\ 5 \text{ oz} & 10 \text{ oz} & 15 \text{ oz} \end{matrix}\right] \end{matrix} \begin{matrix} \text{Cereal A} \\ \text{Cereal B} \end{matrix}$$

(A) Find the amount of protein in mix X by computing the dot product

$$[4 \quad 2] \cdot \begin{bmatrix} 15 \\ 5 \end{bmatrix}$$

(B) Find the amount of fat in mix Z. Set up a dot product as in part A and multiply.

(C) What is the dimension of MN?

(D) Find MN and interpret.

(E) Find $\frac{1}{20}MN$ and interpret.

Social Sciences

49. *Politics.* In a local California election, a group hired a public relations firm to promote its candidate in three ways: telephone calls, house calls, and letters. The cost per contact is given in matrix M:

$$M = \begin{bmatrix} \$0.40 \\ \$0.75 \\ \$0.25 \end{bmatrix} \begin{matrix} \text{Telephone call} \\ \text{House call} \\ \text{Letter} \end{matrix}$$

with the heading "Cost per contact" above the matrix.

The number of contacts of each type made in two adjacent cities is given in matrix N:

$$N = \begin{bmatrix} 1{,}000 & 500 & 5{,}000 \\ 2{,}000 & 800 & 8{,}000 \end{bmatrix} \begin{matrix} \text{Berkeley} \\ \text{Oakland} \end{matrix}$$

with column headings "Telephone call", "House call", "Letter".

(A) Find the total amount spent in Berkeley by computing the dot product

$$[1{,}000 \quad 500 \quad 5{,}000] \cdot \begin{bmatrix} \$0.40 \\ \$0.75 \\ \$0.25 \end{bmatrix}$$

(B) Find the total amount spent in Oakland by computing the dot product of appropriate matrices.

(C) Compute NM and interpret.

(D) Multiply N by the matrix $[1 \quad 1]$ and interpret.

SECTION 5-6 **Inverse of a Square Matrix; Matrix Equations**

◆ IDENTITY MATRIX FOR MULTIPLICATION
◆ INVERSE OF A SQUARE MATRIX
◆ MATRIX EQUATIONS
◆ APPLICATION

◆ **IDENTITY MATRIX FOR MULTIPLICATION**

We know that

$$1a = a1 = a \qquad \text{for all real numbers } a$$

The number 1 is called the **identity** for real number multiplication. Does the set of all matrices of a given dimension have an identity element for multiplication? That is, if M is an arbitrary $m \times n$ matrix, does M have an identity element I such that $IM = MI = M$? The answer, in general, is no. However, the set of all **square** matrices of order **n** (dimension $n \times n$) does have an identity, and it is given as follows:

> The **identity element for multiplication** for the set of all square matrices of order n is the square matrix of order n, denoted by *I*, with 1's along the **principal diagonal** (from the upper left corner to the lower right) and 0's elsewhere.

For example,

$$\begin{bmatrix} 1 & 0 \\ 0 & 1 \end{bmatrix} \quad \text{and} \quad \begin{bmatrix} 1 & 0 & 0 \\ 0 & 1 & 0 \\ 0 & 0 & 1 \end{bmatrix}$$

are the identity matrices for all square matrices of order 2 and 3, respectively.

◆ EXAMPLE 27

(A) $\begin{bmatrix} 1 & 0 & 0 \\ 0 & 1 & 0 \\ 0 & 0 & 1 \end{bmatrix} \begin{bmatrix} a & b & c \\ d & e & f \\ g & h & i \end{bmatrix} = \begin{bmatrix} a & b & c \\ d & e & f \\ g & h & i \end{bmatrix}$

(B) $\begin{bmatrix} a & b & c \\ d & e & f \\ g & h & i \end{bmatrix} \begin{bmatrix} 1 & 0 & 0 \\ 0 & 1 & 0 \\ 0 & 0 & 1 \end{bmatrix} = \begin{bmatrix} a & b & c \\ d & e & f \\ g & h & i \end{bmatrix}$

(C) $\begin{bmatrix} 1 & 0 \\ 0 & 1 \end{bmatrix} \begin{bmatrix} a & b & c \\ d & e & f \end{bmatrix} = \begin{bmatrix} a & b & c \\ d & e & f \end{bmatrix}$

(D) $\begin{bmatrix} a & b & c \\ d & e & f \end{bmatrix} \begin{bmatrix} 1 & 0 & 0 \\ 0 & 1 & 0 \\ 0 & 0 & 1 \end{bmatrix} = \begin{bmatrix} a & b & c \\ d & e & f \end{bmatrix}$

◆

PROBLEM 27 Multiply:

(A) $\begin{bmatrix} 1 & 0 \\ 0 & 1 \end{bmatrix} \begin{bmatrix} 2 & -3 \\ 5 & 7 \end{bmatrix} \quad \text{and} \quad \begin{bmatrix} 2 & -3 \\ 5 & 7 \end{bmatrix} \begin{bmatrix} 1 & 0 \\ 0 & 1 \end{bmatrix}$

(B) $\begin{bmatrix} 1 & 0 & 0 \\ 0 & 1 & 0 \\ 0 & 0 & 1 \end{bmatrix} \begin{bmatrix} 4 & 2 \\ 3 & -5 \\ 6 & 8 \end{bmatrix} \quad \text{and} \quad \begin{bmatrix} 4 & 2 \\ 3 & -5 \\ 6 & 8 \end{bmatrix} \begin{bmatrix} 1 & 0 \\ 0 & 1 \end{bmatrix}$

◆

In general, we can show that if *M* is a square matrix of order *n* and *I* is the identity matrix of order *n*, then

$$IM = MI = M$$

If *M* is an $m \times n$ matrix that is not square ($m \neq n$), it is not possible to find a *single* identity matrix *I* satisfying both $IM = M$ and $MI = M$ (see Examples 27C and 27D). For this reason we restrict our attention in this section to square matrices.

◆ INVERSE OF A SQUARE MATRIX

In the set of real numbers, we know that for each real number a (except 0) there exists a real number a^{-1} such that

$$a^{-1}a = 1$$

The number a^{-1} is called the **inverse** of the number a relative to multiplication, or the **multiplicative inverse** of a. For example, 2^{-1} is the multiplicative inverse of 2, since $2^{-1} \cdot 2 = 1$. We use this idea to define the inverse of a square matrix.

■ Inverse of a Square Matrix

Let M be a square matrix of order n and I be the identity matrix of order n. If there exists a matrix M^{-1} (read "M inverse") such that

$$M^{-1}M = MM^{-1} = I$$

then M^{-1} is called the **multiplicative inverse of M** or, more simply, the **inverse of M.**

The multiplicative inverse of a nonzero real number a can also be written as $1/a$. This notation is not used for matrix inverses.

Let us use the above definition to find M^{-1} for

$$M = \begin{bmatrix} 2 & 3 \\ 1 & 2 \end{bmatrix}$$

We are looking for

$$M^{-1} = \begin{bmatrix} a & c \\ b & d \end{bmatrix}$$

such that

$$MM^{-1} = M^{-1}M = I$$

Thus, we write

$$\overset{M}{\begin{bmatrix} 2 & 3 \\ 1 & 2 \end{bmatrix}} \overset{M^{-1}}{\begin{bmatrix} a & c \\ b & d \end{bmatrix}} = \overset{I}{\begin{bmatrix} 1 & 0 \\ 0 & 1 \end{bmatrix}}$$

and try to find a, b, c, and d so that the product of M and M^{-1} is the identity matrix I. Multiplying M and M^{-1} on the left side, we obtain

$$\begin{bmatrix} (2a + 3b) & (2c + 3d) \\ (a + 2b) & (c + 2d) \end{bmatrix} = \begin{bmatrix} 1 & 0 \\ 0 & 1 \end{bmatrix}$$

which is true only if

$$2a + 3b = 1 \qquad 2c + 3d = 0$$
$$a + 2b = 0 \qquad c + 2d = 1$$

Solving these two systems, we find that $a = 2$, $b = -1$, $c = -3$, and $d = 2$. Thus,

$$M^{-1} = \begin{bmatrix} 2 & -3 \\ -1 & 2 \end{bmatrix}$$

as is easily checked:

$$\overset{M}{\begin{bmatrix} 2 & 3 \\ 1 & 2 \end{bmatrix}} \overset{M^{-1}}{\begin{bmatrix} 2 & -3 \\ -1 & 2 \end{bmatrix}} = \overset{I}{\begin{bmatrix} 1 & 0 \\ 0 & 1 \end{bmatrix}} = \overset{M^{-1}}{\begin{bmatrix} 2 & -3 \\ -1 & 2 \end{bmatrix}} \overset{M}{\begin{bmatrix} 2 & 3 \\ 1 & 2 \end{bmatrix}}$$

Unlike nonzero real numbers, inverses do not always exist for square matrices. For example, if

$$N = \begin{bmatrix} 2 & 1 \\ 4 & 2 \end{bmatrix}$$

then, proceeding as above, we are led to the systems

$$2a + b = 1 \qquad 2c + d = 0$$
$$4a + 2b = 0 \qquad 4c + 2d = 1$$

These are both inconsistent and have no solution. Hence, N^{-1} does not exist.

Being able to find inverses, when they exist, leads to direct and simple solutions to many practical problems. At the end of this section, we shall show how inverses can be used to solve systems of linear equations.

The method outlined above for finding M^{-1}, if it exists, gets very involved for matrices of order larger than 2. Now that we know what we are looking for, we can introduce the idea of the augmented matrix (considered in Sections 5-2 and 5-3) to make the process more efficient.

◆ E X A M P L E 28 Find the inverse, if it exists, of the matrix

$$M = \begin{bmatrix} 1 & -1 & 1 \\ 0 & 2 & -1 \\ 2 & 3 & 0 \end{bmatrix}$$

Solution   We start as before and write

$$
\begin{array}{ccc}
M & M^{-1} & I
\end{array}
$$

$$
\begin{bmatrix} 1 & -1 & 1 \\ 0 & 2 & -1 \\ 2 & 3 & 0 \end{bmatrix}
\begin{bmatrix} a & d & g \\ b & e & h \\ c & f & i \end{bmatrix}
=
\begin{bmatrix} 1 & 0 & 0 \\ 0 & 1 & 0 \\ 0 & 0 & 1 \end{bmatrix}
$$

which is true only if

$$
\begin{array}{lll}
a - b + c = 1 & d - e + f = 0 & g - h + i = 0 \\
2b - c = 0 & 2e - f = 1 & 2h - i = 0 \\
2a + 3b = 0 & 2d + 3e = 0 & 2g + 3h = 1
\end{array}
$$

Now we write augmented matrices for each of the three systems:

$$
\begin{array}{ccc}
\text{First} & \text{Second} & \text{Third}
\end{array}
$$

$$
\left[\begin{array}{ccc|c} 1 & -1 & 1 & 1 \\ 0 & 2 & -1 & 0 \\ 2 & 3 & 0 & 0 \end{array}\right]
\quad
\left[\begin{array}{ccc|c} 1 & -1 & 1 & 0 \\ 0 & 2 & -1 & 1 \\ 2 & 3 & 0 & 0 \end{array}\right]
\quad
\left[\begin{array}{ccc|c} 1 & -1 & 1 & 0 \\ 0 & 2 & -1 & 0 \\ 2 & 3 & 0 & 1 \end{array}\right]
$$

Since each matrix to the left of the vertical bar is the same, exactly the same row operations can be used on each augmented matrix to trānsform it into a reduced form. We can speed up the process substantially by combining all three augmented matrices into the single augmented matrix form below:

$$
\left[\begin{array}{ccc|ccc} 1 & -1 & 1 & 1 & 0 & 0 \\ 0 & 2 & -1 & 0 & 1 & 0 \\ 2 & 3 & 0 & 0 & 0 & 1 \end{array}\right] = [M|I]
\tag{1}
$$

We now try to perform row operations on matrix (1) until we obtain a row-equivalent matrix that looks like matrix (2):

$$
\begin{array}{cc}
I & B
\end{array}
$$

$$
\left[\begin{array}{ccc|ccc} 1 & 0 & 0 & a & d & g \\ 0 & 1 & 0 & b & e & h \\ 0 & 0 & 1 & c & f & i \end{array}\right] = [I|B]
\tag{2}
$$

If this can be done, then the new matrix B to the right of the vertical bar will be M^{-1}! Now let us try to transform (1) into a form like (2). We will follow the same sequence of steps as we did in the solution of linear systems by Gauss–Jordan elimination (see Section 5-3).

$$
\begin{array}{cc}
M & I \\
\end{array}
$$

$$
\left[\begin{array}{ccc|ccc}
1 & -1 & 1 & 1 & 0 & 0 \\
0 & 2 & -1 & 0 & 1 & 0 \\
2 & 3 & 0 & 0 & 0 & 1
\end{array}\right] \quad R_3 + (-2)R_1 \rightarrow R_3
$$

$$
\sim \left[\begin{array}{ccc|ccc}
1 & -1 & 1 & 1 & 0 & 0 \\
0 & 2 & -1 & 0 & 1 & 0 \\
0 & 5 & -2 & -2 & 0 & 1
\end{array}\right] \quad \tfrac{1}{2}R_2 \rightarrow R_2
$$

$$
\sim \left[\begin{array}{ccc|ccc}
1 & -1 & 1 & 1 & 0 & 0 \\
0 & 1 & -\tfrac{1}{2} & 0 & \tfrac{1}{2} & 0 \\
0 & 5 & -2 & -2 & 0 & 1
\end{array}\right] \quad R_3 + (-5)R_2 \rightarrow R_3
$$

$$
\sim \left[\begin{array}{ccc|ccc}
1 & -1 & 1 & 1 & 0 & 0 \\
0 & 1 & -\tfrac{1}{2} & 0 & \tfrac{1}{2} & 0 \\
0 & 0 & \tfrac{1}{2} & -2 & -\tfrac{5}{2} & 1
\end{array}\right] \quad 2R_3 \rightarrow R_3
$$

$$
\sim \left[\begin{array}{ccc|ccc}
1 & -1 & 1 & 1 & 0 & 0 \\
0 & 1 & -\tfrac{1}{2} & 0 & \tfrac{1}{2} & 0 \\
0 & 0 & 1 & -4 & -5 & 2
\end{array}\right] \quad \begin{array}{l} R_1 + (-1)R_3 \rightarrow R_1 \\ R_2 + \tfrac{1}{2}R_3 \rightarrow R_2 \end{array}
$$

$$
\sim \left[\begin{array}{ccc|ccc}
1 & -1 & 0 & 5 & 5 & -2 \\
0 & 1 & 0 & -2 & -2 & 1 \\
0 & 0 & 1 & -4 & -5 & 2
\end{array}\right] \quad R_1 + R_2 \rightarrow R_1
$$

$$
\begin{array}{cc}
I & B \\
\end{array}
$$

$$
\sim \left[\begin{array}{ccc|ccc}
1 & 0 & 0 & 3 & 3 & -1 \\
0 & 1 & 0 & -2 & -2 & 1 \\
0 & 0 & 1 & -4 & -5 & 2
\end{array}\right] = [I|B]
$$

Converting back to systems of equations equivalent to our three original systems (we will not have to do this step in practice), we have

$$
\begin{array}{lll}
a = 3 & d = 3 & g = -1 \\
b = -2 & e = -2 & h = 1 \\
c = -4 & f = -5 & i = 2
\end{array}
$$

And these are just the elements of M^{-1} that we are looking for! Hence,

$$
M^{-1} = \left[\begin{array}{ccc}
3 & 3 & -1 \\
-2 & -2 & 1 \\
-4 & -5 & 2
\end{array}\right]
$$

Note that this is the matrix to the right of the vertical line in the last augmented matrix. That is, $M^{-1} = B$.

Since the definition of matrix inverse requires that

$$
M^{-1}M = I \quad \text{and} \quad MM^{-1} = I \tag{3}
$$

it appears that we must compute both $M^{-1}M$ and MM^{-1} to check our work. However, it can be shown that if one of the equations in (3) is satisfied, then the other is also satisfied. Thus, for checking purposes, it is sufficient to compute either $M^{-1}M$ or MM^{-1}; we do not need to do both.

Check

$$M^{-1}M = \begin{bmatrix} 3 & 3 & -1 \\ -2 & -2 & 1 \\ -4 & -5 & 2 \end{bmatrix} \begin{bmatrix} 1 & -1 & 1 \\ 0 & 2 & -1 \\ 2 & 3 & 0 \end{bmatrix} = \begin{bmatrix} 1 & 0 & 0 \\ 0 & 1 & 0 \\ 0 & 0 & 1 \end{bmatrix} = I \qquad \blacklozenge$$

PROBLEM 28 Let: $M = \begin{bmatrix} 3 & -1 & 1 \\ -1 & 1 & 0 \\ 1 & 0 & 1 \end{bmatrix}$

(A) Form the augmented matrix $[M|I]$.
(B) Use row operations to transform $[M|I]$ into $[I|B]$.
(C) Verify by multiplication that $B = M^{-1}$ (that is, show that $BM = I$). $\qquad \blacklozenge$

The procedure shown in Example 28 can be used to find the inverse of any square matrix, if the inverse exists, and will also indicate when the inverse does not exist. These ideas are summarized in Theorem 5.

THEOREM 5 ■ **Inverse of a Square Matrix M**

If $[M|I]$ is transformed by row operations into $[I|B]$, then the resulting matrix B is M^{-1}. However, if we obtain 0's in one or more rows to the left of the vertical line, then M^{-1} does not exist.

◆ EXAMPLE 29 Find M^{-1}, given: $M = \begin{bmatrix} 3 & -1 \\ -4 & 2 \end{bmatrix}$

Solution

$$\begin{bmatrix} 3 & -1 & | & 1 & 0 \\ -4 & 2 & | & 0 & 1 \end{bmatrix} \qquad \tfrac{1}{3}R_1 \to R_1$$

$$\sim \begin{bmatrix} 1 & -\tfrac{1}{3} & | & \tfrac{1}{3} & 0 \\ -4 & 2 & | & 0 & 1 \end{bmatrix} \qquad R_2 + 4R_1 \to R_2$$

$$\sim \begin{bmatrix} 1 & -\tfrac{1}{3} & | & \tfrac{1}{3} & 0 \\ 0 & \tfrac{2}{3} & | & \tfrac{4}{3} & 1 \end{bmatrix} \qquad \tfrac{3}{2}R_2 \to R_2$$

$$\sim \begin{bmatrix} 1 & -\tfrac{1}{3} & | & \tfrac{1}{3} & 0 \\ 0 & 1 & | & 2 & \tfrac{3}{2} \end{bmatrix} \qquad R_1 + \tfrac{1}{3}R_2 \to R_1$$

$$\sim \begin{bmatrix} 1 & 0 & | & 1 & \tfrac{1}{2} \\ 0 & 1 & | & 2 & \tfrac{3}{2} \end{bmatrix}$$

Thus,

$$M^{-1} = \begin{bmatrix} 1 & \tfrac{1}{2} \\ 2 & \tfrac{3}{2} \end{bmatrix} \qquad \text{Check by showing that } M^{-1}M = I. \qquad \blacklozenge$$

P R O B L E M 29 Find M^{-1}, given: $M = \begin{bmatrix} 2 & -6 \\ 1 & -2 \end{bmatrix}$ ◆

◆ **E X A M P L E 30** Find M^{-1}, given: $M = \begin{bmatrix} 2 & -4 \\ -3 & 6 \end{bmatrix}$

Solution

$$\begin{bmatrix} 2 & -4 & | & 1 & 0 \\ -3 & 6 & | & 0 & 1 \end{bmatrix} \quad \tfrac{1}{2}R_1 \to R_1$$

$$\sim \begin{bmatrix} 1 & -2 & | & \tfrac{1}{2} & 0 \\ -3 & 6 & | & 0 & 1 \end{bmatrix} \quad R_2 + 3R_1 \to R_2$$

$$\sim \begin{bmatrix} 1 & -2 & | & \tfrac{1}{2} & 0 \\ 0 & 0 & | & \tfrac{3}{2} & 1 \end{bmatrix}$$

We have all 0's in the second row to the left of the vertical bar; therefore, the inverse does not exist. ◆

P R O B L E M 30 Find M^{-1}, given: $M = \begin{bmatrix} -6 & 3 \\ -4 & 2 \end{bmatrix}$ ◆

◆ MATRIX EQUATIONS

We will now show how independent systems of linear equations with the same number of variables as equations can be solved using inverses of square matrices.

◆ **E X A M P L E 31** Solve the system

$$\begin{aligned} x_1 - \quad x_2 + x_3 &= k_1 \\ 2x_2 - x_3 &= k_2 \\ 2x_1 + 3x_2 \quad\quad &= k_3 \end{aligned} \qquad (4)$$

for:

(A) $k_1 = 1$, $k_2 = 1$, $k_3 = 1$ (B) $k_1 = 3$, $k_2 = 1$, $k_3 = 4$
(C) $k_1 = -5$, $k_2 = 2$, $k_3 = -3$

Solutions The inverse of the coefficient matrix

$$A = \begin{bmatrix} 1 & -1 & 1 \\ 0 & 2 & -1 \\ 2 & 3 & 0 \end{bmatrix}$$

can be used to solve parts A, B, and C very easily. To see how, we convert system (4) into the following **matrix equation:**

$$\overset{A}{\begin{bmatrix} 1 & -1 & 1 \\ 0 & 2 & -1 \\ 2 & 3 & 0 \end{bmatrix}} \overset{X}{\begin{bmatrix} x_1 \\ x_2 \\ x_3 \end{bmatrix}} = \overset{B}{\begin{bmatrix} k_1 \\ k_2 \\ k_3 \end{bmatrix}} \qquad (5)$$

You should check that matrix equation (5) is equivalent to system (4) by finding the product on the left side and then equating corresponding elements on the left with those on the right. (Now you see another reason for defining matrix multiplication as it was defined.)

We are now interested in finding a column matrix X that will satisfy the matrix equation

$$AX = B$$

To solve this equation, we multiply both sides (on the left) by A^{-1} (assuming it exists) to isolate X on the left side:

$AX = B$	Multiply both sides (on the left) by A^{-1}.
$A^{-1}(AX) = A^{-1}B$	Use the associative property.
$(A^{-1}A)X = A^{-1}B$	$A^{-1}A = I$
$IX = A^{-1}B$	$IX = X$
$X = A^{-1}B$	

The inverse of A was found in Example 28 to be

$$A^{-1} = \begin{bmatrix} 3 & 3 & -1 \\ -2 & -2 & 1 \\ -4 & -5 & 2 \end{bmatrix}$$

Thus,

$$\begin{matrix} X & A^{-1} & B \end{matrix}$$

$$\begin{bmatrix} x_1 \\ x_2 \\ x_3 \end{bmatrix} = \begin{bmatrix} 3 & 3 & -1 \\ -2 & -2 & 1 \\ -4 & -5 & 2 \end{bmatrix} \begin{bmatrix} k_1 \\ k_2 \\ k_3 \end{bmatrix}$$

Now, to solve parts A, B, and C, we simply replace k_1, k_2, and k_3 with the given values and multiply.

(A) $$\begin{bmatrix} x_1 \\ x_2 \\ x_3 \end{bmatrix} = \begin{bmatrix} 3 & 3 & -1 \\ -2 & -2 & 1 \\ -4 & -5 & 2 \end{bmatrix} \begin{bmatrix} 1 \\ 1 \\ 1 \end{bmatrix} = \begin{bmatrix} 5 \\ -3 \\ -7 \end{bmatrix}$$

Thus, $x_1 = 5$, $x_2 = -3$, and $x_3 = -7$.

(B) $$\begin{bmatrix} x_1 \\ x_2 \\ x_3 \end{bmatrix} = \begin{bmatrix} 3 & 3 & -1 \\ -2 & -2 & 1 \\ -4 & -5 & 2 \end{bmatrix} \begin{bmatrix} 3 \\ 1 \\ 4 \end{bmatrix} = \begin{bmatrix} 8 \\ -4 \\ -9 \end{bmatrix}$$

Thus, $x_1 = 8$, $x_2 = -4$, and $x_3 = -9$.

(C) $$\begin{bmatrix} x_1 \\ x_2 \\ x_3 \end{bmatrix} = \begin{bmatrix} 3 & 3 & -1 \\ -2 & -2 & 1 \\ -4 & -5 & 2 \end{bmatrix} \begin{bmatrix} -5 \\ 2 \\ -3 \end{bmatrix} = \begin{bmatrix} -6 \\ 3 \\ 4 \end{bmatrix}$$

Thus, $x_1 = -6$, $x_2 = 3$, and $x_3 = 4$. ◆

PROBLEM 31 Solve the system

$$3x_1 - x_2 + x_3 = k_1$$
$$-x_1 + x_2 \qquad = k_2$$
$$x_1 \qquad + x_3 = k_3$$

for:

(A) $k_1 = 1$, $k_2 = 3$, $k_3 = 2$ (B) $k_1 = 3$, $k_2 = -3$, $k_3 = 2$
(C) $k_1 = -5$, $k_2 = 1$, $k_3 = -4$

[Note: The inverse of the coefficient matrix was found in Problem 28.]   ◆

Computer programs are readily available for finding inverses of square matrices.* A great advantage of using an inverse matrix to solve an independent system of equations is that once the inverse is found, it can be used to solve any new system formed by changing the constant terms. However, this method is not suited for a system where the number of equations and the number of variables is not the same, since the coefficient matrix for such a system is not square.

◆ APPLICATION

The following application will illustrate the usefulness of the inverse matrix method for solving systems of equations.

◆ EXAMPLE 32

Investment Analysis

An investment advisor currently has two types of investments available for clients: a conservative investment A that pays 10% per year and an investment B of higher risk that pays 20% per year. Clients may divide their investments between the two to achieve any total return desired between 10% and 20%. However, the higher the desired return, the higher the risk. How should each client listed in the table invest to achieve the indicated return?

	CLIENT			
	1	2	3	k
TOTAL INVESTMENT	$20,000	$50,000	$10,000	k_1
ANNUAL RETURN DESIRED	$ 2,400	$ 7,500	$ 1,300	k_2
	(12%)	(15%)	(13%)	

---

* See the Preface (under "Student Aids" and "Instructor Aids") for a description of the Interactive Computer Applications Package (ICAP) that accompanies this book.

Solution We will solve the problem for an arbitrary client k by finding an inverse matrix. Then we will apply the result to the three specific clients.

Let

$x_1 =$ Amount invested in A

$x_2 =$ Amount invested in B

Then

$$x_1 + \quad x_2 = k_1 \qquad \text{Total invested}$$
$$0.1x_1 + 0.2x_2 = k_2 \qquad \text{Total annual return desired}$$

Write as a matrix equation:

$$\begin{array}{ccc} A & X & B \end{array}$$
$$\begin{bmatrix} 1 & 1 \\ 0.1 & 0.2 \end{bmatrix} \begin{bmatrix} x_1 \\ x_2 \end{bmatrix} = \begin{bmatrix} k_1 \\ k_2 \end{bmatrix}$$

If A^{-1} exists, then

$$X = A^{-1}B$$

We now find A^{-1} by starting with $[A|I]$ and proceeding as discussed earlier in this section:

$$\begin{bmatrix} 1 & 1 & | & 1 & 0 \\ 0.1 & 0.2 & | & 0 & 1 \end{bmatrix} \quad 10R_2 \rightarrow R_2$$

$$\sim \begin{bmatrix} 1 & 1 & | & 1 & 0 \\ 1 & 2 & | & 0 & 10 \end{bmatrix} \quad R_2 + (-1)R_1 \rightarrow R_2$$

$$\sim \begin{bmatrix} 1 & 1 & | & 1 & 0 \\ 0 & 1 & | & -1 & 10 \end{bmatrix} \quad R_1 + (-1)R_2 \rightarrow R_1$$

$$\sim \begin{bmatrix} 1 & 0 & | & 2 & -10 \\ 0 & 1 & | & -1 & 10 \end{bmatrix}$$

Thus,

$$A^{-1} = \begin{bmatrix} 2 & -10 \\ -1 & 10 \end{bmatrix} \qquad \textit{Check:} \qquad \overset{A^{-1}}{\begin{bmatrix} 2 & -10 \\ -1 & 10 \end{bmatrix}} \overset{A}{\begin{bmatrix} 1 & 1 \\ 0.1 & 0.2 \end{bmatrix}} = \overset{I}{\begin{bmatrix} 1 & 0 \\ 0 & 1 \end{bmatrix}}$$

and

$$\overset{X}{\begin{bmatrix} x_1 \\ x_2 \end{bmatrix}} = \overset{A^{-1}}{\begin{bmatrix} 2 & -10 \\ -1 & 10 \end{bmatrix}} \overset{B}{\begin{bmatrix} k_1 \\ k_2 \end{bmatrix}}$$

To solve each client's investment problem, we replace k_1 and k_2 with appropriate values from the table and multiply by A^{-1}:

Client 1

$$\begin{bmatrix} x_1 \\ x_2 \end{bmatrix} = \begin{bmatrix} 2 & -10 \\ -1 & 10 \end{bmatrix} \begin{bmatrix} 20,000 \\ 2,400 \end{bmatrix} = \begin{bmatrix} 16,000 \\ 4,000 \end{bmatrix}$$

Solution: $x_1 = \$16,000$ in A, $x_2 = \$4,000$ in B

Client 2

$$\begin{bmatrix} x_1 \\ x_2 \end{bmatrix} = \begin{bmatrix} 2 & -10 \\ -1 & 10 \end{bmatrix} \begin{bmatrix} 50,000 \\ 7,500 \end{bmatrix} = \begin{bmatrix} 25,000 \\ 25,000 \end{bmatrix}$$

Solution: $x_1 = \$25,000$ in A, $x_2 = \$25,000$ in B

Client 3

$$\begin{bmatrix} x_1 \\ x_2 \end{bmatrix} = \begin{bmatrix} 2 & -10 \\ -1 & 10 \end{bmatrix} \begin{bmatrix} 10,000 \\ 1,300 \end{bmatrix} = \begin{bmatrix} 7,000 \\ 3,000 \end{bmatrix}$$

Solution: $x_1 = \$7,000$ in A, $x_2 = \$3,000$ in B ◆

PROBLEM 32   Repeat Example 32 with investment A paying 8% and investment B paying 24%.
◆

Answers to Matched Problems   27. (A) $\begin{bmatrix} 2 & -3 \\ 5 & 7 \end{bmatrix}$ (B) $\begin{bmatrix} 4 & 2 \\ 3 & -5 \\ 6 & 8 \end{bmatrix}$

28. (A) $\left[\begin{array}{ccc|ccc} 3 & -1 & 1 & 1 & 0 & 0 \\ -1 & 1 & 0 & 0 & 1 & 0 \\ 1 & 0 & 1 & 0 & 0 & 1 \end{array}\right]$ (B) $\left[\begin{array}{ccc|ccc} 1 & 0 & 0 & 1 & 1 & -1 \\ 0 & 1 & 0 & 1 & 2 & -1 \\ 0 & 0 & 1 & -1 & -1 & 2 \end{array}\right]$

(C) $\begin{bmatrix} 1 & 1 & -1 \\ 1 & 2 & -1 \\ -1 & -1 & 2 \end{bmatrix} \begin{bmatrix} 3 & -1 & 1 \\ -1 & 1 & 0 \\ 1 & 0 & 1 \end{bmatrix} = \begin{bmatrix} 1 & 0 & 0 \\ 0 & 1 & 0 \\ 0 & 0 & 1 \end{bmatrix}$

29. $\begin{bmatrix} -1 & 3 \\ -\frac{1}{2} & 1 \end{bmatrix}$ 30. Does not exist

31. (A) $x_1 = 2$, $x_2 = 5$, $x_3 = 0$ (B) $x_1 = -2$, $x_2 = -5$, $x_3 = 4$
(C) $x_1 = 0$, $x_2 = 1$, $x_3 = -4$

32. $A^{-1} = \begin{bmatrix} 1.5 & -6.25 \\ -0.5 & 6.25 \end{bmatrix}$;   Client 1: \$15,000 in A and \$5,000 in B;
Client 2: \$28,125 in A and \$21,875 in B;
Client 3: \$6,875 in A and \$3,125 in B

EXERCISE 5-6

A   *Perform the indicated operations.*

1. $\begin{bmatrix} 1 & 0 \\ 0 & 1 \end{bmatrix} \begin{bmatrix} 2 & -3 \\ 4 & 5 \end{bmatrix}$ 2. $\begin{bmatrix} 2 & -3 \\ 4 & 5 \end{bmatrix} \begin{bmatrix} 1 & 0 \\ 0 & 1 \end{bmatrix}$

$$
3. \quad
\begin{bmatrix} 1 & 0 & 0 \\ 0 & 1 & 0 \\ 0 & 0 & 1 \end{bmatrix}
\begin{bmatrix} -2 & 1 & 3 \\ 2 & 4 & -2 \\ 5 & 1 & 0 \end{bmatrix}
\qquad
4. \quad
\begin{bmatrix} -2 & 1 & 3 \\ 2 & 4 & -2 \\ 5 & 1 & 0 \end{bmatrix}
\begin{bmatrix} 1 & 0 & 0 \\ 0 & 1 & 0 \\ 0 & 0 & 1 \end{bmatrix}
$$

For each problem, show that the two matrices are inverses of each other by showing that their product is the identity matrix I.

$$
5. \quad
\begin{bmatrix} 3 & -4 \\ -2 & 3 \end{bmatrix};
\begin{bmatrix} 3 & 4 \\ 2 & 3 \end{bmatrix}
\qquad
6. \quad
\begin{bmatrix} 5 & -7 \\ -2 & 3 \end{bmatrix};
\begin{bmatrix} 3 & 7 \\ 2 & 5 \end{bmatrix}
$$

$$
7. \quad
\begin{bmatrix} 1 & -1 & 1 \\ 0 & 2 & -1 \\ 2 & 3 & 0 \end{bmatrix};
\begin{bmatrix} 3 & 3 & -1 \\ -2 & -2 & 1 \\ -4 & -5 & 2 \end{bmatrix}
$$

$$
8. \quad
\begin{bmatrix} 3 & 3 & -1 \\ -2 & -2 & 1 \\ -4 & -5 & 2 \end{bmatrix};
\begin{bmatrix} 1 & -1 & 1 \\ 0 & 2 & -1 \\ 2 & 3 & 0 \end{bmatrix}
$$

Find x_1 and x_2.

$$
9. \quad
\begin{bmatrix} x_1 \\ x_2 \end{bmatrix}
=
\begin{bmatrix} 3 & -2 \\ 1 & 4 \end{bmatrix}
\begin{bmatrix} -2 \\ 1 \end{bmatrix}
\qquad
10. \quad
\begin{bmatrix} x_1 \\ x_2 \end{bmatrix}
=
\begin{bmatrix} -2 & 1 \\ -1 & 2 \end{bmatrix}
\begin{bmatrix} 3 \\ -2 \end{bmatrix}
$$

$$
11. \quad
\begin{bmatrix} x_1 \\ x_2 \end{bmatrix}
=
\begin{bmatrix} -2 & 3 \\ 2 & -1 \end{bmatrix}
\begin{bmatrix} 3 \\ 2 \end{bmatrix}
\qquad
12. \quad
\begin{bmatrix} x_1 \\ x_2 \end{bmatrix}
=
\begin{bmatrix} 3 & -1 \\ 0 & 2 \end{bmatrix}
\begin{bmatrix} -2 \\ 1 \end{bmatrix}
$$

B   *Given M as indicated, find M^{-1} and show that $M^{-1}M = I$.*

$$
13. \quad
\begin{bmatrix} 1 & 2 \\ 1 & 3 \end{bmatrix}
\qquad
14. \quad
\begin{bmatrix} 2 & 1 \\ 5 & 3 \end{bmatrix}
\qquad
15. \quad
\begin{bmatrix} 1 & 3 \\ 2 & 7 \end{bmatrix}
\qquad
16. \quad
\begin{bmatrix} 2 & 1 \\ 1 & 1 \end{bmatrix}
$$

$$
17. \quad
\begin{bmatrix} 1 & -3 & 0 \\ 0 & 3 & 1 \\ 2 & -1 & 2 \end{bmatrix}
\qquad
18. \quad
\begin{bmatrix} 2 & 9 & 0 \\ 1 & 2 & 3 \\ 0 & -1 & 1 \end{bmatrix}
$$

$$
19. \quad
\begin{bmatrix} 1 & 1 & 0 \\ 0 & 3 & -1 \\ 1 & 0 & 1 \end{bmatrix}
\qquad
20. \quad
\begin{bmatrix} 1 & 0 & -1 \\ 2 & -1 & 0 \\ 1 & 1 & 1 \end{bmatrix}
$$

Write each system as a matrix equation and solve by using inverses. [Note: The inverses were found in Problems 13–20.]

21. $x_1 + 2x_2 = k_1$
      $x_1 + 3x_2 = k_2$

      (A) $k_1 = 1, \quad k_2 = 3$
      (B) $k_1 = 3, \quad k_2 = 5$
      (C) $k_1 = -2, \quad k_2 = 1$

22. $2x_1 + x_2 = k_1$
      $5x_1 + 3x_2 = k_2$

      (A) $k_1 = 2, \quad k_2 = 13$
      (B) $k_1 = 2, \quad k_2 = 4$
      (C) $k_1 = 1, \quad k_2 = -3$

23. $x_1 + 3x_2 = k_1$
      $2x_1 + 7x_2 = k_2$

      (A) $k_1 = 2, \quad k_2 = -1$
      (B) $k_1 = 1, \quad k_2 = 0$
      (C) $k_1 = 3, \quad k_2 = -1$

24. $2x_1 + x_2 = k_1$
      $x_1 + x_2 = k_2$

      (A) $k_1 = -1, \quad k_2 = -2$
      (B) $k_1 = 2, \quad k_2 = 3$
      (C) $k_1 = 2, \quad k_2 = 0$

25.
$$\begin{aligned} x_1 - 3x_2 \quad\quad &= k_1 \\ 3x_2 + x_3 &= k_2 \\ 2x_1 - x_2 + 2x_3 &= k_3 \end{aligned}$$

(A) $k_1 = 1$, $k_2 = 0$, $k_3 = 2$
(B) $k_1 = -1$, $k_2 = 1$, $k_3 = 0$
(C) $k_1 = 2$, $k_2 = -2$, $k_3 = 1$

26.
$$\begin{aligned} 2x_1 + 9x_2 \quad\quad &= k_1 \\ x_1 + 2x_2 + 3x_3 &= k_2 \\ - x_2 + x_3 &= k_3 \end{aligned}$$

(A) $k_1 = 0$, $k_2 = 2$, $k_3 = 1$
(B) $k_1 = -2$, $k_2 = 0$, $k_3 = 1$
(C) $k_1 = 3$, $k_2 = 1$, $k_3 = 0$

27.
$$\begin{aligned} x_1 + x_2 \quad\quad &= k_1 \\ 3x_2 - x_3 &= k_2 \\ x_1 \quad\quad + x_3 &= k_3 \end{aligned}$$

(A) $k_1 = 2$, $k_2 = 0$, $k_3 = 4$
(B) $k_1 = 0$, $k_2 = 4$, $k_3 = -2$
(C) $k_1 = 4$, $k_2 = 2$, $k_3 = 0$

28.
$$\begin{aligned} x_1 \quad\quad - x_3 &= k_1 \\ 2x_1 - x_2 \quad\quad &= k_2 \\ x_1 + x_2 + x_3 &= k_3 \end{aligned}$$

(A) $k_1 = 4$, $k_2 = 8$, $k_3 = 0$
(B) $k_1 = 4$, $k_2 = 0$, $k_3 = -4$
(C) $k_1 = 0$, $k_2 = 8$, $k_3 = -8$

C *Find the inverse of each matrix, if it exists.*

29. $\begin{bmatrix} 3 & 9 \\ 2 & 6 \end{bmatrix}$ **30.** $\begin{bmatrix} 2 & -4 \\ -3 & 6 \end{bmatrix}$ **31.** $\begin{bmatrix} 3 & 1 \\ 4 & 2 \end{bmatrix}$ **32.** $\begin{bmatrix} -5 & 3 \\ 2 & -2 \end{bmatrix}$

33. $\begin{bmatrix} -5 & -2 & -2 \\ 2 & 1 & 0 \\ 1 & 0 & 1 \end{bmatrix}$ **34.** $\begin{bmatrix} 2 & 0 & 1 \\ 1 & 1 & 0 \\ 1 & 0 & 1 \end{bmatrix}$ **35.** $\begin{bmatrix} 2 & 1 & 1 \\ 1 & 1 & 0 \\ -1 & -1 & 0 \end{bmatrix}$

36. $\begin{bmatrix} 1 & -1 & 0 \\ 2 & -1 & 1 \\ 0 & 1 & 1 \end{bmatrix}$ **37.** $\begin{bmatrix} -1 & -2 & 2 \\ 4 & 2 & 0 \\ 4 & 0 & 4 \end{bmatrix}$ **38.** $\begin{bmatrix} 3 & 0 & 2 \\ 4 & 2 & 0 \\ 5 & 0 & 5 \end{bmatrix}$

39. Show that $(A^{-1})^{-1} = A$ for: $A = \begin{bmatrix} 3 & 4 \\ 2 & 3 \end{bmatrix}$

40. Show that $(AB)^{-1} = B^{-1}A^{-1}$ for: $A = \begin{bmatrix} 3 & 4 \\ 2 & 3 \end{bmatrix}$ and $B = \begin{bmatrix} 3 & 7 \\ 2 & 5 \end{bmatrix}$

APPLICATIONS

Solve using systems of equations and matrix inverses.

Business & Economics

41. *Resource allocation.* A concert hall has 10,000 seats. If tickets are $4 and $8, how many of each type of ticket should be sold (assuming all seats can be sold) to bring in each of the returns indicated in the table?

| | CONCERT | | |
	1	2	3
TICKETS SOLD	10,000	10,000	10,000
RETURN REQUIRED	$56,000	$60,000	$68,000

GUITAR MODEL	LABOR COST	MATERIAL COST
A	$30	$20
B	$40	$30

42. *Production scheduling.* Labor and material costs for manufacturing two guitar models are given in the table in the margin. If a total of $3,000 a week is allowed for labor and material, how many of each model should be produced each week to use exactly each of the allocations of the $3,000 indicated in the following table?

	WEEKLY ALLOCATION		
	1	2	3
LABOR	$1,800	$1,750	$1,720
MATERIAL	$1,200	$1,250	$1,280

Life Sciences

MIX	PROTEIN (%)	FAT (%)
A	20	2
B	10	6

43. *Diets.* A biologist has available two commercial food mixes containing the percentages of protein and fat given in the table in the margin. How many ounces of each mix should be used to prepare each of the diets listed in the following table?

	DIET		
	1	2	3
PROTEIN	20 oz	10 oz	10 oz
FAT	6 oz	4 oz	6 oz

Important Terms and Symbols

Chapter Review

5-1 *Review: Systems of Linear Equations.* Graphing method; consistent; inconsistent; independent; unique solution; dependent; substitution method; elimination by addition; equivalent systems; parameter; particular solution; equilibrium price; equilibrium quantity; linear equation in two variables; linear equation in three variables; solution of a system; solution set

5-2 *Systems of Linear Equations and Augmented Matrices.* Matrix; element; augmented matrix; column; row; equivalent systems; row-equivalent matrices; row operations; Gauss–Jordan elimination

$$R_i \leftrightarrow R_j; \quad kR_i \to R_i; \quad R_i + kR_j \to R_i$$

5-3 *Gauss–Jordan Elimination.* Reduced form; leftmost variable; submatrix; Gauss–Jordan elimination

5-4 *Matrices—Addition and Multiplication by a Number.* Size or dimension of a matrix; $m \times n$ matrix; square matrix; column matrix; row matrix; equal matrices; sum of two matrices; zero matrix; negative of a matrix M; subtraction of matrices; product of a number k and a matrix M

5-5 *Matrix Multiplication.* Dot product; matrix product; associative property; left and right distributive properties; left and right multiplication properties

5-6 *Inverse of a Square Matrix; Matrix Equations.* Identity element for multiplication; principal diagonal; multiplicative inverse; matrix equation

$$M^{-1}$$

Chapter Review

Work through all the problems in this chapter review and check your answers in the back of the book. (Answers to all review problems are there.) Where weaknesses show up, review appropriate sections in the text.

A
1. Solve the following system by graphing:

$$2x - y = 4$$
$$x - 2y = -4$$

2. Solve the system in Problem 1 by substitution.

In Problems 3–11, perform the operations that are defined, given the following matrices:

$$A = \begin{bmatrix} 1 & 2 \\ 3 & 1 \end{bmatrix} \qquad B = \begin{bmatrix} 2 & 1 \\ 1 & 1 \end{bmatrix} \qquad C = \begin{bmatrix} 2 & 3 \end{bmatrix} \qquad D = \begin{bmatrix} 1 \\ 2 \end{bmatrix}$$

3. $A + B$ 4. $B + D$ 5. $A - 2B$
6. AB 7. AC 8. AD
9. DC 10. $C \cdot D$ 11. $C + D$

12. Find the inverse of the matrix A given below by appropriate row operations on $[A|I]$. Show that $A^{-1}A = I$.

$$A = \begin{bmatrix} 3 & 2 \\ 4 & 3 \end{bmatrix}$$

13. Solve the following system using elimination by addition:

$$3x_1 + 2x_2 = 3$$
$$4x_1 + 3x_2 = 5$$

14. Solve the system in Problem 13 by performing appropriate row operations on the augmented matrix of the system.

15. Solve the system in Problem 13 by writing the system as a matrix equation and using the inverse of the coefficient matrix (see Problem 12). Also, solve the system if the constants 3 and 5 are replaced by 7 and 10, respectively. By 4 and 2, respectively.

B
In Problems 16–21, perform the specified operations, given the following matrices:

$$A = \begin{bmatrix} 2 & -2 \\ 1 & 0 \\ 3 & 2 \end{bmatrix} \qquad B = \begin{bmatrix} -1 \\ 2 \\ 3 \end{bmatrix} \qquad C = \begin{bmatrix} 2 & 1 & 3 \end{bmatrix}$$

$$D = \begin{bmatrix} 3 & -2 & 1 \\ -1 & 1 & 2 \end{bmatrix} \qquad E = \begin{bmatrix} 3 & -4 \\ -1 & 0 \end{bmatrix}$$

16. $A + D$

17. $E + DA$

18. $DA - 3E$

19. $C \cdot B$

20. CB

21. $AD - BC$

22. Find the inverse of the matrix A given below by appropriate row operations on $[A|I]$. Show that $A^{-1}A = I$.

$$A = \begin{bmatrix} 1 & 2 & 3 \\ 2 & 3 & 4 \\ 1 & 2 & 1 \end{bmatrix}$$

23. Solve by Gauss–Jordan elimination:

(A) $x_1 + 2x_2 + 3x_3 = 1$ (B) $x_1 + 2x_2 - x_3 = 2$

     $2x_1 + 3x_2 + 4x_3 = 3$ $2x_1 + 3x_2 + x_3 = -3$

     $x_1 + 2x_2 + x_3 = 3$ $3x_1 + 5x_2 \quad\quad = -1$

24. Solve the system in Problem 23A by writing the system as a matrix equation and using the inverse of the coefficient matrix (see Problem 22). Also, solve the system if the constants 1, 3, and 3 are replaced by 0, 0, and -2, respectively. By -3, -4, and 1, respectively.

C   **25.** Find the inverse of the matrix A given below. Show that $A^{-1}A = I$.

$$A = \begin{bmatrix} 4 & 5 & 6 \\ 4 & 5 & -6 \\ 1 & 1 & 1 \end{bmatrix}$$

26. Solve the system

$$0.04x_1 + 0.05x_2 + 0.06x_3 = 360$$
$$0.04x_1 + 0.05x_2 - 0.06x_3 = 120$$
$$x_1 + x_2 + x_3 = 7,000$$

by writing as a matrix equation and using the inverse of the coefficient matrix. (Before starting, multiply the first two equations by 100 to eliminate decimals. Also, see Problem 25.)

27. Solve Problem 26 by Gauss–Jordan elimination.

APPLICATIONS

Business & Economics

ORE	NICKEL (%)	COPPER (%)
A	1	2
B	2	5

28. *Resource allocation.* A mining company has two mines with ore composition as given in the table. How many tons of each ore should be used to obtain 4.5 tons of nickel and 10 tons of copper? Set up a system of equations and solve using Gauss–Jordan elimination.

29. (A) Set up Problem 28 as a matrix equation and solve using the inverse of the coefficient matrix.

(B) Solve Problem 28 (as in part A) if 2.3 tons of nickel and 5 tons of copper are needed.

30. *Material costs.* A manufacturer wishes to make two different bronze alloys in a metal foundry. The quantities of copper, tin, and zinc needed are indicated in matrix M. The costs for these materials (in dollars per pound) from two suppliers are summarized in matrix N. The company must choose one supplier or the other.

$$M = \begin{array}{c} \\ \\ \end{array} \begin{array}{ccc} \text{Copper} & \text{Tin} & \text{Zinc} \\ \left[\begin{array}{ccc} 4{,}800\text{ lb} & 600\text{ lb} & 300\text{ lb} \\ 6{,}000\text{ lb} & 1{,}400\text{ lb} & 700\text{ lb} \end{array}\right] & & \end{array} \begin{array}{c} \text{Alloy 1} \\ \text{Alloy 2} \end{array}$$

$$N = \begin{array}{c} \text{Supplier } A \quad \text{Supplier } B \\ \left[\begin{array}{cc} \$0.75 & \$0.70 \\ \$6.50 & \$6.70 \\ \$0.40 & \$0.50 \end{array}\right] \end{array} \begin{array}{c} \text{Copper} \\ \text{Tin} \\ \text{Zinc} \end{array}$$

(A) Find MN and interpret.   (B) Find [1 1]MN and interpret.

31. *Labor costs.* A company with manufacturing plants in California and Texas has labor-hour and wage requirements for the manufacture of two inexpensive calculators as given in matrices M and N below:

Labor-hours per calculator

$$M = \begin{array}{c} \\ \\ \end{array} \begin{array}{ccc} \text{Fabricating} & \text{Assembly} & \text{Packaging} \\ \text{department} & \text{department} & \text{department} \\ \left[\begin{array}{ccc} 0.15\text{ hr} & 0.10\text{ hr} & 0.05\text{ hr} \\ 0.25\text{ hr} & 0.20\text{ hr} & 0.05\text{ hr} \end{array}\right] & & \end{array} \begin{array}{c} \text{Model } A \\ \text{Model } B \end{array}$$

Hourly wages

$$N = \begin{array}{c} \text{California} \quad \text{Texas} \\ \text{plant} \qquad \text{plant} \\ \left[\begin{array}{cc} \$15 & \$12 \\ \$12 & \$10 \\ \$\ 4 & \$\ 4 \end{array}\right] \end{array} \begin{array}{l} \text{Fabricating department} \\ \text{Assembly department} \\ \text{Packaging department} \end{array}$$

(A) What is the labor cost for producing one model B calculator in California? Set up a dot product and multiply.

(B) Find MN and interpret.

32. *Investment analysis.* A person has $5,000 to invest, part at 5% and the rest at 10%. How much should be invested at each rate to yield $400 per year? Set up a system of equations and solve using augmented matrix methods.

33. *Investment analysis.* Solve Problem 32 by using a matrix equation and the inverse of the coefficient matrix.

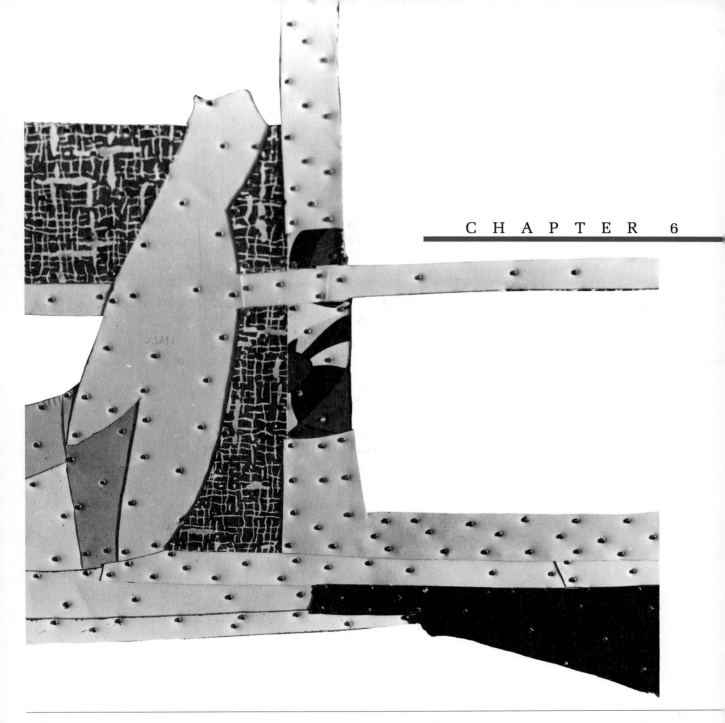

Linear Inequalities and Linear Programming

Contents

In this chapter we will discuss linear inequalities in two and more variables. In addition, we will introduce a relatively new and powerful mathematical tool called *linear programming*, which will be used to solve a variety of interesting practical problems. The row operations on matrices introduced in Chapter 5 will be particularly useful in Section 6-4.

S E C T I O N   6-1   Systems of Linear Inequalities in Two Variables

◆ GRAPHING LINEAR INEQUALITIES IN TWO VARIABLES
◆ SOLVING SYSTEMS OF LINEAR INEQUALITIES GRAPHICALLY
◆ APPLICATIONS

Many applications of mathematics involve systems of inequalities rather than systems of equations. A graph is often the most convenient way to represent the solutions of a system of linear inequalities in two variables. In this section we discuss techniques for graphing both a single linear inequality in two variables and a system of linear inequalities in two variables.

◆ GRAPHING LINEAR INEQUALITIES IN TWO VARIABLES

We know how to graph first-degree equations such as

$$y = 2x - 3 \qquad \text{and} \qquad 2x - 3y = 5$$

but how do we graph first-degree inequalities such as the following?

$$y \leq 2x - 3 \qquad \text{and} \qquad 2x - 3y > 5$$

We will find that graphing these inequalities is almost as easy as graphing the equations, but first, we must discuss some important subsets of a plane with a rectangular coordinate system.

A line divides the plane into two halves called **half-planes.** A vertical line divides it into **left** and **right half-planes;** a nonvertical line divides it into **upper** and **lower half-planes** (Fig. 1).

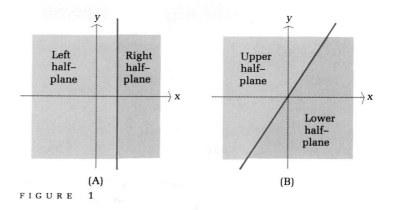

(A)   (B)

FIGURE 1

To investigate the half-planes determined by a linear equation such as $y - x = -2$, we rewrite the equation as $y = x - 2$. For any given value of x, there is exactly one value for y such that (x, y) lies on the line. For example, for $x = 4$, we have $y = 4 - 2 = 2$. For the same x and smaller values of y, the point (x, y) will lie below the line, since $y < x - 2$. Thus, the lower half-plane corresponds to the solution of the inequality $y < x - 2$. Similarly, the upper half-plane corresponds to $y > x - 2$, as shown in Figure 2.

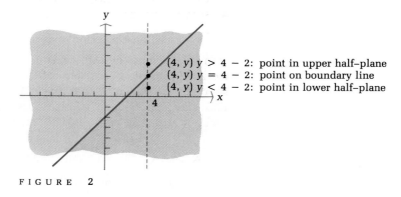

FIGURE 2

The four inequalities formed from $y = x - 2$ by replacing the $=$ sign by $>$, $\geq$, $<$, and $\leq$, respectively, are

$$y > x - 2 \qquad y \geq x - 2 \qquad y < x - 2 \qquad y \leq x - 2$$

The graph of each is a half-plane, excluding the boundary line for $<$ and $>$, and including the boundary line for $\leq$ and $\geq$. In Figure 3 (page 268) the half-planes are indicated with small arrows on the graph of $y = x - 2$ and then graphed as shaded regions. Excluded boundary lines are shown as dashed lines, and included boundary lines are shown as solid lines.

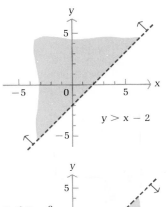

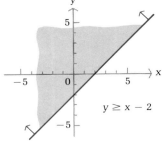

$y > x - 2$

$y \geq x - 2$

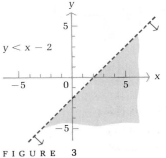

$y < x - 2$

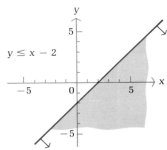

$y \leq x - 2$

FIGURE 3

The preceding discussion suggests the following theorem, which is stated without proof:

THEOREM 1

Graphs of Linear Inequalities

The graph of the linear inequality

$$Ax + By < C \quad \text{or} \quad Ax + By > C$$

with $B \neq 0$, is either the upper half-plane or the lower half-plane (but not both) determined by the line $Ax + By = C$.

If $B = 0$, the graph of

$$Ax < C \quad \text{or} \quad Ax > C$$

is either the left half-plane or the right half-plane (but not both) determined by the line $Ax = C$.

As a consequence of this theorem, we state a simple and fast mechanical procedure for graphing linear inequalities.

Procedure for Graphing Linear Inequalities

Step 1. First graph $Ax + By = C$ as a dashed line if equality is not included in the original statement or as a solid line if equality is included.

Step 2. Choose a test point anywhere in the plane not on the line [the origin $(0, 0)$ often requires the least computation] and substitute the coordinates into the inequality.

Step 3. The graph of the original inequality includes the half-plane containing the test point if the inequality is satisfied by that point or the half-plane not containing the test point if the inequality is not satisfied by that point.

◆ EXAMPLE 1   Graph $2x - 3y \leqslant 6$.

Solution   *Step 1.* Graph $2x - 3y = 6$ as a solid line, since equality is included in the original statement.

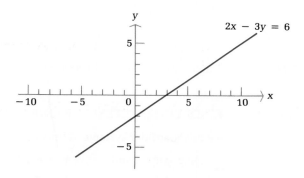

Step 2. Pick a convenient test point above or below the line. The origin $(0, 0)$ requires the least computation, so, substituting $(0, 0)$ into the inequality, we get

$$2x - 3y \leqslant 6$$
$$2(0) - 3(0) = 0 \leqslant 6$$

This is a true statement; therefore, the point $(0, 0)$ is in the solution set.

Step 3. The line $2x - 3y = 6$ and the half-plane containing the origin form the graph of $2x - 3y \leqslant 6$, as shown at the top of the next page.

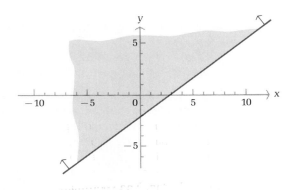

PROBLEM   1   Graph $6x - 3y > 18$.

◆ EXAMPLE   2   Graph: (A) $y > -3$ (B) $2x \leqslant 5$ (C) $x \leqslant 3y$

Solutions   (A)   (B)   (C)

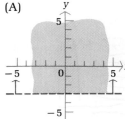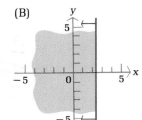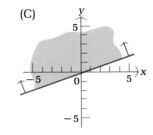

PROBLEM   2   Graph: (A) $y < 4$ (B) $4x \geqslant -9$ (C) $3x \geqslant 2y$

◆ SOLVING SYSTEMS OF LINEAR INEQUALITIES GRAPHICALLY

We now consider systems of linear inequalities such as

$$x + y \geqslant 6 \qquad \text{and} \qquad 2x + y \leqslant 22$$
$$2x - y \geqslant 0 \qquad\qquad\qquad\quad x + y \leqslant 13$$
$$\qquad\qquad\qquad\qquad\qquad\qquad 2x + 5y \leqslant 50$$
$$\qquad\qquad\qquad\qquad\qquad\qquad\quad x \geqslant 0$$
$$\qquad\qquad\qquad\qquad\qquad\qquad\quad y \geqslant 0$$

We wish to **solve** such systems **graphically;** that is, to find the graph of all ordered pairs of real numbers (x, y) that simultaneously satisfy all the inequalities in the system. The graph is called the **solution region** for the system. (In many applications, the solution region is also called the **feasible region.**) To find the solution region, we graph each inequality in the system and then take the intersection of all the graphs. To simplify the discussion that follows, **we will consider only systems of linear inequalities where equality is included in each statement in the system.**

◆ E X A M P L E 3 Solve the following system of linear inequalities graphically:

$$x + y \geq 6$$
$$2x - y \geq 0$$

Solution Graph the line $x + y = 6$ and shade the region that satisfies the inequality $x + y \geq 6$. This region is shaded in gray in figure A. Next, graph the line $2x - y = 0$ and shade the region that satisfies the inequality $2x - y \geq 0$. This region is shaded in color in figure A. The solution region for the system of inequalities is the intersection of these two regions. This is the region shaded in both gray and color in figure A and redrawn in figure B with only the solution region shaded. The coordinates of any point in the shaded region of figure B specify a solution to the system. For example, the points $(2, 4)$, $(6, 3)$, and $(7.43, 8.56)$ are three of infinitely many solutions, as can be easily checked. The intersection point $(2, 4)$ is obtained by solving the equations $x + y = 6$ and $2x - y = 0$ simultaneously.

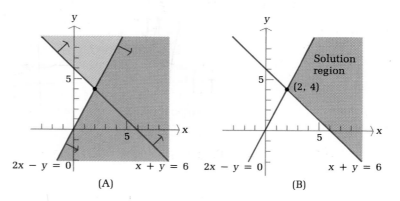

(A)   (B)   ◆

P R O B L E M 3 Solve the following system of linear inequalities graphically:

$$3x + y \leq 21$$
$$x - 2y \leq 0$$

◆

The points of intersection of the lines that form the boundary of a solution region will play a fundamental role in the solution of linear programming problems, which are discussed in the next section.

Corner Point

A **corner point** of a solution region is a point in the solution region that is the intersection of two boundary lines.

For example, the point $(2, 4)$ is the only corner point of the solution region in Example 3.

◆ E X A M P L E   4   Solve the following system of linear inequalities graphically, and find the corner points:

$$2x + y \le 22$$
$$x + y \le 13$$
$$2x + 5y \le 50$$
$$x \ge 0$$
$$y \ge 0$$

Solution   The inequalities $x \ge 0$ and $y \ge 0$ indicate that the solution region will lie in the first quadrant.* Thus, we can restrict our attention to that portion of the plane. First, we graph the lines

$$2x + y = 22$$ Find the x and y intercepts of each line; then sketch the
$$x + y = 13$$ line through these points.
$$2x + 5y = 50$$

Next, choosing (0, 0) as a test point, we see that the graph of each of the first three inequalities in the system consists of its corresponding line and the half-plane lying below the line, as indicated by the small arrows in the figure. Thus, the solution region of the system consists of the points in the first quadrant that simultaneously lie on or below all three of these lines (see the shaded region in the figure).

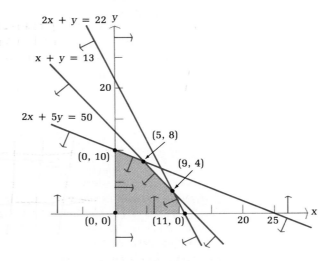

_____

* The inequalities $x \ge 0$ and $y \ge 0$ will occur frequently in applications involving systems of inequalities, since x and y will often represent quantities that cannot be negative (number of units produced, number of hours worked, etc.).

The corner points (0, 0), (0, 10), and (11, 0) can be determined from the graph. The other two corner points are determined as follows:

Solve the system

$$2x + 5y = 50$$
$$x + y = 13$$

to obtain (5, 8).

Solve the system

$$2x + y = 22$$
$$x + y = 13$$

to obtain (9, 4).

Note that the lines $2x + 5y = 50$ and $2x + y = 22$ also intersect, but the intersection point is not part of the solution region, and hence, is not a corner point.

◆

PROBLEM 4   Solve the following system of linear inequalities graphically, and find the corner points:

$$5x + y \geqslant 20$$
$$x + y \geqslant 12$$
$$x + 3y \geqslant 18$$
$$x \geqslant 0$$
$$y \geqslant 0$$

◆

If we compare the solution regions of Example 3 and Example 4, we see that there is a fundamental difference between these two regions. We can draw a circle around the solution region in Example 4; however, it is impossible to include all the points in the solution region in Example 3 in any circle, no matter how large we draw it. This leads to the following definition:

Bounded and Unbounded Solution Regions

A solution region of a system of linear inequalities is **bounded** if it can be enclosed within a circle. If it cannot be enclosed within a circle, then it is **unbounded.**

Thus, the solution region for Example 4 is bounded and the solution region for Example 3 is unbounded. This definition will be important in the next section.

◆ E X A M P L E 5

Medicine

A patient in a hospital is required to have at least 84 units of drug A and 120 units of drug B each day (assume that an overdosage of either drug is harmless). Each gram of substance M contains 10 units of drug A and 8 units of drug B, and each gram of substance N contains 2 units of drug A and 4 units of drug B. How many grams of substances M and N can be mixed to meet the minimum daily requirements?

Solution

To clarify relationships, we summarize the information in the following table:

| | AMOUNT OF DRUG PER GRAM | | MINIMUM DAILY |
	Substance M	Substance N	REQUIREMENT
DRUG A	10 units	2 units	84 units
DRUG B	8 units	4 units	120 units

Let

x = Number of grams of substance M used

y = Number of grams of substance N used

Then

$10x$ = Number of units of drug A in x grams of substance M

$2y$ = Number of units of drug A in y grams of substance N

$8x$ = Number of units of drug B in x grams of substance M

$4y$ = Number of units of drug B in y grams of substance N

The following conditions must be satisfied to meet daily requirements:

$$\begin{pmatrix} \text{Number of units of} \\ \text{drug } A \\ \text{in } x \text{ grams of substance } M \end{pmatrix} + \begin{pmatrix} \text{Number of units of} \\ \text{drug } A \\ \text{in } y \text{ grams of substance } N \end{pmatrix} \geq 84$$

$$\begin{pmatrix} \text{Number of units of} \\ \text{drug } B \\ \text{in } x \text{ grams of substance } M \end{pmatrix} + \begin{pmatrix} \text{Number of units of} \\ \text{drug } B \\ \text{in } y \text{ grams of substance } N \end{pmatrix} \geq 120$$

$$(\text{Number of grams of substance } M \text{ used}) \geq 0$$

$$(\text{Number of grams of substance } N \text{ used}) \geq 0$$

Converting these verbal statements into symbolic statements by using the variables x and y introduced above, we obtain the following system of linear inequalities:

$$10x + 2y \geqslant 84 \qquad \text{Drug } A \text{ restriction}$$
$$8x + 4y \geqslant 120 \qquad \text{Drug } B \text{ restriction}$$
$$x \geqslant 0 \qquad \text{Cannot use a negative amount of } M$$
$$y \geqslant 0 \qquad \text{Cannot use a negative amount of } N$$

Graphing this system of linear inequalities, we obtain the set of feasible solutions, or the feasible region (solution region), as shown in the figure. Thus, any point in the shaded area (including the straight-line boundaries) will meet the daily requirements; any point outside the shaded area will not. For example, 4 units of drug M and 23 units of drug N will meet the daily requirements, but 4 units of drug M and 21 units of drug N will not. (Note that the feasible region is unbounded.)

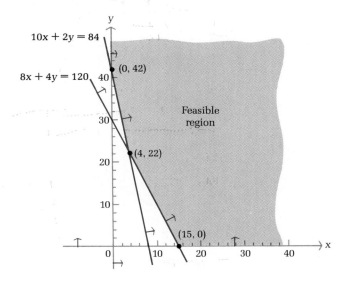

PROBLEM 5

Resource Allocation

A manufacturing plant makes two types of inflatable boats, a two-person boat and a four-person boat. Each two-person boat requires 0.9 labor-hour in the cutting department and 0.8 labor-hour in the assembly department. Each four-person boat requires 1.8 labor-hours in the cutting department and 1.2 labor-hours in the assembly department. The maximum labor-hours available each month in the cutting and assembly departments are 864 and 672, respectively.

(A) Summarize this information in a table.
(B) If x two-person boats and y four-person boats are manufactured each month, write a system of linear inequalities that reflect the conditions indicated. Find the set of feasible solutions graphically. ◆

Answers to Matched Problems

1. Graph $6x - 3y = 18$ as a dashed line (since equality is not included). Choosing the origin $(0, 0)$ as a test point, we see that $6(0) - 3(0) > 18$ is a false statement; thus, the lower half-plane determined by $6x - 3y = 18$ is the graph of $6x - 3y > 18$.

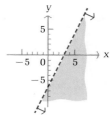

2. (A) 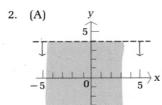 (B) (C)

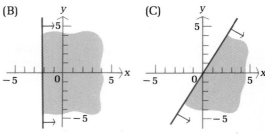

3. 4.

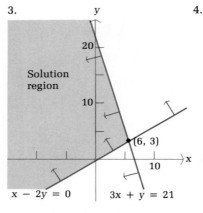

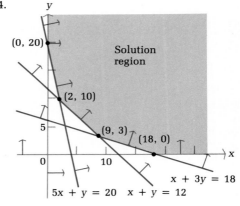

5. (A)

| | LABOR-HOURS REQUIRED | | MAXIMUM LABOR-HOURS AVAILABLE PER MONTH |
	Two-Person Boat	*Four-Person Boat*	
CUTTING DEPARTMENT	0.9	1.8	864
ASSEMBLY DEPARTMENT	0.8	1.2	672

(B) $0.9x + 1.8y \leq 864$
   $0.8x + 1.2y \leq 672$
   $x \geq 0$
   $y \geq 0$

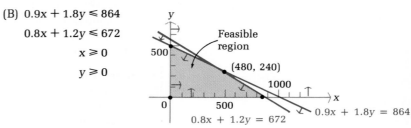

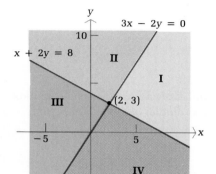

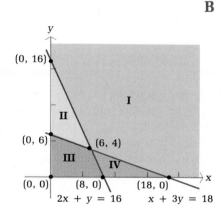

A   *Graph each inequality.*

1. $y \leqslant x - 1$ 2. $y > x + 1$ 3. $3x - 2y > 6$
4. $2x - 5y \leqslant 10$ 5. $x \geqslant -4$ 6. $y < 5$
7. $-4 \leqslant y < 4$ 8. $0 \leqslant x < 6$ 9. $6x + 4y \geqslant 24$
10. $4x + 8y \geqslant 32$ 11. $5x \leqslant -2y$ 12. $6x \geqslant 4y$

In Problems 13–16, match the solution region of each system of linear inequalities with one of the four regions shown in the figure in the margin.

13. $x + 2y \leqslant 8$ 14. $x + 2y \geqslant 8$ 15. $x + 2y \geqslant 8$ 16. $x + 2y \leqslant 8$
    $3x - 2y \geqslant 0$ $3x - 2y \leqslant 0$ $3x - 2y \geqslant 0$ $3x - 2y \leqslant 0$

Solve each system of linear inequalities graphically.

17. $3x + y \geqslant 6$ 18. $3x + 4y \leqslant 12$ 19. $x - 2y \leqslant 12$ 20. $2x + 5y \leqslant 20$
    $x \leqslant 4$ $y \geqslant -3$ $2x + y \geqslant 4$ $x - 5y \geqslant -5$

B   *In Problems 21–24, match the solution region of each system of linear inequalities with one of the four regions shown in the figure in the margin. Identify the corner points of each solution region.*

21. $x + 3y \leqslant 18$ 22. $x + 3y \leqslant 18$ 23. $x + 3y \geqslant 18$
    $2x + y \geqslant 16$ $2x + y \leqslant 16$ $2x + y \geqslant 16$
    $x \geqslant 0$ $x \geqslant 0$ $x \geqslant 0$
    $y \geqslant 0$ $y \geqslant 0$ $y \geqslant 0$

24. $x + 3y \geqslant 18$
    $2x + y \leqslant 16$
    $x \geqslant 0$
    $y \geqslant 0$

Solve the following systems graphically, and indicate whether each solution region is bounded or unbounded. Find the coordinates of each corner point.

25. $2x + 3y \leqslant 12$ 26. $3x + 4y \leqslant 24$ 27. $2x + y \leqslant 10$
    $x \geqslant 0$ $x \geqslant 0$ $x + 2y \leqslant 8$
    $y \geqslant 0$ $y \geqslant 0$ $x \geqslant 0$
                                                             $y \geqslant 0$

28. $6x + 3y \leqslant 24$ 29. $2x + y \geqslant 10$ 30. $4x + 3y \geqslant 24$
    $3x + 6y \leqslant 30$ $x + 2y \geqslant 8$ $3x + 4y \geqslant 8$
    $x \geqslant 0$ $x \geqslant 0$ $x \geqslant 0$
    $y \geqslant 0$ $y \geqslant 0$ $y \geqslant 0$

31. $2x + y \leqslant 10$
    $x + y \leqslant 7$
    $x + 2y \leqslant 12$
    $x \geqslant 0$
    $y \geqslant 0$

32. $3x + y \leqslant 21$
    $x + y \leqslant 9$
    $x + 3y \leqslant 21$
    $x \geqslant 0$
    $y \geqslant 0$

33. $2x + y \geqslant 16$
    $x + y \geqslant 12$
    $x + 2y \geqslant 14$
    $x \geqslant 0$
    $y \geqslant 0$

34. $3x + y \geqslant 24$
    $x + y \geqslant 16$
    $x + 3y \geqslant 30$
    $x \geqslant 0$
    $y \geqslant 0$

C *Solve the following systems graphically, and indicate whether each solution region is bounded or unbounded. Find the coordinates of each corner point.*

35. $x + 4y \leqslant 32$
    $3x + y \leqslant 30$
    $4x + 5y \geqslant 51$

36. $x + y \leqslant 11$
    $x + 5y \geqslant 15$
    $2x + y \geqslant 12$

37. $4x + 3y \leqslant 48$
    $2x + y \geqslant 24$
    $x \leqslant 9$

38. $2x + 3y \geqslant 24$
    $x + 3y \leqslant 15$
    $y \geqslant 4$

39. $x - y \leqslant 0$
    $2x - y \leqslant 4$
    $0 \leqslant x \leqslant 8$

40. $2x + 3y \geqslant 12$
    $-x + 3y \leqslant 3$
    $0 \leqslant y \leqslant 5$

41. $-x + 3y \geqslant 1$
    $5x - y \geqslant 9$
    $x + y \leqslant 9$
    $x \leqslant 5$

42. $x + y \leqslant 10$
    $5x + 3y \geqslant 15$
    $-2x + 3y \leqslant 15$
    $2x - 5y \leqslant 6$

APPLICATIONS

Business & Economics

43. *Manufacturing—resource allocation.* A manufacturing company makes two types of water skis, a trick ski and a slalom ski. The trick ski requires 6 labor-hours for fabricating and 1 labor-hour for finishing. The slalom ski requires 4 labor-hours for fabricating and 1 labor-hour for finishing. The maximum labor-hours available per day for fabricating and finishing are 108 and 24, respectively. If x is the number of trick skis and y is the number of slalom skis produced per day, write a system of linear inequalities that indicates appropriate restraints on x and y. Find the set of feasible solutions graphically for the number of each type of ski that can be produced.

44. *Manufacturing—resource allocation.* A furniture manufacturing company manufactures dining room tables and chairs. A table requires 8 labor-hours for assembling and 2 labor-hours for finishing. A chair requires 2 labor-hours for assembling and 1 labor-hour for finishing. The maximum labor-hours available per day for assembly and finishing are 400 and 120, respectively. If x is the number of tables and y is the number of chairs produced per day, write a system of linear inequalities that indicates appropriate restraints on x and y. Find the set of feasible solutions graphically for the number of tables and chairs that can be produced.

Life Sciences

45. *Nutrition — plants.* A farmer can buy two types of plant food, mix *A* and mix *B*. Each cubic yard of mix *A* contains 20 pounds of phosphoric acid, 30 pounds of nitrogen, and 5 pounds of potash. Each cubic yard of mix *B* contains 10 pounds of phosphoric acid, 30 pounds of nitrogen, and 10 pounds of potash. The minimum monthly requirements are 460 pounds of phosphoric acid, 960 pounds of nitrogen, and 220 pounds of potash. If *x* is the number of cubic yards of mix *A* used and *y* is the number of cubic yards of mix *B* used, write a system of linear inequalities that indicates appropriate restraints on *x* and *y*. Find the set of feasible solutions graphically for the amounts of mix *A* and mix *B* that can be used.

46. *Nutrition — people.* A dietitian in a hospital is to arrange a special diet using two foods. Each ounce of food *M* contains 30 units of calcium, 10 units of iron, and 10 units of vitamin A. Each ounce of food *N* contains 10 units of calcium, 10 units of iron, and 30 units of vitamin A. The minimum requirements in the diet are 360 units of calcium, 160 units of iron, and 240 units of vitamin A. If *x* is the number of ounces of food *M* used and *y* is the number of ounces of food *N* used, write a system of linear inequalities that reflects the conditions indicated. Find the set of feasible solutions graphically for the amount of each kind of food that can be used.

Social Sciences

47. *Psychology.* In an experiment on conditioning, a psychologist uses two types of Skinner (conditioning) boxes with mice and rats. Each mouse spends 10 minutes per day in box *A* and 20 minutes per day in box *B*. Each rat spends 20 minutes per day in box *A* and 10 minutes per day in box *B*. The total maximum time available per day is 800 minutes for box *A* and 640 minutes for box *B*. We are interested in the various numbers of mice and rats that can be used in the experiment under the conditions stated. If we let *x* be the number of mice used and *y* the number of rats used, write a system of linear inequalities that indicates appropriate restrictions on *x* and *y*. Find the set of feasible solutions graphically.

SECTION 6-2

Linear Programming in Two Dimensions — A Geometric Approach

◆ A LINEAR PROGRAMMING PROBLEM
◆ LINEAR PROGRAMMING — A GENERAL DESCRIPTION
◆ GEOMETRIC SOLUTION OF LINEAR PROGRAMMING PROBLEMS
◆ APPLICATIONS

Several problems discussed in the last section are related to a more general type of problem called a *linear programming problem*. Linear programming is a mathematical process that has been developed to help management in decision-making, and it has become one of the most widely used and best-known tools of management science. We will introduce this topic by considering an example in detail, using an intuitive geometric approach. Insight gained from this approach will prove invaluable when we later consider an algebraic approach that is less intuitive but necessary in solving most real-world problems.

> ### ▬ Notation Change
>
> For efficiency of generalization in later sections, we will now change variable notation from letters such as x and y to subscript forms such as x_1 and x_2.

◆ A LINEAR PROGRAMMING PROBLEM

We begin our discussion with a concrete example. The geometric method of solution will suggest two important theorems and a simple general geometric procedure for solving linear programming problems in two variables.

◆ E X A M P L E   6

Production Scheduling

A manufacturer of lightweight mountain tents makes a standard model and an expedition model for national distribution. Each standard tent requires 1 labor-hour from the cutting department and 3 labor-hours from the assembly department. Each expedition tent requires 2 labor-hours from the cutting department and 4 labor-hours from the assembly department. The maximum labor-hours available per day in the cutting department and the assembly department are 32 and 84, respectively. If the company makes a profit of $50 on each standard tent and $80 on each expedition tent, how many tents of each type should be manufactured each day to maximize the total daily profit (assuming all tents can be sold)?

Solution

This is an example of a linear programming problem. To see relationships more clearly, we summarize the manufacturing requirements, objectives, and restrictions in Table 1.

T A B L E   1

| | LABOR-HOURS PER TENT | | MAXIMUM LABOR-HOURS AVAILABLE PER DAY |
	Standard Model	*Expedition Model*	
CUTTING DEPARTMENT	1	2	32
ASSEMBLY DEPARTMENT	3	4	84
PROFIT PER TENT	$50	$80	

We now proceed to formulate a mathematical model for the problem and then to solve it by using geometric methods.

Objective Function

The *objective* of management is to *decide* how many of each tent model should be produced each day so as to maximize profit. Let

x_1 = Number of standard tents produced per day $\left.\right\}$ Decision
x_2 = Number of expedition tents produced per day $\left.\right\}$ variables

The following equation gives the total profit for x_1 standard tents and x_2 expedition tents manufactured each day, assuming all tents manufactured are sold:

$$P = 50x_1 + 80x_2 \qquad \text{Objective function}$$

or $P = 50x + 80y$

Mathematically, the manufacturer needs to decide on values for the **decision variables** (x_1, x_2) that achieve its objective—that is, maximize the **objective function** (profit) $P = 50x_1 + 80x_2$. (Note that P is a function of two independent variables.) It appears that the profit can be made as large as we like by manufacturing more and more tents—or can it?

Constraints

Any manufacturing company, no matter how large or small, has manufacturing limits imposed by available resources, plant capacity, demand, and so forth. These limits are referred to as **problem constraints.**

Cutting Department Constraint

$$\begin{pmatrix} \text{Daily cutting} \\ \text{time for } x_1 \\ \text{standard tents} \end{pmatrix} + \begin{pmatrix} \text{Daily cutting} \\ \text{time for } x_2 \\ \text{expedition tents} \end{pmatrix} \leq \begin{pmatrix} \text{Maximum labor-} \\ \text{hours available} \\ \text{per day} \end{pmatrix}$$

$$1x_1 \quad + \quad 2x_2 \quad \leq \quad 32$$

or $x + 2y \leq 32$

Assembly Department Constraint

$$\begin{pmatrix} \text{Daily assembly} \\ \text{time for } x_1 \\ \text{standard tents} \end{pmatrix} + \begin{pmatrix} \text{Daily assembly} \\ \text{time for } x_2 \\ \text{expedition tents} \end{pmatrix} \leq \begin{pmatrix} \text{Maximum labor-} \\ \text{hours available} \\ \text{per day} \end{pmatrix}$$

$$3x_1 \quad + \quad 4x_2 \quad \leq \quad 84$$

or $3x + 4y \leq 84$

Nonnegative Constraints

It is not possible to manufacture a negative number of tents; thus, we have the **nonnegative constraints**

$$x_1 \geq 0$$
$$x_2 \geq 0$$

which we usually write in the form

$$x_1, x_2 \geq 0$$

Mathematical Model   We now have a **mathematical model** for the problem under consideration:

Maximize $P = 50x_1 + 80x_2$ Objective function

Subject to $\left.\begin{array}{l} x_1 + 2x_2 \leq 32 \\ 3x_1 + 4x_2 \leq 84 \end{array}\right\}$ Problem constraints

$x_1, x_2 \geq 0$ Nonnegative constraints

Graphical Solution   **Solving** the set of linear inequality constraints **graphically** (see the last section), we obtain the feasible region for production schedules (Fig. 4).

Only need to plug in coordinates for points of intersections* to find max. profits.

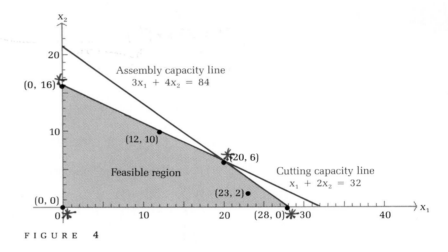

FIGURE 4

By choosing a production schedule (x_1, x_2) from the feasible region, a profit can be determined using the objective function

$P = 50x_1 + 80x_2$

For example, if $x_1 = 12$ and $x_2 = 10$, then the profit for the day would be

$P = 50(12) + 80(10)$
$\quad = \$1{,}400$

Or if $x_1 = 23$ and $x_2 = 2$, then the profit for the day would be

$P = 50(23) + 80(2)$
$\quad = \$1{,}310$

The question is, out of all possible production schedules (x_1, x_2) from the feasible region, which schedule(s) produces the maximum profit? Thus, we have a **maximization problem.** Since point by point checking is impossible (there are infinitely many points to check), we must find another way.

By assigning P in $P = 50x_1 + 80x_2$ a particular value and plotting the resulting equation in Figure 4, we obtain a **constant-profit line (isoprofit line).** Every point in the feasible region on this line represents a production schedule that will produce the same profit. By doing this for a number of values for P, we obtain a family of constant-profit lines (Fig. 5) that are parallel to each other, since they all have the same slope. To see this, we write $P = 50x_1 + 80x_2$ in the slope–intercept form

$$x_2 = -\frac{5}{8} x_1 + \frac{P}{80}$$

and note that for any profit P, the constant-profit line has slope $-\frac{5}{8}$. We also observe that as the profit P increases, the x_2 intercept $(P/80)$ increases, and the line moves away from the origin.

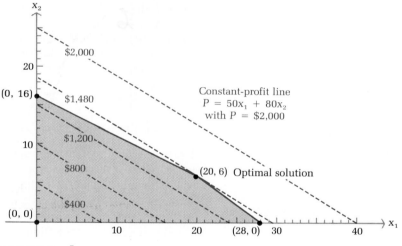

FIGURE 5
Constant-profit lines

Thus, the maximum profit occurs at a point where a constant-profit line is the farthest from the origin but still in contact with the feasible region. In this example, this occurs at (20, 6), as is seen in Figure 5. Thus, if the manufacturer makes 20 standard tents and 6 expedition tents per day, the profit will be maximized at

$$P = 50(20) + 80(6)$$
$$= \$1,480$$

The point (20, 6) is called an **optimal solution** to the problem, because it maximizes the objective (profit) function and is in the feasible region. In general, it appears that a maximum profit occurs at one of the corner points. We also note that the minimum profit ($P = 0$) occurs at the corner point (0, 0). ◆

PROBLEM 6   A manufacturing plant makes two types of inflatable boats, a two-person boat and a four-person boat. Each two-person boat requires 0.9 labor-hour from the cutting department and 0.8 labor-hour from the assembly department. Each four-person boat requires 1.8 labor-hours from the cutting department and 1.2 labor-hours from the assembly department. The maximum labor-hours available per month in the cutting department and the assembly department are 864 and 672, respectively. The company makes a profit of $25 on each two-person boat and $40 on each four-person boat.

(A) Summarize the relevant material in a table similar to Table 1 in Example 6.
(B) Identify the decision variables.
(C) Write the objective function P.
(D) Write the problem constraints and the nonnegative constraints.
(E) Graph the feasible region. Include graphs of the objective function for $P =$ $5,000, $P = $10,000, $P = $15,000, and $P = $21,600.
(F) From the graph and constant-profit lines, determine how many boats should be manufactured each month to maximize the profit. What is the maximum profit? ◆

◆ LINEAR PROGRAMMING — A GENERAL DESCRIPTION

In Example 6 and Problem 6, the optimal solution occurs at a corner point of the feasible region. Is this always the case? The answer is a qualified yes, as will be seen in Theorem 2, stated at the top of the next page. First, we give a few general definitions.

A **linear programming problem** is one that is concerned with finding the **optimal value** (maximum or minimum value) of a linear **objective function** of the form

$$z = c_1x_1 + c_2x_2 + \cdots + c_nx_n$$

where the **decision variables** $x_1, x_2, \ldots, x_n$ are subject to **problem constraints** in the form of linear inequalities and equations. In addition, the decision variables must satisfy the **nonnegative constraints** $x_i \geq 0, i = 1, 2, \ldots, n$. The set of points satisfying both the problem constraints and the nonnegative constraints is called the **feasible region** for the problem. Any point in the feasible region that produces the optimal value of the objective function over the feasible region is called an **optimal solution**.

THEOREM 2

Fundamental Theorem of Linear Programming

If the optimal value of the objective function in a linear programming problem exists, then that value must occur at one (or more) of the corner points of the feasible region.

Theorem 2 provides a simple procedure for solving a linear programming problem, *provided the problem has a solution—not all do.* In order to use Theorem 2, we must know that the problem under consideration has a solution. Theorem 3 provides some conditions that will ensure that a linear programming problem has a solution.

THEOREM 3

Existence of Solutions

Given a linear programming problem with feasible region S and objective function $z = ax_1 + bx_2$:

(A) If S is bounded, then z has both a maximum and a minimum value on S. That is, both

   Maximize z over S and   Minimize z over S

have solutions.

(B) If S is unbounded and $a > 0$ and $b > 0$, then z has a minimum value over S, but no maximum value over S. That is,

   Minimize z over S has a solution

but

   Maximize z over S has no solution

(C) If S is the empty set (that is, there are no points that satisfy all the constraints), then z has neither a maximum value nor a minimum value over S.

Theorem 3 does not cover all possibilities. For example, what happens if S is unbounded and one (or both) of the coefficients of the objective function are negative? Problems of this type require special techniques that we will not discuss. Since virtually all applied problems satisfy one of the conditions listed in Theorem 3, we will consider only problems of this type.

◆ GEOMETRIC SOLUTION OF LINEAR PROGRAMMING PROBLEMS

The discussion above leads to the following procedure for the geometric solution of linear programming problems with two decision variables:

Geometric Solution of a Linear Programming Problem with Two Decision Variables

Step 1. For an applied problem, summarize relevant material in table form (see Table 1 in Example 6).

Step 2. Form a mathematical model for the problem:

(A) Introduce decision variables, and write a linear objective function.
(B) Write problem constraints using linear inequalities and/or equations.
(C) Write nonnegative constraints.

Step 3. Graph the feasible region. Then, if according to Theorem 3 an optimal solution exists, find the coordinates of each corner point.

Step 4. Make a table listing the value of the objective function at each corner point.

Step 5. Determine the optimal solution(s) from the table in step 4.

Step 6. For an applied problem, interpret the optimal solution(s) in terms of the original problem.

Before we consider additional applications, let us use this procedure to solve some linear programming problems where the model has already been determined.

◆ EXAMPLE 7

(A) Minimize and maximize

$$z = 3x_1 + x_2$$

Subject to

$$2x_1 + x_2 \leq 20$$
$$10x_1 + x_2 \geq 36$$
$$2x_1 + 5x_2 \geq 36$$
$$x_1, x_2 \geq 0$$

(B) Minimize and maximize

$$z = 10x_1 + 20x_2$$

Subject to

$$6x_1 + 2x_2 \geq 36$$
$$2x_1 + 4x_2 \geq 32$$
$$x_2 \leq 20$$
$$x_1, x_2 \geq 0$$

Solutions

(A) We begin with step 3, since step 1 does not apply and step 2 has already been done for us.

Step 3. Graph the feasible region. Then, after checking Theorem 3 to determine that an optimal solution exists, find the coordinates of each corner point. Since S is bounded, z will have both a maximum and a minimum on S (Theorem 3A) and these will both occur at corner points (Theorem 2).

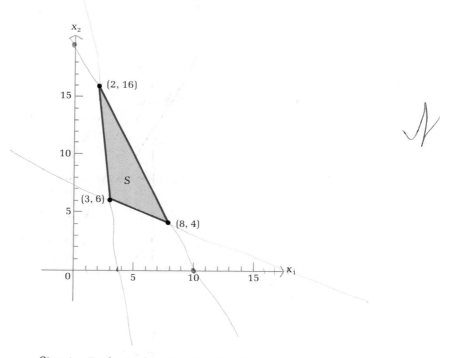

Step 4. Evaluate the objective function at each corner point, as shown in the table in the margin.

Step 5. Determine the optimal solutions from step 4. Examining the values in the table, we see that the minimum value of z is 15 at (3, 6) and the maximum value of z is 28 at (8, 4).

CORNER POINT (x_1, x_2)	$z = 3x_1 + x_2$
(3, 6)	15
(2, 16)	22
(8, 4)	28

(B) Again, we can begin with step 3.

Step 3. Graph the feasible region, as shown on the next page. Then, after checking Theorem 3 to determine that an optimal solution exists, find the coordinates of each corner point. Since S is unbounded and the coefficients of the objective function are positive, z has a minimum value on S but no maximum value (Theorem 3B).

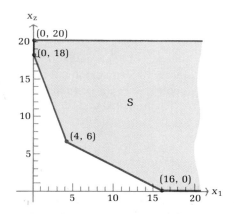

(0, 20)

(0, 18)

S

(4, 6)

(16, 0)

Step 4. Evaluate the objective function at each corner point, as shown in the table in the margin.

Step 5. Determine the optimal solution from step 4. The minimum value of z is 160 at (4, 6) and at (16, 0). ◆

CORNER POINT

(x_1, x_2)	$z = 10x_1 + 20x_2$
(0, 20)	400
(0, 18)	360
(4, 6)	160
(16, 0)	160

This is a **multiple optimal solution.**

In general, if two corner points are both optimal solutions to a linear programming problem, then any point on the line segment joining them is also an optimal solution.

This is the only time that optimal solutions also occur at noncorner points.

PROBLEM 7

(A) Maximize and minimize $z = 4x_1 + 2x_2$ subject to the constraints given in Example 7A.

(B) Maximize and minimize $z = 20x_1 + 5x_2$ subject to the constraints given in Example 7B. ◆

For an illustration of Theorem 3C, consider the following:

Maximize $P = 2x_1 + 3x_2$

Subject to $x + x_2 \geqslant 8$

$x_1 + 2x_2 \leqslant 8$

$2x_1 + x_2 \leqslant 10$

$x_1, x_2 \geqslant 0$

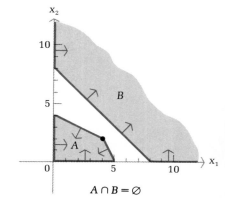

$A \cap B = \emptyset$

The intersection of the graphs of the constraint inequalities is the empty set; hence, the feasible region is empty. If this happens, then the problem should be reexamined to see if it has been formulated properly. If it has, then the management may have to reconsider items such as labor-hours, overtime, budget, and supplies allocated to the project in order to obtain a nonempty feasible region and a solution to the original problem.

◆ APPLICATIONS

◆ **E X A M P L E   8**

Medicine

We now convert Example 5 (in the preceding section) into a linear programming problem. A patient in a hospital is required to have at least 84 units of drug D_1 and 120 units of drug D_2 each day (assume that an overdosage of either drug is harmless). Each gram of substance M contains 10 units of drug D_1 and 8 units of drug D_2, and each gram of substance N contains 2 units of drug D_1 and 4 units of drug D_2. Now, suppose both M and N contain an undesirable drug D_3, 3 units per gram in M and 1 unit per gram in N. How many grams of each of substances M and N should be mixed to meet the minimum daily requirements and at the same time minimize the intake of drug D_3? How many units of the undesirable drug D_3 will be in this mixture?

Solution

Step 1. Summarize relevant material in table form:

	AMOUNT OF DRUG PER GRAM		MINIMUM DAILY
	Substance M	*Substance N*	REQUIREMENT
DRUG D_1	10 units	2 units	84 units
DRUG D_2	8 units	4 units	120 units
DRUG D_3	3 units	1 unit	

Step 2. Form a mathematical model for the problem. Let

$x_1 =$ Number of grams of substance M used ⎱ Decision
$x_2 =$ Number of grams of substance N used ⎰ variables

We form the linear objective function

$$C = 3x_1 + x_2$$

which gives the amount of the undesirable drug D_3 in x_1 grams of M and x_2 grams of N. Proceeding as in Example 5, we formulate the following mathematical model for the problem:

Minimize	$C = 3x_1 + x_2$	Objective function
Subject to	$10x_1 + 2x_2 \geq 84$	Drug D_1 constraint
	$8x_1 + 4x_2 \geq 120$	Drug D_2 constraint
	$x_1, x_2 \geq 0$	Nonnegative constraints

Step 3. Graph the feasible region. Then, after checking Theorem 3 to determine that an optimal solution exists, find the coordinates of each corner point. Solving the system of constraint inequalities graphically, we obtain the feasible region shown in Figure 6. Since the feasible region is unbounded and the coefficients of the objective function are positive, this minimization problem has a solution.

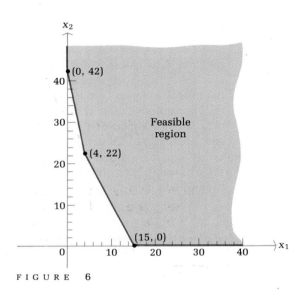

FIGURE 6

Step 4. Evaluate the objective function at each corner point, as shown in the table.

CORNER POINT (x_1, x_2)	$C = 3x_1 + x_2$
(0, 42)	42
(4, 22)	34
(15, 0)	45

Step 5. Determine the optimal solution from step 4. The optimal solution is $C = 34$ at the corner point (4, 22).

Step 6. Interpret the optimal solution in terms of the original problem. If we use 4 grams of substance M and 22 grams of substance N, we will supply the minimum daily requirements for drugs D_1 and D_2 and minimize the intake of the undesirable drug D_3 at 34 units. (Any other combination of M and N from the feasible region will result in a larger amount of the undesirable drug D_3.) ◆

A chicken farmer can buy a special food mix A at 20¢ per pound and a special food mix B at 40¢ per pound. Each pound of mix A contains 3,000 units of nutrient N_1 and 1,000 units of nutrient N_2; each pound of mix B contains 4,000 units of nutrient N_1 and 4,000 units of nutrient N_2. If the minimum daily requirements for the chickens collectively are 36,000 units of nutrient N_1 and 20,000 units of nutrient N_2, how many pounds of each food mix should be used each day to minimize daily food costs while meeting (or exceeding) the minimum daily nutrient requirements? What is the minimum daily cost? ◆

Answers to Matched Problems

6. (A)

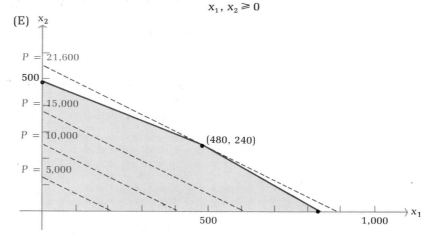

| | LABOR-HOURS REQUIRED | | MAXIMUM |
	Two-Person Boat	Four-Person Boat	LABOR-HOURS AVAILABLE PER MONTH
CUTTING DEPARTMENT	0.9	1.8	864
ASSEMBLY DEPARTMENT	0.8	1.2	672
PROFIT PER BOAT	$25	$40	

(B) $x_1 =$ Number of two-person boats produced each month
$x_2 =$ Number of four-person boats produced each month

(C) $P = 25x_1 + 40x_2$ (D) $0.9x_1 + 1.8x_2 \leqslant 864$
$0.8x_1 + 1.2x_2 \leqslant 672$
$x_1, x_2 \geqslant 0$

(E)

(F) 480 two-person boats, 240 four-person boats; Max $P = \$21,600$ per month

7. (A) Min $z = 24$ at $(3, 6)$; Max $z = 40$ at $(2, 16)$ and $(8, 4)$ (multiple optimal solution)

(B) Min $z = 90$ at $(0, 18)$; no maximum value

8. 8 lb of mix A, 3 lb of mix B; Min $C = \$2.80$ per day

A   Find the maximum value of each objective function over the feasible region S shown in the figure.

1. $z = x + y$
2. $z = 4x + y$
3. $z = 3x + 7y$
4. $z = 9x + 3y$

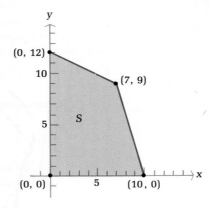

Find the minimum value of each objective function over the feasible region T shown in the figure.

5. $z = 7x + 4y$
6. $z = 7x + 9y$
7. $z = 3x + 8y$
8. $z = 5x + 4y$

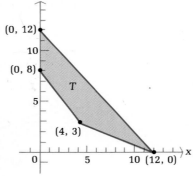

B   Solve the following linear programming problems:

9. Maximize $P = 5x_1 + 5x_2$
   Subject to $2x_1 + x_2 \leq 10$
   $x_1 + 2x_2 \leq 8$
   $x_1, x_2 \geq 0$

10. Maximize $P = 3x_1 + 2x_2$
    Subject to $6x_1 + 3x_2 \leq 24$
    $3x_1 + 6x_2 \leq 30$
    $x_1, x_2 \geq 0$

11. Minimize and maximize

    $z = 2x_1 + 3x_2$

    Subject to $2x_1 + x_2 \geq 10$
    $x_1 + 2x_2 \geq 8$
    $x_1, x_2 \geq 0$

12. Minimize and maximize

    $z = 8x_1 + 7x_2$

    Subject to $4x_1 + 3x_2 \geq 24$
    $3x_1 + 4x_2 \geq 8$
    $x_1, x_2 \geq 0$

13. Maximize $P = 30x_1 + 40x_2$
    Subject to $\quad 2x_1 + x_2 \leqslant 10$
    $x_1 + x_2 \leqslant 7$
    $x_1 + 2x_2 \leqslant 12$
    $x_1, x_2 \geqslant 0$

14. Maximize $P = 20x_1 + 10x_2$
    Subject to $\quad 3x_1 + x_2 \leqslant 21$
    $x_1 + x_2 \leqslant 9$
    $x_1 + 3x_2 \leqslant 21$
    $x_1, x_2 \geqslant 0$

15. Minimize and maximize

    $z = 10x_1 + 30x_2$

    Subject to $\quad 2x_1 + x_2 \geqslant 16$
    $x_1 + x_2 \geqslant 12$
    $x_1 + 2x_2 \geqslant 14$
    $x_1, x_2 \geqslant 0$

16. Minimize and maximize

    $z = 400x_1 + 100x_2$

    Subject to $\quad 3x_1 + x_2 \geqslant 24$
    $x_1 + x_2 \geqslant 16$
    $x_1 + 3x_2 \geqslant 30$
    $x_1, x_2 \geqslant 0$

17. Minimize and maximize

    $P = 30x_1 + 10x_2$

    Subject to $\quad 2x_1 + 2x_2 \geqslant 4$
    $6x_1 + 4x_2 \leqslant 36$
    $2x_1 + x_2 \leqslant 10$
    $x_1, x_2 \geqslant 0$

18. Minimize and maximize

    $P = 2x_1 + x_2$

    Subject to $\quad x_1 + x_2 \geqslant 2$
    $6x_1 + 4x_2 \leqslant 36$
    $4x_1 + 2x_2 \leqslant 20$
    $x_1, x_2 \geqslant 0$

19. Minimize and maximize

    $P = 3x_1 + 5x_2$

    Subject to $\quad x_1 + 2x_2 \leqslant 6$
    $x_1 + x_2 \leqslant 4$
    $2x_1 + 3x_2 \geqslant 12$
    $x_1, x_2 \geqslant 0$

20. Minimize and maximize

    $P = -x_1 + 3x_2$

    Subject to $\quad 2x_1 - x_2 \geqslant 4$
    $-x_1 + 2x_2 \leqslant 4$
    $x_2 \leqslant 6$
    $x_1, x_2 \geqslant 0$

21. Minimize and maximize

    $P = 20x_1 + 10x_2$

    Subject to $\quad 2x_1 + 3x_2 \geqslant 30$
    $2x_1 + x_2 \leqslant 26$
    $-2x_1 + 5x_2 \leqslant 34$
    $x_1, x_2 \geqslant 0$

22. Minimize and maximize

    $P = 12x_1 + 14x_2$

    Subject to $\quad -2x_1 + x_2 \geqslant 6$
    $x_1 + x_2 \leqslant 15$
    $3x_1 - x_2 \geqslant 0$
    $x_1, x_2 \geqslant 0$

23. Maximize $P = 20x_1 + 30x_2$
    Subject to $\quad 0.6x_1 + 1.2x_2 \leqslant 960$
    $0.03x_1 + 0.04x_2 \leqslant 36$
    $0.3x_1 + 0.2x_2 \leqslant 270$
    $x_1, x_2 \geqslant 0$

24. Minimize $C = 30x_1 + 10x_2$
    Subject to $\quad 1.8x_1 + 0.9x_2 \geqslant 270$
    $0.3x_1 + 0.2x_2 \geqslant 54$
    $0.01x_1 + 0.03x_2 \geqslant 3.9$
    $x_1, x_2 \geqslant 0$

C   **25.** The corner points for the bounded feasible region determined by the system of linear inequalities

$$x_1 + 2x_2 \leqslant 10$$
$$3x_1 + x_2 \leqslant 15$$
$$x_1, x_2 \geqslant 0$$

are $O = (10, 0)$, $A = (0, 5)$, $B = (4, 3)$, and $C = (5, 0)$. If $P = ax_1 + bx_2$ and $a, b > 0$, determine conditions on a and b that will ensure that the maximum value of P occurs:

(A) Only at A (B) Only at B (C) Only at C
(D) At both A and B (E) At both B and C

26. The corner points for the feasible region determined by the system of linear inequalities

$$x_1 + 4x_2 \geqslant 30$$
$$3x_1 + x_2 \geqslant 24$$
$$x_1, x_2 \geqslant 0$$

are $A = (0, 24)$, $B = (6, 6)$, $D = (30, 0)$. If $C = ax_1 + bx_2$ and $a, b > 0$, determine conditions on a and b that will ensure that the minimum value of C occurs:

(A) Only at A (B) Only at B (C) Only at D
(D) At both A and B (E) At both B and D

APPLICATIONS

Business & Economics

27. *Manufacturing—resource allocation.* A manufacturing company makes two types of water skis, a trick ski and a slalom ski. The relevant manufacturing data are given in the table. How many of each type of ski should be manufactured each day to realize a maximum profit? What is the maximum profit?

| | LABOR-HOURS PER SKI | | MAXIMUM LABOR-HOURS |
	Trick Ski	Slalom Ski	AVAILABLE PER DAY
FABRICATING DEPARTMENT	6	4	108
FINISHING DEPARTMENT	1	1	24
PROFIT PER SKI	$40	$30	

28. *Manufacturing—resource allocation.* A furniture manufacturing company manufactures dining room tables and chairs. The relevant manufacturing data are given in the accompanying table.

(A) How many tables and chairs should be manufactured each day to realize a maximum profit? What is the maximum profit?

(B) Repeat part A if the marketing department of the company has decided that the number of chairs produced should be at least four times the number of tables produced.

	LABOR-HOURS PER UNIT		MAXIMUM LABOR-HOURS AVAILABLE PER DAY
	Table	Chair	
ASSEMBLY DEPARTMENT	8	2	400
FINISHING DEPARTMENT	2	1	120
PROFIT PER UNIT	$90	$25	

29. *Manufacturing — production scheduling.* A furniture company has two plants that produce the lumber used in manufacturing tables and chairs. In 1 day of operation, plant A can produce the lumber required to manufacture 20 tables and 60 chairs, and plant B can produce the lumber required to manufacture 25 tables and 50 chairs. The company needs enough lumber to manufacture at least 200 tables and 500 chairs.

(A) If it costs $1,000 to operate plant A for 1 day and $900 to operate plant B for 1 day, how many days should each plant be operated in order to produce a sufficient amount of lumber at a minimum cost? What is the minimum cost?

(B) Repeat part A if the daily cost of operating plant A is reduced to $600.

(C) Repeat part A if the daily cost of operating plant B is reduced to $800.

30. *Manufacturing — resource allocation.* An electronics firm manufactures two types of personal computers, a standard model and a portable model. The production of a standard computer requires a capital expenditure of $400 and 40 hours of labor. The production of a portable computer requires a capital expenditure of $250 and 30 hours of labor. The firm has $20,000 capital and 2,160 labor-hours available for production of standard and portable computers.

(A) What is the maximum number of computers the company is capable of producing?

(B) If each standard computer contributes a profit of $320 and each portable computer contributes a profit of $220, how many of each type of computer should the firm produce in order to maximize profit? What is the maximum profit?

31. *Transportation.* The officers of a high school senior class are planning to rent buses and vans for a class trip. Each bus can transport 40 students, requires 3 chaperones, and costs $1,200 to rent. Each van can transport 8 students, requires 1 chaperone, and costs $100 to rent. Since there are 400 students in the senior class that may be eligible to go on the trip, the officers

must plan to accommodate at least 400 students. Since only 36 parents have volunteered to serve as chaperones, the officers must plan to use at most 36 chaperones. How many vehicles of each type should the officers rent in order to minimize the transportation costs? What are the minimal transportation costs?

32. *Investment.* An investor has $24,000 to invest in bonds of AAA and B qualities. The AAA bonds yield on the average 6% and the B bonds yield 10%. The investor's policy requires that she invest at least three times as much money in AAA bonds as in B bonds. How much should she invest in each type of bond to maximize her return? What is the maximum return?

33. *Pollution control.* Because of new federal regulations on pollution, a chemical plant introduced a new, more expensive process to supplement or replace an older process used in the production of a particular chemical. The older process emitted 15 grams of sulfur dioxide and 40 grams of particulate matter into the atmosphere for each gallon of chemical produced. The new process emits 5 grams of sulfur dioxide and 20 grams of particulate matter for each gallon produced. The company makes a profit of 30¢ per gallon and 20¢ per gallon on the old and new processes, respectively. If the government allows the plant to emit no more than 10,500 grams of sulfur dioxide and no more than 30,000 grams of particulate matter daily, how many gallons of the chemical should be produced by each process to maximize daily profit? What is the maximum profit?

34. *Capital expansion.* A fast-food chain plans to expand by opening several new restaurants. The chain operates two types of restaurants, drive-through and full-service. A drive-through restaurant costs $100,000 to construct, requires 5 employees, and has an expected annual revenue of $200,000. A full-service restaurant costs $150,000 to construct, requires 15 employees, and has an expected annual revenue of $500,000. The chain has $2,400,000 in capital available for expansion. Labor contracts require that they hire no more than 210 employees, and licensing restrictions require that they open no more than 20 new restaurants. How many restaurants of each type should the chain open in order to maximize the expected revenue? What is the maximum expected revenue? How much of their capital will they use and how many employees will they hire?

Life Sciences **35.** *Nutrition—plants.* A fruit grower can use two types of fertilizer in his orange grove, brand *A* and brand *B*. The amounts (in pounds) of nitrogen, phosphoric acid, and chlorine in a bag of each brand are given in the accompanying table. Tests indicate that the grove needs at least 1,000 pounds of phosphoric acid and at most 400 pounds of chlorine.

(A) If the grower wants to maximize the amount of nitrogen added to the grove, how many bags of each mix should be used? How much nitrogen will be added?

(B) If the grower wants to minimize the amount of nitrogen added to the grove, how many bags of each mix should be used? How much nitrogen will be added?

| | POUNDS PER BAG | |
	Brand A	Brand B
NITROGEN	8	3
PHOSPHORIC ACID	4	4
CHLORINE	2	1

36. *Nutrition — people.* A dietitian in a hospital is to arrange a special diet composed of two foods, M and N. Each ounce of food M contains 30 units of calcium, 10 units of iron, 10 units of vitamin A, and 8 units of cholesterol. Each ounce of food N contains 10 units of calcium, 10 units of iron, 30 units of vitamin A, and 4 units of cholesterol. If the minimum daily requirements are 360 units of calcium, 160 units of iron, and 240 units of vitamin A, how many ounces of each food should be used to meet the minimum requirements and at the same time minimize the cholesterol intake? What is the minimum cholesterol intake?

37. *Nutrition — plants.* A farmer can buy two types of plant food, mix A and mix B. Each cubic yard of mix A contains 20 pounds of phosphoric acid, 30 pounds of nitrogen, and 5 pounds of potash. Each cubic yard of mix B contains 10 pounds of phosphoric acid, 30 pounds of nitrogen, and 10 pounds of potash. The minimum monthly requirements are 460 pounds of phosphoric acid, 960 pounds of nitrogen, and 220 pounds of potash. If mix A costs $30 per cubic yard and mix B costs $35 per cubic yard, how many cubic yards of each mix should the farmer blend to meet the minimum monthly requirements at a minimal cost? What is this cost?

38. *Nutrition — animals.* A laboratory technician in a medical research center is asked to formulate a diet from two commercially packaged foods, food A and food B, for a group of animals. Each ounce of food A contains 8 units of fat, 16 units of carbohydrate, and 2 units of protein. Each ounce of food B contains 4 units of fat, 32 units of carbohydrate, and 8 units of protein. The minimum daily requirements are 176 units of fat, 1,024 units of carbohydrate, and 384 units of protein. If food A costs 5¢ per ounce and food B costs 5¢ per ounce, how many ounces of each food should be used to meet the minimum daily requirements at the least cost? What is the cost for this amount of food?

Social Sciences 39. *Psychology.* In an experiment on conditioning, a psychologist uses two types of Skinner boxes with mice and rats. The amount of time (in minutes) each mouse and each rat spends in each box per day is given in the table.

What is the maximum number of mice and rats that can be used in this experiment? How many mice and how many rats produce this maximum?

| | TIME | | MAXIMUM TIME |
	Mice	Rats	AVAILABLE PER DAY
SKINNER BOX A	10 min	20 min	800 min
SKINNER BOX B	20 min	10 min	640 min

40. *Sociology.* A city council voted to conduct a study on inner-city community problems. A nearby university was contacted to provide sociologists and research assistants. Allocation of time and costs per week are given in the accompanying table. How many sociologists and how many research assistants should be hired to minimize the cost and meet the weekly labor-hour requirements? What is the minimum weekly cost?

| | LABOR-HOURS | | MINIMUM |
| | | Research | LABOR-HOURS |
	Sociologist	Assistant	NEEDED PER WEEK
FIELDWORK	10	30	180
RESEARCH CENTER	30	10	140
COSTS PER WEEK	$500	$300	

SECTION 6-3 A Geometric Introduction to the Simplex Method

- ◆ STANDARD MAXIMIZATION PROBLEMS IN STANDARD FORM
- ◆ SLACK VARIABLES
- ◆ BASIC AND NONBASIC VARIABLES; BASIC SOLUTIONS AND BASIC FEASIBLE SOLUTIONS
- ◆ BASIC FEASIBLE SOLUTIONS AND THE SIMPLEX METHOD

The geometric method of solving linear programming problems provides us with an overview of the subject and some useful terminology. But, practically speaking, the method is useful only for problems involving two decision variables and relatively few problem constraints. What happens when we need more decision variables and more problem constraints? We use an algebraic method called the *simplex method,* which was developed by George B. Dantzig in 1947 while on assignment to the U.S. Department of the Air Force. Ideally suited to computer use, the method is used routinely on applied problems involving hundreds and even thousands of variables and problem constraints.

The algebraic procedures utilized in the simplex method require the problem constraints to be written as equations rather than inequalities. This new form of the linear programming problem also prompts the use of some new terminology.

We introduce this new form and associated terminology through a simple example and an appropriate geometric interpretation. From this example we can illustrate what the simplex method does geometrically before we immerse ourselves in the algebraic details of the process.

◆ STANDARD MAXIMIZATION PROBLEMS IN STANDARD FORM

We now return to the tent production problem in Example 6 from the last section. Recall the mathematical model for the problem:

$$\begin{aligned}
\text{Maximize} \quad & P = 50x_1 + 80x_2 & & \text{Objective function} \\
\text{Subject to} \quad & x_1 + 2x_2 \leq 32 & & \text{Cutting department constraint} \\
& 3x_1 + 4x_2 \leq 84 & & \text{Assembly department constraint} \\
& x_1, x_2 \geq 0 & & \text{Nonnegative constraints}
\end{aligned} \tag{1}$$

where the decision variables x_1 and x_2 are the number of standard and expedition tents, respectively, produced each day.

Notice that the problem constraints involve $\leq$ inequalities with positive constants to the right of the inequality. Maximization problems that satisfy this condition are called *standard maximization problems*. In this and the next section we will restrict our attention to standard maximization problems.

A Standard Maximization Problem in Standard Form

A linear programming problem is said to be a **standard maximization problem in standard form** if its mathematical model is of the following form:

Maximize the objective function

$$P = c_1x_1 + c_2x_2 + \cdots + c_nx_n$$

Subject to problem constraints of the form

$$a_1x_1 + a_2x_2 + \cdots + a_nx_n \leq b \qquad b \geq 0$$

With nonnegative constraints

$$x_1, x_2, \ldots, x_n \geq 0$$

[*Note:* Mathematical model (1) above is a standard maximization problem in standard form. Also note that the coefficients of the objective function can be any real numbers.]

◆ SLACK VARIABLES

To adapt a linear programming problem to the matrix methods used in the simplex process (as discussed in the next section), we convert the problem constraint inequalities into a system of linear equations by using a simple device called a *slack variable*. In particular, to convert the system of problem constraint inequalities from (1),

$$x_1 + 2x_2 \leq 32 \qquad \text{Cutting department constraint}$$
$$3x_1 + 4x_2 \leq 84 \qquad \text{Assembly department constraint} \tag{2}$$

into a system of equations, we add variables s_1 and s_2 to the left sides of (2) to obtain

$$x_1 + 2x_2 + s_1 \qquad = 32$$
$$3x_1 + 4x_2 \qquad + s_2 = 84 \tag{3}$$

The variables s_1 and s_2 are called **slack variables** because each makes up the difference (takes up the slack) between the left and right sides of the inequalities in (2). For example, if we produced 20 standard tents ($x_1 = 20$) and 5 expedition tents ($x_2 = 5$), then the number of labor-hours used in the cutting department would be $20 + 2(5) = 30$, leaving a slack of 2 unused labor-hours out of the 32 available. Thus, s_1 would have the value of 2.

Notice that if the decision variables x_1 and x_2 satisfy the system of constraint inequalities (2), then the slack variables s_1 and s_2 are nonnegative. We will have more to say about this later in this discussion.

◆ BASIC AND NONBASIC VARIABLES; BASIC SOLUTIONS AND BASIC FEASIBLE SOLUTIONS

Observe that system (3) has infinitely many solutions — just solve for s_1 and s_2 in terms of x_1 and x_2, and then assign x_1 and x_2 arbitrary values. Certain solutions of system (3), called *basic solutions*, are related to the intersection points of the (extended) boundary lines of the feasible region in Figure 7.

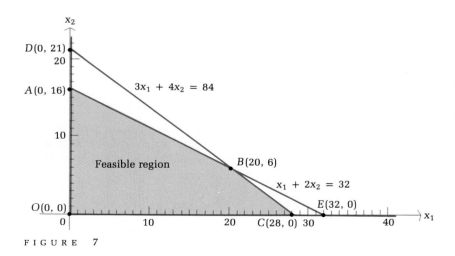

FIGURE 7

How are basic solutions to system (3) determined? System (3) involves four variables and two equations. We divide the four variables into two groups, called *basic variables* and *nonbasic variables*, as follows: Basic variables are selected arbitrarily with the restriction that there be as many basic variables as there are equations. The remaining variables are nonbasic variables.

Since system (3) has two equations, we can select any two of the four variables as basic variables. The remaining two variables are then nonbasic variables. A solution found by setting the two nonbasic variables equal to 0 and solving for the two basic variables is a basic solution. [Note that setting two variables equal to 0 in system (3) results in a system of two equations with two variables, which has (from Chapter 5) exactly one solution, infinitely many solutions, or no solution.]

◆ E X A M P L E 9   (A) Find two basic solutions for system (3) by first selecting s_1 and s_2 as basic variables, and then by selecting x_2 and s_1 as basic variables.

(B) Associate each basic solution found in part A with an intersection point of the (extended) boundary lines of the feasible region in Figure 7, and indicate which boundary lines produce each intersection point.

(C) Indicate which of the intersection points found in part B are in the feasible region.

Solutions   (A) Since s_1 and s_2 were selected as basic variables, x_1 and x_2 are nonbasic variables. A basic solution is found by setting the nonbasic variables equal to 0 and solving for the basic variables. If $x_1 = 0$ and $x_2 = 0$, then system (3) becomes

$$s_1 = 32 \qquad \overset{0}{x_1} + \overset{0}{2x_2} + s_1 \qquad = 32$$
$$s_2 = 84 \qquad 3x_1 + 4x_2 \qquad + s_2 = 84$$

and the basic solution is

$$x_1 = 0, \quad x_2 = 0, \quad s_1 = 32, \quad s_2 = 84 \tag{4}$$

If we select x_2 and s_1 as basic variables, then x_1 and s_2 are nonbasic variables. Setting the nonbasic variables equal to 0, system (3) becomes

$$2x_2 + s_1 = 32 \qquad \overset{0}{x_1} + 2x_2 + s_1 \qquad = 32$$
$$4x_2 = 84 \qquad 3x_1 + 4x_2 \qquad + \overset{0}{s_2} = 84$$

Solving, we see that $x_2 = 21$ and $s_1 = -10$, and the basic solution is

$$x_1 = 0, \quad x_2 = 21, \quad s_1 = -10, \quad s_2 = 0 \tag{5}$$

(B) Basic solution (4)—since $x_1 = 0$ and $x_2 = 0$—corresponds to the origin $O(0, 0)$ in Figure 7, which is the intersection of the boundary lines $x_1 = 0$ and $x_2 = 0$. Basic solution (5)—since $x_1 = 0$ and $x_2 = 21$—corresponds to the intersection point $D(0, 21)$, which is the intersection of the boundary lines $x_1 = 0$ and $3x_1 + 4x_2 = 84$.

(C) The intersection point $O(0, 0)$ is in the feasible region; hence, it is natural to call the corresponding basic solution a *basic feasible solution*. The intersection point $D(0, 21)$ is not in the feasible region; hence, the corresponding basic solution is not feasible. ◆

PROBLEM 9

(A) Find two basic solutions for system (3) by first selecting x_1 and s_1 as basic variables, and then by selecting x_1 and s_2 as basic variables.

(B) Associate each basic solution found in part A with an intersection point of the (extended) boundary lines of the feasible region in Figure 7, and indicate which boundary lines produce each intersection point.

(C) Indicate which of the intersection points found in part B are in the feasible region. ◆

Proceeding systematically as in Example 9 and Problem 9, we can obtain all basic solutions to system (3). The results are summarized in Table 2, which also includes geometric interpretations of the basic solutions relative to Figure 7. Figure 8 summarizes these interpretations. A careful study of Table 2 and Figure 8 is very worthwhile. (Note that to be sure we have listed all possible basic solutions in Table 2, it is convenient to organize the table in terms of the zero values of the nonbasic variables.)

TABLE 2

Basic Solutions

| BASIC SOLUTIONS | | | | INTERSECTION | INTERSECTING | |
x_1	x_2	s_1	s_2	POINT	BOUNDARY LINES	FEASIBLE
0	0	32	84	$O(0, 0)$	$x_1 = 0$ $x_2 = 0$	Yes
0	16	0	20	$A(0, 16)$	$x_1 = 0$ $x_1 + 2x_2 = 32$	Yes
0	21	-10	0	$D(0, 21)$	$x_1 = 0$ $3x_1 + 4x_2 = 84$	No
32	0	0	-12	$E(32, 0)$	$x_2 = 0$ $x_1 + 2x_2 = 32$	No
28	0	4	0	$C(28, 0)$	$x_2 = 0$ $3x_1 + 4x_2 = 84$	Yes
20	6	0	0	$B(20, 6)$	$x_1 + 2x_2 = 32$ $3x_1 + 4x_2 = 84$	Yes

Observations from Table 2 and Figure 8 Important to the Development of the Simplex Method

1. In Table 2, observe that a basic solution that is not feasible includes at least one negative value and that a basic feasible solution does not include any negative values. That is, we can determine the feasibility of a basic solution simply by examining the sign of all the variables in the solution.

2. In Table 2 and Figure 8, observe that basic feasible solutions are associated with the corner points of the feasible region, which include the optimal solution to the original linear programming problem.

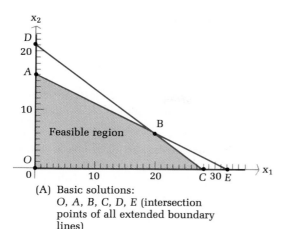

(A) Basic solutions:
O, A, B, C, D, E (intersection
points of all extended boundary
lines)

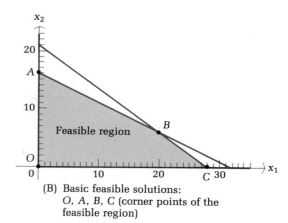

(B) Basic feasible solutions:
O, A, B, C (corner points of the
feasible region)

FIGURE 8

Before proceeding further, let us formalize the definitions alluded to in the discussion above so that they apply to standard maximization problems in general, without any reference to geometric forms.

Basic Variables and Nonbasic Variables; Basic Solutions and Basic Feasible Solutions

Given a system of linear equations associated with a linear programming problem (such a system will always have more variables than equations):

The variables are divided into two (mutually exclusive) groups, as follows: **Basic variables** are selected arbitrarily with the one restriction that there be as many basic variables as there are equations. The remaining variables are called **nonbasic variables.**

A solution found by setting the nonbasic variables equal to 0 and solving for the basic variables is called a **basic solution.** If a basic solution has no negative values, it is a **basic feasible solution.**

◆ E X A M P L E   10   Suppose there is a system of three problem constraint equations with eight (slack and decision) variables associated with a standard maximization problem.

(A) How many basic variables and how many nonbasic variables are associated with the system?
(B) Setting the nonbasic variables equal to 0 will result in a system of how many linear equations with how many variables?
(C) If a basic solution has all nonnegative elements, is it feasible or not feasible?

(A) Since there are three equations in the system, there should be three basic variables and five nonbasic variables.
(B) Three equations with three variables
(C) Feasible ◆

PROBLEM 10 Suppose there is a system of five problem constraint equations with eleven (slack and decision) variables associated with a standard maximization problem.

(A) How many basic variables and how many nonbasic variables are associated with the system?
(B) Setting the nonbasic variables equal to 0 will result in a system of how many linear equations with how many variables?
(C) If a basic solution has one or more negative elements, is it feasible or not feasible? ◆

◆ BASIC FEASIBLE SOLUTIONS AND THE SIMPLEX METHOD

The following important theorem, which is equivalent to the fundamental theorem (Theorem 2 in the preceding section), is stated without proof:

THEOREM 4

> If the optimal value of the objective function in a linear programming problem exists, then that value must occur at one (or more) of the basic feasible solutions.

Now you can understand why the concepts of basic and nonbasic variables and basic solutions and basic feasible solutions are so important — the concepts are central to the process of finding optimal solutions to linear programming problems.

We have taken the first step toward finding a general method of solving linear programming problems involving any number of variables and problem constraints. That is, **we have found a method of identifying all the corner points (basic feasible solutions) of a feasible region without drawing its graph.** This is a critical step if we want to consider problems with more than two decision variables. Unfortunately, the number of corner points increases dramatically as the number of variables and constraints increases. In real-world problems, it is not practical to find all the corner points in order to find the optimal solution. Thus, the next step is to find a method of locating the optimal solution without finding every corner point. The procedure for doing this is the simplex method mentioned at the beginning of this section.

The simplex method, using a special matrix and row operations, automatically moves from one basic feasible solution to another — that is, from one corner point of the feasible region to another — each time getting closer to an optimal solution (if one exists), until an optimal solution is reached. Then the process stops. The **simplex method** is an iterative algorithm (a repetitive procedure) that is shown schematically in Figure 9. A remarkable property of the simplex method is that in large linear programming problems it usually arrives at an optimal solution (if one exists) by testing relatively few of the large number of basic feasible solutions (corner points) available.

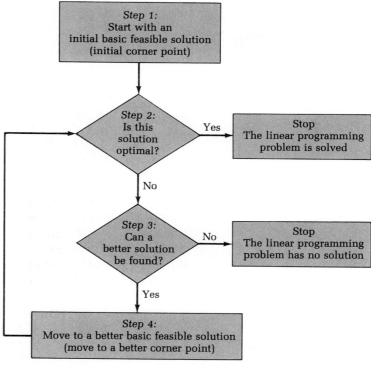

FIGURE 9
Simplex algorithm

With this background, we are now ready to discuss the algebraic details of the simplex method in the next section.

Answers to Matched Problems

9. (A) Basic solution corresponding to basic variables x_1 and s_1: $x_1 = 28$, $x_2 = 0$, $s_1 = 4$, $s_2 = 0$. Basic solution corresponding to basic variables x_1 and s_2: $x_1 = 32$, $x_2 = 0$, $s_1 = 0$, $s_2 = -12$.

(B) The first basic solution corresponds to $C(28, 0)$, which is the intersection of the boundary lines $x_2 = 0$ and $3x_1 + 4x_2 = 84$. The second basic solution corresponds to $E(32, 0)$, which is the intersection of the boundary lines $x_2 = 0$ and $x_1 + 2x_2 = 32$.

(C) $C(28, 0)$ is in the feasible region (hence, the corresponding basic solution is a basic feasible solution); $E(32, 0)$ is not in the feasible region (hence, the corresponding basic solution is not feasible).

10. (A) Five basic variables and six nonbasic variables
(B) Five equations and five variables   (C) Not feasible

E X E R C I S E   6-3

A   1. Associated with a standard maximization problem is a system of problem constraint equations with nine variables, including five slack variables.

(A) How many constraint equations are in the system?

(B) How many decision variables are in the system?

(C) How many basic variables and how many nonbasic variables are associated with the system?

(D) Setting the nonbasic variables equal to 0 will result in a system of how many linear equations with how many variables?

2. Associated with a standard maximization problem is a system of problem constraint equations with ten variables, including four decision variables.

(A) How many slack variables are in the system?

(B) How many constraint equations are in the system?

(C) How many basic variables and how many nonbasic variables are associated with the system?

(D) Setting the nonbasic variables equal to 0 will result in a system of how many linear equations with how many variables?

3. Listed in the table below are all the basic solutions for the system

$$2x_1 + 3x_2 + s_1 \qquad = 24$$
$$4x_1 + 3x_2 \qquad + s_2 = 36$$

For each basic solution, identify the nonbasic variables and the basic variables. Then classify each basic solution as feasible or not feasible.

	x_1	x_2	s_1	s_2
(A)	0	0	24	36
(B)	0	8	0	12
(C)	0	12	-12	0
(D)	12	0	0	-12
(E)	9	0	6	0
(F)	6	4	0	0

4. Repeat Problem 3 for the system

$$2x_1 + x_2 + s_1 \qquad = 30$$
$$x_1 + 5x_2 \qquad + s_2 = 60$$

whose basic solutions are given in the following table:

	x_1	x_2	s_1	s_2
(A)	0	0	30	60
(B)	0	30	0	-90
(C)	0	12	18	0
(D)	15	0	0	45
(E)	60	0	-90	0
(F)	10	10	0	0

5. Listed in the table below are all the possible choices of nonbasic variables for the system

$$2x_1 + x_2 + s_1 \qquad = 50$$
$$x_1 + 2x_2 \qquad + s_2 = 40$$

In each case, find the values of the basic variables and determine whether the basic solution is feasible.

	x_1	x_2	s_1	s_2
(A)	0	0	?	?
(B)	0	?	0	?
(C)	0	?	?	0
(D)	?	0	0	?
(E)	?	0	?	0
(F)	?	?	0	0

6. Repeat Problem 5 for the system

$$x_1 + 2x_2 + s_1 \qquad = 12$$
$$3x_1 + 2x_2 \qquad + s_2 = 24$$

B *Graph the following systems of inequalities. Introduce slack variables to convert each system of inequalities to a system of equations, and find all the basic solutions of the system. Construct a table (similar to Table 2) listing each basic* solution, the corresponding point on the graph, and whether the basic solution is feasible. (You do not need to list the intersecting lines.)

7. $\quad x_1 + x_2 \leqslant 16$
   $\quad 2x_1 + x_2 \leqslant 20$
   $\quad x_1, x_2 \geqslant 0$

8. $\quad 5x_1 + x_2 \leqslant 35$
   $\quad 4x_1 + x_2 \leqslant 32$
   $\quad x_1, x_2 \geqslant 0$

9. $\quad 2x_1 + \quad x_2 \leqslant 22$
   $\quad x_1 + \quad x_2 \leqslant 12$
   $\quad x_1 + 2x_2 \leqslant 20$
   $\quad x_1, x_2 \geqslant 0$

10. $\quad 4x_1 + x_2 \leqslant 28$
    $\quad 2x_1 + x_2 \leqslant 16$
    $\quad x_1 + x_2 \leqslant 13$
    $\quad x_1, x_2 \geqslant 0$

SECTION 6-4 The Simplex Method: Maximization with Problem Constraints of the Form $\leqslant$

◆ INITIAL SYSTEM
◆ THE SIMPLEX TABLEAU
◆ THE PIVOT OPERATION
◆ INTERPRETING THE SIMPLEX PROCESS GEOMETRICALLY
◆ THE SIMPLEX METHOD SUMMARIZED
◆ APPLICATION

We are now ready to develop the simplex method for a standard maximization problem. Specific details in the presentation of the method generally vary from one person to another or from one book to another, even though the underlying process is the same. The presentation developed here emphasizes concept development and understanding.

As was pointed out in the last section, the simplex method is most efficient when used with computers. Consequently, it is not intended that you become expert in manually solving linear programming problems using the simplex method. It is important that you become proficient in setting up linear programming problems so that they can be solved using computers, and it is also important that you develop skill in interpreting the results. The best way to gain this proficiency and interpretive skill is to set up and manually solve a number of fairly simple linear programming problems using the simplex method, and this is the purpose of the remainder of this chapter. (The computer supplement for this book contains a computer program for solving linear programming problems—see the Preface.)

◆ INITIAL SYSTEM

We will introduce the concepts and procedures involved in the simplex method through an example—the tent production example we have discussed earlier.

We restate the problem here in standard form for convenient reference:

Maximize $P = 50x_1 + 80x_2$ Objective function

Subject to $\left.\begin{array}{l} x_1 + 2x_2 \leqslant 32 \\ 3x_1 + 4x_2 \leqslant 84 \end{array}\right\}$ Problem constraints   (1)

$x_1, x_2 \geqslant 0$ Nonnegative constraints

Introducing slack variables s_1 and s_2, we convert the problem constraint inequalities in (1) into the system of problem constraint equations:

$$
\begin{array}{rl}
x_1 + 2x_2 + s_1 &= 32 \\
3x_1 + 4x_2 + s_2 &= 84 \\
x_1, x_2, s_1, s_2 &\geqslant 0
\end{array} \tag{2}
$$

Since a basic solution of (2) is not feasible if it contains any negative values, we have also included the nonnegative constraints for both the decision variables x_1 and x_2 and the slack variables s_1 and s_2. From our earlier discussion (in the last section), we know that out of the infinitely many solutions to system (2), an optimal solution is among the basic feasible solutions, which correspond to the corner points of the feasible region.

As part of the simplex method we add the objective function equation in the form $-50x_1 - 80x_2 + P = 0$ to system (2) to create what is called the **initial system:**

$$
\begin{array}{rl}
x_1 + 2x_2 + s_1 &= 32 \\
3x_1 + 4x_2 + s_2 &= 84 \\
-50x_1 - 80x_2 + P &= 0 \\
x_1, x_2, s_1, s_2 &\geqslant 0
\end{array} \tag{3}
$$

When we add the objective function equation to system (2), we must slightly modify the earlier definitions of basic solution and basic feasible solution so that they apply to the initial system (3).

Basic Solutions and Basic Feasible Solutions for Initial Systems

1. The objective function variable P is always selected as a basic variable and is never selected as a nonbasic variable.
2. Note that a basic solution of system (3) is also a basic solution of system (2) after P is deleted.
3. If a basic solution of system (3) is a basic feasible solution of system (2) after deleting P, then the basic solution of (3) is called a **basic feasible solution** of (3).
4. A basic feasible solution of system (3) can contain a negative number, but only if it is the value of P, the objective function variable.

These changes lead to a small change in Theorem 4:

THEOREM 5

If the optimal value of the objective function in a linear programming problem exists, then that value must occur at one (or more) of the basic feasible solutions of the initial system.

With these adjustments understood, we start the simplex process with a basic feasible solution of the initial system (3), which we will refer to as an **initial basic feasible solution.** An initial basic feasible solution that is easy to find is the one associated with the origin.

Since system (3) has three equations and five variables, it has three basic variables and two nonbasic variables. Looking at the system, we see that x_1 and x_2 appear in all equations and s_1, s_2, and P each appears only once and each in a different equation. A basic solution can be found by inspection by selecting s_1, s_2, and P as the basic variables (remember, P is always selected as a basic variable) and x_1 and x_2 as the nonbasic variables to be set equal to 0. Setting x_1 and x_2 equal to 0 and solving for the basic variables, we obtain the basic solution:

$$x_1 = 0, \quad x_2 = 0, \quad s_1 = 32, \quad s_2 = 84, \quad P = 0$$

This basic solution is feasible since none of the variables (excluding P) are negative. Thus, this is the initial basic feasible solution we seek.

Now you can see why we wanted to add the objective function equation to system (2): A basic feasible solution of (3) not only includes a basic feasible solution of (2), but, in addition, it includes the value of P for that basic feasible solution of (2).

The initial basic feasible solution we just found is associated with the origin. Of course, if we do not produce any tents, we do not expect a profit, so $P = \$0$. Starting with this easily obtained initial basic feasible solution, the simplex process moves through each iteration (each repetition) to another basic feasible solution, each time improving the profit, and the process continues until the maximum profit is reached. Then the process stops.

◆ THE SIMPLEX TABLEAU

To facilitate the search for the optimal solution, we now turn to matrix methods discussed in Chapter 5. Our first step is to write the augmented matrix for the initial system (3). This matrix is called the **initial simplex tableau,** and it is simply a tabulation of the coefficients in system (3).

$$
\begin{array}{c}
 \\
s_1 \\
s_2 \\
P
\end{array}
\begin{array}{ccccc}
x_1 & x_2 & s_1 & s_2 & P \\
\left[\begin{array}{ccccc|c}
1 & 2 & 1 & 0 & 0 & 32 \\
3 & 4 & 0 & 1 & 0 & 84 \\
\hline
-50 & -80 & 0 & 0 & 1 & 0
\end{array}\right]
\end{array}
\qquad \text{Initial simplex tableau} \qquad (4)
$$

In tableau (4), the row below the dashed line always corresponds to the objective function. Each of the basic variables we selected above, s_1, s_2, and P, is also placed on the left of the tableau so that the intersection element in its row and column is not 0. For example, we place the basic variable s_1 on the left so that the intersection element of the s_1 row and the s_1 column is 1 and not 0. The basic variable s_2 is similarly placed. The objective function variable P is always placed at the bottom. The reason for writing the basic variables on the left in this way, is that this placement makes it possible to read certain basic feasible solutions directly from the tableau. If $x_1 = 0$ and $x_2 = 0$, then the basic variables on the left of tableau (4) are lined up with their corresponding values, 32, 84, and 0, to the right of the vertical line.

Looking at tableau (4) relative to the choice of s_1, s_2, and P as basic variables, we see that each basic variable is above a column that has all 0 elements except for a single 1 and that no two such columns contain 1's in the same row. These observations lead to a formalization of the process of selecting basic and nonbasic variables that is an important part of the simplex method:

Selecting Basic and Nonbasic Variables for the Simplex Method

Given a simplex tableau:

Step 1. Determine the number of basic variables and the number of nonbasic variables. These numbers do not change during the simplex process.

Step 2. *Selecting basic variables:* A variable can be selected as a basic variable only if it corresponds to a column in the tableau that has exactly one nonzero element (usually 1) and the nonzero element in the column is not in the same row as the nonzero element in the column of another basic variable. (This procedure always selects P as a basic variable, since the P column never changes during the simplex process.)

Step 3. *Selecting nonbasic variables:* After the basic variables are selected in step 2, the remaining variables are selected as the nonbasic variables. (The tableau columns under the nonbasic variables will usually contain more than one nonzero element.)

The earlier selection of s_1, s_2, and P as basic variables and x_1 and x_2 as nonbasic variables conforms to this prescribed convention of selecting basic and nonbasic variables for the simplex process.

◆ THE PIVOT OPERATION

The simplex method will now switch one of the nonbasic variables, x_1 or x_2, for one of the basic variables, s_1 or s_2 (but not P), as a step toward improving the profit. For a nonbasic variable to be classified as a basic variable we need to perform appropriate row operations on the tableau so that the newly selected basic variable will end up with exactly one nonzero element in its column. In this process, the old basic variable will usually gain additional nonzero elements in its column as it becomes nonbasic.

Which nonbasic variable should we select to become basic? It makes sense to select the nonbasic variable that will increase the profit the most per unit change in that variable. Looking at the objective function

$$P = 50x_1 + 80x_2$$

we see that if x_1 stays a nonbasic variable (set equal to 0) and if x_2 becomes a new basic variable, then

$$P = 50(0) + 80x_2 = 80x_2$$

and for each unit increase in x_2, P will increase \$80. If x_2 stays a nonbasic variable and x_1 becomes a new basic variable, then (reasoning in the same way) for each unit increase in x_1, P will increase only \$50. So, we select the nonbasic variable x_2 to enter the set of basic variables, and call it the **entering variable.** (The basic variable leaving the set of basic variables to become a nonbasic variable is called the **exiting variable.** Exiting variables will be discussed shortly.)

We call the column corresponding to the entering variable the **pivot column.** Looking at the bottom row in tableau (4)—the objective function row below the dashed line—we see that the pivot column is associated with the column to the left of the P column that has the most negative bottom element. In general, the most negative element in the bottom row to the left of the P column *indicates* the variable above it that will produce the greatest increase in P for a unit increase in that variable. For this reason, we call the elements in the bottom row of the tableau to the left of the P column **indicators.**

We illustrate the indicators, the pivot column, the entering variable, and the initial basic feasible solution below:

$$
\begin{array}{c}
\text{Entering} \\
\text{variable} \\
\downarrow
\end{array}
$$

$$
\begin{array}{c}
\begin{array}{ccccc}
x_1 & x_2 & s_1 & s_2 & P
\end{array} \\
\begin{array}{c}
s_1 \\
s_2 \\
P
\end{array}
\left[
\begin{array}{ccccc|c}
1 & 2 & 1 & 0 & 0 & 32 \\
3 & 4 & 0 & 1 & 0 & 84 \\
\hline
-50 & -80 & 0 & 0 & 1 & 0
\end{array}
\right]
\\
\uparrow \\
\text{Pivot} \\
\text{column}
\end{array}
$$

Initial simplex tableau   (5)

Indicators are shown in color.

$$x_1 = 0, \quad x_2 = 0, \quad s_1 = 32, \quad s_2 = 84, \quad P = 0 \qquad \text{Initial basic feasible solution}$$

Now that we have chosen the nonbasic variable x_2 as the entering variable (the nonbasic variable to become basic), which of the two basic variables, s_1 or s_2, should we choose as the exiting variable (the basic variable to become nonbasic)? We saw above that for $x_1 = 0$, each unit increase in the entering variable x_2 results in an increase of $80 for P. Can we increase x_2 without limit? No! A limit is imposed by the nonnegative requirements for s_1 and s_2. (Remember that if any of the basic variables except P become negative, we no longer have a feasible solution.) So we rephrase the question and ask: How much can x_2 be increased when $x_1 = 0$ without causing s_1 or s_2 to become negative? To see how much x_2 can be increased, we refer to tableau (5) or system (3) and write the two problem constraint equations with $x_1 = 0$:

$$2x_2 + s_1 = 32$$
$$4x_2 + s_2 = 84$$

Solving for s_1 and s_2, we have

$$s_1 = 32 - 2x_2$$
$$s_2 = 84 - 4x_2$$

For s_1 and s_2 to be nonnegative, x_2 must be chosen so that both $32 - 2x_2$ and $84 - 4x_2$ are nonnegative. That is, so that

$$32 - 2x_2 \geqslant 0 \qquad \text{and} \qquad 84 - 4x_2 \geqslant 0$$
$$-2x_2 \geqslant -32 \qquad\qquad\qquad -4x_2 \geqslant -84$$
$$x_2 \leqslant \tfrac{32}{2} = 16 \qquad\qquad\qquad x_2 \leqslant \tfrac{84}{4} = 21$$

For both inequalities to be satisfied, x_2 must be less than or equal to the smaller of the values, which is 16. Thus, x_2 can increase to 16 without either s_1 or s_2 becoming negative. Now, observe how each value (16 and 21) can be obtained directly from tableau (6) below:

From (6) we can determine the amount the entering variable can increase by choosing the smallest of the quotients obtained by dividing each element in the last column above the dashed line by the corresponding *positive* element in the pivot column. **If the corresponding element in the pivot column is 0 or nega-**

tive, no value is computed. The row with the smallest quotient is called the **pivot row,** and the variable to the left of the pivot row is the exiting variable. In this case, s_1 will be the exiting variable, and the roles of x_2 and s_1 will be interchanged. The element at the intersection of the pivot column and the pivot row, 2, is called the **pivot element,** and we circle this element for ease of recognition. Tableau (7) illustrates these ideas, which are also summarized in the box below.

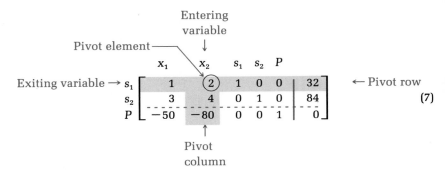

$$(7)$$

Step 1. Locate the most negative indicator in the bottom row of the tableau to the left of the P column (the negative number with the largest absolute value). The column containing this element is the *pivot column.* If there is a tie for the most negative, choose either.

Step 2. Divide each *positive* element in the pivot column above the dashed line into the corresponding element in the last column. The *pivot row* is the row corresponding to the smallest quotient obtained. If there is a tie for the smallest quotient, choose either. If the pivot column above the dashed line has no positive elements, then there is no solution, and we stop.

Step 3. The *pivot* (or *pivot element*) is the element in the intersection of the pivot column and pivot row. [*Note:* The pivot element is always positive and is never in the bottom row.]

[*Remember:* The entering variable is at the top of the pivot column and the exiting variable is at the left of the pivot row.]

In order for x_2 to be classified as a basic variable, we perform row operations on tableau (7) so that the pivot element is transformed into 1 and all other elements in the column into 0's. This procedure for transforming a nonbasic variable into a basic variable is called a *pivot operation,* or *pivoting,* and is summarized in the box.

A **pivot operation,** or **pivoting,** consists of performing row operations as follows:

Step 1.   Multiply the pivot row by the reciprocal of the pivot element to transform the pivot element into a 1. (If the pivot element is already a 1, omit this step.)

Step 2.   Add multiples of the pivot row to other rows in the tableau to transform all other nonzero elements in the pivot column into 0's.

[Note: Rows are not to be interchanged while performing a pivot operation. The only way the (positive) pivot element can be transformed into 1 (if it is not a 1 already) is for the pivot row to be multiplied by the reciprocal of the pivot element.]

Performing a pivot operation has the following effects:

1.   The (entering) nonbasic variable becomes a basic variable.
2.   The (exiting) basic variable becomes a nonbasic variable.
3.   The value of the objective function is increased, or, in some cases, remains the same.

We now carry out the pivot operation on tableau (7). (To facilitate the process, we do not repeat the variables after the first tableau, and we use "Enter" and "Exit" for "Entering variable" and "Exiting variable," respectively.)

$$
\begin{array}{c}
\qquad\qquad\quad \text{Enter} \\
\qquad\qquad\quad \downarrow \\
\begin{array}{c}
\\
\text{Exit} \rightarrow s_1 \\
s_2 \\
P
\end{array}
\begin{array}{cccccc}
x_1 & x_2 & s_1 & s_2 & P & \\
\left[\begin{array}{ccccc|c}
1 & \textcircled{2} & 1 & 0 & 0 & 32 \\
3 & 4 & 0 & 1 & 0 & 84 \\
\hline
-50 & -80 & 0 & 0 & 1 & 0
\end{array}\right] & \tfrac{1}{2}R_1 \rightarrow R_1
\end{array}
\end{array}
$$

$$
\sim
\left[\begin{array}{ccccc|c}
\tfrac{1}{2} & \textcircled{1} & \tfrac{1}{2} & 0 & 0 & 16 \\
3 & 4 & 0 & 1 & 0 & 84 \\
\hline
-50 & -80 & 0 & 0 & 1 & 0
\end{array}\right]
\quad
\begin{array}{l}
R_2 + (-4)R_1 \rightarrow R_2 \\
R_3 + 80R_1 \rightarrow R_3
\end{array}
$$

$$
\sim
\left[\begin{array}{ccccc|c}
\tfrac{1}{2} & 1 & \tfrac{1}{2} & 0 & 0 & 16 \\
1 & 0 & -2 & 1 & 0 & 20 \\
\hline
-10 & 0 & 40 & 0 & 1 & 1{,}280
\end{array}\right]
$$

We have completed the pivot operation, and now we must insert appropriate variables for this new tableau. Observing that x_2, s_2, and P are now the basic variables, we write the new basic feasible solution by setting the nonbasic variables x_1 and s_1 equal to 0 and solving for the basic variables by inspection.

(Remember, the values of the basic variables listed on the left are the corresponding numbers to the right of the vertical line.)

$$
\begin{array}{c}
\begin{array}{ccccc} x_1 & x_2 & s_1 & s_2 & P \end{array} \\
\begin{array}{c} x_2 \\ s_2 \\ P \end{array}
\left[\begin{array}{ccccc|c}
\frac{1}{2} & 1 & \frac{1}{2} & 0 & 0 & 16 \\
1 & 0 & -2 & 1 & 0 & 20 \\
\hline
-10 & 0 & 40 & 0 & 1 & 1{,}280
\end{array}\right]
\end{array}
$$

$$x_1 = 0, \quad x_2 = 16, \quad s_1 = 0, \quad s_2 = 20, \quad P = 1{,}280$$

A profit of \$1,280 is a marked improvement over the \$0 profit produced by the initial basic feasible solution. But we can improve P still further, since a negative indicator still remains in the bottom row. To see why, we write out the objective function:

$$-10x_1 + 40s_1 + P = 1{,}280$$

or

$$P = 10x_1 - 40s_1 + 1{,}280$$

If s_1 stays a nonbasic variable (set equal to 0) and x_1 becomes a new basic variable, then

$$P = 10x_1 - 40(0) + 1{,}280 = 10x_1 + 1{,}280$$

and for each unit increase in x_1, P will increase \$10.

We now go through another iteration of the simplex process (that is, we repeat the above sequence of steps) using another pivot element. The pivot element and the entering and exiting variables are shown in the following tableau:

$$
\begin{array}{cc}
 & \begin{array}{c} \text{Enter} \\ \downarrow \end{array} \\
\end{array}
$$

$$
\begin{array}{c}
\quad\quad\begin{array}{ccccc} x_1 & x_2 & s_1 & s_2 & P \end{array} \\
\begin{array}{c} x_2 \\ \text{Exit} \to s_2 \\ P \end{array}
\left[\begin{array}{ccccc|c}
\frac{1}{2} & 1 & \frac{1}{2} & 0 & 0 & 16 \\
\boxed{1} & 0 & -2 & 1 & 0 & 20 \\
\hline
-10 & 0 & 40 & 0 & 1 & 1{,}280
\end{array}\right]
\begin{array}{l} \frac{16}{1/2} = 32 \\[4pt] \frac{20}{1} = 20 \end{array}
\end{array}
$$

We now pivot on (the circled) 1. That is, we perform a pivot operation using this 1 as the pivot element. Since the pivot element is 1, we do not need to perform the first step in the pivot operation, so we proceed to the second step to get 0's above and below the pivot element 1. As before, to facilitate the process, we omit writing the variables, except for the first tableau.

$$
\begin{array}{cc}
 & \begin{array}{c} \text{Enter} \\ \downarrow \end{array} \\
\end{array}
$$

$$
\begin{array}{c}
\quad\quad\begin{array}{ccccc} x_1 & x_2 & s_1 & s_2 & P \end{array} \\
\begin{array}{c} x_2 \\ \text{Exit} \to s_2 \\ P \end{array}
\left[\begin{array}{ccccc|c}
\frac{1}{2} & 1 & \frac{1}{2} & 0 & 0 & 16 \\
\boxed{1} & 0 & -2 & 1 & 0 & 20 \\
\hline
-10 & 0 & 40 & 0 & 1 & 1{,}280
\end{array}\right]
\begin{array}{l} R_1 + (-\tfrac{1}{2})R_2 \to R_1 \\[8pt] R_3 + 10R_2 \to R_3 \end{array}
\end{array}
$$

$$
\sim
\left[\begin{array}{ccccc|c}
0 & 1 & \frac{3}{2} & -\frac{1}{2} & 0 & 6 \\
1 & 0 & -2 & 1 & 0 & 20 \\
\hline
0 & 0 & 20 & 10 & 1 & 1{,}480
\end{array}\right]
$$

Since there are no more negative indicators in the bottom row, we are through. Let us insert the appropriate variables for this last tableau and write the corresponding basic feasible solution. The basic variables are now x_1, x_2, and P, so to get the corresponding basic feasible solution, we set the nonbasic variables s_1 and s_2 equal to 0 and solve for the basic variables by inspection.

$$\begin{array}{c} \begin{array}{ccccc} x_1 & x_2 & s_1 & s_2 & P \end{array} \\ \begin{array}{c} x_2 \\ x_1 \\ P \end{array} \left[\begin{array}{ccccc|c} 0 & 1 & \frac{3}{2} & -\frac{1}{2} & 0 & 6 \\ 1 & 0 & -2 & 1 & 0 & 20 \\ \hline 0 & 0 & 20 & 10 & 1 & 1{,}480 \end{array} \right] \end{array}$$

$$x_1 = 20, \quad x_2 = 6, \quad s_1 = 0, \quad s_2 = 0, \quad P = 1{,}480$$

To see why this is the maximum, we rewrite the objective function from the bottom row:

$$20s_1 + 10s_2 + P = 1{,}480$$
$$P = 1{,}480 - 20s_1 - 10s_2$$

Since s_1 and s_2 cannot be negative, any increase of either from 0 will make the profit smaller.

Finally, returning to our original problem, we conclude that a production schedule of 20 standard tents and 6 expedition tents will produce a maximum profit of $1,480 per day, which is the same as the geometric solution obtained in Section 6-2. The fact that the slack variables are both 0 means that for this production schedule, the plant will operate at full capacity — there is no slack in either the cutting department or the assembly department.

◆ INTERPRETING THE SIMPLEX PROCESS GEOMETRICALLY

We can now interpret the simplex process just completed geometrically in terms of the feasible region graphed in the preceding section. Table 3 lists the three basic feasible solutions we just found using the simplex method (in the order they were found). The table also includes the corresponding corner points of the feasible region illustrated in Figure 10 (at the top of the next page).

TABLE 3

Basic Feasible Solutions (Obtained Above)

x_1	x_2	s_1	s_2	P	CORNER POINT
0	0	32	84	$ 0	$O(0, 0)$
0	16	0	20	$1,280	$A(0, 16)$
20	6	0	0	$1,480	$B(20, 6)$

Looking at Table 3 and Figure 10, we see that the simplex process started at the origin, moved to the adjacent corner point $A(0, 16)$, then to the optimal

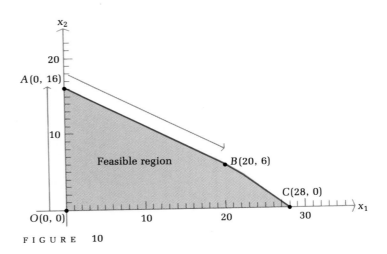

FIGURE 10

solution $B(20, 6)$ at the next adjacent corner point. This is typical of the simplex process.

◆ THE SIMPLEX METHOD SUMMARIZED

Before commencing with further examples, we summarize the important parts of the simplex method:

■ **Simplex Method**

KEY STEPS FOR STANDARD MAXIMIZATION PROBLEMS

(Problem constraints are of the ≤ form with nonnegative constants on the right. The coefficients of the objective function can be any real numbers.)

Step 1. Start with a standard maximization problem written in standard form.

Step 2. Introduce slack variables and write the initial system.

Step 3. Write the simplex tableau associated with the initial system.

Step 4. Determine the pivot element (if it exists) and the entering and exiting variables.

Step 5. Perform the pivot operation.

Step 6. Repeat steps 4 and 5 until all indicators in the bottom row are nonnegative. When this occurs, we stop the process and read the optimal solution.

There are two different reasons for stopping the simplex process:

1. If we cannot select a new pivot column, we stop because the optimal solution has been found (see step 6 above).
2. If we select a new pivot column and then are unable to select a new pivot row, we stop because the problem has no solution (see step 4 above and step 2 on selecting the pivot element, page 314).

◆ E X A M P L E   11   Solve the following linear programming problem using the simplex method:

$$\text{Maximize} \quad P = 10x_1 + 5x_2$$
$$\text{Subject to} \quad 6x_1 + 2x_2 \leqslant 36$$
$$2x_1 + 4x_2 \leqslant 32$$
$$x_1, x_2 \geqslant 0$$

Solution   Introduce slack variables s_1 and s_2 and write the initial system:

$$6x_1 + 2x_2 + s_1 \qquad\qquad = 36$$
$$2x_1 + 4x_2 \qquad + s_2 \qquad = 32$$
$$-10x_1 - 5x_2 \qquad\qquad + P = 0$$
$$x_1, x_2, s_1, s_2 \geqslant 0$$

Write the simplex tableau and identify the first pivot element and the entering and exiting variables:

Enter
↓

$$
\text{Exit} \rightarrow \begin{array}{c} s_1 \\ s_2 \\ P \end{array}
\begin{array}{c} x_1 \quad x_2 \quad s_1 \quad s_2 \quad P \\
\left[\begin{array}{ccccc|c}
⑥ & 2 & 1 & 0 & 0 & 36 \\
2 & 4 & 0 & 1 & 0 & 32 \\
\hline
-10 & -5 & 0 & 0 & 1 & 0
\end{array}\right]
\end{array}
\quad
\begin{array}{l}
\frac{36}{6} = 6 \\
\frac{32}{2} = 16
\end{array}
$$

Perform the pivot operation:

Enter
↓

$$
\text{Exit} \rightarrow \begin{array}{c} s_1 \\ s_2 \\ P \end{array}
\begin{array}{c} x_1 \quad x_2 \quad s_1 \quad s_2 \quad P \\
\left[\begin{array}{ccccc|c}
⑥ & 2 & 1 & 0 & 0 & 36 \\
2 & 4 & 0 & 1 & 0 & 32 \\
\hline
-10 & -5 & 0 & 0 & 1 & 0
\end{array}\right]
\end{array}
\quad \frac{1}{6}R_1 \rightarrow R_1
$$

$$
\sim
\left[\begin{array}{ccccc|c}
① & \frac{1}{3} & \frac{1}{6} & 0 & 0 & 6 \\
2 & 4 & 0 & 1 & 0 & 32 \\
\hline
-10 & -5 & 0 & 0 & 1 & 0
\end{array}\right]
\quad
\begin{array}{l}
R_2 + (-2)R_1 \rightarrow R_2 \\
R_3 + 10R_1 \rightarrow R_3
\end{array}
$$

$$
\sim \begin{array}{c} x_1 \\ s_2 \\ P \end{array}
\left[\begin{array}{ccccc|c}
1 & \frac{1}{3} & \frac{1}{6} & 0 & 0 & 6 \\
0 & \frac{10}{3} & -\frac{1}{3} & 1 & 0 & 20 \\
\hline
0 & -\frac{5}{3} & \frac{5}{3} & 0 & 1 & 60
\end{array}\right]
$$

Since there still is a negative indicator in the last row, we repeat the process by finding a new pivot element:

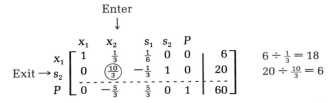

$$
\begin{array}{c} \\ x_1 \\ \text{Exit} \to s_2 \\ P \end{array}
\begin{array}{cccccc}
x_1 & x_2 & s_1 & s_2 & P & \\
1 & \frac{1}{3} & \frac{1}{6} & 0 & 0 & 6 \\
0 & \boxed{\frac{10}{3}} & -\frac{1}{3} & 1 & 0 & 20 \\
\hline
0 & -\frac{5}{3} & \frac{5}{3} & 0 & 1 & 60
\end{array}
\qquad
\begin{array}{l}
6 \div \frac{1}{3} = 18 \\[4pt]
20 \div \frac{10}{3} = 6
\end{array}
$$

Performing the pivot operation, we obtain

Enter
↓

$$
\begin{array}{c} \\ x_1 \\ \text{Exit} \to s_2 \\ P \end{array}
\begin{array}{cccccc}
x_1 & x_2 & s_1 & s_2 & P & \\
1 & \frac{1}{3} & \frac{1}{6} & 0 & 0 & 6 \\
0 & \boxed{\frac{10}{3}} & -\frac{1}{3} & 1 & 0 & 20 \\
\hline
0 & -\frac{5}{3} & \frac{5}{3} & 0 & 1 & 60
\end{array}
\qquad \frac{3}{10}R_2 \to R_2
$$

$$
\sim
\begin{bmatrix}
1 & \frac{1}{3} & \frac{1}{6} & 0 & 0 & 6 \\
0 & \boxed{1} & -\frac{1}{10} & \frac{3}{10} & 0 & 6 \\
\hline
0 & -\frac{5}{3} & \frac{5}{3} & 0 & 1 & 60
\end{bmatrix}
\qquad
\begin{array}{l}
R + (-\frac{1}{3})R_2 \to R_1 \\[6pt]
R_3 + \frac{5}{3}R_2 \to R_3
\end{array}
$$

$$
\begin{array}{c} x_1 \\ \sim x_2 \\ P \end{array}
\begin{bmatrix}
1 & 0 & \frac{1}{5} & -\frac{1}{10} & 0 & 4 \\
0 & 1 & -\frac{1}{10} & \frac{3}{10} & 0 & 6 \\
\hline
0 & 0 & \frac{3}{2} & \frac{1}{2} & 1 & 70
\end{bmatrix}
$$

Since all the indicators in the last row are nonnegative, we stop and read the optimal solution:

Max $P = 70$ at $x_1 = 4$, $x_2 = 6$, $s_1 = 0$, $s_2 = 0$

(If this still is not clear, write the objective function corresponding to the last row to see what happens to P when you try to increase s_1 or s_2.) ◆

PROBLEM   11   Solve the following linear programming problem using the simplex method:

Maximize $P = 2x_1 + x_2$

Subject to $4x_1 + x_2 \le 8$

$2x_1 + 2x_2 \le 10$

$x_1, x_2 \ge 0$ ◆

◆ EXAMPLE   12   Solve using the simplex method:

Maximize $P = 6x_1 + 3x_2$

Subject to $-2x_1 + 3x_2 \le 9$

$-x_1 + 3x_2 \le 12$

$x_1, x_2 \ge 0$

Solution   Write the initial system using the slack variables s_1 and s_2:

$$-2x_1 + 3x_2 + s_1 \qquad\qquad = 9$$
$$-x_1 + 3x_2 \qquad + s_2 \qquad = 12$$
$$-6x_1 - 3x_2 \qquad\qquad + P = 0$$

Write the simplex tableau and identify the first pivot element:

$$
\begin{array}{c}
\quad\;\; x_1 \quad x_2 \quad s_1 \quad s_2 \quad P \\
\begin{array}{c} s_1 \\ s_2 \\ P \end{array}
\left[
\begin{array}{rrrrr|r}
-2 & 3 & 1 & 0 & 0 & 9 \\
-1 & 3 & 0 & 1 & 0 & 12 \\
\hline
-6 & -3 & 0 & 0 & 1 & 0
\end{array}
\right]
\end{array}
$$

$\qquad\qquad\;\; \uparrow$
$\qquad\quad$ Pivot column

Since both elements in the pivot column above the dashed line are negative, we are unable to select a pivot row. We stop and conclude that there is no solution.

$\qquad\qquad\qquad\qquad\qquad\qquad\qquad\qquad\qquad\qquad\qquad\qquad\qquad\quad$ ◆

Notice that we do not try to continue with the simplex method by using a different column for the pivot column. The pivot column must correspond to the most negative indicator in the bottom row.

Once the pivot column has been selected, either there is a pivot row and the simplex method can be continued, or there is no pivot row and the simplex method stops, and we conclude that there is no solution.

P R O B L E M   12   Solve using the simplex method:

$$\text{Maximize} \quad P = 2x_1 + 3x_2$$
$$\text{Subject to} \quad -3x_1 + 4x_2 \leqslant 12$$
$$x_2 \leqslant 9$$
$$x_1, x_2 \geqslant 0$$

$\qquad\qquad\qquad\qquad\qquad\qquad\qquad\qquad\qquad\qquad\qquad\qquad\qquad\qquad\qquad$ ◆

◆ APPLICATION

◆ E X A M P L E   13   A farmer owns a 100 acre farm and plans to plant at most three crops. The seed for crops A, B, and C costs \$40, \$20, and \$30 per acre, respectively. A maximum

Agriculture   of \$3,200 can be spent on seed. Crops A, B, and C require 1, 2, and 1 workdays per acre, respectively, and there are a maximum of 160 workdays available. If the farmer can make a profit of $100 per acre on crop A, $300 per acre on crop B, and $200 per acre on crop C, how many acres of each crop should be planted to maximize profit?

Solution Let

$$x_1 = \text{Number of acres of crop } A$$
$$x_2 = \text{Number of acres of crop } B$$
$$x_3 = \text{Number of acres of crop } C$$
$$P = \text{Total profit}$$

Then we have the following linear programming problem:

Maximize $P = 100x_1 + 300x_2 + 200x_3$ Objective function

Subject to

$$\left. \begin{array}{c} x_1 + x_2 + x_3 \le 100 \\ 40x_1 + 20x_2 + 30x_3 \le 3{,}200 \\ x_1 + 2x_2 + x_3 \le 160 \end{array} \right\} \text{Problem constraints}$$

$$x_1, x_2, x_3 \ge 0 \qquad \text{Nonnegative constraints}$$

Next, we introduce slack variables and form the initial system:

$$\begin{array}{rcl} x_1 + x_2 + x_3 + s_1 & = & 100 \\ 40x_1 + 20x_2 + 30x_3 + s_2 & = & 3{,}200 \\ x_1 + 2x_2 + x_3 + s_3 & = & 160 \\ -100x_1 - 300x_2 - 200x_3 + P & = & 0 \end{array}$$

$$x_1, x_2, x_3, s_1, s_2, s_3 \ge 0$$

Notice that the initial system has $7 - 4 = 3$ nonbasic variables and 4 basic variables. Now we form the simplex tableau and solve by the simplex method:

Enter
↓

	x_1	x_2	x_3	s_1	s_2	s_3	P		
s_1	1	1	1	1	0	0	0	100	
s_2	40	20	30	0	1	0	0	3,200	
Exit → s_3	1	②	1	0	0	1	0	160	$0.5R_3 \rightarrow R_3$
P	−100	−300	−200	0	0	0	1	0	

~	1	1	1	1	0	0	0	100	$R_1 + (-1)R_3 \rightarrow R_1$
	40	20	30	0	1	0	0	3,200	$R_2 + (-20)R_3 \rightarrow R_2$
	0.5	①	0.5	0	0	0.5	0	80	
	−100	−300	−200	0	0	0	1	0	$R_4 + 300R_3 \rightarrow R_4$

$$
\begin{array}{c}
\text{Enter} \\
\downarrow
\end{array}
$$

		x_1	x_2	x_3	s_1	s_2	s_3	P		
Exit →	s_1	0.5	0	(0.5)	1	0	−0.5	0	20	$2R_1 \to R_1$
	s_2	30	0	20	0	1	−10	0	1,600	
	x_2	0.5	1	0.5	0	0	0.5	0	80	
	P	50	0	−50	0	0	150	1	24,000	

	x_1	x_2	x_3	s_1	s_2	s_3	P		
~	1	0	(1)	2	0	−1	0	40	
	30	0	20	0	1	−10	0	1,600	$R_2 + (-20)R_1 \to R_2$
	0.5	1	0.5	0	0	0.5	0	80	$R_3 + (-0.5)R_1 \to R_3$
	50	0	−50	0	0	150	1	24,000	$R_4 + 50R_1 \to R_4$

		x_1	x_2	x_3	s_1	s_2	s_3	P	
~	x_3	1	0	1	2	0	−1	0	40
	s_2	10	0	0	−40	1	10	0	800
	x_2	0	1	0	−1	0	1	0	60
	P	100	0	0	100	0	100	1	26,000

All indicators in the bottom row are nonnegative, and we can now read the optimal solution:

$$x_1 = 0, \quad x_2 = 60, \quad x_3 = 40, \quad s_1 = 0, \quad s_2 = 800, \quad s_3 = 0, \quad P = \$26,000$$

Thus, if the farmer plants 60 acres in crop B, 40 acres in crop C, and no crop A, the maximum profit of \$26,000 will be realized. The fact that $s_2 = 800$ tells us (look at the second row in the equations at the start) that this maximum profit is reached by using only \$2,400 of the \$3,200 available for seed; that is, we have a slack of \$800 that can be used for some other purpose. ◆

PROBLEM 13   Repeat Example 13 modified as follows:

	INVESTMENT PER ACRE			MAXIMUM
	Crop A	Crop B	Crop C	AVAILABLE
SEED COST	\$24	\$40	\$30	\$3,600
WORKDAYS	1	2	2	160
PROFIT	\$140	\$200	\$160	

◆

Remarks

1. It can be shown that the feasible region for the linear programming problem in Example 13 has eight corner points, yet the simplex method found the solution in only two steps. Now you begin to see the power of the simplex method. In larger problems, the difference between the total number of corner points and the number of steps required by the simplex method is

even more dramatic. A feasible region may have hundreds or even thousands of corner points, yet the simplex method will often find the optimal solution in 10 or 15 steps.

2. Refer to the second problem constraint in the model for Example 13:

$$40x_1 + 20x_2 + 30x_3 \leqslant 3{,}200$$

Multiplying both sides of this inequality by $\frac{1}{10}$ before introducing a slack variable simplifies subsequent calculations. However, performing this operation has a side effect — it changes the units of the slack variable from dollars to tens of dollars. Compare the equations

$$40x_1 + 20x_2 + 30x_3 + s_2 = 3{,}200 \qquad s_2 \text{ represents dollars}$$

and

$$4x_1 + 2x_2 + 3x_3 + s_2' = 320 \qquad s_2' \text{ represents tens of dollars}$$

to see why this happens. In general, if you multiply a problem constraint by a number, remember to take this into account when you interpret the value of the slack variable for that constraint.

3. It is important to realize that in order to keep this introduction as simple as possible, we have purposely avoided certain degenerate cases that lead to difficulties. Discussion and resolution of these problems is left to a more advanced treatment of the subject.

Answers to Matched Problems

11. Max $P = 6$ when $x_1 = 1$ and $x_2 = 4$ 12. No solution
13. 40 acres of crop A, 60 acres of crop B, no crop C; Max $P = \$17{,}600$ (since $s_2 = 240$, $\$240$ out of the $\$3{,}600$ will not be spent)

EXERCISE 6-4

A *For the simplex tableaux in Problems 1–4:*

(A) *Identify the basic and nonbasic variables.*
(B) *Find the corresponding basic feasible solution.*
(C) *Determine whether the optimal solution has been found, an additional pivot is required, or the problem has no solution.*

1.

x_1	x_2	s_1	s_2	P	
2	1	0	3	0	12
3	0	1	−2	0	15
−4	0	0	4	1	20

2.

x_1	x_2	s_1	s_2	P	
1	4	−2	0	0	10
0	2	3	1	0	25
0	5	6	0	1	35

3.

	x_1	x_2	x_3	s_1	s_2	s_3	P	
	-2	0	1	3	1	0	0	5
	0	1	0	-2	0	0	0	15
	-1	0	0	4	1	1	0	12
	-4	0	0	2	4	0	1	45

4.

	x_1	x_2	x_3	s_1	s_2	s_3	P	
	0	2	-1	1	4	0	0	5
	0	1	2	0	-2	1	0	2
	1	3	0	0	5	0	0	11
	0	-5	4	0	-3	0	1	27

In Problems 5–8, find the pivot element, identify the entering and exiting variables, and perform one pivot operation.

5.

	x_1	x_2	s_1	s_2	P	
	1	4	1	0	0	4
	3	5	0	1	0	24
	-8	-5	0	0	1	0

6.

	x_1	x_2	s_1	s_2	P	
	1	6	1	0	0	36
	3	1	0	1	0	5
	-1	-2	0	0	1	0

7.

	x_1	x_2	s_1	s_2	s_3	P	
	2	1	1	0	0	0	4
	3	0	1	1	0	0	8
	0	0	2	0	1	0	2
	-4	0	-3	0	0	1	5

8.

	x_1	x_2	s_1	s_2	s_3	P	
	0	0	2	1	1	0	2
	1	0	-4	0	1	0	3
	0	1	5	0	2	0	11
	0	0	-6	0	-5	1	18

In Problems 9–12:

(A) Using slack variables, write the initial system for each linear programming problem.
(B) Write the simplex tableau, circle the first pivot, and identify the entering and exiting variables.
(C) Use the simplex method to solve the problem.

9. Maximize $P = 15x_1 + 10x_2$
   Subject to $2x_1 + x_2 \leqslant 10$
   $x_1 + 2x_2 \leqslant 8$
   $x_1, x_2 \geqslant 0$

10. Maximize $P = 3x_1 + 2x_2$
    Subject to $6x_1 + 3x_2 \leqslant 24$
    $3x_1 + 6x_2 \leqslant 30$
    $x_1, x_2 \geqslant 0$

11. Repeat Problem 9 with the objective function changed to $P = 30x_1 + x_2$.
12. Repeat Problem 10 with the objective function changed to $P = x_1 + 3x_2$.

B *Solve the following linear programming problems using the simplex method:*

13. Maximize $P = 30x_1 + 40x_2$
    Subject to $2x_1 + x_2 \leq 10$
    $x_1 + x_2 \leq 7$
    $x_1 + 2x_2 \leq 12$
    $x_1, x_2 \geq 0$

14. Maximize $P = 20x_1 + 10x_2$
    Subject to $3x_1 + x_2 \leq 21$
    $x_1 + x_2 \leq 9$
    $x_1 + 3x_2 \leq 21$
    $x_1, x_2 \geq 0$

15. Maximize $P = 2x_1 + 3x_2$
    Subject to $-2x_1 + x_2 \leq 2$
    $-x_1 + x_2 \leq 5$
    $x_2 \leq 6$
    $x_1, x_2 \geq 0$

16. Repeat Problem 15 with
    $P = -x_1 + 3x_2$.

17. Maximize $P = -x_1 + 2x_2$
    Subject to $-x_1 + x_2 \leq 2$
    $-x_1 + 3x_2 \leq 12$
    $x_1 - 4x_2 \leq 4$
    $x_1, x_2 \geq 0$

18. Repeat Problem 17 with
    $P = x_1 + 2x_2$.

19. Maximize $P = 5x_1 + 2x_2 - x_3$
    Subject to $x_1 + x_2 - x_3 \leq 10$
    $2x_1 + 4x_2 + x_3 \leq 30$
    $x_1, x_2, x_3 \geq 0$

20. Maximize $P = 4x_1 - 3x_2 + 2x_3$
    Subject to $x_1 + 2x_2 - x_3 \leq 5$
    $3x_1 + 2x_2 + 2x_3 \leq 22$
    $x_1, x_2, x_3 \geq 0$

21. Maximize $P = 2x_1 + 3x_2 + 4x_3$
    Subject to $x_1 + x_3 \leq 4$
    $x_2 + x_3 \leq 3$
    $x_1, x_2, x_3 \geq 0$

22. Maximize $P = x_1 + x_2 + 2x_3$
    Subject to $x_1 - 2x_2 + x_3 \leq 9$
    $2x_1 + x_2 + 2x_3 \leq 28$
    $x_1, x_2, x_3 \geq 0$

23. Maximize $P = 4x_1 + 3x_2 + 2x_3$
    Subject to $3x_1 + 2x_2 + 5x_3 \leq 23$
    $2x_1 + x_2 + x_3 \leq 8$
    $x_1 + x_2 + 2x_3 \leq 7$
    $x_1, x_2, x_3 \geq 0$

24. Maximize $P = 4x_1 + 2x_2 + 3x_3$
    Subject to $x_1 + x_2 + x_3 \leq 11$
    $2x_1 + 3x_2 + x_3 \leq 20$
    $x_1 + 3x_2 + 2x_3 \leq 20$
    $x_1, x_2, x_3 \geq 0$

C

25. Maximize $P = 20x_1 + 30x_2$
    Subject to $0.6x_1 + 1.2x_2 \leq 960$
    $0.03x_1 + 0.04x_2 \leq 36$
    $0.3x_1 + 0.2x_2 \leq 270$
    $x_1, x_2 \geq 0$

26. Repeat Problem 25 with
    $P = 20x_1 + 20x_2$.

27. Maximize $P = x_1 + 2x_2 + 3x_3$
    Subject to $2x_1 + 2x_2 + 8x_3 \leq 600$
    $x_1 + 3x_2 + 2x_3 \leq 600$
    $3x_1 + 2x_2 + x_3 \leq 400$
    $x_1, x_2, x_3 \geq 0$

28. Maximize $P = 10x_1 + 50x_2 + 10x_3$
    Subject to $3x_1 + 3x_2 + 3x_3 \leq 66$
    $6x_1 - 2x_2 + 4x_3 \leq 48$
    $3x_1 + 6x_2 + 9x_3 \leq 108$
    $x_1, x_2, x_3 \geq 0$

In Problems 29 and 30, first solve the linear programming problem by the simplex method, keeping track of the basic feasible solutions at each step. Then graph the feasible region and illustrate the path to the optimal solution determined by the simplex method.

29. Maximize $P = 2x_1 + 5x_2$
    Subject to $x_1 + 2x_2 \leqslant 40$
    $x_1 + 3x_2 \leqslant 48$
    $x_1 + 4x_2 \leqslant 60$
    $x_2 \leqslant 14$
    $x_2, x_2 \geqslant 0$

30. Maximize $P = 5x_1 + 3x_2$
    Subject to $5x_1 + 4x_2 \leqslant 100$
    $2x_1 + x_2 \leqslant 28$
    $4x_1 + x_2 \leqslant 42$
    $x_1 \leqslant 10$
    $x_1, x_2 \geqslant 0$

APPLICATIONS

Formulate each of the following as a linear programming problem. Then solve the problem using the simplex method.

Business & Economics

31. *Manufacturing — resource allocation.* A small company manufactures three different electronic components for computers. Component A requires 2 hours of fabrication and 1 hour of assembly; component B requires 1 hour of fabrication and 2 hours of assembly; and component C requires 2 hours of fabrication and 2 hours of assembly. The company has up to 1,000 labor-hours of fabrication time and 800 labor-hours of assembly time available per week. The profit on each component, A, B, and C, is $7, $9, and $10, respectively. How many components of each type should the company manufacture each week in order to maximize its profit (assuming all components that it manufactures can be sold)? What is the maximum profit?

32. *Manufacturing — resource allocation.* Repeat Problem 31 under the additional assumption that the combined total number of components produced by the company cannot exceed 550.

33. *Investment.* An investor has at most $100,000 to invest in government bonds, mutual funds, and money market funds. The average yields for government bonds, mutual funds, and money market funds are 8%, 13%, and 15%, respectively. The investor's policy requires that the total amount invested in mutual and money market funds not exceed the amount invested in government bonds. How much should be invested in each type of investment in order to maximize the return? What is the maximum return?

34. *Investment.* Repeat Problem 33 under the additional assumption that no more than $30,000 can be invested in money market funds.

35. *Advertising.* A department store chain has up to $20,000 to spend on television advertising for a sale. All ads will be placed with one television station, where a 30-second ad costs $1,000 on daytime TV and is viewed by 14,000 potential customers, $2,000 on prime-time TV and is viewed by 24,000 potential customers, and $1,500 on late-night TV and is viewed by 18,000

potential customers. The television station will not accept a total of more than 15 ads in all three time periods. How many ads should be placed in each time period in order to maximize the number of potential customers who will see the ads? How many potential customers will see the ads? (Ignore repeated viewings of the ad by the same potential customer.)

36. *Advertising.* Repeat Problem 35 if the department store increases its budget to $24,000 and requires that at least half of the ads be placed in prime-time shows.

37. *Construction — resource allocation.* A contractor is planning to build a new housing development consisting of colonial, split-level, and ranch-style houses. A colonial house requires $\frac{1}{2}$ acre of land, $60,000 capital, and 4,000 labor-hours to construct, and returns a profit of $20,000. A split-level house requires $\frac{1}{2}$ acre of land, $60,000 capital, and 3,000 labor-hours to construct, and returns a profit of $18,000. A ranch house requires 1 acre of land, $80,000 capital, and 4,000 labor-hours to construct, and returns a profit of $24,000. The contractor has available 30 acres of land, $3,200,000 capital, and 180,000 labor-hours. How many houses of each type should be constructed to maximize the contractor's profit? What is the maximum profit?

38. *Manufacturing — resource allocation.* A company manufactures three-speed, five-speed, and ten-speed bicycles. Each bicycle passes through three departments, fabrication, painting & plating, and final assembly. The relevant manufacturing data are given in the table. How many bicycles of each type should the company manufacture per day in order to maximize its profit? What is the maximum profit?

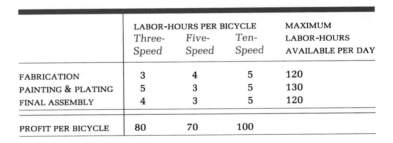

| | LABOR-HOURS PER BICYCLE | | | MAXIMUM |
	Three-Speed	Five-Speed	Ten-Speed	LABOR-HOURS AVAILABLE PER DAY
FABRICATION	3	4	5	120
PAINTING & PLATING	5	3	5	130
FINAL ASSEMBLY	4	3	5	120
PROFIT PER BICYCLE	80	70	100	

39. *Packaging — product mix.* A candy company makes three types of candy, solid-center, fruit-filled, and cream-filled, and packages these candies in three different assortments. A box of assortment I contains 4 solid-center, 4 fruit-filled, and 12 cream-filled candies, and sells for $9.40. A box of assortment II contains 12 solid-center, 4 fruit-filled, and 4 cream-filled candies, and sells for $7.60. A box of assortment III contains 8 solid-center, 8 fruit-filled, and 8 cream-filled candies, and sells for $11.00. The manufacturing costs per piece of candy are $0.20 for solid-center, $0.25 for fruit-filled, and $0.30 for cream-filled. The company can manufacture 4,800 solid-center, 4,000 fruit-filled, and 5,600 cream-filled candies weekly. How many boxes

of each type should they produce each week in order to maximize their profit? What is the maximum profit?

40. *Scheduling — resource allocation.* A small accounting firm prepares tax returns for three types of customers: individual, commercial, and industrial. The tax preparation process begins with a 1 hour interview with the customer. The data collected during this interview are entered into a time-sharing computer system, which produces the customer's tax return. It takes 1 hour to enter the data for an individual customer, 2 hours for a commercial customer, and $1\frac{1}{2}$ hours for an industrial customer. It takes 10 minutes of computer time to process an individual return, 25 minutes to process a commercial return, and 20 minutes to process an industrial return. The firm has one employee who conducts the initial interview and two who enter the data into the computer. The interviewer can work a maximum of 50 hours a week, and each of the data-entry employees can work a maximum of 40 hours a week. The computer is available for a maximum of 1,025 minutes a week. The firm makes a profit of $50 on each individual customer, $65 on each commercial customer, and $60 on each industrial customer. How many customers of each type should the firm schedule each week in order to maximize its profit? What is the maximum profit?

Life Sciences

41. *Nutrition — animals.* The natural diet of a certain animal consists of three foods, A, B, and C. The number of units of calcium, iron, and protein in 1 gram of each food and the average daily intake are given in the table. A scientist wants to investigate the effect of increasing the protein in the animal's diet while not allowing the units of calcium and iron to exceed their average daily intakes. How many grams of each food should be used to maximize the amount of protein in the diet? What is the maximum amount of protein?

| | UNITS PER GRAM | | | AVERAGE DAILY INTAKE (UNITS) |
	Food A	Food B	Food C	
CALCIUM	1	3	2	30
IRON	2	1	1	24
PROTEIN	3	3	5	60

42. *Nutrition — animals.* Repeat Problem 41 if the scientist wants to maximize the daily calcium intake while not allowing the intake of iron or protein to exceed the average daily intake.

Social Sciences

43. *Opinion survey.* A political scientist has received a grant to fund a research project involving voting trends. The budget of the grant includes $1,620 for conducting door-to-door interviews the day before an election. Undergraduate students, graduate students, and faculty members will be hired to conduct the interviews. Each undergraduate student will conduct 18 interviews and be paid $60. Each graduate student will conduct 25 interviews and be paid $90. Each faculty member will conduct 30 interviews and be paid $120. Due to limited transportation facilities, no more than 20 interviewers can be hired. How many undergraduate students, graduate students, and faculty members should be hired in order to maximize the number of interviews that will be conducted? What is the maximum number of interviews?

44. *Opinion survey.* Repeat Problem 43 if one of the requirements of the grant is that at least 50% of the interviewers be undergraduate students.

SECTION 6-5 | **Chapter Review**

Important Terms and Symbols

6-1 *Systems of Linear Inequalities in Two Variables.* Graph of a linear inequality in two variables; upper half-plane; lower half-plane; graphical solution of systems of linear inequalities; solution region; feasible region; corner point; bounded regions; unbounded regions

6-2 *Linear Programming in Two Dimensions—A Geometric Approach.* Linear programming problem; decision variables; objective function; problem constraints; nonnegative constraints; mathematical model; graphical solution; maximization problem; constant-profit line; optimal solution; feasible region; fundamental theorem of linear programming; multiple optimal solution; empty feasible region; unbounded objective function

6-3 *A Geometric Introduction to the Simplex Method.* Standard maximization problem in standard form; slack variables; basic variables; nonbasic variables; basic solution; basic feasible solution; simplex method (algorithm)

6-4 *The Simplex Method: Maximization with Problem Constraints of the Form ≤.* Initial system; basic solutions and basic feasible solutions for initial systems; initial basic feasible solution; simplex tableau; initial simplex tableau; selecting basic and nonbasic variables for the simplex method; pivot operation; entering variable; exiting variable; pivot column; indicators; pivot row; pivot element; pivoting

---

EXERCISE 6-5 | **Chapter Review**

Work through all the problems in this chapter review and check your answers in the back of the book. (Answers to all review problems are there.) Where weaknesses show up, review appropriate sections in the text.

A *Solve the systems in Problems 1 and 2 graphically, and indicate whether each solution region is bounded or unbounded. Find the coordinates of each corner point.*

1. $2x_1 + x_2 \leqslant 8$
   $3x_1 + 9x_2 \leqslant 27$
   $x_1, x_2 \geqslant 0$

2. $3x_1 + x_2 \geqslant 9$
   $2x_1 + 4x_2 \geqslant 16$
   $x_1, x_2 \geqslant 0$

3. Solve the linear programming problem geometrically:

   Maximize $P = 6x_1 + 2x_2$

   Subject to $2x_1 + x_2 \leqslant 8$

   $x_1 + 2x_2 \leqslant 10$

   $x_1, x_2 \geqslant 0$

4. Convert the problem constraints in Problem 3 into a system of equations using slack variables.

5. How many basic variables and how many nonbasic variables are associated with the system in Problem 4?

6. Find all basic solutions for the system in Problem 4 and determine which basic solutions are feasible.

7. Write the simplex tableau for Problem 3 and circle the pivot element. Indicate the entering and exiting variables.

8. Solve Problem 3 using the simplex method.

9. For the simplex tableau below, identify the basic and nonbasic variables. Find the pivot element, the entering and exiting variables, and perform one pivot operation.

$$
\begin{array}{ccccccc|c}
x_1 & x_2 & x_3 & s_1 & s_2 & s_3 & P & \\
2 & 1 & 3 & -1 & 0 & 0 & 0 & 20 \\
3 & 0 & 4 & 1 & 1 & 0 & 0 & 30 \\
2 & 0 & 5 & 2 & 0 & 1 & 0 & 10 \\
-8 & 0 & -5 & 3 & 0 & 0 & 1 & 50
\end{array}
$$

10. Find the basic solution for each tableau. Determine whether the optimal solution has been reached, additional pivoting is required, or the problem has no solution.

(A)
$$
\begin{array}{ccccc|c}
x_1 & x_2 & s_1 & s_2 & P & \\
4 & 1 & 0 & 0 & 0 & 2 \\
2 & 0 & 1 & 1 & 0 & 5 \\
-2 & 0 & 3 & 0 & 1 & 12
\end{array}
$$

(B)
$$
\begin{array}{ccccc|c}
x_1 & x_2 & s_1 & s_2 & P & \\
-1 & 3 & 0 & 1 & 0 & 7 \\
0 & 2 & 1 & 0 & 0 & 0 \\
-2 & 1 & 0 & 0 & 1 & 22
\end{array}
$$

(C)
$$
\begin{array}{ccccc|c}
x_1 & x_2 & s_1 & s_2 & P & \\
1 & -2 & 0 & 4 & 0 & 6 \\
0 & 2 & 1 & 6 & 0 & 15 \\
0 & 3 & 0 & 2 & 1 & 10
\end{array}
$$

B

11. Solve the linear programming problem geometrically:

$$\text{Minimize} \quad C = 5x_1 + 2x_2$$
$$\text{Subject to} \quad x_1 + 3x_2 \geq 15$$
$$2x_1 + x_2 \geq 20$$
$$x_1, x_2 \geq 0$$

12. Solve the linear programming problem geometrically:

$$\text{Maximize} \quad P = 3x_1 + 4x_2$$
$$\text{Subject to} \quad 2x_1 + 4x_2 \leq 24$$
$$3x_1 + 3x_2 \leq 21$$
$$4x_1 + 2x_2 \leq 20$$
$$x_1, x_2 \geq 0$$

13. Solve Problem 12 using the simplex method.

14. Solve the linear programming problem geometrically:

$$\text{Minimize} \quad C = 3x_1 + 8x_2$$
$$\text{Subject to} \quad x_1 + x_2 \geq 10$$
$$x_1 + 2x_2 \geq 15$$
$$x_2 \geq 3$$
$$x_1, x_2 \geq 0$$

Solve the following linear programming problems:

15. Maximize
$$P = 5x_1 + 3x_2 - 3x_3$$
Subject to
$$x_1 - x_2 - 2x_3 \leq 3$$
$$2x_1 + 2x_2 - 5x_3 \leq 10$$
$$x_1, x_2, x_3 \geq 0$$

16. Maximize
$$P = 5x_1 + 3x_2 - 3x_3$$
Subject to
$$x_1 - x_2 - 2x_3 \leq 3$$
$$x_1 + x_2 \leq 5$$
$$x_1, x_2, x_3 \geq 0$$

C

17. Solve the following linear programming problem by the simplex method, keeping track of the obvious basic solution at each step. Then graph the feasible region and illustrate the path to the optimal solution determined by the simplex method.

$$\text{Maximize} \quad P = 2x_1 + 3x_2$$
$$\text{Subject to} \quad x_1 + 2x_2 \leq 22$$
$$2x_1 + x_2 \leq 20$$
$$x_1 \leq 8$$
$$x_2 \leq 10$$
$$x_1, x_2 \geq 0$$

Formulate the following as linear programming problems. (Do not solve.)

Business & Economics

18. *Manufacturing — resource allocation.* South Shore Sail Loft manufactures regular and competition sails. Each regular sail takes 2 hours to cut and 4 hours to sew. Each competition sail takes 3 hours to cut and 9 hours to sew. The Loft makes a profit of $100 on each regular sail and $200 on each competition sail. If there are 150 hours available in the cutting department and 360 hours available in the sewing department, how many sails of each type should the company manufacture in order to maximize their profit?

19. *Investment.* An investor has $150,000 to invest in oil stock, steel stock, and government bonds. The oil stock is expected to yield 12%, the steel stock is expected to yield 9%, and the government bonds are guaranteed to yield 5%. The investor's policy requires that the total amount of money invested in stock cannot exceed the amount invested in bonds and that at least twice as much money should be invested in steel stock as in oil stock. How much should be invested in each alternative in order to maximize the return?

Life Sciences

20. *Nutrition — animals.* A special diet for laboratory animals is to contain at least 300 units of vitamins, 200 units of minerals, and 900 calories. There are two feed mixes available, mix *A* and mix *B*. A gram of mix *A* contains 3 units of vitamins, 2 units of minerals, and 6 calories. A gram of mix *B* contains 4 units of vitamins, 5 units of minerals, and 10 calories. Mix *A* costs $0.02 per gram and mix *B* costs $0.04 per gram. How many grams of each mix should be used to satisfy the requirements of the diet at minimal cost?

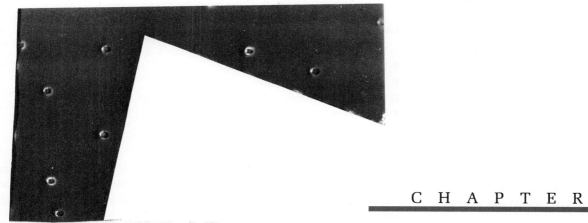

Probability

CHAPTER 7

Contents

Probability, like many branches of mathematics, evolved out of practical considerations. Girolamo Cardano (1501–1576), a gambler and physician, produced some of the best mathematics of his time, including a systematic analysis of gambling problems. In 1654, another gambler, the Chevalier de Méré, plagued with bad luck, approached the well-known French philosopher and mathematician Blaise Pascal (1623–1662) regarding certain dice problems. Pascal became interested in these problems, studied them, and discussed them with Pierre de Fermat (1601–1665), another French mathematician. Thus, out of the gaming rooms of western Europe the study of probability was born.

In spite of this lowly birth, probability has matured into a highly respected and immensely useful branch of mathematics. It is used in practically every field. Probability theory can be thought of as the science of uncertainty. If, for example, a card is drawn from a deck of 52 playing cards, it is uncertain which card will be drawn. But suppose a card is drawn and replaced in the deck and a card is again drawn and replaced, and this action is repeated a large number of times. A particular card, say the ace of spades, will be drawn over the long run with a relative frequency that is approximately predictable. Probability theory is concerned with determining the long-run frequency of the occurrence of a given event.

$$\text{Probability} = \frac{\text{favorable}}{\text{Total \# of possibilities}} =$$

$$\frac{4}{52} = \frac{1}{13}$$

How do we assign probabilities to events? There are two basic approaches to this problem, one theoretical and the other empirical. An example will illustrate the difference between the two approaches.

Suppose you were asked, "What is the probability of obtaining a 2 on a single throw of a die?" Using a *theoretical approach*, we would reason as follows: Since there are 6 *equally likely* ways the die can turn up (assuming the die is fair) and there is only 1 way a 2 can turn up, then the probability of obtaining a 2 is $\frac{1}{6}$. Here, we have arrived at a probability assignment without even rolling a die once; we have used certain assumptions and a reasoning process.

What does the result have to do with reality? We would expect that in the long run (after rolling a die many times), the 2 would appear approximately $\frac{1}{6}$ of the time.

With the *empirical approach*, we make no assumption about the equally likely ways in which the die can turn up. We simply set up an experiment and roll the die a large number of times. Then we compute the percentage of times the 2 appears and use this number as an estimate of the probability of obtaining a 2 on a single roll of the die. Each approach has advantages and drawbacks; these will be discussed in the following sections.

We will first consider the theoretical approach and develop procedures that will lead to the solution of a large variety of interesting problems. These procedures require counting the number of ways certain events can happen, and this is not always easy. However, powerful mathematical tools can assist us in this counting task. The development of these tools is the subject matter of the next section.

Multiplication Principle, Permutations, and Combinations

◆ MULTIPLICATION PRINCIPLE
◆ FACTORIAL
◆ PERMUTATIONS
◆ COMBINATIONS
◆ APPLICATIONS

◆ MULTIPLICATION PRINCIPLE

The best way to start this discussion is with an example.

◆ EXAMPLE 1 Suppose we spin a spinner that can land on four possible numbers, 1, 2, 3, or 4. We then flip a coin that can turn up either heads (H) or tails (T). What are the possible combined outcomes?

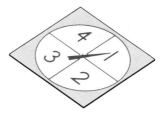

Solution To solve this problem, let us use a **tree diagram:**

Spinner Outcomes	Coin Outcomes	Combined Outcomes
1	H	(1, H)
	T	(1, T)
2	H	(2, H)
	T	(2, T)
3	H	(3, H)
	T	(3, T)
4	H	(4, H)
	T	(4, T)

Start

Thus, there are 8 possible combined outcomes (there are 4 places the spinner can stop, followed by 2 ways the coin can land). The order in the ordered pair representing a combined outcome is important: The first element in the ordered pair represents a spinner outcome and the second element a coin outcome. ◆

PROBLEM 1 Use a tree diagram to determine the possible combined outcomes of flipping a coin followed by spinning the dial in Example 1. ◆

Now suppose you asked, "From the 26 letters in the alphabet, how many ways can 3 letters appear in a row on a license plate if no letter is repeated?" To try to count the possibilities using a tree diagram would be extremely tedious, to say the least. The following **multiplication principle** will enable us to solve this problem easily; in addition, it forms the basis for several other counting devices that are developed later in this section:

▬ Multiplication Principle (for Counting)

1. If two operations O_1 and O_2 are performed in order, with N_1 possible outcomes for the first operation and N_2 possible outcomes for the second operation, then there are

   $$N_1 \cdot N_2$$

   possible combined outcomes of the first operation followed by the second.

2. In general, if n operations $O_1, O_2, \ldots, O_n$ are performed in order, with possible number of outcomes $N_1, N_2, \ldots, N_n$, respectively, then there are

   $$N_1 \cdot N_2 \cdot \cdots \cdot N_n$$

   possible combined outcomes of the operations performed in the given order.

In Example 1, we see that there are 4 possible outcomes of spinning the dial (first operation) and 2 possible outcomes of flipping the coin (second operation); hence, by the multiplication principle, there are $4 \cdot 2 = 8$ possible combined outcomes. Use the multiplication principle to solve Problem 1. [*Answer:* $2 \cdot 4 = 8$.]

To answer the license plate question: There are 26 ways the first letter can be chosen; after a first letter is chosen, there are 25 ways a second letter can be chosen; and after 2 letters are chosen, there are 24 ways a third letter can be chosen. Hence, using the multiplication principle, there are $26 \cdot 25 \cdot 24 = 15{,}600$ possible ways 3 letters can be chosen from the alphabet without repeats.

◆ E X A M P L E　2　Many colleges and universities are now using computer-assisted testing procedures. Suppose a screening test is to consist of 5 questions, and a computer stores
5 comparable questions for the first test question, 8 for the second, 6 for the
third, 5 for the fourth, and 10 for the fifth. How many different 5-question tests
can the computer select? (Two tests are considered different if they differ in one
or more questions.)

Solution　O_1:　Selecting the first question　　　　N_1:　5 ways
　　　　　　O_2:　Selecting the second question　　　N_2:　8 ways
　　　　　　O_3:　Selecting the third question　　　　N_3:　6 ways
　　　　　　O_4:　Selecting the fourth question　　　N_4:　5 ways
　　　　　　O_5:　Selecting the fifth question　　　　N_5:　10 ways

Thus, the computer can generate

$$5 \cdot 8 \cdot 6 \cdot 5 \cdot 10 = 12,000 \text{ different tests}$$　　◆

P R O B L E M　2　Each question on a multiple-choice test has 5 choices. If there are 5 such questions on a test, how many different response sheets are possible if only 1 choice is
marked for each question?　$5 \cdot 5 \cdot 5 \cdot 5 \cdot 5 = 3,125$　　◆

◆ E X A M P L E　3　How many 3-letter code words are possible using the first 8 letters of the alphabet if:

(A) No letter can be repeated?　　　　(B) Letters can be repeated?
(C) Adjacent letters cannot be alike?

Solutions　To form 3-letter code words from the 8 letters available, we select a letter for the
first position, one for the second position, and one for the third position.
Altogether, there are three operations.

(A) No letter can be repeated:

　　　O_1:　Selecting the first letter　　　N_1:　8 ways
　　　O_2:　Selecting the second letter　　N_2:　7 ways　　Since 1 letter has
　　　　　　　　　　　　　　　　　　　　　　　　　　　　　been used
　　　O_3:　Selecting the third letter　　　N_3:　6 ways　　Since 2 letters have
　　　　　　　　　　　　　　　　　　　　　　　　　　　　　been used

Thus, there are

$$8 \cdot 7 \cdot 6 = 336 \text{ possible code words}$$　　Possible combined operations

(B) Letters can be repeated:

　　　O_1:　Selecting the first letter　　　N_1:　8 ways
　　　O_2:　Selecting the second letter　　N_2:　8 ways　　Repeats allowed
　　　O_3:　Selecting the third letter　　　N_3:　8 ways　　Repeats allowed

Thus, there are

$$8 \cdot 8 \cdot 8 = 8^3 = 512 \text{ possible code words}$$

(C) Adjacent letters cannot be alike:

O_1:	Selecting the first letter	N_1:	8 ways	
O_2:	Selecting the second letter	N_2:	7 ways	Cannot be the same as the first
O_3:	Selecting the third letter	N_3:	7 ways	Cannot be the same as the second, but can be the same as the first

Thus, there are

$$8 \cdot 7 \cdot 7 = 392 \text{ possible code words}$$ ◆

PROBLEM 3 How many 4-letter code words are possible using the first 10 letters of the alphabet under the three different conditions stated in Example 3? ◆

The multiplication principle can be used to develop two additional devices for counting that are extremely useful in more complicated counting problems. Both of these devices use a function called a *factorial function*, which we introduce first.

◆ FACTORIAL

When using the multiplication principle, we encountered expressions of the form

$$26 \cdot 25 \cdot 24 \qquad 8 \cdot 7 \cdot 6$$

where each natural number factor is decreased by 1 as we move from left to right. The factors in the following product continue to decrease by 1 until a factor of 1 is reached:

$$5 \cdot 4 \cdot 3 \cdot 2 \cdot 1$$

Products of this type are encountered so frequently in counting problems that it is useful to be able to express them in a concise notation. The product of the first n natural numbers is called **n factorial** and is denoted by **$n!$**. Also, we define **zero factorial, 0!,** to be 1. Symbolically:

■ Factorial

For n a natural number,

$$n! = n(n-1)(n-2) \cdot \cdots \cdot 2 \cdot 1$$

$$0! = 1$$

$$n! = n \cdot (n-1)!$$

[*Note:* The $\boxed{n!}$ key appears on many calculators.]

♦ E X A M P L E   4   (A) $5! = 5 \cdot 4 \cdot 3 \cdot 2 \cdot 1 = 120$

(B) $\dfrac{7!}{6!} = \dfrac{7 \cdot \cancel{6!}}{\cancel{6!}} = 7$

(C) $\dfrac{8!}{5!} = \dfrac{8 \cdot 7 \cdot 6 \cdot \cancel{5!}}{\cancel{5!}} = 8 \cdot 7 \cdot 6 = 336$

(D) $\dfrac{52!}{5!47!} = \dfrac{52 \cdot 51 \cdot 50 \cdot 49 \cdot 48 \cdot \cancel{47!}}{5 \cdot 4 \cdot 3 \cdot 2 \cdot 1 \cdot \cancel{47!}} = 2{,}598{,}960$ ♦

P R O B L E M   4   Find: (A) $6!$ (B) $\dfrac{10!}{9!}$ (C) $\dfrac{10!}{7!}$ (D) $\dfrac{5!}{0!3!}$ (E) $\dfrac{20!}{3!17!}$ ♦

It is interesting and useful to note that $n!$ grows very rapidly. Compare the following:

$5! = 120$

$10! = 3{,}628{,}800$

$15! = 1{,}307{,}674{,}368{,}000$

Try $69!$, $70!$, and $71!$ on your calculator.

♦ PERMUTATIONS

A particular (horizontal) arrangement of a set of paintings on a wall is called a *permutation* of the set of paintings. In general:

■ A Permutation of a Set of Objects

A **permutation** of a set of distinct objects is an arrangement of the objects in a specific order without repetition.

Suppose 4 pictures are to be arranged from left to right on one wall of an art gallery. How many permutations (ordered arrangements) are possible? Using the multiplication principle, there are 4 ways of selecting the first picture; after the first picture is selected, there are 3 ways of selecting the second picture; after the first 2 pictures are selected, there are 2 ways of selecting the third picture; and after the first 3 pictures are selected, there is only 1 way to select the fourth. Thus, the number of permutations (ordered arrangements) of the set of 4 pictures is

$4 \cdot 3 \cdot 2 \cdot 1 = 4! = 24$

In general, how many permutations of a set of n distinct objects are possible? Reasoning as above, there are n ways in which the first object can be chosen, there are $n - 1$ ways in which the second object can be chosen, and so on. Using the multiplication principle, we have the following:

The Number of Permutations of *n* Objects

The number of permutations of n distinct objects without repetition, denoted by $P_{n,n}$, is

$$P_{n,n} = n(n-1) \cdot \cdots \cdot 2 \cdot 1 = n! \qquad \text{n factors}$$

The number of permutations of 7 objects is

$$P_{7,7} = 7 \cdot 6 \cdot 5 \cdot 4 \cdot 3 \cdot 2 \cdot 1 = 7! \qquad \text{7 factors}$$

Now suppose the museum director decides to use only 2 of the 4 available paintings, and they will be arranged on the wall from left to right. We are now talking about a particular arrangement of 2 paintings out of the 4, which is called a *permutation of 4 objects taken 2 at a time*. In general:

A Permutation of *n* Objects Taken *r* at a Time

A permutation of a set of n distinct objects taken r at a time without repetition, denoted by $P_{n,r}$, is an arrangement of the r objects in a specific order.

How many ordered arrangements of 2 pictures can be formed from the 4? That is, how many permutations of 4 objects taken 2 at a time are there? There are 4 ways the first picture can be selected; after selecting the first picture, there are 3 ways the second picture can be selected. Thus, the number of permutations of a set of 4 objects taken 2 at a time, which is denoted by $P_{4,2}$, is given by

$$P_{4,2} = 4 \cdot 3$$

In terms of factorials, multiplying $4 \cdot 3$ by 1 in the form $2!/2!$, we have

$$P_{4,2} = 4 \cdot 3 = \frac{4 \cdot 3 \cdot 2!}{2!} = \frac{4!}{2!}$$

Reasoning in the same way as in the example, we find that the number of permutations of n distinct objects taken r at a time without repetition $(0 \leqslant r \leqslant n)$ is given by

$$P_{n,r} = n(n-1)(n-2) \cdot \cdots \cdot (n-r+1) \qquad \text{r factors}$$
$$P_{9,6} = 9(9-1)(9-2) \cdot \cdots \cdot (9-6+1) \qquad \text{6 factors}$$
$$= 9 \cdot 8 \cdot 7 \cdot 6 \cdot 5 \cdot 4$$

Multiplying the right side by 1 in the form $(n - r)!/(n - r)!$, we obtain a factorial form for $P_{n,r}$:

$$P_{n,r} = n(n - 1)(n - 2) \cdot \cdots \cdot (n - r + 1) \frac{(n - r)!}{(n - r)!}$$

But

$$n(n - 1)(n - 2) \cdot \cdots \cdot (n - r + 1)(n - r)! = n!$$

Hence,

$$P_{n,r} = \frac{n!}{(n - r)!}$$

We summarize these results in the following box:

Number of Permutations of n Objects Taken r at a Time

The number of permutations of n distinct objects taken r at a time without repetition is given by*

$$P_{n,r} = n(n - 1)(n - 2) \cdot \cdots \cdot (n - r + 1) \qquad r \text{ factors}$$
$$P_{5,2} = 5 \cdot 4 \qquad 2 \text{ factors}$$

or

$$P_{n,r} = \frac{n!}{(n - r)!} \qquad 0 \leqslant r \leqslant n$$
$$P_{5,2} = \frac{5!}{(5 - 2)!} = \frac{5!}{3!}$$

[Note: $P_{n,n} = \dfrac{n!}{(n - n)!} = \dfrac{n!}{0!} = n!$ permutations of n objects taken n at a time. Remember, by definition, $0! = 1$.]

---

* In place of the symbol $P_{n,r}$, the symbols P_r^n, $_nP_r$, and $P(n, r)$ are often used.

◆ E X A M P L E   5   Given the set $\{A, B, C\}$, how many permutations are there of this set of 3 objects taken 2 at a time? Answer the question:

(A) Using a tree diagram   (B) Using the multiplication principle
(C) Using the two formulas for $P_{n,r}$

Solutions   (A) Using a tree diagram:

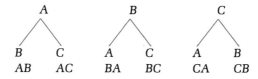

There are 6 permutations of 3 objects taken 2 at a time.

(B) Using the multiplication principle:

O_1: Fill the first position N_1: 3 ways
O_2: Fill the second position N_2: 2 ways

Thus, there are

$3 \cdot 2 = 6$ permutations of 3 objects taken 2 at a time

(C) Using the two formulas for $P_{n,r}$:

2 factors
↓

$$P_{3,2} = 3 \cdot 2 = 6 \qquad \text{or} \qquad P_{3,2} = \frac{3!}{(3-2)!} = \frac{3 \cdot 2 \cdot 1}{1} = 6$$

Thus, there are 6 permutations of 3 objects taken 2 at a time. Of course, all three methods produce the same answer. ◆

PROBLEM 5   Given the set $\{A, B, C, D\}$, how many permutations are there of this set of 4 objects taken 2 at a time? Answer the question:

(A) Using a tree diagram   (B) Using the multiplication principle
(C) Using the two formulas for $P_{n,r}$ ◆

In Example 5 you probably found the multiplication principle the easiest to use and wonder about the usefulness of the factorial formula for $P_{n,r}$. When the numbers get large you will find that the factorial formula is very useful, since most scientific calculators have a key marked $\boxed{n!}$. (In fact, many scientific calculators even have a $\boxed{P_{n,r}}$ key that can evaluate $P_{n,r}$ directly.) Check your calculator and user's manual. The use of a calculator will significantly reduce the burden of computing in this and the following sections. To appreciate this fact, try solving Example 6 without a calculator.

◆ EXAMPLE 6   Find the number of permutations of 13 objects taken 8 at a time. Compute the answer using the $\boxed{n!}$ key on your calculator, if it has one.

Solution   We use the factorial formula for $P_{n,r}$:

$$P_{13,8} = \frac{13!}{(13-8)!} = \frac{13!}{5!} = 51{,}891{,}840$$

Using a tree diagram to solve this problem would involve a monumental effort. Using the multiplication principle would involve multiplying $13 \cdot 12 \cdot 11 \cdot 10 \cdot 9 \cdot 8 \cdot 7 \cdot 6$ (8 factors), which is not too bad, but using the factorial formula is easier if your calculator has the $\boxed{n!}$ key. ◆

PROBLEM 6   Find the number of permutations of 30 objects taken 4 at a time. Compute the answer exactly using a calculator.   ◆

◆ COMBINATIONS

Now suppose that an art museum owns 8 paintings by a given artist and another art museum wishes to borrow 3 of these paintings for a special show. In selecting 3 of the 8 paintings for shipment, the order would not matter, and we would simply be selecting a 3-element subset from the set of 8 paintings. That is, we would be selecting what is called a *combination of 8 objects taken 3 at a time*. In general:

▬ A Combination of *n* Objects Taken *r* at a Time

A **combination** of a set of *n* distinct objects taken *r* at a time without repetition, denoted by $C_{n,r}$, is an r-element subset of the set of n objects. The arrangement of the elements in the subset does not matter.

How many ways can the 3 paintings be selected for shipment out of the 8 available? That is, what is the number of combinations of 8 objects taken 3 at a time? To answer this question, and to get a better insight into the general problem, we return to Example 5.

In Example 5 we were given the set {*A*, *B*, *C*} and found the number of permutations of 3 objects taken 2 at a time using a tree diagram. From this tree diagram we also can determine the number of combinations of 3 objects taken 2 at a time (the number of 2-element subsets from a 3-element set), and compare it with the number of permutations (see Fig. 1).

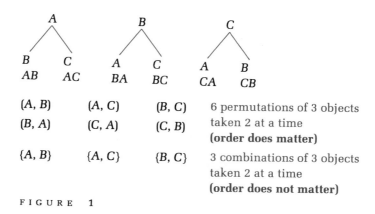

FIGURE 1

There are fewer combinations than permutations, as one would expect. To each subset (combination) there corresponds two ordered pairs (permutations).

We denote the number of combinations in Figure 1 by

$$C_{3,2} \quad \text{or} \quad \binom{3}{2}$$

Our final goal is to find a factorial formula for $C_{n,r}$, the number of combinations of n objects taken r at a time. But first, we will develop a formula for $C_{3,2}$, and then we will generalize from this experience.

We know the number of permutations of 3 objects taken 2 at a time is given by $P_{3,2}$, and we have a formula for computing this number. Now, suppose we think of $P_{3,2}$ in terms of two operations:

O_1: Selecting a subset of 2 elements $\qquad N_1$: $C_{3,2}$ ways

O_2: Arranging the subset in a given order $\qquad N_2$: 2! ways

The combined operation, O_1 followed by O_2, produces a permutation of 3 objects taken 2 at a time. Thus,

$$P_{3,2} = C_{3,2} \cdot 2!$$

or

$$C_{3,2} = \frac{P_{3,2}}{2!}$$

To find $C_{3,2}$, the number of combinations of 3 objects taken 2 at a time, we substitute

$$P_{3,2} = \frac{3!}{(3-2)!}$$

and solve for $C_{3,2}$:

$$C_{3,2} = \frac{3!}{2!(3-2)!} = \frac{3 \cdot 2 \cdot 1}{(2 \cdot 1)(1)} = 3$$

This result agrees with the result we got using a tree diagram. Note that the number of combinations of 3 objects taken 2 at a time is the same as the number of permutations of 3 objects taken 2 at a time divided by the number of permutations of the elements in a 2-element subset. This observation also can be made in Figure 1.

Reasoning the same way as in the example, the number of combinations of n objects taken r at a time $(0 \leqslant r \leqslant n)$ is given by

$$C_{n,r} = \frac{P_{n,r}}{r!}$$

$$= \frac{n!}{r!(n-r)!} \qquad \text{Since } P_{n,r} = \frac{n!}{(n-r)!}$$

In summary:

> ### ■ Number of Combinations of n Objects Taken r at a Time
>
> The number of combinations of n distinct objects taken r at a time without repetition is given by*
>
> $$C_{n,r} = \binom{n}{r} = \frac{P_{n,r}}{r!} = \frac{n!}{r!(n-r)!} \qquad 0 \leq r \leq n \qquad C_{52,5} = \binom{52}{5}$$
>
> $$= \frac{P_{52,5}}{5!} = \frac{52!}{5!(52-5)!}$$

* In place of the symbols $C_{n,r}$ and $\binom{n}{r}$, the symbols C_r^n, $_nC_r$, and $C(n, r)$ are often used.

Now we can answer the question posed earlier in the museum example. There are

$$C_{8,3} = \frac{8!}{3!(8-3)!} = \frac{8!}{3!5!} = \boxed{\frac{8 \cdot 7 \cdot 6 \cdot 5!}{3 \cdot 2 \cdot 1 \cdot 5!}} = 56$$

ways the 3 paintings can be selected for shipment. That is, there are 56 combinations of 8 objects taken 3 at a time.

◆ **EXAMPLE   7**   From a committee of 10 people:

2598960

(A) In how many ways can we choose a chairperson, a vice chairperson, and a secretary, assuming that one person cannot hold more than one position?

(B) In how many ways can we choose a subcommittee of 3 people?

Solutions   Note how parts A and B differ. In part A, order of choice makes a difference in the selection of the officers. In part B, the ordering does not matter in choosing a 3-person subcommittee. Thus, in part A, we are interested in the number of *permutations* of 10 objects taken 3 at a time; and in part B, we are interested in the number of *combinations* of 10 objects taken 3 at a time. These quantities are computed as follows (and since the numbers are not large, we do not need to use a calculator):

(A) $P_{10,3} = \dfrac{10!}{(10-3)!} = \dfrac{10!}{7!} = \boxed{\dfrac{10 \cdot 9 \cdot 8 \cdot 7!}{7!}} = 720$ ways

(B) $C_{10,3} = \dfrac{10!}{3!(10-3)!} = \dfrac{10!}{3!7!} = \boxed{\dfrac{10 \cdot 9 \cdot 8 \cdot 7!}{3 \cdot 2 \cdot 1 \cdot 7!}} = 120$ ways   ◆

PROBLEM   7   From a committee of 12 people:

(A) In how many ways can we choose a chairperson, a vice chairperson, a secretary, and a treasurer, assuming that one person cannot hold more than one position? Permutation

(B) In how many ways can we choose a subcommittee of 4 people?   ◆

Combination

If n and r are other than small numbers (as in Example 7), the $\boxed{n!}$ key on your calculator will simplify the computation. (Many scientific calculators also have a $\boxed{C_{n,r}}$ key that will simplify the computation even further.)

◆ EXAMPLE 8   Find the number of combinations of 13 objects taken 8 at a time. Compute the answer using the $\boxed{n!}$ key on your calculator, if it has one.

Solution $$C_{13,8} = \binom{13}{8} = \frac{13!}{8!(13-8)!} = \frac{13!}{8!5!} = 1{,}287$$ ◆

Compare the result in Example 8 with that obtained in Example 6 and note that $C_{13,8}$ is substantially smaller than $P_{13,8}$.

PROBLEM 8   Find the number of combinations of 30 objects taken 4 at a time. Compute the answer exactly using a calculator.   ◆

Remember

In a permutation, order counts.
In a combination, order does not count.

To determine whether permutations or combinations are involved in a problem, see if rearranging the collection or listing produces a different object. If so, use permutations; if not, use combinations.

◆ APPLICATIONS

We now consider some applications of the concepts discussed above. Several applications in this and the following sections involve a standard 52-card deck of playing cards, which is described below:

Standard 52-Card Deck of Playing Cards

A standard deck of 52 cards has four 13-card suits: diamonds, hearts, clubs, and spades. The diamonds and hearts are red, and the clubs and spades are black. Each 13-card suit contains cards numbered from 2 to 10, a jack, a queen, a king, and an ace. The jack, queen, and king are called *face cards*. Depending on the game, the ace may be counted as the lowest and/or the highest card in the suit.

◆ **E X A M P L E   9**   How many 5-card hands will have 3 aces and 2 kings?

Solution   The solution involves both the multiplication principle and combinations. Think of selecting the 5-card hand in terms of the following two operations:

O_1: Choosing 3 aces out of 4 possible N_1: $C_{4,3}$
        (order is not important)

O_2: Choosing 2 kings out of 4 possible N_2: $C_{4,2}$
        (order is not important)

Using the multiplication principle, we have

Number of hands $= C_{4,3} \cdot C_{4,2}$

$$= \frac{4!}{3!(4-3)!} \cdot \frac{4!}{2!(4-2)!}$$

$$= 4 \cdot 6 = 24 \qquad \qquad \blacklozenge$$

P R O B L E M   9   How many 5-card hands will have 3 hearts and 2 spades?   ◆

◆ **E X A M P L E   10**   Serial numbers for a product are to be made using 2 letters followed by 3 numbers. If the letters are to be taken from the first 8 letters of the alphabet with <u>no repeats</u> and the numbers are to be taken from the 10 digits (0–9) with no repeats, how many serial numbers are possible?

permutation

Solution   The solution involves both the multiplication principle and permutations. Think of selecting a serial number in terms of the following two operations:

O_1: Choosing 2 letters out of 8 available N_1: $P_{8,2}$
        (order is important)

O_2: Choosing 3 numbers out of 10 available N_2: $P_{10,3}$
        (order is important)

Using the multiplication principle, we have

Number of serial numbers $= P_{8,2} \cdot P_{10,3}$

$$= \frac{8!}{(8-2)!} \cdot \frac{10!}{(10-3)!}$$

$$= 56 \cdot 720 = 40,320 \qquad \qquad \blacklozenge$$

P R O B L E M   10   Repeat Example 10 under the same conditions, except the serial numbers are now to have 3 letters followed by 2 digits (no repeats).   ◆

◆ **E X A M P L E   11**   A company has 7 senior and 5 junior officers. An ad hoc legislative committee is to be formed. In how many ways can a 4-officer committee be formed so that it is composed of:

(A) Any 4 officers?   (B) 4 senior officers?
(C) 3 senior officers and 1 junior officer?

(D) 2 senior and 2 junior officers?

(E) At least 2 senior officers?

Solutions

(A) Since there are a total of 12 officers in the company, the number of different 4-member committees is

$$C_{12,4} = \frac{12!}{4!(12-4)!} = \frac{12!}{4!8!} = 495$$

(B) If only senior officers can be on the committee, the number of different committees is

$$C_{7,4} = \frac{7!}{4!(7-4)!} = \frac{7!}{4!3!} = 35$$

(C) The 3 senior officers can be selected in $C_{7,3}$ ways, and the 1 junior officer can be selected in $C_{5,1}$ ways. Applying the multiplication principle, the number of ways that 3 senior officers and 1 junior officer can be selected is

$$C_{7,3} \cdot C_{5,1} = \frac{7!}{3!(7-3)!} \cdot \frac{5!}{1!(5-1)!} = \frac{7!5!}{3!4!1!4!} = 175$$

(D) $$C_{7,2} \cdot C_{5,2} = \frac{7!}{2!(7-2)!} \cdot \frac{5!}{2!(5-2)!} = \frac{7!5!}{2!5!2!3!} = 210$$

(E) The committees with *at least* 2 senior officers can be divided into three disjoint collections:

1. Committees with 4 senior officers and 0 junior officers
2. Committees with 3 senior officers and 1 junior officer
3. Committees with 2 senior officers and 2 junior officers

The number of committees of types 1, 2, and 3 were computed in parts B, C, and D, respectively. The total number of committees of all three types is the sum of these quantities:

$$
\begin{array}{ccc}
\text{Type 1} & \text{Type 2} & \text{Type 3} \\
C_{7,4} & + C_{7,3} \cdot C_{5,1} + C_{7,2} \cdot C_{5,2} = 35 + 175 + 210 = 420
\end{array}
$$

◆

PROBLEM 11

Given the information in Example 11, answer the following questions:

(A) How many 4-officer committees with 1 senior officer and 3 junior officers can be formed?

(B) How many 4-officer committees with 4 junior officers can be formed?

(C) How many 4-officer committees with at least 2 junior officers can be formed?

◆

1.

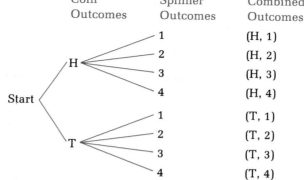

Coin Outcomes	Spinner Outcomes	Combined Outcomes

2. 5^5 or 3,125

3. (A) $10 \cdot 9 \cdot 8 \cdot 7 = 5,040$ (B) $10 \cdot 10 \cdot 10 \cdot 10 = 10,000$
   (C) $10 \cdot 9 \cdot 9 \cdot 9 = 7,290$

4. (A) 720   (B) 10   (C) 720   (D) 20   (E) 1,140

5. (A)

A	B	C	D
B C D	A C D	A B D	A B C
AB AC AD	BA BC BD	CA CB CD	DA DB DC

   12 permutations of 4 objects taken 2 at a time

   (B) O_1: Fill first position N_1: 4 ways
        O_2: Fill second position N_2: 3 ways
        $4 \cdot 3 = 12$

   (C) $P_{4,2} = 4 \cdot 3 = 12$; $P_{4,2} = \dfrac{4!}{(4-2)!} = 12$

6. $P_{30,4} = \dfrac{30!}{(30-4)!} = 657,720$

7. (A) $P_{12,4} = \dfrac{12!}{(12-4)!} = 11,880$ ways   (B) $C_{12,4} = \dfrac{12!}{4!(12-4)!} = 495$ ways

8. $C_{30,4} = \dfrac{30!}{4!(30-4)!} = 27,405$

9. $C_{13,3} \cdot C_{13,2} = 22,308$ 10. $P_{8,3} \cdot P_{10,2} = 30,240$

11. (A) $C_{7,1}C_{5,3} = 70$ (B) $C_{5,4} = 5$ (C) $C_{7,2}C_{5,2} + C_{7,1}C_{5,3} + C_{5,4} = 285$

EXERCISE 7-1

A *Evaluate.*

1. $4! = 24$ 2. $6!$ 3. $\dfrac{9!}{8!} = 9$ 4. $\dfrac{14!}{13!}$

5. $\dfrac{11!}{8!} = 990$ 6. $\dfrac{14!}{12!}$ 7. $\dfrac{5!}{2!3!} = 120$ 8. $\dfrac{6!}{4!2!}$

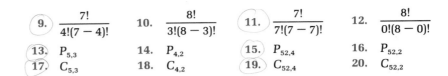

9. $\dfrac{7!}{4!(7-4)!}$ 10. $\dfrac{8!}{3!(8-3)!}$ 11. $\dfrac{7!}{7!(7-7)!}$ 12. $\dfrac{8!}{0!(8-0)!}$

13. $P_{5,3}$ 14. $P_{4,2}$ 15. $P_{52,4}$ 16. $P_{52,2}$

17. $C_{5,3}$ 18. $C_{4,2}$ 19. $C_{52,4}$ 20. $C_{52,2}$

21. A particular new car model is available with 5 choices of color, 3 choices of transmission, 4 types of interior, and 2 types of engine. How many different variations of this model car are possible?

22. A delicatessen serves meat sandwiches with the following options: 3 kinds of bread, 5 kinds of meat, and lettuce or sprouts. How many different sandwiches are possible, assuming one item is used out of each category?

23. In a horse race, how many different finishes among the first 3 places are possible if 10 horses are running? (Exclude ties.)

24. In a long-distance foot race, how many different finishes among the first 5 places are possible if 50 people are running? (Exclude ties.)

25. How many ways can a 3-person subcommittee be selected from a committee of 7 people? How many ways can a president, vice president, and secretary be chosen from a committee of 7 people?

26. Nine cards are numbered with the digits from 1 to 9. A 3-card hand is dealt, 1 card at a time. How many hands are possible where:

(A) Order is taken into consideration?

(B) Order is not taken into consideration?

27. There are 10 teams in a conference. If each team is to play every other team exactly once, how many games must be scheduled?

28. Given 7 points, no 3 of which are on a straight line, how many lines can be drawn joining 2 points at a time?

B 29. How many 4-letter code words are possible from the first 6 letters of the alphabet with no letter repeated? Allowing letters to repeat?

30. How many 5-letter code words are possible from the first 7 letters of the alphabet with no letters repeated? Allowing letters to repeat?

31. A combination lock has 5 wheels, each labeled with the 10 digits from 0 to 9. How many 5-digit opening combinations are possible, assuming no digit is repeated? Assuming digits can be repeated?

32. A small combination lock on a suitcase has 3 wheels, each labeled with the 10 digits from 0 to 9. How many 3-digit combinations are possible, assuming no digit is repeated? Assuming digits can be repeated?

33. From a standard 52-card deck, how many 5-card hands will have all hearts?

34. From a standard 52-card deck, how many 5-card hands will have all face cards? All face cards, but no kings?

35. How many different license plates are possible if each contains 3 letters (out of the 26 letters of the alphabet) followed by 3 digits (from 0 to 9)? How many of these license plates contain no repeated letters and no repeated digits?

36. How many 5-digit ZIP code numbers are possible? How many of these numbers contain no repeated digits?

37. From a standard 52-card deck, how many 7-card hands have exactly 5 spades and 2 hearts?

38. From a standard 52-card deck, how many 5-card hands will have 2 clubs and 3 hearts?

39. A catering service offers 8 appetizers, 10 main courses, and 7 desserts. A banquet committee is to select 3 appetizers, 4 main courses, and 2 desserts. How many ways can this be done?

40. Three departments have 12, 15, and 18 members, respectively. If each department is to select a delegate and an alternate to represent the department at a conference, how many ways can this be done?

C 41. A sporting goods store has 12 pairs of ski gloves of the same size, but of all different brands, in a large bin. In how many ways can a left-hand glove and a right-hand glove be selected that do not match?

42. A sporting goods store has 6 pairs of running shoes of the same size, but all different styles. In how many ways can a left shoe and a right shoe be selected that do not match?

43. Eight distinct points are selected on the circumference of a circle.

(A) How many chords can be drawn by joining the points in all possible ways?

(B) How many triangles can be drawn using these 8 points as vertices?

(C) How many quadrilaterals can be drawn using these 8 points as vertices?

44. Five distinct points are selected on the circumference of a circle.

(A) How many chords can be drawn by joining the points in all possible ways?

(B) How many triangles can be drawn using these 5 points as vertices?

45. How many ways can 2 people be seated in a row of 5 chairs? 3 people? 4 people? 5 people?

46. Each of 2 countries sends 5 delegates to a negotiating conference. A rectangular table is used with 5 chairs on each long side. If each country is assigned a long side of the table (operation 1), how many seating arrangements are possible?

47. A basketball team has 5 distinct positions. Out of 8 players, how many starting teams are possible if:

(A) The distinct positions are taken into consideration?

(B) The distinct positions are not taken into consideration?

(C) The distinct positions are not taken into consideration, but either Mike or Ken (but not both) must start?

48. How many 4-person committees are possible from a group of 9 people if:

(A) There are no restrictions?

(B) Both Jim and Mary must be on the committee?

(C) Either Jim or Mary (but not both) must be on the committee?

Business & Economics

49. *Management selection.* A management selection service classifies its applicants (using tests and interviews) as high-IQ, middle-IQ, or low-IQ and as aggressive or passive. How many combined classifications are possible?

(A) Solve by using a tree diagram.
(B) Solve by using the multiplication principle.

50. *Management selection.* A corporation plans to fill 2 different positions for vice president, V_1 and V_2, from administrative officers in 2 of its manufacturing plants. Plant A has 6 officers and plant B has 8. How many ways can these 2 positions be filled if the V_1 position is to be filled from plant A and the V_2 position from plant B? How many ways can the 2 positions be filled if the selection is made without regard to plant?

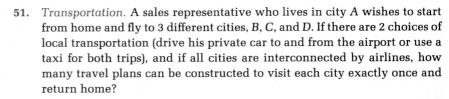

51. *Transportation.* A sales representative who lives in city A wishes to start from home and fly to 3 different cities, B, C, and D. If there are 2 choices of local transportation (drive his private car to and from the airport or use a taxi for both trips), and if all cities are interconnected by airlines, how many travel plans can be constructed to visit each city exactly once and return home?

52. *Transportation.* A manufacturing company in city A wishes to truck its product to four different cities, B, C, D, and E. If the cities are all interconnected by roads, how many different route plans can be constructed so that a single truck, starting from A, will visit each city exactly once, then return home?

53. *Personnel selection.* Suppose 6 female and 5 male applicants have been successfully screened for 5 positions. In how many ways can the following compositions be selected?

(A) 3 females and 2 males   (B) 4 females and 1 male
(C) 5 females   (D) 5 people regardless of sex
(E) At least 4 females

54. *Committee selection.* A 4-person grievance committee is to be selected out of 2 departments, A and B, with 15 and 20 people, respectively. In how many ways can the following committees be selected?

(A) 3 from A and 1 from B (B) 2 from A and 2 from B
(C) All from A (D) 4 people regardless of department
(E) At least 3 from department A

Life Sciences

55. *Medicine.* A medical researcher classifies subjects according to male or female; smoker or nonsmoker; and underweight, average weight, or overweight. How many combined classifications are possible?

(A) Solve using a tree diagram.
(B) Solve using the multiplication principle.

56. *Family planning.* A couple is planning to have 3 children. How many boy–girl combinations are possible? Distinguish between combined outcomes such as (B, B, G), (B, G, B), and (G, B, B).

 (A) Solve by using a tree diagram.
 (B) Solve by using the multiplication principle.

57. *Medicine.* There are 8 standard classifications of blood type. An examination for prospective laboratory technicians consists of having each candidate determine the type for 3 blood samples. How many different examinations can be given if no 2 of the samples provided for the candidate have the same type? If 2 or more samples can have the same type?

58. *Medical research.* Because of limited funds, 5 research centers are to be chosen out of 8 suitable ones for a study on heart disease. How many choices are possible?

Social Sciences

59. *Politics.* A nominating convention is to select a president and vice president from among 4 candidates. Campaign buttons, listing a president and a vice president, are to be designed for each possible outcome before the convention. How many different kinds of buttons should be designed?

60. *Politics.* In how many different ways can 6 candidates for an office be listed on a ballot?

SECTION 7-2 Sample Spaces and Events

◆ EXPERIMENTS
◆ SAMPLE SPACES AND EVENTS
◆ PROBABILITY OF AN EVENT
◆ EQUALLY LIKELY ASSUMPTION

This section provides a relatively brief and informal introduction to probability. More detailed and formal treatments can be found in books and courses devoted entirely to the subject. Probability studies involve many subtle notions, and care must be taken at the beginning to understand the fundamental concepts.

◆ EXPERIMENTS

Our first step in constructing a mathematical model for probability studies is to describe the type of experiments on which probability studies are based. Some experiments do not yield the same results each time they are performed no matter how carefully they are repeated under the same conditions. These experiments are called **random experiments.** Familiar examples of random experiments are flipping coins, rolling dice, observing the frequency of defective items from an assembly line, or observing the frequency of deaths in a certain age group.

Probability theory is a branch of mathematics that has been developed to deal with outcomes of random experiments, both real and conceptual. In the work that follows, we will simply use the word **experiment** to mean a random experiment.

◆ SAMPLE SPACES AND EVENTS

Associated with outcomes of experiments are *sample spaces* and *events*. Our second step in constructing a mathematical model for probability studies is to define these two terms. Set concepts, which are reviewed in Appendix A, are useful in this regard.

Consider the experiment, "A wheel with 18 numerals on the perimeter (Fig. 2) is spun and allowed to come to rest so that a pointer points within a numbered sector."

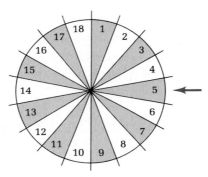

FIGURE 2

What outcomes might we observe? When the wheel stops, we might be interested in which number is next to the pointer, or whether that number is an odd number, or whether that number is divisible by 5, or whether that number is prime, or whether the pointer is in a gray or white sector, and so on. The list of possible outcomes appears endless. In general, there is no unique method of analyzing all possible outcomes of an experiment. Therefore, before conducting an experiment, it is important to decide just what outcomes are of interest.

Suppose we limit our interest to the set of numbers on the wheel and to various subsets of these numbers, such as the set of prime numbers on the wheel or the set of odd numbers. Having decided what to observe, we make a list of outcomes of the experiment, called *simple outcomes* or *simple events*, such that in each trial of the experiment (each spin of the wheel), one and only one of the outcomes on the list will occur. For our stated interests, we choose each number on the wheel as a simple event and form the set

$$S = \{1, 2, 3, \ldots, 17, 18\}$$

The set of simple events S for the experiment is called a *sample space* for the experiment.

Now consider the outcome, "When the wheel comes to rest, the number next to the pointer is divisible by 4." This outcome is not a simple outcome (or simple event), since it is not associated with one and only one element in the sample space S. The outcome will occur whenever any one of the simple events 4, 8, 12, or 16 occurs; that is, whenever an element in the subset

$$E = \{4, 8, 12, 16\}$$

occurs. Subset E is called a *compound event* (and the outcome, a *compound outcome*).

In general:

Sample Spaces and Events

If we formulate a set S of outcomes (events) of an experiment in such a way that in each trial of the experiment one and only one of the outcomes (events) in the set will occur, then we call the set S a **sample space** for the experiment. Each element in S is called a **simple outcome**, or **simple event.**

An **event E** is defined to be any subset of S (including the empty set ∅ and the sample space S). Event E is a **simple event** if it contains only one element and a **compound event** if it contains more than one element. We say that **an event E occurs** if any of the simple events in E occurs.

We use the terms *event* and *outcome of an experiment* interchangeably. Technically, an event is the mathematical counterpart of an outcome of an experiment (as outlined below), but we will not insist on a strict adherence to this distinction in our development of probability.

Real World	Mathematical Model
Experiment (real or conceptual)	Sample space (set S)
Outcome (simple or compound)	Event (subset of S; simple or compound)

◆ E X A M P L E   12   Relative to the number wheel experiment (see Fig. 2) and the sample space

$$S = \{1, 2, 3, \ldots, 17, 18\}$$

what is the event E (subset of the sample space S) that corresponds to each of the following outcomes? Indicate whether the event is a simple event or a compound event.

(A) The outcome is a prime number.   (B) The outcome is the square of 4.

Solutions

(A) The outcome is a prime number if any of the simple events 2, 3, 5, 7, 11, 13, or 17 occurs.* Thus, to say the event "A prime number occurs" is the same as saying the experiment has an outcome in the set

$$E = \{2, 3, 5, 7, 11, 13, 17\}$$

Since event E has more than one element, it is a compound event.

(B) The outcome is the square of 4 if 16 occurs. Thus, to say the event "The square of 4 occurs" is the same as saying the experiment has an outcome in the set

$$E = \{16\}$$

Since E has only one element, it is a simple event. ◆

PROBLEM 12   Repeat Example 12 for:

(A) The outcome is a number divisible by 12.
(B) The outcome is an even number greater than 15. ◆

◆ EXAMPLE 13   A nickel and a dime are tossed. How shall we identify a sample space for this experiment? There are a number of possibilities, depending on our interest. We shall consider three.

(A) If we are interested in whether each coin falls heads (H) or tails (T), then, using a tree diagram, we can easily determine an appropriate sample space for the experiment:

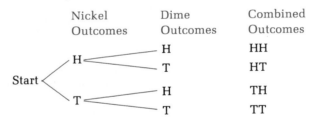

Nickel Outcomes	Dime Outcomes	Combined Outcomes
	H	HH
H	T	HT
	H	TH
T	T	TT

Thus,

$$S_1 = \{HH, HT, TH, TT\}$$

and there are 4 simple events in the sample space.

(B) If we are interested only in the number of heads that appear on a single toss of the two coins, then we can let

$$S_2 = \{0, 1, 2\}$$

and there are 3 simple events in the sample space.

---

* Technically, we should write {2}, {3}, {5}, {7}, {11}, {13}, and {17} for the simple events, since there is a logical distinction between an element of a set and a subset consisting of only that element. But we will just keep this in mind and drop the braces for simple events to simplify the notation.

(C) If we are interested in whether the coins match (M) or do not match (D), then we can let

$$S_3 = \{M, D\}$$

and there are only 2 simple events in the sample space. ◆

In Example 13, which sample space would be appropriate for all three interests? Sample space S_1 contains more information than either S_2 or S_3. If we know which outcome has occurred in S_1, then we know which outcome has occurred in S_2 and S_3. However, the reverse is not true. (Note that the simple events in S_2 and S_3 are compound events in S_1.) In this sense, we say that S_1 is a more **fundamental sample space** than either S_2 or S_3. Thus, we would choose S_1 as an appropriate sample space for all three expressed interests.

Choosing Sample Spaces

There is no one correct sample space for a given experiment. When specifying a sample space for an experiment, we include as much detail as is necessary to answer *all* questions of interest regarding the outcomes of the experiment. When in doubt, choose a sample space with more elements rather than fewer.

PROBLEM 13 An experiment consists of recording the boy–girl composition of 2-child families.

(A) What is an appropriate sample space if we are interested in the sex of each child in the order of their births? Draw a tree diagram.

(B) What is an appropriate sample space if we are interested only in the number of girls in a family?

(C) What is an appropriate sample space if we are interested only in whether the sexes are alike (A) or different (D)?

(D) What is an appropriate sample space for all three interests expressed in parts A–C? ◆

◆ **EXAMPLE 14** Consider an experiment of rolling two dice. A convenient sample space that will enable us to answer many questions about interesting events is shown in Figure 3 on page 360. Let S be the set of all ordered pairs in the figure. The simple event (3, 2) is to be distinguished from the simple event (2, 3). The former indicates that a 3 turned up on the first die and a 2 on the second, while the latter indicates that a 2 turned up on the first die and a 3 on the second.

What is the event (subset of the sample space S) that corresponds to each of the following outcomes?

(A) A sum of 7 turns up.　　　　(B) A sum of 11 turns up.

(C) A sum less than 4 turns up.　　(D) A sum of 12 turns up.

Solutions

(A) By "A sum of 7 turns up," we mean that the sum of all dots on both turned-up faces is 7. This outcome corresponds to the event

$$\{(6, 1), (5, 2), (4, 3), (3, 4), (2, 5), (1, 6)\}$$

(B) "A sum of 11 turns up" corresponds to the event

$$\{(6, 5), (5, 6)\}$$

(C) "A sum less than 4 turns up" corresponds to the event

$$\{(1, 1), (2, 1), (1, 2)\}$$

(D) "A sum of 12 turns up" corresponds to the event

$$\{(6, 6)\}$$

◆

PROBLEM   14   Refer to the sample space shown in Figure 3. What is the event that corresponds to each of the following outcomes?

(A) A sum of 5 turns up.

(B) A sum that is a prime number greater than 7 turns up.   ◆

As indicated earlier, we often use the terms *event* and *outcome of an experiment* interchangeably. Thus, in Example 14 we might say "the event, 'A sum of 11 turns up'" in place of "the outcome, 'A sum of 11 turns up,'" or even write

$$E = \text{A sum of 11 turns up} = \{(6, 5), (5, 6)\}$$

◆ PROBABILITY OF AN EVENT

The next step in developing our mathematical model for probability studies is the introduction of a *probability function*. This is a function that assigns to an arbitrary event associated with a sample space a real number between 0 and 1, inclusive. We start by discussing ways in which probabilities are assigned to simple events in the sample space S.

█ Probabilities for Simple Events

Given a sample space

$$S = \{e_1, e_2, \ldots, e_n\}$$

with n simple events, to each simple event e_i we assign a real number, denoted by $P(e_i)$, called the **probability of the event e_i**. These numbers can be assigned in an arbitrary manner as long as the following two conditions are satisfied:

1. The probability of a simple event is a number between 0 and 1, inclusive. That is,

$$0 \leqslant P(e_i) \leqslant 1$$

2. The sum of the probabilities of all simple events in the sample space is 1. That is,

$$P(e_1) + P(e_2) + \cdots + P(e_n) = 1$$

Any probability assignment that meets conditions 1 and 2 is said to be an **acceptable probability assignment.**

Our mathematical theory does not explain how acceptable probabilities are assigned to simple events. These assignments are generally based on the expected or actual percentage of times a simple event occurs when an experiment is repeated a large number of times. Assignments based on this principle are called **reasonable.**

Let an experiment be the flipping of a single coin, and let us choose a sample space S to be

$$S = \{H, T\}$$

If a coin appears to be fair, we are inclined to assign probabilities to the simple events in S as follows:

$$P(H) = \tfrac{1}{2} \quad \text{and} \quad P(T) = \tfrac{1}{2}$$

These assignments are based on reasoning that, since there are 2 ways a coin can land, in the long run, a head will turn up half the time and a tail will turn up half the time. These probability assignments are acceptable, since both conditions for acceptable probability assignments stated in the box are satisfied:

1. $0 \leqslant P(H) \leqslant 1$, $\quad 0 \leqslant P(T) \leqslant 1$
2. $P(H) + P(T) = \frac{1}{2} + \frac{1}{2} = 1$

But there are other acceptable assignments. Maybe after flipping a coin 1,000 times, we find that the head turns up 376 times and the tail turns up 624 times. With this result, we might suspect that the coin is not fair and assign the simple events in the sample space S the following probabilities, based on our experimental results:

$$P(H) = .376 \quad \text{and} \quad P(T) = .624$$

This is also an acceptable assignment. However, the probability assignment

$$P(H) = 1 \quad \text{and} \quad P(T) = 0$$

though acceptable, is not reasonable (unless the coin has 2 heads). And the assignment

$$P(H) = .6 \quad \text{and} \quad P(T) = .8$$

is not acceptable, since $.6 + .8 = 1.4$, which violates condition 2 in the box. [*Note:* In probability studies, the 0 to the left of the decimal is usually omitted. Thus, we write .6 and .8 instead of 0.6 and 0.8.]

It is important to keep in mind that out of the infinitely many possible acceptable probability assignments to simple events in a sample space, we are generally inclined to choose one assignment over another based on reasoning or experimental results.

Given an acceptable probability assignment for simple events in a sample space S, how do we define the probability of an arbitrary event E associated with S?

▬ Probability of an Event *E*

Given an acceptable probability assignment for the simple events in a sample space S, we define the **probability of an arbitrary event *E*,** denoted by ***P(E)*,** as follows:

(A) If E is the empty set, then $P(E) = 0$.

(B) If E is a simple event, then $P(E)$ has already been assigned.

(C) If E is a compound event, then $P(E)$ is the sum of the probabilities of all the simple events in E.

(D) If E is the sample space S, then $P(E) = P(S) = 1$ (this is a special case of part C).

◆ E X A M P L E   15   Let us return to Example 13, the tossing of a nickel and a dime, and the sample space

$$S = \{HH, HT, TH, TT\}$$

Since there are 4 simple outcomes and the coins are assumed to be fair, it would appear that each outcome would occur 25% of the time, in the long run. Let us assign the same probability of $\frac{1}{4}$ to each simple event in S:

SIMPLE EVENT e_i	HH	HT	TH	TT
$P(e_i)$	$\frac{1}{4}$	$\frac{1}{4}$	$\frac{1}{4}$	$\frac{1}{4}$

This is an acceptable assignment according to conditions 1 and 2, and it is a reasonable assignment for ideal (perfectly balanced) coins or coins close to ideal.

(A) What is the probability of getting 1 head (and 1 tail)?
(B) What is the probability of getting at least 1 head?
(C) What is the probability of getting 1 head or 1 tail?
(D) What is the probability of getting 3 heads?   ◆

Solutions   (A) $E_1 = $ Getting 1 head $= \{HT, TH\}$

Since E_1 is a compound event, we use part C in the box and find $P(E_1)$ by adding the probabilities of the simple events in E_1:

$$P(E_1) = P(HT) + P(TH) = \tfrac{1}{4} + \tfrac{1}{4} = \tfrac{1}{2}$$

(B) $E_2 = $ Getting at least 1 head $= \{HH, HT, TH\}$

$$P(E_2) = P(HH) + P(HT) + P(TH) = \tfrac{1}{4} + \tfrac{1}{4} + \tfrac{1}{4} = \tfrac{3}{4}$$

(C) $E_3 = \{HH, HT, TH, TT\} = S$

$$P(E_3) = P(S) = 1 \qquad\qquad \tfrac{1}{4} + \tfrac{1}{4} + \tfrac{1}{4} + \tfrac{1}{4} = 1$$

(D) $E_4 = $ Getting 3 heads $= \varnothing$ Empty set

$$P(\varnothing) = 0 \qquad\qquad ◆$$

■ Steps for Finding the Probability of an Event E

Step 1.   Set up an appropriate sample space S for the experiment.

Step 2.   Assign acceptable probabilities to the simple events in S.

Step 3.   To obtain the probability of an arbitrary event E, add the probabilities of the simple events in E.

The function P defined in steps 2 and 3 is a **probability function** whose domain is all possible events (subsets) in the sample space S and whose range is a set of real numbers between 0 and 1, inclusive.

PROBLEM 15   Suppose in Example 15 that, after flipping the nickel and dime 1,000 times, we find that HH turns up 273 times, HT turns up 206 times, TH turns up 312 times, and TT turns up 209 times. On the basis of this evidence, we assign probabilities to simple events in S as follows:

SIMPLE EVENT e_i	HH	HT	TH	TT
$P(e_i)$	.273	.206	.312	.209

This is an acceptable and reasonable probability assignment for the simple events in S. What are the probabilities of the following events?

(A) $E_1 =$ Getting at least 1 tail   (B) $E_2 =$ Getting 2 tails
(C) $E_3 =$ Getting either 1 head or 1 tail   ◆

Example 15 and Problem 15 illustrate two important ways in which acceptable and reasonable probability assignments are made for simple events in a sample space S:

1. *Theoretical.* We use assumptions and a deductive reasoning process to assign probabilities to simple events. No experiments are actually conducted. This is what we did in Example 15.
2. *Empirical.* We assign probabilities to simple events based on the results of actual experiments. This is what we did in Problem 15. (As an experiment is repeated without end, the percentage of times an event occurs may get closer and closer to a single fixed number. If so, the single fixed number is generally called the **actual probability of the event.**)

Each approach has its advantages in certain situations. For the rest of this section, we will emphasize the theoretical approach. In the next section, we will consider the empirical approach in more detail.

◆ EQUALLY LIKELY ASSUMPTION

In tossing a nickel and a dime (Example 15), we assigned the same probability, $\frac{1}{4}$, to each simple event in the sample space $S = \{$HH, HT, TH, TT$\}$. By assigning the same probability to each simple event in S, we are actually making the assumption that each simple event is as likely to occur as any other. We refer to this as an **equally likely assumption.** In general, we have the following:

◼ **Probability of a Simple Event under an Equally Likely Assumption**

If, in a sample space

$$S = \{e_1, e_2, \ldots, e_n\}$$

with n elements, we assume each simple event e_i is as likely to occur as any other, then we assign the probability $1/n$ to each. That is,

$$P(e_i) = \frac{1}{n}$$

Under an equally likely assumption, we can develop a very useful formula for finding probabilities of arbitrary events associated with a sample space S. Consider the following example.

If a single die is rolled and we assume each face is as likely to come up as any other, then for the sample space

$$S = \{1, 2, 3, 4, 5, 6\}$$

we assign a probability of $\frac{1}{6}$ to each simple event, since there are 6 simple events. The probability of

$$E = \text{Rolling a prime number} = \{2, 3, 5\}$$

is

$$\begin{array}{c} \text{Number of elements in } E \\ \downarrow \end{array}$$

$$P(E) = P(2) + P(3) + P(5) = \tfrac{1}{6} + \tfrac{1}{6} + \tfrac{1}{6} = \tfrac{3}{6} = \tfrac{1}{2}$$

$$\begin{array}{c} \uparrow \\ \text{Number of elements in } S \end{array}$$

Thus, under the assumption that each simple event is as likely to occur as any other, the computation of the probability of the occurrence of any event E in a sample space S is the number of elements in E divided by the number of elements in S.

Probability of an Arbitrary Event under an Equally Likely Assumption

If we assume each simple event in sample space S is as likely to occur as any other, then the probability of an arbitrary event E in S is given by

$$P(E) = \frac{\text{Number of elements in } E}{\text{Number of elements in } S} = \frac{n(E)}{n(S)}$$

◆ EXAMPLE 16   Let us again consider rolling two dice, and assume each simple event in the sample space shown in Figure 3 (page 360) is as likely as any other. Find the probabilities of the following events:

(A) $E_1 =$ A sum of 7 turns up   (B) $E_2 =$ A sum of 11 turns up
(C) $E_3 =$ A sum less than 4 turns up   (D) $E_4 =$ A sum of 12 turns up

Solutions   Referring to Figure 3 (page 360) and the results found in Example 14, we find:

(A) $P(E_1) = \dfrac{n(E_1)}{n(S)} = \dfrac{6}{36} = \dfrac{1}{6}$ $E_1 = \{(6, 1), (5, 2), (4, 3), (3, 4), (2, 5), (1, 6)\}$

(B) $P(E_2) = \dfrac{n(E_2)}{n(S)} = \dfrac{2}{36} = \dfrac{1}{18}$ $E_2 = \{(6, 5), (5, 6)\}$

(C) $P(E_3) = \dfrac{n(E_3)}{n(S)} = \dfrac{3}{36} = \dfrac{1}{12}$ $E_3 = \{(1, 1), (2, 1), (1, 2)\}$

(D) $P(E_4) = \dfrac{n(E_4)}{n(S)} = \dfrac{1}{36}$ $E_4 = \{(6, 6)\}$ ◆

PROBLEM 16   Under the conditions in Example 16, find the probabilities of the following events (each event refers to the sum of the dots facing up on both dice):

(A) $E_5 =$ A sum of 5 turns up
(B) $E_6 =$ A sum that is a prime number greater than 7 turns up   ◆

We now turn to some examples that make use of the counting techniques developed in the preceding section.

◆ EXAMPLE 17   In drawing 5 cards from a 52-card deck without replacement, what is the probability of getting 5 spades?

Solution   Let the sample space S be the set of all 5-card hands from a 52-card deck. Since the order in a hand does not matter, $n(S) = C_{52,5}$. Let event E be the set of all 5-card hands from 13 spades. Again, the order does not matter and $n(E) = C_{13,5}$.

Thus, assuming each 5-card hand is as likely as any other,

$$P(E) = \frac{n(E)}{n(S)} = \frac{C_{13,5}}{C_{52,5}} = \frac{13!/5!8!}{52!/5!47!} = \frac{13!}{5!8!} \cdot \frac{5!47!}{52!}$$

$$= \frac{13 \cdot 12 \cdot 11 \cdot 10 \cdot 9 \cdot 8!}{8!} \cdot \frac{47!}{52 \cdot 51 \cdot 50 \cdot 49 \cdot 48 \cdot 47!}$$

$$\approx .0005$$

Or, using a calculator to evaluate $C_{13,5}$ and $C_{52,5}$,

$$P(E) = \frac{n(E)}{n(S)} = \frac{C_{13,5}}{C_{52,5}} = \frac{1,287}{2,593,960} \approx .0005$$

◆

PROBLEM 17   In drawing 7 cards from a 52-card deck without replacement, what is the probability of getting 7 hearts?

◆

◆ EXAMPLE 18   The board of regents of a university is made up of 12 men and 16 women. If a committee of 6 is chosen at random, what is the probability that it will contain 3 men and 3 women?

Solution   Let S be the set of all 6-person committees out of 28 people. Then

$$n(S) = C_{28,6}$$

Let E be the set of all 6-person committees with 3 men and 3 women. To find $n(E)$, we use the multiplication principle and the following two operations:

O_1: Select 3 men out of the 12 available N_1: $C_{12,3}$
O_2: Select 3 women out of the 16 available N_2: $C_{16,3}$

Thus,

$$n(E) = N_1 \cdot N_2 = C_{12,3} \cdot C_{16,3}$$

and

$$P(E) = \frac{n(E)}{n(S)} = \frac{C_{12,3} \cdot C_{16,3}}{C_{28,6}} \approx .327$$

◆

PROBLEM 18   What is the probability that the committee in Example 18 will have 4 men and 2 women?

◆

It needs to be pointed out that there are many counting problems for which it is not possible to produce a simple formula that will yield the number of possible cases. In situations of this type, we often revert back to tree diagrams and counting branches.

Answers to Matched Problems

12. (A) $E = \{12\}$; simple event   (B) $E = \{16, 18\}$; compound event
13. (A) $S_1 = \{BB, BG, GB, GG\}$;
    (B) $S_2 = \{0, 1, 2\}$
    (C) $S_3 = \{A, D\}$
    (D) S_1

Sex of First Child	Sex of Second Child	Combined Outcomes
B	B	BB
	G	BG
G	B	GB
	G	GG

14. (A) $\{(4, 1), (3, 2), (2, 3), (1, 4)\}$ (B) $\{(6, 5), (5, 6)\}$
15. (A) .727   (B) .209   (C) 1   16. (A) $P(E_5) = \frac{1}{9}$ (B) $P(E_6) = \frac{1}{18}$
17. $C_{13,7}/C_{52,7} \approx .000\ 013$ 18. $C_{12,4}C_{16,2}/C_{28,6} \approx .158$

EXERCISE 7-2

A

1. How would you interpret $P(E) = 1$?
2. How would you interpret $P(E) = 0$?
3. In a family with 2 children, excluding multiple births, what is the probability of having 2 children of the opposite sex? Assume a girl is as likely as a boy at each birth.
4. In a family with 2 children, excluding multiple births, what is the probability of having 2 girls? Assume a girl is as likely as a boy at each birth.
5. A spinner can land on four different colors: red (R), green (G), yellow (Y), and blue (B). If we do not assume each color is as likely to turn up as any other, which of the following probability assignments have to be rejected, and why?

   (A) $P(R) = .15$, $P(G) = -.35$, $P(Y) = .50$, $P(B) = .70$
   (B) $P(R) = .32$, $P(G) = .28$, $P(Y) = .24$, $P(B) = .30$
   (C) $P(R) = .26$, $P(G) = .14$, $P(Y) = .30$, $P(B) = .30$

6. Using the probability assignments in Problem 5C, what is the probability that the spinner will not land on blue?
7. Using the probability assignments in Problem 5C, what is the probability that the spinner will land on red or yellow?
8. Using the probability assignments in Problem 5C, what is the probability that the spinner will not land on red or yellow?

B

9. In a family with 3 children, excluding multiple births, what is the probability of having 2 boys and 1 girl, in that order? Assume a boy is as likely as a girl at each birth.
10. In a family with 3 children, excluding multiple births, what is the probability of having 2 boys and 1 girl, in any order? Assume a boy is as likely as a girl at each birth.

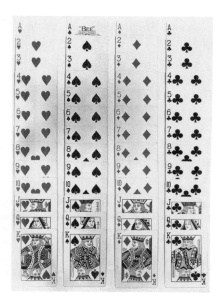

11. A small combination lock on a suitcase has 3 wheels, each labeled with the 10 digits from 0 to 9. If an opening combination is a particular sequence of 3 digits with no repeats, what is the probability of a person guessing the right combination?

12. A combination lock has 5 wheels, each labeled with the 10 digits from 0 to 9. If an opening combination is a particular sequence of 5 digits with no repeats, what is the probability of a person guessing the right combination?

Refer to the description of a standard deck of 52 cards on page 348 and the photograph in the margin. An experiment consists of dealing 5 cards from a standard 52-card deck. In Problems 13–16, what is the probability of being dealt:

13. 5 black cards?   14. 5 hearts?

15. 5 face cards?   16. 5 nonface cards?

17. If 4-digit numbers less than 5,000 are randomly formed from the digits 1, 3, 5, 7, and 9, what is the probability of forming a number divisible by 5? (Digits may be repeated; for example, 1,355 is acceptable.)

18. If 4-letter code words are generated at random using the letters A, B, C, D, E, and F, what is the probability of forming a word without a vowel in it? (Letters may be repeated.)

19. Suppose 5 thank-you notes are written and 5 envelopes are addressed. Accidentally, the notes are randomly inserted into the envelopes and mailed without checking the addresses. What is the probability that all the notes will be inserted into the correct envelopes?

20. Suppose 6 people check their coats in a checkroom. If all claim checks are lost and the 6 coats are randomly returned, what is the probability that all the people will get their own coats back?

An experiment consists of rolling two fair dice and adding the dots on the two sides facing up. Using the sample space shown in Figure 3 (page 360) and assuming each simple event is as likely as any other, find the probability of the sum of the dots indicated in Problems 21–36:

21. Sum is 2.   22. Sum is 10.   23. Sum is 6.   24. Sum is 8.

25. Sum is less than 5.   26. Sum is greater than 8.

27. Sum is not 7 or 11.   28. Sum is not 2, 4, or 6.

29. Sum is 1.   30. Sum is 13.

31. Sum is divisible by 3.   32. Sum is divisible by 4.

33. Sum is 7 or 11 (a "natural").   34. Sum is 2, 3, or 12 ("craps").

35. Sum is divisible by 2 or 3.   36. Sum is divisible by 2 and 3.

An experiment consists of tossing three fair (not weighted) coins, except one of the three coins has a head on both sides. Compute the probability of obtaining the indicated results in Problems 37–42.

37. 1 head   38. 2 heads   39. 3 heads

40. 0 heads   41. More than 1 head   42. More than 1 tail

C An experiment consists of rolling two fair (not weighted) dice and adding the dots on the two sides facing up. Each die has the number 1 on two opposite faces, the number 2 on two opposite faces, and the number 3 on two opposite faces. Compute the probability of obtaining the indicated sums in Problems 43–50.

43. 2   **44.** 3   **45.** 4   **46.** 5

47. 6   **48.** 7   **49.** An odd sum   **50.** An even sum

Refer to the description of a standard deck of 52 cards on page 348. An experiment consists of dealing 5 cards from a standard 52-card deck. In Problems 51–58, what is the probability of being dealt:

51. 5 face cards or aces?   **52.** 5 numbered cards (2 through 10)?

53. 4 aces?   **54.** Four of a kind (4 queens, 4 kings, and so on)?

55. A 10, jack, queen, king, and ace, all in the same suit?

56. A 2, 3, 4, 5, and 6, all in the same suit?

57. 2 aces and 3 queens?   **58.** 2 kings and 3 aces?

APPLICATIONS

Business & Economics

59. *Consumer testing.* Twelve popular brands of beer are to be used in a blind taste study for consumer recognition.

(A) If 4 distinct brands are chosen at random from the 12 and if a consumer is not allowed to repeat any answers, what is the probability that all 4 brands could be identified by just guessing?

(B) If repeats are allowed in the 4 brands chosen at random from the 12 and if the consumer is allowed to repeat answers, what is the probability of correct identification of all 4 by just guessing?

60. *Consumer testing.* Six popular brands of cola are to be used in a blind taste study for consumer recognition.

(A) If 3 distinct brands are chosen at random from the 6 and if a consumer is not allowed to repeat any answers, what is the probability that all 3 brands could be identified by just guessing?

(B) If repeats are allowed in the 3 brands chosen at random from the 6 and if the consumer is allowed to repeat answers, what is the probability of correct identification of all 3 by just guessing?

61. *Personnel selection.* Suppose 6 female and 5 male applicants have been successfully screened for 5 positions. If the 5 positions are selected at random from the 11 finalists, what is the probability of selecting:

(A) 3 females and 2 males?   (B) 4 females and 1 male?

(C) 5 females?   (D) At least 4 females?

62. *Committee selection.* A 4-person grievance committee is to be composed of employees in 2 departments, A and B, with 15 and 20 employees, respectively. If the 4 people are selected at random from the 35 employees, what is the probability of selecting:

(A) 3 from A and 1 from B?   (B) 2 from A and 2 from B?
(C) All from A?   (D) At least 3 from A?

Life Sciences

63. *Medicine.* A prospective laboratory technician is to be tested on identifying blood types from 8 standard classifications.

(A) If 3 distinct samples are chosen at random from the 8 types and if the examinee is not allowed to repeat any answers, what is the probability that all 3 could be correctly identified by just guessing?
(B) If repeats are allowed in the 3 blood types chosen at random from the 8 and if the examinee is allowed to repeat answers, what is the probability of correct identification of all 3 by just guessing?

64. *Medical research.* Because of limited funds, 5 research centers are to be chosen out of 8 suitable ones for a study on heart disease. If the selection is made at random, what is the probability that 5 particular research centers will be chosen?

Social Sciences

65. *Membership selection.* A town council has 11 members, 6 Democrats and 5 Republicans.

(A) If the president and vice president are selected at random, what is the probability that they are both Democrats?
(B) If a 3-person committee is selected at random, what is the probability that Republicans make up the majority?

SECTION 7-3 Empirical Probability

◆ THEORETICAL VERSUS EMPIRICAL PROBABILITY
◆ STATISTICS VERSUS PROBABILITY THEORY
◆ LAW OF LARGE NUMBERS

◆ THEORETICAL VERSUS EMPIRICAL PROBABILITY

In Section 7-2 we indicated that acceptable and reasonable probability assignments are made for events in a sample space in two common ways, theoretical and empirical. Let us look at another example and compare the two approaches.

There are 20,000 students registered in a state university. Students are legally either state residents, out-of-state residents, or foreign residents. What is the probability that a student chosen at random is a state resident? An out-of-state resident? A foreign resident? How do we proceed to find these probabilities?

Theoretical Approach

Suppose resident information is available in the registrar's office and can be obtained from a computer printout. Requesting the printout, we find:

State residents (E_1)	12,000
Out-of-state residents (E_2)	5,000
Foreign residents (E_3)	$\dfrac{3,000}{20,000} = N$

Looking at the total structure, we reason as follows: We choose the total list of registered students (with resident status indicated) as our sample space S. We assume that any one student is as likely to be chosen as any other in a random sample of 1. Thus, we assign the probability $\frac{1}{20,000}$ to each simple event in S. This is an acceptable assignment. Under the equally likely assumption,

$$P(E_1) = \frac{n(E_1)}{n(S)} = \frac{12,000}{20,000} = .60 \qquad P(E_2) = \frac{n(E_2)}{n(S)} = \frac{5,000}{20,000} = .25$$

$$P(E_3) = \frac{n(E_3)}{n(S)} = \frac{3,000}{20,000} = .15$$

Our approach here is analogous to that used in assigning a probability of $\frac{1}{4}$ to the drawing of a heart in a single draw of 1 card from a 52-card deck.

Empirical Approach

Suppose residency status was not recorded during registration and the information is not available through the registrar. Not having the time, inclination, or money to interview each student, we choose a random sample of 200 students and find:

State residents (E_1)	128
Out-of-state residents (E_2)	47
Foreign residents (E_3)	$\dfrac{25}{200} = n$

It would be reasonable to say that

$$P(E_1) \approx \frac{128}{200} = .640 \qquad P(E_2) \approx \frac{47}{200} = .235 \qquad P(E_3) \approx \frac{25}{200} = .125$$

As we increase the sample size, our confidence in the probability assignments would likely increase. We refer to these probability assignments as **approximate empirical probabilities** and use them to approximate the actual probabilities for the total population.

In general, if we conduct an experiment n times and an event E occurs with **frequency f(E)**, then the ratio $f(E)/n$ is called the **relative frequency** of the occurrence of event E in n trials. We define the **empirical probability** of E, denoted by **P(E)**, by the number (if it exists) that the relative frequency $f(E)/n$ approaches as n gets larger and larger. Of course, for any particular n, the relative frequency $f(E)/n$ is generally only approximately equal to P(E). However, as n increases in size, we would expect the approximation to improve.

Empirical Probability Approximation

$$P(E) \approx \frac{\text{Frequency of occurrence of } E}{\text{Total number of trials}} = \frac{f(E)}{n}$$

(The larger n is, the better the approximation.)

If equally likely assumptions used to obtain theoretical probability assignments are actually warranted, then we would also expect corresponding approximate empirical probabilities to approach the theoretical values as the number of trials n of actual experiments becomes very large.

◆ EXAMPLE 19   Two coins are tossed 1,000 times with the following frequencies of outcomes:

2 heads	200
1 head	560
0 heads	240

(A) Compute the approximate empirical probability for each type of outcome.
(B) Compute the theoretical probabilities for each type of outcome.

Solutions   (A) $P(2 \text{ heads}) \approx \dfrac{200}{1,000} = .20$ $P(1 \text{ head}) \approx \dfrac{560}{1,000} = .56$

$P(0 \text{ heads}) \approx \dfrac{240}{1,000} = .24$

(B)   A sample space of equally likely simple events is $S = \{HH, HT, TH, TT\}$. Let

$$E_1 = \{HH\} \qquad E_2 = \{HT, TH\} \qquad E_3 = \{TT\}$$

Then,

$$P(2 \text{ heads}) = .25 \qquad P(1 \text{ head}) = .50 \qquad P(0 \text{ heads}) = .25 \qquad ◆$$

PROBLEM 19   One die is rolled 1,000 times with the following frequencies of outcomes:

1	180	4	138
2	140	5	175
3	152	6	215

(A) Calculate approximate empirical probabilities for each indicated outcome.
(B) Do the indicated outcomes seem equally likely?
(C) Assuming the indicated outcomes are equally likely, compute their theoretical probabilities.   ◆

◆ EXAMPLE 20   An insurance company selected 1,000 drivers at random in a particular city to determine a relationship between age and accidents. The data obtained are

listed in Table 1. Compute the following approximate empirical probabilities for a driver chosen at random in the city:

(A) Of being under 20 years old **and** having 3 accidents in 1 year (E_1)
(B) Of being 30–39 years old **and** having 1 or more accidents in 1 year (E_2)
(C) Of having no accidents in 1 year (E_3)
(D) Of being under 20 years old **or** having 3 accidents in 1 year (E_4)

TABLE 1

| | ACCIDENTS IN 1 YEAR | | | | |
AGE	0	1	2	3	Over 3
Under 20	50	62	53	35	20
20–29	64	93	67	40	36
30–39	82	68	32	14	4
40–49	38	32	20	7	3
Over 49	43	50	35	28	24

Solutions

(A) $P(E_1) \approx \dfrac{35}{1,000} = .035$

(B) $P(E_2) \approx \dfrac{68 + 32 + 14 + 4}{1,000} = .118$

(C) $P(E_3) \approx \dfrac{50 + 64 + 82 + 38 + 43}{1,000} = .277$

(D) $P(E_4) \approx \dfrac{50 + 62 + 53 + 35 + 20 + 40 + 14 + 7 + 28}{1,000} = .309$ ◆

Notice that in this type of problem, which is typical of many realistic problems, approximate empirical probabilities are the only type we can compute.

PROBLEM 20

Referring to Table 1, compute each of the following approximate empirical probabilities for a driver chosen at random in the city:

(A) Of being under 20 years old with no accidents in 1 year (E_1)
(B) Of being 20–29 years old and having fewer than 2 accidents in 1 year (E_2)
(C) Of not being over 49 years old (E_3) ◆

Approximate empirical probabilities are often used to test theoretical probabilities. As we said before, equally likely assumptions may not be justified in reality. In addition to this use, there are many situations in which it is either very difficult or impossible to compute the theoretical probabilities for given events. For example, insurance companies use past experience to establish approximate empirical probabilities to predict the future, baseball teams use batting averages (approximate empirical probabilities based on past experience) to predict the future performance of a player, and pollsters use approximate empirical probabilities to predict outcomes of elections.

◆ STATISTICS VERSUS PROBABILITY THEORY

We are now entering the area of *mathematical statistics,* a subject that we will not pursue too far in this book. Mathematical statistics is a branch of mathematics that draws inferences about certain characteristics of a total population, called **population parameters,** based on corresponding characteristics of a random sample from the population. In general, a **population** is the set containing every element we are describing (all people in a school, all flashbulbs produced by a given company using a particular type of manufacturing process, all flips of a certain coin, all rolls of a certain die, etc.). A **sample** is a subset of a population. The population size, if finite, is denoted by N; the sample size is denoted by n. Except when the sample is a census (the whole population), n is less than N.

Because samples are used to draw inferences about the total population, it is desirable that a sample be **representative** of the population—that is, that various population characteristics are proportionately represented in the sample. **Random samples** are those in which each element of the population has the same probability of being chosen for the sample. Much statistical theory is based on random samples.

Statisticians start with a known sample and proceed to describe certain characteristics of the total population that are not known. For example, in Example 20, the insurance company used the approximate empirical probability .035 (computed from the sample) as an approximation for the actual probability of a person drawn at random from the *total* population being under 20 years old and having 3 accidents in 1 year.

Probability theory, on the other hand, starts with a known composition of a population and from this deduces the probable composition of a sample. For example, knowing the composition of a standard deck of 52 cards, we can (assuming each 5-card hand has the same probability of being dealt as any other) deduce that the probability of being dealt 5 cards of the same suit (a "flush") is given by $4C_{13,5}/C_{52,5} = .001\ 98$. In short, statistics proceeds from a sample to the population, while probability theory proceeds from a population to a sample.

◆ LAW OF LARGE NUMBERS

How does the approximate empirical probability of an event determined from a sample relate to the actual probability of the event relative to the total population? In mathematical statistics an important theorem called the **law of large numbers** (or the **law of averages**) is proved. Informally, it states that the approximate empirical probability can be made as close to the actual probability as we please by making the sample sufficiently large.

For example, if we roll a fair die a large number of times, we would expect to get each number about (not exactly) $\frac{1}{6}$ of the time. The law of large numbers states (informally) that the greater the number of times we roll a fair die, the closer the relative frequency of the occurrence of a given number will be to $\frac{1}{6}$ [or, if the die is not fair (and no die can be absolutely fair), then the closer the relative frequency of the occurrence of a given number will be to the actual probability of the occurrence of that number].

19. (A) $P(1) \approx .180, P(2) \approx .140, P(3) \approx .152, P(4) \approx .138, P(5) \approx .175, P(6) \approx .215$
    (B) No   (C) $\frac{1}{6} \approx .167$ for each
20. (A) $P(E_1) \approx .05$ (B) $P(E_2) \approx .157$ (C) $P(E_3) \approx .82$

E X E R C I S E   7-3

A

1. A ski jumper has jumped over 300 feet in 25 out of 250 jumps. What is the approximate empirical probability of the next jump being over 300 feet?
2. In a certain city there are 4,000 youths between 16 and 20 years old who drive cars. If 560 of them were involved in accidents last year, what is the approximate empirical probability of a youth in this age group being involved in an accident this year?
3. Out of 420 times at bat, a baseball player gets 189 hits. What is the approximate empirical probability that the player will get a hit next time at bat?
4. In a medical experiment, a new drug is found to help 2,400 out of 3,000 people. If a doctor prescribes the drug for a particular patient, what is the approximate empirical probability that the patient will be helped?
5. A thumbtack is tossed 1,000 times with the following outcome frequencies:

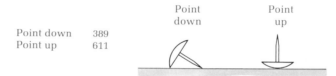

		Point down	Point up
Point down	389		
Point up	611		

Compute the approximate empirical probability for each outcome. Does each outcome appear to be equally likely?
6. Toss a thumbtack 100 times and let it fall to the floor. Count the number of times it lands point down. What is the approximate empirical probability of the tack landing point down? Point up? (Actually, you can toss 10 tacks at a time and count the total number pointing down in 10 throws.)

B

7. A random sample of 10,000 families with 2 children, excluding those with twins, produced the following frequencies:

   2,351 families with 2 girls

   5,435 families with 1 girl

   2,214 families with 0 girls

   (A) Compute the approximate empirical probability for each outcome.
   (B) Compute the theoretical probability for each outcome assuming a boy is as likely as a girl at each birth.

8. If we multiply the probability of the occurrence of an event E by the total number of trials n, we obtain the **expected frequency** of the occurrence of E in n trials. Using the theoretical probabilities found in Problem 7B, compute the expected frequency of each outcome in Problem 7 from the sample of 10,000.

9. Three coins are flipped 1,000 times with the following frequencies of outcomes:

3 heads	132	1 head	380
2 heads	368	0 heads	120

(A) Compute the approximate empirical probability for each outcome.
(B) Compute the theoretical probability for each outcome, assuming fair coins.
(C) Using the theoretical probabilities computed in part B, compute the expected frequency in 1,000 trials for each outcome. (See Problem 8 above for a definition of expected frequency.)

10. Toss 3 coins 50 times and compute the approximate empirical probabilities for 3 heads, 2 heads, 1 head, and 0 heads, respectively.

C 11. If 4 fair coins are tossed 80 times, what is the expected frequency of 4 heads turning up? 3 heads? 2 heads? 1 head? 0 heads? (See Problem 8 for a definition of expected frequency.)

12. Actually toss 4 coins 80 times and tabulate the frequencies of the outcomes indicated in Problem 11. What are the approximate empirical probabilities for these outcomes?

APPLICATIONS

Business & Economics

13. *Market analysis.* A company selected 1,000 households at random and surveyed them to determine a relationship between income level and the number of television sets in a home. The information gathered is listed in the table:

	TELEVISIONS PER HOUSEHOLD				
YEARLY INCOME	0	1	2	3	Above 3
Less then $12,000	0	40	51	11	0
$12,000–19,999	0	70	80	15	1
$20,000–39,999	2	112	130	80	12
$40,000–59,999	10	90	80	60	21
$60,000 or more	30	32	28	25	20

Compute the approximate empirical probabilities:

(A) Of a household earning $12,000–19,999 per year **and** owning exactly 3 television sets

(B) Of a household earning $20,000–39,999 per year **and** owning more than 1 television set

(C) Of a household earning $60,000 or more per year **or** owning more than 3 television sets

(D) Of a household not owning 0 television sets

14. *Market analysis.* Use the sample results in Problem 13 to compute the approximate empirical probabilities:

(A) Of a household earning $40,000–59,999 per year **and** owning 0 television sets

(B) Of a household earning $12,000–39,999 per year **and** owning more than 2 television sets

(C) Of a household earning less than $20,000 per year **or** owning exactly 2 television sets

(D) Of a household not owning more than 3 television sets

Life Sciences

GENES	FLOWERS
RR	Red
RW or WR	Pink
WW	White

15. *Genetics.* A particular type of flowering plant has the possible colors listed in the table in the margin. If 2 pink plants are crossed, the theoretical probabilities associated with each possible flower color are determined by the following table:

		PINK-FLOWERED PLANT	
		R	W
PINK-FLOWERED PLANT	R	RR	RW
	W	WR	WW

$P(\text{Red}) = \frac{1}{4}$
$P(\text{Pink}) = \frac{1}{2}$
$P(\text{White}) = \frac{1}{4}$

In an experiment, 1,000 crosses were made with pink-flowered plants with the following results:

Red	300
Pink	440
White	260

(A) What is the approximate empirical probability for each color?

(B) What is the expected number of plants with each color in the experiment, based on the theoretical probabilities?

Social Sciences

16. *Sociology.* One thousand women between the ages of 50 and 60, who had been married at least once, were chosen at random. They were surveyed to determine a relationship between the age at which they were first married and the total number of marriages they had had to date. The results are

given in the table:

FIRST MARRIAGE Age	NUMBER OF MARRIAGES				
	1	2	3	4	Above 4
Under 18	44	88	25	12	7
18–20	82	70	30	14	8
21–25	130	110	30	10	4
26–30	95	84	12	6	3
Over 30	56	48	25	5	2

Compute the approximate empirical probabilities:

(A) Of a woman being 21–25 years old on her first marriage **and** having a total of 3 marriages

(B) Of a woman being 18–20 years old on her first marriage **and** having more than 1 marriage

(C) Of a woman being under 18 on her first marriage **or** having 2 marriages

(D) Of a woman not being over 30 on her first marriage

SECTION 7-4

Random Variable, Probability Distribution, and Expectation

◆ RANDOM VARIABLE; PROBABILITY DISTRIBUTION
◆ EXPECTED VALUE OF A RANDOM VARIABLE
◆ DECISION-MAKING AND EXPECTED VALUE

◆ RANDOM VARIABLE; PROBABILITY DISTRIBUTION

When performing a random experiment, a sample space S is selected in such a way that all probability problems of interest relative to the experiment can be solved. In many situations we may not be interested in each simple event in the sample space S but in some numerical value associated with the event. For example, if 3 coins are tossed, we may be interested in the number of heads that turn up rather than in the particular pattern that turns up. Or, in selecting a random sample of students, we may be interested in the proportion that are women rather than which particular students are women. In the same way, a "craps" player is usually interested in the sum of the dots on the showing faces rather than the pattern of dots on each face.

In each of these examples, we have a rule that assigns to each simple event in S a single real number. Mathematically speaking, we are dealing with a function (see Section 2-4). Historically, this particular type of function has been called a "random variable."

> A **random variable** is a function that assigns a numerical value to each simple event in a sample space S.

The term *random variable* is an unfortunate choice, since it is neither random nor a variable—it is a function with a numerical value and it is defined on a sample space. But the terminology has stuck and is now standard, so we shall have to live with it. Capital letters, such as X, are used to represent random variables.

Let us return to the experiment of tossing 3 coins. A sample space S of equally likely simple events is indicated in Table 2. Suppose we are interested in the number of heads (0, 1, 2, or 3) appearing on each toss of the 3 coins and the probability of each of these events. We introduce a random variable X (a function) that indicates the number of heads for each simple event in S (see the second column in Table 2). For example, $X(e_1) = 0$, $X(e_2) = 1$, and so on. The random variable X assigns a numerical value to each simple event in the sample space S.

We are interested in the probability of the occurrence of each image value of X; that is, in the probability of the occurrence of 0 heads, 1 head, 2 heads, or 3 heads in the single toss of 3 coins. We indicate this probability by

$$p(x) \quad \text{where} \quad x \in \{0, 1, 2, 3\}$$

The function p is called the **probability function* of the random variable X.**

What is $p(2)$, the probability of getting exactly 2 heads on the single toss of 3 coins? "Exactly 2 heads occur" is the event

$$E = \{THH, HTH, HHT\}$$

Thus,

$$p(2) = \frac{n(E)}{n(S)} = \frac{3}{8}$$

Proceeding similarly for $p(0)$, $p(1)$, and $p(3)$, we obtain the results in Table 3. This table is called a **probability distribution for the random variable X.** Probability distributions are also represented graphically, as shown in Figure 4. The graph of a probability distribution is often called a **histogram.**

Note from Table 3 or Figure 4 that

1. $0 \leqslant p(x) \leqslant 1$, $x \in \{0, 1, 2, 3\}$
2. $p(0) + p(1) + p(2) + p(3) = \frac{1}{8} + \frac{3}{8} + \frac{3}{8} + \frac{1}{8} = 1$

---

* Formally, the probability function p of the random variable X is defined by $p(x) = P(\{e_i \in S | X(e_i) = x\})$, which, because of its cumbersome nature, is usually simplified to $p(x) = P(X = x)$ or, simply, $p(x)$. We will use the simplified notation.

TABLE 2

Number of Heads in the Toss of 3 Coins

SAMPLE SPACE S	NUMBER OF HEADS $X(e_i)$
e_1: TTT	0
e_2: TTH	1
e_3: THT	1
e_4: HTT	1
e_5: THH	2
e_6: HTH	2
e_7: HHT	2
e_8: HHH	3

TABLE 3

Probability Distribution

NUMBER OF HEADS x	0	1	2	3
PROBABILITY $p(x)$	$\frac{1}{8}$	$\frac{3}{8}$	$\frac{3}{8}$	$\frac{1}{8}$

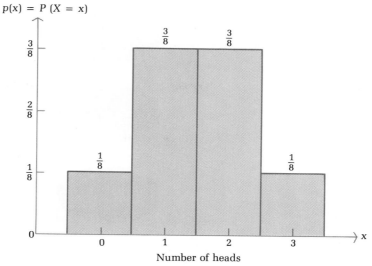

FIGURE 4
Histogram for a probability distribution

These are general properties that any probability distribution of a random variable X associated with a finite sample space must have.

Probability Distribution of a Random Variable X

A probability function $P(X = x) = p(x)$ is a **probability distribution of the random variable X** if

1. $0 \leqslant p(x) \leqslant 1$, $x \in \{x_1, x_2, \ldots, x_n\}$
2. $p(x_1) + p(x_2) + \cdots + p(x_n) = 1$

where $\{x_1, x_2, \ldots, x_n\}$ are the (range) values of X (see Fig. 5).

Figure 5 (at the top of the next page) illustrates the process of forming a probability distribution of a random variable.

◆ EXPECTED VALUE OF A RANDOM VARIABLE

Suppose the experiment of tossing 3 coins was repeated a large number of times. What would be the average number of heads per toss (the total number of heads in all tosses divided by the total number of tosses)? Consulting the probability distribution in Table 3 or Figure 4, we see that we would expect to toss 0 heads

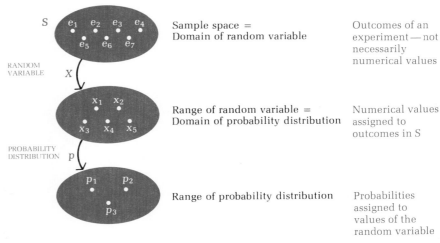

S	e_1 e_2 e_3 e_4 e_5 e_6 e_7	Sample space = Domain of random variable
RANDOM VARIABLE X		
	x_1 x_2 x_3 x_4 x_5	Range of random variable = Domain of probability distribution
PROBABILITY DISTRIBUTION p		
	p_1 p_2 p_3	Range of probability distribution

Sample space = Domain of random variable — Outcomes of an experiment—not necessarily numerical values

Range of random variable = Domain of probability distribution — Numerical values assigned to outcomes in S

Range of probability distribution — Probabilities assigned to values of the random variable

FIGURE 5
Probability distribution of a random variable for a finite sample space

$\frac{1}{8}$ of the time, 1 head $\frac{3}{8}$ of the time, 2 heads $\frac{3}{8}$ of the time, and 3 heads $\frac{1}{8}$ of the time. Thus, in the long run, we would expect the average number of heads per toss of the 3 coins, or the *expected value* $E(X)$, to be given by

$$E(X) = 0\left(\tfrac{1}{8}\right) + 1\left(\tfrac{3}{8}\right) + 2\left(\tfrac{3}{8}\right) + 3\left(\tfrac{1}{8}\right) = \tfrac{12}{8} = 1.5$$

It is important to note that the expected value is not a value that will necessarily occur in a single experiment (1.5 heads cannot occur in the toss of 3 coins), but it is an average of what occurs over a large number of experiments. Sometimes we will toss more than 1.5 heads and sometimes less, but if the experiment is repeated many times, the average number of heads per experiment should be close to 1.5.

We now make the above discussion more precise through the following definition of expected value:

Expected Value of a Random Variable X

Given the probability distribution for the random variable X,

x_i	x_1	x_2	$\cdots$	x_m
p_i	p_1	p_2	$\cdots$	p_m

where $p_i = p(x_i)$, we define the **expected value of X,** denoted $E(X)$, by the formula

$$E(X) = x_1 p_1 + x_2 p_2 + \cdots + x_m p_m$$

We again emphasize that the expected value is not to be expected to occur in a single experiment; it is a long-run average of repeated experiments — it is the weighted average of the possible outcomes, each weighted by its probability.

> **Steps for Computing the Expected Value of a Random Variable X**
>
> *Step 1.* Form the probability distribution of the random variable X.
>
> *Step 2.* Multiply each image value of X, x_i, by its corresponding probability of occurrence p_i; then add the results.

◆ **EXAMPLE 21** What is the expected value (long-run average) of the number of dots facing up for the roll of a single die?

Solution If we choose

$$S = \{1, 2, 3, 4, 5, 6\}$$

as our sample space, then each simple event is a numerical outcome reflecting our interest, and each is equally likely. The random variable X in this case is just the identity function (each number is associated with itself). Thus, the probability distribution for X is

x_i	1	2	3	4	5	6
p_i	$\frac{1}{6}$	$\frac{1}{6}$	$\frac{1}{6}$	$\frac{1}{6}$	$\frac{1}{6}$	$\frac{1}{6}$

Hence,

$$E(X) = 1\left(\tfrac{1}{6}\right) + 2\left(\tfrac{1}{6}\right) + 3\left(\tfrac{1}{6}\right) + 4\left(\tfrac{1}{6}\right) + 5\left(\tfrac{1}{6}\right) + 6\left(\tfrac{1}{6}\right)$$
$$= \tfrac{21}{6} = 3.5$$

◆

PROBLEM 21 Suppose the die in Example 21 is not fair and we obtain (empirically) the following probability distribution for X:

x_i	1	2	3	4	5	6	
p_i	.14	.13	.18	.20	.11	.24	[*Note:* Sum $= 1$.]

What is the expected value of X?

◆

◆ E X A M P L E   22   A carton of 20 calculator batteries contains 2 dead ones. A random sample of 3 is selected from the 20 and tested. Let X be the random variable associated with the number of dead batteries found in a sample.

(A) Find the probability distribution of X.
(B) Find the expected number of dead batteries in a sample.

Solutions   (A) The number of ways of selecting a sample of 3 from 20 (order is not important) is $C_{20,3}$. This is the number of simple events in the experiment, each as likely as the other. A sample will have either 0, 1, or 2 dead batteries. These are the values of the random variable in which we are interested. The probability distribution is computed as follows:

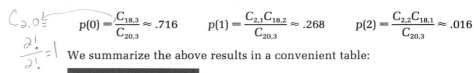

$$p(0) = \frac{C_{18,3}}{C_{20,3}} \approx .716 \qquad p(1) = \frac{C_{2,1}C_{18,2}}{C_{20,3}} \approx .268 \qquad p(2) = \frac{C_{2,2}C_{18,1}}{C_{20,3}} \approx .016$$

$C_{2,0} =$
$\frac{2!}{2!} = 1$

We summarize the above results in a convenient table:

x_i	0	1	2
p_i	.716	.268	.016

[Note: .716 + .268 + .016 = 1.]

(B) The expected number of dead batteries in a sample is readily computed as follows:

$$E(X) = (0)(.716) + (1)(.268) + (2)(.016) = .3$$

The expected value is not one of the random variable values; rather, it is a number that the average number of dead batteries in a sample would approach as the experiment is repeated without end.   ◆

P R O B L E M   22   Repeat Example 22 using a random sample of 4.   ◆

◆ E X A M P L E   23   A spinner device is numbered from 0 to 5, and each of the 6 numbers is as likely to come up as any other. A player who bets $1 on any given number wins $4 (and gets the bet back) if the pointer comes to rest on the chosen number; otherwise, the $1 bet is lost. What is the expected value of the game (long-run average gain or loss per game)?

Solution   The sample space of equally likely events is

$$S = \{0, 1, 2, .3, 4, 5\}$$

Each sample point occurs with a probability of $\frac{1}{6}$. The random variable X assigns $4 to the winning number and $-$1 to each of the remaining numbers. Thus, the probability distribution for X, called a **payoff table,** is as shown in the margin. The probability of winning $4 is $\frac{1}{6}$ and of losing $1 is $\frac{5}{6}$. We can now compute the expected value of the game:

Payoff Table (Probability Distribution for X)

x_i	$4	$-$1
p_i	$\frac{1}{6}$	$\frac{5}{6}$

$$E(X) = \$4(\tfrac{1}{6}) + (-\$1)(\tfrac{5}{6}) = -\$\tfrac{1}{6} \approx -\$0.1667 \approx -17\text{¢ per game}$$

Thus, in the long run the player will lose an average of about 17¢ per game. ◆

In general, a game is said to be **fair** if $E(X) = 0$. The game in Example 23 is not fair — the "house" has an advantage, on the average, of about 17¢ per game.

PROBLEM 23   Repeat Example 23 with the player winning $5 instead of $4 if the chosen number turns up. The loss is still $1 if any other number turns up. Is this now a fair game?

◆

EXAMPLE 24   Suppose you are interested in insuring a car stereo system for $500 against theft. An insurance company charges a premium of $60 for coverage for 1 year, claiming an empirically determined probability of .1 that the stereo will be stolen some time during the year. What is your expected return from the insurance company if you take out this insurance?

Solution   This is actually a game of chance in which your stake is $60. You have a .1 chance of receiving $440 from the insurance company ($500 minus your stake of $60) and a .9 chance of losing your stake of $60. What is the expected value of this "game"? We form a payoff table (the probability distribution for X) as shown in the margin. Then we compute the expected value as follows:

Payoff Table

x_i	$440	$-$60
p_i	.1	.9

$$E(X) = (\$440)(.1) + (-\$60)(.9) = -\$10$$

This means that if you insure with this company over many years and the circumstances remain the same, you would have an average net loss to the insurance company of $10 per year.

◆

PROBLEM 24   Find the expected value in Example 24 from the insurance company's point of view.

◆

◆ DECISION-MAKING AND EXPECTED VALUE

We conclude this section with an example in decision-making.

EXAMPLE 25

Decision Analysis

An outdoor concert featuring a very popular musical group is scheduled for a Sunday afternoon in a large open stadium. The promoter, worrying about being rained out, contacts a long-range weather forecaster who predicts the chance of rain on that Sunday to be .24. If it does not rain, the promoter is certain to net $100,000; if it does rain, the promoter estimates that the net will be only $10,000. An insurance company agrees to insure the concert for $100,000 against rain at a premium of $20,000. Should the promoter buy the insurance?

Solution   The promoter has a choice between two courses of action; A_1: Insure and A_2: Do not insure. As an aid in making a decision, the expected value is computed for each course of action. Probability distributions are indicated in the following payoff table (read vertically):

Payoff Table

p_i	A_1: INSURE x_i	A_2: DO NOT INSURE x_i
.24 (rain)	$90,000	$ 10,000
.76 (no rain)	$80,000	$100,000

Note that the $90,000 entry comes from the insurance company's payoff ($100,000) minus the premium ($20,000) plus gate receipts ($10,000). The reasons for the other entries should be clear. The expected value for each course of action is computed as follows:

A_1: Insure

$$E(X) = x_1 p_1 + x_2 p_2$$
$$= (\$90,000)(.24) + (\$80,000)(.76)$$
$$= \$82,400$$

A_2: Do Not Insure

$$E(X) = (\$10,000)(.24) + (\$100,000)(.76)$$
$$= \$78,400$$

It appears that the promoter's best course of action is to buy the insurance at $20,000. ◆

PROBLEM 25

In Example 25, what is the insurance company's expected value if it writes the policy? ◆

Answers to Matched Problems

21. $E(X) = 3.73$
22. (A)

x_i	0	1	2
p_i	.632	.337	.032

(B) .4

23. $E(X) = \$0$; the game is fair
24. $E(X) = (-\$440)(.1) + (\$60)(.9) = \$10$ (This amount, of course, is necessary to cover expenses and profit.)
25. $E(X) = (-\$80,000)(.24) + (\$20,000)(.76) = -\$4,000$ (This means the insurance company had other information regarding the weather than the promoter; otherwise, the company would not have written this policy.)

EXERCISE 7-4

Where possible, construct a probability distribution or payoff table for a suitable random variable X; then complete the problem.

A 1. If the probability distribution for the random variable X is given in the table, what is the expected value of X?

x_i	−3	0	4
p_i	.3	.5	.2

2. If the probability distribution for the random variable X is given in the table, what is the expected value of X?

x_i	-2	-1	0	1	2
p_i	.1	.2	.4	.2	.1

3. In tossing 2 fair coins, what is the expected number of heads?
4. In a 2-child family, excluding multiple births and assuming a boy is as likely as a girl at each birth, what is the expected number of boys?
5. A fair coin is flipped. If a head turns up, you win $1. If a tail turns up, you lose $1. What is the expected value of the game? Is the game fair?
6. Repeat Problem 5, assuming an unfair coin with the probability of a head being .55 and a tail being .45.

B

7. After paying $4 to play, a single fair die is rolled and you are paid back the number of dollars corresponding to the number of dots facing up. For example, if a 5 turns up, $5 is returned to you for a net gain, or payoff, of $1; if a 1 turns up, $1 is returned for a net gain of $-$3; and so on. What is the expected value of the game? Is the game fair?
8. Repeat Problem 7 with the same game costing $3.50 for each play.
9. Two coins are flipped. You win $2 if either 2 heads or 2 tails turn up; you lose $3 if a head and a tail turn up. What is the expected value of the game?
10. In Problem 9, for the game to be fair, how much *should* you lose if a head and a tail turn up?
11. A friend offers the following game: She wins $1 from you if on 4 rolls of a single die, a 6 turns up at least once; otherwise, you win $1 from her. What is the expected value of the game to you? To her?
12. On 3 rolls of a single die, you will lose $10 if a 5 turns up at least once, and you will win $7 otherwise. What is the expected value of the game?
13. A pair of dice is rolled once. Suppose you lose $10 if a 7 turns up and win $11 if 11 or 12 turn up. How much should you win or lose if any other number turns up in order for the game to be fair?
14. A coin is tossed 3 times. Suppose you lose $3 if 3 heads appear, lose $2 if 2 heads appear, and win $3 if 0 heads appear. How much should you win or lose if 1 head appears in order for the game to be fair?
15. The payoff table for two courses of action, A_1 or A_2, is given in the margin. Which of the two courses, A_1 or A_2, will produce the largest expected value? What is it?

p_i	A_1 x_i	A_2 x_i
.1	$-$200	$-$100
.2	$100	$200
.4	$400	$300
.3	$100	$200

16. The payoff table for three possible courses of action is given as follows:

p_i	A_1 x_i	A_2 x_i	A_3 x_i
.2	$ 500	$ 400	$ 300
.4	1,200	1,100	1,000
.3	1,200	1,800	1,700
.1	1,200	1,800	2,400

Which of the three courses, A_1, A_2, or A_3, will produce the largest expected value? What is it?

17. Roulette wheels in Nevada generally have 38 equally spaced slots numbered 00, 0, 1, 2, . . . , 36. A player who bets $1 on any given number wins $35 (and gets the bet back) if the ball comes to rest on the chosen number; otherwise, the $1 bet is lost. What is the expected value of this game?

18. In roulette (see Problem 17) the numbers from 1 to 36 are evenly divided between red and black. A player who bets $1 on black wins $1 (and gets the bet back) if the ball comes to rest on black; otherwise (if the ball lands on red, 0, or 00), the $1 bet is lost. What is the expected value of the game?

C 19. Five thousand tickets are sold at $1 each for a charity raffle. Tickets are to be drawn at random and monetary prizes awarded as follows: 1 prize of $500, 3 prizes of $100, 5 prizes of $20, and 20 prizes of $5. What is the expected value of this raffle if you buy 1 ticket?

20. A player rolls 2 dice and receives a number of dollars equal to the number of dots showing on both faces. Assuming the dice are fair, what should the player pay each time for the game to be fair?

21. A box of 10 flashbulbs contains 3 defective bulbs. A random sample of 2 is selected and tested. Let X be the random variable associated with the number of defective bulbs in the sample.

(A) Find the probability distribution of X.
(B) Find the expected number of defective bulbs in a sample.

22. A box of 8 flashbulbs contains 3 defective bulbs. A random sample of 2 is selected and tested. Let X be the random variable associated with the number of defective bulbs in a sample.

(A) Find the probability distribution of X.
(B) Find the expected number of defective bulbs in a sample.

23. One thousand raffle tickets are sold at $1 each. Three tickets will be drawn at random (without replacement) and each will pay $200. Suppose you buy 5 tickets.

(A) Create a payoff table for 0, 1, 2, and 3 winning tickets among the 5 tickets you purchased. (If you do not have any winning tickets, you lose $5; if you have 1 winning ticket, you net $195, since your initial $5 will not be returned to you; and so on.)
(B) What is the expected value of the raffle (to you)?

24. Repeat Problem 23 with the purchase of 10 tickets.

Business & Economics

25. *Insurance.* The annual premium for a $5,000 insurance policy against the theft of a painting is $150. If the (empirical) probability that the painting will be stolen during the year is .01, what is your expected return from the insurance company if you take out this insurance?

26. *Insurance.* Repeat Problem 25 from the point of view of the insurance company.

27. *Decision analysis.* An oil company, after careful testing and analysis, is considering drilling in two different sites. It is estimated that site *A* will net $30 million if successful (probability .2) and lose $3 million if not (probability .8); site *B* will net $70 million if successful (probability .1) and lose $4 million if not (probability .9). Which site should the company choose according to the expected return from each site?

28. *Decision analysis.* Repeat Problem 27, assuming additional analysis caused the estimated probability of success in field *B* to be changed from .1 to .11.

Life Sciences

29. *Genetics.* Suppose that, at each birth, having a girl is not as likely as having a boy. The probability assignments for the number of boys in a 3-child family are approximated empirically from past records and are given in the table. What is the expected number of boys in a 3-child family?

NUMBER OF BOYS x_i	p_i
0	.12
1	.36
2	.38
3	.14

NUMBER OF W GENES PRESENT x_i	p_i
0	.25
1	.50
2	.25

30. *Genetics.* A pink-flowering plant is of genotype RW. If two such plants are crossed, we obtain a red plant (RR) with probability .25, a pink plant (RW or WR) with probability .50, and a white plant (WW) with probability .25, as shown in the table in the margin. What is the expected number of W genes present in a crossing of this type?

Social Sciences

31. *Politics.* A money drive is organized by a campaign committee for a candidate running for public office. Two approaches are considered:

A_1: A general mailing with a followup mailing

A_2: Door-to-door solicitation with followup telephone calls

From campaign records of previous committees, average donations and their corresponding probabilities are estimated to be:

A_1		A_2	
x_i (Return per Person)	p_i	x_i (Return per Person)	p_i
$10	.3	$15	.3
5	.2	3	.1
0	.5	0	.6
	1.0		1.0

What are the expected returns? Which course of action should be taken according to the expected returns?

SECTION 7-5   Chapter Review

Important Terms and Symbols

7-1 *Multiplication Principle, Permutations, and Combinations.* Tree diagram; multiplication principle; n factorial; zero factorial; permutation; permutation of n objects; permutation of n objects taken r at a time; combination; combination of n objects taken r at a time

$$n! = n(n-1)(n-2) \cdot \cdots \cdot 2 \cdot 1; \quad 0! = 1; \quad P_{n,n} = n!;$$

$$P_{n,r} = \frac{n!}{(n-r)!}; \quad C_{n,r} = \binom{n}{r} = \frac{P_{n,r}}{r!} = \frac{n!}{r!(n-r)!}, \quad 0 \le r \le n$$

7-2 *Sample Spaces and Events.* Random experiment; sample space; event; simple outcome; simple event; compound event; fundamental sample space; probability of an event; acceptable probability assignment; reasonable probability assignment; probability function; equally likely assumption

$$P(E); \quad n(E)$$

7-3 *Empirical Probability.* Approximate empirical probability; empirical probability; relative frequency; population parameters; population; sample; representative sample; random sample; law of large numbers (or law of averages); expected frequency

$$P(E) = \frac{f(E)}{n}$$

7-4 *Random Variable, Probability Distribution, and Expectation.* Random variable; probability function of a random variable X; probability distribution for a random variable; histogram; expected value of a random variable; payoff table; fair game

$$E(X) = x_1 p_1 + x_2 p_2 + \cdots + x_n p_n$$

Chapter Review

Work through all the problems in this chapter review and check your answers in the back of the book. (Answers to all review problems are there.) Where weaknesses show up, review appropriate sections in the text.

A

1. A single die is rolled and a coin is flipped. How many combined outcomes are possible? Solve:

    (A) By using a tree diagram   (B) By using the multiplication principle

2. Evaluate $C_{6,2}$ and $P_{6,2}$.

3. How many seating arrangements are possible with 6 people and 6 chairs in a row? Solve by using the multiplication principle.

4. Solve Problem 3 using permutations or combinations, whichever is applicable.

5. In a single deal of 5 cards from a standard 52-card deck, what is the probability of being dealt 5 clubs?

6. Betty and Bill are members of a 15-person ski club. If the president and treasurer are selected by lottery, what is the probability that Betty will be president and Bill will be treasurer? (A person cannot hold more than one office.)

7. Each of the first 10 letters of the alphabet is printed on a separate card. What is the probability of drawing 3 cards and getting the code word *dig* by drawing *d* on the first draw, *i* on the second draw, and *g* on the third draw? What is the probability of being dealt a 3-card hand containing the letters *d*, *i*, and *g* in any order?

8. A drug has side effects for 50 out of 1,000 people in a test. What is the approximate empirical probability that a person using the drug will have side effects?

9. A spinning device has 5 numbers, 1, 2, 3, 4, and 5, each as likely to turn up as the other. A person pays $3 and then receives back the dollar amount corresponding to the number turning up on a single spin. What is the expected value of the game? Is the game fair?

B

10. A person tells you that the following approximate empirical probabilities apply to the sample space $\{e_1, e_2, e_3, e_4\}$: $P(e_1) \approx .1$, $P(e_2) \approx -.2$, $P(e_3) \approx .6$, $P(e_4) \approx 2$. There are three reasons why P cannot be a probability function. Name them.

11. Six distinct points are selected on the circumference of a circle. How many triangles can be formed using these points as vertices?

12. How many 3-letter code words are possible using the first 8 letters of the alphabet if no letter can be repeated? If letters can be repeated? If adjacent letters cannot be alike?

13. Solve the following problems using $P_{n,r}$ or $C_{n,r}$:

    (A) How many 3-digit opening combinations are possible on a combination lock with 6 digits if the digits cannot be repeated?

(B) Five tennis players have made the finals. If each of the 5 players is to play every other player exactly once, how many games must be scheduled?

14. Two coins are flipped 1,000 times with the following frequencies:

2 heads   210
1 head   480
0 heads   310

(A) Compute the empirical probability for each outcome.
(B) Compute the theoretical probability for each outcome.
(C) Using the theoretical probabilities computed in part B, compute the expected frequency of each outcome, assuming fair coins.

15. From a standard deck of 52 cards, what is the probability of obtaining a 5-card hand:

(A) Of all diamonds?   (B) Of 3 diamonds and 2 spades?

Write answers in terms of $C_{n,r}$ or $P_{n,r}$; do not evaluate.

16. A group of 10 people includes one married couple. If 4 people are selected at random, what is the probability that the married couple is selected?

17. A player tosses 2 coins and receives $5 if 2 heads turn up, loses $4 if 1 head turns up, and wins $2 if 0 heads turn up. (Would you play this game?) Compute the expected value of the game. Is the game fair?

18. A spinning device has 3 numbers, 1, 2, and 3, each as likely to turn up as the other. If the device is spun twice, what is the probability that:

(A) The same number turns up both times?
(B) The sum of the numbers turning up is 5?

19. An experiment consists of rolling a pair of fair dice. Let X be the random variable associated with the sum of the values that turn up.

(A) Find the probability distribution for X.
(B) Find the expected value of X.

C 20. How many different 5-child families are possible where the sex of each child in the order of their birth is taken into consideration [that is, birth sequences such as (B, G, G, B, B) and (G, B, G, B, B) produce different families]? How many families are possible if the order pattern is not taken into account?

21. If 3 people are selected from a group of 7 men and 3 women, what is the probability that at least 1 woman is selected?

22. A software development department consists of 6 women and 4 men.

(A) How many ways can they select a chief programmer, a backup programmer, and a programming librarian?
(B) If the positions in part A are selected by lottery, what is the probability that women are selected for all 3 positions?
(C) How many ways can they select a team of 3 programmers to work on a particular project?

(D) If the selections in part C are made by lottery, what is the probability that a majority of the team members will be women?

23. How many ways can 2 people be seated in a row of 4 chairs?

24. Three fair coins are tossed 1,000 times with the following frequencies of outcomes:

NUMBER OF HEADS	0	1	2	3
FREQUENCY	120	360	350	170

(A) What is the approximate empirical probability of obtaining 2 heads?
(B) What is the theoretical probability of obtaining 2 heads?
(C) What is the expected frequency of obtaining 2 heads?

25. You bet a friend $1 that you will get 1 or more double 6's on 24 rolls of a pair of fair dice. What is your expectation for this game? What is your friend's expectation? Is the game fair?

26. Two fair (not weighted) dice are each numbered with a 3 on one side, a 2 on two sides, and a 1 on three sides. The dice are rolled and the numbers on the two up faces are added. If X is the random variable associated with the sample space $S = \{2, 3, 4, 5, 6\}$:

(A) Find the probability distribution of X.
(B) Find the expected value of X.

27. If you pay $3.50 to play the game in Problem 26 (the dice are rolled once) and you are returned the dollar amount corresponding to the sum on the faces, what is the expected value of the game? Is the game fair?

APPLICATIONS

Business & Economics

28. *Transportation.* A distribution center A wishes to distribute its products to five different retail stores, B, C, D, E, and F, in a city. How many different route plans can be constructed so that a single truck, starting from A, will deliver to each store exactly once, and then return to the center?

29. *Market analysis.* A clothing company selected 1,000 persons at random and surveyed them to determine a relationship between age of purchaser and annual purchases of jeans. The results are given in the table:

AGE	JEANS PURCHASED ANNUALLY				Totals
	0	*1*	*2*	*Above 2*	
Under 12	60	70	30	10	170
12–18	30	100	100	60	290
19–25	70	110	120	30	330
Over 25	100	50	40	20	210
Totals	260	330	290	120	1,000

Find the empirical probability that a person selected at random:

(A) Is over 25 **and** buys 2 pairs of jeans annually

(B) Is 12–18 years old **and** buys more than 1 pair of jeans annually

(C) Is 12–18 years old **or** buys more than 1 pair of jeans annually

30. *Decision analysis.* A company sales manager, after careful analysis, presents two sales plans. It is estimated that plan A will net $10 million if successful (probability .8) and lose $2 million if not (probability .2); plan B will net $12 million if successful (probability .7) and lose $2 million if not (probability .3). What is the expected return for each plan? Which plan should be chosen based on the expected return?

31. *Insurance.* A $300 bicycle is insured against theft for an annual premium of $30. If the probability that the bicycle will be stolen during the year is .08 (empirically determined), what is the expected value of the policy?

32. *Quality control.* Twelve precision parts, including 2 that are substandard, are sent to an assembly plant. The plant will select 4 at random and will return the whole shipment if 1 or more of the sample are found to be substandard. What is the probability that the shipment will be returned?

33. *Quality control.* A dozen computer circuit boards, including 2 that are defective, are sent to a computer service center. A random sample of 3 is selected and tested. Let X be the random variable associated with the number of circuit boards in a sample that is defective.

(A) Find the probability distribution of X.

(B) Find the expected number of defective boards in a sample.

Additional Topics in Probability

Contents

SECTION 8-1

Union, Intersection, and Complement of Events; Odds

- ◆ UNION AND INTERSECTION
- ◆ COMPLEMENT OF AN EVENT
- ◆ ODDS
- ◆ APPLICATIONS TO EMPIRICAL PROBABILITY

Recall that in Section 7-2 we said that given a sample space

$$S = \{e_1, e_2, \ldots, e_n\}$$

any function P defined on S such that

$$0 \leq P(e_i) \leq 1 \qquad i = 1, 2, \ldots, n$$

and

$$P(e_1) + P(e_2) + \cdots + P(e_n) = 1$$

is called a *probability function*. In addition, we said that any subset of S is called an *event E*, and we defined the probability of E to be the sum of the probabilities of the simple events in E.

◆ UNION AND INTERSECTION

Since events are subsets of a sample space, the union and intersection of events are simply the union and intersection of sets as defined in Appendix A and in the following box. In this section we will concentrate on the union of events and consider only simple cases of intersection. The latter will be investigated in more detail in the next section.

Union and Intersection of Events

If A and B are two events in a sample space S, then the **union** of A and B, denoted by $A \cup B$, and the **intersection** of A and B, denoted by $A \cap B$, are defined as follows:

$$A \cup B = \{e \in S | e \in A \text{ or } e \in B\} \qquad A \cap B = \{e \in S | e \in A \text{ and } e \in B\}$$

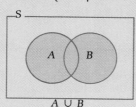

$A \cup B$

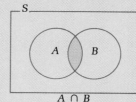

$A \cap B$

Furthermore, we define:

The **event A or B** to be $A \cup B$

The **event A and B** to be $A \cap B$

◆ E X A M P L E   1   Consider the sample space of equally likely events for the rolling of a single fair die:

$$S = \{1, 2, 3, 4, 5, 6\}$$

(A) What is the probability of rolling a number that is odd **and** exactly divisible by 3?

(B) What is the probability of rolling a number that is odd **or** exactly divisible by 3?

Solutions   (A) Let A be the event of rolling an odd number, B the event of rolling a number divisible by 3, and F the event of rolling a number that is odd **and** divisible by 3. Then

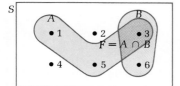

$$A = \{1, 3, 5\} \qquad B = \{3, 6\} \qquad F = A \cap B = \{3\}$$

Thus, the probability of rolling a number that is odd **and** exactly divisible by 3 is

$$P(F) = P(A \cap B) = \frac{n(A \cap B)}{n(S)} = \frac{1}{6}$$

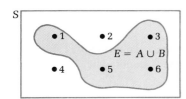

(B) Let A and B be the same events as in part A and let E be the event of rolling a number that is odd **or** divisible by 3. Then

$$A = \{1, 3, 5\} \qquad B = \{3, 6\} \qquad E = A \cup B = \{1, 3, 5, 6\}$$

Thus, the probability of rolling a number that is odd **or** exactly divisible by 3 is

$$P(E) = P(A \cup B) = \frac{n(A \cup B)}{n(E)} = \frac{4}{6} = \frac{2}{3}$$

◆

PROBLEM 1   Use the sample space in Example 1 to answer the following:

(A) What is the probability of rolling an odd number **and** a prime number?

(B) What is the probability of rolling an odd number **or** a prime number?   ◆

Suppose

$$E = A \cup B$$

Can we find $P(E)$ in terms of A and B? The answer is almost yes, but we must be careful. It would be nice if

$$P(A \cup B) = P(A) + P(B) \tag{1}$$

FIGURE 1
Mutually exclusive
$A \cap B = \varnothing$

This turns out to be true if events A and B are **mutually exclusive (disjoint);** that is, if $A \cap B = \varnothing$ (Fig. 1). In this case, $P(A \cup B)$ is the sum of all the probabilities of simple events in A added to the sum of all the probabilities of simple events in B. But what happens if events A and B are not mutually exclusive; that is, if $A \cap B \neq \varnothing$ (see Fig. 2)? If we simply add the probabilities of the elements in A to the probabilities of the elements in B, we are including some of the probabilities twice, namely those for elements that are in both A and B. To compensate for this double counting, we subtract $P(A \cap B)$ from equation (1) to obtain

$$P(A \cup B) = P(A) + P(B) - P(A \cap B) \tag{2}$$

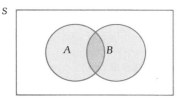

FIGURE 2
Not mutually exclusive
$A \cap B \neq \varnothing$

We notice that (2) holds for both cases, $A \cap B \neq \varnothing$ and $A \cap B = \varnothing$, since (2) reduces to (1) for the latter case $[P(A \cap B) = P(\varnothing) = 0]$. It is better to use (2) if there is any doubt that A and B are mutually exclusive. We summarize this discussion in the box for convenient reference.

Probability of a Union of Two Events

For any events A and B,

$$P(A \cup B) = P(A) + P(B) - P(A \cap B) \tag{2}$$

If A and B are mutually exclusive, then

$$P(A \cup B) = P(A) + P(B) \tag{1}$$

◆ EXAMPLE 2   Suppose two fair dice are rolled:

(A) What is the probability that a sum of 7 or 11 turns up?
(B) What is the probability that both dice turn up the same or that a sum less than 5 turns up?

Solutions   (A) If A is the event that a sum of 7 turns up and B is the event that a sum of 11 turns up, then (see Fig. 3) the event that a sum of 7 or 11 turns up is $A \cup B$, where

$$A = \{(1, 6), (2, 5), (3, 4), (4, 3), (5, 2), (6, 1)\}$$

$$B = \{(5, 6), (6, 5)\}$$

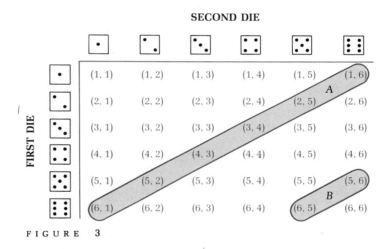

SECOND DIE

FIRST DIE

FIGURE 3

Since events A and B are mutually exclusive, we can use equation (1) to calculate $P(A \cup B)$:

$$P(A \cup B) = P(A) + P(B)$$
$$= \tfrac{6}{36} + \tfrac{2}{36} \qquad \text{In this equally likely sample space,}$$
$$= \tfrac{8}{36} = \tfrac{2}{9} \qquad n(A) = 6, \, n(B) = 2, \text{ and } n(S) = 36.$$

(B) If A is the event that both dice turn up the same and B is the event that the sum is less than 5, then (see Fig. 4 at the top of the next page) the event that both dice turn up the same or the sum is less than 5 is $A \cup B$, where

$$A = \{(1, 1), (2, 2), (3, 3), (4, 4), (5, 5), (6, 6)\}$$

$$B = \{(1, 1), (1, 2), (1, 3), (2, 1), (2, 2), (3, 1)\}$$

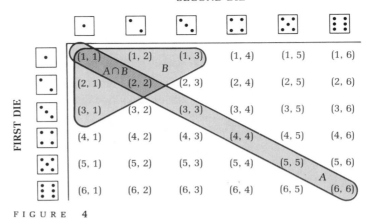

SECOND DIE

	$\boxed{\cdot}$	$\boxed{\because}$	$\boxed{\because}$	$\boxed{::}$	$\boxed{\therefore\cdot}$	$\boxed{:::}$
$\boxed{\cdot}$	(1, 1)	(1, 2)	(1, 3)	(1, 4)	(1, 5)	(1, 6)
$\boxed{\because}$	(2, 1)	(2, 2)	(2, 3)	(2, 4)	(2, 5)	(2, 6)
$\boxed{\because}$	(3, 1)	(3, 2)	(3, 3)	(3, 4)	(3, 5)	(3, 6)
$\boxed{::}$	(4, 1)	(4, 2)	(4, 3)	(4, 4)	(4, 5)	(4, 6)
$\boxed{\therefore\cdot}$	(5, 1)	(5, 2)	(5, 3)	(5, 4)	(5, 5)	(5, 6)
$\boxed{:::}$	(6, 1)	(6, 2)	(6, 3)	(6, 4)	(6, 5)	(6, 6)

FIRST DIE

FIGURE 4

Since $A \cap B = \{(1, 1), (2, 2)\}$, A and B are not mutually exclusive and we use equation (2) to calculate $P(A \cup B)$:

$$P(A \cup B) = P(A) + P(B) - P(A \cap B)$$
$$= \tfrac{6}{36} + \tfrac{6}{36} - \tfrac{2}{36}$$
$$= \tfrac{10}{36} = \tfrac{5}{18}$$

◆

PROBLEM 2   Use the sample space in Example 2 to answer the following:

(A) What is the probability that a sum of 2 or 3 turns up?

(B) What is the probability that both dice turn up the same or that a sum greater than 8 turns up? ◆

You no doubt noticed in Example 2 that we actually did not have to use either formula (1) or (2). We could have proceeded as in Example 1 and simply counted sample points in $A \cup B$. The following example illustrates the use of formula (2) in a situation where visual representation of sample points is not practical.

◆ EXAMPLE 3   What is the probability that a number selected at random from the first 500 positive integers is (exactly) divisible by 3 or 4?

Solution   Let A be the event that a drawn integer is divisible by 3 and B the event that a drawn integer is divisible by 4. Note that events A and B are not mutually exclusive (Why?). Since each of the positive integers from 1 to 500 is as likely to be drawn as any other, we can use $n(A)$, $n(B)$, and $n(A \cap B)$ to determine $P(A \cup B)$, where (think about this)

$$n(A) = \text{The largest integer less than or equal to } \tfrac{500}{3} = 166$$
$$n(B) = \text{The largest integer less than or equal to } \tfrac{500}{4} = 125$$
$$n(A \cap B) = \text{The largest integer less than or equal to } \tfrac{500}{12} = 41$$

[*Note*: A positive integer is divisible by 3 and 4 if and only if it is divisible by 12, the least common multiple of 3 and 4.]

Now we can compute $P(A \cup B)$:

$$P(A \cup B) = P(A) + P(B) - P(A \cap B)$$
$$= \frac{n(A)}{n(S)} + \frac{n(B)}{n(S)} - \frac{n(A \cap B)}{n(S)}$$
$$= \frac{166}{500} + \frac{125}{500} - \frac{41}{500} = \frac{250}{500} = .5 \qquad \blacklozenge$$

PROBLEM 3   What is the probability that a number selected at random from the first 140 positive integers is (exactly) divisible by 4 or 6? $\blacklozenge$

◆ COMPLEMENT OF AN EVENT

Suppose we divide a finite sample space

$$S = \{e_1, \ldots, e_n\}$$

into two subsets E and E' such that

$$E \cap E' = \varnothing$$

that is, E and E' are mutually exclusive, and

$$E \cup E' = S$$

Then E' is called the **complement of E** relative to S. Thus, E' contains all the elements of S that are not in E (Fig. 5). Furthermore,

$$P(S) = P(E \cup E')$$
$$= P(E) + P(E') = 1$$

Hence, we have:

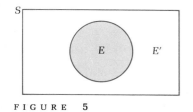

FIGURE 5

Complements	
$P(E) = 1 - P(E')$ $P(E') = 1 - P(E)$	(3)

If the probability of rain is .67, then the probability of no rain is $1 - .67 = .33$; if the probability of striking oil is .01, then the probability of not striking oil is .99. If the probability of having at least 1 boy in a 2-child family is .75, what is the probability of having 2 girls? [*Answer*: .25.]

In looking for $P(E)$, there are situations in which it is easier to find $P(E')$ first, and then use equations (3) to find $P(E)$. The next two examples illustrate two such situations.

◆ **E X A M P L E   4**

Quality Control

A shipment of 45 precision parts, including 9 that are defective, is sent to an assembly plant. The quality control division selects 10 at random for testing and rejects the whole shipment if 1 or more in the sample are found defective. What is the probability that the shipment will be rejected?

Solution

If E is the event that 1 or more parts in a random sample of 10 are defective, then E', the complement of E, is the event that no parts in a random sample of 10 are defective. It is easier to compute $P(E')$ than to compute $P(E)$ directly (try the latter to see why). Once $P(E')$ is found, we will use $P(E) = 1 - P(E')$ to find $P(E)$.

The sample space S for this experiment is the set of all subsets of 10 elements from the set of 45 parts shipped. Thus, since there are $45 - 9 = 36$ nondefective parts,

$$P(E') = \frac{n(E')}{n(S)} = \frac{C_{36,10}}{C_{45,10}} \approx .08$$

and

$$P(E) = 1 - P(E') \approx 1 - .08 = .92$$

◆

P R O B L E M   4

A shipment of 40 precision parts, including 8 that are defective, is sent to an assembly plant. The quality control division selects 10 at random for testing and rejects the whole shipment if 1 or more in the sample are found defective. What is the probability that the shipment will be rejected? ◆

◆ **E X A M P L E   5**

Birthday Problem

In a group of n people, what is the probability that at least 2 people have the same birthday (the same month and day, excluding leap year)? (Make a guess for a class of 40 people, and check your guess with the conclusion of this example.)

Solution

If we form a list of the birthdays of all the people in the group, then we have a simple event in the sample space

$S =$ Set of all lists of n birthdays

For any person in the group, we will assume that any birthday is as likely as any other, so that the simple events in S are equally likely. How many simple events are in the set S? Since any person could have any one of 365 birthdays (excluding leap year), the multiplication principle implies that the number of simple events in S is

$$
\begin{array}{ccccccc}
\text{1st} & & \text{2nd} & & \text{3rd} & & \text{nth} \\
\text{person} & & \text{person} & & \text{person} & & \text{person} \\
n(S) = 365 & \cdot & 365 & \cdot & 365 & \cdots\cdots & 365 \\
= 365^n
\end{array}
$$

Now, let E be the event that at least 2 people in the group have the same birthday. Then E' is the event that no 2 people have the same birthday. The multiplication principle also can be used to determine the number of simple events in E':

<table>
<tr><td></td><td>1st
person</td><td></td><td>2nd
person</td><td></td><td>3rd
person</td><td></td><td>nth
person</td></tr>
<tr><td>$n(E') =$</td><td>365</td><td>·</td><td>364</td><td>·</td><td>363</td><td>· · · · ·</td><td>$(366 - n)$</td></tr>
</table>

$$n(E') = \frac{[365 \cdot 364 \cdot 363 \cdot \cdots \cdot (366 - n)](365 - n)!}{(365 - n)!}$$

Multiply numerator and denominator by $(365 - n)!$

$$= \frac{365!}{(365 - n)!}$$

Since we have assumed that S is an equally likely sample space,

$$P(E') = \frac{n(E')}{n(S)} = \frac{\dfrac{365!}{(365 - n)!}}{365^n} = \frac{365!}{365^n(365 - n)!}$$

Thus,

$$P(E) = 1 - P(E')$$

$$= 1 - \frac{365!}{365^n(365 - n)!} \tag{4}$$

Equation (4) is valid for any n satisfying $1 \le n \le 365$. [What is $P(E)$ if $n > 365$?] For example, in a group of 6 people,

$$P(E) = 1 - \frac{365!}{(365)^6 359!}$$

$$= 1 - \frac{\cancel{365} \cdot 364 \cdot 363 \cdot 362 \cdot 361 \cdot 360 \cdot \cancel{359!}}{\cancel{365} \cdot 365 \cdot 365 \cdot 365 \cdot 365 \cdot 365 \cdot \cancel{359!}}$$

$$= .04$$

It is interesting to note that as the size of the group increases, $P(E)$ increases more rapidly than you might expect. Table 1 gives the value of $P(E)$ for selected values of n. Notice that for a group of only 23 people, the probability that 2 or more have the same birthday is greater than $\frac{1}{2}$. ◆

TABLE 1

The Birthday Problem

NUMBER OF PEOPLE IN GROUP n	PROBABILITY THAT 2 OR MORE HAVE SAME BIRTHDAY $P(E)$
5	.027
10	.117
15	.253
20	.411
23	.507
30	.706
40	.891
50	.970
60	.994
70	.999

PROBLEM 5

Use equation (4) to evaluate $P(E)$ for $n = 4$. ◆

◆ ODDS

When the probability of an event E is known, it is often customary (particularly in gaming situations) to speak of *odds* for (or against) the event E, rather than the *probability* of the event E. The relationship between these two designations is outlined below:

From Probability to Odds

If $P(E)$ is the probability of the event E, then we define:

(A) Odds for $E = \dfrac{P(E)}{1 - P(E)} = \dfrac{P(E)}{P(E')}$ $\qquad P(E) \neq 1$

(B) Odds against $E = \dfrac{P(E')}{P(E)}$ $\qquad P(E) \neq 0$

[*Note:* When possible, odds are expressed as ratios of whole numbers.]

◆ E X A M P L E   6   If you roll a fair die once, the probability of rolling a 4 is $\frac{1}{6}$, whereas the odds in favor of rolling a 4 are

$$\frac{P(E)}{P(E')} = \frac{\frac{1}{6}}{\frac{5}{6}} = \frac{1}{5} \qquad \text{Read "1 to 5" and also written "1:5"}$$

and

$$\text{Odds against rolling a 4} = \frac{5}{1}$$

In terms of a fair game, if you bet $1 on a 4 turning up, you would lose $1 to the house if any number other than 4 turns up and would be paid $5 by the house (and in addition your bet of $1 returned) if a 4 turns up. (An experienced gambler would say that the house pays 5 to 1 on a 4 turning up on a single roll of a die.) Note that, under these rules, the expected value of the game is

$$E(X) = x_1 p_1 + x_2 p_2$$
$$= (\$5)(\tfrac{1}{6}) + (-\$1)(\tfrac{5}{6}) = 0$$

The game is fair (and you would not likely see it in a casino).   ◆

P R O B L E M   6   (A) What are the odds for rolling a sum of 8 in a single roll of two fair dice?
(B) If you bet $5 that a sum of 8 will turn up, what should the house pay (plus returning your $5 bet) for the game to be fair?   ◆

Now we will go in the other direction: If we are given the odds for an event, what is the probability of the event? (The verification of the following formula is left to Problem 53 in Exercise 8-1.)

If the odds for event E are a/b, then the probability of E is

$$P(E) = \frac{a}{a+b}$$

♦ EXAMPLE 7

If in repeated rolls of two fair dice the odds for rolling a 5 before rolling a 7 are 2 to 3, then the probability of rolling a 5 before rolling a 7 is

$$P(E) = \frac{a}{a+b} = \frac{2}{2+3} = \frac{2}{5}$$

♦

PROBLEM 7

If in repeated rolls of two fair dice the odds against rolling a 6 before rolling a 7 are 6 to 5, then what is the probability of rolling a 6 before rolling a 7? (Be careful! Read the problem again.)

♦

♦ APPLICATIONS TO EMPIRICAL PROBABILITY

The following example illustrates the application of the concepts discussed in this section to problems involving data from surveys of a randomly selected sample from a total population. In this situation, the distinction between theoretical and empirical probabilities is a subtle one. If we use the data to assign probabilities to events in the sample population, we are dealing with theoretical probabilities. If we use the same data to assign probabilities to events in the total population, then we are working with empirical probabilities. (See the discussion at the beginning of Section 7-3.) Fortunately, the procedures for computing the probabilities are the same in either case, and all we must do is be careful to use the correct terminology. In the following discussions, we will use *empirical probability* to mean the probability of an event determined by a sample that is used to approximate the probability of the corresponding event in the total population.

♦ EXAMPLE 8

Market Research

From a survey involving 1,000 people in a certain city, it was found that 500 people had tried a certain brand of diet cola, 600 had tried a certain brand of regular cola, and 200 had tried both brands. If a resident of the city is selected at random, what is the (empirical) probability that:

(A) The resident has tried the diet or the regular cola? What are the (empirical) odds for this event?

(B) The resident has tried one of the colas but not both? What are the (empirical) odds against this event?

Solutions Let D be the event that a person has tried the diet cola and R the event that a person has tried the regular cola. The events D and R can be used to partition the residents of the city into four mutually exclusive subsets (a collection of subsets is mutually exclusive if the intersection of any two of them is the empty set):

$D \cap R$ = Set of people who have tried both colas

$D \cap R'$ = Set of people who have tried the diet cola but not the regular cola

$D' \cap R$ = Set of people who have tried the regular cola but not the diet cola

$D' \cap R'$ = Set of people who have not tried either cola

These sets are displayed in the Venn diagram in Figure 6.

The sample population of 1,000 residents is also partitioned into four mutually exclusive sets, with $n(D) = 500$, $n(R) = 600$, and $n(D \cap R) = 200$. By using a Venn diagram (Fig. 7), we can determine the number of sample points in the sets $D \cap R'$, $D' \cap R$, and $D' \cap R'$ (see Example 5 in Appendix A).

These frequencies can be conveniently displayed in a table:

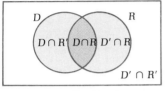

FIGURE 6
Total population

		REGULAR R	NO REGULAR R'	Totals
DIET	D	200	300	500
NO DIET	D'	400	100	500
	Totals	600	400	1,000

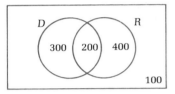

FIGURE 7
Sample population

Assuming that each sample point is equally likely, we form a probability table by dividing each entry in this table by 1,000, the total number surveyed. These are theoretical probabilities for the sample population, which we can use as empirical probabilities to approximate the corresponding probabilities for the total population.

		REGULAR R	NO REGULAR R'	Totals
DIET	D	.2	.3	.5
NO DIET	D'	.4	.1	.5
	Totals	.6	.4	1.0

Now we are ready to compute the required probabilities.

(A) The event that a person has tried the diet or the regular cola is $E = D \cup R$.

We compute $P(E)$ two ways.

Method 1. Directly:

$$P(E) = P(D \cup R)$$
$$= P(D) + P(R) - P(D \cap R)$$
$$= .5 + .6 - .2 = .9$$

Method 2. Using the complement of E:

$$P(E) = 1 - P(E')$$
$$= 1 - P(D' \cap R') \qquad E' = (D \cup R)' = D' \cap R' \text{ (see Fig. 6)}$$
$$= 1 - .1 = .9$$

In either case,

$$\text{Odds for } E = \frac{P(E)}{P(E')} = \frac{.9}{.1} = \frac{9}{1}$$

(B) The event that a person has tried one cola but not both is the event that the person has tried diet and not regular cola or has tried regular and not diet cola. In terms of sets, this is event $E = (D \cap R') \cup (D' \cap R)$. Since $D \cap R'$ and $D' \cap R$ are mutually exclusive (look at the Venn diagrams in Fig. 6),

$$P(E) = P[(D \cap R') \cup (D' \cap R)]$$
$$= P(D \cap R') + P(D' \cap R)$$
$$= .3 + .4 = .7$$

$$\text{Odds against } E = \frac{P(E')}{P(E)} = \frac{.3}{.7} = \frac{3}{7}$$

◆

PROBLEM   8   If a resident from the city in Example 8 is selected at random, what is the (empirical) probability that:

(A) The resident has not tried either cola? What are the (empirical) odds for this event?
(B) The resident has tried the diet cola or has not tried the regular cola? What are the (empirical) odds against this event?

◆

Answers to Matched Problems

1. (A) $\frac{1}{3}$ (B) $\frac{2}{3}$ 2. (A) $\frac{1}{12}$ (B) $\frac{7}{18}$ 3. $\frac{47}{140} \approx .336$ 4. .92
5. .016   6. (A) $5:31$ (B) $\$31$ 7. $\frac{5}{11} \approx .455$
8. (A) $P(D' \cap R') = .1$; odds for $D' \cap R' = \frac{1}{9}$
   (B) $P(D \cup R') = .6$; odds against $D \cup R' = \frac{2}{3}$

E X E R C I S E   8-1

A   1.   If a manufactured item has the probability of .003 of failing within 90 days, what is the probability that the item will not fail in that time period?
2.   In a particular cross of two plants the probability that the flowers will be red is .25. What is the probability that they will not be red?

A spinner is numbered from 1 through 10, and each number is as likely to come up as any other. Use equation (1) or (2), indicating which is used, to compute the probability that in a single spin the spinner will stop at:

3. A number less than 3 or larger than 7
4. A 2 or a number larger than 6
5. An even number or a number divisible by 3
6. An odd number or a number divisible by 3

Problems 7–18 refer to the Venn diagram for events A and B in an equally likely sample space S shown in the margin. Find each of the following probabilities:

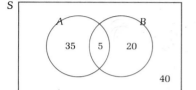

7. $P(A)$ 8. $P(A')$ 9. $P(B)$ 10. $P(B')$
11. $P(A \cap B)$ 12. $P(A \cap B')$ 13. $P(A' \cap B)$, ?   14. $P(A' \cap B') = .40$
15. $P(A \cup B)$ 16. $P(A \cup B')$ 17. $P(A' \cup B)$ 18. $P(A' \cup B')$

$35 + 40 = 70$

In Problems 19–22, use the equally likely sample space in Example 2 and equation (1) or (2), indicating which is used, to compute the probability of the following events:

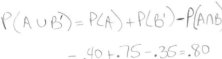

$P(A \cup B') = P(A) + P(B') - P(A \cap B)$

$= .40 + .75 - .35 = .80$

19. A sum of 5 or 6   20. A sum of 9 or 10
21. The number on the first die is a 1 or the number on the second die is a 1.
22. The number on the first die is a 1 or the number on the second die is less than 3.

23. Given the following probabilities for an event E, find the odds for and against E:
    (A) $\frac{3}{8}$ (B) $\frac{1}{4}$ (C) .4   (D) .55

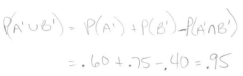

$P(A' \cup B') = P(A') + P(B') - P(A' \cap B')$

$= .60 + .75 - .40 = .95$

24. Given the following probabilities for an event E, find the odds for and against E:
    (A) $\frac{3}{5}$ (B) $\frac{1}{7}$ (C) .6   (D) .35

25. Compute the probability of event E if the odds in favor of E are:
    (A) $\frac{3}{8}$ (B) $\frac{11}{7}$ (C) $\frac{4}{1}$ (D) $\frac{49}{51}$

26. Compute the probability of event E if the odds in favor of E are:
    (A) $\frac{5}{9}$ (B) $\frac{4}{3}$ (C) $\frac{3}{7}$ (D) $\frac{23}{77}$

B In Problems 27–30, compute the odds in favor of obtaining:

27. A head in a single toss of a coin
28. A number divisible by 3 in a single roll of a die
29. At least 1 head when a single coin is tossed 3 times
30. 1 head when a single coin is tossed twice

In Problems 31–34, compute the odds against obtaining:

31. A number greater than 4 in a single roll of a die
32. 2 heads when a single coin is tossed twice
33. A 3 or an even number in a single roll of a die
34. An odd number or a number divisible by 3 in a single roll of a die

35. (A) What are the odds for rolling a sum of 5 in a single roll of two fair dice?
    (B) If you bet $1 that a sum of 5 will turn up, what should the house pay (plus returning your $1 bet) for the game to be fair?
36. (A) What are the odds for rolling a sum of 10 in a single roll of two fair dice?
    (B) If you bet $1 that a sum of 10 will turn up, what should the house pay (plus returning your $1 bet) for the game to be fair?

A pair of dice are rolled 1,000 times with the following frequencies of outcomes:

SUM	2	3	4	5	6	7	8	9	10	11	12
FREQUENCY	10	30	50	70	110	150	170	140	120	80	70

Use these frequencies to calculate the approximate empirical probabilities and odds for the events in Problems 37 and 38.

37. (A) The sum is less than 4 or greater than 9.
    (B) The sum is even or exactly divisible by 5.
38. (A) The sum is a prime number or is exactly divisible by 4.
    (B) The sum is an odd number or exactly divisible by 3.

In Problems 39–42, a single card is drawn from a standard 52-card deck. Calculate the probability of and odds for each event.

39. A face card or a club is drawn
40. A king or a heart is drawn
41. A black card or an ace is drawn
42. A heart or a number less than 7 (count an ace as 1) is drawn

43. What is the probability of getting at least 1 diamond in a 5-card hand dealt from a standard 52-card deck?
44. What is the probability of getting at least 1 black card in a 7-card hand dealt from a standard 52-card deck?
45. What is the probability that a number selected at random from the first 1,000 positive integers is (exactly) divisible by 6 or 8?
46. What is the probability that a number selected at random from the first 600 positive integers is (exactly) divisible by 6 or 9?
47. Five integers are selected at random from the first 50 positive integers, without replacement. What is the probability that there is at least 1 number divisible by 3 among the 5 drawn?
48. Four integers are selected at random from the first 60 positive integers, without replacement. What is the probability that there is at least 1 number divisible by 7 among the 4 drawn?
49. On a roulette wheel (with a 0 and 00), the true odds for landing on red are 9 to 10. What is the probability (to four decimal places) of landing on red? If the house pays $1 for a $1 winning bet on red (plus returning the $1 bet),

what is the expected value of the game? [*Moral:* Nobody said that casinos play fair.]

50. In a game of craps, if a shooter gets an 8 on the first roll of the dice, then in order to win, the shooter, on subsequent rolls of the dice, must roll an 8 before a 7. The shooter loses if a 7 turns up. If the true odds for obtaining an 8 before a 7 are 5 to 6, what is the probability of getting an 8 before a 7? If the house pays $1 for a $1 winning side bet on 8 (plus returning the $1 bet), what is the expected value of the side bet game? (Now you know why casinos make so much money.)

C 51. In a group of n people (n ⩽ 12), what is the probability that at least 2 of them have the same birth month? (Assume any birth month is as likely as any other.)

52. In a group of n people (n ⩽ 100), each person is asked to select a number between 1 and 100, write the number on a slip of paper, and place the slip in a hat. What is the probability that at least 2 of the slips in the hat have the same number written on them?

53. If the odds in favor of an event E occurring are a to b, show that

$$P(E) = \frac{a}{a+b}$$

[*Hint:* Solve the equation $P(E)/P(E') = a/b$ for $P(E)$.]

54. If $P(E) = c/d$, show that the odds in favor of E occurring are c to $d - c$.

APPLICATIONS

Business & Economics

55. *Market research.* From a survey involving 1,000 students at a large university, a market research company found that 750 students owned stereos, 450 owned cars, and 350 owned cars and stereos. If a student at the university is selected at random, what is the (empirical) probability that:

(A) The student owns either a car or a stereo?
(B) The student owns neither a car nor a stereo?

56. *Market research.* If a student at the university in Problem 55 is selected at random, what is the (empirical) probability that:

(A) The student does not own a car?
(B) The student owns a car but not a stereo?

57. *Insurance.* By examining the past driving records of drivers in a certain city, an insurance company has determined the (empirical) probabilities in

the table:

| | | MILES DRIVEN PER YEAR | | | |
		Less than 10,000, M_1	10,000–15,000 Inclusive, M_2	More than 15,000, M_3	Totals
ACCIDENT	A	.05	.1	.15	.3
NO ACCIDENT	A'	.15	.2	.35	.7
	Totals	.2	.3	.5	1.0

If a driver in this city is selected at random, what is the probability that:

(A) He or she drives less than 10,000 miles per year or has an accident?

(B) He or she drives 10,000 or more miles per year and has no accidents?

58. *Insurance.* Use the (empirical) probabilities in Problem 57 to find the probability that a driver in the city selected at random:

(A) Drives more than 15,000 miles per year or has an accident

(B) Drives 15,000 or fewer miles per year and has an accident

59. *Manufacturing.* Manufacturers of a portable computer provide a 90-day limited warranty covering only the keyboard and the disk drive. Their records indicate that during the warranty period, 6% of their computers are returned because they have defective keyboards, 5% are returned because they have defective disk drives, and 1% are returned because both the keyboard and the disk drive are defective. What is the (empirical) probability that a computer will not be returned during the warranty period?

60. *Product testing.* In order to test a new car, an automobile manufacturer wants to select 4 employees to test drive the car for 1 year. If 12 management and 8 union employees volunteer to be test drivers and the selection is made at random, what is the probability that at least 1 union employee is selected?

61. *Quality control.* A shipment of 60 inexpensive digital watches, including 9 that are defective, are sent to a department store. The receiving department selects 10 at random for testing and rejects the whole shipment if 1 or more in the sample are found defective. What is the probability that the shipment will be rejected?

62. *Quality control.* An automated manufacturing process produces 40 computer circuit boards, including 7 that are defective. The quality control department selects 10 at random (from the 40 produced) for testing and will shut down the plant for trouble shooting if 1 or more in the sample are found defective. What is the probability that the plant will be shut down?

63. *Medicine.* In order to test a new drug for adverse reactions, the drug was administered to 1,000 test subjects with the following results: 60 subjects reported that their only adverse reaction was a loss of appetite, 90 subjects reported that their only adverse reaction was a loss of sleep, and 800 subjects reported no adverse reactions at all. If this drug is released for general use, what is the (empirical) probability that a person using the drug will suffer both a loss of appetite and a loss of sleep?

64. *Medicine.* Thirty animals are to be used in a medical experiment on diet deficiency: 3 male and 7 female rhesus monkeys, 6 male and 4 female chimpanzees, and 2 male and 8 female dogs. If one animal is selected at random, what is the probability of getting:

(A) A chimpanzee or a dog?   (B) A chimpanzee or a male?

(C) An animal other than a female monkey?

Social Sciences

Problems 65 and 66 refer to the data in the table below, obtained from a random survey of 1,000 residents of a state. The participants were asked their political affiliations and their preferences in an upcoming gubernatorial election.

		DEMOCRAT D	REPUBLICAN R	UNAFFILIATED U	Totals
CANDIDATE A	A	200	100	85	385
CANDIDATE B	B	250	230	50	530
NO PREFERENCE	N	50	20	15	85
	Totals	500	350	150	1,000

65. *Politics.* If a resident of the state is selected at random, what is the (empirical) probability that the resident is:

(A) Not affiliated with a political party or has no preference? What are the odds for this event?

(B) Affiliated with a political party and prefers candidate *A*? What are the odds against this event?

66. *Politics.* If a resident of the state is selected at random, what is the (empirical) probability that the resident is:

(A) A Democrat or prefers candidate *B*? What are the odds for this event?

(B) Not a Democrat and has no preference? What are the odds against this event?

67. *Sociology.* A group of 5 Blacks, 5 Asians, 5 Latinos, and 5 Whites was used in a study on racial influence in small group dynamics. If 3 people are chosen at random, what is the probability that at least 1 is Black?

Conditional Probability, Intersection, and Independence

- ◆ CONDITIONAL PROBABILITY
- ◆ INTERSECTION OF EVENTS — PRODUCT RULE
- ◆ PROBABILITY TREES
- ◆ INDEPENDENT EVENTS

In the previous section we asked if the probability of the union of two events could be expressed in terms of the probabilities of the individual events, and we found the answer to be a qualified yes (see page 398). Now we ask the same question for the intersection of two events; that is, can the probability of the intersection of two events be represented in terms of the probabilities of the individual events? The answer again is a qualified yes. But before we find out how and under what conditions, we must investigate a related concept called *conditional probability*.

◆ CONDITIONAL PROBABILITY

The probability of an event may change if we are told of the occurrence of another event. For example, if an adult (a person 21 years or older) is selected at random from all adults in the United States, the probability of that person having lung cancer would not be too high. However, if we are told that the person is also a heavy smoker, then we would certainly want to revise the probability upward.

In general, the probability of the occurrence of an event A, given the occurrence of another event B, is called a **conditional probability** and is denoted by $P(A|B)$.

In the above illustration, events A and B would be

$A =$ An adult has lung cancer

$B =$ An adult is a heavy smoker

And $P(A|B)$ would represent the probability of an adult having lung cancer, given that he or she is a heavy smoker.

Our objective now is to try to formulate a precise definition of $P(A|B)$. It is helpful to start with a relatively simple problem, solve it intuitively, and then generalize from this experience.

What is the probability of rolling a prime number (2, 3, or 5) in a single roll of a fair die? Let

$S = \{1, 2, 3, 4, 5, 6\}$

Then the event of rolling a prime number is (see Fig. 8)

$A = \{2, 3, 5\}$

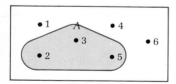

FIGURE 8

Thus, since we assume each simple event in the sample space is equally likely,

$$P(A) = \frac{n(A)}{n(S)} = \frac{3}{6} = \frac{1}{2}$$

Now suppose you are asked, "In a single roll of a fair die, what is the probability that a prime number has turned up if we are given the additional information that an odd number has turned up?" The additional knowledge that another event has occurred, namely,

$$B = \text{An odd number turns up}$$

puts the problem in a new light. We are now interested only in the part of event A (rolling a prime number) that is in event B (rolling an odd number). Event B, since we know it has occurred, becomes the new sample space. The Venn diagrams in Figure 9 illustrate the various relationships. Thus, the probability of A given B is the number of A elements in B divided by the total number of elements in B. Symbolically,

$$P(A|B) = \frac{n(A \cap B)}{n(B)} = \frac{2}{3}$$

Dividing the numerator and denominator of $n(A \cap B)/n(B)$ by $n(S)$, the number of elements in the original sample space, we can express $P(A|B)$ in terms of $P(A \cap B)$ and $P(B)$:*

$$P(A|B) = \frac{n(A \cap B)}{n(B)} = \frac{\dfrac{n(A \cap B)}{n(S)}}{\dfrac{n(B)}{n(S)}} = \frac{P(A \cap B)}{P(B)}$$

Using the right side to compute $P(A|B)$ for the example above, we obtain the same result (as we should):

$$P(A|B) = \frac{P(A \cap B)}{P(B)} = \frac{\frac{2}{6}}{\frac{3}{6}} = \frac{2}{3}$$

We use this latter form for $P(A|B)$ to generalize the concept of conditional

FIGURE   9
B is the new sample space

---

* Note that $P(A|B)$ is a probability based on the new sample space B, while $P(A \cap B)$ and $P(B)$ are both probabilities based on the original sample space S.

probability to arbitrary sample spaces (spaces where simple events are not necessarily equally likely):

Conditional Probability

For events A and B in an arbitrary sample space S, we define the conditional probability of A given B by

$$P(A|B) = \frac{P(A \cap B)}{P(B)} \qquad P(B) \neq 0 \tag{1}$$

◆ **EXAMPLE 9**

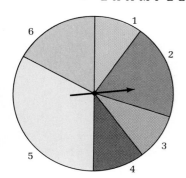

A pointer is spun once on the circular spinner shown in the margin. The probability assigned to the pointer landing on a given integer (from 1 to 6) is the ratio of the area of the corresponding circular sector to the area of the whole circle, as given in the table:

e_i	1	2	3	4	5	6
p_i	.1	.2	.1	.1	.3	.2

(A) What is the probability of the pointer landing on a prime number?
(B) What is the probability of the pointer landing on a prime number, given that it landed on an odd number?

Solutions   Let the events E and F be defined as follows:

$E =$ The pointer lands on a prime number $= \{2, 3, 5\}$
$F =$ The pointer lands on an odd number $= \{1, 3, 5\}$

(A) $P(E) = p(2) + p(3) + p(5)$
$= .2 + .1 + .3 = .6$

(B) First note that $E \cap F = \{3, 5\}$.

$$P(E|F) = \frac{P(E \cap F)}{P(F)} = \frac{p(3) + p(5)}{p(1) + p(3) + p(5)}$$

$$= \frac{.1 + .3}{.1 + .1 + .3} = \frac{.4}{.5} = .8$$

◆

PROBLEM 9   Refer to the spinner and table in Example 9:

(A) What is the probability of the pointer landing on a number greater than 4?
(B) What is the probability of the pointer landing on a number greater than 4, given that it landed on an even number?   ◆

◆ E X A M P L E 10 Suppose past records in a large city produced the following probability data on a driver being in an accident on the last day of a Memorial Day weekend:

		RAIN R	NO RAIN R′	Totals
ACCIDENT	A	.025	.015	.040
NO ACCIDENT	A′	.335	.625	.960
	Totals	.360	.640	1.000

(A) The probability of an accident (rain or shine) is

$P(A) = .040$

(B) The probability of an accident and rain is

$P(A \cap R) = .025$

(C) The probability of rain is

$P(R) = .360$

(D) The probability of an accident, given rain is [using equation (1)]

$$P(A|R) = \frac{P(A \cap R)}{P(R)} = \frac{.025}{.360} = .069$$

(Compare this result with part A.) ◆

P R O B L E M 10 Referring to the table in Example 10, determine the following:

(A) The probability of no rain
(B) The probability of an accident and no rain
(C) The probability of an accident, given no rain [use formula (1) and the results of parts A and B] ◆

◆ INTERSECTION OF EVENTS — PRODUCT RULE

We now return to the original problem of this section; that is, representing the probability of an intersection of two events in terms of the probabilities of the individual events. If $P(A) \neq 0$ and $P(B) \neq 0$, then using (1) we can write:

$$P(A|B) = \frac{P(A \cap B)}{P(B)} \quad \text{and} \quad P(B|A) = \frac{P(B \cap A)}{P(A)}$$

Solving the first equation for $P(A \cap B)$ and the second equation for $P(B \cap A)$, we have

$$P(A \cap B) = P(B)P(A|B) \quad \text{and} \quad P(B \cap A) = P(A)P(B|A)$$

Since $A \cap B = B \cap A$ for any sets A and B, it follows that

$$P(A \cap B) = P(B)P(A|B) = P(A)P(B|A)$$

In summary, we have the **product rule:**

Product Rule

For events A and B with nonzero probabilities in a sample space S,

$$P(A \cap B) = P(A)P(B|A) = P(B)P(A|B) \tag{2}$$

and we can use either $P(A)P(B|A)$ or $P(B)P(A|B)$ to compute $P(A \cap B)$.

◆ EXAMPLE 11 If 60% of a department store's customers are female and 75% of the female customers have charge accounts at the store, what is the probability that a customer selected at random is a female and has a charge account?

Solution Let

$F =$ Female customer

$C =$ Customer with a charge account

If 60% of the customers are female, then the probability that a customer selected at random is a female is

$$P(F) = .60$$

Since 75% of the female customers have charge accounts, the probability that a customer has a charge account, given that the customer is a female, is

$$P(C|F) = .75$$

Using equation (2), the probability that a customer is a female and has a charge account is

$$P(F \cap C) = P(F)P(C|F) = (.60)(.75) = .45 \qquad \blacklozenge$$

PROBLEM 11 If 80% of the male customers of the department store in Example 11 have charge accounts, what is the probability that a customer selected at random is a male and has a charge account? $\qquad \blacklozenge$

◆ PROBABILITY TREES

We used tree diagrams in Section 7-1 to help us count the number of combined outcomes in a sequence of experiments. In much the same way we will now use probability trees to help us compute the probabilities of combined outcomes in a sequence of experiments. An example will help make the process of forming and using probability trees clear.

◆ EXAMPLE 12   Two balls are drawn in succession, without replacement, from a box containing 3 red and 2 white balls. What is the probability of drawing a white ball on the second draw?

Solution   We start with a tree diagram showing the combined outcomes of the two experiments (first draw and second draw):

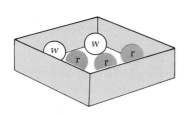

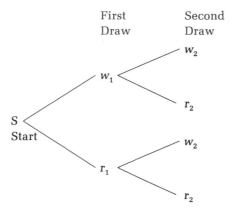

We now assign a probability to each branch on the tree. For example, we assign the probability $\frac{2}{5}$ to the branch Sw_1, since this is the probability of drawing a white ball on the first draw (there are 2 white balls and 3 red balls in the box). What probability should be assigned to the branch w_1w_2? This is the conditional probability $P(w_2|w_1)$; that is, the probability of drawing a white ball on the second draw given that a white ball was drawn on the first draw and not replaced. Since the box now contains 1 white ball and 3 red balls, the probability is $\frac{1}{4}$. Continuing in the same way, we assign probabilities to the other branches of the tree and obtain the following probability tree:

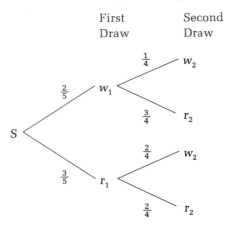

What is the probability of the combined outcome $w_1 \cap w_2$; that is, of drawing a white ball on the first draw and a white ball on the second draw? Using the product rule (2), we have

$$P(w_1 \cap w_2) = P(w_1)P(w_2|w_1)$$
$$= (\tfrac{2}{5})(\tfrac{1}{4}) = \tfrac{1}{10}$$

The combined outcome $w_1 \cap w_2$ corresponds to the unique path Sw_1w_2 in the tree diagram, and we see that the probability of reaching w_2 along this path is just the product of the probabilities assigned to the branches on the path. Reasoning in the same way, we obtain the probability of each remaining combined outcome by multiplying the probabilities assigned to the branches on the path corresponding to the given combined outcome. These probabilities are often written at the ends of the paths to which they correspond.

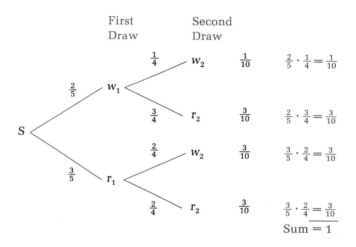

Now it is an easy matter to complete the problem. A white ball drawn on the second draw corresponds to either the combined outcome $w_1 \cap w_2$ or $r_1 \cap w_2$ occurring. Thus, since these combined outcomes are mutually exclusive:

$$P(w_2) = P(w_1 \cap w_2) + P(r_1 \cap w_2)$$
$$= \tfrac{1}{10} + \tfrac{3}{10} = \tfrac{4}{10} = \tfrac{2}{5}$$

which is just the sum of the probabilities listed at the ends of the two paths terminating in w_2. ◆

PROBLEM   12   Two balls are drawn in succession without replacement from a box containing 4 red and 2 white balls. What is the probability of drawing a red ball on the second draw? ◆

The sequence of two experiments in Example 12 is an example of a *stochastic process*. In general, a **stochastic process** involves a sequence of experiments where the outcome of each experiment is not certain. Our interest is in making predictions about the process as a whole. The analysis in Example 12 generalizes to stochastic processes involving any finite sequence of experiments. We summarize the procedures used in Example 12 for general application:

■ Constructing Probability Trees

Step 1.   Draw a tree diagram corresponding to all combined outcomes of the sequence of experiments.

Step 2.   Assign a probability to each tree branch. (This is the probability of the occurrence of the event on the right end of the branch subject to the occurrence of all events on the path leading to the event on the right end of the branch. The probability of the occurrence of a combined outcome that corresponds to a path through the tree is the product of all branch probabilities on the path.*)

Step 3.   Use the results in steps 1 and 2 to answer various questions related to the sequence of experiments as a whole.

◆ E X A M P L E   13

Product Defects

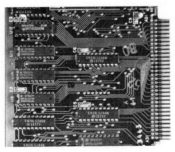

A large computer company A subcontracts the manufacturing of its circuit boards to two companies, 40% to company B and 60% to company C. Company B in turn subcontracts 70% of the orders it receives from company A to company D and the remaining 30% to company E, both subsidiaries of company B. When the boards are completed by companies D, E, and C, they are shipped to company A to be used in various computer models. It has been found that 1.5%, 1%, and .5% of the boards from D, E, and C, respectively, prove defective during the 90-day warranty period after a computer is first sold. What is the probability that a given board in a computer will be defective during the 90-day warranty period?

---

* If we form a sample space S such that each simple event in S corresponds to one path through the tree, and if the probability assigned to each simple event in S is the product of the branch probabilities on the corresponding path, then it can be shown that this is not only an acceptable assignment (all probabilities for the simple events in S are nonnegative and their sum is 1), but it is the only assignment consistent with the method used to assign branch probabilities within the tree.

Solution   Draw a tree diagram and assign probabilities to each branch:

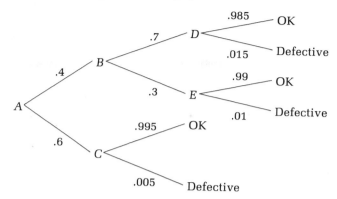

There are three paths leading to defective (the board will be defective within the 90-day warranty period). We multiply the branch probabilities on each path and add the three products:

$$P(\text{Defective}) = (.4)(.7)(.015) + (.4)(.3)(.01) + (.6)(.005)$$
$$= .0084 \qquad \blacklozenge$$

PROBLEM 13   In Example 13, what is the probability that a circuit board in a completed computer came from company E or C? $\qquad \blacklozenge$

◆ INDEPENDENT EVENTS

Suppose events A and B are subsets of a sample space S with $P(A) \neq 0$ and $P(B) \neq 0$. Both $P(A|B)$ and $P(A)$ are probabilities of the event A, the first with respect to a new sample space B and the second with respect to the original sample space S. [*Note:* $P(A)$ is simply an abbreviation for $P(A|S)$.] In general, $P(A|B)$ and $P(A)$ are not equal. However, there is an important class of events for which they are. If

$$P(A|B) = P(A) \qquad (3)$$

then the knowledge of B occurring does not change the probability of A, and we say that A *is independent of B*. Similarly, if

$$P(B|A) = P(B) \qquad (4)$$

we say that B *is independent of A*.

◆ EXAMPLE 14   A single die is rolled and the number of dots facing up are counted. The sample space

$$S = \{1, 2, 3, 4, 5, 6\}$$

is chosen for this experiment. Consider the following four events:

A = The number is divisible by $3 = \{3, 6\}$
B = The number is even $= \{2, 4, 6\}$
C = The number is prime $= \{2, 3, 5\}$
D = The number is odd $= \{1, 3, 5\}$

(A) Determine whether A is independent of B by computing $P(A|B)$ and $P(A)$.
(B) Determine whether C is independent of D by computing $P(C|D)$ and $P(C)$.

Solutions

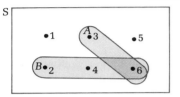

(A) To visualize relationships among S, A, and B, we draw the Venn diagram shown in the margin. Then

$$P(A|B) = \tfrac{1}{3} \qquad P(A) = \tfrac{2}{6} = \tfrac{1}{3} \qquad P(A) \text{ means } P(A|S)$$

Since $P(A|B) = P(A)$, event A is independent of event B.

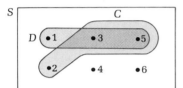

(B) To visualize relationships among S, C, and D, we draw the Venn diagram in the margin. Then

$$P(C|D) = \tfrac{2}{3} \qquad P(C) = \tfrac{3}{6} = \tfrac{1}{2} \qquad P(C) \text{ means } P(C|S)$$

Since $P(C|D) \neq P(C)$, event C is not independent of D. ◆

PROBLEM 14 Referring to Example 14:

(A) Determine whether B is independent of A by computing $P(B|A)$ and $P(B)$.
(B) Determine whether D is independent of C by computing $P(D|C)$ and $P(D)$. ◆

Independence is a reciprocal relationship. That is, if event A is independent of event B, then it can be shown that event B is independent of event A, and we just say that *events A and B are independent*. If events A and B are independent, then the product rule (2) takes on a particularly simple form. Returning to the conditional probability formula (1),

$$P(A|B) = \frac{P(A \cap B)}{P(B)}$$

we replace $P(A|B)$ with $P(A)$ — since events A and B are independent — to obtain

$$P(A) = \frac{P(A \cap B)}{P(B)}$$

or

$$P(A \cap B) = P(A)P(B)$$

We will use this last equation for a general definition of independence of two events:

■■ **Independence**

If A and B are any events in a sample space S, we say that **A and B are independent** if and only if

$$P(A \cap B) = P(A)P(B) \tag{5}$$

Otherwise, A and B are said to be **dependent**.

It is important to note that the general definition of independence allows either event A or event B (or both) to have 0 probability — a situation we rarely encounter in practice, but an important consideration in a more advanced treatment of the subject. Equations (3), (4), and (5) are equivalent (in that either all three are true or all three are false) whenever events A and B have nonzero probabilities.

The definition of independence is used in two ways: (A) if we know that two events are independent, then we can use (5) in place of the product rule (2) to find the probability of the intersection of the two events;* and (B) if we do not know whether two events are independent or dependent, we can use (5) to decide which.

In practice, one often has correct feelings about independence. For example, if one tosses a coin twice, the second toss is independent of the first (a coin has no memory); if a card is drawn from a deck twice, with replacement, the second draw is independent of the first (the deck has no memory); if one rolls a pair of dice twice, the second roll is independent of the first (dice have no memory); and so on. However, there are pairs of events that are not obviously independent or dependent (see Example 14), so we need equation (5) in the definition to decide which. Example 15 considers two events that are obviously independent and Example 16 considers events whose independence or dependence is not obvious.

◆ E X A M P L E   15   In two tosses of a single fair coin show that the events "A head on the first toss" and "A head on the second toss" are independent.

Solution   Consider the sample space of equally likely outcomes for the tossing of a fair coin twice,

$S = \{HH, HT, TH, TT\}$

---

* It is important to keep in mind that equation (5) is a special case of the product rule (2); that is, (2) applies to both dependent and independent events, while (5) only applies to independent events. If independence is in doubt, use the product rule (2).

and the two events,

$$A = \text{A head on the first toss} = \{HH, HT\}$$
$$B = \text{A head on the second toss} = \{HH, TH\}$$

Then

$$P(A) = \tfrac{2}{4} = \tfrac{1}{2} \qquad P(B) = \tfrac{2}{4} = \tfrac{1}{2} \qquad P(A \cap B) = \tfrac{1}{4}$$

Thus,

$$P(A \cap B) = \tfrac{1}{4} = \tfrac{1}{2} \cdot \tfrac{1}{2} = P(A)P(B)$$

and the two events are independent. (The theory agrees with our intuition—a comforting thought!)   ◆

PROBLEM   15   In Example 15, compute $P(B|A)$ and compare with $P(B)$.   ◆

◆ EXAMPLE   16   A single card is drawn from a standard 52-card deck. Test the following events for independence (try guessing the answer to each part before looking at the solution):

(A) $E = $ The drawn card is a spade.
      $F = $ The drawn card is a face card.
(B) $G = $ The drawn card is a club.
      $H = $ The drawn card is a heart.

Solutions   (A) To test E and F for independence, we compute $P(E \cap F)$ and $P(E)P(F)$. If they are equal, then events E and F are independent; if they are not equal, then events E and F are dependent.

$$P(E \cap F) = \tfrac{3}{52} \qquad P(E)P(F) = (\tfrac{13}{52})(\tfrac{12}{52}) = \tfrac{3}{52}$$

Events E and F are independent. (Did you guess this?)
(B) Proceeding as in part A, we see that

$$P(G \cap H) = P(\varnothing) = 0 \qquad P(G)P(H) = (\tfrac{13}{52})(\tfrac{13}{52}) = \tfrac{1}{16}$$

Events G and H are dependent. (Did you guess this?)   ◆

Students often confuse *mutually exclusive (disjoint) events* with *independent events*. One does not necessarily imply the other. In fact, it is not difficult to show (see Problem 45, Exercise 8-2) that any two mutually exclusive events A and B, with nonzero probabilities, are always dependent!

PROBLEM   16   A single card is drawn from a standard 52-card deck. Test the following events for independence:

(A) $E = $ The drawn card is a red card.
      $F = $ The drawn card's number is divisible by 5 (face cards are not assigned values).
(B) $G = $ The drawn card is a king.
      $H = $ The drawn card is a queen.   ◆

◆ E X A M P L E 17   Using the data from Example 10, which are reproduced in the table below, test the indicated events for independence:

(A) Events A and R (B) Events A and R'

		RAIN R	NO RAIN R'	Totals
ACCIDENT	A	.025	.015	.040
NO ACCIDENT	A'	.335	.625	.960
	Totals	.360	.640	1.000

Solutions   (A) $P(A \cap R) = .025$ $P(A)P(R) = (.040)(.360) = .0144$

Events A and R are dependent.

(B) $P(A \cap R') = .015$ $P(A)P(R') = (.040)(.640) = .0256$

Events A and R' are dependent.   ◆

P R O B L E M 17   Referring to the table in Example 17, test the following events for independence:

(A) Events A' and R (B) Events A' and R' ◆

The notion of independence can be extended to more than two events:

▰ Independent Set of Events

A **finite set of events** is said to be **independent** if the probability of each possible intersection of events in the set is the product of the probabilities of the events in the intersection.

For example, events A, B, and C are independent if and only if

$$P(A \cap B) = P(A)P(B) \qquad P(A \cap C) = P(A)P(C) \qquad P(B \cap C) = P(B)P(C)$$
$$P(A \cap B \cap C) = P(A)P(B)P(C)$$

◆ E X A M P L E 18   A space shuttle has four independent computer control systems. If the probability of failure (during flight) of any one system is .001, what is the probability of failure of all four systems?

Solution   Let

E_1 = Failure of system 1 E_3 = Failure of system 3

E_2 = Failure of system 2 E_4 = Failure of system 4

Then, since events E_1, E_2, E_3, and E_4 are given to be independent,

$$P(E_1 \cap E_2 \cap E_3 \cap E_4) = P(E_1)P(E_2)P(E_3)P(E_4)$$
$$= (.001)^4$$
$$= .000\ 000\ 000\ 001 \qquad \blacklozenge$$

PROBLEM   18   A single die is rolled 6 times. What is the probability of getting the sequence, 1, 2, 3, 4, 5, 6? $\blacklozenge$

Answers to Matched Problems

9. (A) .5   (B) .4

10. (A) $P(R') = .640$ (B) $P(A \cap R') = .015$ (C) $P(A|R') = \dfrac{P(A \cap R')}{P(R)} = .023$

11. $P(M \cap C) = P(M)P(C|M) = .32$ 12. $\frac{2}{3}$ 13. .72

14. (A) $P(B|A) = \frac{1}{2} = P(B)$; B is independent of A
    (B) $P(D|C) = \frac{2}{3}$, $P(D) = \frac{1}{2}$; D is not independent of C

15. $P(B|A) = \dfrac{P(A \cap B)}{P(A)} = \dfrac{\frac{1}{4}}{\frac{1}{2}} = \dfrac{1}{2} = P(B)$

16. (A) E and F are independent   (B) G and H are dependent

17. (A) A' and R are dependent   (B) A' and R' are dependent

18. $(\frac{1}{6})^6 \approx .000\ 021\ 4$

EXERCISE   8-2

A   *Given the probabilities in the table in the margin for events in a sample space S, solve Problems 1–16 relative to these probabilities.*

Read the following directly from the table:

	A	B	C	Totals
D	.10	.04	.06	.20
E	.40	.26	.14	.80
Totals	.50	.30	.20	1.00

1. $P(A)$ 2. $P(C)$ 3. $P(D)$ 4. $P(E)$
5. $P(A \cap D)$ 6. $P(A \cap E)$ 7. $P(C \cap D)$ 8. $P(C \cap E)$

Compute the following using equation (1) and appropriate table values:

9. $P(A|D)$ 10. $P(A|E)$ 11. $P(C|D)$ 12. $P(C|E)$

Test the following pairs of events for independence:

13. A and $D \cdot$ 14. A and E 15. C and D 16. C and E

17. A fair coin is tossed 8 times.

   (A) What is the probability of tossing a head on the 8th toss, given that the preceding 7 tosses were heads?

   (B) What is the probability of getting 8 heads or 8 tails?

18. A fair die is rolled 5 times.

   (A) What is the probability of getting a 6 on the 5th roll, given that a 6 turned up on the preceding 4 rolls?

   (B) What is the probability that the same number turns up every time?

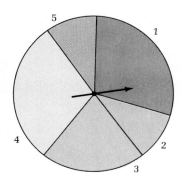

19. A pointer is spun once on the circular spinner shown in the margin. The probability assigned to the pointer landing on a given integer (from 1 to 5) is the ratio of the area of the corresponding circular sector to the area of the whole circle, as given in the table:

e_i	1	2	3	4	5
p_i	.3	.1	.2	.3	.1

Given the events:

E = The pointer lands on an even number.

F = The pointer lands on a number less than 4.

(A) Find $P(F|E)$.   (B) Test events E and F for independence.

20. Repeat Problem 19 with the following events:

E = The pointer lands on an odd number.

F = The pointer lands on a prime number.

Compute the indicated probabilities in Problems 21 and 22 by referring to the following probability tree:

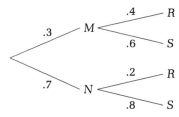

21. (A) $P(M \cap S)$ (B) $P(R)$ 22. (A) $P(N \cap R)$ (B) $P(S)$

B 23. A fair coin is tossed twice. Given S = {HH, HT, TH, TT} as the sample space of equally likely sample points and given the events

E_1 = A head on the first toss

E_2 = A tail on the first toss

E_3 = A tail on the second toss

(A) Are E_1 and E_3 independent?   (B) Are E_1 and E_3 mutually exclusive?

24. For the events in Problem 23:

(A) Are E_2 and E_3 independent?   (B) Are E_2 and E_3 mutually exclusive?
(C) Are E_1 and E_2 mutually exclusive?

25. In 2 throws of a fair die, what is the probability that you will get an even number on each throw? An even number on the first or second throw?

26. In 2 throws of a fair die, what is the probability that you will get at least 5 on each throw? At least 5 on the first or second throw?

27. Two cards are drawn in succession from a standard 52-card deck. What is the probability that the first card is a club and the second card is a heart:

(A) If the cards are drawn without replacement?

(B) If the cards are drawn with replacement?

28. Two cards are drawn in succession from a standard 52-card deck. What is the probability that both cards are red:

(A) If the cards are drawn without replacement?

(B) If the cards are drawn with replacement?

29. A card is drawn at random from a standard 52-card deck. Events G and H are:

$G =$ The drawn card is black.

$H =$ The drawn card is divisible by 3 (face cards are not valued).

(A) Find $P(H|G)$.   (B) Test H and G for independence.

30. A card is drawn at random from a standard 52-card deck. Events M and N are:

$M =$ The drawn card is a diamond.

$N =$ The drawn card is even (face cards are not valued).

(A) Find $P(N|M)$.   (B) Test M and N for independence.

31. Let A be the event that all of a family's children are the same sex, and let B be the event that the family has at most 1 boy. Assuming the probability of having a girl is the same as the probability of having a boy (both .5), test events A and B for independence if:

(A) The family has 2 children.   (B) The family has 3 children.

32. An experiment consists of tossing n coins. Let A be the event that at least 2 heads turn up, and let B be the event that all the coins turn up the same. Test A and B for independence if:

(A) 2 coins are tossed.   (B) 3 coins are tossed.

Problems 33 – 36 refer to the following experiment; 2 balls are drawn in succession out of a box containing 2 red and 5 white balls.

33. Construct a probability tree for this experiment and find the probability of each of the events $R_1 \cap R_2$, $R_1 \cap W_2$, $W_1 \cap R_2$, and $W_1 \cap W_2$, given that the first ball drawn was:

(A) Replaced before the second draw

(B) Not replaced before the second draw

34. Find the probability that the second ball was red, given that the first ball was:

(A) Replaced before the second draw

(B) Not replaced before the second draw

35. Find the probability that at least 1 ball was red, given that the first ball was:

(A) Replaced before the second draw

(B) Not replaced before the second draw

36. Find the probability that both balls were the same color, given that the first ball was:

(A) Replaced before the second draw

(B) Not replaced before the second draw

37. Find the probability of drawing exactly 1 ace if 2 cards are drawn from a standard 52-card deck and the first card is:

(A) Replaced before the second card is drawn

(B) Not replaced before the second card is drawn

38. Find the probability of drawing exactly 1 heart if 2 cards are drawn from a standard 52-card deck and the first card is:

(A) Replaced before the second card is drawn

(B) Not replaced before the second card is drawn

C 39. A box contains 2 red, 3 white, and 4 green balls. Two balls are drawn out of the box in succession without replacement. What is the probability that both balls are the same color?

40. For the experiment in Problem 39, what is the probability that no white balls are drawn?

41. An urn contains 2 one-dollar bills, 1 five-dollar bill, and 1 ten-dollar bill. A player draws bills one at a time without replacement from the urn until a ten-dollar bill is drawn. Then the game stops. All bills are kept by the player.

(A) What is the probability of winning $16?

(B) What is the probability of winning all bills in the urn?

(C) What is the probability of the game stopping at the second draw?

(D) What should the player pay to play the game if the game is to be fair?

42. Ann and Barbara are playing a tennis match. The first player to win 2 sets wins the match. For any given set, the probability that Ann wins that set is $\frac{2}{3}$. Find the probability that:

(A) Ann wins the match.   (B) 3 sets are played.

(C) The player who wins the first set goes on to win the match.

43. Show that $P(A|A) = 1$, when $P(A) \neq 0$.

44. Show that $P(A|B) + P(A'|B) = 1$.

45. Show that A and B are dependent if A and B are mutually exclusive and $P(A) \neq 0$, $P(B) \neq 0$.

46. Show that $P(A|B) = 1$ if B is a subset of A and $P(B) \neq 0$.

Business & Economics

47. *Labor relations.* In a study to determine employee voting patterns in a recent strike election, 1,000 employees were selected at random and the following tabulation was made:

| | | SALARY CLASSIFICATION | | | |
		Hourly (H)	Salary (S)	Salary + Bonus (B)	Totals
TO STRIKE	Yes (Y)	400	180	20	600
	No (N)	150	120	130	400
	Totals	550	300	150	1,000

(A) Convert this table to a probability table by dividing each entry by 1,000.

(B) What is the probability of an employee voting to strike, given the person is paid hourly?

(C) What is the probability of an employee voting to strike, given the person receives a salary plus bonus?

(D) What is the probability of an employee being on straight salary (S)? Of being on straight salary given he or she voted in favor of striking?

(E) What is the probability of an employee being paid hourly? Of being paid hourly given he or she voted in favor of striking?

(F) What is the probability of an employee being in a salary plus bonus position and voting against striking?

(G) Are events S and Y independent?

(H) Are events H and Y independent?

(I) Are events B and N independent?

48. *Quality control.* An automobile manufacturer produces 37% of its cars at plant A. If 5% of the cars manufactured at plant A have defective emission control devices, what is the probability that one of this manufacturer's cars was manufactured at plant A and has a defective emission control device?

49. *Bonus incentives.* If a salesperson has gross sales of over $600,000 in a year, he or she is eligible to play the company's bonus game: A black box contains 1 twenty-dollar bill, 2 five-dollar bills, and 1 one-dollar bill. Bills are drawn out of the box one at a time without replacement until a twenty-dollar bill is drawn. Then the game stops. The salesperson's bonus is 1,000 times the value of the bills drawn.

(A) What is the probability of winning a $26,000 bonus?

(B) What is the probability of winning the maximum bonus, $31,000, by drawing out all bills in the box?

(C) What is the probability of the game stopping at the third draw?

(D) If the salesperson does not like to gamble and requests the expected value of the game for a bonus, what would the bonus be?

50. *Personnel selection.* To transfer into a particular technical department, a company requires an employee to pass a screening test. A maximum of 3 attempts are allowed at 6 month intervals between trials. From past records it is found that 40% pass on the first trial; of those that fail the first trial and take the test a second time, 60% pass; and of those that fail on the second trial and take the test a third time, 20% pass. For an employee wishing to transfer:

(A) What is the probability of passing the test on the first or second try?

(B) What is the probability of failing on the first 2 trials and passing on the third?

(C) What is the probability of failing on all 3 attempts?

Life Sciences

51. *Medicine.* In order to test a new drug for adverse reactions, the drug was administered to 1,000 test subjects with the following results: 60 subjects reported that their only adverse reaction was a loss of appetite, 90 subjects reported that their only adverse reaction was a loss of sleep, and 800 students reported no adverse reactions at all.

(A) If a randomly selected test subject suffered a loss of appetite, what is the probability that the subject also suffered a loss of sleep?

(B) If a randomly selected test subject suffered a loss of sleep, what is the probability that the subject also suffered a loss of appetite?

(C) If a randomly selected test subject did not suffer a loss of appetite, what is the probability that the subject suffered a loss of sleep?

(D) If a randomly selected test subject did not suffer a loss of sleep, what is the probability that the subject suffered a loss of appetite?

52. *Genetics.* In a study to determine frequency and dependency of color-blindness relative to females and males, 1,000 people were chosen at random and the following results were recorded:

		FEMALE F	MALE F′	*Totals*
COLOR-BLIND	C	2	24	26
NORMAL	C′	518	456	974
	Totals	520	480	1,000

(A) Convert this table to a probability table by dividing each entry by 1,000.

(B) What is the probability that a person is a woman, given that the person is color-blind?

(C) What is the probability that a person is color-blind, given that the person is a male?

(D) Are the events color-blindness and male independent?

(E) Are the events color-blindness and female independent?

53. *Psychology.* In a study to determine the frequency and dependency of IQ ranges relative to males and females, 1,000 people were chosen at random and the following results were recorded:

		IQ Below 90 (A)	IQ 90–120 (B)	IQ Above 120 (C)	Totals
FEMALE	F	130	286	104	520
MALE	F'	120	264	96	480
	Totals	250	550	200	1,000

(A) Convert this table to a probability table by dividing each entry by 1,000.

(B) What is the probability of a person having an IQ below 90, given that the person is a female? A male?

(C) What is the probability of a person having an IQ above 120, given that the person is a female? A male?

(D) What is the probability of a person having an IQ below 90?

(E) What is the probability of a person having an IQ between 90 and 120? Of a person having an IQ between 90 and 120, given that the person is a male?

(F) What is the probability of a person being female and having an IQ above 120?

(G) Are any of the events A, B, or C dependent relative to F or F'?

54. *Voting patterns.* A survey of the residents of a precinct in a large city revealed that 55% of the residents were members of the Democratic party and that 60% of the Democratic party members voted in the last election. What is the probability that a person selected at random from the residents of this precinct is a member of the Democratic party and voted in the last election?

S E C T I O N 8-3 **Bayes' Formula**

In the preceding section we discussed the conditional probability of the occurrence of an event, given the occurrence of an earlier event (see Example 9). Now we are going to reverse the problem and try to find the probability of an earlier event conditioned on the occurrence of a later event. As you will see before the discussion is over, a number of practical problems are of this form. First, let us consider a relatively simple problem that will provide the basis for a generalization.

◆ E X A M P L E 19 One urn has 3 red and 2 white balls; a second urn has 1 red and 3 white balls. A single fair die is rolled and if 1 or 2 comes up, a ball is drawn out of the first urn;

otherwise, a ball is drawn out of the second urn. If the drawn ball is red, what is the probability that it came out of the first urn? Out of the second urn?

Solution We form a probability tree, letting U_1 represent urn 1, U_2 urn 2, R a red ball, and W a white ball. Then, on the various outcome branches we assign appropriate probabilities. For example, $P(U_1) = \frac{1}{3}$, $P(R|U_1) = \frac{3}{5}$, and so on:

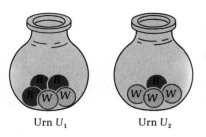

Urn U_1 Urn U_2

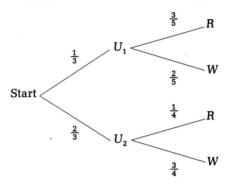

Now we are interested in finding $P(U_1|R)$; that is, the probability that the ball came out of urn 1, given the drawn ball is red. Using equation (1) from Section 8-2, we can write

$$P(U_1|R) = \frac{P(U_1 \cap R)}{P(R)} \tag{1}$$

If we look at the tree diagram, we can see that R is at the end of two different branches; thus,

$$P(R) = P(U_1 \cap R) + P(U_2 \cap R) \tag{2}$$

After substituting equation (2) into equation (1), we get

$$P(U_1|R) = \frac{P(U_1 \cap R)}{P(U_1 \cap R) + P(U_2 \cap R)} \qquad P(A \cap B) = P(A)P(B|A)$$

$$= \frac{P(U_1)P(R|U_1)}{P(U_1)P(R|U_1) + P(U_2)P(R|U_2)}$$

$$= \frac{P(R|U_1)P(U_1)}{P(R|U_1)P(U_1) + P(R|U_2)P(U_2)} \tag{3}$$

Formula (3) is really a lot simpler to use than it looks. You do not need to memorize it; you simply need to understand its form relative to the probability tree above. Referring to the probability tree, we see that

$P(R|U_1)P(U_1) = $ Product of branch probabilities leading to R through U_1
$\qquad\qquad = (\frac{3}{5})(\frac{1}{3}) \qquad$ We usually start at R and work back through U_1.

$P(R|U_2)P(U_2) = $ Product of branch probabilities leading to R through U_2
$\qquad\qquad = (\frac{1}{4})(\frac{2}{3}) \qquad$ We usually start at R and work back through U_2.

Equation (3) now can be interpreted in terms of the probability tree as follows:

$$P(U_1|R) = \frac{\text{Product of branch probabilities leading to } R \text{ through } U_1}{\text{Sum of all branch products leading to } R}$$

$$= \frac{(\frac{3}{5})(\frac{1}{3})}{(\frac{3}{5})(\frac{1}{3}) + (\frac{1}{4})(\frac{2}{3})} = \frac{6}{11} \approx .55$$

Similarly,

$$P(U_2|R) = \frac{\text{Product of branch probabilities leading to } R \text{ through } U_2}{\text{Sum of all branch products leading to } R}$$

$$= \frac{(\frac{1}{4})(\frac{2}{3})}{(\frac{3}{5})(\frac{1}{3}) + (\frac{1}{4})(\frac{2}{3})} = \frac{5}{11} \approx .45$$

[Note: We also could have obtained $P(U_2|R)$ by subtracting $P(U_1|R)$ from 1. Why?] ◆

PROBLEM 19   Repeat Example 19, but find $P(U_1|W)$ and $P(U_2|W)$. ◆

In generalizing the results in Example 19, it is helpful to look at its structure in terms of the Venn diagram shown in Figure 10. We note that U_1 and U_2 are mutually exclusive (disjoint) and their union forms S. The following two equations can now be interpreted in terms of this diagram:

$$P(U_1|R) = \frac{P(U_1 \cap R)}{P(R)} = \frac{P(U_1 \cap R)}{P(U_1 \cap R) + P(U_2 \cap R)}$$

$$P(U_2|R) = \frac{P(U_2 \cap R)}{P(R)} = \frac{P(U_2 \cap R)}{P(U_1 \cap R) + P(U_2 \cap R)}$$

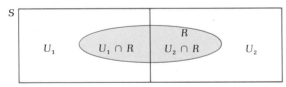

FIGURE 10

Look over the equations and the diagram carefully.

Of course, there is no reason to stop here. Suppose U_1, U_2, and U_3 are three mutually exclusive events whose union is the whose sample space S. Then, for an arbitrary event E in S, with $P(E) \neq 0$, the corresponding Venn diagram looks like Figure 11, and

$$P(U_1|E) = \frac{P(U_1 \cap E)}{P(E)} = \frac{P(U_1 \cap E)}{P(U_1 \cap E) + P(U_2 \cap E) + P(U_3 \cap E)}$$

Similar results hold for U_2 and U_3.

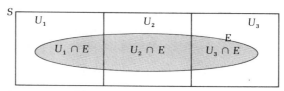

S
U_1 U_2 U_3
$U_1 \cap E$ $U_2 \cap E$ $U_3 \cap E$ E

FIGURE 11

Reasoning in the same way, we arrive at the following famous theorem, which was first stated by the Presbyterian minister Thomas Bayes (1702–1763):

THEOREM 1

Bayes' Formula

Let $U_1, U_2, \ldots, U_n$ be n mutually exclusive events whose union is the sample space S. Let E be an arbitrary event in S such that $P(E) \neq 0$. Then

$$P(U_1|E) = \frac{P(U_1 \cap E)}{P(E)}$$

$$= \frac{P(U_1 \cap E)}{P(U_1 \cap E) + P(U_2 \cap E) + \cdots + P(U_n \cap E)}$$

$$= \frac{P(E|U_1)P(U_1)}{P(E|U_1)P(U_1) + P(E|U_2)P(U_2) + \cdots + P(E|U_n)P(U_n)}$$

Similar results hold for $U_2, U_3, \ldots, U_n$.

You do not need to memorize Bayes' formula. In practice, it is often easier to draw a probability tree and use the following:

Bayes' Formula and Probability Trees

$$P(U_1|E) = \frac{\text{Product of branch probabilities leading to } E \text{ through } U_1}{\text{Sum of all branch products leading to } E}$$

Similar results hold for $U_2, U_3, \ldots, U_n$.

◆ EXAMPLE 20

Tuberculosis Screening

A new, inexpensive skin test is devised for detecting tuberculosis. To evaluate the test before it is put into use, a medical researcher randomly selects 1,000 people. Using precise but more expensive methods already available, it is found that 8% of the 1,000 people tested have tuberculosis. Now each of the 1,000 subjects is given the new skin test and the following results are recorded: The test indicates tuberculosis in 96% of those who have it and in 2% of those who do not. Based on these results, what is the probability of a randomly chosen person having tuberculosis, given that the skin test indicates the disease? What is the probability of a person not having tuberculosis, given that the skin test indicates the disease? (That is, what is the probability of the skin test giving a *false positive result?*)

Solution

Now we will see the power of Bayes' formula in an important application. To start, we form a tree diagram and place appropriate probabilities on each branch:

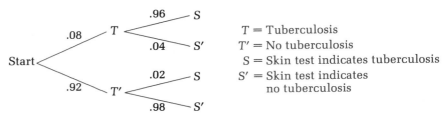

We are interested in finding $P(T|S)$; that is, the probability of a person having tuberculosis given that the skin test indicates the disease. Bayes' formula for this case is

$$P(T|S) = \frac{\text{Product of branch probabilities leading to } S \text{ through } T}{\text{Sum of all branch products leading to } S}$$

Substituting appropriate values from the probability tree, we obtain

$$P(T|S) = \frac{(.08)(.96)}{(.08)(.96) + (.92)(.02)} = .81$$

The probability of a person not having tuberculosis, given that the skin test indicates the disease, denoted by $P(T'|S)$, is

$$P(T'|S) = 1 - P(T|S) = 1 - .81 = .19 \qquad P(T|S) + P(T'|S) = 1 \qquad \blacklozenge$$

Other important questions that need to be answered are indicated in Problem 20.

PROBLEM   20

What is the probability that a person has tuberculosis, given that the test indicates no tuberculosis is present? (That is, what is the probability of the skin test giving a *false negative result?*) What is the probability that a person does not have tuberculosis, given that the test indicates no tuberculosis is present? ◆

◆ EXAMPLE   21

Product Defects

A company produces 1,000 refrigerators a week at three plants. Plant A produces 350 refrigerators a week, plant B produces 250 refrigerators a week, and plant C produces 400 refrigerators a week. Production records indicate that 5% of the refrigerators produced at plant A will be defective, 3% of those produced at plant B will be defective, and 7% of those produced at plant C will be defective. All the refrigerators are shipped to a central warehouse. If a refrigerator at the warehouse is found to be defective, what is the probability that it was produced at plant A?

Solution We begin by constructing a tree diagram:

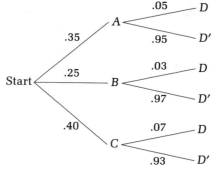

A = Produced at plant A
B = Produced at plant B
C = Produced at plant C
D = Defective

The probability that a defective refrigerator was produced at plant A is $P(A|D)$. Bayes' formula for this case is

$$P(A|D) = \frac{\text{Product of branch probabilities leading to } D \text{ through } A}{\text{Sum of all branch products leading to } D}$$

Using the values from the probability tree, we have

$$P(A|D) = \frac{(.35)(.05)}{(.35)(.05) + (.25)(.03) + (.4)(.07)}$$

$$\approx .33 \qquad \blacklozenge$$

PROBLEM 21 In Example 21, what is the probability that a defective refrigerator in the warehouse was produced at plant B? At plant C? $\qquad \blacklozenge$

Answers to Matched Problems

19. $P(U_1|W) = \frac{4}{19} \approx .21$; $P(U_2|W) = \frac{15}{19} \approx .79$
20. $P(T|S') = .004$; $P(T'|S') = .996$
21. $P(B|D) \approx .14$; $P(C|D) \approx .53$

EXERCISE 8-3

A

Find the probabilities in Problems 1–6 by referring to the tree diagram in the margin.

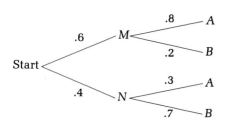

1. $P(M \cap A) = P(M)P(A|M)$
2. $P(N \cap B) = P(N)P(B|N)$
3. $P(A) = P(M \cap A) + P(N \cap A)$
4. $P(B) = P(M \cap B) + P(N \cap B)$
5. $P(M|A) = \dfrac{P(M \cap A)}{P(M \cap A) + P(N \cap A)}$
6. $P(N|B) = \dfrac{P(N \cap B)}{P(N \cap B) + P(M \cap B)}$

Find the probabilities in Problems 7–10 by referring to the following Venn diagram and using Bayes' formula (assume that the simple events in S are equally likely):

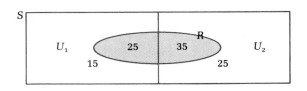

7. $P(U_1|R)$ 8. $P(U_2|R)$ 9. $P(U_1|R')$ 10. $P(U_2|R')$

B   *Find the probabilities in Problems 11–16 by referring to the following tree diagram and using Bayes' formula:*

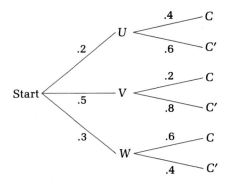

11. $P(U|C)$ 12. $P(V|C')$ 13. $P(W|C)$
14. $P(U|C')$ 15. $P(V|C)$ 16. $P(W|C')$

Find the probabilities in Problems 17–22 by referring to the following Venn diagram and using Bayes' formula (assume that the simple events in S are equally likely):

S

U_1	U_2	U_3
5	15	20
10	20	30

R

17. $P(U_1|R)$ 18. $P(U_2|R')$ 19. $P(U_3|R)$
20. $P(U_1|R')$ 21. $P(U_2|R)$ 22. $P(U_3|R')$

In Problems 23 and 24, use the probabilities on the tree diagram on the left to find the probability of each branch of the tree diagram on the right.

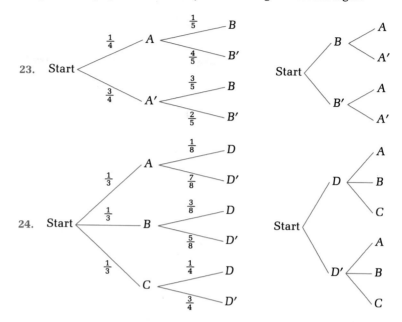

One of two urns is chosen at random with one as likely to be chosen as the other. Then a ball is withdrawn from the chosen urn. Urn 1 contains 1 white and 4 red balls, and urn 2 has 3 white and 2 red balls.

25. If a white ball is drawn, what is the probability that it came from urn 1?
26. If a white ball is drawn, what is the probability that it came from urn 2?
27. If a red ball is drawn, what is the probability that it came from urn 2?
28. If a red ball is drawn, what is the probability that it came from urn 1?

An urn contains 4 red and 5 white balls. Two balls are drawn in succession without replacement.

29. If the second ball is white, what is the probability that the first ball was white?
30. If the second ball is red, what is the probability that the first ball was red?

Urn 1 contains 7 red and 3 white balls. Urn 2 contains 4 red and 5 white balls. A ball is drawn from urn 1 and placed in urn 2. Then a ball is drawn from urn 2.

31. If the ball drawn from urn 2 is red, what is the probability that the ball drawn from urn 1 was red?
32. If the ball drawn from urn 2 is white, what is the probability that the ball drawn from urn 1 was white?

C 33. If 2 cards are drawn in succession from a standard 52-card deck without replacement and the second card is a heart, what is the probability that the first card is a heart?

34. A box contains 10 balls numbered 1 through 10. Two balls are drawn in succession without replacement. If the second ball drawn has the number 4 on it, what is the probability that the first ball had a smaller number on it? An even number on it?

35. Show that $P(U_1|R) + P(U_1'|R) = 1$.

36. If U_1 and U_2 are two mutually exclusive events whose union is the equally likely sample space S and if E is an arbitrary event in S such that $P(E) \neq 0$, show that

$$P(U_1|E) = \frac{n(U_1 \cap E)}{n(U_1 \cap E) + n(U_2 \cap E)}$$

A P P L I C A T I O N S

In the following applications, the word "probability" is often understood to mean "approximate empirical probability."

Business & Economics

37. *Employee screening.* The management of a company finds that 30% of the secretaries hired are unsatisfactory. The personnel director is instructed to devise a test that will improve the situation. One hundred employed secretaries are chosen at random and are given a newly constructed test. Out of these, 90% of the successful secretaries pass the test and 20% of the unsuccessful secretaries pass. Based on these results, if a person applies for a secretarial job, takes the test, and passes it, what is the probability that he or she is a good secretary? If the applicant fails the test, what is the probability that he or she is a good secretary?

38. *Employee rating.* A company has rated 75% of its employees as satisfactory and 25% as unsatisfactory. Personnel records indicate that 80% of the satisfactory workers had previous work experience, while only 40% of the unsatisfactory workers had any previous work experience. If a person with previous work experience is hired, what is the probability that this person will be a satisfactory employee? If a person with no previous work experience is hired, what is the probability that this person will be a satisfactory employee?

39. *Product defects.* A manufacturer obtains clock–radios from three different subcontractors: 20% from A, 40% from B, and 40% from C. The defective rates for these subcontractors are 1%, 3%, and 2%, respectively. If a defective clock–radio is returned by a customer, what is the probability that it came from subcontractor A? From B? From C?

40. *Product defects.* A computer store sells three types of microcomputers, brand A, brand B, and brand C. Of the computers they sell, 60% are brand A, 25% are brand B, and 15% are brand C. They have found that 20% of the

brand *A* computers, 15% of the brand *B* computers, and 5% of the brand *C* computers are returned for service during the warranty period. If a computer is returned for service during the warranty period, what is the probability that it is a brand *A* computer? A brand *B* computer? A brand *C* computer?

Life Sciences

41. *Cancer screening.* A new, simple test has been developed to detect a particular type of cancer. The test must be evaluated before it is put into use. A medical researcher selects a random sample of 1,000 adults and finds (by other means) that 2% have this type of cancer. Each of the 1,000 adults is given the test, and it is found that the test indicates cancer in 98% of those who have it and in 1% of those who do not. Based on these results, what is the probability of a randomly chosen person having cancer given that the test indicates cancer? Of a person having cancer given that the test does not indicate cancer?

42. *Pregnancy testing.* In a random sample of 200 women who suspect that they are pregnant, 100 turn out to be pregnant. A new pregnancy test given to these women indicated pregnancy in 92 of the 100 pregnant women and in 12 of the 100 nonpregnant women. If a woman suspects she is pregnant and this test indicates that she is pregnant, what is the probability that she is pregnant? If the test indicates that she is not pregnant, what is the probability that she is not pregnant?

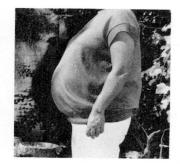

43. *Medical research.* In a random sample of 1,000 people, it is found that 7% have a liver ailment. Of those who have a liver ailment, 40% are heavy drinkers, 50% are moderate drinkers, and 10% are nondrinkers. Of those who do not have a liver ailment, 10% are heavy drinkers, 70% are moderate drinkers, and 20% are nondrinkers. If a person is chosen at random and it is found that he or she is a heavy drinker, what is the probability of that person having a liver ailment? What is the probability for a nondrinker?

44. *Tuberculosis screening.* A test for tuberculosis was given to 1,000 subjects, 8% of whom were known to have tuberculosis. For the subjects who had tuberculosis, the test indicated tuberculosis in 90% of the subjects, was inconclusive for 7%, and indicated no tuberculosis in 3%. For the subjects who did not have tuberculosis, the test indicated tuberculosis in 5% of the subjects, was inconclusive for 10%, and indicated no tuberculosis in the remaining 85%. What is the probability of a randomly selected person having tuberculosis given that the test indicates tuberculosis? Of not having tuberculosis given that the test was inconclusive?

Social Sciences

45. *Police science.* A new lie-detector test has been devised and must be tested before it is put into use. One hundred people are selected at random, and each person draws and keeps a card from a box of 100 cards. Half the cards instruct the person to lie and the others instruct the person to tell the truth. The test indicates lying in 80% of those who lied and in 5% of those who did not. What is the probability that a randomly chosen subject will have lied given that the test indicates lying? That the subject will not have lied given that the test indicates lying?

46. *Politics.* In a given county, records show that of the registered voters, 45% are Democrats, 35% are Republicans, and 20% are independents. In an election, 70% of the Democrats, 40% of the Republicans, and 80% of the independents voted in favor of a parks and recreation bond proposal. If a registered voter chosen at random is found to have voted in favor of the bond, what is the probability that the voter is a Republican? An independent? A Democrat?

SECTION 8-4   Chapter Review

Important Terms and Symbols

8-1 *Union, Intersection, and Complement of Events; Odds.* Union; intersection; event A **or** event B; event A **and** event B; mutually exclusive; disjoint; complement of an event; odds

$$A \cup B; \quad A \cap B; \quad P(A \cup B) = P(A) + P(B) - P(A \cap B);$$
$$P(A \cup B) = P(A) + P(B) \quad \text{if } A \cap B = \varnothing; \quad A'; \quad P(A') = 1 - P(A);$$

$$\frac{P(E)}{P(E')} \text{ (odds for } E)$$

8-2 *Conditional Probability, Intersection, and Independence.* **Conditional** probability of A given B; product rule; stochastic process; independent events; dependent events

$$P(A|B) = \frac{P(A \cap B)}{P(B)} \quad \text{when } P(B) \neq 0; \quad P(A \cap B) = P(A)P(B|A);$$

$$P(A \cap B) = P(A)P(B) \quad \text{if and only if } A \text{ and } B \text{ are independent}$$

8-3 *Bayes' Formula.*

$$P(U_1|E) = \frac{P(E|U_1)P(U_1)}{P(E|U_1)P(U_1) + P(E|U_2)P(U_2) + \cdots + P(E|U_n)P(U_n)}$$

$$= \frac{\text{Product of branch probabilities leading to } E \text{ through } U_1}{\text{Sum of all branch products leading to } E}$$

Similar results hold for $U_2, U_3, \ldots, U_n$.

EXERCISE 8-4   Chapter Review

Work through all the problems in this chapter review and check your answers in the back of the book. (Answers to all review problems are there.) Where weaknesses show up, review appropriate sections in the text.

A   **1.** If A and B are events in a sample space S and $P(A) = .3$, $P(B) = .4$, and $P(A \cap B) = .1$, find:

(A) $P(A')$ (B) $P(A \cup B)$

2. A spinner lands on R with probability .3, on G with probability .5, and on B with probability .2. Find the probability and odds for the spinner landing on either R or G.

3. If in repeated rolls of two fair dice the odds for rolling a sum of 8 before rolling a sum of 7 are 5 to 6, then what is the probability of rolling a sum of 8 before rolling a sum of 7?

Answer Problems 4–12 using the table of probabilities shown in the margin.

	X	Y	Z	Totals
S	.10	.25	.15	.50
T	.05	.20	.02	.27
R	.05	.15	.03	.23
Totals	.20	.60	.20	1.00

4. Find $P(T)$.
5. Find $P(Z)$.
6. Find $P(T \cap Z)$.
7. Find $P(R \cap Z)$.
8. Find $P(R|Z)$.
9. Find $P(Z|R)$.
10. Find $P(T|Z)$.
11. Are T and Z independent?
12. Are S and X independent?

Answer Problems 13–20 using the following probability tree:

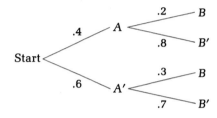

13. $P(A)$
14. $P(B|A)$
15. $P(B|A')$
16. $P(A \cap B)$
17. $P(A' \cap B)$
18. $P(B)$
19. $P(A|B)$
20. $P(A|B')$

B

21. In a single draw from a standard 52-card deck, what are the probability and odds for drawing:

    (A) A jack or a queen?
    (B) A jack or a spade?
    (C) A card other than an ace?

22. (A) What are the odds for rolling a sum of 5 on the single roll of two fair dice?

    (B) If you bet $1 that a sum of 5 will turn up, what should the house pay (plus return your $1 bet) for the game to be fair?

23. Two dice are rolled. The sample space is chosen as the set of all ordered pairs of integers taken from {1, 2, 3, 4, 5, 6}. What is the event A that corresponds to the sum being divisible by 4? What is the event B that corresponds to the sum being divisible by 6? What are $P(A)$, $P(B)$, $P(A \cap B)$, and $P(A \cup B)$?

24. A pointer is spun on a circular spinner. The probabilities of the pointer landing on the integers from 1 to 5 are given in the table in the margin.

    (A) What is the probability of the pointer landing on an odd number?
    (B) What is the probability of the pointer landing on a number less than 4, given that it landed on an odd number?

e_i	1	2	3	4	5
p_i	.2	.1	.3	.3	.1

25. A card is drawn at random from a standard 52-card deck. If E is the event "The drawn card is red" and F is the event "The drawn card is an ace," then:

(A) Find $P(F|E)$.   (B) Test E abnd F for independence.

In Problems 26–30, urn U_1 contains 2 white balls and 3 red balls; urn U_2 contains 2 white balls and 1 red ball.

26. Two balls are drawn out of urn U_1 in succession. What is the probability of drawing a white ball followed by a red ball if the first ball is:

(A) Replaced?   (B) Not replaced?

27. Which of the two parts in Problem 26 involve dependent events?

28. In Problem 26, what is the expected number of red balls if the first ball is:

(A) Replaced?   (B) Not replaced?

29. An urn is selected at random by flipping a fair coin; then a ball is drawn from the urn. Compute:

(A) $P(R|U_1)$ (B) $P(R|U_2)$ (C) $P(R)$
(D) $P(U_1|R)$ (E) $P(U_2|W)$ (F) $P(U_1 \cap R)$

30. In Problem 29, are the events "Selecting urn U_1" and "Drawing a red ball" independent?

31. What is the probability that a number selected at random from the first 200 positive integers is (exactly) divisible by 3 or 5?

32. Three integers are selected at random, without replacement, from the first 30 positive integers. What is the probability that there is at least 1 number divisible by 4 among the 3 drawn?

C Two cards are drawn in succession without replacement from a standard 52-card deck. In Problems 33 and 34, compute the indicated probabilities.

33. The second card is a heart given the first card is a heart.

34. The first card is a heart given the second card is a heart.

35. Suppose 3 white balls and 1 black ball are placed in a box. Balls are drawn in succession without replacement until a black ball is drawn, and then the game is over. You win if the black ball is drawn on the fourth draw.

(A) What are the probability and odds for winning?

(B) If you bet $1, what should the house pay you for winning (plus return your $1 bet) if the game is to be fair?

36. If each of 5 people is asked to identify his or her favorite book from a list of 10 best-sellers, what is the probability that at least 2 of them identify the same book?

Business & Economics

37. *Market analysis.* From a survey of 100 students in a school, it was found that 70 played video games at home, 60 played video games in an arcade, and 40 played video games both at home and in arcades. If a student in the school is selected at random, what is the (empirical) probability that:

    (A) The student plays video games at home or in arcades?
    (B) The student plays video games only at home?

38. *Market analysis.* From a survey of 100 residents of a city, it was found that 40 read the daily morning paper, 70 read the daily evening paper, and 30 read both papers. What is the (empirical) probability that a resident selected at random:

    (A) Reads a daily paper?   (B) Does not read a daily paper?
    (C) Reads exactly one daily paper?

39. *Quality control.* A shipment of 30 parts, including 5 that are defective, is sent to an assembly plant. The quality control division of the plant selects 10 at random for testing and rejects the whole shipment if 1 or more in the sample are found defective. What is the probability that the shipment will be rejected?

40. *Market research.* A market research firm has determined that 40% of the people in a certain area have seen the advertising for a new product and that 85% of those who have seen the advertising have purchased the product. What is the probability that a person in this area has seen the advertising and purchased the product?

Life Sciences

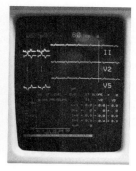

41. *Medicine — cardiogram test.* By testing a large number of individuals, it has been determined that 82% of the population have normal hearts, 11% have some minor heart problems, and 7% have severe heart problems. Ninety-five percent of the persons with normal hearts, 30% of those with minor problems, and 5% of those with severe problems will pass a cardiogram test. What is the probability that a person who passes the cardiogram test has a normal heart?

42. *Genetics.* Six men in 100 and 1 woman in 100 are color-blind. A person is selected at random and is found to be color-blind. What is the probability that this person is a man? (Assume the total population contains the same number of women as men.)

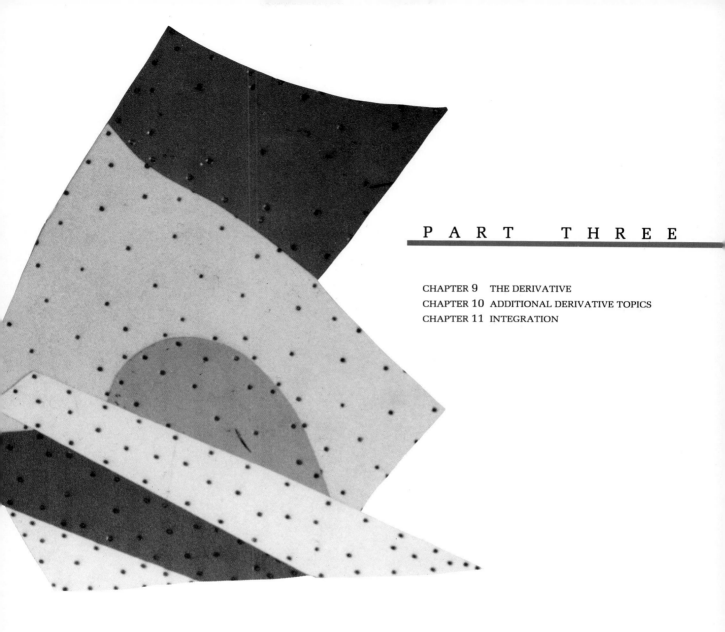

Calculus

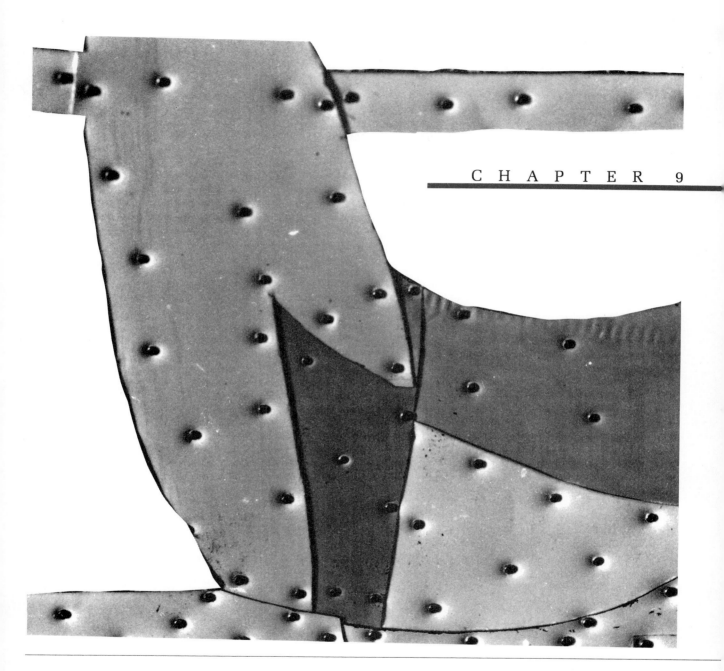

The Derivative

Contents

How do algebra and calculus differ? The two words *static* and *dynamic* probably come as close as any in expressing the difference between the two disciplines. In algebra, we solve equations for a particular value of a variable—a static notion. In calculus, we are interested in how a change in one variable affects another variable —a dynamic notion.

Figure 1 illustrates three basic problems in calculus. It may surprise you to learn that all three problems—as different as they appear—are mathematically related. The solutions to these problems and the discovery of their relationship required the creation of a new kind of mathematics. Isaac Newton (1642–1727) of England and Gottfried Wilhelm von Leibniz (1646–1716) of Germany simultaneously and independently developed this new mathematics, called **the calculus**—it was an idea whose time had come.

In addition to solving the problems described in Figure 1, calculus will enable us to solve many other important problems. Until fairly recently, calculus was used primarily in the physical sciences, but now people in many other disciplines are finding it a useful tool.

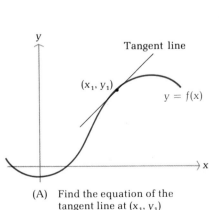

(A) Find the equation of the tangent line at (x_1, y_1) given $y = f(x)$

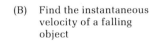

(B) Find the instantaneous velocity of a falling object

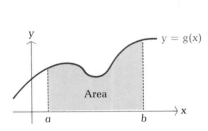

(C) Find the indicated area bounded by $y = g(x)$, $x = a$, $x = b$, and the x axis

FIGURE 1

Limits and Continuity — A Geometric Introduction

◆ FUNCTIONS AND GRAPHS — A BRIEF REVIEW
◆ LIMITS
◆ CONTINUITY
◆ CONTINUITY PROPERTIES
◆ APPLICATION

Basic to the study of calculus is the concept of *limit*. This concept helps us to describe in a precise way the behavior of $f(x)$ when x is close to, but not equal to, a particular number c, or when x increases or decreases without bound. Our discussion here will concentrate on concept development and understanding rather than on formal mathematical detail.

◆ FUNCTIONS AND GRAPHS — A BRIEF REVIEW

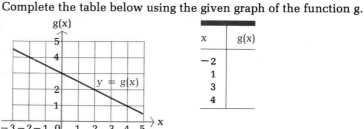

The graph of the function $y = f(x) = x + 2$ is the graph of the set of all ordered pairs $(x, f(x))$. For example, if $x = 2$, then $f(2) = 4$ and $(2, f(2)) = (2, 4)$ is a point on the graph of f. Figure 2 shows $(-1, f(-1))$, $(1, f(1))$, and $(2, f(2))$ plotted on the graph of f. Notice that the domain values -1, 1, and 2 are associated with the x axis, and the range values $f(-1) = 1$, $f(1) = 3$, and $f(2) = 4$ are associated with the y axis.

Given x, it is sometimes useful to be able to read $f(x)$ directly from the graph of f. Example 1 reviews this process.

FIGURE 2

◆ EXAMPLE 1

Complete the table below using the given graph of the function g.

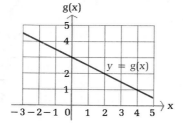

x	g(x)
−2	
1	
3	
4	

Solution

To determine g(x), proceed vertically from the x value on the x axis to the graph of g, then horizontally to the corresponding y value, g(x), on the y axis (as indicated by the dashed lines):

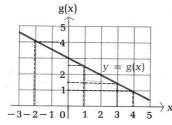

x	g(x)
−2	4.0
1	2.5
3	1.5
4	1.0

PROBLEM 1   Complete the table below using the given graph of the function h.

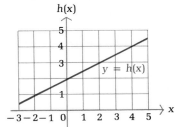

x	h(x)
−2	
−1	
0	
1	
2	
3	
4	

◆

◆ LIMITS

We introduce the important notion of *limit* through two examples, after which the limit concept will be defined.

◆ EXAMPLE 2   Let $f(x) = x + 2$. What happens to $f(x)$ when x is chosen closer and closer to 2, but not equal to 2?

Solution   We construct a table of values of $f(x)$ for some values of x close to 2 and on either side of 2:

x approaches 2 from the left → 2 ← x approaches 2 from the right

x	1.5	1.8	1.9	1.99	1.999 → 2 ← 2.001	2.01	2.1	2.2	2.5
f(x)	3.5	3.8	3.9	3.99	3.999 → ? ← 4.001	4.01	4.1	4.2	4.5

f(x) approaches 4 → 4 ← f(x) approaches 4

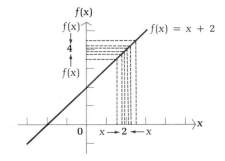

We also draw a graph of f for x near 2, as shown in the margin. Referring to the table and the graph, we see that $f(x)$ approaches 4 as x approaches 2 from either side of 2. We say that "the limit of $x + 2$ as x approaches 2 is 4" and write

$$\lim_{x \to 2} (x + 2) = 4 \quad \text{or} \quad x + 2 \to 4 \quad \text{as} \quad x \to 2$$

This means that we can make $f(x) = x + 2$ as close to 4 as we like by restricting x to a sufficiently small interval about 2 but excluding 2.

Also note that $f(2) = 4$. Thus, the value of the function at 2 and the limit of the function at 2 are the same. That is,

$$\lim_{x \to 2} (x + 2) = f(2)$$

Geometrically, this means there is no break, or hole, in the graph of f at $x = 2$.

◆

PROBLEM 2 Let $f(x) = x + 1$.

(A) Complete the following table:

x	0.9	0.99	0.999 → 1 ← 1.001	1.01	1.1
f(x)	?	?	? → ? ← ?	?	?

(B) Referring to the table in part A, find

$$\lim_{x \to 1} (x + 1)$$

(That is, what value does $x + 1$ approach as x approaches 1 from either side of 1, but is not equal to 1?)
(C) Graph $f(x) = x + 1$ for $-1 \leqslant x \leqslant 4$.
(D) Referring to the graph in part C, find

$$\lim_{x \to 0} (x + 1) \qquad \text{and} \qquad \lim_{x \to 3} (x + 1) \qquad \blacklozenge$$

Example 2 and Problem 2 were fairly obvious. The next example is a little less obvious.

♦ EXAMPLE 3 Let

$$g(x) = \frac{x^2 - 4}{x - 2} \qquad x \neq 2$$

Even though the function is not defined when $x = 2$ (both the numerator and denominator are 0), we can still ask how g(x) behaves when x is near 2, but not equal to 2. Can you guess what happens to g(x) as x approaches 2? The numerator tending to 0 is a force pushing the fraction toward 0. The denominator tending to 0 is another force pushing the fraction toward larger values. How do these two forces balance out?

Solution We could proceed by constructing a table of values to the left and to the right of 2, as in Example 2. Instead, we take advantage of some algebraic simplification using factoring:

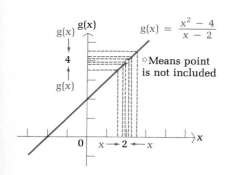

$$g(x) = \frac{x^2 - 4}{x - 2} = \frac{(x - 2)(x + 2)}{x - 2} = x + 2 \qquad x \neq 2$$

Thus, we see that the graph of function g is the same as the graph of function f in Example 2, except that the graph of g has a hole at the point with coordinates (2, 4), as shown in the margin.

Since the behavior of $(x^2 - 4)/(x - 2)$ for x near 2, but not equal to 2, is the same as the behavior of $x + 2$ for x near 2, but not equal to 2, we have

$$\lim_{x \to 2} \frac{x^2 - 4}{x - 2} = \lim_{x \to 2} (x + 2) = 4$$

And we see that the limit of the function g at 2 exists even though the function is not defined there (the graph has a hole at $x = 2$).   ◆

P R O B L E M   3   Let

$$g(x) = \frac{x^2 - 1}{x - 1} \qquad x \ne 1$$

[Before proceeding with parts A and B, try to guess what g(x) approaches as x approaches 1, but does not equal 1. Experiment a little with a calculator.]

(A) Graph g for $-1 \le x \le 4$. [*Hint:*   Proceed as in Example 3.]
(B) Using the graph in part A, find

$$\lim_{x \to 1} \frac{x^2 - 1}{x - 1}$$ ◆

We now present an informal definition of the important concept of limit:

▬ Limit

We write

$$\lim_{x \to c} f(x) = L \qquad \text{or} \qquad f(x) \to L \quad \text{as} \quad x \to c$$

if the functional value $f(x)$ is close to the single real number L whenever x is close to, but not equal to, c (on either side of c).

[*Note:*   The existence of a limit at c has nothing to do with the value of the function at c. In fact, c may not even be in the domain of f (see Examples 2 and 3). However, the function must be defined on both sides of c.]

The next example involves the **absolute value function:**

$$|x| = \begin{cases} x & \text{if } x > 0 \\ 0 & \text{if } x = 0 \\ -x & \text{if } x < 0 \end{cases} \qquad \begin{aligned} |3| &= 3 \\ |0| &= 0 \\ |-2| &= -(-2) = 2 \end{aligned}$$

◆ E X A M P L E   4   Let $h(x) = |x|/x$. Explore the behavior of h(x) for x near 0, but not equal to 0, using a table and a graph. Find $\lim_{x \to 0} h(x)$, if it exists.*

---

* "$\lim_{x \to 0} h(x)$" is another way of writing "$\lim_{x \to 0} h(x)$."

Solution   The function h is defined for all real numbers except 0. For example,

$$h(-2) = \frac{|-2|}{-2} = \frac{2}{-2} = -1$$

$$h(0) = \frac{|0|}{0} = \frac{0}{0} \qquad \text{Not defined}$$

$$h(2) = \frac{|2|}{2} = \frac{2}{2} = 1$$

In general, $h(x)$ is -1 for all negative x and 1 for all positive x. The following table and graph illustrate the behavior of $h(x)$ for x near 0:

	x approaches 0 from the left →					0	← x approaches 0 from the right				
x	-2	-1	-0.1	-0.01	$-0.001 \rightarrow$	0	$\leftarrow 0.001$	0.01	0.1	1	2
$h(x)$	-1	-1	-1	-1	$-1 \;\rightarrow -1$		$1 \leftarrow 1$	1	1	1	1

$$h(x) = -1 \rightarrow -1 \qquad\qquad 1 \leftarrow h(x) = 1$$

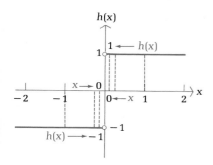

When x is near 0 (on either side of 0), is $h(x)$ near one specific number? The answer is "No," because $h(x)$ is -1 for $x < 0$ and 1 for $x > 0$. Consequently, we say that

$$\lim_{x \to 0} \frac{|x|}{x} \quad \text{does not exist}$$

Thus, neither $h(x)$ nor the limit of $h(x)$ exist at $x = 0$. However, the limit from the left and the limit from the right both exist at 0, but they are not equal. (We will discuss this further below.)   ◆

PROBLEM   4   Graph

$$h(x) = \frac{x - 2}{|x - 2|} \qquad -1 \leqslant x \leqslant 5$$

and find $\lim_{x \to 2} h(x)$, if it exists.   ◆

In Example 4, we mentioned the "limit from the left" and the "limit from the right." These phrases suggest the useful notion of **one-sided limits,** which we now define.

We write

$$\lim_{x \to c^-} f(x) = K \qquad c^- \text{ means } x < c \;\; (c^- \neq -c)$$

and call K the **limit from the left** (or **left-hand limit**) if $f(x)$ is close to K whenever x is close to c, but to the left of c on the real number line. We write

$$\lim_{x \to c^+} f(x) = L \qquad c^+ \text{ means } x > c \;\; (c^+ \neq +c)$$

and call L the **limit from the right** (or **right-hand limit**) if $f(x)$ is close to L whenever x is close to c, but to the right of c on the real number line.

We now make the following important observation:

> ### On the Existence of a Limit
>
> In order for a limit to exist, the limit from the left and the limit from the right must exist and be equal.

In Example 4,

$$\lim_{x \to 0^-} \frac{|x|}{x} = -1 \quad \text{and} \quad \lim_{x \to 0^+} \frac{|x|}{x} = 1$$

Since the left- and right-hand limits are not the same,

$$\lim_{x \to 0} \frac{|x|}{x} \quad \text{does not exist}$$

◆ EXAMPLE 5

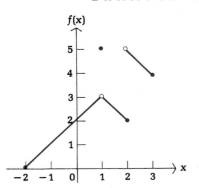

Given the graph of the function f shown in the margin, we discuss the behavior of $f(x)$ for x near -1, 1, 2, and 3:

(A) Behavior of $f(x)$ for x near -1:

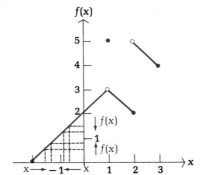

$$\lim_{x \to -1^-} f(x) = 1$$

$$\lim_{x \to -1^+} f(x) = 1$$

$$\lim_{x \to -1} f(x) = 1$$

$$f(-1) = 1$$

(B) Behavior of $f(x)$ for x near 1:

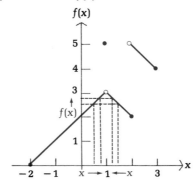

$$\lim_{x \to 1^-} f(x) = 3$$

$$\lim_{x \to 1^+} f(x) = 3$$

$$\lim_{x \to 1} f(x) = 3$$

$$f(1) = 5$$

(C) Behavior of $f(x)$ for x near 2:

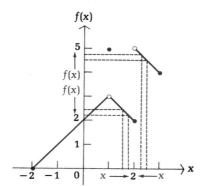

$$\lim_{x \to 2^-} f(x) = 2$$

$$\lim_{x \to 2^+} f(x) = 5$$

$$\lim_{x \to 2} f(x) \quad \text{does not exist}$$

$$f(2) = 2$$

(D) Behavior of $f(x)$ for x near 3:

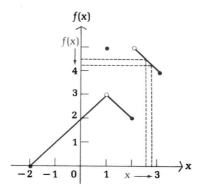

$$\lim_{x \to 3^-} f(x) = 4$$

$$\lim_{x \to 3^+} f(x) \quad \text{does not exist}$$

f is not defined for $x > 3$

$$\lim_{x \to 3} f(x) \quad \text{does not exist}$$

$$f(3) = 4$$

◆

PROBLEM 5

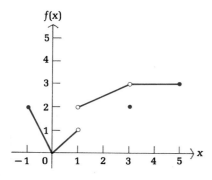

Given the graph of the function f shown in the margin, discuss the following, as we did in Example 5:

(A) Behavior of $f(x)$ for x near -1
(B) Behavior of $f(x)$ for x near 0
(C) Behavior of $f(x)$ for x near 1
(D) Behavior of $f(x)$ for x near 3

◆

◆ CONTINUITY

Compare the graphs from Examples 2–4, which are repeated in Figure 3 (at the top of the next page). Notice that two of the graphs are broken; that is, they cannot be drawn without lifting a pen off the paper. Informally, a function is *continuous over an interval* if its graph over the interval can be drawn without removing a pen from the paper. A function whose graph is broken (disconnected) at $x = c$ is said to be *discontinuous* at $x = c$. Function f (Figure 3A) is continuous for all x. Function g (Figure 3B) is discontinuous at $x = 2$, but is continuous over any interval that does not include 2. Function h (Figure 3C) is discontinuous at $x = 0$, but is continuous over any interval that does not include 0.

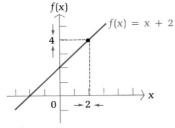

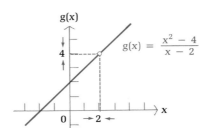

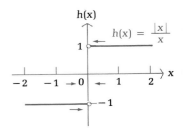

(A) $\lim\limits_{x \to 2} f(x) = 4$
    $f(2) = 4$

(B) $\lim\limits_{x \to 2} g(x) = 4$
    $g(2)$ is not defined

(C) $\lim\limits_{x \to 0} h(x)$ does not exist
    $h(0)$ is not defined

FIGURE 3

Most graphs of natural phenomena are continuous, whereas many graphs in business and economics applications have discontinuities. Figure 4A illustrates temperature variation over a 24 hour period — a continuous phenomenon. Figure 4B illustrates warehouse inventory over a 1 week period — a discontinuous phenomenon.

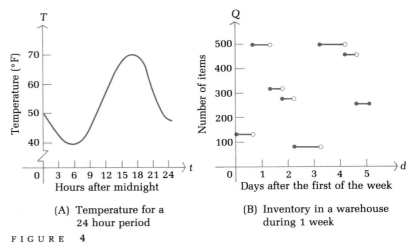

(A) Temperature for a
    24 hour period

(B) Inventory in a warehouse
    during 1 week

FIGURE 4

The preceding discussion, examples, and figures lead to the following formal definition of continuity:

Continuity

A function f is **continuous at the point $x = c$** if

1. $\lim\limits_{x \to c} f(x)$ exists   2. $f(c)$ exists   3. $\lim\limits_{x \to c} f(x) = f(c)$

A function is **continuous on the open interval (a, b)** if it is continuous at each point on the interval.

If one or more of the three conditions in the definition fails, then the function is **discontinuous** at $x = c$.

◆ E X A M P L E   6   Using the definition of continuity, discuss the continuity of each function at the indicated point(s).

(A) $f(x) = x + 2$ at $x = 2$ (B) $g(x) = \dfrac{x^2 - 4}{x - 2}$ at $x = 2$

(C) $h(x) = \dfrac{|x|}{x}$ at $x = 0$ and at $x = 1$

Solutions   (A) f is continuous at $x = 2$, since

$$\lim_{x \to 2} f(x) = 4 = f(2)$$

(See Example 2 and Figure 3A.)

(B) g is not continuous at $x = 2$, since $g(2) = 0/0$ is not defined. (See Example 3 and Figure 3B.)

(C) h is not continuous at $x = 0$, since $h(0) = |0|/0$ is not defined; also, $\lim_{x \to 0} h(x)$ does not exist.

h is continuous at $x = 1$, since

$$\lim_{x \to 1} \frac{|x|}{x} = 1 = h(1)$$

(See Example 4 and Figure 3C.)   ◆

P R O B L E M   6   Using the definition of continuity, discuss the continuity of each function at the indicated point(s). (Compare with Problems 2–4.)

(A) $f(x) = x + 1$ at $x = 1$ (B) $g(x) = \dfrac{x^2 - 1}{x - 1}$ at $x = 1$

(C) $h(x) = \dfrac{x - 2}{|x - 2|}$ at $x = 2$ and at $x = 0$ ◆

We can also talk about one-sided continuity, just as we talked about one-sided limits. For example, a function is said to be **continuous on the right** at $x = c$ if $\lim_{x \to c^+} f(x) = f(c)$ and **continuous on the left** at $x = c$ if $\lim_{x \to c^-} f(x) = f(c)$. A function is **continuous on the closed interval [a, b]** if it is continuous on the open interval (a, b) and is continuous on the right at a and continuous on the left at b.

Figure 5A (at the top of the next page) illustrates a function that is continuous on the closed interval $[-1, 1]$. Figure 5B illustrates a function that is continuous on a half-closed interval $[0, \infty)$.

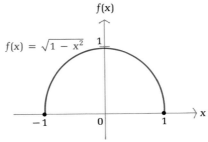

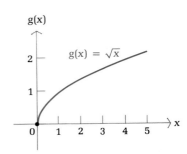

(A) f is continuous on the
closed interval $[-1, 1]$

(B) g is continuous on the
half–closed interval $[0, \infty)$

FIGURE 5
Continuity on closed and half-closed intervals

◆ CONTINUITY PROPERTIES

Functions have useful continuity properties:

If two functions are continuous on the same interval, then their sum, difference, product, and quotient are continuous on the same interval, except for values of x that make a denominator 0.

These properties, along with Theorem 1 below, enable us to determine intervals of continuity for some important classes of functions without having to look at their graphs or use the three conditions in the definition.

THEOREM 1 | **Continuity Properties of Some Specific Functions**

(A) A constant function $f(x) = k$, where k is a constant, is continuous for all x.

$f(x) = 7$ is continuous for all x.

(B) For n a positive integer, $f(x) = x^n$ is continuous for all x.

$f(x) = x^5$ is continuous for all x.

(C) A polynomial function is continuous for all x.

$2x^3 - 3x^2 + x - 5$ is continuous for all x.

(D) A rational function is continuous for all x except those values that make a denominator 0.

$\dfrac{x^2 + 1}{x - 1}$ is continuous for all x except $x = 1$, a value that makes the denominator 0.

(E) For n an odd positive integer greater than 1, $\sqrt[n]{f(x)}$ is continuous wherever $f(x)$ is continuous.

$\sqrt[3]{x^2}$ is continuous for all x.

(F) For n an even positive integer, $\sqrt[n]{f(x)}$ is continuous wherever $f(x)$ is continuous and nonnegative.

$\sqrt[4]{x}$ is continuous on the interval $[0, \infty)$.

Notice that Theorem 1C follows from parts A, B, and the general continuity properties stated above. Also, note that part D follows from part C and the general continuity properties, since a rational function is a function that can be expressed as the quotient of two polynomials.

◆ E X A M P L E   7   Using Theorem 1 and the general properties of continuity, determine where each function is continuous.

(A) $f(x) = x^2 - 2x + 1$

(B) $f(x) = \dfrac{x}{(x + 2)(x - 3)}$

(C) $f(x) = \sqrt[3]{x^2 - 4}$

(D) $f(x) = \sqrt{x - 2}$

Solutions   (A) Since f is a polynomial function, f is continuous for all x.

(B) Since f is a rational function, f is continuous for all x except -2 and 3 (values that make the denominator 0).

(C) The polynomial function $x^2 - 4$ is continuous for all x. Since $n = 3$ is odd, f is continuous for all x.

(D) The polynomial function $x - 2$ is continuous for all x and nonnegative for $x \geqslant 2$. Since $n = 2$ is even, f is continuous for $x \geqslant 2$, or on the interval $[2, \infty)$.

◆

P R O B L E M   7   Using Theorem 1 and the general properties of continuity, determine where each function is continuous.

(A) $f(x) = x^4 + 2x^2 + 1$

(B) $f(x) = \dfrac{x^2}{(x + 1)(x - 4)}$

(C) $f(x) = \sqrt{x - 4}$

(D) $f(x) = \sqrt[3]{x^3 + 1}$ ◆

◆ APPLICATION

A bicycle messenger service in a downtown financial district uses the weight of a package to determine the charge for a delivery. The charge is $12 for the first pound (or any fraction thereof) and $1 for each additional pound (or fraction thereof) up to 10 pounds. If $C(x)$ is the charge for delivering a package weighing x pounds, then

$$C(x) = \begin{cases} \$12 & \text{if } 0 < x \leqslant 1 \\ \$13 & \text{if } 1 < x \leqslant 2 \\ \$14 & \text{if } 2 < x \leqslant 3 \\ \text{and so on} \end{cases}$$

The function C is graphed for $0 < x \leqslant 3$ in Figure 6 (at the top of the next page).

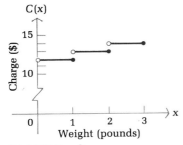

FIGURE 6
Delivery cost function

The following are a few observations about limits and continuity relative to the cost function C over the interval $0 < x \leqslant 3$:

1. $\lim\limits_{x \to 1.5^-} C(x) = 13$

   $\lim\limits_{x \to 1.5^+} C(x) = 13$

   $\lim\limits_{x \to 1.5} C(x) = 13$

   $C(1.5) = 13$

2. $\lim\limits_{x \to 1^-} C(x) = 12$

   $\lim\limits_{x \to 1^+} C(x) = 13$

   $\lim\limits_{x \to 1} C(x)$ does not exist

   $C(1) = 12$

3. $\lim\limits_{x \to 2^-} C(x) = 13$

   $\lim\limits_{x \to 2^+} C(x) = 14$

   $\lim\limits_{x \to 2} C(x)$ does not exist

   $C(2) = 13$

4. C is continuous at $x = 1.5$, since

   $$\lim\limits_{x \to 1.5} C(x) = C(1.5) = 13$$

5. C is discontinuous at 1 and 2, since

   $\lim\limits_{x \to 1} C(x)$ does not exist

   $\lim\limits_{x \to 2} C(x)$ does not exist

6. C is continuous from the left, but not from the right, at 1:

   $\lim\limits_{x \to 1^-} C(x) = 12 = C(1)$

   $\lim\limits_{x \to 1^+} C(x) = 13 \neq C(1) = 12$

7. C is continuous on the intervals $(0, 1]$, $(1, 2]$, and $(2, 3]$.

Answers to Matched Problems

1.

x	−2	−1	0	1	2	3	4
h(x)	1.0	1.5	2.0	2.5	3.0	3.5	4.0

2. (A)

x	0.9	0.99	0.999 → 1 ← 1.001	1.01	1.1
f(x)	1.9	1.99	1.999 → 2 ← 2.001	2.01	2.1

(B) $\lim\limits_{x \to 1} (x + 1) = 2$

(D) $\lim\limits_{x \to 0} (x + 1) = 1$;

   $\lim\limits_{x \to 3} (x + 1) = 4$

(C)

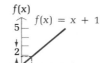

3. (A) $g(x)$

$$g(x) = \frac{x^2 - 1}{x - 1}$$

(Note that 1 is not in the domain of g.)

(B) $\displaystyle\lim_{x \to 1} \frac{x^2 - 1}{x - 1} = \lim_{x \to 1} (x + 1) = 2$

4. $h(x)$

$$h(x) = \frac{x - 2}{|x - 2|}$$

$\displaystyle\lim_{x \to 2} \frac{x - 2}{|x - 2|}$ does not exist

5. (A) $\displaystyle\lim_{x \to -1^-} f(x)$ does not exist   (B) $\displaystyle\lim_{x \to 0^-} f(x) = 0$

$\displaystyle\lim_{x \to -1^+} f(x) = 2$ $\displaystyle\lim_{x \to 0^+} f(x) = 0$

$\displaystyle\lim_{x \to -1} f(x)$ does not exist $\displaystyle\lim_{x \to 0} f(x) = 0$

$f(-1) = 2$ $f(0) = 0$

(C) $\displaystyle\lim_{x \to 1^-} f(x) = 1$ (D) $\displaystyle\lim_{x \to 3^-} f(x) = 3$

$\displaystyle\lim_{x \to 1^+} f(x) = 2$ $\displaystyle\lim_{x \to 3^+} f(x) = 3$

$\displaystyle\lim_{x \to 1} f(x)$ does not exist $\displaystyle\lim_{x \to 3} f(x) = 3$

$f(1)$ not defined $f(3) = 2$

6. (A) f is continuous at $x = 1$, since $\lim_{x \to 1} f(x) = 2 = f(1)$. (See answer to Problem 2 above.)

(B) g is not continuous at $x = 1$, since $g(1) = 0/0$ is not defined. (See answer to Problem 3 above.)

(C) h is not continuous at $x = 2$, since $h(2) = 0/|0| = 0/0$ is not defined; also, $\lim_{x \to 2} h(x)$ does not exist. h is continuous at $x = 0$, since $\lim_{x \to 0} h(x) = -1 = h(0)$. (See answer to Problem 4 above.)

7. (A) Since f is a polynomial function, it is continuous for all x.

(B) Since f is a rational function, f is continuous for all x except -1 and 4 (values that make the denominator 0).

(C) The polynomial function $x - 4$ is continuous for all x and nonnegative for $x \geq 4$. Since $n = 2$ is even, f is continuous for $x \geq 4$, or on the interval $[4, \infty)$.

(D) The polynomial function $x^3 + 1$ is continuous for all x. Since $n = 3$ is odd, f is continuous for all x.

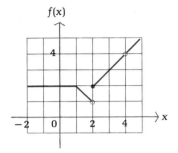

$f(x)$

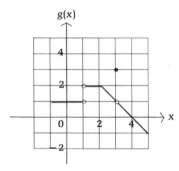

$g(x)$

A

1. Use the graph of the function f shown in the margin to estimate each limit, if it exists.
   (A) $\lim_{x \to 0} f(x)$
   (B) $\lim_{x \to 1} f(x)$
   (C) $\lim_{x \to 2} f(x)$
   (D) $\lim_{x \to 4} f(x)$

2. Use the graph of the function g shown in the margin to estimate each limit, if it exists.
   (A) $\lim_{x \to 0} g(x)$
   (B) $\lim_{x \to 1} g(x)$
   (C) $\lim_{x \to 2} g(x)$
   (D) $\lim_{x \to 3} g(x)$

3. Use the graph of the function f in Problem 1 to estimate each value, if it is defined:
   (A) $f(0)$　　(B) $f(1)$　　(C) $f(2)$　　(D) $f(4)$

4. Use the graph of the function g in Problem 2 to estimate each value, if it is defined:
   (A) $g(0)$　　(B) $g(1)$　　(C) $g(2)$　　(D) $g(3)$

5. Referring to the graph of function f in Problem 1, for which of the values $c = 0, 1, 2,$ or 4 does $\lim_{x \to c} f(x) = f(c)$?

6. Referring to the graph of function g in Problem 2, for which of the values $c = 0, 1, 2,$ or 3 does $\lim_{x \to c} g(x) = g(c)$?

7. For which of the following values of x is the function f in Problem 1 discontinuous?

   0,　1,　2,　4

8. For which of the following values of x is the function g in Problem 2 discontinuous?

   0,　1,　2,　3

Use Theorem 1 to determine where each function in Problems 9–14 is continuous.

9. $f(x) = 2x - 3$ 　　　10. $g(x) = 3 - 5x$ 　　　11. $h(x) = \dfrac{2}{x - 5}$

12. $k(x) = \dfrac{x}{x + 3}$ 　　13. $g(x) = \dfrac{x - 5}{(x - 3)(x + 2)}$ 　　14. $F(x) = \dfrac{1}{x(x + 7)}$

Problems 15–28 refer to the function f shown in the following graph. Use the graph to estimate limits.

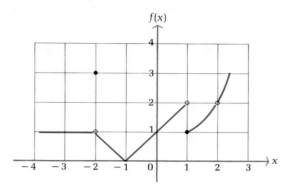

15. (A) $\lim_{x \to 0^-} f(x)$ (B) $\lim_{x \to 0^+} f(x)$ (C) $\lim_{x \to 0} f(x)$

16. (A) $\lim_{x \to -1^-} f(x)$ (B) $\lim_{x \to -1^+} f(x)$ (C) $\lim_{x \to -1} f(x)$

17. (A) $\lim_{x \to 1^-} f(x)$ (B) $\lim_{x \to 1^+} f(x)$ (C) $\lim_{x \to 1} f(x)$

18. (A) $\lim_{x \to 2^-} f(x)$ (B) $\lim_{x \to 2^+} f(x)$ (C) $\lim_{x \to 2} f(x)$

19. (A) $\lim_{x \to -2^-} f(x)$ (B) $\lim_{x \to -2^+} f(x)$ (C) $\lim_{x \to -2} f(x)$

20. (A) $\lim_{x \to 0.5^-} f(x)$ (B) $\lim_{x \to 0.5^+} f(x)$ (C) $\lim_{x \to 0.5} f(x)$

21. (A) $\lim_{x \to 0} f(x)$ (B) $f(0) = ?$ (C) Is f continuous at x = 0?

22. (A) $\lim_{x \to -1} f(x)$ (B) $f(-1) = ?$ (C) Is f continuous at x = −1?

23. (A) $\lim_{x \to 1} f(x)$ (B) $f(1) = ?$ (C) Is f continuous at x = 1?

24. (A) $\lim_{x \to 2} f(x)$ (B) $f(2) = ?$ (C) Is f continuous at x = 2?

25. (A) $\lim_{x \to -2} f(x)$ (B) $f(-2) = ?$ (C) Is f continuous at x = −2?

26. (A) $\lim_{x \to 0.5} f(x)$ (B) $f(0.5) = ?$ (C) Is f continuous at x = 0.5?

27. Function f is discontinuous at which of the following values of x?

   −3, −2, −1

28. Function f is discontinuous at which of the following values of x?

   0, 1, 2

29. Given the following function f:

$$f(x) = \begin{cases} 2 & \text{if x is an integer} \\ 1 & \text{if x is not an integer} \end{cases}$$

(A) Graph f.   (B) $\lim_{x \to 2} f(x) = ?$ (C) $f(2) = ?$

(D) Is f continuous at x = 2?   (E) Where is f discontinuous?

30. Given the following function g:

$$g(x) = \begin{cases} -1 & \text{if x is an even integer} \\ 1 & \text{if x is not an even integer} \end{cases}$$

(A) Graph g.   (B) $\lim\limits_{x \to 1} g(x) = ?$ (C) $g(1) = ?$

(D) Is g continuous at $x = 1$?

(E) Where is g discontinuous?

In Problems 31–38, find all limits that exist using intuition, graphing, or a little algebra, as needed.

31. $\lim\limits_{x \to 2} (2x + 1)$ 32. $\lim\limits_{x \to 1} (3x - 2)$ 33. $\lim\limits_{x \to 2} 7$

34. $\lim\limits_{x \to 5} 9$ 35. $\lim\limits_{x \to -3} \dfrac{x^2 - 9}{x + 3}$ 36. $\lim\limits_{x \to -5} \dfrac{x^2 - 25}{x + 5}$

37. $\lim\limits_{x \to 1^+} \dfrac{|x - 1|}{x - 1}$ 38. $\lim\limits_{x \to 3^-} \dfrac{x - 3}{|x - 3|}$

Use Theorem 1 to determine where each function in Problems 39–46 is continuous. Express the answer in interval notation.

39. $F(x) = 2x^8 - 3x^4 + 5$ 40. $h(x) = \dfrac{x^4 - 3x + 5}{x^2 + 2x}$

41. $g(x) = \sqrt{x - 5}$ 42. $f(x) = \sqrt{3 - x}$

43. $K(x) = \sqrt[3]{x - 5}$ 44. $H(x) = \sqrt[3]{3 - x}$

45. $f(x) = \dfrac{x^2 - 1}{x^2 - 3x + 2}$ 46. $k(x) = \dfrac{x^2 - 4}{x^2 + x - 2}$

Complete the following table for each function in Problems 47–50:

x	0.9	0.99	0.999 → 1 ← 1.001	1.01	1.1
f(x)			→ ? ←		

From the completed table, guess the following (a calculator will be helpful for some):

(A) $\lim\limits_{x \to 1^-} f(x)$ (B) $\lim\limits_{x \to 1^+} f(x)$ (C) $\lim\limits_{x \to 1} f(x)$

47. $f(x) = \dfrac{|x - 1|}{x - 1}$ 48. $f(x) = \dfrac{x - 1}{|x| - 1}$

49. $f(x) = \dfrac{x^3 - 1}{x - 1}$ 50. $f(x) = \dfrac{x^4 - 1}{x - 1}$

In Problems 51–56, graph f and locate all points of discontinuity.

51. $f(x) = \begin{cases} 1 + x & \text{if } x < 1 \\ 5 - x & \text{if } x \geq 1 \end{cases}$

52. $f(x) = \begin{cases} x^2 & \text{if } x \leq 1 \\ 2x & \text{if } x > 1 \end{cases}$

53. $f(x) = \begin{cases} 1 + x & \text{if } x \leq 2 \\ 5 - x & \text{if } x > 2 \end{cases}$

54. $f(x) = \begin{cases} x^2 & \text{if } x \leq 2 \\ 2x & \text{if } x > 2 \end{cases}$

55. $f(x) = \begin{cases} -x & \text{if } x < 0 \\ 1 & \text{if } x = 0 \\ x & \text{if } x > 0 \end{cases}$

56. $f(x) = \begin{cases} 1 & \text{if } x < 0 \\ 0 & \text{if } x = 0 \\ 1 + x & \text{if } x > 0 \end{cases}$

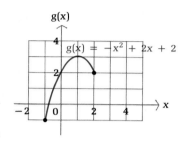

g(x)

$g(x) = -x^2 + 2x + 2$

C

57. Use the graph of the function g shown in the margin to answer the following questions:
   (A) Is g continuous on the open interval $(-1, 2)$?
   (B) Is g continuous from the right at $x = -1$? That is, does $\lim_{x \to -1^+} g(x) = g(-1)$?
   (C) Is g continuous from the left at $x = 2$? That is, does $\lim_{x \to 2^-} g(x) = g(2)$?
   (D) Is g continuous on the closed interval $[-1, 2]$?

58. Use the graph of the function f shown in the margin to answer the following questions:
   (A) Is f continuous on the open interval $(0, 3)$?
   (B) Is f continuous from the right at $x = 0$? That is, does $\lim_{x \to 0^+} f(x) = f(0)$?
   (C) Is f continuous from the left at $x = 3$? That is, does $\lim_{x \to 3^-} f(x) = f(3)$?
   (D) Is f continuous on the closed interval $[0, 3]$?

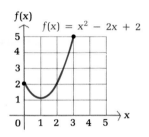

f(x)

$f(x) = x^2 - 2x + 2$

Problems 59 and 60 refer to the **greatest integer function,** which is denoted by $[\![x]\!]$ and is defined as follows:

   $[\![x]\!]$ = Greatest integer $\leq x$

For example,

$$[\![-3.6]\!] = \text{Greatest integer} \leq -3.6 = -4$$
$$[\![2]\!] = \text{Greatest integer} \leq 2 = 2$$
$$[\![2.5]\!] = \text{Greatest integer} \leq 2.5 = 2$$

The graph of $f(x) = [\![x]\!]$ is shown in the margin. There, we can see that

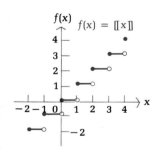

f(x) $f(x) = [\![x]\!]$

$[\![x]\!] = -2$	for	$-2 \leq x < -1$
$[\![x]\!] = -1$	for	$-1 \leq x < 0$
$[\![x]\!] = 0$	for	$0 \leq x < 1$
$[\![x]\!] = 1$	for	$1 \leq x < 2$
$[\![x]\!] = 2$	for	$2 \leq x < 3$

and so on.

59. (A) Is f continuous from the right at $x = 0$?
   (B) Is f continuous from the left at $x = 0$?
   (C) Is f continuous on the open interval $(0, 1)$?
   (D) Is f continuous on the closed interval $[0, 1]$?
   (E) Is f continuous on the half-closed interval $[0, 1)$?

60. (A) Is f continuous from the right at $x = 2$?
  (B) Is f continuous from the left at $x = 2$?
  (C) Is f continuous on the open interval $(1, 2)$?
  (D) Is f continuous on the closed interval $[1, 2]$?
  (E) Is f continuous on the half-closed interval $[1, 2)$?

Use your intuition and current knowledge about continuous functions to answer Problems 61 and 62. (The ideas in these two problems will be generalized in later sections.)

61. If a function f is continuous and never equal to 0 on the interval $(1, 5)$, and if $f(2) = 3$, can $f(4)$ be negative? Can $f(x)$ be negative for any x on the interval $(1, 5)$?

62. If a function g is continuous and never equal to 0 on the interval $(-1, 8)$, and if $g(0) = -3$, can $g(1)$ be positive? Is $g(x)$ negative for all x on the interval $(-1, 8)$?

APPLICATIONS

Business & Economics

63. *Postal rates.* First-class postage in 1989 was \$0.25 for the first ounce (or any fraction thereof) and \$0.20 for each additional ounce (or fraction thereof) up to 11 ounces. If $P(x)$ is the amount of postage for a letter weighing x ounces, then we can write

$$P(x) = \begin{cases} \$0.25 & \text{if } 0 < x \leq 1 \\ \$0.45 & \text{if } 1 < x \leq 2 \\ \$0.65 & \text{if } 2 < x \leq 3 \\ \text{and so on} \end{cases}$$

(A) Graph P for $0 < x \leq 5$.
(B) Find $\lim_{x \to 4.5} P(x)$ and $P(4.5)$.
(C) Find $\lim_{x \to 4} P(x)$ and $P(4)$.
(D) Is P continuous at $x = 4.5$? At $x = 4$?

64. *Telephone rates.* A person placing a station-to-station call on Saturday from San Francisco to New York is charged \$0.30 for the first minute (or any fraction thereof) and \$0.20 for each additional minute (or fraction thereof). If the length of a call is x minutes, then the long-distance charge $R(x)$ is

$$R(x) = \begin{cases} \$0.30 & \text{if } 0 < x \leq 1 \\ \$0.50 & \text{if } 1 < x \leq 2 \\ \$0.70 & \text{if } 2 < x \leq 3 \\ \text{and so on} \end{cases}$$

(A) Graph R for $0 < x \le 6$.

(B) Find $\lim_{x \to 2.5} R(x)$ and $R(2.5)$.

(C) Find $\lim_{x \to 2} R(x)$ and $R(2)$.

(D) Is R continuous at $x = 2.5$? At $x = 2$?

65. *Income.* A personal computer salesperson receives a base salary of \$1,000 per month and a commission of 5% of all sales over \$10,000 during the month. If the monthly sales are \$20,000 or more, the salesperson is given an additional \$500 bonus. Let $E(s)$ represent the person's earnings during the month as a function of the monthly sales s.

(A) Graph $E(s)$ for $0 \le s \le 30,000$.

(B) Find $\lim_{s \to 10,000} E(s)$ and $E(10,000)$.

(C) Find $\lim_{s \to 20,000} E(s)$ and $E(20,000)$.

(D) Is E continuous at $s = 10,000$? At $s = 20,000$?

66. *Equipment rental.* An office equipment rental and leasing company rents electric typewriters for \$10 per day (and any fraction thereof) or for \$50 per 7 day week. Let $C(x)$ be the cost of renting a typewriter for x days.

(A) Graph $C(x)$ for $0 \le x \le 10$.

(B) Find $\lim_{x \to 4.5} C(x)$ and $C(4.5)$.

(C) Find $\lim_{x \to 8} C(x)$ and $C(8)$.

(D) Is C continuous at $x = 4.5$? At $x = 8$?

Life Sciences

67. *Animal supply.* A medical laboratory raises its own rabbits. The number of rabbits $N(t)$ available at any time t depends on the number of births and deaths. When a birth or death occurs, the function N generally has a discontinuity, as shown in the figure.

(A) Where is the function N discontinuous?

(B) $\lim_{t \to t_5} N(t) = ?$; $N(t_5) = ?$ (C) $\lim_{t \to t_3} N(t) = ?$; $N(t_3) = ?$

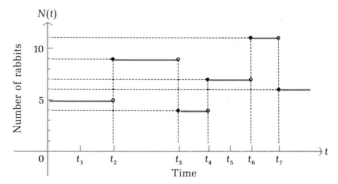

Social Sciences

68. *Learning.* The graph shown on the next page might represent the history of a particular person learning the material on limits and continuity in this book. At time t_2, the student's mind goes blank during a quiz. At time t_4, the instructor explains a concept particularly well, and suddenly, a big jump in understanding takes place.

(A) Where is the function p discontinuous?
(B) $\lim_{t \to t_1} p(t) = ?; \quad p(t_1) = ?$
(C) $\lim_{t \to t_2} p(t) = ?; \quad p(t_2) = ?$
(D) $\lim_{t \to t_4} p(t) = ?; \quad p(t_4) = ?$

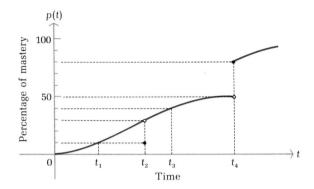

Computation of Limits

◆ LIMITS AT POINTS OF CONTINUITY
◆ LIMITS AND INFINITY
◆ LIMIT PROPERTIES
◆ LIMITS AT INFINITY
◆ SUMMARY OF LIMIT FORMS

To introduce the concept of limit in the preceding section, we relied heavily on tables, graphs, and intuition. But we will find it very useful to be able to compute limits of functions without having to resort to graphs or tables. In this section, we will introduce and use various properties of limits to find limits of a wide variety of functions, in the same way we used continuity properties earlier. We will also discuss the use of the infinity symbol, ∞, relative to limits.

◆ LIMITS AT POINTS OF CONTINUITY

If we know that a function is continuous at a point, then, using the definition of continuity at a point from the preceding section, it is easy to find the limit of the function at that point—we simply evaluate the function at that point.

If a function f is continuous at $x = c$, then

$$\lim_{x \to c} f(x) = f(c)$$

For example,

$$\lim_{x \to 2} (x^2 - x + 1) = 2^2 - 2 + 1 = 3$$

(since a polynomial function is continuous for all x).

EXAMPLE 8 Find the following limits:

(A) $\lim_{x \to 3} \dfrac{x^2 - 3x}{x + 7}$ (B) $\lim_{x \to 5} \sqrt{x - 2}$ (C) $\lim_{x \to 3} 5$

Solutions   (A) Since $(x^2 - 3x)/(x + 7)$ is continuous for all x except $x = -7$, it is continuous at $x = 3$. Thus,

$$\lim_{x \to 3} \frac{x^2 - 3x}{x + 7} = \frac{3^2 - 3(3)}{3 + 7} = \frac{9 - 9}{10} = \frac{0}{10} = 0$$

(B) Since $x - 2$ is continuous for all x and nonnegative for $x \geq 2$, $\sqrt{x - 2}$ is continuous for all $x \geq 2$. In particular, $\sqrt{x - 2}$ is continuous for $x = 5$. Thus,

$$\lim_{x \to 5} \sqrt{x - 2} = \sqrt{5 - 2} = \sqrt{3}$$

(C) Since $f(x) = 5$ is a constant function, it is continuous for all x. In particular, it is continuous when $x = 3$. Thus,

$$\lim_{x \to 3} f(x) = f(3) = 5$$

That is, $\lim_{x \to 3} 5 = 5$.

PROBLEM 8 Find the following limits:

(A) $\lim_{x \to 4} (x + \sqrt{x})$ (B) $\lim_{x \to -7} \sqrt[3]{1 - x}$ (C) $\lim_{x \to -2} 10$

◆ **LIMITS AND INFINITY**

How does $f(x) = 1/x$ behave for x near 0? When a number gets small (close to 0), the reciprocal of its absolute value gets large. For example, when $x = 0.001$, $1/x = 1,000$; when $x = -0.001$, $1/x = -1,000$. Thus, as x approaches 0 from the right, $f(x)$ is positive and increases without bound. We indicate this behavior by writing

$$\lim_{x \to 0^+} \frac{1}{x} = \infty$$

Similarly, as x approaches 0 from the left, $f(x)$ is negative and decreases without bound. We indicate this behavior by writing

$$\lim_{x \to 0^-} \frac{1}{x} = -\infty$$

In neither case does the limit exist, since neither ∞ nor $-\infty$ are numbers. Nevertheless, the notation tells us something about the behavior of $f(x)$ for x near 0. The graph of f is illustrated in Figure 7A. Figure 7B illustrates the behavior of $g(x) = 1/x^2$ for x near 0. Note that for $g(x)$ we can write

$$\lim_{x \to 0} \frac{1}{x^2} = \infty$$

since $g(x)$ remains positive and increases without bound as x approaches 0 from either side.

For both functions f and g, the line $x = 0$ (the vertical axis) is called a *vertical asymptote*. In general, a line $x = a$ is a **vertical asymptote** for the graph of $y = f(x)$ if $f(x)$ either increases or decreases without bound as x approaches a

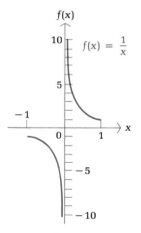

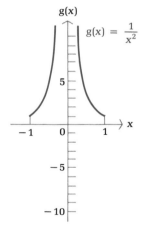

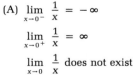

(A) $\lim\limits_{x \to 0^-} \dfrac{1}{x} = -\infty$

$\lim\limits_{x \to 0^+} \dfrac{1}{x} = \infty$

$\lim\limits_{x \to 0} \dfrac{1}{x}$ does not exist

(B) $\lim\limits_{x \to 0^-} \dfrac{1}{x^2} = \infty$

$\lim\limits_{x \to 0^+} \dfrac{1}{x^2} = \infty$

$\lim\limits_{x \to 0} \dfrac{1}{x^2} = \infty$

FIGURE 7

from either the left or the right. Symbolically, $x = a$ is a vertical asymptote if any of the following hold:

$$\lim_{x \to a^-} f(x) = -\infty \text{ (or } \infty)$$

$$\lim_{x \to a^+} f(x) = -\infty \text{ (or } \infty)$$

$$\lim_{x \to a} f(x) = -\infty \text{ (or } \infty)$$

◆ LIMIT PROPERTIES

We now turn to some basic properties of limits that will enable us to evaluate many limits when either the function involved is not continuous at the point or it is not certain whether the function is continuous at the point. [Remember, if a function f is continuous at $x = c$, then we can find $\lim_{x \to c} f(x)$ by evaluating the function f at $x = c$.]

THEOREM 2

Properties of Limits

Let f and g be two functions, and assume that

$$\lim_{x \to c} f(x) = L \qquad \lim_{x \to c} g(x) = M$$

where L and M are real numbers (both limits exist). Then,

1. $\lim_{x \to c} [f(x) + g(x)] = \lim_{x \to c} f(x) + \lim_{x \to c} g(x) = L + M$

2. $\lim_{x \to c} [f(x) - g(x)] = \lim_{x \to c} f(x) - \lim_{x \to c} g(x) = L - M$

3. $\lim_{x \to c} kf(x) = k \lim_{x \to c} f(x) = kL \qquad$ for any constant k

4. $\lim_{x \to c} [f(x) \cdot g(x)] = \left[\lim_{x \to c} f(x) \right]\left[\lim_{x \to c} g(x) \right] = LM$

5. $\lim_{x \to c} \dfrac{f(x)}{g(x)} = \dfrac{\lim\limits_{x \to c} f(x)}{\lim\limits_{x \to c} g(x)} = \dfrac{L}{M} \qquad$ if $M \neq 0$

6. $\lim_{x \to c} \sqrt[n]{f(x)} = \sqrt[n]{\lim\limits_{x \to c} f(x)} = \sqrt[n]{L} \qquad L > 0$ for n even

Suppose

$$\lim_{x \to 2} f(x) = 3 \qquad \text{and} \qquad \lim_{x \to 2} g(x) = 7$$

then no other information is needed about the functions f and g to determine that

$$\lim_{x \to 2} [f(x) - g(x)] = \lim_{x \to 2} f(x) - \lim_{x \to 2} g(x) = 3 - 7 = -4$$

$$\lim_{x \to 2} 9f(x) = 9 \lim_{x \to 2} f(x) = 9 \cdot 3 = 27$$

$$\lim_{x \to 2} \frac{f(x)}{g(x)} = \frac{\lim_{x \to 2} f(x)}{\lim_{x \to 2} g(x)} = \frac{3}{7}$$

$$\lim_{x \to 2} \sqrt{f(x)} = \sqrt{\lim_{x \to 2} f(x)} = \sqrt{3}$$

Now, suppose

$$\lim_{x \to c} f(x) = 0 \quad \text{and} \quad \lim_{x \to c} g(x) = 0$$

then finding

$$\lim_{x \to c} \frac{f(x)}{g(x)} \tag{1}$$

may present some difficulties, since limit property 5 (the limit of a quotient) does not apply when $\lim_{x \to c} g(x) = 0$. We often have to use algebraic manipulation or other devices to determine the outcome. Recall from Examples 3 and 4 (in Section 9-1) that

$$\lim_{x \to 2} \frac{x^2 - 4}{x - 2} = \lim_{x \to 2} \frac{(x - 2)(x + 2)}{x - 2} = \lim_{x \to 2} (x + 2) = 4$$

and

$$\lim_{x \to 0} \frac{|x|}{x} \quad \text{does not exist}$$

From these two examples, it is clear that knowing only that $\lim_{x \to c} f(x) = 0$ and $\lim_{x \to c} g(x) = 0$ is not enough to determine limit (1). Depending on the choice of functions f and g, the limit (1) may or may not exist. Consequently, if we are given (1) and $\lim_{x \to c} f(x) = 0$ and $\lim_{x \to c} g(x) = 0$, then (1) is said to be **indeterminate,** or, more specifically, a **0/0 indeterminate form.**

The 0/0 indeterminate form may seem unusual at first—a form one might not encounter too often—but it turns out that this is not the case. This form turns up in a number of very important developments in calculus. Example 9 provides a preview of a form that will be considered in detail in the following sections.

◆ E X A M P L E 9   Find $\lim_{h \to 0} \dfrac{f(2 + h) - f(2)}{h}$ for:

(A) $f(x) = 3x - 4$ (B) $f(x) = \sqrt{x}$ (C) $f(x) = |x - 2|$

Solutions   (A) $\lim_{h \to 0} \dfrac{f(2 + h) - f(2)}{h}$ $f(x) = 3x - 4$

$$= \lim_{h \to 0} \frac{[3(2 + h) - 4] - [3(2) - 4]}{h}$$

Since this is a 0/0 indeterminate form and property 5 in Theorem 2 does not apply, we proceed with algebraic simplification.

$$= \lim_{h \to 0} \frac{6 + 3h - 4 - 6 + 4}{h}$$

$$= \lim_{h \to 0} \frac{3h}{h} = \lim_{h \to 0} 3 = 3 \qquad h \neq 0$$

(B) $\lim_{h \to 0} \dfrac{f(2 + h) - f(2)}{h}$ $\qquad\qquad f(x) = \sqrt{x}$

$$= \lim_{h \to 0} \frac{\sqrt{2 + h} - \sqrt{2}}{h}$$

This is a $0/0$ indeterminate form, so property 5 in Theorem 2 does not apply. Rationalizing the numerator will be of help.

$$= \lim_{h \to 0} \frac{\sqrt{2 + h} - \sqrt{2}}{h} \cdot \frac{\sqrt{2 + h} + \sqrt{2}}{\sqrt{2 + h} + \sqrt{2}}$$

$(A - B)(A + B) = A^2 - B^2$

$$= \lim_{h \to 0} \frac{2 + h - 2}{h(\sqrt{2 + h} + \sqrt{2})}$$

$$= \lim_{h \to 0} \frac{1}{\sqrt{2 + h} + \sqrt{2}}$$

Use properties 1, 5, and 6 ($h \neq 0$).

$$= \frac{1}{\sqrt{2} + \sqrt{2}} = \frac{1}{2\sqrt{2}}$$

(C) $\lim_{h \to 0} \dfrac{f(2 + h) - f(2)}{h}$ $\qquad\qquad f(x) = |x - 2|$

$$= \lim_{h \to 0} \frac{|(2 + h) - 2| - |2 - 2|}{h}$$

Since this is a $0/0$ indeterminate form and property 5 in Theorem 2 does not apply, we proceed with algebraic simplification.

$$= \lim_{h \to 0} \frac{|h|}{h} \quad \text{does not exist}$$

See Example 4. ◆

PROBLEM 9 Find $\lim\limits_{h \to 0} \dfrac{g(1 + h) - g(1)}{h}$ for:

(A) $g(x) = 1 - x$ \qquad (B) $g(x) = \sqrt{x}$ \qquad (C) $g(x) = |x - 1|$ ◆

◆ LIMITS AT INFINITY

Earlier in this section, we discussed the behavior of $f(x) = 1/x$ for x near 0. We now investigate the behavior of $f(x)$ as x increases or decreases without bound. When the absolute value of a number increases, its reciprocal gets smaller. For example, if $x = 10,000$, then $1/x = 0.0001$; if $x = -10,000$, then $1/x = -0.0001$. Thus, as x increases without bound, indicated by $x \to \infty$, $f(x)$ approaches 0. We indicate this behavior by writing

$$\lim_{x \to \infty} \frac{1}{x} = 0$$

Similarly, as x decreases without bound, indicated by $x \to -\infty$, $f(x)$ approaches 0. We indicate this behavior by writing

$$\lim_{x \to -\infty} \frac{1}{x} = 0$$

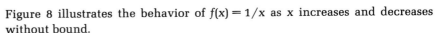

$f(x) = \frac{1}{x}$ $\lim_{x \to \infty} \frac{1}{x} = 0$

$\lim_{x \to -\infty} \frac{1}{x} = 0$

FIGURE 8

Figure 8 illustrates the behavior of $f(x) = 1/x$ as x increases and decreases without bound.

It is important to understand that the symbol ∞ does not represent an actual number that x is approaching, but is used to indicate only that the value of x is increasing with no upper limit on its size. In particular, the statement "$x = \infty$" is meaningless, since ∞ is not a symbol for a real number.

For the function f in Figure 8, the line $y = 0$ (the x axis) is called a *horizontal asymptote*. In general, a line $y = b$ is a **horizontal asymptote** for the graph of $y = f(x)$ if $f(x)$ approaches b as x either increases or decreases without bound. Symbolically, $y = b$ is a horizontal asymptote if either

$$\lim_{x \to -\infty} f(x) = b \quad \text{or} \quad \lim_{x \to \infty} f(x) = b$$

In the first case, the graph of f will be close to the horizontal line $y = b$ for large (in absolute value) negative x. In the second case, the graph will be close to the horizontal line $y = b$ for large positive x.

How can we evaluate limits of the form $\lim_{x \to -\infty} f(x)$ and $\lim_{x \to \infty} f(x)$ without drawing graphs or performing complicated calculations? Fortunately, **limit properties 1–6, listed earlier in this section in Theorem 2, are valid if we replace the statement $x \to c$ with $x \to -\infty$ or $x \to \infty$.** These properties, together with Theorem 3, enable us to evaluate limits at infinity for many functions.

THEOREM 3

Two Special Limits at Infinity

If p is a positive real number, and k is any real constant, then

$$\lim_{x \to -\infty} \frac{k}{x^p} = 0 \quad \text{and} \quad \lim_{x \to \infty} \frac{k}{x^p} = 0$$

provided that x^p names a real number for negative values of x.

Figure 9 illustrates Theorem 3 for several values of p. (Note that the x axis is a horizontal asymptote in all three cases.)

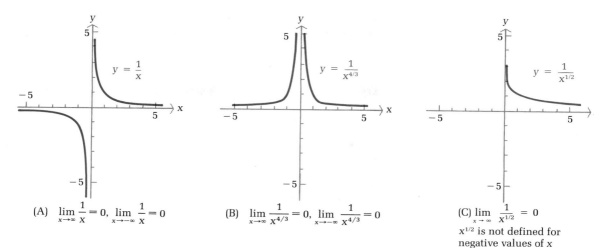

(A) $\lim\limits_{x\to\infty} \dfrac{1}{x} = 0, \ \lim\limits_{x\to-\infty} \dfrac{1}{x} = 0$

(B) $\lim\limits_{x\to\infty} \dfrac{1}{x^{4/3}} = 0, \ \lim\limits_{x\to-\infty} \dfrac{1}{x^{4/3}} = 0$

(C) $\lim\limits_{x\to\infty} \dfrac{1}{x^{1/2}} = 0$

$x^{1/2}$ is not defined for negative values of x

FIGURE 9

Now, consider the following limit of a polynomial:

$$\lim_{x\to\infty} (2x^3 - 4x^2 - 9x - 100)$$

At first glance, it is not clear which part of the polynomial, $-4x^2 - 9x - 100$ or $2x^3$, dominates as x increases without bound. To help resolve the conflict, we factor out the highest power of x (x^3):

$$\lim_{x\to\infty} x^3 \left(2 - \frac{4}{x} - \frac{9}{x^2} - \frac{100}{x^3} \right)$$

Looking at the factors separately, we have

$$\lim_{x\to\infty} x^3 = \infty$$

and (using Theorems 2 and 3)

$$\lim_{x\to\infty} \left(2 - \frac{4}{x} - \frac{9}{x^2} - \frac{100}{x^3} \right) \quad \boxed{= 2 - 0 - 0 - 0} = 2$$

Now, intuitively, we would like to write

$$\lim_{x\to\infty} x^3 \left(2 - \frac{4}{x} - \frac{9}{x^2} - \frac{100}{x^3} \right) = \infty$$

because if one factor is increasing without bound and at the same time the other factor is approaching a positive constant, it seems reasonable to think that their product must increase without bound. This reasoning turns out to be correct; that is:

If $\quad \lim\limits_{x\to\infty} f(x) = \infty \quad$ and $\quad \lim\limits_{x\to\infty} g(x) = L > 0$

then $\quad \lim\limits_{x\to\infty} [f(x) \cdot g(x)] = \infty$

This statement also holds if $x \to \infty$ is replaced with $x \to -\infty$ or $x \to c$, for c any real number. (Other variations of this statement are also possible.)

We are now ready to turn our attention to the limits of rational functions as x increases without bound.

◆ E X A M P L E   10   Find the following limits, if they exist.

(A) $\lim\limits_{x\to\infty} \dfrac{5x^2 + 3}{3x^2 - 2}$ (B) $\lim\limits_{x\to\infty} \dfrac{3x^4 - x^2 + 1}{8x^6 - 10}$ (C) $\lim\limits_{x\to\infty} \dfrac{2x^5 - x^3 - 1}{6x^3 + 2x^2 - 7}$

Solutions   In all three cases, we begin (as in the polynomial example above) by factoring out the highest power of x from the numerator and the highest power of x from the denominator.

(A) $\lim\limits_{x\to\infty} \dfrac{5x^2 + 3}{3x^2 - 2} = \lim\limits_{x\to\infty} \dfrac{x^2\left(5 + \dfrac{3}{x^2}\right)}{x^2\left(3 - \dfrac{2}{x^2}\right)} = \lim\limits_{x\to\infty} \dfrac{5 + \dfrac{3}{x^2}}{3 - \dfrac{2}{x^2}}$ Use Theorems 2 and 3.

$= \dfrac{5 + 0}{3 - 0} = \dfrac{5}{3}$

(B) $\lim\limits_{x\to\infty} \dfrac{3x^4 - x^2 + 1}{8x^6 - 10} = \lim\limits_{x\to\infty} \dfrac{x^4\left(3 - \dfrac{1}{x^2} + \dfrac{1}{x^4}\right)}{x^6\left(8 - \dfrac{10}{x^6}\right)}$ Use Theorems 2 and 3.

$= \lim\limits_{x\to\infty} \dfrac{1}{x^2} \cdot \lim\limits_{x\to\infty} \dfrac{3 - \dfrac{1}{x^2} + \dfrac{1}{x^4}}{8 - \dfrac{10}{x^6}} = 0 \cdot \dfrac{3 - 0 + 0}{8 - 0} = 0 \cdot \dfrac{3}{8} = 0$

(C) $\lim\limits_{x\to\infty} \dfrac{2x^5 - x^3 - 1}{6x^3 + 2x^2 - 7} = \lim\limits_{x\to\infty} \dfrac{x^5\left(2 - \dfrac{1}{x^2} - \dfrac{1}{x^5}\right)}{x^3\left(6 + \dfrac{2}{x} - \dfrac{7}{x^3}\right)}$

$= \lim\limits_{x\to\infty} x^2 \left(\dfrac{2 - \dfrac{1}{x^2} - \dfrac{1}{x^5}}{6 + \dfrac{2}{x} - \dfrac{7}{x^3}}\right)$

$= \infty$ Since $\lim\limits_{x\to\infty} x^2 = \infty$ and $\lim\limits_{x\to\infty} \dfrac{2 - \dfrac{1}{x^2} - \dfrac{1}{x^5}}{6 + \dfrac{2}{x} - \dfrac{7}{x^3}} = \dfrac{2}{6}$ ◆

P R O B L E M   10   Find the following limits, if they exist:

(A) $\lim\limits_{x\to\infty} \dfrac{4x^3 - 5x + 8}{2x^4 - 7}$ (B) $\lim\limits_{x\to\infty} \dfrac{5x^6 + 3x}{2x^5 - x - 5}$ (C) $\lim\limits_{x\to\infty} \dfrac{2x^3 - x + 7}{4x^3 + 3x^2 - 100}$ ◆

◆ SUMMARY OF LIMIT FORMS

Table 1 provides a summary of some of the limit forms we have considered.

TABLE 1

Summary of Some Important Limit Forms

NOTATION	DESCRIPTION	EXAMPLE
$\lim_{x \to c} f(x) = L$	$f(x)$ approaches L as x approaches c from either side of c, but is not equal to c	$\lim_{x \to 2} \dfrac{x^2 - 4}{x - 2} = 4$
$\lim_{x \to c^-} f(x) = L$	$f(x)$ approaches L as x approaches c from the left	$\lim_{x \to 0^-} \dfrac{|x|}{x} = -1$
$\lim_{x \to c^+} f(x) = L$	$f(x)$ approaches L as x approaches c from the right	$\lim_{x \to 0^+} \dfrac{|x|}{x} = 1$
$\lim_{x \to c^-} f(x) = -\infty$	$f(x)$ decreases without bound as x approaches c from the left	$\lim_{x \to 0^-} \dfrac{1}{x} = -\infty$
$\lim_{x \to c^+} f(x) = \infty$	$f(x)$ increases without bound as x approaches c from the right	$\lim_{x \to 0^+} \dfrac{1}{x} = \infty$
$\lim_{x \to c} f(x) = \infty$	$f(x)$ increases without bound as x approaches c from either side of c, but is not equal to c	$\lim_{x \to 0} \dfrac{1}{x^2} = \infty$
$\lim_{x \to \infty} f(x) = L$	$f(x)$ approaches L as x increases without bound	$\lim_{x \to \infty} \dfrac{2x^2 + 1}{3x^2 - 1} = \dfrac{2}{3}$
$\lim_{x \to \infty} f(x) = \infty$	$f(x)$ increases without bound as x increases without bound	$\lim_{x \to \infty} \dfrac{2x^3}{x^2 - 7} = \infty$

Answers to Matched Problems

8. (A) $\lim_{x \to 4} (x + \sqrt{x}) = 4 + \sqrt{4} = 6$ (B) $\lim_{x \to -7} \sqrt[3]{1 - x} = \sqrt[3]{1 - (-7)} = 2$

(C) $\lim_{x \to -2} 10 = 10$

9. (A) -1 (B) $\frac{1}{2}$ (C) Does not exist

10. (A) 0 (B) ∞ (does not exist) (C) $\frac{1}{2}$

EXERCISE 9-2

A Given $\lim_{x \to 3} f(x) = 5$ and $\lim_{x \to 3} g(x) = 9$, find the indicated limits in Problems 1–10.

1. $\lim_{x \to 3} [f(x) - g(x)]$

2. $\lim_{x \to 3} [f(x) + g(x)]$

3. $\lim_{x \to 3} 4g(x)$

4. $\lim_{x \to 3} (-2)f(x)$

5. $\lim_{x \to 3} \dfrac{f(x)}{g(x)}$

6. $\lim_{x \to 3} [f(x) \cdot g(x)]$

7. $\lim_{x \to 3} \sqrt{f(x)}$

8. $\lim_{x \to 3} \sqrt{g(x)}$

9. $\lim_{x \to 3} \dfrac{f(x) + g(x)}{2f(x)}$

10. $\lim_{x \to 3} \dfrac{g(x) - f(x)}{3g(x)}$

Find each limit in Problems 11–26, if it exists. (Algebraic manipulation is useful in some cases.)

11. $\lim\limits_{x \to 5} (2x^2 - 3)$

12. $\lim\limits_{x \to 2} (x^2 - 8x + 2)$

13. $\lim\limits_{x \to 4} (x^2 - 5x)$

14. $\lim\limits_{x \to -2} (3x^3 - 9)$

15. $\lim\limits_{x \to 2} \dfrac{5x}{2 + x^2}$

16. $\lim\limits_{x \to 10} \dfrac{2x + 5}{3x - 5}$

17. $\lim\limits_{x \to 2} (x + 1)^3 (2x - 1)^2$

18. $\lim\limits_{x \to 3} (x + 2)^2 (2x - 4)$

19. $\lim\limits_{x \to 0} \dfrac{x^2 - 3x}{x}$

20. $\lim\limits_{x \to 0} \dfrac{2x^2 + 5x}{x}$

21. $\lim\limits_{x \to \infty} \dfrac{3}{x^2}$

22. $\lim\limits_{x \to \infty} \dfrac{6}{x^4}$

23. $\lim\limits_{x \to \infty} \left(5 - \dfrac{3}{x} + \dfrac{2}{x^2} \right)$

24. $\lim\limits_{x \to \infty} \left(4 + \dfrac{1}{x^2} - \dfrac{3}{x^4} \right)$

25. $\lim\limits_{h \to 0} \dfrac{2(3 + h) - 2(3)}{h}$

26. $\lim\limits_{h \to 0} \dfrac{[(4 + h) - 2] - (4 - 2)}{h}$

B   Find each limit in Problems 27–54, if it exists. (Use algebraic manipulation where necessary. Also, use $-\infty$ or ∞ where appropriate.)

27. $\lim\limits_{x \to 1} (3x^4 - 2x^2 + x - 2)$

28. $\lim\limits_{x \to -1} (5x^3 - 3x^2 - 5x + 3)$

29. $\lim\limits_{x \to 1} \dfrac{x - 2}{x^2 - 2x}$

30. $\lim\limits_{x \to 1} \dfrac{x + 3}{x^2 + 3x}$

31. $\lim\limits_{x \to 2} \dfrac{x - 2}{x^2 - 2x}$

32. $\lim\limits_{x \to -3} \dfrac{x + 3}{x^2 + 3x}$

33. $\lim\limits_{x \to \infty} \dfrac{x - 2}{x^2 - 2x}$

34. $\lim\limits_{x \to \infty} \dfrac{x + 3}{x^2 + 3x}$

35. $\lim\limits_{x \to 2} \dfrac{x^2 - x - 6}{x + 2}$

36. $\lim\limits_{x \to 3} \dfrac{x^2 + x - 6}{x + 3}$

37. $\lim\limits_{x \to -2} \dfrac{x^2 - x - 6}{x + 2}$

38. $\lim\limits_{x \to -3} \dfrac{x^2 + x - 6}{x + 3}$

39. $\lim\limits_{x \to \infty} \dfrac{x^2 - x - 6}{x + 2}$

40. $\lim\limits_{x \to \infty} \dfrac{x^2 + x - 6}{x + 3}$

41. $\lim\limits_{x \to \infty} \dfrac{2x + 4}{x}$

42. $\lim\limits_{x \to \infty} \dfrac{3x^2 + 5}{x^2}$

43. $\lim\limits_{x \to \infty} \dfrac{3x^3 - x + 1}{5x^3 - 7}$

44. $\lim\limits_{x \to \infty} \dfrac{3x^4 - x^3 + 5}{2x^4 - 10}$

45. $\lim\limits_{h \to 0} \dfrac{(2 + h)^2 - 2^2}{h}$

46. $\lim\limits_{h \to 0} \dfrac{(3 + h)^2 - 3^2}{h}$

47. $\lim\limits_{x \to 3} \left(\dfrac{x}{x + 3} + \dfrac{x - 3}{x^2 - 9} \right)$

48. $\lim\limits_{x \to 2} \left(\dfrac{1}{x + 2} + \dfrac{x - 2}{x^2 - 4} \right)$

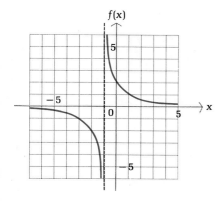

49. $\lim\limits_{x \to 0^-} \dfrac{x-2}{x^2-2x}$

50. $\lim\limits_{x \to 0^-} \dfrac{x+3}{x^2+3x}$

51. $\lim\limits_{x \to 0^+} \dfrac{x-2}{x^2-2x}$

52. $\lim\limits_{x \to 0^+} \dfrac{x+3}{x^2+3x}$

53. $\lim\limits_{x \to 0} \dfrac{x-2}{x^2-2x}$

54. $\lim\limits_{x \to 0} \dfrac{x+3}{x^2+3x}$

Use the graph of the function f shown in the margin to answer Problems 55 and 56.

55. (A) $\lim\limits_{x \to -\infty} f(x) = ?$ (B) $\lim\limits_{x \to \infty} f(x) = ?$

(C) Write the equation of any horizontal asymptote.

56. (A) $\lim\limits_{x \to -1^-} f(x) = ?$ (B) $\lim\limits_{x \to -1^+} f(x) = ?$ (C) $\lim\limits_{x \to -1} f(x) = ?$

(D) Write the equation of any vertical asymptote.

Use the graph of the function g shown in the margin to answer Problems 57 and 58.

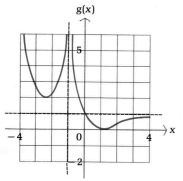

57. (A) $\lim\limits_{x \to -1^-} g(x) = ?$ (B) $\lim\limits_{x \to -1^+} g(x) = ?$ (C) $\lim\limits_{x \to -1} g(x) = ?$

(D) Write the equation of any vertical asymptote.

58. (A) $\lim\limits_{x \to \infty} g(x) = ?$ (B) $\lim\limits_{x \to -\infty} g(x) = ?$

(C) Write the equation of any horizontal asymptote.

Compute the following limit for each function in Problems 59–68:

$$\lim_{h \to 0} \frac{f(2+h)-f(2)}{h}$$

59. $f(x) = 3x + 1$

60. $f(x) = 5x - 1$

61. $f(x) = x^2 + 1$

62. $f(x) = x^2 - 2$

63. $f(x) = 5$

64. $f(x) = -2$

65. $f(x) = \sqrt{x} - 2$

66. $f(x) = 1 + \sqrt{x}$

67. $f(x) = |x - 2| - 3$

68. $f(x) = 2 + |x - 2|$

C   Find each limit in Problems 69–82, if it exists. (Use algebraic manipulation where necessary and $-\infty$ or ∞ where appropriate.) [Recall: $a^3 - b^3 = (a - b)(a^2 + ab + b^2)$ and $a^3 + b^3 = (a + b)(a^2 - ab + b^2)$.]

69. $\lim\limits_{x \to 1} \sqrt{x^2 + 2x}$

70. $\lim\limits_{x \to 1} \sqrt{2x^4 + 1}$

71. $\lim\limits_{x \to 4} \sqrt[3]{x^2 - 3x}$

72. $\lim\limits_{x \to 2} \sqrt[3]{x^3 - 2x}$

73. $\lim\limits_{x \to 2} \dfrac{4}{(x-2)^2}$

74. $\lim\limits_{x \to 0} \dfrac{3}{|x|}$

75. $\lim\limits_{x \to \infty} \left(\dfrac{1}{x^2} + \dfrac{1}{\sqrt{x}} \right)$

76. $\lim\limits_{x \to \infty} \left(\dfrac{1}{\sqrt[3]{x}} - \dfrac{2}{x^3} \right)$

77. $\lim\limits_{x \to 0} \left(\sqrt{x^2 + 9} - \dfrac{x^2 + 3x}{x} \right)$

78. $\lim\limits_{x \to 1} \left(\dfrac{x^2 - 1}{x - 1} + \sqrt{x^2 + 3} \right)$

79. $\displaystyle\lim_{x \to 4} \frac{\sqrt{x} - 2}{x - 4}$

80. $\displaystyle\lim_{x \to 0} \frac{\sqrt{x + 4} - 2}{x}$

81. $\displaystyle\lim_{x \to 2} \frac{x^3 - 8}{x - 2}$

82. $\displaystyle\lim_{x \to -1} \frac{x^2 - 1}{x^3 + 1}$

83. Find each limit, if it exists. (Use $-\infty$ or ∞, as appropriate.)

(A) $\displaystyle\lim_{x \to -2^-} \frac{2}{x + 2}$ (B) $\displaystyle\lim_{x \to -2^+} \frac{2}{x + 2}$ (C) $\displaystyle\lim_{x \to -2} \frac{2}{x + 2}$

(D) Is $x = -2$ a vertical asymptote?

84. Find each limit, if it exists. (Use $-\infty$ or ∞, as appropriate.)

(A) $\displaystyle\lim_{x \to 1^-} \frac{1}{x - 1}$ (B) $\displaystyle\lim_{x \to 1^+} \frac{1}{x - 1}$ (C) $\displaystyle\lim_{x \to 1} \frac{1}{x - 1}$

(D) Is $x = 1$ a vertical asymptote?

Find each limit in Problems 85–88, where a is a real constant.

85. $\displaystyle\lim_{h \to 0} \frac{(a + h)^2 - a^2}{h}$

86. $\displaystyle\lim_{h \to 0} \frac{[3(a + h) - 2] - (3a - 2)}{h}$

87. $\displaystyle\lim_{h \to 0} \frac{\sqrt{a + h} - \sqrt{a}}{h}, \quad a > 0$

88. $\displaystyle\lim_{h \to 0} \frac{\dfrac{1}{a + h} - \dfrac{1}{a}}{h}, \quad a \neq 0$

APPLICATIONS

Business & Economics

89. *Average cost.* The cost equation for manufacturing a particular compact disk album is

$$C(x) = 20,000 + 3x$$

where x is the number of disks produced. The average cost per disk, denoted by $\overline{C}(x)$, is found by dividing $C(x)$ by x:

$$\overline{C}(x) = \frac{C(x)}{x} = \frac{20,000 + 3x}{x}$$

If only 10 disks were manufactured, for example, the average cost per disk would be $2,003. Find:

(A) $\overline{C}(1,000)$ (B) $\overline{C}(100,000)$ (C) $\displaystyle\lim_{x \to 10,000} \overline{C}(x)$ (D) $\displaystyle\lim_{x \to \infty} \overline{C}(x)$

90. *Employee training.* A company producing computer components has established that on the average, a new employee can assemble $N(t)$ components per day after t days of on-the-job training, as given by

$$N(t) = \frac{100t}{t + 9}$$

(See the figure below.) Find:

(A) $N(1)$ (B) $N(11)$ (C) $\lim_{t \to 11} N(t)$ (D) $\lim_{t \to \infty} N(t)$

91. *Compound interest.* If $100 is invested at 8% compounded n times per year, then the amount in the account $A(n)$ at the end of 1 year is given by

$$A(n) = 100\left(1 + \frac{0.08}{n}\right)^n$$

(A) Use a calculator with a y^x key to complete the table in the margin. (Give each entry to the nearest cent.)

(B) Using the results of part A, guess the following limit:

$$\lim_{n \to \infty} A(n) = ?$$

(This problem leads to the important concept of compounding continuously, which will be discussed in detail in Section 10-5.)

COMPOUNDED	n	$A(n)$
Annually	1	$108.00
Semiannually	2	$108.16
Quarterly	4	$108.24
Monthly	12	
Weekly	52	
Daily	365	
Hourly	8,760	

92. *Pollution.* In Silicon Valley (in California), a number of computer-related manufacturing firms were found to be contaminating underground water supplies with toxic chemicals stored in leaking underground containers. A water quality control agency ordered the companies to take immediate corrective action and to contribute to a monetary pool for testing and cleanup of the underground contamination. Suppose the required monetary pool (in millions of dollars) for the testing and cleanup is estimated by

$$P(x) = \frac{2x}{1 - x}$$

where x is the percentage (expressed as a decimal fraction) of the total contaminant removed.

(A) Complete the table in the margin.

(B) Find $\lim_{x \to 0.80} P(x)$.

(C) What happens to the required monetary pool as the desired percentage of contaminant removed approaches 100% (x approaches 1 from the left)?

PERCENTAGE REMOVED	POOL REQUIRED
0.50 (50%)	$2 million
0.60 (60%)	$3 million
0.70 (70%)	
0.80 (80%)	
0.90 (90%)	
0.95 (95%)	
0.99 (99%)	

Life Sciences

93. *Medicine.* A drug is injected into the bloodstream of a patient through her right arm. The concentration of the drug in the bloodstream of the left arm t hours after the injection is given by

$$C(t) = \frac{0.14t}{t^2 + 1}$$

(See the figure below.) Find:

(A) $C(0.5)$ (B) $C(1)$ (C) $\lim_{t \to 1} C(t)$ (D) $\lim_{t \to \infty} C(t)$

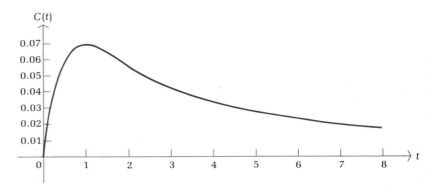

94. *Physiology.* In a study on the speed of muscle contraction in frogs under various loads, researchers W. O. Fems and J. Marsh found that the speed of contraction decreases with increasing loads. More precisely, they found that the relationship between speed of contraction S (in centimeters per second) and load w (in grams) is given approximately by

$$S(w) = \frac{26 + 0.06w}{w} \qquad w > 5$$

Find:

(A) $S(10)$ (B) $S(50)$ (C) $\lim_{w \to 50} S(w)$ (D) $\lim_{w \to \infty} S(w)$

Social Sciences

95. *Psychology — learning theory.* In 1917, L. L. Thurstone, a pioneer in quantitative learning theory, proposed the function

$$f(x) = \frac{a(x + c)}{(x + c) + b}$$

to describe the number of successful acts per unit time that a person could accomplish after x practice sessions. Suppose that for a particular person enrolling in a typing school,

$$f(x) = \frac{60(x + 1)}{x + 5}$$

where $f(x)$ is the number of words per minute the person is able to type after x weeks of lessons. (See the figure below.) Find:

(A) $f(3)$ (B) $f(10)$ (C) $\lim\limits_{x \to 10} f(x)$ (D) $\lim\limits_{x \to \infty} f(x)$

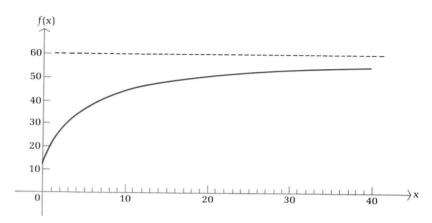

96. *Psychology — retention.* An experiment on retention is conducted in a psychology class. Each student in the class is given 1 day to memorize the same list of 30 special characters. The lists are turned in at the end of the day, and for each succeeding day for 30 days each student is asked to turn in a list of as many of the symbols as can be recalled. Averages are taken, and it is found that

$$N(t) = \frac{5t + 20}{t} \qquad t \geq 1$$

provides a good approximation of the average number of symbols, $N(t)$, retained after t days. (See the figure below.) Find:

(A) $N(2)$ (B) $N(10)$ (C) $\lim\limits_{t \to 10} N(t)$ (D) $\lim\limits_{t \to \infty} N(t)$

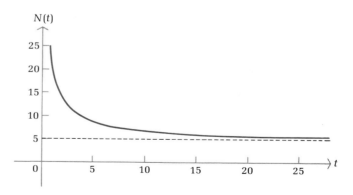

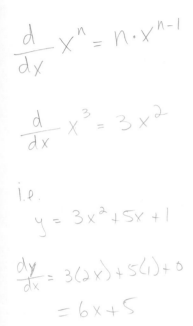

$$\frac{d}{dx} x^n = n \cdot x^{n-1}$$

$$\frac{d}{dx} x^3 = 3x^2$$

i.e.

$$y = 3x^2 + 5x + 1$$

$$\frac{dy}{dx} = 3(2x) + 5(1) + 0$$

$$= 6x + 5$$

We now use the concept of limit to solve two of the important problems illustrated in Figure 1 at the beginning of this chapter. These two problems are repeated in Figure 10. The solution of these two apparently unrelated problems involves a common concept called *the derivative*, which we will be able to define after we have solved the problems illustrated in Figure 10.

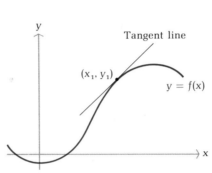

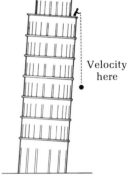

(A) Find the equation of the tangent line at (x_1, y_1) given $y = f(x)$

(B) Find the instantaneous velocity of a falling object

FIGURE 10

◆ TANGENT LINES

From plane geometry, we know that a tangent to a circle is a line that passes through one and only one point on the circle. This definition is not satisfactory for graphs of functions in general, as shown in Figure 11.

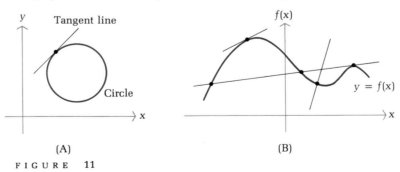

(A)

(B)

FIGURE 11

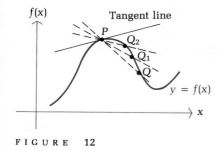

$f(x)$

Tangent line

P

Q_2

Q_1

Q

$y = f(x)$

x

FIGURE 12

To define a tangent line for the graph of a function f at a point P on its graph, we proceed as follows. Select another point Q on the graph of f, and draw a line through P and Q. This line is called a **secant line**. Let Q approach P along the graph, as indicated by Q_1 and Q_2 in Figure 12. If the slopes of the corresponding secant lines have a limit m as Q approaches P, then m is defined to be the *slope of the tangent line* at P and the *slope of the graph of f* at P. (These definitions will be made more precise below.) If we know the slope of the tangent line and the coordinates of P, we can write the equation of the tangent line at P using the point–slope formula for a line (from Section 2-3).

To find the slope of the tangent line at a particular point $P(x_1, f(x_1))$ on the graph of f, we locate any other point Q on the graph, as indicated in Figure 13. Since h can be positive or negative (but not equal to 0), Q can be on the right or on the left of P (but cannot coincide with P). Both cases are illustrated in Figure 13.

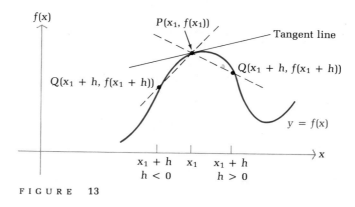

$f(x)$

$P(x_1, f(x_1))$

Tangent line

$Q(x_1 + h, f(x_1 + h))$

$Q(x_1 + h, f(x_1 + h))$

$y = f(x)$

x

$x_1 + h$ x_1 $x_1 + h$

$h < 0$ $h > 0$

FIGURE 13

We now write the slope of the secant line through P and Q, using the slope formula, $m = (y_2 - y_1)/(x_2 - x_1)$, from Section 2-3:

$$\text{Slope of secant line} = \frac{f(x_1 + h) - f(x_1)}{(x_1 + h) - x_1}$$

$$= \frac{f(x_1 + h) - f(x_1)}{h} \tag{1}$$

We refer to (1) as a **difference quotient.** Now, as h approaches 0 from either side of 0, but not equal to 0, Q will approach P along the graph of f (assuming f is continuous at $x = x_1$). If the limit of the difference quotient (1) exists as $h \to 0$, we define this limit to be the slope of the tangent line to the graph of f at the point P.

It is useful to note that the limit used to define the slope of a tangent line (and the slope of a graph) is usually a $0/0$ indeterminate form — one of the reasons we placed special emphasis on this form earlier.

Tangent Line

Given the graph of $y = f(x)$, the **tangent line** at $x = x_1$ is the line that passes through the point $(x_1, f(x_1))$ with slope

$$\text{Slope of tangent line} = \lim_{h \to 0} \frac{f(x_1 + h) - f(x_1)}{h}$$

if the limit exists. The slope of the tangent line is also referred to as the **slope of the graph** at $(x_1, f(x_1))$.

◆ E X A M P L E   11   Given $f(x) = x^2$:

(A) Find the slope of the tangent line at $x = 1$.
(B) Find the equation of the tangent line at $x = 1$; that is, through $(1, f(1))$.
(C) Sketch the graph of f, the tangent line at $(1, f(1))$, and the secant line passing through $(1, f(1))$ and $(2, f(2))$.

Solutions   (A) Slope of the tangent line:

Step 1.   Write the difference quotient and simplify:

$$\frac{f(1 + h) - f(1)}{h} = \frac{(1 + h)^2 - 1^2}{h}$$

This is the slope of a secant line passing through $(1, f(1))$ and $(1 + h, f(1 + h))$.

$$= \frac{1 + 2h + h^2 - 1}{h}$$

$$= \frac{2h + h^2}{h} = \frac{h(2 + h)}{h} = 2 + h \qquad h \neq 0$$

Step 2.   Find the limit of the difference quotient:

$$\text{Slope of tangent line} = \lim_{h \to 0} \frac{f(1 + h) - f(1)}{h}$$

$$= \lim_{h \to 0} (2 + h) = 2$$

This is also the slope of the graph of $f(x) = x^2$ at $(1, f(1))$.

(B) Equation of the tangent line: The tangent line passes through $(1, f(1)) = (1, 1)$ with slope $m = 2$ (from part A). The point–slope formula from Section 2-3 gives its equation:

$$y - y_1 = m(x - x_1) \qquad \text{Point–slope formula}$$
$$y - 1 = 2(x - 1)$$
$$y = 2x - 1 \qquad \text{Tangent line equation}$$

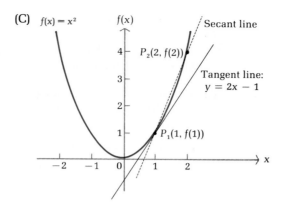

(C) $f(x) = x^2$

$P_2(2, f(2))$

Secant line

Tangent line:
$y = 2x - 1$

$P_1(1, f(1))$

◆

PROBLEM   11   Find the slope of the tangent line for the graph of $f(x) = x^2$ at $x = 2$, and write the equation of the tangent line in the form $y = mx + b$.   ◆

◆ AVERAGE AND INSTANTANEOUS RATES OF CHANGE

We now turn to the second problem stated at the beginning of this section.

◆ EXAMPLE   12   A small steel ball dropped from a tower will fall a distance of y feet in x seconds, as given approximately by the formula (from physics)

$$y = f(x) = 16x^2$$

Figure 14 shows the position of the ball on a coordinate line (positive direction down) at the end of 0, 1, 2, and 3 seconds. Note that the distances fallen between these times increases with time; thus, the velocity of the ball is increasing as it falls. Our main objective in this example is to find the velocity of the ball at a given instant, say, at the end of 2 seconds. (*Velocity* is the rate of change of distance with respect to time.)

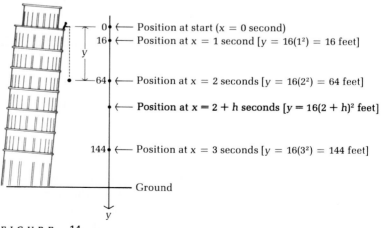

0 ←— Position at start ($x = 0$ second)
16 ←— Position at $x = 1$ second [$y = 16(1^2) = 16$ feet]
64 ←— Position at $x = 2$ seconds [$y = 16(2^2) = 64$ feet]
←— Position at $x = 2 + h$ seconds [$y = 16(2 + h)^2$ feet]
144 ←— Position at $x = 3$ seconds [$y = 16(3^2) = 144$ feet]
Ground

FIGURE   14
Note: Positive y direction is down.

(A) Find the average velocity from $x = 2$ seconds to $x = 3$ seconds.

(B) Find the average velocity from $x = 2$ seconds to $x = 2 + h$ seconds, $h \neq 0$.

(C) Find the limit of the expression from part B as $h \to 0$, if it exists. (What do you think the limit represents?)

Solutions

(A) Average velocity from $x = 2$ seconds to $x = 3$ seconds: Recall the formula $d = rt$, which can be written in the form

$$r = \frac{d}{t} = \frac{\text{Distance covered}}{\text{Elapsed time}} = \text{Average rate}$$

For example, if a person drives from San Francisco to Los Angeles (a distance of about 420 miles) in 7 hours, then the average velocity is

$$r = \frac{d}{t} = \frac{420}{7} = 60 \text{ miles per hour}$$

Sometimes the person will be traveling faster and sometimes slower, but the average velocity is 60 miles per hour. In our present problem, the average velocity of the steel ball is given by

$$\text{Average velocity} = \frac{\text{Distance covered}}{\text{Elapsed time}}$$

$$= \frac{f(3) - f(2)}{3 - 2}$$

$$= \frac{16 \cdot 3^2 - 16 \cdot 2^2}{1} = 80 \text{ feet per second}$$

(B) Average velocity from $x = 2$ seconds to $x = 2 + h$ seconds, $h \neq 0$: Proceeding as in part A,

$$\text{Average velocity} = \frac{\text{Distance covered}}{\text{Elapsed time}}$$

$$= \frac{f(2 + h) - f(2)}{h} \qquad \text{Difference quotient}$$

$$= \frac{16(2 + h)^2 - 16(2)^2}{h} \qquad \begin{array}{l} 0/0 \text{ indeterminate form} \\ \text{requires simplification} \\ \text{(for part C).} \end{array}$$

$$= \frac{64 + 64h + 16h^2 - 64}{h}$$

$$= \frac{h(64 + 16h)}{h} = 64 + 16h \qquad h \neq 0$$

Note that if $h = 1$, the average velocity is 80 feet per second (our result in part A); if $h = 0.5$, the average velocity is 72 feet per second; if $h = -0.01$, the average velocity is 63.84 feet per second; and so on. The closer to 0 that h gets (on either side of 0), the closer the average velocity is to 64 feet per second.

(C) Limit of the expression from part B as $h \to 0$:

$$\lim_{h \to 0} \frac{f(2 + h) - f(2)}{h} = \lim_{h \to 0} (64 + 16h)$$
$$= 64 \text{ feet per second} \qquad \blacklozenge$$

We call the answer to Example 12C, 64 feet per second, the *instantaneous velocity* (or the *instantaneous rate of change*) at $x = 2$ seconds, and we have solved the second basic problem stated at the beginning of this chapter!

Average and Instantaneous Rates of Change

For $y = f(x)$, the **average rate of change from $x = x_1$ to $x = x_1 + h$** is

$$\frac{f(x_1 + h) - f(x_1)}{h} \qquad h \neq 0$$

and the **instantaneous rate of change at $x = x_1$** is

$$\lim_{h \to 0} \frac{f(x_1 + h) - f(x_1)}{h}$$

if the limit exists.

Note that the limit used to define instantaneous rate of change is usually a $0/0$ indeterminate form — another reason we placed special emphasis on this form earlier.

PROBLEM   12   For the falling steel ball in Example 12, find:

(A) The average velocity from $x = 1$ second to $x = 2$ seconds.
(B) The average velocity from $x = 1$ second to $x = 1 + h$ seconds, $h \neq 0$.
(C) The instantaneous velocity (instantaneous rate of change) at $x = 1$. $\qquad \blacklozenge$

◆ THE DERIVATIVE

In the last two examples, we found that the special limit

$$\lim_{h \to 0} \frac{f(x_1 + h) - f(x_1)}{h} \qquad \qquad (2)$$

if it exists, gives us the slope of the tangent line to the graph of $y = f(x)$ at $(x_1, f(x_1))$ and also the instantaneous rate of change of y with respect to x at $x = x_1$. Many other applications give rise to this form. In fact, the limit (2) is of such basic importance to calculus and to the applications of calculus that we give it a special name and study it in detail. To keep form (2) simple and general, we drop the subscript on x_1 and think of the difference quotient

$$\frac{f(x + h) - f(x)}{h}$$

as a function of h, with x held fixed as we let h tend to 0. We are now ready to define one of the basic concepts in calculus, *the derivative*:

The Derivative

For $y = f(x)$, we define the **derivative of f at x,** denoted by $f'(x)$, to be

$$f'(x) = \lim_{h \to 0} \frac{f(x + h) - f(x)}{h} \qquad \text{if the limit exists}$$

If $f'(x)$ exists for each x in the open interval (a, b), then f is said to be **differentiable** over (a, b).

(Differentiability from the left or from the right is defined using $h \to 0^-$ or $h \to 0^+$, respectively, in place of $h \to 0$ in the above definition.)

The process of finding the derivative of a function is called **differentiation.** That is, the derivative of a function is obtained by **differentiating** the function.

The Function $f'(x)$

The derivative of a function f is a new function f' that gives us, among other things, the slope of the tangent line to the graph of $y = f(x)$ for each x and the instantaneous rate of change of $y = f(x)$ with respect to x. The domain of f' is a subset of the domain of f.

◆ E X A M P L E   13   Find $f'(x)$, the derivative of f at x, for $f(x) = 4x - x^2$.

Solution   To find $f'(x)$, we use a two-step process:

Step 1.   Form the difference quotient and simplify:

$$\frac{f(x + h) - f(x)}{h} = \frac{[4(x + h) - (x + h)^2] - (4x - x^2)}{h}$$

$$= \frac{4x + 4h - x^2 - 2xh - h^2 - 4x + x^2}{h}$$

$$= \frac{4h - 2xh - h^2}{h}$$

$$= \frac{h(4 - 2x - h)}{h}$$

$$= 4 - 2x - h \qquad h \neq 0$$

Step 2. Find the limit of the difference quotient:

$$f'(x) = \lim_{h \to 0} \frac{f(x + h) - f(x)}{h}$$

$$= \lim_{h \to 0} (4 - 2x - h) = 4 - 2x$$

Thus, if $f(x) = 4x - x^2$, then $f'(x) = 4 - 2x$. The derivative f' is a new function derived from the function f. ◆

PROBLEM 13   Find $f'(x)$, the derivative of f at x, for $f(x) = 8x - 2x^2$. ◆

◆ EXAMPLE 14   In Example 13, we started with the function specified by $f(x) = 4x - x^2$ and found the derivative of f at x to be $f'(x) = 4 - 2x$. Thus, the slope of a tangent line to the graph of f at any point $(x, f(x))$ on the graph of f is

$$m = f'(x) = 4 - 2x$$

(A) Find the slope of the graph of f at $x = 0$, 2, and 3.
(B) Graph $y = f(x) = 4x - x^2$, and use the slopes found in part A to make a rough sketch of the tangent lines to the graph at $x = 0$, 2, and 3.

Solutions   (A) Using $f'(x) = 4 - 2x$, we have

$$m = f'(0) = 4 - 2(0) = 4 \qquad \text{Slope at } x = 0$$
$$m = f'(2) = 4 - 2(2) = 0 \qquad \text{Slope at } x = 2$$
$$m = f'(3) = 4 - 2(3) = -2 \qquad \text{Slope at } x = 3$$

(B)

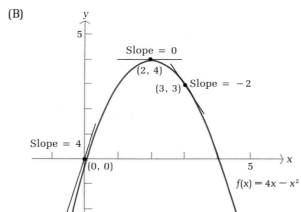

◆

PROBLEM 14   In Problem 13, we started with the function specified by $f(x) = 8x - 2x^2$. Using the derivative found there:

(A) Find the slope of the graph of f at $x = 1$, 2, and 4.
(B) Graph $y = f(x) = 8x - 2x^2$, and use the slopes from part A to make a rough sketch of the tangent lines to the graph at $x = 1$, 2, and 4. ◆

♦ E X A M P L E   15   Find $f'(x)$, the derivative of f at x, for $f(x) = \sqrt{x} + 2$.

Solution   To find $f'(x)$, we find

$$\lim_{h \to 0} \frac{f(x + h) - f(x)}{h}$$

using the two-step process.

Step 1.   Form the difference quotient and simplify:

$$\frac{f(x + h) - f(x)}{h} = \frac{(\sqrt{x + h} + 2) - (\sqrt{x} + 2)}{h}$$

$$= \frac{\sqrt{x + h} - \sqrt{x}}{h}$$

Since this is a $0/0$ indeterminate form, we change the form by rationalizing the numerator:

$$\frac{\sqrt{x + h} - \sqrt{x}}{h} \cdot \frac{\sqrt{x + h} + \sqrt{x}}{\sqrt{x + h} + \sqrt{x}} = \frac{x + h - x}{h(\sqrt{x + h} + \sqrt{x})}$$

$$= \frac{h}{h(\sqrt{x + h} + \sqrt{x})}$$

$$= \frac{1}{\sqrt{x + h} + \sqrt{x}} \qquad h \neq 0$$

Step 2.   Find the limit of the difference quotient:

$$f'(x) = \lim_{h \to 0} \frac{f(x + h) - f(x)}{h}$$

$$= \lim_{h \to 0} \frac{1}{\sqrt{x + h} + \sqrt{x}}$$

$$= \frac{1}{\sqrt{x} + \sqrt{x}} = \frac{1}{2\sqrt{x}} \qquad x > 0$$

Thus, the derivative of $f(x) = \sqrt{x} + 2$ is $f'(x) = 1/(2\sqrt{x})$, a new function. The domain of f is $[0, \infty)$. Since $f'(0)$ is not defined, the domain of f' is $(0, \infty)$, a subset of the domain of f.   ♦

P R O B L E M   15   Find $f'(x)$ for $f(x) = x^{-1}$.   ♦

♦ NONEXISTENCE OF THE DERIVATIVE

The existence of a derivative at $x = a$ depends on the existence of a limit at $x = a$; that is, on the existence of

$$f'(a) = \lim_{h \to 0} \frac{f(a + h) - f(a)}{h} \tag{3}$$

If the limit does not exist at $x = a$, we say that the function f is **nondifferentiable at $x = a$, or $f'(a)$ does not exist.**

How can we recognize the points on the graph of f where $f'(a)$ does not exist? It is impossible to describe all the ways that the limit in (3) can fail to exist. However, we can illustrate some common situations where $f'(a)$ does fail to exist:

1.   If f is not continuous at $x = a$, then $f'(a)$ does not exist (Figure 15A). Or, equivalently, **if f is differentiable at $x = a$, then f must be continuous at $x = a$.**
2.   If the graph of f has a sharp corner at $x = a$, then $f'(a)$ does not exist and the graph has no tangent line at $x = a$ (Figure 15B). (In Figure 15B, the left- and right-hand derivatives exist but are not equal.)
3.   If the graph of f has a vertical tangent line at $x = a$, then $f'(a)$ does not exist (Figure 15C and D).

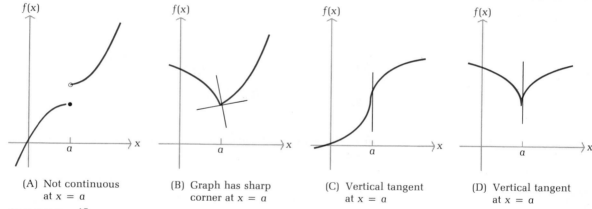

(A) Not continuous
       at $x = a$

(B) Graph has sharp
       corner at $x = a$

(C) Vertical tangent
       at $x = a$

(D) Vertical tangent
       at $x = a$

FIGURE   15
The function f is nondifferentiable at $x = a$.

◆ SUMMARY

The concept of the derivative is a very powerful mathematical idea, and its applications are many and varied. In the next three sections we will develop formulas and general properties of derivatives that will enable us to find the derivatives of many functions without having to go through the two-step limiting process each time.

11. $y = 4x - 4$

12. (A) **48 ft/sec**   (B) $32 + 16h$ (C) **32 ft/sec**

13. $f'(x) = 8 - 4x$

14. (A) $f'(1) = 4, f'(2) = 0, f'(4) = -8$ (B)

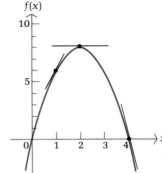

15. $f'(x) = -1/x^2$, or $-x^{-2}$

E X E R C I S E   9-3

A   Use $f(x) = 3x - 2$ in Problems 1 and 2 to find the given difference quotients and limits.

1. (A) $\dfrac{f(1 + h) - f(1)}{h}$ (B) $\lim\limits_{h \to 0} \dfrac{f(1 + h) - f(1)}{h}$

2. (A) $\dfrac{f(2 + h) - f(2)}{h}$ (B) $\lim\limits_{h \to 0} \dfrac{f(2 + h) - f(2)}{h}$

Use $f(x) = 2x^2$ in Problems 3 and 4 to find the given difference quotients and limits.

3. (A) $\dfrac{f(2 + h) - f(2)}{h}$ (B) $\lim\limits_{h \to 0} \dfrac{f(2 + h) - f(2)}{h}$

4. (A) $\dfrac{f(3 + h) - f(3)}{h}$ (B) $\lim\limits_{h \to 0} \dfrac{f(3 + h) - f(3)}{h}$

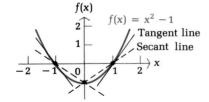

Solve Problems 5 and 6 for the graph of $y = f(x) = x^2 - 1$ (see the figure in the margin).

5. (A) Find the slope of the secant line joining $(0, f(0))$ and $(1, f(1))$.
   (B) Find the slope of the secant line joining $(1, f(1))$ and $(1 + h, f(1 + h))$.
   (C) Find the slope of the tangent line at $x = 1$.

6. (A) Find the slope of the secant line joining $(-1, f(-1))$ and $(0, f(0))$.
   (B) Find the slope of the secant line joining $(-1, f(-1))$ and $(-1 + h, f(-1 + h))$.
   (C) Find the slope of the tangent line at $x = -1$.

In Problems 7–10, find $f'(x)$ using the two-step process:

Step 1.   Simplify: $\dfrac{f(x+h) - f(x)}{h}$

Step 2.   Evaluate: $\displaystyle\lim_{h \to 0} \dfrac{f(x+h) - f(x)}{h}$

Then find $f'(1)$, $f'(2)$, and $f'(3)$.

7. $f(x) = 2x - 3$
9. $f(x) = 2 - x^2$

8. $f(x) = 4x + 3$
10. $f(x) = 2x^2 + 5$

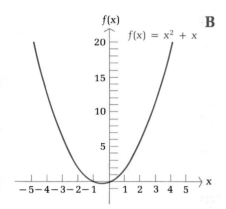

B   Problems 11 and 12 refer to the graph of $y = f(x) = x^2 + x$ shown in the margin.

11. (A) Find the slope of the secant line joining $(1, f(1))$ and $(3, f(3))$.
     (B) Find the slope of the secant line joining $(1, f(1))$ and $(1 + h, f(1 + h))$.
     (C) Find the slope of the tangent line at $(1, f(1))$.
     (D) Find the equation of the tangent line at $(1, f(1))$.
12. (A) Find the slope of the secant line joining $(2, f(2))$ and $(4, f(4))$.
     (B) Find the slope of the secant line joining $(2, f(2))$ and $(2 + h, f(2 + h))$.
     (C) Find the slope of the tangent line at $(2, f(2))$.
     (D) Find the equation of the tangent line at $(2, f(2))$.

In Problems 13 and 14, suppose an object moves along the y axis so that its location is $y = f(x) = x^2 + x$ at time x (y is in meters and x is in seconds). Find:

13. (A) The average velocity (the average rate of change of y with respect to x) for x changing from 1 to 3 seconds
     (B) The average velocity for x changing from 1 to $1 + h$ seconds
     (C) The instantaneous velocity at $x = 1$ second
14. (A) The average velocity (the average rate of change of y with respect to x) for x changing from 2 to 4 seconds
     (B) The average velocity for x changing from 2 to $2 + h$ seconds
     (C) The instantaneous velocity at $x = 2$ seconds

In Problems 15–20, find $f'(x)$ using the two-step limiting process. Then find $f'(1)$, $f'(2)$, and $f'(3)$.

15. $f(x) = 6x - x^2$
17. $f(x) = \sqrt{x} - 3$
19. $f(x) = \dfrac{-1}{x}$

16. $f(x) = 2x - 3x^2$
18. $f(x) = 2 - \sqrt{x}$
20. $f(x) = \dfrac{1}{x + 1}$

Problems 21–28 refer to the function F in the graph below. Use the graph to determine whether F'(x) exists at the indicated value of x.

21. x = a   22. x = b   23. x = c   24. x = d
25. x = e   26. x = f   27. x = g   28. x = h

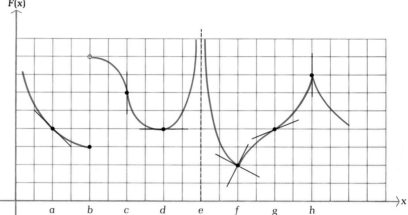

29. Given f(x) = x² − 4x:
    (A) Find f'(x).
    (B) Find the slopes of the tangent lines to the graph of f at x = 0, 2, and 4.
    (C) Graph f, and sketch in the tangent lines at x = 0, 2, and 4.

30. Given f(x) = x² + 2x:
    (A) Find f'(x).
    (B) Find the slopes of the tangent lines to the graph of f at x = −2, −1, and 1.
    (C) Graph f, and sketch in the tangent lines at x = −2, −1, and 1.

31. If an object moves along a line so that it is at y = f(x) = 4x² − 2x at time x (in seconds), find the instantaneous velocity function v = f'(x), and find the velocity at times x = 1, 3, and 5 seconds (y is measured in feet).

32. Repeat Problem 31 with f(x) = 8x² − 4x.

C In Problems 33 and 34, sketch the graph of f and determine where f is nondifferentiable.

33. $f(x) = \begin{cases} 2x & \text{if } x < 1 \\ 2 & \text{if } x \geqslant 1 \end{cases}$

34. $f(x) = \begin{cases} 2x & \text{if } x < 2 \\ 6 - x & \text{if } x \geqslant 2 \end{cases}$

In Problems 35–38, determine whether f is differentiable at $x = 0$ by considering

$$\lim_{h \to 0} \frac{f(0 + h) - f(0)}{h}$$

35. $f(x) = |x|$ **36.** $f(x) = 1 - |x|$ **37.** $f(x) = x^{1/3}$ **38.** $f(x) = x^{2/3}$

39. Show that $f(x) = 2x - x^2$ is differentiable over the closed interval $[0, 2]$ by showing that each of the following limits exists:

(A) $\lim\limits_{h \to 0} \dfrac{f(x + h) - f(x)}{h}, \quad 0 < x < 2$ (B) $\lim\limits_{h \to 0^+} \dfrac{f(0 + h) - f(0)}{h}, \quad x = 0$

(C) $\lim\limits_{h \to 0^-} \dfrac{f(2 + h) - f(2)}{h}, \quad x = 2$

40. Show that $f(x) = \sqrt{x}$ is differentiable over the open interval $(0, \infty)$ but not over the half-closed interval $[0, \infty)$ by considering

$$\lim_{h \to 0} \frac{f(x + h) - f(x)}{h} \qquad 0 < x < \infty$$

and

$$\lim_{h \to 0^+} \frac{f(0 + h) - f(0)}{h} \qquad x = 0$$

APPLICATIONS

Business & Economics

Problems 41 and 42 provide a first look at the concept of "marginal analysis," an important topic in business and economics that is developed further in succeeding sections and in detail in Section 9-7.

41. *Cost analysis.* The total cost per day, $C(x)$ (in hundreds of dollars), for manufacturing x windsurfing boards is given by

$$C(x) = 3 + 10x - x^2 \qquad 0 \le x \le 5$$

(A) Find: $\dfrac{C(4) - C(3)}{4 - 3}$

(This is the increase in cost for a 1 unit increase in production at the level of production of 3 boards per day.)

(B) Find: $\dfrac{C(3 + h) - C(3)}{h}$

(C) Find: $C'(3)$
(This is called the *marginal cost* at the level of production of 3 boards per day; it is the instantaneous rate of change of cost relative to production and approximates the result found in part A.)

(D) Find $C'(x)$, the *marginal cost function.*

(E) Using $C'(x)$, find $C'(1)$, $C'(2)$, $C'(3)$, and $C'(4)$, and interpret. [*Note:* As production levels increase, the rate of change of cost relative to production (marginal cost) decreases.]

42. *Cost analysis.* Repeat Problem 41 for

$$C(x) = 5 + 12x - x^2 \qquad 0 \leqslant x \leqslant 5$$

SECTION   9-4

Derivatives of Constants, Power Forms, and Sums

♦ DERIVATIVE OF A CONSTANT
♦ POWER RULE
♦ DERIVATIVE OF A CONSTANT TIMES A FUNCTION
♦ DERIVATIVES OF SUMS AND DIFFERENCES
♦ APPLICATIONS

In the preceding section, we defined the derivative of f at x as

$$f'(x) = \lim_{h \to 0} \frac{f(x + h) - f(x)}{h}$$

if the limit exists, and we used this definition and a two-step process to find the derivatives of several functions. In this and the next two sections, we will develop some rules based on this definition that will enable us to determine the derivatives of a rather large class of functions without having to go through the two-step process each time.

Before starting on these rules, we list some symbols that are widely used to represent derivatives:

Derivative Notation

Given $y = f(x)$, then

$$f'(x) \qquad y' \qquad \frac{dy}{dx} \qquad D_x\, f(x)$$

all represent the derivative of f at x.

Each of these symbols for the derivative has its particular advantage in certain situations. All of them will become familiar to you after a little experience.

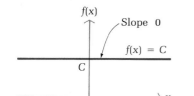

FIGURE 16

◆ DERIVATIVE OF A CONSTANT

Suppose

$$f(x) = C \qquad C \text{ a constant} \qquad \text{A constant function}$$

Geometrically, the graph of $f(x) = C$ is a horizontal straight line with slope 0 (see Figure 16); hence, we would expect $f'(x) = 0$. We will show that this is actually the case using the definition of the derivative and the two-step process introduced earlier. We want to find

$$f'(x) = \lim_{h \to 0} \frac{f(x + h) - f(x)}{h} \qquad \text{Definition of } f'(x)$$

Step 1. $\dfrac{f(x + h) - f(x)}{h} = \dfrac{C - C}{h} = \dfrac{0}{h} = 0 \qquad h \neq 0$

Step 2. $\lim_{h \to 0} 0 = 0$

Thus,

$$f'(x) = 0$$

We conclude that:

The derivative of any constant is 0.

■ Derivative of a Constant

If $y = f(x) = C$, then

$$f'(x) = 0$$

Also, $y' = 0$, $dy/dx = 0$, and $D_x C = 0$.

[*Note:* When we write $D_x C = 0$, we mean $D_x f(x) = 0$, where $f(x) = C$.]

◆ E X A M P L E   16   (A) If $f(x) = 3$, then $f'(x) = 0$.   (B) If $y = -1.4$, then $y' = 0$.
(C) If $y = \pi$, then $dy/dx = 0$.   (D) $D_x (23) = 0$ ◆

P R O B L E M   16   Find:

(A) $f'(x)$ for $f(x) = -24$ (B) y' for $y = 12$
(C) dy/dx for $y = -\sqrt{7}$ (D) $D_x(-\pi)$ ◆

◆ POWER RULE

Using the definition of derivative and the two-step process introduced in the preceding section, we can show that:

If $f(x) = x$, then $f'(x) = 1$.

If $f(x) = x^2$, then $f'(x) = 2x$.

If $f(x) = x^3$, then $f'(x) = 3x^2$.

If $f(x) = x^4$, then $f'(x) = 4x^3$.

In general, for any positive integer n:

$$\text{If } f(x) = x^n, \quad \text{then} \quad f'(x) = nx^{n-1}. \tag{1}$$

In fact, more advanced techniques can be used to show that (1) holds for *any* real number n. We will assume this general result for the remainder of this book.

> **■ Power Rule**
>
> If $y = f(x) = x^n$, where n is a real number, then
>
> $$f'(x) = nx^{n-1}$$

◆ **EXAMPLE 17**   (A) If $f(x) = x^5$, then $f'(x) = 5x^{5-1} = 5x^4$.
(B) If $y = x^{25}$, then $y' = 25x^{25-1} = 25x^{24}$.
(C) If $y = x^{-3}$, then $dy/dx = -3x^{-3-1} = -3x^{-4}$.
(D) $D_x x^{5/3} = \frac{5}{3}x^{(5/3)-1} = \frac{5}{3}x^{2/3}$ ◆

PROBLEM 17   Find:

(A) $f'(x)$ for $f(x) = x^6$ (B) y' for $y = x^{30}$
(C) dy/dx for $y = x^{-2}$ (D) $D_x x^{3/2}$ ◆

In some cases, properties of exponents must be used to rewrite an expression before the power rule is applied.

◆ **EXAMPLE 18**   (A) If $f(x) = 1/x^4$, then we can write $f(x) = x^{-4}$ and

$$f'(x) = -4x^{-4-1} = -4x^{-5} \quad \text{or} \quad \frac{-4}{x^5}$$

(B) If $y = \sqrt{x}$, then we can write $y = x^{1/2}$ and

$$y' = \frac{1}{2}x^{(1/2)-1} = \frac{1}{2}x^{-1/2} \quad \text{or} \quad \frac{1}{2\sqrt{x}}$$

(C) $D_x \frac{1}{\sqrt[3]{x}} = D_x x^{-1/3} = -\frac{1}{3}x^{(-1/3)-1} = -\frac{1}{3}x^{-4/3} \quad \text{or} \quad \frac{-1}{3\sqrt[3]{x^4}}$ ◆

Find:

$$\text{(A)} \quad f'(x) \quad \text{for} \quad f(x) = \frac{1}{x} \qquad \text{(B)} \quad y' \quad \text{for} \quad y = \sqrt[3]{x^2} \qquad \text{(C)} \quad D_x \frac{1}{\sqrt{x}} \qquad \blacklozenge$$

◆ DERIVATIVE OF A CONSTANT TIMES A FUNCTION

Let $f(x) = ku(x)$; where k is a constant and u is differentiable at x. Then, using the two-step process, we have the following:

Step 1. $\quad \dfrac{f(x + h) - f(x)}{h} = \dfrac{ku(x + h) - ku(x)}{h} = k\left[\dfrac{u(x + h) - u(x)}{h}\right]$

Step 2. $\quad \displaystyle\lim_{h \to 0} \dfrac{f(x + h) - f(x)}{h} = \lim_{h \to 0} k\left[\dfrac{u(x + h) - u(x)}{h}\right] \qquad \displaystyle\lim_{x \to c} kg(x) = k \lim_{x \to c} g(x)$

$$= k \lim_{h \to 0} \left[\dfrac{u(x + h) - u(x)}{h}\right] \qquad \text{Definition of } u'(x)$$

$$= ku'(x)$$

Thus:

The derivative of a constant times a differentiable function is the constant times the derivative of the function.

Constant Times a Function Rule

If $y = f(x) = ku(x)$, then

$\quad f'(x) = ku'(x)$

Also, $y' = ku'$, $dy/dx = k\, du/dx$, and $D_x\, ku(x) = k\, D_x\, u(x)$.

◆ E X A M P L E   19 (A) If $f(x) = 3x^2$, then $f'(x) \;\boxed{= 3 \cdot 2x^{2-1}}\; = 6x.$

(B) If $y = \dfrac{x^3}{6} = \dfrac{1}{6}x^3$, then $\dfrac{dy}{dx} \;\boxed{= \dfrac{1}{6} \cdot 3x^{3-1}}\; = \dfrac{1}{2}x^2.$

(C) If $y = \dfrac{1}{2x^4} = \dfrac{1}{2}x^{-4}$, then $y' \;\boxed{= \dfrac{1}{2}(-4x^{-4-1})}\; = -2x^{-5} \quad \text{or} \quad \dfrac{-2}{x^5}.$

(D) $D_x \dfrac{4}{\sqrt{x^3}} = D_x \dfrac{4}{x^{3/2}} = D_x\, 4x^{-3/2} \;\boxed{= 4\left[-\dfrac{3}{2}x^{(-3/2)-1}\right]}$

$$= -6x^{-5/2} \quad \text{or} \quad -\dfrac{6}{\sqrt{x^5}} \qquad \blacklozenge$$

PROBLEM 19   Find:

(A) $f'(x)$ for $f(x) = 4x^5$ (B) $\dfrac{dy}{dx}$ for $y = \dfrac{x^4}{12}$

(C) y' for $y = \dfrac{1}{3x^3}$ (D) $D_x \dfrac{9}{\sqrt[3]{x}}$ ◆

◆ DERIVATIVES OF SUMS AND DIFFERENCES

Let $f(x) = u(x) + v(x)$, where $u'(x)$ and $v'(x)$ exist. Then, using the two-step process, we have the following:

Step 1. $\dfrac{f(x+h) - f(x)}{h} = \dfrac{[u(x+h) + v(x+h)] - [u(x) + v(x)]}{h}$

$$= \dfrac{u(x+h) + v(x+h) - u(x) - v(x)}{h}$$

$$= \dfrac{u(x+h) - u(x)}{h} + \dfrac{v(x+h) - v(x)}{h}$$

Step 2. $\lim\limits_{h \to 0} \dfrac{f(x+h) - f(x)}{h} = \lim\limits_{h \to 0} \left[\dfrac{u(x+h) - u(x)}{h} + \dfrac{v(x+h) - v(x)}{h} \right]$

$$\lim\limits_{x \to c} [g(x) + h(x)] = \lim\limits_{x \to c} g(x) + \lim\limits_{x \to c} h(x)$$

$$= \lim\limits_{h \to 0} \dfrac{u(x+h) - u(x)}{h} + \lim\limits_{h \to 0} \dfrac{v(x+h) - v(x)}{h}$$

$$= u'(x) + v'(x)$$

Thus:

> The derivative of the sum of two differentiable functions is the sum of the derivatives.

Similarly, we can show that:

> The derivative of the difference of two differentiable functions is the difference of the derivatives.

Together, we then have the **sum and difference rule** for differentiation:

Sum and Difference Rule

If $y = f(x) = u(x) \pm v(x)$, then

$\quad f'(x) = u'(x) \pm v'(x)$

[*Note:* This rule generalizes to the sum and difference of any given number of functions.]

With this and the other rules stated previously, we will be able to compute the derivatives of all polynomials and a variety of other functions.

◆ E X A M P L E   20   (A)   If $f(x) = 3x^2 + 2x$, then

$$f'(x) \boxed{= (3x^2)' + (2x)' = 3(2x) + 2(1)} = 6x + 2$$

(B)   If $y = 4 + 2x^3 - 3x^{-1}$, then

$$y' \boxed{= (4)' + (2x^3)' - (3x^{-1})' = 0 + 2(3x^2) - 3(-1)x^{-2}} = 6x^2 + 3x^{-2}$$

(C)   If $y = \sqrt[3]{x} - 3x$, then

$$\frac{dy}{dx} = \frac{d}{dx}\, x^{1/3} - \frac{d}{dx}\, 3x = \frac{1}{3}\, x^{-2/3} - 3$$

(D) $D_x\left(\dfrac{5}{3x^2} - \dfrac{2}{x^4} + \dfrac{x^3}{9}\right) \boxed{= D_x\,\dfrac{5}{3}\,x^{-2} - D_x\, 2x^{-4} + D_x\,\dfrac{1}{9}\,x^3}$

$$= \frac{5}{3}(-2)x^{-3} - 2(-4)x^{-5} + \frac{1}{9}\cdot 3x^2 = -\frac{10}{3x^3} + \frac{8}{x^5} + \frac{1}{3}\, x^2 \qquad ◆$$

P R O B L E M   20   Find:

(A) $f'(x)$ for $f(x) = 3x^4 - 2x^3 + x^2 - 5x + 7$

(B) y' for $y = 3 - 7x^{-2}$ (C) $\dfrac{dy}{dx}$ for $y = 5x^3 - \sqrt[4]{x}$

(D) $D_x\left(-\dfrac{3}{4x} + \dfrac{4}{x^3} - \dfrac{x^4}{8}\right)$ ◆

◆ APPLICATIONS

◆ E X A M P L E   21
Instantaneous Velocity

An object moves along the y axis (marked in feet) so that its position at time x (in seconds) is

$$f(x) = x^3 - 6x^2 + 9x$$

(A) Find the instantaneous velocity function v.
(B) Find the velocity at $x = 2$ and $x = 5$ seconds.
(C) Find the time(s) when the velocity is 0.

Solutions   (A) $v = f'(x) \boxed{= (x^3)' - (6x^2)' + (9x)'} = 3x^2 - 12x + 9$

(B) $f'(2) = 3(2)^2 - 12(2) + 9 = -3$ feet per second
    $f'(5) = 3(5)^2 - 12(5) + 9 = 24$ feet per second
(C) $v = f'(x) = 3x^2 - 12x + 9 = 0$
    $3(x^2 - 4x + 3) = 0$
    $3(x - 1)(x - 3) = 0$
    $x = 1, 3$
    Thus, $v = 0$ at $x = 1$ and $x = 3$ seconds. ◆

Repeat Example 21 for $f(x) = x^3 - 15x^2 + 72x$.   ◆

◆   E X A M P L E   22   Let $f(x) = x^4 - 8x^2 + 10$.

Tangents   (A) Find $f'(x)$.
(B) Find the equation of the tangent line at $x = 1$.
(C) Find the values of x where the tangent line is horizontal.

Solutions   (A) $f'(x) = (x^4)' - (8x^2)' + (10)'$

$$= 4x^3 - 16x$$

(B) $y - y_1 = m(x - x_1)$ $y_1 = f(x_1) = f(1) = (1)^4 - 8(1)^2 + 10 = 3$
    $y - 3 = -12(x - 1)$ $m = f'(x_1) = f'(1) = 4(1)^3 - 16(1) = -12$
      $y = -12x + 15$ Tangent line at $x = 1$

(C) Since a horizontal line has 0 slope, we must solve $f'(x) = 0$ for x:

$$f'(x) = 4x^3 - 16x = 0$$
$$4x(x^2 - 4) = 0$$
$$4x(x - 2)(x + 2) = 0$$
$$x = 0, 2, -2$$

Thus, the tangent line to the graph of f will be horizontal at $x = -2$, $x = 0$, and $x = 2$. (In the next chapter, we will see how this information is used to help sketch the graph of f.)   ◆

P R O B L E M   22   Repeat Example 22 for $f(x) = x^4 - 4x^3 + 7$.   ◆

In business and economics, one is often interested in the rate at which something is taking place. A manufacturer, for example, is not only interested in the total cost $C(x)$ at certain production levels x, but is also interested in the rate of change of costs at various production levels.

In economics, the word **marginal** refers to a rate of change; that is, to a derivative. Thus, if

$C(x) = $ Total cost of producing x units during some unit of time

then

$C'(x) = $ Marginal cost
      = Rate of change in cost per unit change in production at an output level of x units

Just as with instantaneous velocity, marginal cost $C'(x)$ is an instantaneous rate of change of total costs $C(x)$ with respect to production at a production level of x units. If this rate remains constant as production is increased by 1 unit, then it represents the change in cost for a 1 unit change in production. If the rate does not remain constant, then the instantaneous rate is an approximation of what actually happens during the next unit change in production.

The Marginal Cost Function

If the marginal cost function $C'(x)$ is a constant function, then $C'(x)$ represents the cost of producing 1 more unit at any production level x. If $C'(x)$ is not a constant function, then $C'(x)$ approximates the cost of producing 1 more unit at a production level of x units.

Example 23 should help to clarify these ideas.

◆ EXAMPLE 23
Marginal Cost

Suppose the total cost $C(x)$, in thousands of dollars, for manufacturing x sailboats per year is given by the function

$$C(x) = 575 + 25x - \frac{x^2}{4} \qquad 0 \leqslant x \leqslant 50$$

shown in the figure below.

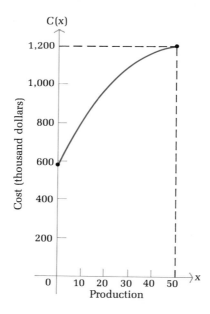

(A) Find the marginal cost at a production level of x boats.

(B) Find the marginal cost at a production level of 40 boats, and interpret the result.

(C) Find the actual cost of producing the 41st boat, and compare this cost with the result found in part B.

(D) Find $C'(30)$, and interpret the result.

(A) The marginal cost at a production level of x boats is

$$C'(x) = (575)' + (25x)' - \left(\frac{x^2}{4}\right)' = 25 - \frac{x}{2}$$

(B) The marginal cost at a production level of 40 boats is

$$C'(40) = 25 - \frac{40}{2} = 5 \quad \text{or} \quad \$5{,}000 \text{ per boat}$$

At a production level of 40 boats, the rate of change of total cost relative to production is $5,000 per boat. Thus, the cost of producing 1 more boat at a production level of 40 boats is approximately $5,000.

(C) The actual cost of producing the 41st boat is

$$\left(\begin{array}{c} \text{Total cost of} \\ \text{producing} \\ \text{41 boats} \end{array}\right) - \left(\begin{array}{c} \text{Total cost of} \\ \text{producing} \\ \text{40 boats} \end{array}\right)$$
$$= C(41) \quad - \quad C(40)$$
$$= 1{,}179.75 - 1{,}175.00 = 4.75 \quad \text{or} \quad \$4{,}750$$

The marginal cost of $5,000 per boat found in part B is a close approximation to this value.

(D) $\quad C'(30) = 25 - \dfrac{30}{2} = 10 \quad \text{or} \quad \$10{,}000 \text{ per boat}$

At a production level of 30 boats, the rate of change of total cost relative to production is $10,000 per boat. Thus, the cost of producing 1 more boat at this level of production is approximately $10,000. ◆

In Example 23, we observe that as production goes up, the marginal cost goes down.

P R O B L E M   23   Suppose the total cost $C(x)$, in thousands of dollars, for manufacturing x sailboats per year is given by the function

$$C(x) = 500 + 24x - \frac{x^2}{5} \qquad 0 \leqslant x \leqslant 50$$

(A) Find the marginal cost at a production level of x boats.
(B) Find the marginal cost at a production level of 35 boats, and interpret the result.
(C) Find the actual cost of producing the 36th boat, and compare this cost with the result found in part B.
(D) Find $C'(40)$, and interpret the result. ◆

16. All are 0.   17. (A) $6x^5$ (B) $30x^{29}$ (C) $-2x^{-3}$ (D) $\frac{3}{2}x^{1/2}$

18. (A) $-x^{-2}$ or $-1/x^2$ (B) $\frac{2}{3}x^{-1/3}$ or $2/(3\sqrt[3]{x})$
    (C) $-\frac{1}{2}x^{-3/2}$ or $-1/(2\sqrt{x^3})$

19. (A) $20x^4$ (B) $x^3/3$ (C) $-x^{-4}$ or $-1/x^4$ (D) $-3x^{-4/3}$ or $-3/\sqrt[3]{x^4}$

20. (A) $12x^3 - 6x^2 + 2x - 5$ (B) $14x^{-3}$
    (C) $15x^2 - \frac{1}{4}x^{-3/4}$ (D) $3/(4x^2) - (12/x^4) - (x^3/2)$

21. (A) $v = 3x^2 - 30x + 72$ (B) $f'(2) = 24$ ft/sec; $f'(5) = -3$ ft/sec
    (C) $x = 4$ and $x = 6$ sec

22. (A) $f'(x) = 4x^3 - 12x^2$ (B) $y = -8x + 12$ (C) $x = 0$ and $x = 3$

23. (A) $C'(x) = 24 - (2x/5)$
    (B) $C'(35) = 10$ or $10,000 per boat; at a production level of 35 boats, the rate of change of total cost relative to production is $10,000 per boat; thus, the cost of producing 1 more boat at this level of production is approx. $10,000
    (C) $C(36) - C(35) = 9.8$ or $9,800; the marginal cost of $10,000 per boat found in part B is a close approximation to this value
    (D) $C'(40) = 8$ or $8,000 per boat; at a production level of 40 boats, the rate of change of total cost relative to production is $8,000 per boat; thus, the cost of producing 1 more boat at this level of production is approx. $8,000

EXERCISE   9-4

Find each of the following:

A

1. $f'(x)$ for $f(x) = 12$ 2. $\dfrac{dy}{dx}$ for $y = -\sqrt{3}$ 3. $D_x\, 23$

4. y' for $y = \pi$ 5. $\dfrac{dy}{dx}$ for $y = x^{12}$ 6. $D_x\, x^5$

7. $f'(x)$ for $f(x) = x$ 8. y' for $y = x^7$ 9. y' for $y = x^{-7}$

10. $f'(x)$ for $f(x) = x^{-11}$ 11. $\dfrac{dy}{dx}$ for $y = x^{5/2}$

12. $D_x\, x^{7/3}$ 13. $D_x\, \dfrac{1}{x^5}$ 14. $f'(x)$ for $f(x) = \dfrac{1}{x^9}$

15. $f'(x)$ for $f(x) = 2x^4$ 16. $\dfrac{dy}{dx}$ for $y = -3x$ 17. $D_x(\frac{1}{3}x^6)$

18. y' for $y = \frac{1}{2}x^4$ 19. $\dfrac{dy}{dx}$ for $y = \dfrac{x^5}{15}$ 20. $f'(x)$ for $f(x) = \dfrac{x^6}{24}$

B

21. $D_x(2x^{-5})$ 22. y' for $y = -4x^{-1}$

23. $f'(x)$ for $f(x) = \dfrac{4}{x^4}$ 24. $\dfrac{dy}{dx}$ for $y = \dfrac{-3}{x^6}$

25. $D_x \dfrac{-1}{2x^2}$

26. y' for $y = \dfrac{1}{6x^3}$

27. $f'(x)$ for $f(x) = -3x^{1/3}$

28. $\dfrac{dy}{dx}$ for $y = -8x^{1/4}$

29. $D_x(2x^2 - 3x + 4)$

30. y' for $y = 3x^2 + 4x - 7$

31. $\dfrac{dy}{dx}$ for $y = 3x^5 - 2x^3 + 5$

32. $f'(x)$ for $f(x) = 2x^3 - 6x + 5$

33. $D_x(3x^{-4} + 2x^{-2})$

34. y' for $y = 2x^{-3} - 4x^{-1}$

35. $\dfrac{dy}{dx}$ for $y = \dfrac{1}{2x} - \dfrac{2}{3x^3}$

36. $f'(x)$ for $f(x) = \dfrac{3}{4x^3} + \dfrac{1}{2x^5}$

37. $D_x(3x^{2/3} - 5x^{1/3})$

38. $D_x(8x^{3/4} + 4x^{-1/4})$

39. $D_x\left(\dfrac{3}{x^{3/5}} - \dfrac{6}{x^{1/2}}\right)$

40. $D_x\left(\dfrac{5}{x^{1/5}} - \dfrac{8}{x^{3/2}}\right)$

41. $D_x \dfrac{1}{\sqrt[3]{x}}$

42. y' for $y = \dfrac{10}{\sqrt[5]{x}}$

43. $\dfrac{dy}{dx}$ for $y = \dfrac{12}{\sqrt{x}} - 3x^{-2} + x$

44. $f'(x)$ for $f(x) = 2x^{-3} - \dfrac{6}{\sqrt[3]{x^2}} + 7$

For Problems 45–48, find:

(A) $f'(x)$
(B) The slope of the graph of f at $x = 2$ and $x = 4$.
(C) The equations of the tangent lines at $x = 2$ and $x = 4$.
(D) The value(s) of x where the tangent line is horizontal.

45. $f(x) = 6x - x^2$

46. $f(x) = 2x^2 + 8x$

47. $f(x) = 3x^4 - 6x^2 - 7$

48. $f(x) = x^4 - 32x^2 + 10$

If an object moves along the y axis (marked in feet) so that its position at time x (in seconds) is given by the indicated function in Problems 49–52, find:

(A) The instantaneous velocity function $v = f'(x)$.
(B) The velocity when $x = 0$ and $x = 3$ seconds.
(C) The time(s) when $v = 0$.

49. $f(x) = 176x - 16x^2$

50. $f(x) = 80x - 10x^2$

51. $f(x) = x^3 - 9x^2 + 15x$

52. $f(x) = x^3 - 9x^2 + 24x$

C In Problems 53–56, find each derivative.

53. $f'(x)$ for $f(x) = \dfrac{10x + 20}{x}$

54. $\dfrac{dy}{dx}$ for $y = \dfrac{x^2 + 25}{x^2}$

55. $D_x \dfrac{x^4 - 3x^3 + 5}{x^2}$

56. y' for $y = \dfrac{2x^5 - 4x^3 + 2x}{x^3}$

In Problems 57 and 58, use the definition of derivative and the two-step process to verify each statement.

57. $D_x x^3 = 3x^2$

58. $D_x x^4 = 4x^3$

APPLICATIONS

Business & Economics

59. *Marginal cost.* The total cost of producing x tennis rackets per day is

$$C(x) = 800 + 60x - \frac{x^2}{4} \qquad 0 \leqslant x \leqslant 120$$

(A) Find the marginal cost at a production level of x rackets.
(B) Find the marginal cost at a production level of 60 rackets, and interpret the result.
(C) Find the actual cost of producing the 61st racket, and compare this cost with the result found in part B.
(D) Find $C'(80)$, and interpret the result.

60. **Marginal cost.** The total cost of producing x portable radios per day is

$$C(x) = 1,000 + 100x - \frac{x^2}{2} \qquad 0 \leqslant x \leqslant 100$$

(A) Find the marginal cost at a production level of x radios.
(B) Find the marginal cost at a production level of 80 radios, and interpret the result.
(C) Find the actual cost of producing the 81st radio, and compare this cost with the result found in part B.
(D) Find $C'(50)$, and interpret the result.

61. *Advertising.* Using past records, it is estimated that a company will sell $N(x)$ units of a product after spending $\$x$ thousand on advertising, as given by

$$N(x) = 60x - x^2 \qquad 5 \leqslant x \leqslant 30$$

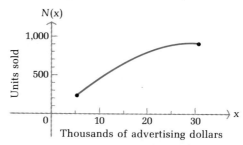

(A) Find $N'(x)$, the instantaneous rate of change of sales with respect to the amount of money spent on advertising at the $\$x$ thousand level.
(B) Find $N'(10)$ and $N'(20)$, and interpret the results.

62. *Demand function.* Suppose that in a given gourmet food store, people are willing to buy $D(x)$ pounds of chocolate candy per day at $\$x$ per quarter pound, as given by the demand function

$$D(x) = 100 - x^2 \qquad 1 \leqslant x \leqslant 10$$

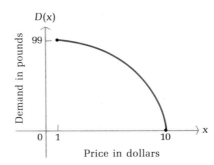

(A) Find $D'(x)$, the instantaneous rate of change of demand with respect to price at the $\$x$ price level.

(B) Find $D'(2)$ and $D'(8)$, and interpret the results.

Life Sciences

63. *Medicine.* A person x inches tall has a pulse rate of y beats per minute, as given approximately by

$$y = 590x^{-1/2} \qquad 30 \leqslant x \leqslant 75$$

What is the instantaneous rate of change of pulse rate at the:

(A) 36 inch level?   (B) 64 inch level?

64. *Ecology.* A coal-burning electrical generating plant emits sulfur dioxide into the surrounding air. The concentration $C(x)$, in parts per million, is given approximately by

$$C(x) = \frac{0.1}{x^2}$$

where x is the distance from the plant in miles. Find the instantaneous rate of change of concentration at:

(A) $x = 1$ mile   (B) $x = 2$ miles

Social Sciences

65. *Learning.* Suppose a person learns y items in x hours, as given by

$$y = 50\sqrt{x} \qquad 0 \leqslant x \leqslant 9$$

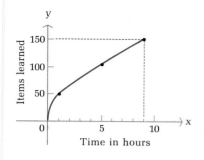

Items learned / Time in hours

(see the figure in the margin). Find the rate of learning at the end of:

(A) 1 hour   (B) 9 hours

66. *Learning.* If a person learns y items in x hours, as given by

$$y = 21\sqrt[3]{x^2} \qquad 0 \leqslant x \leqslant 8$$

find the rate of learning at the end of:

(A) 1 hour   (B) 8 hours

SECTION 9-5

Derivatives of Products and Quotients

◆ DERIVATIVES OF PRODUCTS
◆ DERIVATIVES OF QUOTIENTS

The derivative rules discussed in the preceding section added substantially to our ability to compute and apply derivatives to many practical problems. In this and the next section, we will add a few more rules that will increase this ability even further.

◆ DERIVATIVES OF PRODUCTS

In Section 9-4, we found that the derivative of a sum is the sum of the derivatives. Is the derivative of a product the product of the derivatives? Let us take a look at a simple example. Consider

$$f(x) = u(x)v(x) = (x^2 - 3x)(2x^3 - 1) \tag{1}$$

where $u(x) = x^2 - 3x$ and $v(x) = 2x^3 - 1$. The product of the derivatives is

$$u'(x)v'(x) = (2x - 3)6x^2 = 12x^3 - 18x^2 \tag{2}$$

To see if this is equal to the derivative of the product, we multiply the right side of (1) and use the derivative formulas we already know:

$$f(x) = (x^2 - 3x)(2x^3 - 1) = 2x^5 - 6x^4 - x^2 + 3x$$

Thus,

$$f'(x) = 10x^4 - 24x^3 - 2x + 3 \tag{3}$$

Since (2) and (3) are not equal, we conclude that the derivative of a product is *not* the product of the derivatives. There is a product rule for derivatives, but it is slightly more complicated than you might expect.

Using the definition of derivative and the two-step process, we can show that:

The derivative of the product of two functions is the first function times the derivative of the second function plus the second function times the derivative of the first function.

That is:

■ Product Rule

If

$$y = f(x) = F(x)S(x)$$

and if $F'(x)$ and $S'(x)$ exist, then

$$f'(x) = F(x)S'(x) + S(x)F'(x)$$

Also,

$$y' = FS' + SF' \qquad \frac{dy}{dx} = F\frac{dS}{dx} + S\frac{dF}{dx}$$

$$D_x[F(x)S(x)] = F(x)\,D_x\,S(x) + S(x)\,D_x\,F(x)$$

◆ **EXAMPLE 24**   Use two different methods to find $f'(x)$ for $f(x) = 2x^2(3x^4 - 2)$.

Solution   *Method 1.*   Use the product rule:

$$\begin{aligned}
f'(x) &= 2x^2(3x^4 - 2)' + (3x^4 - 2)(2x^2)' \qquad \text{First times derivative of second} \\
&= 2x^2(12x^3) + (3x^4 - 2)(4x) \qquad\qquad \text{plus second times derivative of first} \\
&= 24x^5 + 12x^5 - 8x \\
&= 36x^5 - 8x
\end{aligned}$$

Method 2.   Multiply first; then take derivatives:

$$\begin{aligned}
f(x) &= 2x^2(3x^4 - 2) = 6x^6 - 4x^2 \\
f'(x) &= 36x^5 - 8x
\end{aligned}$$

◆

PROBLEM 24   Use two different methods to find $f'(x)$ for $f(x) = 3x^3(2x^2 - 3x + 1)$.   ◆

At this point, all the products we will encounter can be differentiated by either of the methods illustrated in Example 24. In the next and later sections, we will see that there are situations where the product rule must be used. Unless instructed otherwise, you should use the product rule to differentiate all products in this section to gain experience with the use of this important differentiation rule.

◆ EXAMPLE 25 Let $f(x) = (2x - 9)(x^2 + 6)$.

(A) Find the equation of the line tangent to the graph of $f(x)$ at $x = 3$.
(B) Find the value(s) of x where the tangent line is horizontal.

Solutions (A) First, find $f'(x)$:

$$f'(x) = (2x - 9)(x^2 + 6)' + (x^2 + 6)(2x - 9)'$$
$$= (2x - 9)(2x) + (x^2 + 6)(2)$$

Now, find the equation of the tangent line at $x = 3$:

$$\begin{array}{ll} y - y_1 = m(x - x_1) & y_1 = f(x_1) = f(3) = -45 \\ y - (-45) = 12(x - 3) & m = f'(x_1) = f'(3) = 12 \\ y = 12x - 81 & \text{Tangent line at } x = 3 \end{array}$$

(B) The tangent line is horizontal at any value of x such that $f'(x) = 0$, so

$$f'(x) = (2x - 9)2x + (x^2 + 6)2 = 0$$
$$6x^2 - 18x + 12 = 0$$
$$x^2 - 3x + 2 = 0$$
$$(x - 1)(x - 2) = 0$$
$$x = 1, 2$$

The tangent line is horizontal at $x = 1$ and at $x = 2$. ◆

PROBLEM 25 Repeat Example 25 for $f(x) = (2x + 9)(x^2 - 12)$. ◆

As Example 25 illustrates, the way we write $f'(x)$ depends on what we want to do with it. If we are interested only in evaluating $f'(x)$ at specified values of x, the form in part A is sufficient. However, if we want to solve $f'(x) = 0$, we must multiply and collect like terms, as we did in part B.

◆ DERIVATIVES OF QUOTIENTS

As in the case with a product, the derivative of a quotient of two functions is not the quotient of the derivatives of the two functions.

Let

$$f(x) = \frac{T(x)}{B(x)} \qquad \text{where } T'(x) \text{ and } B'(x) \text{ exist}$$

Starting with the definition of a derivative, it can be shown that

$$f'(x) = \frac{B(x)T'(x) - T(x)B'(x)}{[B(x)]^2}$$

Thus:

The derivative of the quotient of two functions is the bottom function times the derivative of the top function minus the top function times the derivative of the bottom function, all over the bottom function squared.

Quotient Rule

If

$$y = f(x) = \frac{T(x)}{B(x)}$$

and if $T'(x)$ and $B'(x)$ exist, then

$$f'(x) = \frac{B(x)T'(x) - T(x)B'(x)}{[B(x)]^2}$$

Also,

$$y' = \frac{BT' - TB'}{B^2} \qquad \frac{dy}{dx} = \frac{B\dfrac{dT}{dx} - T\dfrac{dB}{dx}}{B^2}$$

$$D_x \frac{T(x)}{B(x)} = \frac{B(x)\, D_x\, T(x) - T(x)\, D_x\, B(x)}{[B(x)]^2}$$

◆ E X A M P L E   26   (A) If $f(x) = \dfrac{x^2}{2x - 1}$, find $f'(x)$.

(B) Find: $D_x \dfrac{x^2 - x}{x^3 + 1}$

(C) Find $D_x \dfrac{x^2 - 3}{x^2}$ by using the quotient rule and also by splitting the fraction into two fractions.

Solutions   (A) $f'(x) = \dfrac{(2x - 1)(x^2)' - x^2(2x - 1)'}{(2x - 1)^2}$ The bottom times the derivative of the top minus the top times the derivative of the bottom, all over the square of the bottom

$$= \frac{(2x - 1)(2x) - x^2(2)}{(2x - 1)^2}$$

$$= \frac{4x^2 - 2x - 2x^2}{(2x - 1)^2}$$

$$= \frac{2x^2 - 2x}{(2x - 1)^2}$$

(B) $D_x \dfrac{x^2 - x}{x^3 + 1} = \dfrac{(x^3 + 1)\, D_x(x^2 - x) - (x^2 - x)\, D_x(x^3 + 1)}{(x^3 + 1)^2}$

$$= \frac{(x^3 + 1)(2x - 1) - (x^2 - x)(3x^2)}{(x^3 + 1)^2}$$

$$= \frac{2x^4 - x^3 + 2x - 1 - 3x^4 + 3x^3}{(x^3 + 1)^2}$$

$$= \frac{-x^4 + 2x^3 + 2x - 1}{(x^3 + 1)^2}$$

(C) *Method 1.* Use the quotient rule:

$$D_x \frac{x^2 - 3}{x^2} = \frac{x^2 D_x(x^2 - 3) - (x^2 - 3) D_x x^2}{(x^2)^2}$$

$$= \frac{x^2(2x) - (x^2 - 3)2x}{x^4}$$

$$= \frac{2x^3 - 2x^3 + 6x}{x^4} = \frac{6x}{x^4} = \frac{6}{x^3}$$

Method 2. Split into two fractions:

$$\frac{x^2 - 3}{x^2} = \frac{x^2}{x^2} - \frac{3}{x^2} = 1 - 3x^{-2}$$

$$D_x(1 - 3x^{-2}) = 0 - 3(-2)x^{-3} = \frac{6}{x^3}$$

Comparing methods 1 and 2, we see that it often pays to change an expression algebraically before blindly using a differentiation formula. ◆

PROBLEM 26 Find:

(A) $f'(x)$ for $f(x) = \dfrac{2x}{x^2 + 3}$ (B) y' for $y = \dfrac{x^3 - 3x}{x^2 - 4}$

(C) $D_x \dfrac{2 + x^3}{x^3}$ two ways ◆

◆ **EXAMPLE 27**

Sales Analysis

When a successful home video game is first introduced, the monthly sales generally increase rapidly for a period of time, and then begin to decrease. Suppose that the monthly sales $S(t)$, in thousands of games, t months after the game is introduced are given by

$$S(t) = \frac{200t}{t^2 + 100}$$

(A) Find $S'(t)$.   (B) Find $S(5)$ and $S'(5)$, and interpret the results.
(C) Find $S(30)$ and $S'(30)$, and interpret the results.

Solutions

(A) $S'(t) = \dfrac{(t^2 + 100)(200t)' - 200t(t^2 + 100)'}{(t^2 + 100)^2}$

$$= \frac{(t^2 + 100)200 - 200t(2t)}{(t^2 + 100)^2}$$

$$= \frac{200t^2 + 20,000 - 400t^2}{(t^2 + 100)^2}$$

$$= \frac{20,000 - 200t^2}{(t^2 + 100)^2}$$

(B) $S(5) = \dfrac{200(5)}{5^2 + 100} = 8$ and $S'(5) = \dfrac{20,000 - 200(5)^2}{(5^2 + 100)^2} = 0.96$

The sales for the 5th month are 8,000 units. At this point in time, sales are increasing at the rate of 0.96(1,000) = 960 units per month.

(C) $S(30) = \dfrac{200(30)}{30^2 + 100} = 6$ and $S'(30) = \dfrac{20,000 - 200(30)^2}{(30^2 + 100)^2} = -0.16$

The sales for the 30th month are 6,000 units. At this point in time, sales are decreasing at the rate of 0.16(1,000) = 160 units per month.   ◆

The function S(t) in Example 27 is graphed in Figure 17. Notice that the maximum monthly sales seem to occur during the 10th month. In the next chapter, we will see how the derivative S'(t) is used to help sketch the graph of S(t) and to find the maximum monthly sales.

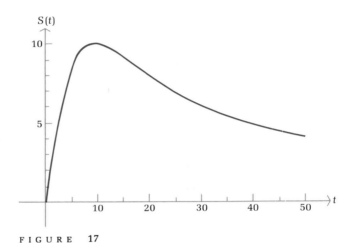

FIGURE   17

PROBLEM   27   Refer to Example 27. Suppose that the monthly sales S(t), in thousands of games, t months after the game is introduced are given by

$$S(t) = \dfrac{200t}{t^2 + 64}$$

(A) Find S'(t).   (B) Find S(4) and S'(4), and interpret the results.
(C) Find S(24) and S'(24), and interpret the results.   ◆

Answers to Matched Problems   24. $30x^4 - 36x^3 + 9x^2$ 25.   (A) $y = 84x - 297$ (B) $x = -4, x = 1$

26.   (A) $\dfrac{(x^2 + 3)2 - (2x)(2x)}{(x^2 + 3)^2} = \dfrac{6 - 2x^2}{(x^2 + 3)^2}$

(B) $\dfrac{(x^2 - 4)(3x^2 - 3) - (x^3 - 3x)(2x)}{(x^2 - 4)^2} = \dfrac{x^4 - 9x^2 + 12}{(x^2 - 4)^2}$ (C) $-\dfrac{6}{x^4}$

27. (A) $S'(t) = \dfrac{12{,}800 - 200t^2}{(t^2 + 64)^2}$

(B) $S(4) = 10$; $S'(4) = 1.5$; at $t = 4$ months, monthly sales are 10,000 and increasing at 1,500 games per month

(C) $S(24) = 7.5$; $S'(24) = -0.25$; at $t = 24$ months, monthly sales are 7,500 and decreasing at 250 games per month

A   *For $f(x)$ as given, find $f'(x)$ and simplify.*

1. $f(x) = 2x^3(x^2 - 2)$

2. $f(x) = 5x^2(x^3 + 2)$

3. $f(x) = (x - 3)(2x - 1)$

4. $f(x) = (3x + 2)(4x - 5)$

5. $f(x) = \dfrac{x}{x - 3}$

6. $f(x) = \dfrac{3x}{2x + 1}$

7. $f(x) = \dfrac{2x + 3}{x - 2}$

8. $f(x) = \dfrac{3x - 4}{2x + 3}$

9. $f(x) = (x^2 + 1)(2x - 3)$

10. $f(x) = (3x + 5)(x^2 - 3)$

11. $f(x) = \dfrac{x^2 + 1}{2x - 3}$

12. $f(x) = \dfrac{3x + 5}{x^2 - 3}$

13. $f(x) = (x^2 + 2)(x^2 - 3)$

14. $f(x) = (x^2 - 4)(x^2 + 5)$

15. $f(x) = \dfrac{x^2 + 2}{x^2 - 3}$

16. $f(x) = \dfrac{x^2 - 4}{x^2 + 5}$

B   *Find each of the following and simplify:*

17. $f'(x)$ for $f(x) = (2x + 1)(x^2 - 3x)$

18. y' for $y = (x^3 + 2x^2)(3x - 1)$

19. $\dfrac{dy}{dx}$ for $y = (2x - x^2)(5x + 2)$

20. $D_x[(3 - x^3)(x^2 - x)]$

21. y' for $y = \dfrac{5x - 3}{x^2 + 2x}$

22. $f'(x)$ for $f(x) = \dfrac{3x^2}{2x - 1}$

23. $D_x \dfrac{x^2 - 3x + 1}{x^2 - 1}$

24. $\dfrac{dy}{dx}$ for $y = \dfrac{x^4 - x^3}{3x - 1}$

In Problems 25–28, find $f'(x)$ and find the equation of the line tangent to the graph of f at $x = 2$.

25. $f(x) = (1 + 3x)(5 - 2x)$

26. $f(x) = (7 - 3x)(1 + 2x)$

27. $f(x) = \dfrac{x - 8}{3x - 4}$

28. $f(x) = \dfrac{2x - 5}{2x - 3}$

In Problems 29–32, find $f'(x)$ and find the value(s) of x where $f'(x) = 0$.

29. $f(x) = (2x - 15)(x^2 + 18)$

30. $f(x) = (2x - 3)(x^2 - 6)$

31. $f(x) = \dfrac{x}{x^2 + 1}$

32. $f(x) = \dfrac{x}{x^2 + 9}$

In Problems 33–36, find $f'(x)$ two ways; by using the product or quotient rule and by simplifying first.

33. $f(x) = x^3(x^4 - 1)$

34. $f(x) = x^4(x^3 - 1)$

35. $f(x) = \dfrac{x^3 + 9}{x^3}$

36. $f(x) = \dfrac{x^4 + 4}{x^4}$

C   Find each of the following. Do not simplify.

37. $f'(x)$ for $f(x) = (2x^4 - 3x^3 + x)(x^2 - x + 5)$

38. $\dfrac{dy}{dx}$ for $y = (x^2 - 3x + 1)(x^3 + 2x^2 - x)$

39. $D_x \dfrac{3x^2 - 2x + 3}{4x^2 + 5x - 1}$

40. y' for $y = \dfrac{x^3 - 3x + 4}{2x^2 + 3x - 2}$

41. $\dfrac{dy}{dx}$ for $y = 9x^{1/3}(x^3 + 5)$

42. $D_x[(4x^{1/2} - 1)(3x^{1/3} + 2)]$

43. $f'(x)$ for $f(x) = \dfrac{6\sqrt[3]{x}}{x^2 - 3}$

44. y' for $y = \dfrac{2\sqrt{x}}{x^2 - 3x + 1}$

45. $D_x \dfrac{x^3 - 2x^2}{\sqrt[3]{x^2}}$

46. $\dfrac{dy}{dx}$ for $y = \dfrac{x^2 - 3x + 1}{\sqrt[4]{x}}$

47. $f'(x)$ for $f(x) = \dfrac{(2x^2 - 1)(x^2 + 3)}{x^2 + 1}$

48. y' for $y = \dfrac{2x - 1}{(x^3 + 2)(x^2 - 3)}$

APPLICATIONS

Business & Economics

49. *Sales analysis.* The monthly sales S (in thousands) for a record album are given by

$$S(t) = \frac{200t}{t^2 + 36}$$

where t is the number of months since the release of the album.

(A) Find $S'(t)$, the instantaneous rate of change of monthly sales with respect to time.

(B) Find $S(2)$ and $S'(2)$, and interpret the results.

(C) Find $S(8)$ and $S'(8)$, and interpret the results.

50. *Sales analysis.* A communications company has installed a cable television system in a city. The total number N (in thousands) of subscribers t months after the installation of the system is given by

$$N(t) = \frac{200t}{t+5}$$

(A) Find $N'(t)$, the instantaneous rate of change of the total number of subscribers with respect to time.

(B) Find $N(5)$ and $N'(5)$, and interpret the results.

(C) Find $N(15)$ and $N'(15)$, and interpret the results.

51. *Price–demand function.* According to classical economic theory, the demand $d(x)$ for a commodity in a free market decreases as the price x increases. Suppose that the number $d(x)$ of transistor radios people are willing to buy per week in a given city at a price $\$x$ is given by

$$d(x) = \frac{50,000}{x^2 + 10x + 25} \qquad 4 \leqslant x \leqslant 15$$

(A) Find $d'(x)$, the instantaneous rate of change of demand with respect to price change.

(B) Find $d'(5)$ and $d'(10)$, and interpret the results.

52. *Employee training.* A company producing computer components has established that on the average, a new employee can assemble $N(t)$ components per day after t days of on-the-job training, as given by

$$N(t) = \frac{100t}{t+9}$$

(A) Find $N'(t)$, the instantaneous rate of change of units assembled with respect to time.

(B) Find $N'(1)$ and $N'(11)$, and interpret the results.

Life Sciences   **53.** *Medicine.* A drug is injected into the bloodstream of a patient through her right arm. The concentration of the drug in the bloodstream of the left arm t hours after the injection is given by

$$C(t) = \frac{0.14t}{t^2 + 1}$$

(A) Find $C'(t)$, the instantaneous rate of change of drug concentration with respect to time.

(B) Find $C'(0.5)$ and $C'(3)$, and interpret the results.

54. *Drug sensitivity.* One hour after x milligrams of a particular drug are given to a person, the change in body temperature $T(x)$, in degrees Fahrenheit, is given approximately by

$$T(x) = x^2 \left(1 - \frac{x}{9} \right) \qquad 0 \leqslant x \leqslant 7$$

The rate at which T changes with respect to the size of the dosage x, $T'(x)$, is called the *sensitivity* of the body to the dosage.

(A) Find $T'(x)$, using the product rule.

(B) Find $T'(1)$, $T'(3)$, and $T'(6)$.

Social Sciences   55. *Learning.* In the early days of quantitative learning theory (around 1917), L. L. Thurstone found that a given person successfully accomplished $N(x)$ acts after x practice acts, as given by

$$N(x) = \frac{100x + 200}{x + 32}$$

(A) Find the instantaneous rate of change of learning, $N'(x)$, with respect to the number of practice acts x.

(B) Find $N'(4)$ and $N'(68)$.

S E C T I O N   9-6   # Chain Rule: Power Form

◆   CHAIN RULE: POWER FORM
◆   COMBINING RULES OF DIFFERENTIATION

◆ CHAIN RULE: POWER FORM

We have already made extensive use of the power rule,

$$D_x x^n = nx^{n-1} \qquad \text{n any real number} \qquad (1)$$

Now we want to generalize this rule so that we can differentiate functions of the form $[u(x)]^n$. Is rule (1) still valid if we replace x with a function $u(x)$? We begin by considering a simple example. Let $u(x) = 2x$ and $n = 4$. Then

$$[u(x)]^n = (2x)^4 = 2^4 x^4 = 16x^4$$

and

$$D_x[u(x)]^n = D_x 16x^4 = 64x^3 \qquad (2)$$

But

$$n[u(x)]^{n-1} = 4(2x)^3 = 32x^3 \qquad (3)$$

Comparing (2) and (3), we see that

$$D_x[u(x)]^n \neq n[u(x)]^{n-1}$$

for this particular choice of $u(x)$ and n. (In fact, it can be shown that the only time this last equation is valid is if $u(x) = x + k$, k a constant.) Thus, we cannot generalize the power rule by simply substituting $u(x)$ for x in rule (1).

How can we find a formula for $D_x[u(x)]^n$ where $u(x)$ is an arbitrary differentiable function? Let us first find $D_x[u(x)]^2$ and $D_x[u(x)]^3$ to see if a general pattern emerges. Since $[u(x)]^2 = u(x)u(x)$, we use the product rule with $F(x) = u(x)$ and $S(x) = u(x)$ to write

$$D_x[u(x)]^2 = D_x[u(x)u(x)] = u(x)u'(x) + u(x)u'(x)$$
$$= 2u(x)u'(x) \qquad (4)$$

Since $[u(x)]^3 = [u(x)]^2u(x)$, we now use the product rule with $F(x) = [u(x)]^2$ and (4) to write

$$D_x[u(x)]^3 = D_x[u(x)]^2u(x) = [u(x)]^2 D_x u(x) + u(x) D_x[u(x)]^2$$
$$= [u(x)]^2u'(x) + u(x)[2u(x)u'(x)]$$
$$= 3[u(x)]^2u'(x)$$

Continuing in this fashion, it can be shown that

$$D_x[u(x)]^n = n[u(x)]^{n-1}u'(x) \qquad n \text{ a positive integer} \qquad (5)$$

Using more advanced techniques, formula (5) can be established for all real numbers n. Thus, we have the **general power rule.**

■ **General Power Rule**

If n is any real number, then

$$D_x[u(x)]^n = n[u(x)]^{n-1}u'(x)$$

provided $u'(x)$ exists. This rule is often written more compactly as

$$D_x u^n = nu^{n-1} \frac{du}{dx} \qquad u = u(x)$$

The general power rule is a special case of a very important and useful differentiation rule called the **chain rule.** In essence, the chain rule will enable us to differentiate a composition form $f[g(x)]$ if we know how to differentiate $f(x)$ and $g(x)$. We defer a complete discussion of the chain rule until Chapter 10.

◆ E X A M P L E   28   Find $f'(x)$:

(A) $f(x) = (3x + 1)^4$ (B) $f(x) = (x^3 + 4)^7$

(C) $f(x) = \dfrac{1}{(x^2 + x + 4)^3}$ (D) $f(x) = \sqrt{3 - x}$

Solutions (A) $f(x) = (3x + 1)^4$ Let $u = 3x + 1$, $n = 4$.

$$f'(x) \;\boxed{= 4(3x + 1)^3 \, D_x(3x + 1)} \qquad nu^{n-1} \frac{du}{dx}$$

$$= 4(3x + 1)^3 3 \qquad \frac{du}{dx} = 3$$

$$= 12(3x + 1)^3$$

(B) $f(x) = (x^3 + 4)^7$ Let $u = (x^3 + 4)$, $n = 7$.

$$f'(x) \;\boxed{= 7(x^3 + 4)^6 \, D_x(x^3 + 4)} \qquad nu^{n-1} \frac{du}{dx}$$

$$= 7(x^3 + 4)^6 3x^2 \qquad \frac{du}{dx} = 3x^2$$

$$= 21x^2(x^3 + 4)^6$$

(C) $f(x) = \dfrac{1}{(x^2 + x + 4)^3} = (x^2 + x + 4)^{-3}$ Let $u = x^2 + x + 4$, $n = -3$.

$$f'(x) \;\boxed{= -3(x^2 + x + 4)^{-4} \, D_x(x^2 + x + 4)} \qquad nu^{n-1} \frac{du}{dx}$$

$$= -3(x^2 + x + 4)^{-4}(2x + 1) \qquad \frac{du}{dx} = 2x + 1$$

$$= \frac{-3(2x + 1)}{(x^2 + x + 4)^4}$$

(D) $f(x) = \sqrt{3 - x} = (3 - x)^{1/2}$ Let $u = 3 - x$, $n = \frac{1}{2}$.

$$f'(x) \;\boxed{= \frac{1}{2}(3 - x)^{-1/2} \, D_x(3 - x)} \qquad nu^{n-1} \frac{du}{dx}$$

$$= \frac{1}{2}(3 - x)^{-1/2}(-1) \qquad \frac{du}{dx} = -1$$

$$= -\frac{1}{2(3 - x)^{1/2}} \quad \text{or} \quad -\frac{1}{2\sqrt{3 - x}} \qquad \blacklozenge$$

PROBLEM 28   Find $f'(x)$:

(A) $f(x) = (5x + 2)^3$ (B) $f(x) = (x^4 - 5)^5$

(C) $f(x) = \dfrac{1}{(x^2 + 4)^2}$ (D) $f(x) = \sqrt{4 - x}$ $\blacklozenge$

Notice that we used two steps to differentiate each function in Example 28. First, we applied the general power rule; then we found du/dx. As you gain experience with the general power rule, you may want to combine these two steps. If you do this, be certain to multiply by du/dx. For example,

$$D_x(x^5 + 1)^4 = 4(x^5 + 1)^3 5x^4 \qquad \text{Correct}$$

$$D_x(x^5 + 1)^4 \neq 4(x^5 + 1)^3 \qquad du/dx = 5x^4 \text{ is missing}$$

If we let $u(x) = x$, then $du/dx = 1$, and the **general power rule reduces to the (ordinary) power rule** discussed in Section 9-4. Compare the following:

$$D_x\, x^n = nx^{n-1} \qquad \text{Yes—power rule}$$

$$D_x\, u^n = nu^{n-1}\frac{du}{dx} \qquad \text{Yes—general power rule}$$

$$D_x\, u^n \neq nu^{n-1} \qquad \text{Unless } u(x) = x + k \text{ so that } du/dx = 1$$

◆ COMBINING RULES OF DIFFERENTIATION

The following examples illustrate the use of the general power rule in combination with other rules of differentiation.

◆ **EXAMPLE   29**   Find the equation of the line tangent to the graph of f at $x = 2$ for $f(x) = x^2\sqrt{2x + 12}$.

Solution

$$f(x) = x^2\sqrt{2x + 12}$$
$$= x^2(2x + 12)^{1/2}$$
Apply the product rule with $F(x) = x^2$ and $S(x) = (2x + 12)^{1/2}$.

$$f'(x) = x^2\, D_x(2x + 12)^{1/2} + (2x + 12)^{1/2}\, D_x\, x^2$$
$$= x^2[\tfrac{1}{2}(2x + 12)^{-1/2}](2) + (2x + 12)^{1/2}(2x)$$
$$= \frac{x^2}{\sqrt{2x + 12}} + 2x\sqrt{2x + 12}$$
Use the general power rule to differentiate $(2x + 12)^{1/2}$ and the ordinary power rule to differentiate x^2.

$$f'(2) = \frac{4}{\sqrt{16}} + 4\sqrt{16} = 1 + 16 = 17$$

$$f(2) = 4\sqrt{16} = 16$$

$$(x_1, y_1) = (2, f(2)) = (2, 16) \qquad \text{Point}$$
$$m = f'(2) = 17 \qquad \text{Slope}$$
$$y - 16 = 17(x - 2) \qquad y - y_1 = m(x - x_1)$$
$$y = 17x - 18 \qquad \text{Tangent line} \qquad ◆$$

PROBLEM   29   Find the equation of the line tangent to the graph of f at $x = 3$ for $f(x) = x\sqrt{15 - 2x}$.   ◆

◆ E X A M P L E   30   Find the value(s) of x where the tangent line is horizontal for

$$f(x) = \frac{x^3}{(2 - 3x)^5}$$

Solution   Use the quotient rule with $T(x) = x^3$ and $B(x) = (2 - 3x)^5$:

$$f'(x) = \frac{(2 - 3x)^5 \, D_x \, x^3 - x^3 \, D_x(2 - 3x)^5}{[(2 - 3x)^5]^2}$$

Use the ordinary power rule to differentiate x^3 and the general power rule to differentiate $(2 - 3x)^5$.

$$= \frac{(2 - 3x)^5 3x^2 - x^3 5(2 - 3x)^4(-3)}{(2 - 3x)^{10}}$$

$$= \frac{(2 - 3x)^4 3x^2[(2 - 3x) + 5x]}{(2 - 3x)^{10}}$$

$$= \frac{3x^2(2 + 2x)}{(2 - 3x)^6} = \frac{6x^2(x + 1)}{(2 - 3x)^6}$$

Since a fraction is 0 when the numerator is 0 and the denominator is not, we see that $f'(x) = 0$ at $x = -1$ and $x = 0$. Thus, the graph of f will have horizontal tangent lines at $x = -1$ and $x = 0$.   ◆

P R O B L E M   30   Find the value(s) of x where the tangent line is horizontal for

$$f(x) = \frac{x^3}{(3x - 2)^2}$$ ◆

◆ E X A M P L E   31   Starting with the function f in Example 30, write f as a product and then differentiate.

Solution $$f(x) = \frac{x^3}{(2 - 3x)^5} = x^3(2 - 3x)^{-5}$$

$$f'(x) = x^3 \, D_x(2 - 3x)^{-5} + (2 - 3x)^{-5} \, D_x \, x^3$$
$$= x^3(-5)(2 - 3x)^{-6}(-3) + (2 - 3x)^{-5} 3x^2$$
$$= 15x^3(2 - 3x)^{-6} + 3x^2(2 - 3x)^{-5}$$

At this point, we have an unsimplified form for $f'(x)$. This may be satisfactory for some purposes, but not for others. For example, if we need to solve the equation $f'(x) = 0$, we must simplify algebraically:

$$f'(x) = \frac{15x^3}{(2 - 3x)^6} + \frac{3x^2}{(2 - 3x)^5} = \frac{15x^3}{(2 - 3x)^6} + \frac{3x^2(2 - 3x)}{(2 - 3x)^6}$$

$$= \frac{15x^3 + 3x^2(2 - 3x)}{(2 - 3x)^6} = \frac{3x^2(5x + 2 - 3x)}{(2 - 3x)^6}$$

$$= \frac{3x^2(2 + 2x)}{(2 - 3x)^6} = \frac{6x^2(1 + x)}{(2 - 3x)^6}$$ ◆

Refer to the function f in Problem 30, above. Write f as a product and then differentiate. Do not simplify. ◆

As Example 31 illustrates, any quotient can be converted to a product and differentiated by the product rule. However, if the derivative must be simplified, it is usually easier to use the quotient rule. (Compare the algebraic simplifications in Example 31 with those in Example 30.) There is one special case where using negative exponents is the preferred method—a fraction whose numerator is a constant.

◆ EXAMPLE 32   Find $f'(x)$ two ways for: $f(x) = \dfrac{4}{(x^2 + 9)^3}$

Solution   *Method 1.*   Use the quotient rule:

$$f'(x) = \frac{(x^2 + 9)^3 \, D_x \, 4 - 4 \, D_x (x^2 + 9)^3}{[(x^2 + 9)^3]^2}$$

$$= \frac{(x^2 + 9)^3 (0) - 4[3(x^2 + 9)^2 (2x)]}{(x^2 + 9)^6}$$

$$= \frac{-24x(x^2 + 9)^2}{(x^2 + 9)^6} = \frac{-24x}{(x^2 + 9)^4}$$

Method 2.   Rewrite as a product, and use the general power rule:

$$f(x) = \frac{4}{(x^2 + 9)^3} = 4(x^2 + 9)^{-3}$$

$$f'(x) = 4(-3)(x^2 + 9)^{-4}(2x)$$

$$= \frac{-24x}{(x^2 + 9)^4}$$

Which method do you prefer? ◆

PROBLEM 32   Find $f'(x)$ two ways for: $f(x) = \dfrac{5}{(x^3 + 1)^2}$ ◆

Answers to Matched Problems   28. (A) $15(5x + 2)^2$ (B) $20x^3(x^4 - 5)^4$ (C) $-4x/(x^2 + 4)^3$
(D) $-1/(2\sqrt{4 - x})$
29. $y = 2x + 3$ 30. $x = 0, x = 2$
31. $-6x^3(3x - 2)^{-3} + 3x^2(3x - 2)^{-2}$ 32. $-30x^2/(x^3 + 1)^3$

EXERCISE 9-6

A   In Problems 1–12, find $f'(x)$ using the general power rule and simplify.

1. $f(x) = (2x + 5)^3$ 2. $f(x) = (3x - 7)^5$ 3. $f(x) = (5 - 2x)^4$
4. $f(x) = (9 - 5x)^2$ 5. $f(x) = (3x^2 + 5)^5$ 6. $f(x) = (5x^2 - 3)^6$

7. $f(x) = (x^3 - 2x^2 + 2)^8$ 8. $f(x) = (2x^2 + x + 1)^7$ 9. $f(x) = (2x - 5)^{1/2}$
10. $f(x) = (4x + 3)^{1/2}$ 11. $f(x) = (x^4 + 1)^{-2}$ 12. $f(x) = (x^5 + 2)^{-3}$

In Problems 13–16, find $f'(x)$ and the equation of the line tangent to the graph of f at the indicated value of x. Find the value(s) of x where the tangent line is horizontal.

13. $f(x) = (2x - 1)^3$; $x = 1$ 14. $f(x) = (3x - 1)^4$; $x = 1$
15. $f(x) = (4x - 3)^{1/2}$; $x = 3$ 16. $f(x) = (2x + 8)^{1/2}$; $x = 4$

B In Problems 17–34, find dy/dx using the general power rule.

17. $y = 3(x^2 - 2)^4$ 18. $y = 2(x^3 + 6)^5$
19. $y = 2(x^2 + 3x)^{-3}$ 20. $y = 3(x^3 + x^2)^{-2}$
21. $y = \sqrt{x^2 + 8}$ 22. $y = \sqrt[3]{3x - 7}$
23. $y = \sqrt[3]{3x + 4}$ 24. $y = \sqrt{2x - 5}$
25. $y = (x^2 - 4x + 2)^{1/2}$ 26. $y = (2x^2 + 2x - 3)^{1/2}$

27. $y = \dfrac{1}{2x + 4}$ 28. $y = \dfrac{1}{3x - 7}$

29. $y = \dfrac{1}{(x^3 + 4)^5}$ 30. $y = \dfrac{1}{(x^2 - 3)^6}$

31. $y = \dfrac{1}{4x^2 - 4x + 1}$ 32. $y = \dfrac{1}{2x^2 - 3x + 1}$

33. $y = \dfrac{4}{\sqrt{x^2 - 3x}}$ 34. $y = \dfrac{3}{\sqrt[3]{x - x^2}}$

In Problems 35–40, find $f'(x)$, and find the equation of the line tangent to the graph of f at the indicated value of x.

35. $f(x) = x(4 - x)^3$; $x = 2$ 36. $f(x) = x^2(1 - x)^4$; $x = 2$

37. $f(x) = \dfrac{x}{(2x - 5)^3}$; $x = 3$ 38. $f(x) = \dfrac{x^4}{(3x - 8)^2}$; $x = 4$

39. $f(x) = x\sqrt{2x + 2}$; $x = 1$ 40. $f(x) = x\sqrt{x - 6}$; $x = 7$

In Problems 41–46, find $f'(x)$, and find the value(s) of x where the tangent line is horizontal.

41. $f(x) = x^2(x - 5)^3$ 42. $f(x) = x^3(x - 7)^4$

43. $f(x) = \dfrac{x}{(2x + 5)^2}$ 44. $f(x) = \dfrac{x - 1}{(x - 3)^3}$

45. $f(x) = \sqrt{x^2 - 8x + 20}$ 46. $f(x) = \sqrt{x^2 + 4x + 5}$

C In Problems 47–58, find each derivative and simplify.

47. $D_x[3x(x^2 + 1)^3]$ 48. $D_x[2x^2(x^3 - 3)^4]$

49. $D_x \dfrac{(x^3 - 7)^4}{2x^3}$ 50. $D_x \dfrac{3x^2}{(x^2 + 5)^3}$

51. $D_x[(2x - 3)^2(2x^2 + 1)^3]$ 52. $D_x[(x^2 - 1)^3(x^2 - 2)^2]$
53. $D_x(4x^2\sqrt{x^2 - 1})$ 54. $D_x(3x\sqrt{2x^2 + 3})$

55. $D_x \dfrac{2x}{\sqrt{x-3}}$

56. $D_x \dfrac{x^2}{\sqrt{x^2+1}}$

57. $D_x \sqrt{(2x-1)^3(x^2+3)^4}$

58. $D_x \sqrt{\dfrac{4x+1}{2x^2+1}}$

APPLICATIONS

Business & Economics

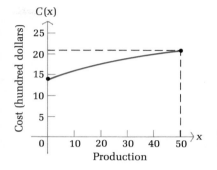

59. *Marginal cost.* The total cost (in hundreds of dollars) of producing x calculators per day is

$$C(x) = 10 + \sqrt{2x+16} \qquad 0 \leqslant x \leqslant 50$$

(see the figure in the margin).

(A) Find the marginal cost at a production level of x calculators.

(B) Find $C'(24)$ and $C'(42)$, and interpret the results.

60. *Marginal cost.* The total cost (in thousands of dollars) of producing x cameras per week is

$$C(x) = 6 + \sqrt{4x+4} \qquad 0 \leqslant x \leqslant 30$$

(A) Find the marginal cost at a production level of x cameras.

(B) Find $C'(15)$ and $C'(24)$, and interpret the results.

61. *Compound interest.* If $1,000 is invested at an interest rate of i compounded monthly, the amount in the account at the end of 4 years is given by

$$A = 1,000(1 + \tfrac{1}{12}i)^{48}$$

Find dA/di.

62. *Compound interest.* If $100 is invested at an interest rate of i compounded semiannually, the amount in the account at the end of 5 years is given by

$$A = 100(1 + \tfrac{1}{2}i)^{10}$$

Find dA/di.

Life Sciences

63. *Bacteria growth.* The number y of bacteria in a certain colony after x days is given approximately by

$$y = (3 \times 10^6) \left[1 - \frac{1}{\sqrt[3]{(x^2-1)^2}} \right]$$

Find dy/dx.

64. *Pollution.* A small lake in a resort area became contaminated with harmful bacteria because of excessive septic tank seepage. After treating the lake with a bactericide, the Department of Public Health estimated the bacteria concentration (number per cubic centimeter) after t days to be given by

$$C(t) = 500(8 - t)^2 \qquad 0 \leqslant t \leqslant 7$$

(A) Find $C'(t)$ using the general power rule.

(B) Find $C'(1)$ and $C'(6)$, and interpret the results.

Social Sciences

65. Learning. In 1930, L. L. Thurstone developed the following formula to indicate how learning time T depends on the length of a list n:

$$T = f(n) = \frac{c}{k} \ n\sqrt{n-a}$$

where a, c, and k are empirical constants. Suppose that for a particular person, time T (in minutes) for learning a list of length n is

$$T = f(n) = 2n\sqrt{n-2}$$

(A) Find dT/dn, the instantaneous rate of change in time with respect to n.
(B) Find $f'(11)$ and $f'(27)$, and interpret the results.

SECTION 9-7 Marginal Analysis in Business and Economics

◆ MARGINAL COST, REVENUE, AND PROFIT
◆ APPLICATION
◆ MARGINAL AVERAGE COST, REVENUE, AND PROFIT

◆ MARGINAL COST, REVENUE, AND PROFIT

One important use of calculus in business and economics is in *marginal analysis*. We introduced the concept of *marginal cost* earlier. There is no reason to stop there. Economists also talk about *marginal revenue* and *marginal profit*. Recall that the word "marginal" refers to an instantaneous rate of change — that is, a derivative. Thus, we define the following:

Marginal Cost, Revenue, and Profit

If x is the number of units of a product produced in some time interval, then

$$\text{Total cost} = C(x) \qquad \text{Total revenue} = R(x)$$
$$\textbf{Marginal cost} = C'(x) \qquad \textbf{Marginal revenue} = R'(x)$$

$$\text{Total profit} = P(x) = R(x) - C(x)$$
$$\textbf{Marginal profit} = P'(x) = R'(x) - C'(x)$$
$$= (\text{Marginal revenue}) - (\text{Marginal cost})$$

Marginal cost (or revenue or profit) is the instantaneous rate of change of cost (or revenue or profit) relative to production at a given production level.

Marginal functions have several important economic interpretations. Earlier in this chapter, we discussed interpretations of marginal cost, which we summarize here. Similar interpretations can be made for marginal revenue and marginal profit.

Returning to the definition of a derivative, we observe the following (assuming the limit exists):

$$C'(x) = \lim_{h \to 0} \frac{C(x + h) - C(x)}{h} \qquad \text{Marginal cost}$$

$$C'(x) \approx \frac{C(x + h) - C(x)}{h} \qquad h \neq 0$$

$$C'(x) \approx \frac{C(x + 1) - C(x)}{1}$$

$$= C(x + 1) - C(x)$$

= Exact change in total cost for 1 unit change in production at the x level of production

= Exact cost of producing the (x + 1)st item at the x level of production

Thus, at a production level of x units:

Marginal cost $C'(x)$ approximates the change in total cost that results from a 1 unit change in production.

In other words,

Marginal cost $C'(x)$ approximates the cost of producing the (x + 1)st item.

These observations are illustrated in Figure 18. Note that if $h = 1$ as shown in the figure, then $C(x + 1) - C(x)$ is the exact change in total cost per unit change in production at a production level of x units. The marginal cost, $C'(x)$, is the slope of the tangent line, and is approximately equal to the change in total cost C per unit change in production at a production level of x units.

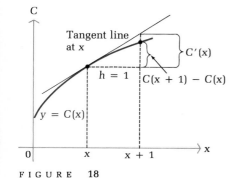

FIGURE 18
$C'(x) \approx C(x + 1) - C(x)$

◆ APPLICATION

We now present an example in market research to show how marginal cost, revenue, and profit are tied together.

◆ EXAMPLE 33

Production Strategy

The market research department of a company recommends that the company manufacture and market a new transistor radio. After suitable test marketing, the research department presents the following **demand equation:**

$$x = 10,000 - 1,000p \qquad \text{x is demand at \$p per radio} \tag{1}$$

or

$$p = 10 - \frac{x}{1,000} \tag{2}$$

where x is the number of radios retailers are likely to buy per week at $p per radio. Equation (2) is simply equation (1) solved for p in terms of x. Notice that as price goes up, demand goes down.

The financial department provides the following **cost equation:**

$$C(x) = 7,000 + 2x \tag{3}$$

where $7,000 is the estimated fixed costs (tooling and overhead), and $2 is the estimated variable costs (cost per unit for materials, labor, marketing, transportation, storage, etc.).

The **marginal cost** is

$$C'(x) = 2$$

Since this is a constant, it costs an additional $2 to produce 1 more radio at all production levels.

The **revenue** (the amount of money R received by the company for manufacturing and selling x units at $p per unit) is

$$R = (\text{Number of units sold})(\text{Price per unit}) = xp$$

In general, the revenue R can be expressed in terms of p by using equation (1) or in terms of x by using equation (2). In marginal analysis (problems involving marginal cost, marginal revenue, or marginal profit), cost, revenue, and profit must be expressed in terms of the number of units x. Thus, the **revenue equation** in terms of x is

$$R(x) = xp = x\left(10 - \frac{x}{1,000}\right) \qquad \text{Using equation (2)} \tag{4}$$

$$= 10x - \frac{x^2}{1,000}$$

The **marginal revenue** is

$$R'(x) = 10 - \frac{x}{500}$$

For production levels of $x = 2,000$, $5,000$, and $7,000$, we have

$$R'(2,000) = 6 \qquad R'(5,000) = 0 \qquad R'(7,000) = -4$$

This means that at production levels of 2,000, 5,000, and 7,000, the respective approximate changes in revenue per unit change in production are $6, $0, and −$4. That is, at the 2,000 output level, revenue increases as production increases; at the 5,000 output level, revenue does not change with a "small" change in production; and at the 7,000 output level, revenue decreases with an increase in production.

When we graph $R(x)$ and $C(x)$ in the same coordinate system, we obtain Figure 19.

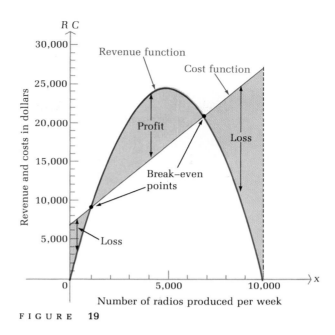

R C

30,000

25,000

20,000

15,000

10,000

5,000

Revenue and costs in dollars

Revenue function

Cost function

Profit

Loss

Break–even
points

Loss

0   5,000   10,000   x

Number of radios produced per week

FIGURE   19

The **break-even points** (the points where revenue equals cost) are obtained as follows:

$$C(x) = R(x)$$

$$7{,}000 + 2x = 10x - \frac{x^2}{1{,}000}$$

$$\frac{x^2}{1{,}000} - 8x + 7{,}000 = 0$$

$$x^2 - 8{,}000x + 7{,}000{,}000 = 0 \qquad \text{Solve using the quadratic formula (see Section 2-2).}$$

$$x = \frac{8{,}000 \pm \sqrt{8{,}000^2 - 4(7{,}000{,}000)}}{2}$$

$$= \frac{8{,}000 \pm \sqrt{36{,}000{,}000}}{2}$$

$$= \frac{8{,}000 \pm 6{,}000}{2}$$

$$= 1{,}000, \quad 7{,}000$$

$$R(1{,}000) = 10(1{,}000) - \frac{1{,}000^2}{1{,}000} = 9{,}000$$

$$C(1{,}000) = 7{,}000 + 2(1{,}000) = 9{,}000$$

$$R(7,000) = 10(7,000) - \frac{7,000^2}{1,000} = 21,000$$

$$C(7,000) = 7,000 + 2(7,000) = 21,000$$

Thus, the break-even points are (1,000, 9,000) and (7,000, 21,000), as shown in Figure 19.

The **profit equation** is

$$P(x) = R(x) - C(x)$$

$$= \left(10x - \frac{x^2}{1,000}\right) - (7,000 + 2x)$$

$$= -\frac{x^2}{1,000} + 8x - 7,000$$

The graph in Figure 19 also provides some useful information concerning the profit equation. At a production level of 1,000 or 7,000, revenue equals cost; hence, profit is 0 and the company will break even. For any production level between 1,000 and 7,000, revenue is greater than cost; hence, $P(x)$ is positive and the company will make a profit. For production levels less than 1,000 or greater than 7,000, revenue is less than cost; hence, $P(x)$ is negative and the company will have a loss.

The **marginal profit** is

$$P'(x) = -\frac{x}{500} + 8$$

For production levels of 1,000, 4,000, and 6,000, we have

$$P'(1,000) = 6 \qquad P'(4,000) = 0 \qquad P'(6,000) = -4$$

This means that at production levels of 1,000, 4,000, and 6,000, the respective approximate changes in profit per unit change in production are $6, $0, and −$4. That is, at the 1,000 output level, profit will be increased if production is increased; at the 4,000 output level, profit does not change for "small" changes in production; and at the 6,000 output level, profits will decrease if production is increased. It seems the best production level to produce a maximum profit is 4,000.

Example 33 warrants careful study, since a number of important ideas in economics and calculus are involved. In the next chapter, we will develop a systematic procedure for finding the production level (and, using the demand equation, the selling price) that will maximize profit. ◆

P R O B L E M   33   Refer to the revenue and profit equations in Example 33.

(A) Find $R'(3,000)$ and $R'(6,000)$, and interpret the results.
(B) Find $P'(2,000)$ and $P'(7,000)$, and interpret the results. ◆

◆ MARGINAL AVERAGE COST, REVENUE, AND PROFIT

Sometimes, it is desirable to carry out marginal analysis relative to **average cost (cost per unit), average revenue (revenue per unit), and average profit (profit per unit).** The relevant definitions are summarized in the following box:

Marginal Average Cost, Revenue, and Profit

If x is the number of units of a product produced in some time interval, then

COST PER UNIT:

$$\text{Average cost} = \overline{C}(x) = \frac{C(x)}{x}$$

$$\text{Marginal average cost} = \overline{C}'(x) = D_x\,\overline{C}(x)$$

REVENUE PER UNIT:

$$\text{Average revenue} = \overline{R}(x) = \frac{R(x)}{x}$$

$$\text{Marginal average revenue} = \overline{R}'(x) = D_x\,\overline{R}(x)$$

PROFIT PER UNIT:

$$\text{Average profit} = \overline{P}(x) = \frac{P(x)}{x}$$

$$\text{Marginal average profit} = \overline{P}'(x) = D_x\,\overline{P}(x)$$

As was the case with marginal cost:

The marginal average cost approximates the change in average cost that results from a unit increase in production.

Similar statements can be made for marginal average revenue and marginal average profit.

◆ E X A M P L E 34

Cost Analysis

A small machine shop manufactures drill bits used in the petroleum industry. The shop manager estimates that the total daily cost (in dollars) of producing x bits is

$$C(x) = 1{,}000 + 25x - \frac{x^2}{10} \qquad \text{Total cost function}$$

Thus,

$$\overline{C}(x) = \frac{C(x)}{x} = \frac{1{,}000}{x} + 25 - \frac{x}{10} \qquad \text{Average cost function}$$

$$\overline{C}'(x) = D_x\,\overline{C}(x) = -\frac{1{,}000}{x^2} - \frac{1}{10} \qquad \text{Marginal average cost function}$$

$$\overline{C}(10) = \frac{1,000}{10} + 25 - \frac{10}{10}$$ Average cost per unit if 10 units are produced

$$= \$124$$

$$\overline{C}'(10) = -\frac{1,000}{100} - \frac{1}{10}$$ A unit increase in production will decrease the average cost per unit by approximately \$10.10 at a production level of 10 units.

$$= -\$10.10$$

PROBLEM   34   Consider the cost function $C(x) = 7,000 + 2x$ from Example 33.

(A) Find $\overline{C}(x)$ and $\overline{C}'(x)$.
(B) Find $\overline{C}(1,000)$ and $\overline{C}'(1,000)$, and interpret the results.   ◆

Answers to Matched Problems

33. (A) $R'(3,000) = 4$ (at a production level of 3,000, a unit increase in production will increase revenue by approx. \$4); $R'(6,000) = -2$ (at a production level of 6,000, a unit increase in production will decrease revenue by approx. \$2)

(B) $P'(2,000) = 4$ (at a production level of 2,000, a unit increase in production will increase profit by approx. \$4); $P'(7,000) = -6$ (at a production level of 7,000, a unit increase in production will decrease profit by approx. \$6)

34. (A) $\overline{C}(x) = \frac{7,000}{x} + 2, \quad \overline{C}'(x) = -\frac{7,000}{x^2}$

(B) $\overline{C}(1,000) = 9$ (at a production level of 1,000, the average cost per unit is \$9); $\overline{C}'(1,000) = -0.007$ (at a production level of 1,000, a unit increase in production will decrease the average cost per unit by approx. 0.7¢)

E X E R C I S E   9-7

A P P L I C A T I O N S

Business & Economics

1. *Cost analysis.* The total cost (in dollars) of producing x food processors is

$$C(x) = 2,000 + 50x - \frac{x^2}{2}$$

(A) Find the exact cost of producing the 21st food processor.
(B) Use the marginal cost to approximate the cost of producing the 21st food processor.

2. *Cost analysis.* The total cost (in dollars) of producing x electric guitars is

$$C(x) = 1{,}000 + 100x - \frac{x^2}{4}$$

(A) Find the exact cost of producing the 51st guitar.
(B) Use the marginal cost to approximate the cost of producing the 51st guitar.

3. *Cost analysis.* The total cost (in dollars) of manufacturing x auto body frames is

$$C(x) = 60{,}000 + 300x$$

(A) Find the average cost per unit if 500 frames are produced.
(B) Find the marginal average cost at a production level of 500 units, and interpret the results.

4. *Cost analysis.* The total cost (in dollars) of printing x dictionaries is

$$C(x) = 20{,}000 + 10x$$

(A) Find the average cost per unit if 1,000 dictionaries are produced.
(B) Find the marginal average cost at a production level of 1,000 units, and interpret the results.

5. *Revenue analysis.* The total revenue (in dollars) from the sale of x clock radios is

$$R(x) = 100x - \frac{x^2}{40}$$

Evaluate the marginal revenue at the given values of x, and interpret the results.

(A) x = 1,600   (B) x = 2,500

6. *Revenue analysis.* The total revenue (in dollars) from the sale of x steam irons is

$$R(x) = 50x - \frac{x^2}{20}$$

Evaluate the marginal revenue at the given values of x, and interpret the results.

(A) x = 400   (B) x = 650

7. *Profit analysis.* The total profit (in dollars) from the sale of x skateboards is

$$P(x) = 30x - \frac{x^2}{2} - 250$$

(A) Find the exact profit from the sale of the 26th skateboard.
(B) Use the marginal profit to approximate the profit from the sale of the 26th skateboard.

8. *Profit analysis.* The total profit (in dollars) from the sale of x portable stereos is

$$P(x) = 22x - \frac{x^2}{10} - 400$$

(A) Find the exact profit from the sale of the 41st stereo.
(B) Use the marginal profit to approximate the profit from the sale of the 41st stereo.

9. *Profit analysis.* The total profit (in dollars) from the sale of x video cassettes is

$$P(x) = 5x - \frac{x^2}{200} - 450$$

Evaluate the marginal profit at the given values of x, and interpret the results.

(A) $x = 450$ (B) $x = 750$

10. *Profit analysis.* The total profit (in dollars) from the sale of x cameras is

$$P(x) = 12x - \frac{x^2}{50} - 1,000$$

Evaluate the marginal profit at the given values of x, and interpret the results.

(A) $x = 200$ (B) $x = 350$

11. *Profit analysis.* Refer to the profit equation in Problem 9.

(A) Find the average profit per unit if 150 cassettes are produced.
(B) Find the marginal average profit at a production level of 150 units, and interpret the results.

12. *Profit analysis.* Refer to the profit equation in Problem 10.

(A) Find the average profit per unit if 200 cameras are produced.
(B) Find the marginal average profit at a production level of 200 units, and interpret the results.

13. *Revenue, cost, and profit.* In Example 33, suppose we have the demand equation

$$x = 6,000 - 30p \quad \text{or} \quad p = 200 - \frac{x}{30}$$

and the cost equation

$$C(x) = 72,000 + 60x$$

(A) Find the marginal cost.
(B) Find the revenue equation in terms of x.
(C) Find the marginal revenue.
(D) Find $R'(1,500)$ and $R'(4,500)$, and interpret the results.

(E) Graph the cost function and the revenue function on the same coordinate system for $0 \leqslant x \leqslant 6{,}000$. Find the break-even points, and indicate regions of loss and profit.

(F) Find the profit equation in terms of x.

(G) Find the marginal profit.

(H) Find $P'(1{,}500)$ and $P'(3{,}000)$, and interpret the results.

14. *Revenue, cost, and profit.* In Example 33, suppose we have the demand equation

$$x = 9{,}000 - 30p \quad \text{or} \quad p = 300 - \frac{x}{30}$$

and the cost equation

$$C(x) = 150{,}000 + 30x$$

(A) Find the marginal cost.

(B) Find the revenue equation in terms of x.

(C) Find the marginal revenue.

(D) Find $R'(3{,}000)$ and $R'(6{,}000)$, and interpret the results.

(E) Graph the cost function and the revenue function on the same coordinate system for $0 \leqslant x \leqslant 9{,}000$. Find the break-even points, and indicate regions of loss and profit.

(F) Find the profit equation in terms of x.

(G) Find the marginal profit.

(H) Find $P'(1{,}500)$ and $P'(4{,}500)$, and interpret the results.

15. *Revenue, cost, and profit.* A company is planning to manufacture and market a new two-slice electric toaster. After conducting extensive market surveys, the research department provides the following estimates: a weekly demand of 200 toasters at a price of \$16 per toaster and a weekly demand of 300 toasters at a price of \$14 per toaster. The financial department estimates that weekly fixed costs will be \$1,400 and variable costs (cost per unit) will be \$4.

(A) Assume that the demand equation is linear. Use the research department's estimates to find the demand equation.

(B) Find the revenue equation in terms of x.

(C) Assume that the cost equation is linear. Use the financial department's estimates to find the cost equation.

(D) Graph the cost function and the revenue function on the same coordinate system for $0 \leqslant x \leqslant 1{,}000$. Find the break-even points, and indicate regions of loss and profit.

(E) Find the profit equation in terms of x.

(F) Evaluate the marginal profit at $x = 250$ and $x = 475$, and interpret the results.

16. *Revenue, cost, and profit.* The company in Problem 15 is also planning to manufacture and market a four-slice toaster. For this toaster, the research department's estimates are a weekly demand of 300 toasters at a price of $25 per toaster and a weekly demand of 400 toasters at a price of $20. The financial department's estimates are fixed weekly costs of $5,000 and variable costs of $5 per toaster. Assume the demand and cost equations are linear (see Problem 15, parts A and C).

(A) Use the research department's estimates to find the demand equation.
(B) Find the revenue equation in terms of x.
(C) Use the financial department's estimates to find the cost equation in terms of x.
(D) Graph the cost function and the revenue function on the same coordinate system for $0 \leq x \leq 800$. Find the break-even points, and indicate regions of loss and profit.
(E) Find the profit equation in terms of x.
(F) Evaluate the marginal profit at $x = 325$ and $x = 425$, and interpret the results.

Chapter Review

Important Terms and Symbols

9-1   *Limits and Continuity — A Geometric Introduction.* Limit as x approaches c; one-sided limits; limit as x approaches c from the left; left-hand limit; limit as x approaches c from the right; right-hand limit; continuous curve; continuity at a point; discontinuity at a point; continuity on an open interval; continuity on a closed or half-closed interval; continuity properties

$$\lim_{x \to c} f(x); \quad \lim_{x \to c^-} f(x); \quad \lim_{x \to c^+} f(x)$$

9-2   *Computation of Limits.* Limits at points of continuity; limits and infinity; vertical asymptote; limit properties; 0/0 indeterminate form; limits at infinity; horizontal asymptote

$$\lim_{x \to c} f(x) = f(c) \quad \text{if } f \text{ is continuous at } x = c;$$

$$\lim_{x \to c} f(x) = \infty; \quad \lim_{x \to c} f(x) = -\infty; \quad \lim_{x \to \infty} f(x); \quad \lim_{x \to -\infty} f(x)$$

9-3   *The Derivative.* Secant lines; tangent lines; difference quotient; slope of a tangent line; slope of a graph; average rate of change; instantaneous rate of change; the derivative; differentiable at a point; nondifferentiable at a point; differentiable over an open interval; differentiable over a closed or half-closed interval; differentiation

$$f'(x) = \lim_{h \to 0} \frac{f(x + h) - f(x)}{h}$$

9-4　*Derivatives of Constants, Power Forms, and Sums.* **Derivative notation; derivative of a constant; power rule; constant times a function rule; sum and difference rule; marginal cost**

$$f'(x); \quad y'; \quad \frac{dy}{dx}; \quad D_x f(x)$$

9-5　*Derivatives of Products and Quotients.* **Product rule; quotient rule**

9-6　*Chain Rule: Power Form.* **General power rule; chain rule; combining rules of differentiation**

9-7　*Marginal Analysis in Business and Economics.* **Marginal cost; marginal revenue; marginal profit; demand equation; cost equation; revenue equation; break-even points; profit equation; average cost; marginal average cost; average revenue; marginal average revenue; average profit; marginal average profit**

$$C'(x); \quad \overline{C}(x); \quad \overline{C}'(x); \quad R'(x); \quad \overline{R}(x); \quad \overline{R}'(x); \quad P'(x); \quad \overline{P}(x); \quad \overline{P}'(x)$$

Summary of Rules of Differentiation

$$D_x k = 0$$
$$D_x x^n = nx^{n-1}$$
$$D_x kf(x) = kf'(x)$$
$$D_x[u(x) \pm v(x)] = u'(x) \pm v'(x)$$

$$D_x[F(x)S(x)] = F(x)S'(x) + S(x)F'(x)$$
$$D_x \frac{T(x)}{B(x)} = \frac{B(x)T'(x) - T(x)B'(x)}{[B(x)]^2}$$
$$D_x[u(x)]^n = n[u(x)]^{n-1}u'(x)$$

EXERCISE 9-8

Chapter Review

Work through all the problems in this chapter review and check your answers in the back of the book. (Answers to all review problems are there.) Where weaknesses show up, review appropriate sections in the text.

A　In Problems 1–10, find $f'(x)$ for $f(x)$ as given.

1.　$f(x) = 3x^4 - 2x^2 + 1$

2.　$f(x) = 2x^{1/2} - 3x$

3.　$f(x) = 5$

4.　$f(x) = \dfrac{1}{2x^2} + \dfrac{x^2}{2}$

5.　$f(x) = (2x - 1)(3x + 2)$

6.　$f(x) = (x^2 - 1)(x^3 - 3)$

7.　$f(x) = \dfrac{2x}{x^2 + 2}$

8.　$f(x) = \dfrac{1}{3x + 2}$

9.　$f(x) = (2x - 3)^3$

10.　$f(x) = (x^2 + 2)^{-2}$

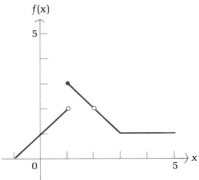

In Problems 11–13, use the graph of the function f shown in the margin to answer each question.

11. (A) $\lim_{x \to 1} f(x) = ?$ (B) $f(1) = ?$
    (C) Is f continuous at x = 1?

12. (A) $\lim_{x \to 2} f(x) = ?$ (B) $f(2) = ?$
    (C) Is f continuous at x = 2?

13. (A) $\lim_{x \to 3} f(x) = ?$ (B) $f(3) = ?$
    (C) Is f continuous at x = 3?

B   In Problems 14–21, find the indicated derivative.

14. $\dfrac{dy}{dx}$ for $y = 3x^4 - 2x^{-3} + 5$

15. y' for $y = (2x^2 - 3x + 2)(x^2 + 2x - 1)$

16. $f'(x)$ for $f(x) = \dfrac{2x - 3}{(x - 1)^2}$ 17. y' for $y = 2\sqrt{x} + \dfrac{4}{\sqrt{x}}$

18. $D_x[(x^2 - 1)(2x + 1)^2]$ 19. $D_x \sqrt[3]{x^3 - 5}$

20. $\dfrac{dy}{dx}$ for $y = \dfrac{3x^2 + 4}{x^2}$ 21. $D_x \dfrac{(x^2 + 2)^4}{2x - 3}$

22. For $y = f(x) = x^2 + 4$, find:
    (A) The slope of the graph at x = 1
    (B) The equation of the tangent line at x = 1 in the form $y = mx + b$

23. Repeat Problem 22 for $f(x) = x^3(x + 1)^2$.

In Problems 24–27, find the value(s) of x where the tangent line is horizontal.

24. $f(x) = 10x - x^2$ 25. $f(x) = (x + 3)(x^2 - 45)$

26. $f(x) = \dfrac{x}{x^2 + 4}$ 27. $f(x) = x^2(2x - 15)^3$

28. If an object moves along the y axis (scale in feet) so that it is at $y = f(x) = 16x^2 - 4x$ at time x (in seconds), find:
    (A) The instantaneous velocity function
    (B) The velocity at time x = 3 seconds

29. An object moves along the y axis (scale in feet) so that at time x (in seconds) it is at $y = f(x) = 96x - 16x^2$. Find:
    (A) The instantaneous velocity function
    (B) The time(s) when the velocity is 0

Problems 30 and 31 refer to the function f described in the figure in the margin.

30. (A) $\lim_{x \to 2^-} f(x) = ?$ (B) $\lim_{x \to 2^+} f(x) = ?$ (C) $\lim_{x \to 2} f(x) = ?$
    (D) $f(2) = ?$ (E) Is f continuous at x = 2?

f(x)

$f(x) = \begin{cases} x^2 & 0 \leq x < 2 \\ 8 - x & x \geq 2 \end{cases}$

31. (A) $\lim\limits_{x \to 5^-} f(x) = ?$ (B) $\lim\limits_{x \to 5^+} f(x) = ?$ (C) $\lim\limits_{x \to 5} f(x) = ?$

(D) $f(5) = ?$ (E) Is f continuous at $x = 5$?

32. Find each limit. (Use $-\infty$ or ∞, if appropriate.)

(A) $\lim\limits_{x \to 0^-} \dfrac{1}{|x|}$ (B) $\lim\limits_{x \to 0^+} \dfrac{1}{|x|}$ (C) $\lim\limits_{x \to 0} \dfrac{1}{|x|}$

33. Find each limit. (Use $-\infty$ or ∞, if appropriate.)

(A) $\lim\limits_{x \to 1^-} \dfrac{1}{x - 1}$ (B) $\lim\limits_{x \to 1^+} \dfrac{1}{x - 1}$ (C) $\lim\limits_{x \to 1} \dfrac{1}{x - 1}$

In Problems 34–38, determine where f is continuous. Express the answer in interval notation.

34. $f(x) = 2x^2 - 3x + 1$ **35.** $f(x) = \dfrac{1}{x + 5}$ **36.** $f(x) = \dfrac{x - 3}{x^2 - x - 6}$

37. $f(x) = \sqrt{x - 3}$ **38.** $f(x) = \sqrt[3]{1 - x^2}$

In Problems 39–52, find each limit, if it exists. (Use $-\infty$ or ∞, if appropriate.)

39. $\lim\limits_{x \to 3} \dfrac{2x - 3}{x + 5}$ **40.** $\lim\limits_{x \to 3} (2x^2 - x + 1)$

41. $\lim\limits_{x \to 0} \dfrac{2x}{3x^2 - 2x}$ **42.** $\lim\limits_{h \to 0} \dfrac{[(2 + h)^2 - 1] - [2^2 - 1]}{h}$

43. $\lim\limits_{h \to 0} \dfrac{f(2 + h) - f(2)}{h}$ for $f(x) = x^2 + 4$

44. $\lim\limits_{x \to 3} \dfrac{x - 3}{x^2 - 9}$ **45.** $\lim\limits_{x \to -3} \dfrac{x - 3}{x^2 - 9}$

46. $\lim\limits_{x \to 7} \dfrac{\sqrt{x} - \sqrt{7}}{x - 7}$ **47.** $\lim\limits_{x \to -2} \sqrt{\dfrac{x^2 + 4}{2 - x}}$

48. $\lim\limits_{x \to \infty} \left(3 + \dfrac{1}{x^{1/3}} + \dfrac{2}{x^3} \right)$ **49.** $\lim\limits_{x \to \infty} (3x^3 - 2x^2 - 10x - 100)$

50. $\lim\limits_{x \to \infty} \dfrac{2x^2 + 3}{3x^2 + 2}$ **51.** $\lim\limits_{x \to \infty} \dfrac{2x + 3}{3x^2 + 2}$

52. $\lim\limits_{x \to \infty} \dfrac{2x^2 + 3}{3x + 2}$

In Problems 53 and 54, use the definition of the derivative to find $f'(x)$.

53. $f(x) = x^2 - x$ **54.** $f(x) = \sqrt{x} - 3$

C Problems *55–58* refer to the function *f* in the figure. Determine whether *f* is differentiable at the indicated value of *x*.

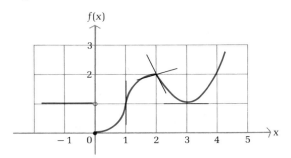

f(x)

55. $x = 0$ 56. $x = 1$ 57. $x = 2$ 58. $x = 3$

In Problems 59 and 60, graph f and find all discontinuities.

59. $f(x) = \begin{cases} 4 - x^2 & \text{if } x < 0 \\ 2 + x^2 & \text{if } x \geq 0 \end{cases}$ 60. $f(x) = \begin{cases} 4 - x^2 & \text{if } x < 1 \\ 3x & \text{if } x \geq 1 \end{cases}$

In Problems 61–64, find f'(x) and simplify.

61. $f(x) = (x - 4)^4(x + 3)^3$ 62. $f(x) = \dfrac{x^5}{(2x + 1)^4}$

63. $f(x) = \dfrac{\sqrt{x^2 - 1}}{x}$ 64. $f(x) = \dfrac{x}{\sqrt{x^2 + 4}}$

Answer the questions in Problems 65–67 for the function f shown in the figure in the margin and given below:

$$f(x) = 1 - |x - 1| \qquad 0 \leq x \leq 2$$

65. (A) $\lim\limits_{x \to 1^-} f(x) = ?$ (B) $\lim\limits_{x \to 1^+} f(x) = ?$

     (C) $\lim\limits_{x \to 1} f(x) = ?$ (D) Is *f* continuous at $x = 1$?

66. (A) Is *f* continuous on the open interval (0, 2)?
     (B) Is *f* continuous from the right at $x = 0$?
     (C) Is *f* continuous from the left at $x = 2$?
     (D) Is *f* continuous on the closed interval [0, 2]?

67. (A) $\lim\limits_{h \to 0^-} \dfrac{f(1 + h) - f(1)}{h} = ?$ (B) $\lim\limits_{h \to 0^+} \dfrac{f(1 + h) - f(1)}{h} = ?$

     (C) $\lim\limits_{h \to 0} \dfrac{f(1 + h) - f(1)}{h} = ?$ (D) Does $f'(1)$ exist?

APPLICATIONS

Business & Economics   68. *Cost analysis.* The total cost (in dollars) of producing *x* television sets is

$$C(x) = 10{,}000 + 200x - 0.1x^2$$

(A) Find the exact cost of producing the 101st television set.

(B) Use the marginal cost to approximate the cost of producing the 101st television set.

69. *Marginal analysis.* Let

$$p = 20 - x \quad \text{and} \quad C(x) = 2x + 56 \quad 0 \leqslant x \leqslant 20$$

be the demand equation and the cost function, respectively, for a certain commodity.

(A) Find the marginal cost, average cost, and marginal average cost functions.

(B) Express the revenue in terms of x, and find the marginal revenue, average revenue, and marginal average revenue functions.

(C) Find the profit, marginal profit, average profit, and marginal average profit functions.

(D) Find the break-even point(s).

(E) Evaluate the marginal profit at $x = 7$, 9, and 11, and interpret the results.

(F) Graph $R = R(x)$ and $C = C(x)$ on the same coordinate system, and locate regions of profit and loss.

70. *Employee training.* A company producing computer components has established that on the average, a new employee can assemble $N(t)$ components per day after t days of on-the-job training, as given by

$$N(t) = \frac{40t}{t + 2}$$

(A) Find the average rate of change of $N(t)$ from 3 days to 6 days.

(B) Find the instantaneous rate of change of $N(t)$ at 3 days.

(C) Find $\lim_{t \to \infty} N(t)$.

Life Sciences

71. *Pollution.* A sewage treatment plant disposes of its effluent through a pipeline that extends 1 mile toward the center of a large lake. The concentration of effluent $C(x)$, in parts per million, x meters from the end of the pipe is given approximately by

$$C(x) = 500(x + 1)^{-2}$$

What is the instantaneous rate of change of concentration at 9 meters? At 99 meters?

Social Sciences

72. *Learning.* If a person learns N items in t hours, as given by

$$N(t) = 20\sqrt{t}$$

find the rate of learning after:

(A) 1 hour   (B) 4 hours

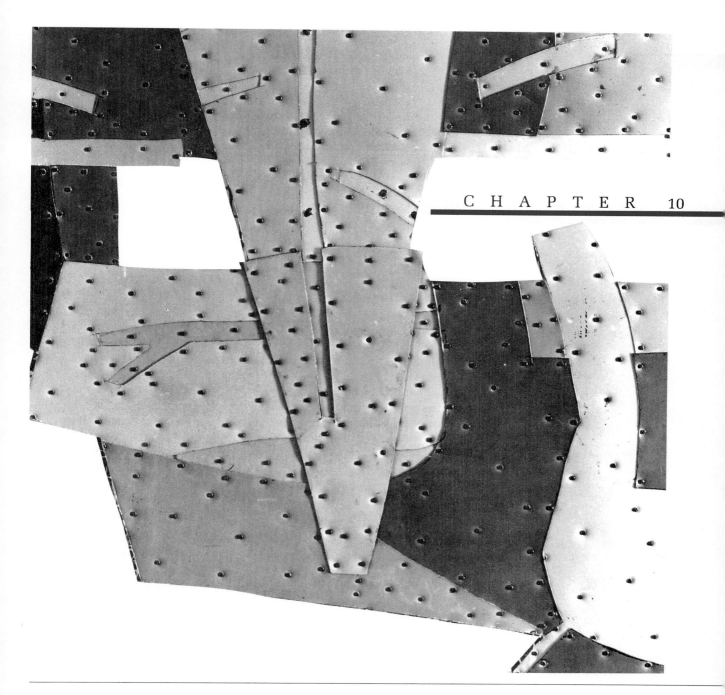

Additional Derivative Topics

Contents

CHAPTER 10

SECTION 10-1

First Derivative and Graphs

- ◆ SOLVING INEQUALITIES USING CONTINUITY PROPERTIES
- ◆ INCREASING AND DECREASING FUNCTIONS
- ◆ CRITICAL VALUES AND LOCAL EXTREMA
- ◆ FIRST-DERIVATIVE TEST

Since the derivative is associated with the slope of the graph of a function at a point, we might expect that it is also associated with other properties of a graph. As we will see in this and the next section, the derivative can tell us a great deal about the shape of the graph of a function. In addition, this investigation will lead to methods for finding absolute maximum and minimum values for functions that do not require graphing. Manufacturing companies can use these methods to find production levels that will minimize cost or maximize profit. Pharmacologists can use them to find levels of drug dosages that will produce maximum sensitivity to a drug. And so on.

We digress for a moment to discuss the use of continuity and *sign charts* in solving inequalities, a process that will see frequent use in this chapter.

◆ SOLVING INEQUALITIES USING CONTINUITY PROPERTIES

In our informal discussion of continuity in Section 9-1, we said that a function is continuous over an interval if we can draw its graph over the interval without lifting a pencil from the paper. Suppose a function f is continuous over the interval $(1, 8)$ and $f(x) \neq 0$ for any x in $(1, 8)$. Also suppose $f(2) = 5$, a positive number. Is it possible for $f(x)$ to be negative for any x in $(1, 8)$? The answer is "no." If $f(7)$ were -3, for example, as shown in Figure 1, how would it be possible to join the points $(2, 5)$ and $(7, -3)$ with the graph of a continuous function without crossing the x axis between 1 and 8 at least once? [Crossing the x axis would violate our assumption that $f(x) \neq 0$ for any x in $(1, 8)$.] Thus, we conclude that $f(x)$ must be positive for all x in $(1, 8)$. If $f(2)$ were negative, then, using the same type of reasoning, $f(x)$ would have to be negative over the whole interval $(1, 8)$.

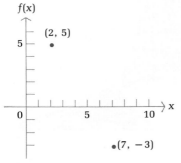

FIGURE 1

In general, **if f is continuous and $f(x) \neq 0$ on the interval (a, b), then $f(x)$ cannot change sign on (a, b).** This is the essence of Theorem 1.

T H E O R E M   1

Sign Properties on an Interval (a, b)

If f is continuous on (a,b) and $f(x) \neq 0$ for all x in (a, b), then either $f(x) > 0$ for all x in (a,b) or $f(x) < 0$ for all x in (a,b).

Theorem 1 provides the basis for an effective method of solving many types of inequalities. Example 1 illustrates the process.

◆ **E X A M P L E   1**   Solve: $\dfrac{x+1}{x-2} > 0$

Solution   We start by using the left side of the inequality to form the function f:

$$f(x) = \frac{x+1}{x-2}$$

The rational function f is discontinuous at $x = 2$, and $f(x) = 0$ for $x = -1$ (a fraction is 0 when the numerator is 0 and the denominator is not 0). We plot $x = 2$ and $x = -1$, which we call *partition numbers*, on a real number line:

(Note that the dot at 2 is open, because the function is not defined at $x = 2$.) The partition numbers 2 and -1 determine three open intervals: $(-\infty, -1)$, $(-1, 2)$, and $(2, \infty)$. The function f is continuous and nonzero on each of these intervals. From Theorem 1 we know that $f(x)$ does not change sign on any of these intervals. Thus, we can find the sign of $f(x)$ on each of these intervals by selecting a **test number** in each interval and evaluating $f(x)$ at that number. Since any number in each subinterval will do, we choose test numbers that are easy to evaluate: -2, 0, and 3. The table in the margin shows the results.

The sign of $f(x)$ at each test number is the same as the sign of $f(x)$ over the interval containing that test number. Using this information, we construct a **sign chart** for $f(x)$:

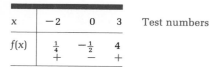

x	-2	0	3	Test numbers
$f(x)$	$\frac{1}{4}$ $+$	$-\frac{1}{2}$ $-$	4 $+$	

Test numbers

Now using the sign chart, we can easily write the solution for the given nonlinear inequality: $f(x) > 0$ for

$x < -1$ or $x > 2$	Inequality notation
$(-\infty, -1) \cup (2, \infty)$	Interval notation

Most of the inequalities we will encounter will involve strict inequalities ($>$ or $<$). If it is necessary to solve inequalities of the form $\geq$ or $\leq$, we simply include the end point of any interval if it is a zero of f [that is, if it is a value of x such that $f(x) = 0$]. For example, referring to the sign chart in Example 1, the solution of the inequality

$$\frac{x+1}{x-2} \geq 0 \quad \text{is} \quad \begin{array}{ll} x \leq -1 \text{ or } x > 2 & \text{Inequality notation} \\ (-\infty, -1] \cup (2, \infty) & \text{Interval notation} \end{array}$$

In general, given a function f, we will call all values x such that f is discontinuous at x or $f(x) = 0$ **partition numbers. Partition numbers determine open intervals where $f(x)$ does not change sign.** By using a test number from each interval, we can construct a sign chart for $f(x)$ on the real number line. It is then an easy matter to determine where $f(x) < 0$ or $f(x) > 0$; that is, to solve the inequality $f(x) < 0$ or $f(x) > 0$.

We summarize the procedure for constructing sign charts in the following box:

Constructing Sign Charts

Given a function f:

Step 1. Find all partition numbers. That is:

(A) Find all numbers where f is discontinuous. (Rational functions are discontinuous for values of x that make a denominator 0.)

(B) Find all numbers where $f(x) = 0$. (For a rational function, this occurs where the numerator is 0 and the denominator is not 0.)

Step 2. Plot the numbers found in step 1 on a real number line, dividing the number line into intervals.

Step 3. Select a test number in each open interval determined in step 2, and evaluate $f(x)$ at each test number to determine whether $f(x)$ is positive ($+$) or negative ($-$) in each interval.

Step 4. Construct a sign chart using the real number line in step 2. This will show the sign of $f(x)$ on each open interval

[*Note:* From the sign chart, it is easy to find the solution for the inequality $f(x) < 0$ or $f(x) > 0$.]

PROBLEM 1 Solve: $\dfrac{x^2 - 1}{x - 3} < 0$

◆ INCREASING AND DECREASING FUNCTIONS

Graphs of functions generally have *rising* or *falling* sections as we scan the graphs from left to right. It would be an aid to graphing if we could determine where these sections occur. Suppose the graph of a function f is as indicated in Figure 2. As we look from left to right, we see that on the interval (a, b) the graph of f is *rising*, $f(x)$ is *increasing*,* and the slope of the graph is positive $[f'(x) > 0]$. On the other hand, on the interval (b, c) the graph of f is *falling*, $f(x)$ is *decreasing*, and the slope of the graph is negative $[f'(x) < 0]$. At $x = b$, the graph of f changes direction (from rising to falling), $f(x)$ changes from increasing to decreasing, the slope of the graph is 0 $[f'(b) = 0]$, and the tangent line is horizontal.

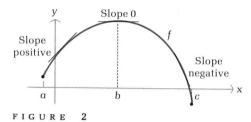

FIGURE 2

In general, if $f'(x) > 0$ (is positive) on the interval (a, b), then $f(x)$ increases (↗) and the graph of f rises as we move from left to right over the interval; if $f'(x) < 0$ (is negative) on an interval (a, b), then $f(x)$ decreases (↘) and the graph of f falls as we move from left to right over the interval. We summarize these important results in the box.

Increasing and Decreasing Functions

For the interval (a, b):

$f'(x)$	$f(x)$	GRAPH OF f	EXAMPLES
+	Increases ↗	Rises ↗	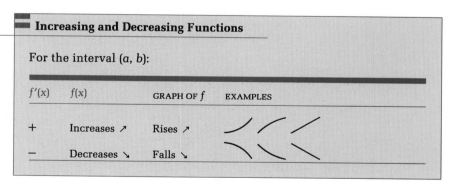
−	Decreases ↘	Falls ↘	

◆ E X A M P L E 2   Given $f(x) = 8x - x^2$:

(A) Which values of x correspond to horizontal tangent lines?

(B) For which values of x is $f(x)$ increasing? Decreasing?

(C) Sketch a graph of f. Add horizontal tangent lines.

---

* Formally, we say that $f(x)$ is **increasing** on an interval (a, b) if $f(x_2) > f(x_1)$ whenever $a < x_1 < x_2 < b$; f is **decreasing** on (a, b) if $f(x_2) < f(x_1)$ whenever $a < x_1 < x_2 < b$.

Solutions

(A) $f'(x) = 8 - 2x = 0$

$$x = 4$$

Thus, a horizontal tangent line exists at $x = 4$ only.

(B) As in Example 1, we will construct a sign chart for $f'(x)$ to determine which values of x make $f'(x) > 0$ and which values make $f'(x) < 0$. To do this, we must first determine the partition numbers (values of x for which $f'(x) = 0$ or for which $f'(x)$ is discontinuous). From part A we know that $f'(x) = 8 - 2x = 0$ at $x = 4$. Since $f'(x) = 8 - 2x$ is a polynomial, it is continuous for all x. Thus, 4 is the only partition number. We construct a sign chart for the intervals $(-\infty, 4)$ and $(4, \infty)$, using test numbers 3 and 5:

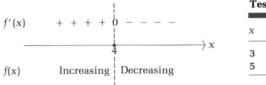

Test Numbers

x	$f'(x)$
3	2 (+)
5	−2 (−)

Thus, $f(x)$ is increasing on $(-\infty, 4)$ and decreasing on $(4, \infty)$.

x	$f(x)$
0	0
2	12
4	16
6	12
8	0

(C)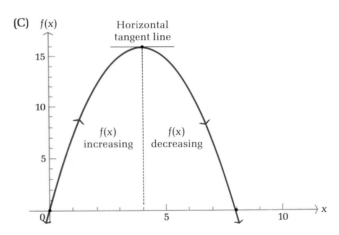

PROBLEM   2   Repeat Example 2 for $f(x) = x^2 - 6x + 10$.

◆ EXAMPLE   3   Determine the intervals where f is increasing and those where f is decreasing for:

(A) $f(x) = 1 + x^3$ (B) $f(x) = (1 - x)^{1/3}$ (C) $f(x) = \dfrac{1}{x - 2}$

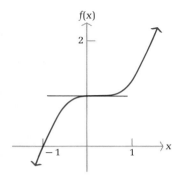

Solutions   (A) $f(x) = 1 + x^3$ $f'(x) = 3x^2 = 0$

$$x = 0$$

Sign chart for $f'(x) = 3x^2$ (partition number is 0):

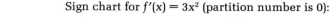

| $f'(x)$ | $+ + + + \ 0 \ + + + +$ |
| | |

$f(x)$ Increasing $\vdots$ Increasing

Test Numbers

x	$f'(x)$	
-1	3	$(+)$
1	3	$(+)$

The sign chart indicates that $f(x)$ is increasing on $(-\infty, 0)$ and $(0, \infty)$. Since f is continuous at $x = 0$, it follows that $f(x)$ is increasing for all x. The graph of f is shown in the margin.

(B) $f(x) = (1 - x)^{1/3}$ $f'(x) = -\dfrac{1}{3}(1 - x)^{-2/3} = \dfrac{-1}{3(1 - x)^{2/3}}$

To find partition numbers for $f'(x)$, we note that f' is continuous for all x except for values of x for which the denominator is 0; that is, f' is discontinuous at $x = 1$. Since the numerator is the constant -1, $f'(x) \neq 0$ for any value of x. Thus, we plot the partition number $x = 1$ on the real number line and use the abbreviation ND to note the fact that $f'(x)$ is *not defined* at $x = 1$.

Sign chart for $f'(x) = -1/[3(1 - x)^{2/3}]$ (partition number is 1):

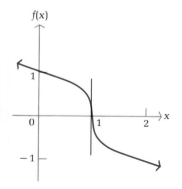

| $f'(x)$ | $- - - - \ \text{ND} \ - - - -$ |
| | |

$f(x)$ Decreasing $\vdots$ Decreasing

Test Numbers

x	$f'(x)$	
0	$-\frac{1}{3}$	$(-)$
2	$-\frac{1}{3}$	$(-)$

The sign chart indicates that f is decreasing on $(-\infty, 1)$ and $(1, \infty)$. Since f is continuous at $x = 1$, it follows that $f(x)$ is decreasing for all x. Thus, **a continuous function can be decreasing (or increasing) on an interval containing values of x where $f'(x)$ does not exist.** The graph of f is shown in the margin. Notice that the undefined derivative at $x = 1$ results in a vertical tangent line at $x = 1$. In general, **a vertical tangent will occur at $x = c$ if f is continuous at $x = c$ and $|f'(x)|$ becomes larger and larger as x approaches c.**

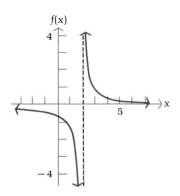

(C) $f(x) = \dfrac{1}{x-2}$ $f'(x) = \dfrac{-1}{(x-2)^2}$

Sign chart for $f'(x) = -1/(x-2)^2$ (partition number is 2):

$f'(x)$ $\quad - - - -$ ND $- - - -$

$\xrightarrow{\hspace{4cm}}$ x

$\qquad\qquad 2$

$f(x)$ $\quad$ Decreasing $|$ Decreasing

Test Numbers

x	$f'(x)$	
1	-1	$(-)$
3	-1	$(-)$

Thus, f is decreasing on $(-\infty, 2)$ and $(2, \infty)$. See the graph of f in the margin. ◆

The values where a function is increasing or decreasing must always be expressed in terms of open intervals that are subsets of the domain of the function.

PROBLEM 3 Determine the intervals where f is increasing and those where f is decreasing for:

(A) $f(x) = 1 - x^3$ (B) $f(x) = (1+x)^{1/3}$ (C) $f(x) = \dfrac{1}{x}$ ◆

◆ CRITICAL VALUES AND LOCAL EXTREMA

When the graph of a continuous function changes from rising to falling, a high point, or *local maximum*, occurs; and when the graph changes from falling to rising, a low point, or *local minimum*, occurs. In Figure 3, high points occur at c_3 and c_6, and low points occur at c_2 and c_4. In general, we call $f(c)$ a **local maximum** if there exists an interval (m, n) containing c such that

$$f(x) \leq f(c)$$

for all x in (m, n).

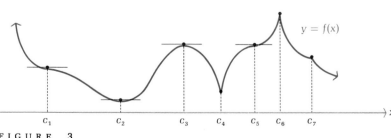

FIGURE 3

The quantity $f(c)$ is called a **local minimum** if there exists an interval (m, n) containing c such that

$f(x) \geq f(c)$

for all x in (m, n). The quantity $f(c)$ is called a **local extremum** if it is either a local maximum or a local minimum. Thus, in Figure 3, we see that local maxima occur at c_3 and c_6, local minima occur at c_2 and c_4, and all four of these points are local extrema.

How can we locate local maxima and minima if we are given the equation for a function and not its graph? Figure 3 suggests an approach. It appears that local maxima and minima occur among those values of x such that $f'(x) = 0$ or $f'(x)$ does not exist; that is, among the values c_1, c_2, c_3, c_4, c_5, c_6, and c_7. [Recall from Section 9-3 that $f'(x)$ is not defined at points on the graph of f where there is a sharp corner or a vertical tangent line.]

Critical Values of f

The values of x in the domain of f where $f'(x) = 0$ or $f'(x)$ does not exist are called the **critical values** of f.

It is possible to prove the following theorem:

THEOREM 2

Existence of Local Extrema

If f is continuous on the interval (a, b) and $f(c)$ is a local extremum, then either $f'(c) = 0$ or $f'(c)$ does not exist (is not defined).

Theorem 2 states that a local extremum can occur only at a critical value, but it does not imply that every critical value produces a local extremum. In Figure 3, c_1 and c_5 are critical values (the slope is 0), but the function does not have a local maximum or local minimum at either of these values.

Our strategy for finding local extrema is now clear. We find all critical values for f and test each one to see if it produces a local maximum, a local minimum, or neither.

◆ FIRST-DERIVATIVE TEST

If $f'(x)$ exists on both sides of a critical value c, then the sign of $f'(x)$ can be used to determine whether the point $(c, f(c))$ is a local maximum, a local minimum, or neither. The various possibilities are summarized in the box and illustrated in Figure 4.

First-Derivative Test for Local Extrema

Let c be a critical value of f [$f(c)$ defined and either $f'(c) = 0$ or $f'(c)$ not defined]. Construct a sign chart for $f'(x)$ close to and on either side of c.

SIGN CHART	$f(c)$
$f'(x)$ $- \; - \; - \mid + \; + \; +$ $\longleftarrow\!\!\!\!(\xrightarrow{\;\;\;m\;\;\;\;\;\;\;\;c\;\;\;\;\;\;\;\;n\;\;\;}) x$ $f(x)$ Decreasing $\mid$ Increasing	$f(c)$ is a local minimum. If $f'(x)$ changes from negative to positive at c, then $f(c)$ is a local minimum.
$f'(x)$ $+ \; + \; + \mid - \; - \; -$ $\longleftarrow\!\!\!\!(\xrightarrow{\;\;\;m\;\;\;\;\;\;\;\;c\;\;\;\;\;\;\;\;n\;\;\;}) x$ $f(x)$ Increasing $\mid$ Decreasing	$f(c)$ is a local maximum. If $f'(x)$ changes from positive to negative at c, then $f(c)$ is a local maximum.
$f'(x)$ $- \; - \; - \mid - \; - \; -$ $\longleftarrow\!\!\!\!(\xrightarrow{\;\;\;m\;\;\;\;\;\;\;\;c\;\;\;\;\;\;\;\;n\;\;\;}) x$ $f(x)$ Decreasing $\mid$ Decreasing	$f(c)$ is not a local extremum. If $f'(x)$ does not change sign at c, then $f(c)$ is neither a local maximum nor a local minimum.
$f'(x)$ $+ \; + \; + \mid + \; + \; +$ $\longleftarrow\!\!\!\!(\xrightarrow{\;\;\;m\;\;\;\;\;\;\;\;c\;\;\;\;\;\;\;\;n\;\;\;}) x$ $f(x)$ Increasing $\mid$ Increasing	$f(c)$ is not a local extremum. If $f'(x)$ does not change sign at c, then $f(c)$ is neither a local maximum nor a local minimum.

◆ E X A M P L E   4   Given $f(x) = x^3 - 6x^2 + 9x + 1$:

(A) Find the critical values of f.   (B) Find the local maxima and minima.
(C) Sketch the graph of f.

Solutions   (A) Find all numbers x in the domain of f where $f'(x) = 0$ or $f'(x)$ does not exist.

$$f'(x) = 3x^2 - 12x + 9 = 0$$
$$3(x^2 - 4x + 3) = 0$$
$$3(x - 1)(x - 3) = 0$$
$$x = 1 \quad \text{or} \quad x = 3$$

$f'(x)$ exists for all x; the critical values are $x = 1$ and $x = 3$.

$f'(c) = 0$
Horizontal tangent

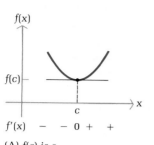

$f'(x)$ $\quad - \quad - \quad 0 \ + \quad +$

(A) $f(c)$ is a
local minimum.

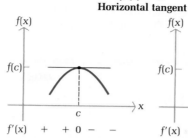

$f'(x)$ $\quad + \quad + \ 0 \ - \quad -$

(B) $f(c)$ is a local
maximum.

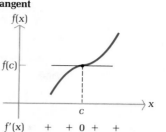

$f'(x)$ $\quad + \quad + \ 0 \ + \quad +$

(C) $f(c)$ is neither
a local maximum
nor a local minimum.

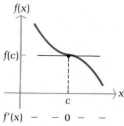

$f'(x)$ $\ - \quad - \quad 0 \ - \quad -$

(D) $f(c)$ is neither a
local maximum nor
a local minimum.

$f'(c)$ is not defined
but $f(c)$ is defined

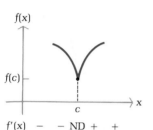

$f'(x)$ $\quad - \quad - \ \text{ND} \ + \quad +$

(E) $f(c)$ is a local
minimum.

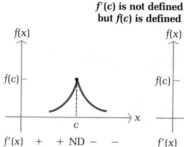

$f'(x)$ $\quad + \quad + \ \text{ND} \ - \quad -$

(F) $f(c)$ is a local
maximum.

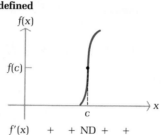

$f'(x)$ $\quad + \quad + \ \text{ND} \ + \quad +$

(G) $f(c)$ is neither a
local maximum nor
a local minimum.

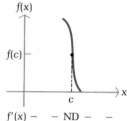

$f'(x)$ $\ - \quad - \ \text{ND} \ - \quad -$

(H) $f(c)$ is neither a
local maximum nor
a local minimum.

FIGURE 4
Local extrema

(B) The easiest way to apply the first-derivative test for local maxima and minima is to construct a sign chart for $f'(x)$ for all x. Partition numbers for $f'(x)$ are $x = 1$ and $x = 3$ (which also happen to be critical values for f).

Sign chart for $f'(x) = 3(x - 1)(x - 3)$:

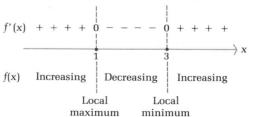

Test Numbers

x	$f'(x)$	
0	9	(+)
2	-3	(−)
4	9	(+)

The sign chart indicates that f increases on $(-\infty, 1)$, has a local maximum at $x = 1$, decreases on $(1, 3)$, has a local minimum at $x = 3$, and increases on $(3, \infty)$. These facts are summarized in the following table:

x	$f'(x)$	$f(x)$	GRAPH OF f
$(-\infty, 1)$	$+$	Increasing	Rising
$x = 1$	0	Local maximum	Horizontal tangent
$(1, 3)$	$-$	Decreasing	Falling
$x = 3$	0	Local minimum	Horizontal tangent
$(3, \infty)$	$+$	Increasing	Rising

(C) We sketch a graph of f using the information from part B and point-by-point plotting.

x	$f(x)$
0	1
1	5
2	3
3	1
4	5

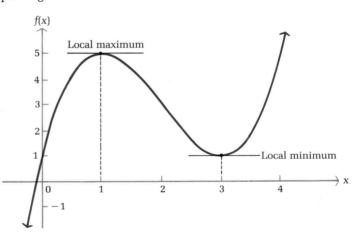

PROBLEM 4   Given $f(x) = x^3 - 9x^2 + 24x - 10$:

(A) Find the critical values of f.
(B) Find the local maxima and minima.
(C) Sketch a graph of f. ◆

In Example 4, the function f had local extrema at both of its critical values. However, as was noted earlier, not every critical value of a function will produce a local extremum. For example, consider the function discussed in Example 3B:

$$f(x) = (1 - x)^{1/3} \qquad \text{and} \qquad f'(x) = \frac{-1}{3(1 - x)^{2/3}}$$

Since $f(1)$ exists and $f'(1)$ does not exist, $x = 1$ is a critical value for this function. However, the sign chart for $f'(x)$ shows that $f'(x)$ does not change sign at $x = 1$:

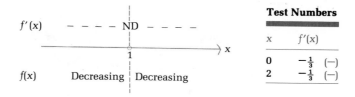

Test Numbers	
x	$f'(x)$
0	$-\frac{1}{3}$ $(-)$
2	$-\frac{1}{3}$ $(-)$

Thus, f does not have a local maximum nor a local minimum at $x = 1$.

Finally, it is important to remember that a critical value must be in the domain of the function. Refer to the function discussed in Example 3C:

$$f(x) = \frac{1}{x - 2} \quad \text{and} \quad f'(x) = \frac{-1}{(x - 2)^2}$$

The derivative is not defined at $x = 2$, but neither is the function. Thus, $x = 2$ is not a critical value for f (in fact, this function does not have any critical values). Nevertheless, $x = 2$ is a partition number for $f'(x)$ and must be included in the sign chart for $f'(x)$.

In general, every critical value for $f(x)$ is a partition number for $f'(x)$, but some partition numbers for $f'(x)$ may not be critical values for f. [Critical values have the added requirement that $f(x)$ must be defined at that value.]

Answers to Matched Problems

1. $-\infty < x < -1$ or $1 < x < 3$; $(-\infty, -1) \cup (1, 3)$
2. (A) Horizontal tangent line at $x = 3$.
    (B) Decreasing on $(-\infty, 3)$; increasing on $(3, \infty)$
    (C)

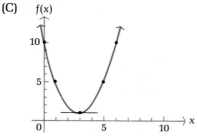

3. (A) Decreasing for all x (B) Increasing for all x
    (C) Decreasing on $(-\infty, 0)$ and $(0, \infty)$

4. (A) Critical values: $x = 2$, $x = 4$
(B) Local maximum at $x = 2$; local minimum at $x = 4$
(C)

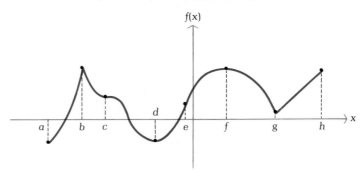

A   *Problems 1–6 refer to the following graph of* $y = f(x)$:

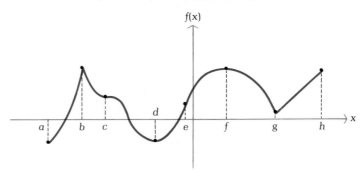

1. Identify the intervals over which $f(x)$ is increasing.
2. Identify the intervals over which $f(x)$ is decreasing.
3. Identify the points where $f'(x) = 0$.
4. Identify the points where $f'(x)$ does not exist.
5. Identify the points where f has a local maximum.
6. Identify the points where f has a local minimum.

B   *Solve each inequality using a sign chart. Express answers in inequality and interval notation.*

7. $x^2 - x - 12 < 0$ 8. $x^2 - 2x - 8 < 0$
9. $x^2 + 21 > 10x$ 10. $x^2 + 7x > -10$

11. $\dfrac{x^2 + 5x}{x - 3} > 0$ 12. $\dfrac{x - 4}{x^2 + 2x} < 0$

Find the intervals where $f(x)$ is increasing, the intervals where $f(x)$ is decreasing, and the local extrema.

13. $f(x) = x^2 - 16x + 12$

14. $f(x) = x^2 + 6x + 7$

15. $f(x) = 4 + 10x - x^2$

16. $f(x) = 5 + 8x - 2x^2$

17. $f(x) = 2x^3 + 4$

18. $f(x) = 2 - 3x^3$

19. $f(x) = 2 - 6x - 2x^3$

20. $f(x) = x^3 + 9x + 7$

21. $f(x) = x^3 - 12x + 8$

22. $f(x) = 3x - x^3$

23. $f(x) = x^3 - 3x^2 - 24x + 7$

24. $f(x) = x^3 + 3x^2 - 9x + 5$

25. $f(x) = 2x^2 - x^4$

26. $f(x) = x^4 - 8x^2 + 3$

Find the intervals where $f(x)$ is increasing, the intervals where $f(x)$ is decreasing, and sketch the graph. Add horizontal tangent lines.

27. $f(x) = 4 + 8x - x^2$

28. $f(x) = 2x^2 - 8x + 9$

29. $f(x) = x^3 - 3x + 1$

30. $f(x) = x^3 - 12x + 2$

31. $f(x) = 10 - 12x + 6x^2 - x^3$

32. $f(x) = x^3 + 3x^2 + 3x$

C Find the critical values, the intervals where $f(x)$ is increasing, the intervals where $f(x)$ is decreasing, and the local extrema. Do not graph.

33. $f(x) = \dfrac{x-1}{x+2}$

34. $f(x) = \dfrac{x+2}{x-3}$

35. $f(x) = x + \dfrac{4}{x}$

36. $f(x) = \dfrac{9}{x} + x$

37. $f(x) = 1 + \dfrac{1}{x} + \dfrac{1}{x^2}$

38. $f(x) = 3 - \dfrac{4}{x} - \dfrac{2}{x^2}$

39. $f(x) = \dfrac{x^2}{x-2}$

40. $f(x) = \dfrac{x^2}{x+1}$

41. $f(x) = x^4(x-6)^2$

42. $f(x) = x^3(x-5)^2$

43. $f(x) = 3(x-2)^{2/3} + 4$

44. $f(x) = 6(4-x)^{2/3} + 4$

45. $f(x) = 2\sqrt{x} - x, \quad x > 0$

46. $f(x) = x - 4\sqrt{x}, \quad x > 0$

APPLICATIONS

Business & Economics

47. *Average cost.* A manufacturer has the following costs in producing x toasters in one day for $0 < x < 150$: fixed costs, \$320; unit production cost, \$20 per toaster; equipment maintenance and repairs, $x^2/20$ dollars. Thus, the cost of manufacturing x toasters in one day is given by

$$C(x) = \frac{x^2}{20} + 20x + 320 \qquad 0 < x < 150$$

and the average cost per toaster is given by

$$\overline{C}(x) = \frac{C(x)}{x} = \frac{x}{20} + 20 + \frac{320}{x} \qquad 0 < x < 150$$

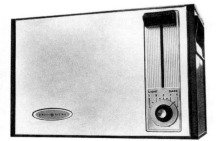

Find the critical values for $\overline{C}(x)$, the intervals where the average cost per toaster is decreasing, the intervals where the average cost per toaster is increasing, and the local extrema. Do not graph.

48. *Average cost.* A manufacturer has the following costs in producing x blenders in one day for $0 < x < 200$: fixed costs, \$450; unit production cost, \$60 per blender; equipment maintenance and repairs, $x^2/18$ dollars.

(A) What is the average cost $\overline{C}(x)$ per blender if x blenders are produced in one day?

(B) Find the critical values for $\overline{C}(x)$, the intervals where the average cost per blender is decreasing, the intervals where the average cost per blender is increasing, and the local extrema. Do not graph.

49. *Marginal analysis.* Show that profit will be increasing over production intervals (a, b) for which marginal revenue is greater than marginal cost. [*Hint:* $P(x) = R(x) - C(x)$]

50. *Marginal analysis.* Show that profit will be decreasing over production intervals (a, b) for which marginal revenue is less than marginal cost.

Life Sciences

51. *Medicine.* A drug is injected into the bloodstream of a patient through the right arm. The concentration of the drug in the bloodstream of the left arm t hours after the injection is approximated by

$$C(t) = \frac{0.14t}{t^2 + 1} \qquad 0 < t < 24$$

Find the critical values for $C(t)$, the intervals where the concentration of the drug is increasing, the intervals where the concentration of the drug is decreasing, and the local extrema. Do not graph.

52. *Medicine.* The concentration $C(t)$, in milligrams per cubic centimeter, of a particular drug in a patient's bloodstream is given by

$$C(t) = \frac{0.16t}{t^2 + 4t + 4} \qquad 0 < t < 12$$

where t is the number of hours after the drug is taken orally. Find the critical values for $C(t)$, the intervals where the concentration of the drug is increasing, the intervals where the concentration of the drug is decreasing, and the local extrema. Do not graph.

Social Sciences

53. *Politics.* Public awareness of a Congressional candidate before and after a successful campaign was approximated by

$$P(t) = \frac{8.4t}{t^2 + 49} + 0.1 \qquad 0 < t < 24$$

where t is time (in months) after the campaign started and $P(t)$ is the fraction of people in the Congressional district who could recall the candidate's (and later, Congressman's) name. Find the critical values for $P(t)$, the time intervals where the fraction is increasing, the time intervals where the fraction is decreasing, and the local extrema. Do not graph.

Second Derivative and Graphs

◆ CONCAVITY
◆ INFLECTION POINTS
◆ SECOND-DERIVATIVE TEST
◆ APPLICATION

In the preceding section, we saw that the derivative can be used to determine when a graph is rising and falling. Now we want to see what the *second derivative* (the derivative of the derivative) can tell us about the shape of a graph.

◆ CONCAVITY

Consider the functions

$$f(x) = x^2 \quad \text{and} \quad g(x) = \sqrt{x}$$

for x in the interval $(0, \infty)$. Since

$$f'(x) = 2x > 0 \qquad \text{for } 0 < x < \infty$$

and

$$g'(x) = \frac{1}{2\sqrt{x}} > 0 \qquad \text{for } 0 < x < \infty$$

both functions are increasing on $(0, \infty)$.

Notice the different shapes of the graphs of f and g, shown in Figure 5. Even though the graph of each function is rising and each graph starts at (0, 0) and goes through (1, 1), the graphs are quite dissimilar. The graph of f opens upward, while the graph of g opens downward. We say that the graph of f is *concave upward*, and the graph of g is *concave downward*. It will help us draw a graph if we can determine the concavity of the graph before we draw it. How can we find a mathematical formulation of concavity?

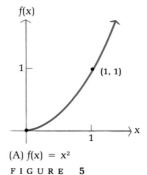

(A) $f(x) = x^2$

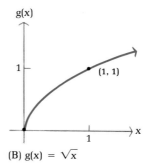

(B) $g(x) = \sqrt{x}$

F I G U R E   5

It will be instructive to examine the slopes of f and g at various points on their graphs (see Figure 6). We can make two observations about each graph. Looking at the graph of f in Figure 6A, we see that $f'(x)$ (the slope of the tangent line) is *increasing* and that the graph lies *above* each tangent line. Looking at Figure 6B, we see that $g'(x)$ is *decreasing* and that the graph lies *below* each tangent line.

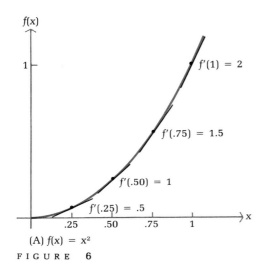

(A) $f(x) = x^2$

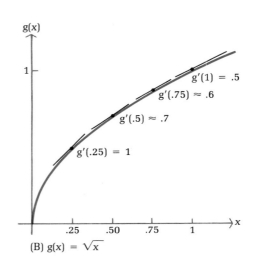

(B) $g(x) = \sqrt{x}$

FIGURE 6

With these ideas in mind, we state the general definition of concavity:

The graph of a function f is concave upward on the interval (a, b) if $f'(x)$ is *increasing* on (a, b) and is concave downward on the interval (a, b) if $f'(x)$ is *decreasing* on (a, b).

Geometrically, the graph is concave upward on (a, b) if it lies above its tangent lines in (a, b) and is concave downward on (a, b) if it lies below its tangent lines in (a, b).

How can we determine when $f'(x)$ is increasing or decreasing? In the preceding section, we used the derivative of a function to determine when that function is increasing or decreasing. Thus, to determine when the function $f'(x)$ is increasing or decreasing, we use the derivative of $f'(x)$. The derivative of the derivative of a function is called the *second derivative* of the function. Various notations for the second derivative are given in the following box:

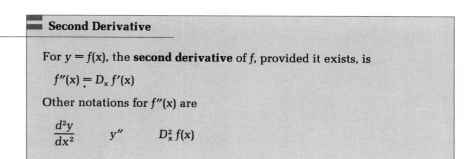

Second Derivative

For $y = f(x)$, the **second derivative** of f, provided it exists, is

$$f''(x) = D_x f'(x)$$

Other notations for $f''(x)$ are

$$\frac{d^2y}{dx^2} \qquad y'' \qquad D_x^2 f(x)$$

Returning to the functions f and g discussed at the beginning of this section, we have

$$f(x) = x^2 \qquad\qquad g(x) = \sqrt{x} = x^{1/2}$$

$$f'(x) = 2x \qquad\qquad g'(x) = \frac{1}{2}x^{-1/2} = \frac{1}{2\sqrt{x}}$$

$$f''(x) = D_x\, 2x = 2 \qquad g''(x) = D_x \frac{1}{2}x^{-1/2} = -\frac{1}{4}x^{-3/2} = -\frac{1}{4\sqrt{x^3}}$$

For $x > 0$, we see that $f''(x) > 0$; thus, $f'(x)$ is increasing and the graph of f is concave upward (see Fig. 6A). For $x > 0$, we also see that $g''(x) < 0$; thus, $g'(x)$ is decreasing and the graph of g is concave downward (see Fig. 6B). These ideas are summarized in the following box:

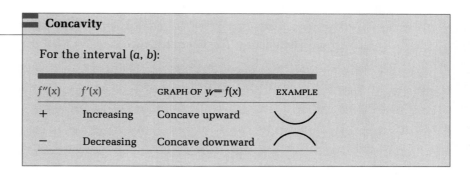

Concavity

For the interval (a, b):

$f''(x)$	$f'(x)$	GRAPH OF $y = f(x)$	EXAMPLE
+	Increasing	Concave upward	⌣
−	Decreasing	Concave downward	⌢

Be careful not to confuse concavity with falling and rising. As Figure 7 (on the next page) illustrates, a graph that is concave upward on an interval may be falling, rising, or both falling and rising on that interval. A similar statement holds for a graph that is concave downward.

$$f''(x) > 0 \text{ over } (a, b)$$
Concave upward

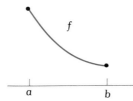

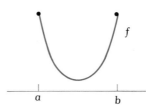

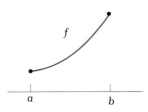

(A) $f'(x)$ is negative and increasing. Graph of f is falling.

(B) $f'(x)$ increases from negative to positive. Graph of f falls, then rises.

(C) $f'(x)$ is positive and increasing. Graph of f is rising.

$$f''(x) < 0 \text{ over } (a, b)$$
Concave downward

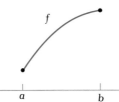

(D) $f'(x)$ is positive and decreasing. Graph of f is rising.

(E) $f'(x)$ decreases from positive to negative. Graph of f rises, then falls.

(F) $f'(x)$ is negative and decreasing. Graph of f is falling.

FIGURE 7
Concavity

◆ EXAMPLE 5 Let $f(x) = x^3$. Find the intervals where the graph of f is concave upward and the intervals where the graph of f is concave downward. Sketch a graph of f.

Solution To determine concavity, we must determine the sign of $f''(x)$.

$$f(x) = x^3 \qquad f'(x) = 3x^2 \qquad f''(x) = 6x$$

Sign chart for $f''(x) = 6x$ (partition number is 0):

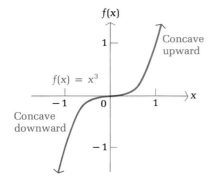

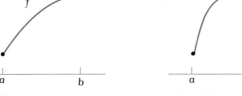

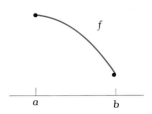

Test Numbers

x	$f''(x)$	
-1	-6	$(-)$
1	6	$(+)$

Thus, the graph of f is concave downward on $(-\infty, 0)$ and concave upward on $(0, \infty)$. The graph of f (without going through other graphing details) is shown in the figure in the margin. ◆

PROBLEM 5 Repeat Example 5 for $f(x) = 1 - x^3$. ◆

The graph in Example 5 changes from concave downward to concave upward at the point (0, 0). This point is called an *inflection point*.

◆ INFLECTION POINTS

In general, an **inflection point** is a point on the graph of a function where the concavity changes (from upward to downward or from downward to upward). In order for the concavity to change at a point, $f''(x)$ must change sign at that point. Reasoning as we did in the previous section, we conclude that the inflection points must occur at points where $f''(x) = 0$ or $f''(x)$ does not exist [but $f(x)$ must exist]. Figure 8 illustrates several typical cases.

If $f'(c)$ exists and $f''(x)$ changes sign at $x = c$, then the tangent line at an inflection point $(c, f(c))$ will always lie below the graph on the side that is concave upward and above the graph on the side that is concave downward (see Figs. 8A, B, and C).

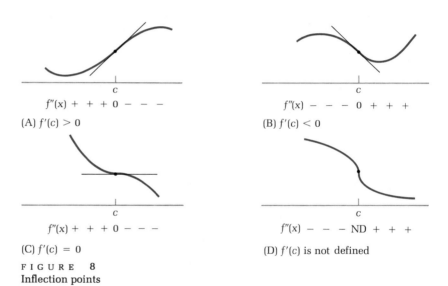

$f''(x) + + + 0 - - -$

(A) $f'(c) > 0$

$f''(x) - - - 0 + + +$

(B) $f'(c) < 0$

$f''(x) + + + 0 - - -$

(C) $f'(c) = 0$

$f''(x) - - - \text{ND} + + +$

(D) $f'(c)$ is not defined

FIGURE 8
Inflection points

◆ EXAMPLE 6 Find the inflection points of $f(x) = x^3 - 6x^2 + 9x + 1$.

Solution Since inflection points occur at values of x where $f''(x)$ changes sign, we construct a sign chart for $f''(x)$.

$$f(x) = x^3 - 6x^2 + 9x + 1$$
$$f'(x) = 3x^2 - 12x + 9$$
$$f''(x) = 6x - 12 = 6(x - 2)$$

Sign chart for $f''(x) = 6(x - 2)$ (partition number is 2):

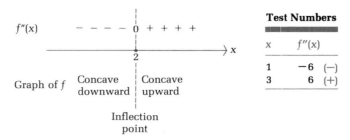

From the sign chart, we see that the graph of f has an inflection point at $x = 2$. The graph of f is shown in the figure. (See Example 4 in Section 10-1.)

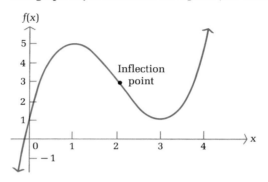

PROBLEM 6   Find the inflection points of $f(x) = x^3 - 9x^2 + 24x - 10$. (See Problem 4 in Section 10-1 for the graph of f.) ◆

It is important to remember that the values of x where $f''(x) = 0$ [or $f''(x)$ does not exist] are only candidates for inflection points. The second derivative must change sign at $x = c$ in order for the graph of f to have an inflection point at $x = c$. For example, consider

$$f(x) = x^4 \qquad f'(x) = 4x^3 \qquad f''(x) = 12x^2$$

The second derivative is 0 at $x = 0$, but $f''(x) > 0$ for all other values of x. Since $f''(x)$ does not change sign at $x = 0$, the graph of f does not have an inflection point at $x = 0$, as illustrated in Figure 9.

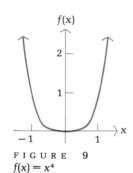

FIGURE 9
$f(x) = x^4$

◆ SECOND-DERIVATIVE TEST

Now we want to see how the second derivative can be used to find local extrema. Suppose f is a function satisfying $f'(c) = 0$ and $f''(c) > 0$. First, note that if $f''(c) > 0$, then it follows from the properties of limits* that $f''(x) > 0$ in some

---

* Actually, we are assuming that $f''(x)$ is continuous in an interval containing c. It is very unlikely that we will encounter a function for which $f''(c)$ exists, but $f''(x)$ is not continuous in an interval containing c.

interval (m, n) containing c. Thus, the graph of f must be concave upward in this interval. But this implies that $f'(x)$ is increasing in this interval. Since $f'(c) = 0$, $f'(x)$ must change from negative to positive at $x = c$ and $f(c)$ is a local minimum (see Figure 10). Reasoning in the same fashion, we conclude that if $f'(c) = 0$ and $f''(c) < 0$, then $f(c)$ is a local maximum. Of course, it is possible that both $f'(c) = 0$ and $f''(c) = 0$. In this case, the second derivative cannot be used to determine the shape of the graph around $x = c$; $f(c)$ may be a local minimum, a local maximum, or neither.

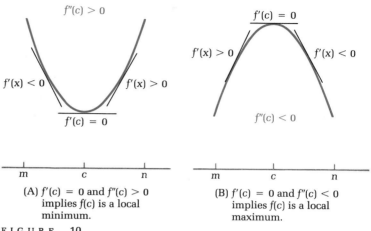

(A) $f'(c) = 0$ and $f''(c) > 0$ implies $f(c)$ is a local minimum.

(B) $f'(c) = 0$ and $f''(c) < 0$ implies $f(c)$ is a local maximum.

FIGURE 10
The second derivative and local extrema

The sign of the second derivative thus provides a simple test for identifying local maxima and minima. This test is most useful when we do not want to draw the graph of the function. If we are interested in drawing the graph and have already constructed the sign chart for $f'(x)$, then the first-derivative test can be used to identify the local extrema.

Second-Derivative Test for Local Maxima and Minima

Let c be a critical value for $f(x)$.

$f'(c)$	$f''(c)$	GRAPH OF f IS	$f(c)$	EXAMPLE
0	+	Concave upward	Local minimum	
0	−	Concave downward	Local maximum	
0	0	?	Test fails	

The first-derivative test must be used whenever $f''(c) = 0$ or $f''(c)$ does not exist.

◆ E X A M P L E   7   Find the local maxima and minima for each function. Use the second-derivative test when it applies.

(A) $f(x) = x^3 - 6x^2 + 9x + 1$ (B) $f(x) = \frac{1}{6}x^6 - 4x^5 + 25x^4$

Solutions   (A) Take first and second derivatives and find critical values:

$$f(x) = x^3 - 6x^2 + 9x + 1$$
$$f'(x) = 3x^2 - 12x + 9 = 3(x - 1)(x - 3)$$
$$f''(x) = 6x - 12 = 6(x - 2)$$

Critical values are $x = 1$ and $x = 3$.

$f''(1) = -6 < 0$ f has a local maximum at $x = 1$.
$f''(3) = 6 > 0$ f has a local minimum at $x = 3$.

(B) $f(x) = \frac{1}{6}x^6 - 4x^5 + 25x^4$
$$f'(x) = x^5 - 20x^4 + 100x^3 = x^3(x - 10)^2$$
$$f''(x) = 5x^4 - 80x^3 + 300x^2$$

Critical values are $x = 0$ and $x = 10$.

$f''(0) = 0$ The second-derivative test fails at both critical values, so
$f''(10) = 0$ the first-derivative test must be used.

Sign chart for $f'(x) = x^3(x - 10)^2$ (partition numbers are 0 and 10):

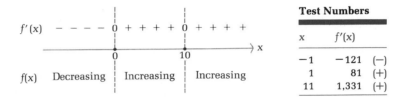

	Test Numbers	
x	$f'(x)$	
−1	−121	(−)
1	81	(+)
11	1,331	(+)

From the sign chart, we see that $f(x)$ has a local minimum at $x = 0$ and does not have a local extremum at $x = 10$.   ◆

P R O B L E M   7   Find the local maxima and minima for each function. Use the second-derivative test when it applies.

(A) $f(x) = x^3 - 9x^2 + 24x - 10$ (B) $f(x) = 10x^6 - 24x^5 + 15x^4$ ◆

A common error is to assume that $f''(c) = 0$ implies that $f(c)$ is not a local extreme point. As Example 7B illustrates, if $f''(c) = 0$, then $f(c)$ may or may not be a local extreme point. **The first-derivative test *must* be used whenever $f''(c) = 0$ or $f''(c)$ does not exist.**

◆ E X A M P L E   8

Maximum Rate of Change

Using past records, a company estimates that it will sell $N(x)$ units of a product after spending $\$x$ thousand on advertising, as given by

$$N(x) = 2{,}000 - 2x^3 + 60x^2 - 450x \qquad 5 \leqslant x \leqslant 15$$

When is the rate of change of sales per unit (thousand dollars) change in advertising increasing? Decreasing? What is the maximum rate of change? Graph N and N' on the same coordinate system and interpret.

Solution

The rate of change of sales per unit (thousand dollars) change in advertising expenditure is

$$N'(x) = -6x^2 + 120x - 450 = -6(x - 5)(x - 15)$$

To determine when this rate is increasing and decreasing, we find $N''(x)$, the derivative of $N'(x)$:

$$N''(x) = -12x + 120 = 12(10 - x)$$

The information obtained by analyzing the signs of $N'(x)$ and $N''(x)$ is summarized in the table (sign charts are omitted).

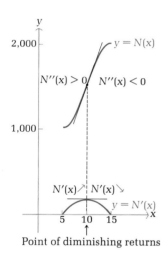

x	$N''(x)$	$N'(x)$	$N'(x)$	$N(x)$
$5 < x < 10$	$+$	$+$	Increasing	Increasing, concave upward
$x = 10$	0	$+$	Local maximum	Inflection point
$10 < x < 15$	$-$	$+$	Decreasing	Increasing, concave downward

Thus, we see that $N'(x)$, the rate of change of sales, is increasing on $(5, 10)$ and decreasing on $(10, 15)$. Both N and N' are graphed in the figure in the margin. An examination of the graph of $N'(x)$ shows that the maximum rate of change is $N'(10) = 150$. Notice that $N'(x)$ has a local maximum and $N(x)$ has an inflection point at $x = 10$. This value of x is referred to as the **point of diminishing returns,** since the rate of change of sales begins to decrease at this point.   ◆

P R O B L E M   8

Repeat Example 8 for $N(x) = 5{,}000 - x^3 + 60x^2 - 900x$, $10 \leqslant x \leqslant 30$.   ◆

Answers to Matched Problems

5. Concave upward on $(-\infty, 0)$; concave downward on $(0, \infty)$

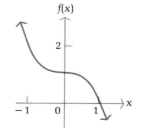

6. Inflection point at x = 3
7. (A) f(2) is a local maximum; f(4) is a local minimum
   (B) f(0) is a local minimum; no local extremum at x = 1
8. N'(x) is increasing on (10, 20), decreasing on (20, 30); maximum rate of change is N'(20) = 300; x = 20 is point of diminishing returns

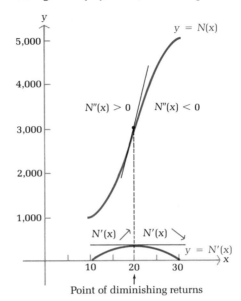

Point of diminishing returns

A   *Problems 1–4 refer to the following graph of y = f(x):*

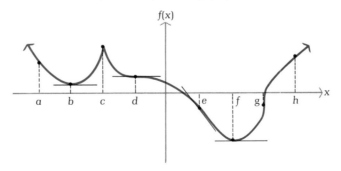

1. Identify intervals over which the graph of f is concave upward.
2. Identify intervals over which the graph of f is concave downward.
3. Identify inflection points.
4. Identify local extrema.

Find the indicated derivative for each function.

5. $f''(x)$ for $f(x) = x^3 - 2x^2 - 1$
6. $g''(x)$ for $g(x) = x^4 - 3x^2 + 5$
7. d^2y/dx^2 for $y = 2x^5 - 3$
8. d^2y/dx^2 for $y = 3x^4 - 7x$
9. $D_x^2(1 - 2x + x^3)$
10. $D_x^2(3x^2 - x^3)$
11. y'' for $y = (x^2 - 1)^3$
12. y'' for $y = (x^2 + 4)^4$
13. $f''(x)$ for $f(x) = 3x^{-1} + 2x^{-2} + 5$
14. $f''(x)$ for $f(x) = x^2 - x^{1/3}$

B *Find all local maxima and minima using the second-derivative test whenever it applies (do not graph). If the second-derivative test fails, use the first-derivative test.*

15. $f(x) = 2x^2 - 8x + 6$
16. $f(x) = 6x - x^2 + 4$
17. $f(x) = 2x^3 - 3x^2 - 12x - 5$
18. $f(x) = 2x^3 + 3x^2 - 12x - 1$
19. $f(x) = 3 - x^3 + 3x^2 - 3x$
20. $f(x) = x^3 + 6x^2 + 12x + 2$
21. $f(x) = x^4 - 8x^2 + 10$
22. $f(x) = x^4 - 18x^2 + 50$
23. $f(x) = x^6 + 3x^4 + 2$
24. $f(x) = 4 - x^6 - 6x^4$

25. $f(x) = x + \dfrac{16}{x}$
26. $f(x) = x + \dfrac{25}{x}$

Find the intervals where the graph of f is concave upward, the intervals where the graph is concave downward, and the inflection points.

27. $f(x) = x^2 - 4x + 5$
28. $f(x) = 9 + 3x - 4x^2$
29. $f(x) = x^3 - 18x^2 + 10x - 11$
30. $f(x) = x^3 + 24x^2 + 15x - 12$
31. $f(x) = x^4 - 24x^2 + 10x - 5$
32. $f(x) = x^4 + 6x^2 + 9x + 11$
33. $f(x) = -x^4 + 4x^3 + 3x + 7$
34. $f(x) = -x^4 - 2x^3 + 12x^2 + 15$

Find local maxima, local minima, and inflection points. Sketch the graph of each function. Include tangent lines at each local extreme point and inflection point.

35. $f(x) = x^3 - 6x^2 + 16$
36. $f(x) = x^3 - 9x^2 + 15x + 10$
37. $f(x) = x^3 + x + 2$
38. $f(x) = 1 - 3x - x^3$
39. $f(x) = (2 - x)^3 + 1$
40. $f(x) = (1 + x)^3 - 1$
41. $f(x) = x^3 - 12x$
42. $f(x) = 27x - x^3$

C *Find the inflection points. Do not graph.*

43. $f(x) = \dfrac{1}{x^2 + 12}$
44. $f(x) = \dfrac{x^2}{x^2 + 12}$

45. $f(x) = \dfrac{x}{x^2 + 12}$
46. $f(x) = \dfrac{x^3}{x^2 + 12}$

APPLICATIONS

Business & Economics

47. *Revenue.* The marketing research department for a computer company used a large city to test market their new product. They found that the relationship between price p (dollars per unit) and the demand x (units per

week) was given approximately by

$$p = 1{,}296 - 0.12x^2 \qquad 0 < x < 80$$

Thus, the weekly revenue can be approximated by

$$R(x) = xp = 1{,}296x - 0.12x^3 \qquad 0 < x < 80$$

(A) Find the local extrema for the revenue function.
(B) Over which intervals is the graph of the revenue function concave upward? Concave downward?

48. *Profit.* Suppose the cost equation for the company in Problem 47 is

$$C(x) = 830 + 396x$$

(A) Find the local extrema for the profit function.
(B) Over which intervals is the graph of the profit function concave upward? Concave downward?

49. *Advertising.* A company estimates that it will sell $N(x)$ units of a product after spending $\$x$ thousand on advertising, as given by

$$N(x) = -3x^3 + 225x^2 - 3{,}600x + 17{,}000 \qquad 10 \leqslant x \leqslant 40$$

(A) When is the rate of change of sales $N'(x)$ increasing? Decreasing?
(B) Find the inflection points for the graph of N.
(C) Graph N and N' on the same coordinate system.
(D) What is the maximum rate of change of sales?

50. *Advertising.* A company estimates that it will sell $N(x)$ units of a product after spending $\$x$ thousand on advertising, as given by

$$N(x) = -2x^3 + 90x^2 - 750x + 2{,}000 \qquad 5 \leqslant x \leqslant 25$$

(A) When is the rate of change of sales $N'(x)$ increasing? Decreasing?
(B) Find the inflection points for the graph of N.
(C) Graph N and N' on the same coordinate system.
(D) What is the maximum rate of change of sales?

Life Sciences 51. *Population growth—bacteria.* A drug that stimulates reproduction is introduced into a colony of bacteria. After t minutes, the number of bacteria is given approximately by

$$N(t) = 1{,}000 + 30t^2 - t^3 \qquad 0 \leqslant t \leqslant 20$$

(A) When is the rate of growth $N'(t)$ increasing? Decreasing?
(B) Find the inflection points for the graph of N.
(C) Sketch the graphs of N and N' on the same coordinate system.
(D) What is the maximum rate of growth?

52. *Drug sensitivity.* One hour after x milligrams of a particular drug are given to a person, the change in body temperature $T(x)$, in degrees Fahrenheit, is given by

$$T(x) = x^2 \left(1 - \frac{x}{9}\right) \qquad 0 \leqslant x \leqslant 6$$

The rate at which $T(x)$ changes with respect to the size of the dosage x, $T'(x)$, is called the *sensitivity* of the body to the dosage.

(A) When is $T'(x)$ increasing? Decreasing?
(B) Where does the graph of T have inflection points?
(C) Sketch the graphs of T and T' on the same coordinate system.
(D) What is the maximum value of $T'(x)$?

Social Sciences   53.   *Learning.* The time T (in minutes) it takes a person to learn a list of length n is

$$T(n) = \tfrac{2}{25}n^3 - \tfrac{6}{5}n^2 + 6n \qquad n \geqslant 0$$

(A) When is the rate of change of T with respect to the length of the list increasing? Decreasing?
(B) Where does the graph of T have inflection points? Graph T and T' on the same coordinate system.
(C) What is the minimum value of $T'(n)$?

S E C T I O N   10-3

Curve Sketching Techniques: Unified and Extended

◆ ASYMPTOTES
◆ GRAPHING STRATEGY
◆ USING THE STRATEGY
◆ APPLICATION

In this section we will apply, in a systematic way, all the graphing concepts discussed in Sections 10-1 and 10-2. Before outlining a graphing strategy and considering the graphs of specific functions, we need to discuss in more detail the concept of *asymptotes*, which we introduced in Section 9-2.

◆ ASYMPTOTES

To refresh your memory concerning horizontal and vertical asymptotes, refer to Figure 11 on the next page. The lines $y = 2$ and $y = -2$ are horizontal asymptotes, since $\lim_{x \to \infty} f(x) = 2$ and $\lim_{x \to -\infty} f(x) = -2$. The line $x = -4$ is a vertical asymptote, since $\lim_{x \to -4} f(x) = \infty$. The line $x = 4$ is also a vertical asymptote, since $\lim_{x \to 4^-} f(x) = -\infty$ and $\lim_{x \to 4^+} f(x) = \infty$.

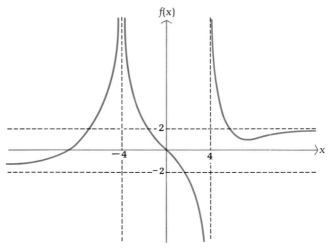

FIGURE 11

We restate the definitions of horizontal and vertical asymptotes in the box for convenient reference.

Horizontal and Vertical Asymptotes

A line $y = b$ is a **horizontal asymptote** for the graph of $y = f(x)$ if

$$\lim_{x \to -\infty} f(x) = b \quad \text{or} \quad \lim_{x \to \infty} f(x) = b$$

A line $x = a$ is a **vertical asymptote** for the graph of $y = f(x)$ if

$$\lim_{x \to a^-} f(x) = \infty \ (\text{or} -\infty), \ \lim_{x \to a^+} f(x) = \infty \ (\text{or} -\infty), \ \text{or} \lim_{x \to a} f(x) = \infty \ (\text{or} -\infty)$$

We are interested in locating any horizontal and vertical asymptotes as an aid to graphing a function. Consequently, we are interested in locating horizontal and vertical asymptotes from an equation that defines a function rather than from its graph. Consider the rational function given by

$$f(x) = \frac{2x - 3}{x - 1}$$

To locate horizontal asymptotes, we compute $\lim_{x \to -\infty} f(x)$ and $\lim_{x \to \infty} f(x)$ as follows:

$$\lim_{x \to -\infty} \frac{2x - 3}{x - 1} = \lim_{x \to -\infty} \frac{x\left(2 - \dfrac{3}{x}\right)}{x\left(1 - \dfrac{1}{x}\right)} = 2 \qquad \lim_{x \to \infty} \frac{2x - 3}{x - 1} = \lim_{x \to \infty} \frac{x\left(2 - \dfrac{3}{x}\right)}{x\left(1 - \dfrac{1}{x}\right)} = 2$$

Thus, $y = 2$ is the only horizontal asymptote, and it could have been obtained using either limit. In general, **a rational function has at most one horizontal asymptote.** (Notice that this implies that the function graphed in Figure 11 is not a rational function, since it has two horizontal asymptotes.)

Since the given function f is continuous for all values of x except those that make the denominator 0, the only candidate for a vertical asymptote is $x = 1$. How can we tell if $x = 1$ is a vertical asymptote? We could try to compute the limits in the definition of a vertical asymptote above—a difficult task in many problems. But Theorem 3, which we state without proof, eliminates the need to compute any of the limits in the definition, and it applies to f, as well as a wide variety of other functions.

THEOREM   3

On Vertical Asymptotes

Let $f(x) = N(x)/D(x)$, where both N and D are continuous at $x = c$. If at $x = c$ the denominator $D(x)$ is 0 and the numerator $N(x)$ is not 0, then the line $x = c$ is a vertical asymptote for the graph of f.

[*Note:*   Since a rational function is the ratio of two polynomial functions, and polynomial functions are continuous for all real numbers, this theorem includes rational functions as a special case.]

Returning to the example above, $x = 1$ is a vertical asymptote for the graph of $f(x) = (2x - 3)/(x - 1)$, since the denominator is 0 at $x = 1$ and the numerator is not.

◆ E X A M P L E   9   Find horizontal and vertical asymptotes, if they exist, for:

(A) $f(x) = \dfrac{6x + 5}{2x - 4}$ (B) $f(x) = \dfrac{x}{x^2 + 1}$ (C) $f(x) = \dfrac{x^2 - 4}{x}$

Solutions   (A) *Horizontal asymptotes:*

$$\lim_{x \to \infty} \frac{6x + 5}{2x - 4} = \lim_{x \to \infty} \frac{x\left(6 + \dfrac{5}{x}\right)}{x\left(2 - \dfrac{4}{x}\right)} = \frac{6}{2} = 3$$

The line $y = 3$ is a horizontal asymptote. [Since a rational function cannot have more than one horizontal asymptote, there is no need to compute $\lim_{x \to -\infty} f(x)$.]

Vertical asymptotes:

$$f(x) = \frac{6x+5}{2x-4} = \frac{6x+5}{2(x-2)}$$

Using Theorem 3, we search for values of x that make the denominator 0 without making the numerator 0 at the same time. The denominator is 0 for $x = 2$, and we see that the line $x = 2$ is a vertical asymptote, since the numerator is not 0 for $x = 2$.

(B) *Horizontal asymptotes:*

$$\lim_{x \to \infty} \frac{x}{x^2+1} = \lim_{x \to \infty} \frac{x}{x^2\left(1 + \dfrac{1}{x^2}\right)} = \lim_{x \to \infty} \left[\frac{1}{x} \cdot \frac{1}{\left(1 + \dfrac{1}{x^2}\right)} \right] = 0$$

The line $y = 0$ (the x axis) is a horizontal asymptote.

Vertical asymptotes: Since the denominator, $x^2 + 1$, is never 0, there are no vertical asymptotes.

(C) *Horizontal asymptotes:*

$$\lim_{x \to \infty} \frac{x^2-4}{x} = \lim_{x \to \infty} \frac{x^2\left(1 - \dfrac{4}{x^2}\right)}{x} = \lim_{x \to \infty} x\left(1 - \frac{4}{x^2}\right) = \infty$$

There are no horizontal asymptotes.

Vertical asymptotes: The line $x = 0$ (the y axis) is a vertical asymptote, since for $x = 0$, the denominator is 0 and the numerator is not 0. ◆

PROBLEM 9   Find horizontal and vertical asymptotes, if they exist, for:

(A) $f(x) = \dfrac{3x+5}{x+2}$ (B) $f(x) = \dfrac{x+1}{x^2}$ (C) $f(x) = \dfrac{x^3}{x^2+4}$ ◆

◆ GRAPHING STRATEGY

We now have powerful tools to determine the shape of a graph of a function, even before we plot any points. We can accurately sketch the graphs of many functions using these tools and point-by-point plotting as needed (often, very little point-by-point plotting is necessary). We organize these tools in the graphing strategy summarized in the box.

■ A Graphing Strategy for $y = f(x)$

Omit any of the following steps if procedures involved appear to be too difficult or impossible (what may seem too difficult now, will become less so with a little practice).

Step 1. Use $f(x)$:

(A) Find the domain of f. [The domain of f is the set of all real numbers x that produce real values for $f(x)$.]

(B) Find intercepts. [The y intercept is $f(0)$, if it exists; the x intercepts are the solutions to $f(x) = 0$, if they exist.]

(C) Find asymptotes. [Find any horizontal asymptotes by calculating $\lim_{x \to \pm\infty} f(x)$. Find any vertical asymptotes by using Theorem 3.]

Step 2. Use $f'(x)$: Find any critical values for $f(x)$ and any partition numbers for $f'(x)$. [Remember, every critical value for $f(x)$ is also a partition number for $f'(x)$, but some partition numbers for $f'(x)$ may not be critical values for $f(x)$.] Construct a sign chart for $f'(x)$, determine the intervals where $f(x)$ is increasing and decreasing, and find local maxima and minima.

Step 3. Use $f''(x)$: Construct a sign chart for $f''(x)$, determine where the graph of f is concave upward and concave downward, and find any inflection points.

Step 4. Sketch the graph of f: Draw asymptotes and locate intercepts, local maxima and minima, and inflection points. Sketch in what you know from steps 1–3. In regions of uncertainty, use point-by-point plotting to complete the graph.

◆ USING THE STRATEGY

Some examples will illustrate the use of the graphing strategy.

◆ E X A M P L E   10   Graph $f(x) = x^4 - 2x^3$ using the graphing strategy.

Solution   Step 1.   Use $f(x)$: $f(x) = x^4 - 2x^3$

(A) Domain: All real x

(B) Intercepts:

y intercept: $f(0) = 0$

x intercepts: $f(x) = 0$
$$x^4 - 2x^3 = 0$$
$$x^3(x - 2) = 0$$
$$x = 0, 2$$

(C) Asymptotes: Since f is a polynomial, there are no horizontal or vertical asymptotes.

Step 2. Use $f'(x)$: $f'(x) = 4x^3 - 6x^2 = 4x^2(x - \frac{3}{2})$

Critical values: 0 and $\frac{3}{2}$

Partition numbers: 0 and $\frac{3}{2}$

Sign chart for $f'(x)$:

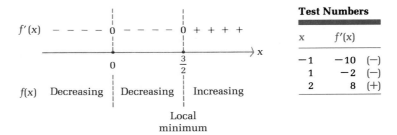

Test Numbers		
x	$f'(x)$	
-1	-10	$(-)$
1	-2	$(-)$
2	8	$(+)$

Thus, $f(x)$ is decreasing on $(-\infty, \frac{3}{2})$, increasing on $(\frac{3}{2}, \infty)$, and has a local minimum at $x = \frac{3}{2}$.

Step 3. Use $f''(x)$: $f''(x) = 12x^2 - 12x = 12x(x - 1)$

Partition numbers for $f''(x)$: 0 and 1

Sign chart for $f''(x)$:

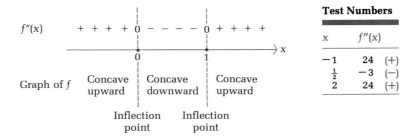

Test Numbers		
x	$f''(x)$	
-1	24	$(+)$
$\frac{1}{2}$	-3	$(-)$
2	24	$(+)$

Thus, the graph of f is concave upward on $(-\infty, 0)$ and $(1, \infty)$, concave downward on $(0, 1)$, and has inflection points at $x = 0$ and $x = 1$.

Step 4. *Sketch the graph of f:*

x	f(x)
0	0
1	-1
$\frac{3}{2}$	$-\frac{27}{16}$
2	0

◆

P R O B L E M   10   Graph $f(x) = x^4 + 4x^3$ using the graphing strategy.

◆

◆ **E X A M P L E   11**   Graph $f(x) = \dfrac{x-1}{x-2}$ using the graphing strategy.

Solution   *Step 1.* Use $f(x)$: $\quad f(x) = \dfrac{x-1}{x-2}$

(A) Domain: All real x, except $x = 2$

(B) Intercepts:

$$y \text{ intercept:} \quad f(0) = \frac{0-1}{0-2} = \frac{1}{2}$$

x intercepts: Since a fraction is 0 when its numerator is 0 and the denominator is not 0, the x intercept is $x = 1$.

(C) Asymptotes:

$$\text{Horizontal asymptote:} \quad \lim_{x \to \infty} \frac{x-1}{x-2} = \lim_{x \to \infty} \frac{x\left(1 - \dfrac{1}{x}\right)}{x\left(1 - \dfrac{2}{x}\right)} = 1$$

Thus, the line $y = 1$ is a horizontal asymptote.

Vertical asymptote: The denominator is 0 for $x = 2$, and the numerator is not 0 for this value. Therefore, the line $x = 2$ is a vertical asymptote.

Step 2.   Use $f'(x)$: $f'(x) = \dfrac{(x-2)(1) - (x-1)(1)}{(x-2)^2} = \dfrac{-1}{(x-2)^2}$

Critical values: None

Partition number: $x = 2$

Sign chart for $f'(x)$:

$f'(x)$ $-\ -\ -\ -\ $ ND $\ -\ -\ -\ -$

$\xrightarrow{\hspace{5cm}} x$

2

$f(x)$ Decreasing $\vert$ Decreasing

Test Numbers

x	$f'(x)$	
1	-1	$(-)$
3	-1	$(-)$

Thus, $f(x)$ is decreasing on $(-\infty, 2)$ and $(2, \infty)$. There are no local extrema.

Step 3.   Use $f''(x)$: $f''(x) = \dfrac{2}{(x-2)^3}$

Partition number for $f''(x)$:   2

Sign chart for $f''(x)$:

$f''(x)$ $-\ -\ -\ -\ $ ND $+\ +\ +\ +$

$\xrightarrow{\hspace{5cm}} x$

2

Graph of f Concave $\vert$ Concave
downward $\vert$ upward

Test Numbers

x	$f''(x)$	
1	-2	$(-)$
3	2	$(+)$

Thus, the graph of f is concave downward on $(-\infty, 2)$ and concave upward on $(2, \infty)$. Since $f(2)$ is not defined, there is no inflection point at $x = 2$, even though $f''(x)$ changes sign at $x = 2$.

Step 4.   Sketch a graph of f: Insert intercepts and asymptotes, and plot a few additional points (for functions with asymptotes, plotting additional points is often helpful). Then sketch the graph.

x	$f(x)$
-2	$\frac{3}{4}$
0	$\frac{1}{2}$
1	0
$\frac{3}{2}$	-1
$\frac{5}{2}$	3
3	2
4	$\frac{3}{2}$

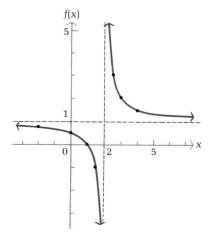

PROBLEM 11   Graph $f(x) = \dfrac{2x}{1-x}$ using the graphing strategy.   ◆

◆ APPLICATION

◆ EXAMPLE 12   Given the cost function $C(x) = 5{,}000 + \frac{1}{2}x^2$, where x is the number of units produced, graph the average cost function and the marginal cost function on the same set of coordinate axes.

Average Cost

Solution   The average cost function is

$$\overline{C}(x) = \frac{C(x)}{x} = \frac{5{,}000}{x} + \frac{x}{2} = \frac{10{,}000 + x^2}{2x}$$

Graph $\overline{C}$ using the graphing strategy.

Step 1.   Use $\overline{C}(x)$:

(A) Domain: Since negative values of x do not make sense and $\overline{C}(0)$ is not defined, the domain is the set of all positive real numbers.
(B) Intercepts: None
(C) Asymptotes:

Horizontal asymptote:

$$\lim_{x \to \infty} \frac{10{,}000 + x^2}{2x} = \lim_{x \to \infty} \frac{x^2\left(\dfrac{10{,}000}{x^2} + 1\right)}{2x} = \lim_{x \to \infty} \frac{x}{2}\left(\frac{10{,}000}{x^2} + 1\right) = \infty$$

Thus, there is no horizontal asymptote.

Vertical asymptote: The line $x = 0$ (the y axis) is a vertical asymptote, since the denominator is 0 and the numerator is not 0 for $x = 0$.

Oblique asymptotes: Some graphs have asymptotes that are neither vertical nor horizontal. These are called **oblique asymptotes.** If we look at

$$\overline{C}(x) = \frac{5{,}000}{x} + \frac{x}{2}$$

we can see that for x near and to the right of 0, $\overline{C}(x)$ is approximated by $5{,}000/x$. On the other hand, as x increases without bound, $\overline{C}(x)$ approaches $x/2$; that is,

$$\lim_{x \to \infty}\left[\overline{C}(x) - \frac{x}{2}\right] = \lim_{x \to \infty} \frac{5{,}000}{x} = 0$$

This implies that the graph of $y = \overline{C}(x)$ approaches the line $y = x/2$ as x approaches ∞. This line is an oblique asymptote for the graph of $y = \overline{C}(x)$.

Step 2. *Use* $\overline{C}'(x)$: $\overline{C}'(x) = -\dfrac{5{,}000}{x^2} + \dfrac{1}{2} = \dfrac{x^2 - 10{,}000}{2x^2}$

Critical value: 100

Partition numbers: 0 and 100

Sign chart for $\overline{C}'(x)$:

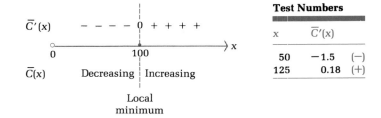

Test Numbers		
x	$\overline{C}'(x)$	
50	−1.5	(−)
125	0.18	(+)

Thus, $\overline{C}(x)$ is decreasing on (0, 100), increasing on (100, ∞), and has a local minimum at $x = 100$.

Step 3. *Use* $\overline{C}''(x)$: $\overline{C}''(x) = \dfrac{10{,}000}{x^3}$

$\overline{C}''(x)$ is positive for all positive x; therefore, the graph of $y = \overline{C}(x)$ is concave upward on (0, ∞).

Step 4. *Sketch the graph of* $\overline{C}$: The graph of $\overline{C}$ is shown in the figure. The marginal cost function is $C'(x) = x$. The graph of this linear function is also shown in the figure.

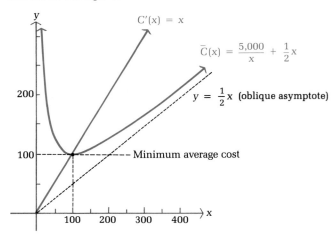

The graph in Example 12 illustrates an important principle in economics.:

The minimum average cost occurs when the average cost is equal to the marginal cost.

P R O B L E M   12   Given the cost function $C(x) = 1{,}600 + \frac{1}{4}x^2$, where x is the number of units produced:

(A) Graph the average cost function (using the graphing strategy) and the marginal cost function on the same set of coordinate axes. Include any oblique asymptotes.

(B) Find the minimum average cost. ◆

Answers to Matched Problems   9. (A) Horizontal asymptote: $y = 3$;   vertical asymptote: $x = -2$
   (B) Horizontal asymptote: $y = 0$ (x axis);   vertical asymptote: $x = 0$ (y axis)
   (C) No horizontal or vertical asymptotes

10. Domain: $(-\infty, \infty)$
   y intercept: $f(0) = 0$; x intercepts: $-4, 0$
   Asymptotes: No horizontal or vertical asymptotes
   Decreasing on $(-\infty, -3)$; increasing on $(-3, \infty)$; local minimum at $x = -3$
   Concave upward on $(-\infty, -2)$ and $(0, \infty)$; concave downward on $(-2, 0)$
   Inflection points at $x = -2$ and $x = 0$

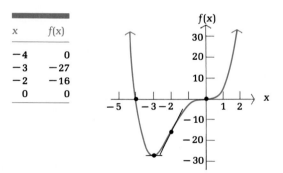

x	$f(x)$
-4	0
-3	-27
-2	-16
0	0

11. Domain: All real x, except $x = 1$
   y intercept: $f(0) = 0$;
   x intercept: 0
   Horizontal asymptote: $y = -2$;
   vertical asymptote: $x = 1$

Increasing on $(-\infty, 1)$ and $(1, \infty)$
Concave upward on $(-\infty, 1)$; concave downward on $(1, \infty)$

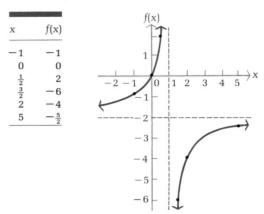

x	f(x)
−1	−1
0	0
$\frac{1}{2}$	2
$\frac{3}{2}$	−6
2	−4
5	$-\frac{5}{2}$

12. (A)

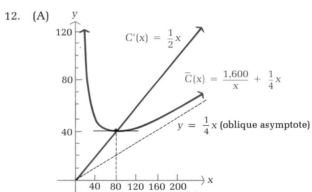

(B) Minimum average cost is 40 at $x = 80$.

A *Problems 1–10 refer to the graph of $y = f(x)$ at the top of the next page:*

1. Identify the intervals over which $f(x)$ is increasing.
2. Identify the intervals over which $f(x)$ is decreasing.
3. Identify the points where $f(x)$ has a local maximum.
4. Identify the points where $f(x)$ has a local minimum.
5. Identify the intervals over which the graph of f is concave upward.
6. Identify the intervals over which the graph of f is concave downward.
7. Identify the inflection points.
8. Identify the horizontal asymptotes.
9. Identify the vertical asymptotes.
10. Identify the x and y intercepts.

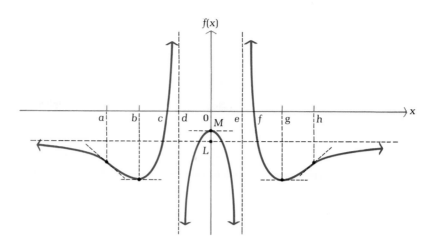

B Find any horizontal and vertical asymptotes.

11. $f(x) = \dfrac{2x}{x+2}$

12. $f(x) = \dfrac{3x+2}{x-4}$

13. $f(x) = \dfrac{x^2+1}{x^2-1}$

14. $f(x) = \dfrac{x^2-1}{x^2+2}$

15. $f(x) = \dfrac{x^3}{x^2+6}$

16. $f(x) = \dfrac{x}{x^2-4}$

17. $f(x) = \dfrac{x}{x^2+4}$

18. $f(x) = \dfrac{x^2+9}{x}$

19. $f(x) = \dfrac{x^2}{x-3}$

20. $f(x) = \dfrac{x+5}{x^2}$

Sketch a graph of $y = f(x)$ using the graphing strategy.

21. $f(x) = x^2 - 6x + 5$

22. $f(x) = 3 + 2x - x^2$

23. $f(x) = x^3 - 6x^2$

24. $f(x) = 3x^2 - x^3$

25. $f(x) = (x+4)(x-2)^2$

26. $f(x) = (2-x)(x+1)^2$

27. $f(x) = 8x^3 - 2x^4$

28. $f(x) = x^4 - 4x^3$

29. $f(x) = \dfrac{x+3}{x-3}$

30. $f(x) = \dfrac{2x-4}{x+2}$

31. $f(x) = \dfrac{x}{x-2}$

32. $f(x) = \dfrac{2+x}{3-x}$

C In Problems 33 and 34, show that the line $y = x$ is an oblique asymptote for the graph of $y = f(x)$, and then use the graphing strategy to sketch a graph of $y = f(x)$.

33. $f(x) = x + \dfrac{1}{x}$

34. $f(x) = x - \dfrac{1}{x}$

Sketch a graph of $y = f(x)$ using the graphing strategy.

35. $f(x) = x^3 - x$

36. $f(x) = x^3 + x$

37. $f(x) = (x^2 + 3)(9 - x^2)$

38. $f(x) = (x^2 + 3)(x^2 - 1)$

39. $f(x) = (x^2 - 4)^2$

40. $f(x) = (x^2 - 1)(x^2 - 5)$

41. $f(x) = 2x^6 - 3x^5$

42. $f(x) = 3x^5 - 5x^4$

43. $f(x) = \dfrac{x}{x^2 - 4}$

44. $f(x) = \dfrac{1}{x^2 - 4}$

45. $f(x) = \dfrac{1}{1 + x^2}$

46. $f(x) = \dfrac{x^2}{1 + x^2}$

APPLICATIONS

Business & Economics

47. *Revenue.* The marketing research department for a computer company used a large city to test market their new product. They found that the relationship between price p (dollars per unit) and the demand x (units per week) was given approximately by

$$p = 1{,}296 - 0.12x^2 \qquad 0 < x < 80$$

Thus, the weekly revenue can be approximated by

$$R(x) = xp = 1{,}296x - 0.12x^3 \qquad 0 < x < 80$$

Graph the revenue function R.

48. *Profit.* Suppose the cost equation for the company in Problem 47 is

$$C(x) = 830 + 396x$$

(A) Write an equation for the profit $P(x)$.

(B) Graph the profit function P.

49. *Pollution.* In Silicon Valley (California), a number of computer-related manufacturing firms were found to be contaminating underground water supplies with toxic chemicals stored in leaking underground containers. A water quality control agency ordered the companies to take immediate corrective action and to contribute to a monetary pool for testing and cleanup of the underground contamination. Suppose the required monetary pool (in millions of dollars) for the testing and cleanup is estimated to be given by

$$P(x) = \dfrac{2x}{1 - x} \qquad 0 \leqslant x < 1$$

where x is the percentage (expressed as a decimal fraction) of the total contaminant removed.

(A) Where is $P(x)$ increasing? Decreasing?

(B) Where is the graph of P concave upward? Downward?

(C) Find any horizontal and vertical asymptotes.

(D) Find the x and y intercepts.

(E) Sketch a graph of P.

50. *Employee training.* A company producing computer components has established that on the average a new employee can assemble $N(t)$ components per day after t days of on-the-job training, as given by

$$N(t) = \frac{100t}{t+9} \qquad t \geq 0$$

(A) Where is $N(t)$ increasing? Decreasing?
(B) Where is the graph of N concave upward? Downward?
(C) Find any horizontal and vertical asymptotes.
(D) Find the intercepts.
(E) Sketch a graph of N.

51. *Replacement time.* An office copier has an initial price of $3,200. A maintenance/service contract costs $300 for the first year and increases $100 per year thereafter. It can be shown that the total cost of the copier after n years is given by

$$C(n) = 3,200 + 250n + 50n^2$$

(A) Write an expression for the average cost per year, $\overline{C}(n)$, for n years.
(B) Graph the average cost function found in part A.
(C) When is the average cost per year minimum? (This is frequently referred to as the **replacement time** for this piece of equipment.)

52. *Construction costs.* The management of a manufacturing plant wishes to add a fenced-in rectangular storage yard of 20,000 square feet, using the plant building as one side of the yard (see the figure in the margin). If x is the distance from the building to the fence parallel to the building, then show that the length of the fence required for the yard is given by

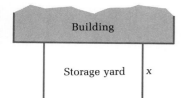

$$L(x) = 2x + \frac{20,000}{x} \qquad x > 0$$

(A) Graph L.
(B) What are the dimensions of the rectangle requiring the least amount of fencing?

53. *Average and marginal costs.* The cost of producing x units of a certain product is given by

$$C(x) = 1,000 + 5x + \tfrac{1}{10}x^2$$

(A) Sketch the graphs of the average cost function and the marginal cost function on the same set of coordinate axes. Include any oblique asymptotes.
(B) Find the minimum average cost.

54. Repeat Problem 53 for $C(x) = 500 + 2x + \tfrac{1}{5}x^2$.

Life Sciences   **55.** *Medicine.* A drug is injected into the bloodstream of a patient through her right arm. The concentration of the drug in the bloodstream of the left arm

t hours after the injection is given by

$$C(t) = \frac{0.14t}{t^2 + 1}$$

Graph C.

56. *Physiology.* In a study on the speed of muscle contraction in frogs under various loads, researchers W. O. Fems and J. Marsh found that the speed of contraction decreases with increasing loads. More precisely, they found that the relationship between speed of contraction S (in centimeters per second) and load w (in grams) is given approximately by

$$S(w) = \frac{26 + 0.06w}{w} \qquad w \geqslant 5$$

Graph S.

Social Sciences

57. *Psychology — retention.* An experiment on retention is conducted in a psychology class. Each student in the class is given 1 day to memorize the same list of 30 special characters. The lists are turned in at the end of the day, and for each succeeding day for 30 days each student is asked to turn in a list of as many of the symbols as can be recalled. Averages are taken, and it is found that

$$N(t) = \frac{5t + 20}{t} \qquad t \geqslant 1$$

provides a good approximation of the average number of symbols, $N(t)$, retained after t days. Graph N.

SECTION 10-4

Optimization; Absolute Maxima and Minima

◆ ABSOLUTE MAXIMA AND MINIMA
◆ APPLICATIONS

We are now ready to consider one of the most important applications of the derivative, namely, the use of derivatives to find the *absolute maximum* or *minimum* value of a function. As we mentioned earlier, an economist may be interested in the price or production level of a commodity that will bring a maximum profit; a doctor may be interested in the time it takes for a drug to reach its maximum concentration in the bloodstream after an injection; and a city planner might be interested in the location of heavy industry in a city to produce minimum pollution in residential and business areas. Before we launch an attack on problems of this type, which are called *optimization* problems, we have to say a few words about the procedures needed to find absolute maximum and absolute minimum values of functions. We have most of the tools we need from the previous sections.

◆ ABSOLUTE MAXIMA AND MINIMA

First, what do we mean by *absolute maximum* and *absolute minimum*? We say that $f(c)$ is an **absolute maximum** of f if

$$f(c) \geqslant f(x)$$

for all x in the domain of f. Similarly, $f(c)$ is called an **absolute minimum** of f if

$$f(c) \leqslant f(x)$$

for all x in the domain of f. Figure 12 illustrates several typical examples.

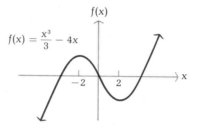

(A) No absolute maximum or minimum
One local maximum at $x = -2$
One local minimum at $x = 2$

(B) Absolute maximum at $x = 0$
No absolute minimum

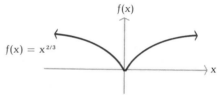

(C) Absolute minimum at $x = 0$
No absolute maximum

FIGURE 12

In many practical problems, the domain of a function is restricted because of practical or physical considerations. If the domain is restricted to some closed interval, as is often the case, then Theorem 4 can be proved.

THEOREM 4

A function f continuous on a closed interval $[a, b]$ (see Section 9-1) assumes both an absolute maximum and an absolute minimum on that interval.

It is important to understand that the absolute maximum and minimum depend on both the function f and the interval $[a, b]$. Figure 13 (at the top of the next page) illustrates four cases.

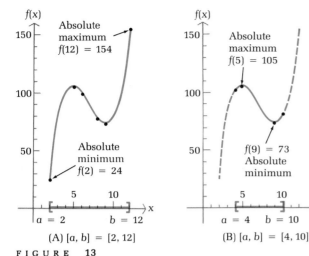

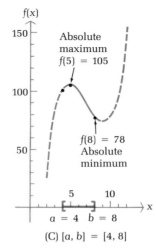

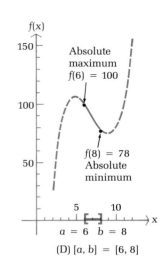

(A) $[a, b] = [2, 12]$ (B) $[a, b] = [4, 10]$ (C) $[a, b] = [4, 8]$ (D) $[a, b] = [6, 8]$

FIGURE 13
Absolute extrema for $f(x) = x^3 - 21x^2 + 135x - 170$ for various closed intervals

In all four cases illustrated in Figure 13, the absolute maximum and absolute minimum both occur at a critical value or an end point. In general:

Absolute extrema (if they exist) must always occur at critical values or at end points.

Thus, to find the absolute maximum or minimum value of a continuous function on a closed interval, we simply identify the end points and the critical values in the interval, evaluate each, and then choose the largest and smallest values out of this group.

Steps in Finding Absolute Maximum and Minimum Values of a Continuous Function f

Step 1.   Check to make certain that f is continuous over $[a, b]$.

Step 2.   Find the critical values in the interval (a, b).

Step 3.   Evaluate f at the end points a and b and at the critical values found in step 2.

Step 4.   The absolute maximum $f(x)$ on $[a, b]$ is the largest of the values found in step 3.

Step 5.   The absolute minimum $f(x)$ on $[a, b]$ is the smallest of the values found in step 3.

◆ E X A M P L E   13   Find the absolute maximum and absolute minimum values of

$$f(x) = x^3 + 3x^2 - 9x - 7$$

on each of the following intervals:

(A) $[-6, 4]$ (B) $[-4, 2]$ (C) $[-2, 2]$

Solutions   (A) The function is continuous for all values of x.

$$f'(x) = 3x^2 + 6x - 9 = 3(x - 1)(x + 3)$$

Thus, $x = -3$ and $x = 1$ are critical values in the interval $(-6, 4)$. Evaluate f at the end points and critical values, -6, -3, 1, and 4, and choose the maximum and minimum from these:

$$f(-6) = -61 \quad \text{Absolute minimum}$$
$$f(-3) = 20$$
$$f(1) = -12$$
$$f(4) = 69 \quad \text{Absolute maximum}$$

(B) Interval: $[-4, 2]$

x	f(x)	
-4	13	
-3	20	Absolute maximum
1	-12	Absolute minimum
2	-5	

(C) Interval: $[-2, 2]$

x	f(x)	
-2	15	Absolute maximum
1	-12	Absolute minimum
2	-5	

The critical value $x = -3$ is not included in this table, because it is not in the interval $[-2, 2]$.   ◆

P R O B L E M   13   Find the absolute maximum and absolute minimum values of

$$f(x) = x^3 - 12x$$

on each of the following intervals:

(A) $[-5, 5]$ (B) $[-3, 3]$ (C) $[-3, 1]$ ◆

Now, suppose we want to find the absolute maximum or minimum value of a function that is continuous on an interval that is not closed. Since Theorem 4 no longer applies, we cannot be certain that the absolute maximum or minimum value exists. Figure 14 (at the top of the next page) illustrates several ways that functions can fail to have absolute extrema.

In general, the best procedure to follow when the interval is not a closed interval (that is, not of the form $[a, b]$) is to sketch the graph of the function. However, one special case that occurs frequently in applications can be ana-

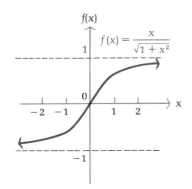

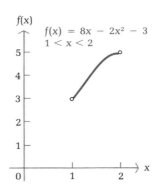

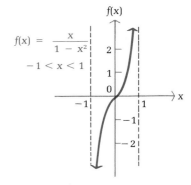

(A) No absolute extrema on $(-\infty, \infty)$:
$-1 < f(x) < 1$ for all x
$[f(x) \neq 1$ or -1 for any $x]$

(B) No absolute extrema on $(1, 2)$:
$3 < f(x) < 5$ for $x \in (1, 2)$
$[f(x) \neq 3$ or 5 for any $x \in (1, 2)]$

(C) No absolute extrema on $(-1, 1)$:
Graph has vertical
asymptotes at $x = -1$ and $x = 1$.

FIGURE 14
Functions with no absolute extrema

lyzed without drawing a graph. It often happens that f is continuous on an interval I and has only one critical value c in the interval I (here, I can be any type of interval—open, closed, or half-closed). If this is the case and if $f''(c)$ exists, then we have the second-derivative test for absolute extrema given in the box below.

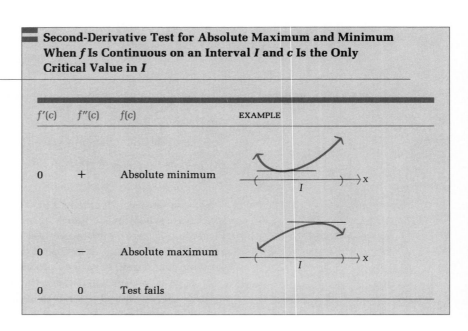

Second-Derivative Test for Absolute Maximum and Minimum When f Is Continuous on an Interval I and c Is the Only Critical Value in I

$f'(c)$	$f''(c)$	$f(c)$	EXAMPLE
0	+	Absolute minimum	
0	−	Absolute maximum	
0	0	Test fails	

◆ E X A M P L E   14   Find the absolute minimum value of $f(x) = x + \dfrac{4}{x}$ on the interval $(0, \infty)$.

Solution $f'(x) = 1 - \dfrac{4}{x^2} = \dfrac{x^2 - 4}{x^2} = \dfrac{(x-2)(x+2)}{x^2}$ $f''(x) = \dfrac{8}{x^3}$

The only critical value in the interval $(0, \infty)$ is $x = 2$. Since $f''(2) = 1 > 0, f(2) = 4$ is the absolute minimum value of f on $(0, \infty)$.   ◆

P R O B L E M   14   Find the absolute maximum value of $f(x) = 12 - x - \dfrac{9}{x}$ on the interval $(0, \infty)$.

◆

◆ APPLICATIONS

Now we want to solve some applied problems that involve absolute extrema. Before beginning, we outline in the next box the steps to follow in solving this type of problem. The first step is the most difficult one. The techniques used to solve optimization problems are best illustrated through a series of examples.

■ A Strategy for Solving Applied Optimization Problems

Step 1. Introduce variables and a function f, including the domain I of f, and then construct a mathematical model of the form

   Maximize (or minimize) $f(x)$ on the interval I

Step 2. Find the absolute maximum (or minimum) value of $f(x)$ on the interval I and the value(s) of x where this occurs.

Step 3. Use the solution to the mathematical model to answer the questions asked in the problem.

◆ E X A M P L E   15

Cost – Demand

A company manufactures and sells x transistor radios per week. If the weekly cost and demand equations are

$C(x) = 5{,}000 + 2x$

$p = 10 - \dfrac{x}{1{,}000} \qquad 0 \le x \le 8{,}000$

find for each week:

(A) The maximum revenue
(B) The maximum profit, the production level that will realize the maximum profit, and the price that the company should charge for each radio to realize the maximum profit

Solutions   (A) The revenue received for selling x radios at $p per radio is

$$R(x) = xp$$

$$= x\left(10 - \frac{x}{1,000}\right)$$

$$= 10x - \frac{x^2}{1,000}$$

Thus, the mathematical model is

$$\text{Maximize}\quad R(x) = 10x - \frac{x^2}{1,000}\qquad 0 \le x \le 8,000$$

$$R'(x) = 10 - \frac{x}{500}$$

$$10 - \frac{x}{500} = 0$$

$$x = 5,000\qquad \text{Only critical value}$$

Use the second-derivative test for absolute extrema:

$$R''(x) = -\frac{1}{500} < 0 \qquad \text{for all } x$$

Thus, the maximum revenue is

$$\text{Max } R(x) = R(5,000) = \$25,000$$

(B) Profit = Revenue − Cost

$$P(x) = R(x) - C(x)$$

$$= 10x - \frac{x^2}{1,000} - 5,000 - 2x$$

$$= 8x - \frac{x^2}{1,000} - 5,000$$

The mathematical model is

$$\text{Maximize}\quad P(x) = 8x - \frac{x^2}{1,000} - 5,000 \qquad 0 \le x \le 8,000$$

$$P'(x) = 8 - \frac{x}{500}$$

$$8 - \frac{x}{500} = 0$$

$$x = 4,000$$

$$P''(x) = -\frac{1}{500} < 0 \qquad \text{for all } x$$

Since x = 4,000 is the only critical value and P''(x) < 0,

$$\text{Max } P(x) = P(4,000) = \$11,000$$

Using the price–demand equation with $x = 4,000$, we find

$$p = 10 - \frac{4,000}{1,000} = \$6$$

Thus, a maximum profit of $11,000 per week is realized when 4,000 radios are produced weekly and sold for $6 each. Notice that this is not the same level of production that produces the maximum revenue. ◆

All the results in Example 15 are illustrated in Figure 15. We also note that profit is maximum when

$$P'(x) = R'(x) - C'(x) = 0$$

that is, when the marginal revenue is equal to the marginal cost (the rate of increase in revenue is the same as the rate of increase in cost at the 4,000 output level — notice that the slopes of the two curves are the same at this point).

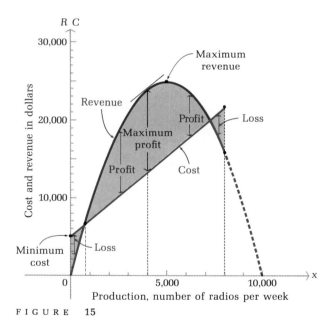

FIGURE   15

PROBLEM   15   Repeat Example 15 for

$$C(x) = 90,000 + 30x$$

$$p = 300 - \frac{x}{30} \qquad 0 \leqslant x \leqslant 9,000$$

◆

◆ E X A M P L E   16

Profit

In Example 15 the government has decided to tax the company $2 for each radio produced. Taking into account this additional cost, how many radios should the company manufacture each week in order to maximize its weekly profit? What is the maximum weekly profit? How much should it charge for the radios to realize the maximum weekly profit?

Solution

The tax of $2 per unit changes the company's cost equation:

$$C(x) = \text{Original cost} + \text{Tax}$$
$$= 5,000 + 2x + 2x$$
$$= 5,000 + 4x$$

The new profit function is

$$P(x) = R(x) - C(x)$$
$$= 10x - \frac{x^2}{1,000} - 5,000 - 4x$$
$$= 6x - \frac{x^2}{1,000} - 5,000$$

Thus, we must solve the following:

$$\text{Maximize} \quad P(x) = 6x - \frac{x^2}{1,000} - 5,000 \qquad 0 \leqslant x \leqslant 8,000$$

$$P'(x) = 6 - \frac{x}{500}$$

$$6 - \frac{x}{500} = 0$$

$$x = 3,000$$

$$P''(x) = -\frac{1}{500} < 0 \qquad \text{for all } x$$

$$\text{Max } P(x) = P(3,000) = \$4,000$$

Using the price–demand equation with $x = 3,000$, we find

$$p = 10 - \frac{3,000}{1,000} = \$7$$

Thus, the company's maximum profit is $4,000 when 3,000 radios are produced and sold weekly at a price of $7.

Even though the tax caused the company's cost to increase by $2 per radio, the price that the company should charge to maximize its profit increases by only $1. The company must absorb the other $1 with a resulting decrease of $7,000 in maximum profit.   ◆

Repeat Example 16 if

$$C(x) = 90,000 + 30x$$

$$p = 300 - \frac{x}{30} \qquad 0 \leqslant x \leqslant 9,000$$

and the government decides to tax the company $20 for each unit produced. Compare the results with the results in Problem 15B. ◆

◆ EXAMPLE 17

Maximize Yield

A walnut grower estimates from past records that if 20 trees are planted per acre, each tree will average 60 pounds of nuts per year. If for each additional tree planted per acre (up to 15) the average yield per tree drops 2 pounds, how many trees should be planted to maximize the yield per acre? What is the maximum yield?

Solution

Let x be the number of additional trees planted per acre. Then

$$20 + x = \text{Total number of trees per acre}$$
$$60 - 2x = \text{Yield per tree}$$

Yield per acre = (Total number of trees per acre)(Yield per tree)

$$Y(x) = (20 + x)(60 - 2x)$$
$$= 1,200 + 20x - 2x^2 \qquad 0 \leqslant x \leqslant 15$$

Thus, we must solve the following:

Maximize $Y(x) = 1,200 + 20x - 2x^2 \qquad 0 \leqslant x \leqslant 15$

$$Y'(x) = 20 - 4x$$
$$20 - 4x = 0$$
$$x = 5$$
$$Y''(x) = -4 < 0 \qquad \text{for all } x$$

Hence,

Max $Y(x) = Y(5) = 1,250$ pounds per acre

Thus, a maximum yield of 1,250 pounds of nuts per acre is realized if 25 trees are planted per acre. ◆

PROBLEM 17

Repeat Example 17 starting with 30 trees per acre and a reduction of 1 pound per tree for each additional tree planted. ◆

◆ EXAMPLE 18

Maximize Area

A farmer wants to construct a rectangular pen next to a barn 60 feet long, using all of the barn as part of one side of the pen. Find the dimensions of the pen with the largest area that the farmer can build if:

(A) 160 feet of fencing material is available
(B) 250 feet of fencing material is available

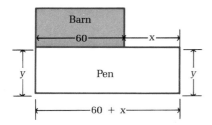

Barn

60

x

y

Pen

y

60 + x

Solutions

(A) We begin by constructing and labeling the figure in the margin. The area of the pen is

$$A = (x + 60)y$$

Before we can maximize the area, we must determine a relationship between x and y in order to express A as a function of one variable. In this case, x and y are related to the total amount of available fencing material:

$$x + y + 60 + x + y = 160$$
$$2x + 2y = 100$$
$$y = 50 - x$$

Thus,

$$A(x) = (x + 60)(50 - x)$$

Now we need to determine the permissible values of x; that is, the domain of the function A. Since the farmer wants to use all of the barn as part of one side of the pen, x cannot be negative. Since y is the other dimension of the pen, y cannot be negative. Thus,

$$y = 50 - x \geq 0$$
$$50 \geq x$$

The domain of A is [0, 50]. Thus, we must solve the following:

$$\text{Maximize} \quad A(x) = (x + 60)(50 - x) \qquad 0 \leq x \leq 50$$
$$A(x) = 3{,}000 - 10x - x^2$$
$$A'(x) = -10 - 2x$$
$$-10 - 2x = 0$$
$$x = -5$$

Since x = −5 is not in the interval [0, 50], there are no critical values in the interval. A(x) is continuous on [0, 50], so the absolute maximum must occur at one of the end points.

$$A(0) = 3{,}000 \qquad \text{Maximum area}$$
$$A(50) = 0$$

If x = 0, then y = 50. Thus, the dimensions of the pen with largest area are 60 feet by 50 feet.

(B) If 250 feet of fencing material is available, then

$$x + y + 60 + x + y = 250$$
$$2x + 2y = 190$$
$$y = 95 - x$$

60 ft

Barn

Pen   50 ft

60 ft

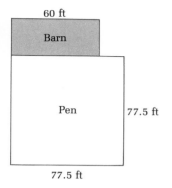

60 ft

Barn

Pen | 77.5 ft

77.5 ft

The model becomes

$$\text{Maximize} \quad A(x) = (x + 60)(95 - x) \qquad 0 \leqslant x \leqslant 95$$

$$A(x) = 5{,}700 + 35x - x^2$$

$$A'(x) = 35 - 2x$$

$$35 - 2x = 0$$

$$x = \tfrac{35}{2} = 17.5 \qquad \text{The only critical value}$$

$$A''(x) = -2 < 0 \qquad \text{for all } x$$

$$\text{Max } A(x) = A(17.5) = 6{,}006.25$$

$$y = 95 - 17.5 = 77.5$$

This time, the dimensions of the pen with the largest area are 77.5 feet by 77.5 feet. ◆

PROBLEM 18   Repeat Example 18 if the barn is 80 feet long. ◆

◆ EXAMPLE 19

Inventory Control

A record company anticipates that there will be a demand for 20,000 copies of a certain album during the following year. It costs the company $0.50 to store a record for 1 year. Each time it must press additional records, it costs $200 to set up the equipment. How many records should the company press during each production run in order to minimize its total storage and set-up costs?

Solution

This type of problem is called an **inventory control problem.** One of the basic assumptions made in such problems is that the demand is uniform. For example, if there are 250 working days in a year, then the daily demand would be $20{,}000/250 = 800$ records. The company could decide to produce all 20,000 records at the beginning of the year. This would certainly minimize the set-up costs, but would result in very large storage costs. At the other extreme, it could produce 800 records each day. This would minimize the storage costs, but would result in very large set-up costs. Somewhere between these two extremes is the optimal solution that will minimize the total storage and set-up costs. Let

$x =$ Number of records pressed during each production run

$y =$ Number of production runs

It is easy to see that the total set-up cost for the year is $200y$, but what is the total storage cost? If the demand is uniform, then the number of records in storage between production runs will decrease from x to 0, and the average number in storage each day is $x/2$. This result is illustrated in the figure on page 602.

Since it costs $0.50 to store a record for 1 year, the total storage cost is $0.5(x/2) = 0.25x$ and the total cost is

$$\text{Total cost} = \text{Set-up cost} + \text{Storage cost}$$

$$C = 200y + 0.25x$$

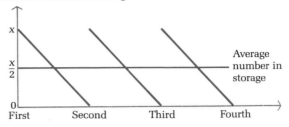

Number of records in storage

Average number in storage

First Second Third Fourth

Production run

In order to write the total cost C as a function of one variable, we must find a relationship between x and y. If the company produces x records in each of y production runs, then the total number of records produced is xy. Thus,

$$xy = 20{,}000$$

$$y = \frac{20{,}000}{x}$$

Certainly, x must be at least 1 and cannot exceed 20,000. Thus, we must solve the following:

$$\text{Minimize} \quad C(x) = 200\left(\frac{20{,}000}{x}\right) + 0.25x \qquad 1 \leq x \leq 20{,}000$$

$$C(x) = \frac{4{,}000{,}000}{x} + 0.25x$$

$$C'(x) = -\frac{4{,}000{,}000}{x^2} + 0.25$$

$$-\frac{4{,}000{,}000}{x^2} + 0.25 = 0$$

$$x^2 = \frac{4{,}000{,}000}{0.25}$$

$$x^2 = 16{,}000{,}000$$

$$x = 4{,}000 \qquad -4{,}000 \text{ is not a critical value}$$
$$\text{since } 1 \leq x \leq 20{,}000$$

$$C''(x) = \frac{8{,}000{,}000}{x^3} > 0 \qquad \text{for } x \in (1,\ 20{,}000)$$

Thus,

$$\text{Min } C(x) = C(4{,}000) = 2{,}000$$

$$y = \frac{20{,}000}{4{,}000} = 5$$

The company will minimize its total cost by pressing 4,000 records five times during the year. ◆

Repeat Example 19 if it costs $250 to set up a production run and $0.40 to store a record for 1 year.

◆

Answers to Matched Problems

13. (A) Absolute maximum: $f(5) = 65$; absolute minimum: $f(-5) = -65$
    (B) Absolute maximum: $f(-2) = 16$; absolute minimum: $f(2) = -16$
    (C) Absolute maximum: $f(-2) = 16$; absolute minimum: $f(1) = -11$
14. $f(3) = 6$
15. (A) Max $R(x) = R(4,500) = \$675,000$
    (B) Max $P(x) = P(4,050) = \$456,750$; $p = \$165$
16. Max $P(x) = P(3,750) = \$378,750$; $p = \$175$; price increases $10, profit decreases $78,000
17. Max $Y(x) = Y(15) = 2,025$ lb/acre
18. (A) 80 ft by 40 ft   (B) 82.5 ft by 82.5 ft
19. Press 5,000 records four times during the year

EXERCISE 10-4

A *Find the absolute maximum and minimum, if either exists, for each function.*

1. $f(x) = x^2 - 4x + 5$
2. $f(x) = x^2 + 6x + 7$
3. $f(x) = 10 + 8x - x^2$
4. $f(x) = 6 - 8x - x^2$
5. $f(x) = 1 - x^3$
6. $f(x) = 1 - x^4$

B *Find the indicated extremum of each function.*

7. Absolute maximum value of $f(x) = 24 - 2x - \dfrac{8}{x}$, $x > 0$

8. Absolute minimum value of $f(x) = 3x + \dfrac{27}{x}$, $x > 0$

9. Absolute minimum value of $f(x) = 5 + 3x + \dfrac{12}{x^2}$, $x > 0$

10. Absolute maximum value of $f(x) = 10 - 2x - \dfrac{27}{x^2}$, $x > 0$

Find the absolute maximum and minimum, if either exists, for each function on the indicated intervals.

11. $f(x) = x^3 - 6x^2 + 9x - 6$
    (A) $[-1, 5]$ (B) $[-1, 3]$ (C) $[2, 5]$
12. $f(x) = 2x^3 - 3x^2 - 12x + 24$
    (A) $[-3, 4]$ (B) $[-2, 3]$ (C) $[-2, 1]$
13. $f(x) = (x - 1)(x - 5)^3 + 1$
    (A) $[0, 3]$ (B) $[1, 7]$ (C) $[3, 6]$

14. $f(x) = x^4 - 8x^2 + 16$

    (A) $[-1, 3]$ (B) $[0, 2]$ (C) $[-3, 4]$

Preliminary word problems:

C   **15.** How would you divide a 10 inch line so that the product of the two lengths is a maximum?

     16. What quantity should be added to 5 and subtracted from 5 in order to produce the maximum product of the results?

     17. Find two numbers whose difference is 30 and whose product is a minimum.

     18. Find two positive numbers whose sum is 60 and whose product is a maximum.

     19. Find the dimensions of a rectangle with perimeter 100 centimeters that has maximum area. Find the maximum area.

     20. Find the dimensions of a rectangle of area 225 square centimeters that has the least perimeter. What is the perimeter?

APPLICATIONS

Business & Economics

21. *Average costs.* If the average manufacturing cost (in dollars) per pair of sunglasses is given by

$$\overline{C}(x) = x^2 - 6x + 12 \qquad 0 \leqslant x \leqslant 6$$

where x is the number (in thousands) of pairs manufactured, how many pairs of glasses should be manufactured to minimize the average cost per pair? What is the minimum average cost per pair?

22. *Maximum revenue and profit.* A company manufactures and sells x television sets per month. The monthly cost and demand equations are

$$C(x) = 72,000 + 60x$$

$$p = 200 - \frac{x}{30} \qquad 0 \leqslant x \leqslant 6,000$$

    (A) Find the maximum revenue.

    (B) Find the maximum profit, the production level that will realize the maximum profit, and the price the company should charge for each television set.

    (C) If the government decides to tax the company $5 for each set it produces, how many sets should the company manufacture each month in order to maximize its profit? What is the maximum profit? What should the company charge for each set?

23. *Car rental.* A car rental agency rents 200 cars per day at a rate of $30 per day. For each $1 increase in rate, 5 fewer cars are rented. At what rate should the cars be rented to produce the maximum income? What is the maximum income?

24. *Rental income.* A 300 room hotel in Las Vegas is filled to capacity every night at $80 a room. For each $1 increase in rent, 3 fewer rooms are rented. If each rented room costs $10 to service per day, how much should the management charge for each room to maximize gross profit? What is the maximum gross profit?

25. *Agriculture.* A commercial cherry grower estimates from past records that if 30 trees are planted per acre, each tree will yield an average of 50 pounds of cherries per season. If for each additional tree planted per acre (up to 20), the average yield per tree is reduced by 1 pound, how many trees should be planted per acre to obtain the maximum yield per acre? What is the maximum yield?

26. *Agriculture.* A commercial pear grower must decide on the optimum time to have fruit picked and sold. If the pears are picked now, they will bring 30¢ per pound, with each tree yielding an average of 60 pounds of salable pears. If the average yield per tree increases 6 pounds per tree per week for the next 4 weeks, but the price drops 2¢ per pound per week, when should the pears be picked to realize the maximum return per tree? What is the maximum return?

27. *Manufacturing.* A candy box is to be made out of a piece of cardboard that measures 8 by 12 inches. Squares of equal size will be cut out of each corner, and then the ends and sides will be folded up to form a rectangular box. What size square should be cut from each corner to obtain a maximum volume?

28. *Packaging.* A parcel delivery service will deliver a package only if the length plus girth (distance around) does not exceed 108 inches.

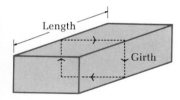

(A) Find the dimensions of a rectangular box with square ends that satisfies the delivery service's restriction and has maximum volume. What is the maximum volume?

(B) Find the dimensions (radius and height) of a cylindrical container that meets the delivery service's requirement and has maximum volume. What is the maximum volume?

29. *Construction costs.* A fence is to be built to enclose a rectangular area of 800 square feet. The fence along three sides is to be made of material that costs $6 per foot. The material for the fourth side costs $18 per foot. Find the dimensions of the rectangle that will allow the most economical fence to be built.

30. *Construction costs.* The owner of a retail lumber store wants to construct a fence to enclose an outdoor storage area adjacent to the store, as indicated in the figure in the margin. Find the dimensions that will enclose the largest area if:

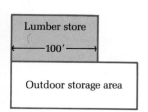

(A) 240 feet of fencing material is used.

(B) 400 feet of fencing material is used.

31. Inventory control. A publishing company sells 50,000 copies of a certain book each year. It costs the company $1.00 to store a book for 1 year. Each time it must print additional copies, it costs the company $1,000 to set up the presses. How many books should the company produce during each printing in order to minimize its total storage and set-up costs?

32. Operational costs. The cost per hour for fuel to run a train is $v^2/4$ dollars, where v is the speed of the train in miles per hour. (Note that the cost goes up as the square of the speed.) Other costs, including labor, are $300 per hour. How fast should the train travel on a 360 mile trip to minimize the total cost for the trip?

33. Construction costs. A freshwater pipeline is to be run from a source on the edge of a lake to a small resort community on an island 5 miles off-shore, as indicated in the figure.

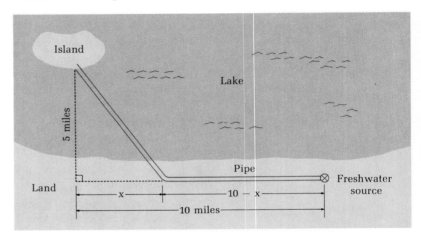

(A) If it costs 1.4 times as much to lay the pipe in the lake as it does on land, what should x be (in miles) to minimize the total cost of the project?

(B) If it costs only 1.1 times as much to lay the pipe in the lake as it does on land, what should x be to minimize the total cost of the project? [Note: Compare with Problem 38.]

34. Manufacturing costs. A manufacturer wants to produce cans that will hold 12 ounces (approximately 22 cubic inches) in the form of a right circular cylinder. Find the dimensions (radius of an end and height) of the can that will use the smallest amount of material. Assume the circular ends are cut out of squares, with the corner portions wasted, and the sides are made from rectangles, with no waste.

Life Sciences

35. Bacteria control. A recreational swimming lake is treated periodically to control harmful bacteria growth. Suppose t days after a treatment, the concentration of bacteria per cubic centimeter is given by

$$C(t) = 30t^2 - 240t + 500 \qquad 0 \leq t \leq 8$$

How many days after a treatment will the concentration be minimal? What is the minimum concentration?

36. *Drug concentration.* The concentration $C(t)$, in milligrams per cubic centimeter, of a particular drug in a patient's bloodstream is given by

$$C(t) = \frac{0.16t}{t^2 + 4t + 4}$$

where t is the number of hours after the drug is taken. How many hours after the drug is given will the concentration be maximum? What is the maximum concentration?

37. *Laboratory management.* A laboratory uses 500 white mice each year for experimental purposes. It costs $4.00 to feed a mouse for 1 year. Each time mice are ordered from a supplier, there is a service charge of $10 for processing the order. How many mice should be ordered each time in order to minimize the total cost of feeding the mice and of placing the orders for the mice?

38. *Bird flights.* Some birds tend to avoid flights over large bodies of water during daylight hours. (It is speculated that more energy is required to fly over water than land, because air generally rises over land and falls over water during the day.) Suppose an adult bird with this tendency is taken from its nesting area on the edge of a large lake to an island 5 miles off-shore and is then released (see the accompanying figure).

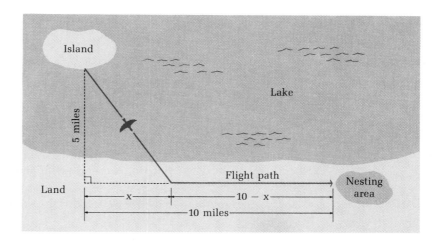

(A) If it takes 1.4 times as much energy to fly over water as land, how far up-shore (x, in miles) should the bird head in order to minimize the total energy expended in returning to the nesting area?

(B) It it takes only 1.1 times as much energy to fly over water as land, how far up-shore should the bird head in order to minimize the total energy expended in returning to the nesting area? [*Note:* Compare with Problem 33.]

39. *Botany.* If it is known from past experiments that the height (in feet) of a given plant after t months is given approximately by

$$H(t) = 4t^{1/2} - 2t \qquad 0 \leqslant t \leqslant 2$$

how long, on the average, will it take a plant to reach its maximum height? What is the maximum height?

40. *Pollution.* Two heavy industrial areas are located 10 miles apart, as indicated in the figure. If the concentration of particulate matter (in parts per million) decreases as the reciprocal of the square of the distance from the source, and area A_1 emits eight times the particulate matter as A_2, then the concentration of particulate matter at any point between the two areas is given by

$$C(x) = \frac{8k}{x^2} + \frac{k}{(10 - x)^2} \qquad 0.5 \leqslant x \leqslant 9.5, \quad k > 0$$

How far from A_1 will the concentration of particulate matter be at a minimum?

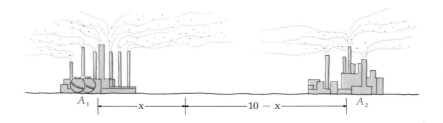

41. *Politics.* In a newly incorporated city, it is estimated that the voting population (in thousands) will increase according to

$$N(t) = 30 + 12t^2 - t^3 \qquad 0 \leqslant t \leqslant 8$$

where t is time in years. When will the rate of increase be most rapid?

42. *Learning.* A large grocery chain found that, on the average, a checker can memorize $P\%$ of a given price list in x continuous hours, as given approxi-

Social Sciences

mately by

$$P(x) = 96x - 24x^2 \qquad 0 \leqslant x \leqslant 3$$

How long should a checker plan to take to memorize the maximum percentage? What is the maximum?

SECTION 10-5 The Constant *e* and Continuous Compound Interest

◆ THE CONSTANT *e*
◆ CONTINUOUS COMPOUND INTEREST

In Chapter 3, both the exponential function with base *e* and continuous compound interest were introduced informally. Now, with limit concepts at our disposal, we can give precise definitions of *e* and continuous compound interest.

◆ THE CONSTANT *e*

The special irrational number *e* is a particularly suitable base for both exponential and logarithmic functions. The reasons for choosing this number as a base will become clear as we develop differentiation formulas for the exponential function e^x and the natural logarithmic function ln x.

In precalculus treatments (Chapter 3), the number *e* is informally defined as an irrational number that can be approximated by the expression $[1 + (1/n)]^n$ by taking n sufficiently large. Now we will use the limit concept to formally define *e* as either of the following two limits:

The Number *e*

$$e = \lim_{n \to \infty} \left(1 + \frac{1}{n}\right)^n \qquad \text{or, alternately,} \qquad e = \lim_{s \to 0}(1 + s)^{1/s}$$

$$e = 2.718\ 281\ 828\ 459. \ . \ .$$

We will use both these limit forms. [*Note:* If s = 1/n, then as n → ∞, s → 0.]

The proof that the indicated limits exist and represent an irrational number between 2 and 3 is not easy and is omitted here. Many people reason (incorrectly) that the limits are 1, since "$(1 + s)$ approaches 1 as $s \to 0$, and 1 to any power is 1." A little experimentation with a calculator can convince you otherwise. Consider the table of values for s and $f(s) = (1 + s)^{1/s}$ and the graph shown in Figure 16 for s close to 0.

	s approaches 0 from the left $\to 0 \leftarrow s$ approaches 0 from the right							
s	-0.5	-0.2	-0.1	$-0.01 \to 0 \leftarrow 0.01$	0.1	0.2	0.5
$(1 + s)^{1/s}$	4.0000	3.0518	2.8680	$2.7320 \to e \leftarrow 2.7048$	2.5937	2.4883	2.2500

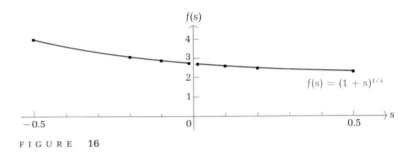

FIGURE 16

Compute some of the table values with a calculator yourself, and also try several values of s even closer to 0. Note that the function is discontinuous at $s = 0$.

Exactly who discovered e is still being debated. It is named after the great mathematician Leonard Euler (1707–1783), who computed e to twenty-three decimal places using $[1 + (1/n)]^n$.

◆ CONTINUOUS COMPOUND INTEREST

Now we will see how e appears quite naturally in the important application of compound interest. Let us start with simple interest, move on to compound interest, and then on to continuous compound interest.

If a principal P is borrowed at an annual rate of r,* then after t years at simple interest the borrower will owe the lender an amount A given by

$$A = P + Prt = P(1 + rt) \qquad \text{Simple interest} \tag{1}$$

---

* If r is the interest rate written as a decimal, then $100r\%$ is the rate using %. For example, if $r = 0.12$, then we have $100r\% = 100(0.12)\% = 12\%$. The expressions 0.12 and 12% are therefore equivalent. Unless stated otherwise, all formulas in this book use r as a decimal.

On the other hand, if interest is compounded n times a year, then the borrower will owe the lender an amount A given by

$$A = P\left(1 + \frac{r}{n}\right)^{nt} \qquad \text{Compound interest} \tag{2}$$

where r/n is the interest rate per compounding period and nt is the number of compounding periods. Suppose P, r, and t in (2) are held fixed and n is increased. Will the amount A increase without bound, or will it tend to approach some limiting value?

Let us perform a calculator experiment before we attack the general limit problem. If $P = \$100$, $r = 0.06$, and $t = 2$ years, then

$$A = 100\left(1 + \frac{0.06}{n}\right)^{2n}$$

We compute A for several values of n in Table 1. The biggest gain appears in the first step; then the gains slow down as n increases. In fact, it appears that A might be tending to approach $\$112.75$ as n gets larger and larger.

TABLE 1

COMPOUNDING FREQUENCY	n	$A = 100\left(1 + \dfrac{0.06}{n}\right)^{2n}$
Annually	1	$112.3600
Semiannually	2	112.5509
Quarterly	4	112.6493
Weekly	52	112.7419
Daily	365	112.7486
Hourly	8,760	112.7496

Now we turn back to the general problem for a moment. Keeping P, r, and t fixed in equation (2), we compute the following limit and observe an interesting and useful result.

$$\lim_{n \to \infty} P\left(1 + \frac{r}{n}\right)^{nt} = P \lim_{n \to \infty}\left(1 + \frac{r}{n}\right)^{(n/r)rt}$$

Insert r/r in the exponent and let $s = r/n$. Note that $n \to \infty$ implies $s \to 0$.

$$= P \lim_{s \to 0}[(1 + s)^{1/s}]^{rt}$$

Use the limit property given in the footnote below.*

$$= P[\lim_{s \to 0}(1 + s)^{1/s}]^{rt}$$

$$\lim_{s \to 0}(1 + s)^{1/s} = e$$

$$= Pe^{rt}$$

---

* The following new limit property is used: If $\lim_{x \to c} f(x)$ exists, then $\lim_{x \to c}[f(x)]^p = [\lim_{x \to c} f(x)]^p$, provided the last expression names a real number.

The resulting formula is called the **continuous compound interest formula,** a very important and widely used formula in business and economics.

▬ Continuous Compound Interest

$$A = Pe^{rt}$$

where

    P = Principal

    r = Annual nominal interest rate compounded continuously

    t = Time in years

    A = Amount at time t

◆ E X A M P L E   20   If $100 is invested at an annual nominal rate of 6% compounded continuously, what amount will be in the account after 2 years?

Solution $A = Pe^{rt}$

         $= 100e^{(0.06)(2)}$ 6% is equivalent to $r = 0.06$.

         $\approx \$112.7497$

(Compare this result with the values calculated in Table 1.)   ◆

P R O B L E M   20   What amount (to the nearest cent) will an account have after 5 years if $100 is invested at an annual nominal rate of 8% compounded annually? Semiannually? Continuously?   ◆

◆ E X A M P L E   21   If $100 is invested at 12% compounded continuously,* graph the amount in the account relative to time for a period of 10 years.

Solution   We want to graph

         $A = 100e^{0.12t}$ $0 \leqslant t \leqslant 10$

---

* Following common usage, we will often write the form "at 12% compounded continuously," understanding that this means "at an annual nominal rate of 12% compounded continuously."

We construct a table of values using a calculator, graph the points from the table, and join the points with a smooth curve.

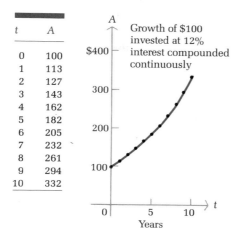

t	A
0	100
1	113
2	127
3	143
4	162
5	182
6	205
7	232
8	261
9	294
10	332

◆

PROBLEM 21 If $5,000 is invested at 20% compounded continuously, graph the amount in the account relative to time for a period of 10 years. ◆

◆ **EXAMPLE 22** How long will it take an investment of $5,000 to grow to $8,000 if it is invested at 12% compounded continuously?

Solution Starting with the continuous compound interest formula $A = Pe^{rt}$, we must solve for t:

$$A = Pe^{rt}$$

$$8{,}000 = 5{,}000e^{0.12t} \quad \text{Divide both sides by 5,000 and reverse the equation.}$$

$$e^{0.12t} = 1.6 \quad \text{Take the natural logarithm of both sides — recall}$$

$$\ln e^{0.12t} = \ln 1.6 \quad \text{that } \log_b b^x = x.$$

$$0.12t = \ln 1.6$$

$$t = \frac{\ln 1.6}{0.12}$$

$$t \approx 3.92 \text{ years}$$ ◆

PROBLEM 22 How long will it take an investment of $10,000 to grow to $15,000 if it is invested at 9% compounded continuously? ◆

◆ **EXAMPLE 23** How long will it take money to double if it is invested at 18% compounded continuously?

Solution   Starting with the continuous compound interest formula $A = Pe^{rt}$, we must solve for t given $A = 2P$ and $r = 0.18$:

$$2P = Pe^{0.18t} \qquad \text{Divide both sides by } P \text{ and reverse the equation.}$$
$$e^{0.18t} = 2 \qquad \text{Take the natural logarithm of both sides.}$$
$$\ln e^{0.18t} = \ln 2$$
$$0.18t = \ln 2$$
$$t = \frac{\ln 2}{0.18}$$
$$t \approx 3.85 \text{ years}$$

◆

P R O B L E M   23   How long will it take money to triple if it is invested at 12% compounded continuously?

◆

Answers to Matched Problems

20. $146.93; $148.02; $149.18

21. $A = 5{,}000e^{0.2t}$

t	A
0	5,000
1	6,107
2	7,459
3	9,111
4	11,128
5	13,591
6	16,601
7	20,276
8	24,765
9	30,248
10	36,945

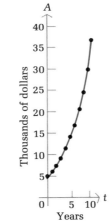

22. 4.51 yr

23. 9.16 yr

A   *Use a calculator or table to evaluate A to the nearest cent in Problems 1 and 2.*

1. $A = \$1{,}000e^{0.1t}$ for $t = 2, 5,$ and 8
2. $A = \$5{,}000e^{0.08t}$ for $t = 1, 4,$ and 10

B In Problems 3–8, solve for t or r to two decimal places.

3. $2 = e^{0.06t}$ 4. $2 = e^{0.03t}$ 5. $3 = e^{0.1t}$

6. $3 = e^{0.25t}$ 7. $2 = e^{5r}$ 8. $3 = e^{10r}$

C In Problems 9 and 10, complete each table to five decimal places using a calculator.

9.

n	$[1 + (1/n)]^n$
10	2.593 74
100	
1,000	
10,000	
100,000	
1,000,000	
10,000,000	
$\downarrow$	$\downarrow$
∞	$e = 2.718\ 281\ 828\ 459...$

10.

s	$(1 + s)^{1/s}$
0.01	2.704 81
-0.01	
0.001	
-0.001	
0.000 1	
-0.000 1	
0.000 01	
-0.000 01	
$\downarrow$	$\downarrow$
0	$e = 2.718\ 281\ 828\ 459...$

APPLICATIONS

Business & Economics

11. *Continuous compound interest.* If $20,000 is invested at an annual nominal rate of 12% compounded continuously, how much will it be worth in 8.5 years?

12. *Continuous compound interest.* Assume $1 had been invested at an annual nominal rate of 4% compounded continuously, at the time of the birth of Christ. What would be the value of the account in solid gold Earths in the year 2000? Assume that the Earth weighs approximately 2.11×10^{26} ounces and that gold will be worth $1,000 an ounce in the year 2000. What would be the value of the account in dollars at simple interest?

13. *Present value.* A note will pay $20,000 at maturity 10 years from now. How much should you be willing to pay for the note now if money is worth 7% compounded continuously?

14. *Present value.* A note will pay $50,000 at maturity 5 years from now. How much should you be willing to pay for the note now if money is worth 8% compounded continuously?

15. *Continuous compound interest.* An investor bought stock for $20,000. Four years later, the stock was sold for $30,000. If interest is compounded continuously, what annual nominal rate of interest did the original $20,000 investment earn?

16. *Continuous compound interest.* A family paid $40,000 cash for a house. Ten years later, they sold the house for $100,000. If interest is compounded continuously, what annual nominal rate of interest did the original $40,000 investment earn?

17. *Present value.* Solving $A = Pe^{rt}$ for P, we obtain

$$P = Ae^{-rt}$$

which is the present value of the amount A due in t years if money earns interest at an annual nominal rate r compounded continuously.

(A) Graph $P = 10,000e^{-0.08t}$, $0 \leqslant t \leqslant 50$.
(B) $\lim_{t \to \infty} 10,000e^{-0.08t} = ?$ (Guess, using part A.)

[*Conclusion:* The longer the duration of time until the amount A is due, the smaller its present value, as we would expect.]

18. *Present value.* Referring to Problem 17, in how many years will the $10,000 have to be due in order for its present value to be $5,000?

19. *Doubling time.* How long will it take money to double if it is invested at 25% compounded continuously?

20. *Doubling time.* How long will it take money to double if it is invested at 5% compounded continuously?

21. *Doubling rate.* At what nominal rate compounded continuously must money be invested to double in 5 years?

22. *Doubling rate.* At what nominal rate compounded continuously must money be invested to double in 3 years?

23. *Doubling time.* It is instructive to look at doubling times for money invested at various nominal rates of interest compounded continuously. Show that doubling time t at an annual rate r compounded continuously is given by

$$t = \frac{\ln 2}{r}$$

24. *Doubling time.* Graph the doubling-time equation from Problem 23 for $0 < r < 1.00$. Identify vertical and horizontal asymptotes.

Life Sciences

25. *World population.* A mathematical model for world population growth over short periods of time is given by

$$P = P_0 e^{rt}$$

where

P_0 = Population at time $t = 0$

r = Continuous compound rate of growth

t = Time in years

P = Population at time t

How long will it take the world population to double if it continues to grow at its current continuous compound rate of 2% per year?

26. *World population.* Repeat Problem 25 under the assumption that the world population is growing at a continuous compound rate of 1% per year.

27. *Population growth.* Some underdeveloped nations have population doubling times of 20 years. At what continuous compound rate is the population growing? (Use the population growth model in Problem 25.)

28. *Population growth.* Some developed nations have population doubling times of 120 years. At what continuous compound rate is the population growing? (Use the population growth model in Problem 25.)

29. *Radioactive decay.* A mathematical model for the decay of radioactive substances is given by

$$Q = Q_0 e^{rt}$$

where

Q_0 = Amount of the substance at time $t = 0$

r = Continuous compound rate of decay

t = Time in years

Q = Amount of the substance at time t

If the continuous compound rate of decay of radium per year is $r = -0.000\ 433\ 2$, how long will it take an amount of radium to decay to half the original amount? (This period of time is the half-life of the substance.)

30. *Radioactive decay.* The continuous compound rate of decay of carbon-14 per year is $r = -0.000\ 123\ 8$. How long will it take an amount of carbon-14 to decay to half the original amount? (Use the radioactive decay model in Problem 29.)

31. *Radioactive decay.* A cesium isotope has a half-life of 30 years. What is the continuous compound rate of decay? (Use the radioactive decay model in Problem 29.)

32. *Radioactive decay.* A strontium isotope has a half-life of 90 years. What is the continuous compound rate of decay? (Use the radioactive decay model in Problem 29.)

33. *World population.* If the world population is now 5 billion (5×10^9) people and if it continues to grow at a continuous compound rate of 2% per year, how long will it be before there is only 1 square yard of land per person? (The Earth has approximately 1.68×10^{14} square yards of land.)

Derivatives of Logarithmic and Exponential Functions

♦ DERIVATIVE FORMULAS FOR ln x AND e^x
♦ GRAPHING TECHNIQUES
♦ APPLICATION

In this section, we discuss derivative formulas for ln x and e^x. Out of all the possible choices for bases for the logarithmic and exponential functions, $\log_b x$ and b^x, it turns out (as we will see in this and the next section) that the simplest derivative formulas occur when the base b is chosen to be e.

♦ DERIVATIVE FORMULAS FOR ln x AND e^x

We are now ready to derive a formula for the derivative of

$$f(x) = \ln x = \log_e x \qquad x > 0$$

using the definition of the derivative

$$f'(x) = \lim_{h \to 0} \frac{f(x + h) - f(x)}{h}$$

and the two-step process discussed in Section 9-3:

Step 1. Simplify the difference quotient first:

$$\frac{f(x + h) - f(x)}{h} = \frac{\ln(x + h) - \ln x}{h}$$

$$= \frac{1}{h}[\ln(x + h) - \ln x] \qquad \text{Use } \ln A - \ln B = \ln \frac{A}{B}.$$

$$= \frac{1}{h} \ln \frac{x + h}{x} \qquad \text{Multiply by } 1 = x/x \text{ to change form.}$$

$$= \frac{x}{x} \cdot \frac{1}{h} \ln \frac{x + h}{x}$$

$$= \frac{1}{x} \left[\frac{x}{h} \ln \left(1 + \frac{h}{x} \right) \right] \qquad \text{Use } p \ln A = \ln A^p.$$

$$= \frac{1}{x} \ln \left(1 + \frac{h}{x} \right)^{x/h}$$

Step 2. Find the limit. Let $s = h/x$. For x fixed, if $h \to 0$, then $s \to 0$. Thus,

$$D_x \ln x = \lim_{h \to 0} \frac{f(x + h) - f(x)}{h}$$

$$= \lim_{h \to 0} \left[\frac{1}{x} \ln \left(1 + \frac{h}{x} \right)^{x/h} \right] \qquad \text{Let } s = h/x. \text{ Note that } h \to 0 \text{ implies } s \to 0.$$

$$= \frac{1}{x} \lim_{s \to 0} [\ln(1 + s)^{1/s}] \qquad \text{Use the new limit property given in the footnote below.}^*$$

$$= \frac{1}{x} \ln[\lim_{s \to 0} (1 + s)^{1/s}] \qquad \text{Use the definition of } e.$$

$$= \frac{1}{x} \ln e \qquad \text{ln } e = \log_e e = 1$$

$$= \frac{1}{x}$$

Thus,

$$D_x \ln x = \frac{1}{x}$$

In the next section, we will show that, in general,

$$D_x \log_b x = \frac{1}{\ln b} \left(\frac{1}{x} \right)$$

which is a somewhat more complicated result than the above — unless $b = e$.

We now apply the two-step process to the exponential function $f(x) = e^x$. In the process, we will use (without proof) the fact that

$$\lim_{h \to 0} \left(\frac{e^h - 1}{h} \right) = 1$$

[Try computing $(e^h - 1)/h$ for values of h closer and closer to 0 and on either side of 0 to convince yourself of the reasonableness of this limit.]

Step 1. Simplify the difference quotient first:

$$\frac{f(x + h) - f(x)}{h} = \frac{e^{x+h} - e^x}{h} \qquad \text{Use } e^{a+b} = e^a e^b.$$

$$= \frac{e^x e^h - e^x}{h} \qquad \text{Factor out } e^x.$$

$$= e^x \left(\frac{e^h - 1}{h} \right)$$

---

* The following new limit property is used: If $\lim_{x \to c} f(x)$ exists and is positive, then $\lim_{x \to c} [\ln f(x)] = \ln[\lim_{x \to c} f(x)]$.

Step 2. Compute the limit of the result in step 1:

$$D_x\, e^x = \lim_{h\to 0} \frac{f(x+h) - f(x)}{h}$$

$$= \lim_{h\to 0} e^x\!\left(\frac{e^h - 1}{h}\right)$$

$$= e^x \lim_{h\to 0}\left(\frac{e^h - 1}{h}\right) \qquad \text{Use the assumed limit given above.}$$

$$= e^x \cdot 1 = e^x$$

Thus,

$$\boldsymbol{D_x\, e^x = e^x}$$

In the next section, we will show that

$$D_x\, b^x = b^x \ln b$$

which is, again, a somewhat more complicated result than the above — unless $b = e$.

The two results just obtained explain why e^x is so widely used that it is sometimes referred to as *the* exponential function. These two new and important derivative formulas are restated in the box for reference.

Derivatives of the Natural Logarithmic and Exponential Functions

$$D_x \ln x = \frac{1}{x} \qquad\qquad D_x\, e^x = e^x$$

These new derivative formulas can be combined with the rules of differentiation discussed in Chapter 9 to differentiate a wide variety of functions.

◆ **E X A M P L E   24**   Find $f'(x)$ for:

(A) $f(x) = 2e^x + 3 \ln x$ (B) $f(x) = \dfrac{e^x}{x^3}$

(C) $f(x) = (\ln x)^4$ (D) $f(x) = \ln x^4$

Solutions   (A) $f'(x) = \boxed{2\,D_x\, e^x + 3\,D_x \ln x}$

$$= 2e^x + 3\!\left(\frac{1}{x}\right) = 2e^x + \frac{3}{x}$$

(B) $f'(x) = \boxed{\dfrac{x^3\,D_x\, e^x - e^x\,D_x\, x^3}{(x^3)^2}}$ Quotient rule

$$= \frac{x^3 e^x - e^x 3x^2}{x^6} = \frac{x^2 e^x(x-3)}{x^6} = \frac{e^x(x-3)}{x^4}$$

(C) $D_x(\ln x)^4 = 4(\ln x)^3 D_x \ln x$ Power rule for functions

$$= 4(\ln x)^3 \left(\frac{1}{x}\right) = \frac{4(\ln x)^3}{x}$$

(D) $D_x \ln x^4 = D_x(4 \ln x)$ Property of logarithms

$$= 4 \left(\frac{1}{x}\right) = \frac{4}{x}$$

◆

PROBLEM   24   Find $f'(x)$ for:

(A) $f(x) = 4 \ln x - 5e^x$ (B) $f(x) = x^2 e^x$
(C) $f(x) = \ln x^3$ (D) $f(x) = (\ln x)^3$

◆

Common Error

$$D_x e^x \neq xe^{x-1} \qquad D_x e^x = e^x$$

The power rule cannot be used to differentiate the exponential function. The power rule applies to exponential forms x^n where the exponent is a constant and the base is a variable. In the exponential form e^x, the base is a constant and the exponent is a variable.

◆ GRAPHING TECHNIQUES

Using the techniques discussed earlier in this chapter, we can use first and second derivatives to gain useful information about the graphs of $y = \ln x$ and $y = e^x$. Using the derivative formulas given above, we can construct Table 2.

TABLE   2

$\ln x$		e^x	
$y = \ln x$	$x > 0$	$y = e^x$	$-\infty < x < \infty$
$y' = 1/x > 0$	$x > 0$	$y' = e^x > 0$	$-\infty < x < \infty$
$y'' = -1/x^2 < 0$	$x > 0$	$y'' = e^x > 0$	$-\infty < x < \infty$

From the table, we can see that both functions are increasing throughout their respective domains, the graph of $y = \ln x$ is always concave downward, and the graph of $y = e^x$ is always concave upward. It can be shown that the y axis is a vertical asymptote for the graph of $y = \ln x$ ($\lim_{x \to 0^+} \ln x = -\infty$), and the x axis is a horizontal asymptote for the graph of $y = e^x$ ($\lim_{x \to -\infty} e^x = 0$). Both equations are graphed in Figure 17 at the top of the next page.

Notice that if we fold the page along the dashed line $y = x$, the two graphs match exactly (see Section 3-3). Also notice that both graphs are unbounded as $x \to \infty$. Comparing each graph with the graph of $y = x$ (the dashed line), we

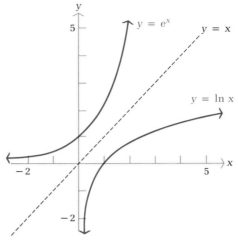

FIGURE 17

e^x is continuous on $(-\infty, \infty)$

ln x is continuous on $(0, \infty)$

conclude that e^x grows more rapidly than x and ln x grows more slowly than x. In fact, the following limits can be established:

$$\lim_{x \to \infty} \frac{x^p}{e^x} = 0, \quad p > 0 \qquad \text{and} \qquad \lim_{x \to \infty} \frac{\ln x}{x^p} = 0, \quad p > 0$$

These limits indicate that e^x grows more rapidly than any positive power of x, and ln x grows more slowly than any positive power of x.

Now we will apply graphing techniques to a slightly more complicated function.

◆ E X A M P L E 25   Sketch the graph of $f(x) = xe^x$ using the graphing strategy discussed in Section 10-3.

Solution   *Step 1.*   Use $f(x)$: $f(x) = xe^x$

(A) Domain: All real numbers
(B) Intercepts:
     y intercept: $f(0) = 0$
     x intercept: $xe^x = 0$ for $x = 0$ only, since $e^x > 0$ for all x (see Figure 17).
(C) Asymptotes:
     Vertical asymptotes: None
     Horizontal asymptotes: We have not developed limit techniques for functions of this type to determine the behavior of $f(x)$ as $x \to -\infty$ and $x \to \infty$.

However, the following tables of values suggest the nature of the graph of f as $x \to -\infty$ and $x \to \infty$:

x	1	5	10	$\to \infty$
$f(x)$	2.72	742.07	220,264.66	$\to \infty$

x	-1	-5	-10	$\to -\infty$
$f(x)$	-0.37	-0.03	$-0.000\ 45$	$\to 0$

Step 2. Use $f'(x)$:

$$f'(x) = x\, D_x\, e^x + e^x\, D_x\, x$$
$$= xe^x + e^x = e^x(x + 1)$$

Critical value: -1
Partition number: -1
Sign chart for $f'(x)$:

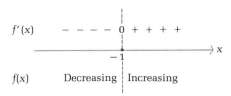

Test Numbers

x	$f'(x)$	
-2	$-e^{-2}$	$(-)$
0	1	$(+)$

Thus, $f(x)$ decreases on $(-\infty, -1)$, has a local minimum at $x = -1$, and increases on $(-1, \infty)$. [Since $e^x > 0$ for all x, we do not have to evaluate e^{-2} to conclude that $-e^{-2} < 0$ when using the test number -2.]

Step 3. Use $f''(x)$:

$$f''(x) = e^x\, D_x(x + 1) + (x + 1)\, D_x\, e^x$$
$$= e^x + (x + 1)e^x = e^x(x + 2)$$

Sign chart for $f''(x)$ (partition number is -2):

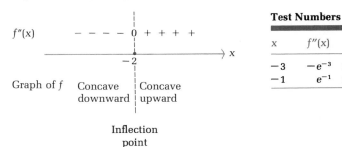

Test Numbers

x	$f''(x)$	
-3	$-e^{-3}$	$(-)$
-1	e^{-1}	$(+)$

Thus, the graph of f is concave downward on $(-\infty, -2)$, has an inflection point at $x = -2$, and is concave upward on $(-2, \infty)$.

Step 4. Sketch the graph of f using the information from steps 1–3:

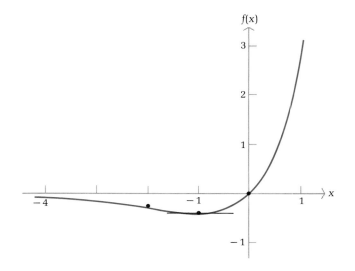

x	f(x)
−2	−0.27
−1	−0.37
0	0

◆

PROBLEM 25   Sketch the graph of $f(x) = x \ln x$.   ◆

◆ APPLICATION

◆ EXAMPLE 26

Maximum profit

The market research department of a chain of pet stores test marketed their aquarium pumps (as well as other items) in several of their stores in a test city. They found that the weekly demand for aquarium pumps is given approximately by

$$p = 12 - 2 \ln x \qquad 0 < x < 90$$

where x is the number of pumps sold each week and $\$p$ is the price of one pump. If each pump costs the chain $\$3$, how should it be priced in order to maximize the weekly profit?

Solution

Although we want to find the price that maximizes the weekly profit, it will be easier to first find the number of pumps that will maximize the weekly profit. The revenue equation is

$$R(x) = xp = 12x - 2x \ln x$$

The cost equation is

$$C(x) = 3x$$

and the profit equation is

$$P(x) = R(x) - C(x)$$
$$= 12x - 2x \ln x - 3x$$
$$= 9x - 2x \ln x$$

Thus, we must solve the following:

Maximize $\quad P(x) = 9x - 2x \ln x \qquad 0 < x < 90$

$$P'(x) = 9 - 2x \left(\frac{1}{x}\right) - 2 \ln x$$

$$= 7 - 2 \ln x = 0$$

$$2 \ln x = 7$$

$$\ln x = 3.5$$

$$x = e^{3.5}$$

$$P''(x) = -2 \left(\frac{1}{x}\right) = -\frac{2}{x}$$

Since $x = e^{3.5}$ is the only critical value and $P''(e^{3.5}) < 0$, the maximum weekly profit occurs when $x = e^{3.5} \approx 33$ and $p = 12 - 2 \ln e^{3.5} = \$5$. ◆

PROBLEM   26   Repeat Example 26 if each pump costs the chain \$3.50. ◆

Answers to Matched Problems

24. (A) $(4/x) - 5e^x$ (B) $xe^x(x + 2)$ (C) $3/x$ (D) $3(\ln x)^2/x$
25. Domain: $(0, \infty)$
    y intercept: None [$f(0)$ is not defined]
    x intercept: $x = 1$
    Increasing on (e^{-1}, ∞)
    Decreasing on $(0, e^{-1})$
    Local minimum at $x = e^{-1} \approx 0.368$
    Concave upward on $(0, \infty)$

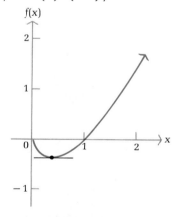

x	5	10	100	$\to \infty$
$f(x)$	8.05	23.03	460.52	$\to \infty$

x	0.1	0.01	0.001	0.000 1	$\to 0$
$f(x)$	-0.23	-0.046	$-0.006\ 9$	$-0.000\ 92$	$\to 0$

26. Maximum profit occurs for $x = e^{3.25} \approx 26$ and $p = \$5.50$.

A   *Find f'(x).*

1. $f(x) = 6e^x - 7 \ln x$ 2. $f(x) = 4e^x + 5 \ln x$ 3. $f(x) = 2x^e + 3e^x$
4. $f(x) = 4e^x - ex^e$ 5. $f(x) = \ln x^5$ 6. $f(x) = (\ln x)^5$
7. $f(x) = (\ln x)^2$ 8. $f(x) = \ln x^2$

B   9. $f(x) = x^4 \ln x$ 10. $f(x) = x^3 \ln x$

11. $f(x) = x^3 e^x$ 12. $f(x) = x^4 e^x$

13. $f(x) = \dfrac{e^x}{x^2 + 9}$ 14. $f(x) = \dfrac{e^x}{x^2 + 4}$

15. $f(x) = \dfrac{\ln x}{x^4}$ 16. $f(x) = \dfrac{\ln x}{x^3}$

17. $f(x) = (x + 2)^3 \ln x$ 18. $f(x) = (x - 1)^2 \ln x$
19. $f(x) = (x + 1)^3 e^x$ 20. $f(x) = (x - 2)^3 e^x$

21. $f(x) = \dfrac{x^2 + 1}{e^x}$ 22. $f(x) = \dfrac{x + 1}{e^x}$

23. $f(x) = x(\ln x)^3$ 24. $f(x) = x(\ln x)^2$
25. $f(x) = (4 - 5e^x)^3$ 26. $f(x) = (5 - \ln x)^4$
27. $f(x) = \sqrt{1 + \ln x}$ 28. $f(x) = \sqrt{1 + e^x}$
29. $f(x) = xe^x - e^x$ 30. $f(x) = x \ln x - x$
31. $f(x) = 2x^2 \ln x - x^2$ 32. $f(x) = x^2 e^x - 2xe^x + 2e^x$

Find the equation of the line tangent to the graph of $y = f(x)$ at the indicated value of x.

33. $f(x) = e^x; \quad x = 1$ 34. $f(x) = e^x; \quad x = 2$
35. $f(x) = \ln x; \quad x = e$ 36. $f(x) = \ln x; \quad x = 1$

C   *Find the indicated extremum of each function for x > 0.*

37. Absolute maximum value of: $f(x) = 4x - x \ln x$
38. Absolute minimum value of: $f(x) = x \ln x - 3x$

39. Absolute minimum value of: $f(x) = \dfrac{e^x}{x}$

40. Absolute maximum value of: $f(x) = \dfrac{x^2}{e^x}$

41. Absolute maximum value of: $f(x) = \dfrac{1 + 2 \ln x}{x}$

42. Absolute minimum value of: $f(x) = \dfrac{1 - 5 \ln x}{x}$

Sketch the graph of $y = f(x)$.

43. $f(x) = 1 - e^x$ 44. $f(x) = 1 - \ln x$ 45. $f(x) = x - \ln x$

46. $f(x) = e^x - x$ 47. $f(x) = (3 - x)e^x$ 48. $f(x) = (x - 2)e^x$

49. $f(x) = x^2 \ln x$ 50. $f(x) = \dfrac{\ln x}{x}$

APPLICATIONS

Business & Economics

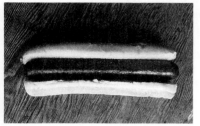

51. *Maximum profit.* A national food service runs food concessions for sporting events throughout the country. Their marketing research department chose a particular football stadium to test market a new jumbo hot dog. It was found that the demand for the new hot dog is given approximately by

$$p = 5 - \ln x \qquad 5 \leqslant x \leqslant 50$$

where x is the number of hot dogs (in thousands) that can be sold during one game at a price of $p. If the concessionaire pays $1 for each hot dog, how should the hot dogs be priced to maximize the profit per game?

52. *Maximum profit.* On a national tour of a rock band, the demand for T-shirts is given by

$$p = 15 - 4 \ln x \qquad 1 \leqslant x \leqslant 40$$

where x is the number of T-shirts (in thousands) that can be sold during a single concert at a price of $p. If the shirts cost the band $5 each, how should they be priced in order to maximize the profit per concert?

53. *Minimum average cost.* The cost of producing x units of a product is given by

$$C(x) = 600 + 100x - 100 \ln x \qquad x \geqslant 1$$

Find the minimum average cost.

54. *Minimum average cost.* The cost of producing x units of a product is given by

$$C(x) = 1{,}000 + 200x - 200 \ln x \qquad x \geqslant 1$$

Find the minimum average cost.

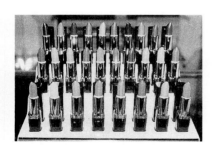

55. *Maximizing revenue.* A cosmetic company is planning the introduction and promotion of a new lipstick line. The marketing research department, after test marketing the new line in a carefully selected large city, found that the demand in that city is given approximately by

$$p = 10e^{-x} \qquad 0 \leqslant x \leqslant 2$$

where x thousand lipsticks were sold per week at a price of p dollars each.

(A) At what price will the weekly revenue $R(x) = xp$ be maximum? What is the maximum weekly revenue in the test city?

(B) Graph R for $0 \leqslant x \leqslant 2$.

56. *Maximizing revenue.* Repeat Problem 55 using the demand equation $p = 12e^{-x}$, $0 \leqslant x \leqslant 2$.

57. *Blood pressure.* An experiment was set up to find a relationship between weight and systolic blood pressure in normal children. Using hospital records for 5,000 normal children, it was found that the systolic blood pressure was given approximately by

$$P(x) = 17.5(1 + \ln x) \qquad 10 \leqslant x \leqslant 100$$

where $P(x)$ is measured in millimeters of mercury and x is measured in pounds. What is the rate of change of blood pressure with respect to weight at the 40 pound weight level? At the 90 pound weight level?

58. *Blood pressure.* Graph the systolic blood pressure equation in Problem 57.

59. *Drug concentration.* The concentration of a drug in the bloodstream t hours after injection is given approximately by

$$C(t) = 4.35e^{-t} \qquad 0 \leqslant t \leqslant 5$$

where $C(t)$ is concentration in milligrams per milliliter.

(A) What is the rate of change of concentration after 1 hour? After 4 hours?

(B) Graph C.

60. *Water pollution.* The use of iodine crystals is a popular way of making small quantities of nonpotable water safe to drink. Crystals placed in a 1 ounce bottle of water will dissolve until the solution is saturated. After saturation, half of this solution is poured into a quart container of nonpotable water, and after about an hour, the water is usually safe to drink. The half empty 1 ounce bottle is then refilled to be used again in the same way. Suppose the concentration of iodine in the 1 ounce bottle t minutes after the crystals are introduced can be approximated by

$$C(t) = 250(1 - e^{-t}) \qquad t \geqslant 0$$

where $C(t)$ is the concentration of iodine in micrograms per milliliter.

(A) What is the rate of change of the concentration after 1 minute? After 4 minutes?

(B) Graph C for $0 \leqslant t \leqslant 5$.

61. *Psychology—stimulus/response.* In psychology, the Weber–Fechner law for stimulus response is

$$R = k \ln\left(\frac{S}{S_0}\right)$$

where R is the response, S is the stimulus, and S_0 is the lowest level of stimulus that can be detected. Find dR/dS.

62. Psychology — learning. A mathematical model for the average of a group of people learning to type is given by

$$N(t) = 10 + 6 \ln t \qquad t \geq 1$$

where $N(t)$ is the number of words per minute typed after t hours of instruction and practice (2 hours per day, 5 days per week). What is the rate of learning after 10 hours of instruction and practice? After 100 hours?

S E C T I O N   10-7

Chain Rule: General Form

◆   COMPOSITE FUNCTIONS
◆   CHAIN RULE
◆   GENERALIZED DERIVATIVE RULES
◆   OTHER LOGARITHMIC AND EXPONENTIAL FUNCTIONS

In Section 9-6, we introduced the power form of the chain rule:

$$D_x[u(x)]^n = n[u(x)]^{n-1} u'(x)$$

For example,
$$D_x(x^2 - 3)^5 = 5(x^2 - 3)^4 D_x(x^2 - 3)$$
$$= 10x(x^2 - 3)^4$$

This general power rule is a special case of one of the most important derivative rules of all — the *chain rule* — which will enable us to determine the derivatives of some fairly complicated functions in terms of derivatives of more elementary functions.

Suppose you were asked to find the derivative of

$$m(x) = \ln(2x + 1) \qquad \text{or} \qquad n(x) = e^{3x^2 - 1}$$

We have formulas for computing derivatives of $\ln x$ and e^x, and polynomial functions in general, but not in the indicated combinations. The chain rule is used to compute derivatives of functions that are *compositions* of more elementary functions whose derivatives are known. We start the section with a brief review of *composite functions*.

◆ COMPOSITE FUNCTIONS

Let us look at function m more closely:

$$m(x) = \ln(2x + 1)$$

The function m is a combination of the natural logarithm function and a linear function. To see this more clearly, let

$$y = f(u) = \ln u \qquad \text{and} \qquad u = g(x) = 2x + 1$$

Then we can express y as a function of x as follows:

$$y = f(u) = f[g(x)] = \ln(2x + 1) = m(x)$$

The function m is said to be the *composite* of the two simpler functions f and g. (Loosely speaking, we can think of m as a function of a function.) In general, we have the following:

Composite Functions

A function m is a **composite** of functions f and g if

$$m(x) = f[g(x)]$$

The domain of m is the set of all numbers x such that x is in the domain of g and $g(x)$ is the domain of f.

◆ E X A M P L E 27   Let $f(u) = e^u$, $g(x) = 3x^2 + 1$, and $m(v) = v^{3/2}$. Find:

(A) $f[g(x)]$ (B) $g[f(u)]$ (C) $m[g(x)]$

Solutions   (A) $f[g(x)] = e^{g(x)} = e^{3x^2+1}$
(B) $g[f(u)] = 3[f(u)]^2 + 1 = 3(e^u)^2 + 1 = 3e^{2u} + 1$
(C) $m[g(x)] = [g(x)]^{3/2} = (3x^2 + 1)^{3/2}$ ◆

P R O B L E M 27   Let $f(u) = \ln u$, $g(x) = 2x^3 + 4$, and $m(v) = v^{-5}$. Find:

(A) $f[g(x)]$ (B) $g[f(u)]$ (C) $m[g(x)]$ ◆

◆ E X A M P L E 28   Write each function as a composition of the natural logarithm or exponential function and a polynomial.

(A) $y = \ln(x^3 - 2x^2 + 1)$ (B) $y = e^{x^2+4}$

Solutions   (A) Let

$$y = f(u) = \ln u$$
$$u = g(x) = x^3 - 2x^2 + 1$$

Check: $y = f[g(x)] = \ln[g(x)] = \ln(x^3 - 2x^2 + 1)$
(B) Let

$$y = f(u) = e^u$$
$$u = g(x) = x^2 + 4$$

Check: $y = f[g(x)] = e^{g(x)} = e^{x^2+4}$ ◆

P R O B L E M 28   Repeat Example 28 for:

(A) $y = e^{2x^3+7}$ (B) $y = \ln(x^4 + 10)$ ◆

◆ CHAIN RULE

The word "chain" in the name *chain rule* comes from the fact that a function formed by composition (such as those in Example 27) involves a chain of functions—that is, a function of a function. The *chain rule* will enable us to compute the derivative of a composite function in terms of the derivatives of the functions making up the composition.

Suppose

$$y = m(x) = f[g(x)]$$

is a composite of f and g, where

$$y = f(u) \qquad \text{and} \qquad u = g(x)$$

We would like to express the derivative dy/dx in terms of the derivatives of f and g. From the definition of a derivative (see Section 9-3), we have

$$\frac{dy}{dx} = \lim_{h \to 0} \frac{m(x + h) - m(x)}{h} \qquad \begin{array}{l} \text{Substitute} \\ m(x + h) = f(g(x + h)) \\ \text{and } m(x) = f(g(x)) \end{array}$$

$$= \lim_{h \to 0} \frac{f(g(x + h)) - f(g(x))}{h} \qquad \begin{array}{l} \text{Multiply by} \\ 1 = \dfrac{g(x + h) - g(x)}{g(x + h) - g(x)} \end{array}$$

$$= \lim_{h \to 0} \left[\frac{f(g(x + h)) - f(g(x))}{h} \cdot \frac{g(x + h) - g(x)}{g(x + h) - g(x)} \right]$$

$$= \lim_{h \to 0} \left[\frac{f(g(x + h)) - f(g(x))}{g(x + h) - g(x)} \cdot \frac{g(x + h) - g(x)}{h} \right] \qquad (1)$$

We recognize the second factor in (1) as the difference quotient for $g(x)$. In order to interpret the first factor as the difference quotient for $f(u)$, we let $k = g(x + h) - g(x)$. Since $u = g(x)$, we can write

$$u + k \overset{\vdots}{=} g(x) + g(x + h) - g(x) \overset{\vdots}{=} g(x + h)$$

Substituting in (1), we now have

$$\frac{dy}{dx} = \lim_{h \to 0} \left[\frac{f(u + k) - f(u)}{k} \cdot \frac{g(x + h) - g(x)}{h} \right] \qquad (2)$$

If we assume that $k = [g(x + h) - g(x)] \to 0$ as $h \to 0$, then we can find the limit of each difference quotient in (2):

$$\frac{dy}{dx} = \left[\lim_{k \to 0} \frac{f(u + k) - f(u)}{k} \right] \left[\lim_{h \to 0} \frac{g(x + h) - g(x)}{h} \right]$$

$$= f'(u)g'(x)$$

$$= \frac{dy}{du} \frac{du}{dx}$$

The result is correct under rather general conditions, and is called the *chain rule*, but our "derivation" is superficial, because it ignores a number of hidden problems. Since a formal proof of the chain rule is beyond the scope of this book, we simply state it as follows:

Chain Rule

If $y = f(u)$ and $u = g(x)$, define the composite function

$$y = m(x) = f[g(x)]$$

Then

$$\frac{dy}{dx} = \frac{dy}{du}\frac{du}{dx} \qquad \text{provided } \frac{dy}{du} \text{ and } \frac{du}{dx} \text{ exist}$$

◆ EXAMPLE 29 — Find dy/dx, given:

(A) $y = \ln(x^2 - 4x + 2)$ (B) $y = e^{2x^3+5}$ (C) $y = (3x^2 + 1)^{3/2}$

Solutions — (A) Let $y = \ln u$ and $u = x^2 - 4x + 2$. Then

$$\frac{dy}{dx} = \frac{dy}{du}\frac{du}{dx} \qquad *$$

$$= \frac{1}{u}(2x - 4)$$

$$= \frac{1}{x^2 - 4x + 2}(2x - 4) \qquad \text{Since } u = x^2 - 4x + 2$$

$$= \frac{2x - 4}{x^2 - 4x + 2}$$

(B) Let $y = e^u$ and $u = 2x^3 + 5$. Then

$$\frac{dy}{dx} = \frac{dy}{du}\frac{du}{dx}$$

$$= e^u(6x^2)$$

$$= 6x^2 e^{2x^3+5} \qquad \text{Since } u = 2x^3 + 5$$

---

* After some experience with the chain rule, the steps in the dashed boxes are usually done mentally.

(C) We have two methods:

Method 1. Chain rule—general form: Let $y = u^{3/2}$ and $u = 3x^2 + 1$. Then

$$\frac{dy}{dx} = \frac{dy}{du}\frac{du}{dx}$$

$$= \tfrac{3}{2}u^{1/2}(6x)$$

$$= \tfrac{3}{2}(3x^2 + 1)^{1/2}(6x) \qquad \text{Since } u = 3x^2 + 1$$

$$= 9x(3x^2 + 1)^{1/2} \quad \text{or} \quad 9x\sqrt{3x^2 + 1}$$

Method 2. Chain rule—power form (general power rule):

$$D_x(3x^2 + 1)^{3/2} = \tfrac{3}{2}(3x^2 + 1)^{1/2} D_x(3x^2 + 1) \qquad D_x[u(x)]^n = n[u(x)]^{n-1} D_x\, u(x)$$

$$= \tfrac{3}{2}(3x^2 + 1)^{1/2}(6x)$$

$$= 9x(3x^2 + 1)^{1/2} \quad \text{or} \quad 9x\sqrt{3x^2 + 1} \qquad \blacklozenge$$

The general power rule can be derived using the chain rule as follows: Given $y = [u(x)]^n$, let $y = v^n$ and $v = u(x)$. Then

$$\frac{dy}{dx} = \frac{dy}{dv}\frac{dv}{dx}$$

$$= nv^{n-1} D_x\, u(x)$$

$$= n[u(x)]^{n-1} D_x\, u(x) \qquad \text{Since } v = u(x)$$

PROBLEM 29   Find dy/dx, given:

(A) $y = e^{3x^4 + 6}$ (B) $y = \ln(x^2 + 9x + 4)$ (C) $y = (2x^3 + 4)^{-5}$
(Use two methods.) $\blacklozenge$

The chain rule can be extended to compositions of three or more functions. For example, if $y = f(w)$, $w = g(u)$, and $u = h(x)$, then

$$\frac{dy}{dx} = \frac{dy}{dw}\frac{dw}{du}\frac{du}{dx}$$

$\blacklozenge$ EXAMPLE 30   For $y = m(x) = e^{1 + (\ln x)^2}$, find dy/dx.

Solution   Note that m is of the form $y = e^w$, where $w = 1 + u^2$ and $u = \ln x$. Thus,

$$\frac{dy}{dx} = \frac{dy}{dw}\frac{dw}{du}\frac{du}{dx}$$

$$= e^w(2u)\left(\frac{1}{x}\right)$$

$$= e^{1 + u^2}(2u)\left(\frac{1}{x}\right) \qquad \text{Since } w = 1 + u^2$$

$$= e^{1 + (\ln x)^2}(2 \ln x)\left(\frac{1}{x}\right) \qquad \text{Since } u = \ln x$$

$$= \frac{2}{x}(\ln x)e^{1 + (\ln x)^2}$$

$\blacklozenge$

For $y = m(x) = [\ln(1 + e^x)]^3$, find dy/dx.   ◆

◆ GENERALIZED DERIVATIVE RULES

In practice, it is not necessary to introduce additional variables when using the chain rule, as we did in Examples 29 and 30. Instead, the chain rule can be used to extend the derivative rules for specific functions to general derivative rules for compositions. This is what we did above when we showed that the general power rule is a consequence of the chain rule. The same technique can be applied to functions of the form $y = e^{f(x)}$ and $y = \ln[f(x)]$ (see Problems 59 and 60 at the end of this section). The results are summarized in the following box:

General Derivative Rules	
$D_x[f(x)]^n = n[f(x)]^{n-1}f'(x)$	(3)
$D_x \ln[f(x)] = \dfrac{1}{f(x)} f'(x)$	(4)
$D_x e^{f(x)} = e^{f(x)}f'(x)$	(5)

For power, natural logarithm, or exponential forms, we can either use the chain rule discussed earlier or these special differentiation formulas based on the chain rule. Use whichever is easier for you. In Example 31, we will use the general derivative rules.

◆ **E X A M P L E   31**   (A) $D_x e^{2x} = e^{2x} D_x 2x$ Using (5)
$$= e^{2x}(2) = 2e^{2x}$$

(B) $D_x \ln(x^2 + 9) = \dfrac{1}{x^2 + 9} D_x(x^2 + 9)$ Using (4)

$$= \dfrac{1}{x^2 + 9} 2x = \dfrac{2x}{x^2 + 9}$$

(C) $D_x(1 + e^{x^2})^3 = 3(1 + e^{x^2})^2 D_x(1 + e^{x^2})$ Using (3)
$$= 3(1 + e^{x^2})^2 e^{x^2} D_x x^2$$ Using (5)
$$= 3(1 + e^{x^2})^2 e^{x^2}(2x)$$
$$= 6xe^{x^2}(1 + e^{x^2})^2$$ ◆

P R O B L E M   31   Find:

(A) $D_x \ln(x^3 + 2x)$ (B) $D_x e^{3x^2+2}$ (C) $D_x(2 + e^{-x^2})^4$ ◆

◆ OTHER LOGARITHMIC AND EXPONENTIAL FUNCTIONS

In most applications involving logarithmic or exponential functions, the number e is the preferred base. However, there are situations where it is convenient to use a base other than e. Derivatives of $y = \log_b x$ and $y = b^x$ can be obtained by expressing these functions in terms of the natural logarithmic and exponential functions. We begin by finding a relationship between $\log_b x$ and $\ln x$ for any base b, $b > 0$ and $b \neq 1$.

$$y = \log_b x \qquad \text{Change to exponential form.}$$
$$b^y = x \qquad \text{Take the natural logarithm of both sides.}$$
$$\ln b^y = \ln x \qquad \text{Recall that } \ln b^y = y \ln b.$$
$$y \ln b = \ln x \qquad \text{Solve for } y.$$

$$y = \frac{1}{\ln b} \ln x$$

Thus,

$$\log_b x = \frac{1}{\ln b} \ln x \qquad \text{Change-of-base formula*} \tag{6}$$

Differentiating both sides of (6), we have

$$D_x \log_b x = \frac{1}{\ln b} D_x \ln x = \frac{1}{\ln b}\left(\frac{1}{x}\right)$$

◆ E X A M P L E 32

Find $f'(x)$ for:

(A) $f(x) = \log_2 x$ (B) $f(x) = \log(1 + x^3)$

Solutions

(A) $f(x) = \log_2 x = \dfrac{1}{\ln 2} \ln x$ Using (6)

$$f'(x) = \frac{1}{\ln 2}\left(\frac{1}{x}\right)$$

(B) $f(x) = \log(1 + x^3)$ Recall that $\log r = \log_{10} r$.

$$= \frac{1}{\ln 10} \ln(1 + x^3) \qquad \text{Using (6)}$$

$$f'(x) = \frac{1}{\ln 10}\left(\frac{1}{1 + x^3}\, 3x^2\right) = \frac{1}{\ln 10}\left(\frac{3x^2}{1 + x^3}\right)$$

◆

P R O B L E M 32

Find $f'(x)$ for:

(A) $f(x) = \log x$ (B) $f(x) = \log_3(x + x^2)$

◆

---

* Equation (6) is a special case of the **general change-of-base formula** for logarithms (which can be derived in the same way): $\log_b x = (\log_a x)/(\log_a b)$.

Now we want to find a relationship between b^x and e^x for any base b, $b > 0$ and $b \neq 1$.

$$y = b^x \qquad \text{Take the natural logarithm of both sides.}$$

$$\ln y = \ln b^x$$
$$\quad\ = x \ln b \qquad \text{If } \ln A = B, \text{ then } A = e^B.$$
$$y = e^{x\ln b}$$

Thus,

$$b^x = e^{x\ln b} \tag{7}$$

Differentiating both sides of (7), we have

$$D_x\, b^x = e^{x\ln b}\, \ln b = b^x \ln b$$

◆ EXAMPLE 33   Find $f'(x)$ for:

(A) $f(x) = 2^x$ (B) $f(x) = 10^{x^5+x}$

Solutions   (A) $f(x) = 2^x = e^{x\ln 2}$ Using (7)
              $f'(x) = e^{x\ln 2}\ln 2 = 2^x \ln 2$
         (B) $f(x) = 10^{x^5+x} = e^{(x^5+x)\ln 10}$ Using (7)
              $f'(x) = e^{(x^5+x)\ln 10}(5x^4 + 1)\ln 10$
                  $= 10^{x^5+x}(5x^4 + 1)\ln 10$ ◆

PROBLEM 33   Find $f'(x)$ for:

(A) $f(x) = 5^x$ (B) $f(x) = 4^{x^2+3x}$ ◆

Answers to Matched Problems   27. (A) $\ln(2x^3 + 4)$ (B) $2(\ln u)^3 + 4$ (C) $(2x^3 + 4)^{-5}$
                             28. (A) $y = f(u) = e^u$; $u = g(x) = 2x^3 + 7$
                                  (B) $y = f(u) = \ln u$; $u = g(x) = x^4 + 10$

29. (A) $12x^3 e^{3x^4+6}$ (B) $\dfrac{2x + 9}{x^2 + 9x + 4}$ (C) $-30x^2(2x^3 + 4)^{-6}$

30. $\dfrac{3e^x[\ln(1 + e^x)]^2}{1 + e^x}$

31. (A) $\dfrac{3x^2 + 2}{x^3 + 2x}$ (B) $6xe^{3x^2+2}$ (C) $-8xe^{-x^2}(2 + e^{-x^2})^3$

32. (A) $\dfrac{1}{\ln 10}\left(\dfrac{1}{x}\right)$ (B) $\dfrac{1}{\ln 3}\left(\dfrac{1 + 2x}{x + x^2}\right)$

33. (A) $5^x \ln 5$ (B) $4^{x^2+3x}(2x + 3)\ln 4$

A *Write each composite function in the form $y = f(u)$ and $u = g(x)$.*

1. $y = (2x + 5)^3$ **2.** $y = (3x - 7)^5$ **3.** $y = \ln(2x^2 + 7)$
4. $y = \ln(x^2 - 2x + 5)$ **5.** $y = e^{x^2 - 2}$ **6.** $y = e^{3x^3 + 5x}$

Express y in terms of x. Use the chain rule to find dy/dx, and then express dy/dx in terms of x.

7. $y = u^2$; $u = 2 + e^x$ **8.** $y = u^3$; $u = 3 - \ln x$
9. $y = e^u$; $u = 2 - x^4$ **10.** $y = e^u$; $u = x^6 + 5x^2$
11. $y = \ln u$; $u = 4x^5 - 7$ **12.** $y = \ln u$; $u = 2 + 3x^4$

Find each derivative.

13. $D_x \ln(x - 3)$ **14.** $D_w \ln(w + 100)$ **15.** $D_t \ln(3 - 2t)$
16. $D_y \ln(4 - 5y)$ **17.** $D_x 3e^{2x}$ **18.** $D_y 2e^{3y}$
19. $D_t 2e^{-4t}$ **20.** $D_r 6e^{-3r}$

B **21.** $D_x 100e^{-0.03x}$ **22.** $D_t 1{,}000e^{0.06t}$ **23.** $D_x \ln(x + 1)^4$
24. $D_x \ln(x + 1)^{-3}$ **25.** $D_x(2e^{2x} - 3e^x + 5)$ **26.** $D_t(1 + e^{-t} - e^{-2t})$
27. $D_x e^{3x^2 - 2x}$ **28.** $D_x e^{x^3 - 3x^2 + 1}$ **29.** $D_t \ln(t^2 + 3t)$
30. $D_x \ln(x^3 - 3x^2)$ **31.** $D_x \ln(x^2 + 1)^{1/2}$ **32.** $D_x \ln(x^4 + 5)^{3/2}$
33. $D_t[\ln(t^2 + 1)]^4$ **34.** $D_w [\ln(w^3 - 1)]^2$ **35.** $D_x(e^{2x} - 1)^4$

36. $D_x(e^{x^2} + 3)^5$ **37.** $D_x \dfrac{e^{2x}}{x^2 + 1}$ **38.** $D_x \dfrac{e^{x+1}}{x + 1}$

39. $D_x(x^2 + 1)e^{-x}$ **40.** $D_x(1 - x)e^{2x}$ **41.** $D_x(e^{-x} \ln x)$

42. $D_x \dfrac{\ln x}{e^x + 1}$ **43.** $D_x \dfrac{1}{\ln(1 + x^2)}$ **44.** $D_x \dfrac{1}{\ln(1 - x^3)}$

45. $D_x \sqrt[3]{\ln(1 - x^2)}$ **46.** $D_t \sqrt[5]{\ln(1 - t^5)}$

C *Sketch the graph of $y = f(x)$.*

47. $f(x) = 1 - e^{-x}$ **48.** $f(x) = 2 - 3e^{-2x}$ **49.** $f(x) = \ln(1 - x)$
50. $f(x) = \ln(2x + 4)$ **51.** $f(x) = e^{-(1/2)x^2}$ **52.** $f(x) = \ln(x^2 + 4)$

Express y in terms of x. Use the chain rule to find dy/dx, and express dy/dx in terms of x.

53. $y = 1 + w^2$; $w = \ln u$; $u = 2 + e^x$
54. $y = \ln w$; $w = 1 + e^u$; $u = x^2$

Find each derivative.

55. $D_x \log_2(3x^2 - 1)$ **56.** $D_x \log(x^3 - 1)$
57. $D_x 10^{x^2 + x}$ **58.** $D_x 8^{1 - 2x^2}$

59. Use the chain rule to derive the formula: $D_x \ln[f(x)] = \dfrac{1}{f(x)} f'(x)$

60. Use the chain rule to derive the formula: $D_x e^{f(x)} = e^{f(x)} f'(x)$

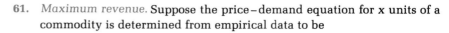

Business & Economics

61. *Maximum revenue.* Suppose the price–demand equation for x units of a commodity is determined from empirical data to be

$$p = 100e^{-0.05x}$$

where x units are sold per day at a price of $p each. Find the production level and price that maximize revenue. What is the maximum revenue?

62. *Maximum revenue.* Repeat Problem 61 using the price–demand equation

$$p = 10e^{-0.04x}$$

63. *Salvage value.* The salvage value S (in dollars) of a company airplane after t years is estimated to be given by

$$S(t) = 300,000e^{-0.1t}$$

What is the rate of depreciation (in dollars per year) after 1 year? 5 years? 10 years?

64. *Resale value.* The resale value R (in dollars) of a company car after t years is estimated to be given by

$$R(t) = 20,000e^{-0.15t}$$

What is the rate of depreciation (in dollars per year) after 1 year? 2 years? 3 years?

65. *Promotion and maximum profit.* A recording company has produced a new compact disk featuring a very popular recording group. Before launching a national sales campaign, the marketing research department chose to test market the disk in a bellwether city. Their interest is in determining the length of a sales campaign that will maximize total profits. From empirical data, the research department estimates that the proportion of a target group of 50,000 persons buying the disk after t days of television promotion is given by $1 - e^{-0.03t}$. If $4 is received for each disk sold, then the total revenue after t days of promotion will be approximated by

$$R(t) = (4)(50,000)(1 - e^{-0.03t}) \qquad t \geq 0$$

Television promotion costs are

$$C(t) = 4,000 + 3,000t \qquad t \geq 0$$

(A) How many days of television promotion should be used to maximize total profit? What is the maximum total profit? What percentage of the target market will have purchased the disk when the maximum profit is reached?

(B) Graph the profit function.

66. *Promotion and maximum profit.* Repeat Problem 65 using the revenue equation

$$R(t) = (3)(60{,}000)(1 - e^{-0.04t})$$

Life Sciences

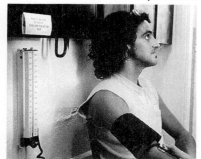

67. *Blood pressure and age.* A research group using hospital records developed the following approximate mathematical model relating systolic blood pressure and age:

$$P(x) = 40 + 25 \ln(x + 1) \qquad 0 \leqslant x \leqslant 65$$

where $P(x)$ is pressure measured in millimeters of mercury and x is age in years. What is the rate of change of pressure at the end of 10 years? At the end of 30 years? At the end of 60 years?

68. *Biology.* A yeast culture at room temperature (68°F) is placed in a refrigerator maintaining a constant temperature of 38°F. After t hours, the temperature T of the culture is given approximately by

$$T = 30e^{-0.58t} + 38 \qquad t \geqslant 0$$

What is the rate of change of temperature of the culture at the end of 1 hour? At the end of 4 hours?

69. *Bacterial growth.* A single cholera bacterium divides every 0.5 hour to produce two complete cholera bacteria. If we start with a colony of 5,000 bacteria, then after t hours there will be

$$A(t) = 5{,}000 \cdot 2^{2t}$$

bacteria. Find $A'(t)$, $A'(1)$, and $A'(5)$, and interpret the results.

70. *Bacterial growth.* Repeat Problem 69 for a starting colony of 1,000 bacteria where a single bacterium divides every 0.25 hour.

Social Sciences

71. *Sociology.* Daniel Lowenthal, a sociologist at Columbia University, made a 5 year study on the sale of popular records relative to their position in the top 20. He found that the average number of sales $N(n)$ of the nth ranking record was given approximately by

$$N(n) = N_1 e^{-0.09(n-1)} \qquad 1 \leqslant n \leqslant 20$$

where N_1 was the number of sales of the number one record on the list at a given time. Graph N for $N_1 = 1{,}000{,}000$ records.

72. *Political science.* Thomas W. Casstevens, a political scientist at Oakland University, has studied legislative turnover. He (with others) found that the number $N(t)$ of continuously serving members of an elected legislative body remaining t years after an election is given approximately by a function of the form

$$N(t) = N_0 e^{-ct}$$

In particular, for the 1965 election for the U.S. House of Representatives, it was found that

$$N(t) = 434e^{-0.0866t}$$

What is the rate of change after 2 years? After 10 years?

SECTION 10-8 # Chapter Review

Important Terms and Symbols

10-1 *First Derivative and Graphs.* Solving inequalities using continuity properties; sign chart; partition numbers; increasing and decreasing functions; rising and falling graphs; local extremum; local maximum; local minimum; critical values; first-derivative test for local extrema

10-2 *Second Derivative and Graphs.* Concave upward; concave downward; second derivative; concavity and the second derivative; inflection point; second-derivative test for local maxima and minima

$$f''(x); \quad \frac{d^2y}{dx^2}; \quad y''; \quad D_x^2 f(x)$$

10-3 *Curve Sketching Techniques: Unified and Extended.* Horizontal asymptote; vertical asymptote; graphing strategy: use $f(x)$ to find the domain, intercepts, horizontal and vertical asymptotes; use $f'(x)$ to find increasing and decreasing regions, and local extrema; use $f''(x)$ to find concave upward and downward regions, and inflection points; oblique asymptote

10-4 *Optimization; Absolute Maxima and Minima.* Absolute maximum; absolute minimum; absolute extrema of a function continuous on a closed interval; second-derivative test for absolute maximum and minimum; optimization problems

10-5 *The Constant e and Continuous Compound Interest.* Definition of e; continuous compound interest

10-6 *Derivatives of Logarithmic and Exponential Functions.* Derivative formulas for the natural logarithmic and exponential functions; graph properties of $y = \ln x$ and $y = e^x$

10-7 *Chain Rule: General Form.* Composite functions; chain rule; general derivative formulas; derivative formulas for $y = \log_b x$ and $y = b^x$

Additional Rules of Differentiation

$$D_x \ln x = \frac{1}{x} \qquad\qquad D_x e^x = e^x$$

$$D_x \ln[f(x)] = \frac{1}{f(x)} f'(x) \qquad D_x e^{f(x)} = e^{f(x)}f'(x)$$

$$D_x \log_b x = D_x \frac{1}{\ln b} \ln x = \frac{1}{\ln b}\left(\frac{1}{x}\right)$$

$$D_x b^x = D_x e^{x \ln b} = e^{x \ln b} \ln b = b^x \ln b$$

$$D_x[f(x)]^n = n[f(x)]^{n-1}f'(x)$$

$$\frac{dy}{dx} = \frac{dy}{du}\frac{du}{dx}, \quad \frac{dy}{dx} = \frac{dy}{dw}\frac{dw}{du}\frac{du}{dx}, \quad \text{and so on}$$

Chapter Review

Work through all the problems in this chapter review and check your answers in the back of the book. (Answers to all review problems are there.) Where weaknesses show up, review appropriate sections in the text.

A   *Problems 1–8 refer to the following graph of $y = f(x)$:*

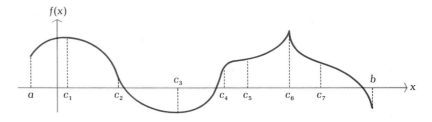

Identify the points or intervals on the x axis that produce the indicated behavior.

1. $f(x)$ is increasing
2. $f'(x) < 0$
3. Graph of f is concave downward
4. Local minima
5. Absolute maxima
6. $f'(x)$ appears to be 0
7. $f'(x)$ does not exist
8. Inflection points

9. Find $f''(x)$ for $f(x) = x^4 + 5x^3$.
10. Find y'' for $y = 3x + 4/x$.
11. Use a calculator to evaluate $A = 2{,}000e^{0.09t}$ to the nearest cent for $t = 5, 10$, and 20.

Find the indicated derivatives in Problems 12–14.

12. $D_x(2 \ln x + 3e^x)$
13. $D_x e^{2x-3}$
14. y' for $y = \ln(2x + 7)$

15. Let $y = \ln u$ and $u = 3 + e^x$.

    (A) Express y in terms of x.

    (B) Use the chain rule to find dy/dx, and then express dy/dx in terms of x.

B   *In Problems 16 and 17, solve each inequality.*

16. $x^2 - x < 12$ 17. $\dfrac{x - 5}{x^2 + 3x} > 0$

Problems 18–21 refer to the function: $f(x) = x^3 - 18x^2 + 81x$

18.   Using $f(x)$:

    (A) Determine the domain of f.

    (B) Find any intercepts for the graph of f.

    (C) Find any horizontal or vertical asymptotes for the graph of f.

19.   Using $f'(x)$:

    (A) Find critical values for $f(x)$.

    (B) Find partition numbers for $f'(x)$.

    (C) Find intervals over which $f(x)$ is increasing; decreasing.

    (D) Find any local maxima and minima.

20.   Using $f''(x)$:

    (A) Find intervals over which the graph of f is concave upward; concave downward.

    (B) Find any inflection points.

21.   Use the results of Problems 18–20 to graph f.

Problems 22–25 refer to the function: $y = f(x) = \dfrac{3x}{x + 2}$

22.   Using $f(x)$:

    (A) Determine the domain of f.

    (B) Find any intercepts for the graph of f.

    (C) Find any horizontal or vertical asymptotes for the graph of f.

23.   Using $f'(x)$:

    (A) Find critical values for $f(x)$.

    (B) Find partition numbers for $f'(x)$.

    (C) Find intervals over which $f(x)$ is increasing; decreasing.

    (D) Find any local maxima and minima.

24.   Using $f''(x)$:

    (A) Find intervals over which the graph of f is concave upward; concave downward.

    (B) Find any inflection points.

25. Use the results of Problems 22–24 to graph f.

26. Find the absolute maximum and absolute minimum for
$$y = f(x) = x^3 - 12x + 12 \qquad -3 \leqslant x \leqslant 5$$

27. Find the absolute minimum for
$$y = f(x) = x^2 + \frac{16}{x^2} \qquad x > 0$$

Find horizontal and vertical asymptotes (if they exist) in Problems 28 and 29.

28. $f(x) = \dfrac{x}{x^2 + 9}$

29. $f(x) = \dfrac{x^3}{x^2 - 9}$

30. Graph $y = 100e^{-0.1x}$.

Find the indicated derivatives in Problems 31–36.

31. $D_z[(\ln z)^7 + \ln z^7]$

32. $D_x(x^6 \ln x)$

33. $D_x \dfrac{e^x}{x^6}$

34. y' for $y = \ln(2x^3 - 3x)$

35. $f'(x)$ for $f(x) = e^{x^3 - x^2}$

36. dy/dx for $y = e^{-2x} \ln 5x$

C   37. Find the absolute maximum for $f'(x)$ if
$$f(x) = 6x^2 - x^3 + 8$$

Graph f and f' on the same coordinate system.

38. Find two positive numbers whose product is 400 and whose sum is a minimum. What is the minimum sum?

39. Sketch the graph of $f(x) = (x - 1)^3(x + 3)$ using the graphing strategy discussed in Section 10-3.

In Problems 40 and 41, find the absolute maximum value of $f(x)$ for $x > 0$.

40. $f(x) = 11x - 2x \ln x$

41. $f(x) = 10xe^{-2x}$

Sketch the graph of $y = f(x)$ in Problems 42 and 43.

42. $f(x) = 5 - 5e^{-x}$

43. $f(x) = x^3 \ln x$

44. Let $y = w^3$, $w = \ln u$, and $u = 4 - e^x$.

(A) Express y in terms of x.

(B) Use the chain rule to find dy/dx, and then express dy/dx in terms of x.

Find the indicated derivatives in Problems 45–47.

45. y' for $y = 5^{x^2-1}$

46. $D_x \log_5(x^2 - x)$

47. $D_x \sqrt{\ln(x^2 + x)}$

Business & Economics

48. *Profit.* The profit for a company manufacturing and selling x units per month is given by

$$P(x) = 150x - \frac{x^2}{40} - 50{,}000 \qquad 0 \leqslant x \leqslant 5{,}000$$

What production level will produce the maximum profit? What is the maximum profit?

49. *Average cost.* The total cost of producing x units per month is given by

$$C(x) = 4{,}000 + 10x + \tfrac{1}{10}x^2$$

Find the minimum average cost. Graph the average cost and the marginal cost functions on the same coordinate system. Include any oblique asymptotes.

50. *Rental income.* A 200 room hotel in Fresno is filled to capacity every night at a rate of $40 per room. For each $1 increase in the nightly rate, 4 fewer rooms are rented. If each rented room costs $8 a day to service, how much should the management charge per room in order to maximize gross profit? What is the maximum gross profit?

51. *Inventory control.* A computer store sells 7,200 boxes of floppy disks annually. It costs the store $0.20 to store a box of disks for 1 year. Each time it reorders disks, the store must pay a $5.00 service charge for processing the order. How many times during the year should the store order disks in order to minimize the total storage and reorder costs?

52. *Doubling time.* How long will it take money to double if it is invested at 5% interest compounded:

(A) Annually?   (B) Continuously?

53. *Continuous compound interest.* If $100 is invested at 10% interest compounded continuously, the amount (in dollars) at the end of t years is given by

$$A = 100e^{0.1t}$$

Find $A'(t)$, $A'(1)$, and $A'(10)$.

54. *Marginal analysis.* If the price–demand equation for x units of a commodity is

$$p(x) = 1{,}000e^{-0.02x}$$

find the marginal revenue equation.

55. *Maximum revenue.* For the price–demand equation in Problem 54, find the production level and price per unit that produces the maximum revenue. What is the maximum revenue?

56. *Maximum revenue.* Graph the revenue function from Problems 54 and 55 for $0 \leqslant x \leqslant 100$.

57. *Minimum average cost.* The cost of producing x units of a product is given by

$$C(x) = 200 + 50x - 50 \ln x \qquad x \geqslant 1$$

Find the minimum average cost.

Life Sciences

58. *Bacteria control.* If t days after a treatment, the bacteria count per cubic centimeter in a body of water is given by

$$C(t) = 20t^2 - 120t + 800 \qquad 0 \leqslant t \leqslant 9$$

in how many days will the count be a minimum?

59. *Drug concentration.* The concentration of a drug in the bloodstream t hours after injection is given approximately by

$$C(t) = 5e^{-0.3t}$$

where $C(t)$ is concentration in milligrams per milliliter. What is the rate of change of concentration after 1 hour? After 5 hours?

Social Sciences

60. *Politics.* In a new suburb, it is estimated that the number of registered voters will grow according to

$$N = 10 + 6t^2 - t^3 \qquad 0 \leqslant t \leqslant 5$$

where t is time in years and N is in thousands. When will the rate of increase be maximum?

61. *Psychology — learning.* In a computer assembly plant, a new employee, on the average, is able to assemble

$$N(t) = 10(1 - e^{-0.4t})$$

units after t days of on-the-job training.

(A) What is the rate of learning after 1 day? After 5 days?
(B) Graph N for $0 \leqslant t \leqslant 10$.

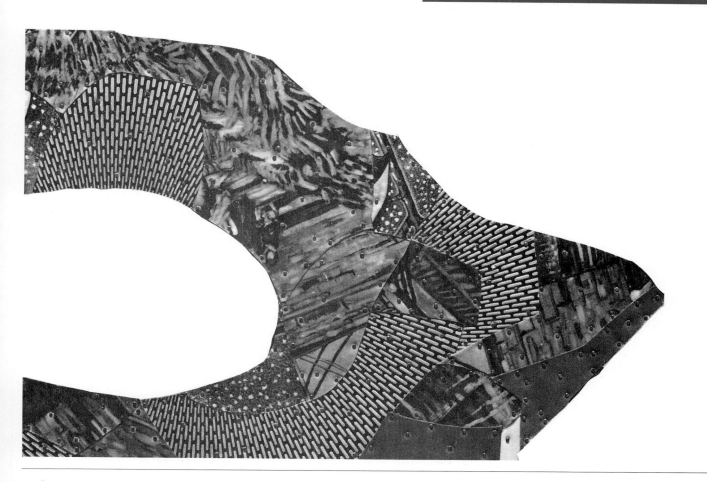

Integration

Contents

The last two chapters dealt with differential calculus. We now begin the development of the second main part of calculus, called *integral calculus*. Two types of integrals will be introduced, the *indefinite integral* and the *definite integral*. Each is quite different from the other, but both are intimately related to differentiation.

SECTION 11-1

Antiderivatives and Indefinite Integrals

- ◆ ANTIDERIVATIVES
- ◆ INDEFINITE INTEGRALS
- ◆ INDEFINITE INTEGRALS INVOLVING ALGEBRAIC FUNCTIONS
- ◆ INDEFINITE INTEGRALS INVOLVING EXPONENTIAL AND LOGARITHMIC FUNCTIONS
- ◆ APPLICATIONS

◆ ANTIDERIVATIVES

Many operations in mathematics have reverses — compare addition and subtraction, multiplication and division, and powers and roots. The function $f(x) = \frac{1}{3}x^3$ has the derivative $f'(x) = x^2$. Reversing this process is referred to as **antidifferentiation.** Thus,

$$\frac{x^3}{3} \quad \text{is an antiderivative of} \quad x^2$$

since

$$D_x\left(\frac{x^3}{3}\right) = x^2$$

In general, we say that $F(x)$ is an **antiderivative** of $f(x)$ if

$$F'(x) = f(x)$$

Note that

$$D_x\left(\frac{x^3}{3} + 2\right) = x^2 \qquad D_x\left(\frac{x^3}{3} - \pi\right) = x^2 \qquad D_x\left(\frac{x^3}{3} + \sqrt{5}\right) = x^2$$

Hence,

$$\frac{x^3}{3} + 2 \qquad \frac{x^3}{3} - \pi \qquad \frac{x^3}{3} + \sqrt{5}$$

are also antiderivatives of x^2, since each has x^2 as a derivative. In fact, it appears that

$$\frac{x^3}{3} + C$$

for any real number C, is an antiderivative of x^2, since

$$D_x\left(\frac{x^3}{3} + C\right) = x^2$$

Thus, antidifferentiation of a given function does not lead to a unique function, but to a whole set of functions.

Does the expression

$$\frac{x^3}{3} + C$$

with C any real number, include all antiderivatives of x^2? Theorem 1 (which we state without proof) indicates that the answer is yes.

THEOREM 1

> If F and G are differentiable functions on the interval (a, b) and $F'(x) = G'(x)$, then $F(x) = G(x) + k$ for some constant k.

◆ INDEFINITE INTEGRALS

In words, Theorem 1 states that **if the derivatives of two functions are equal, then the functions differ by at most a constant.** We use the symbol

$$\int f(x)\, dx$$

called the **indefinite integral,** to represent the set of all antiderivatives of $f(x)$, and we write

$$\int f(x)\, dx = F(x) + C \qquad \text{if} \qquad F'(x) = f(x)$$

The symbol $\int$ is called an **integral sign,** and the function $f(x)$ is called the **integrand.** The symbol dx indicates that the antidifferentiation is performed with respect to the variable x. (We will have more to say about the symbol dx

later.) The arbitrary constant C is called the **constant of integration.** Referring to the preceding discussion, we can write

$$\int x^2 \, dx = \frac{x^3}{3} + C \qquad \text{since} \qquad D_x \left(\frac{x^3}{3} + C \right) = x^2$$

Of course, variables other than x can be used in indefinite integrals. Thus, we also can write

$$\int t^2 \, dt = \frac{t^3}{3} + C \qquad \text{since} \qquad D_t \left(\frac{t^3}{3} + C \right) = t^2$$

or

$$\int u^2 \, du = \frac{u^3}{3} + C \qquad \text{since} \qquad D_u \left(\frac{u^3}{3} + C \right) = u^2$$

The fact that indefinite integration and differentiation are reverse operations, except for the addition of the constant of integration, can be expressed symbolically as

$$D_x \left[\int f(x) \, dx \right] = f(x) \qquad \text{The derivative of the indefinite integral of } f(x) \text{ is } f(x).$$

and

$$\int F'(x) \, dx = F(x) + C \qquad \text{The indefinite integral of the derivative of } F(x) \text{ is } F(x) + C.$$

◆ INDEFINITE INTEGRALS INVOLVING ALGEBRAIC FUNCTIONS

Just as with differentiation, we can develop formulas and special properties that will enable us to find indefinite integrals of many frequently encountered functions. To start, we list some formulas that can be established using the definitions of antiderivative and indefinite integral, and the properties of derivatives considered in Chapter 9.

Indefinite Integral Formulas and Properties

For k and C constants:

1. $\displaystyle\int k \, dx = kx + C$

2. $\displaystyle\int x^n \, dx = \frac{x^{n+1}}{n+1} + C \qquad n \neq -1$

3. $\displaystyle\int kf(x) \, dx = k \int f(x) \, dx$

4. $\displaystyle\int [f(x) \pm g(x)] \, dx = \int f(x) \, dx \pm \int g(x) \, dx$

We will establish formula 2 and property 3 here (the others may be shown to be true in a similar manner). To establish formula 2, we simply differentiate the right side to obtain the integrand on the left side. Thus,

$$D_x\left(\frac{x^{n+1}}{n+1} + C\right) = \frac{(n+1)x^n}{n+1} + 0 = x^n \qquad n \neq -1$$

(Notice that formula 2 cannot be used when $n = -1$; that is, when the integrand is x^{-1} or $1/x$. The indefinite integral of $x^{-1} = 1/x$ will be considered later in this section.)

To establish property 3, let F be a function such that $F'(x) = f(x)$. Then

$$k \int f(x)\, dx = k \int F'(x)\, dx = k[F(x) + C_1] = kF(x) + kC_1$$

and since $(kF(x))' = kF'(x) = kf(x)$, we have

$$\int kf(x)\, dx = \int kF'(x)\, dx = kF(x) + C_2$$

But $kF(x) + kC_1$ and $kF(x) + C_2$ describe the same set of functions, since C_1 and C_2 are arbitrary real numbers. Thus, property 3 is established.

It is important to remember that property 3 states that **a constant factor can be moved across an integral sign; a variable factor cannot be moved across an integral sign:**

Constant Factor $\qquad\qquad$ Variable Factor

$$\int 5x^{1/2}\, dx = 5 \int x^{1/2}\, dx \qquad \int xx^{1/2}\, dx \neq x \int x^{1/2}\, dx$$

Now let us put the formulas and properties to use.

◆ E X A M P L E 1 (A) $\displaystyle\int 5\, dx = 5x + C$

(B) $\displaystyle\int x^4\, dx = \frac{x^{4+1}}{4+1} + C = \frac{x^5}{5} + C$

(C) $\displaystyle\int 5t^7\, dt = 5 \int t^7\, dt = 5\frac{t^8}{8} + C = \frac{5}{8}t^8 + C$

(D) $\displaystyle\int (4x^3 + 2x - 1)\, dx = \int 4x^3\, dx + \int 2x\, dx - \int dx$

$$= 4 \int x^3\, dx + 2 \int x\, dx - \int dx$$

$$= \frac{4x^4}{4} + \frac{2x^2}{2} - x + C$$

$$= x^4 + x^2 - x + C$$

Property 4 can be extended to the sum and difference of an arbitrary number of functions.

(E) $\displaystyle\int \frac{3\,dx}{x^2} = \int 3x^{-2}\,dx = \frac{3x^{-2+1}}{-2+1} + C = -3x^{-1} + C$

(F) $\displaystyle\int 5\sqrt[3]{u^2}\,du = 5\int u^{2/3}\,du = 5\,\frac{u^{(2/3)+1}}{\frac{2}{3}+1} + C$

$\displaystyle\qquad\qquad = 5\,\frac{u^{5/3}}{\frac{5}{3}} + C = 3u^{5/3} + C$ ◆

To check any of the results in Example 1, we differentiate the final result to obtain the integrand in the original indefinite integral. When you evaluate an indefinite integral, do not forget to include the arbitrary constant C.

P R O B L E M   1   Find each of the following:

(A) $\displaystyle\int dx$ (B) $\displaystyle\int 3t^4\,dt$ (C) $\displaystyle\int (2x^5 - 3x^2 + 1)\,dx$

(D) $\displaystyle\int 4\sqrt[5]{w^3}\,dw$ (E) $\displaystyle\int \left(2x^{2/3} - \frac{3}{x^4}\right)dx$ ◆

◆ E X A M P L E   2   (A) $\displaystyle\int \frac{x^3 - 3}{x^2}\,dx = \int \left(\frac{x^3}{x^2} - \frac{3}{x^2}\right)dx$

$\displaystyle\qquad\qquad\qquad = \int (x - 3x^{-2})\,dx$

$\displaystyle\qquad\qquad\qquad = \int x\,dx - 3\int x^{-2}\,dx$

$\displaystyle\qquad\qquad\qquad = \frac{x^{1+1}}{1+1} - 3\,\frac{x^{-2+1}}{-2+1} + C$

$\displaystyle\qquad\qquad\qquad = \tfrac{1}{2}x^2 + 3x^{-1} + C$

(B) $\displaystyle\int \left(\frac{2}{\sqrt[3]{x}} - 6\sqrt{x}\right)dx = \int (2x^{-1/3} - 6x^{1/2})\,dx$

$\displaystyle\qquad\qquad\qquad = 2\int x^{-1/3}\,dx - 6\int x^{1/2}\,dx$

$\displaystyle\qquad\qquad\qquad = 2\,\frac{x^{(-1/3)+1}}{-\frac{1}{3}+1} - 6\,\frac{x^{(1/2)+1}}{\frac{1}{2}+1} + C$

$\displaystyle\qquad\qquad\qquad = 2\,\frac{x^{2/3}}{\frac{2}{3}} - 6\,\frac{x^{3/2}}{\frac{3}{2}} + C$

$\displaystyle\qquad\qquad\qquad = 3x^{2/3} - 4x^{3/2} + C$ ◆

P R O B L E M   2   Find each indefinite integral.

(A) $\displaystyle\int \frac{x^4 - 8x^3}{x^2}\,dx$ (B) $\displaystyle\int \left(8\sqrt[3]{x} - \frac{6}{\sqrt{x}}\right)dx$ ◆

◆ INDEFINITE INTEGRALS INVOLVING EXPONENTIAL AND LOGARITHMIC FUNCTIONS

We now give indefinite integral formulas for e^x and $1/x$. (Recall that the form $x^{-1} = 1/x$ is not covered by formula 2, given earlier.)

■ Indefinite Integral Formulas

5. $\displaystyle\int e^x \, dx = e^x + C$ 6. $\displaystyle\int \frac{1}{x} \, dx = \ln|x| + C$ $x \neq 0$

Formula 5 follows immediately from the derivative formula for the exponential function discussed in the last chapter. Because of the absolute value, formula 6 does not follow directly from the derivative formula for the natural logarithm function. Let us show that

$$D_x \ln|x| = \frac{1}{x} \qquad x \neq 0$$

We consider two cases, $x > 0$ and $x < 0$:

Case 1. $x > 0$:

$$D_x \ln|x| = D_x \ln x \qquad \text{Since } |x| = x \text{ for } x > 0$$

$$= \frac{1}{x}$$

Case 2. $x < 0$:

$$D_x \ln|x| = D_x \ln(-x) \qquad \text{Since } |x| = -x \text{ for } x < 0$$

$$= \frac{1}{-x} D_x(-x)$$

$$= \frac{-1}{-x} = \frac{1}{x}$$

Thus,

$$D_x \ln|x| = \frac{1}{x} \qquad x \neq 0$$

and hence,

$$\int \frac{1}{x} \, dx = \ln|x| + C \qquad x \neq 0$$

◆ E X A M P L E   3 $\int \left(2e^x + \dfrac{3}{x}\right) dx = 2 \int e^x \, dx + 3 \int \dfrac{1}{x} \, dx$

$$= 2e^x + 3 \ln|x| + C$$ ◆

P R O B L E M   3   Find: $\int \left(\dfrac{5}{x} - 4e^x\right) dx$ ◆

◆ APPLICATIONS

Let us now consider some applications of the indefinite integral to see why we are interested in finding antiderivatives of functions.

◆ E X A M P L E   4   Find the equation of the curve that passes through (2, 5) if its slope is given by
Curves $dy/dx = 2x$ at any point x.

Solution   We are interested in finding a function $y = f(x)$ such that

$$\frac{dy}{dx} = 2x \tag{1}$$

and

$$y = 5 \quad \text{when} \quad x = 2 \tag{2}$$

If $dy/dx = 2x$, then

$$y = \int 2x \, dx$$
$$= x^2 + C \tag{3}$$

Since $y = 5$ when $x = 2$, we determine the *particular value of C* so that

$$5 = 2^2 + C$$

Thus, $C = 1$, and

$$y = x^2 + 1$$

is the *particular antiderivative* out of all those possible from (3) that satisfies both (1) and (2). See Figure 1.

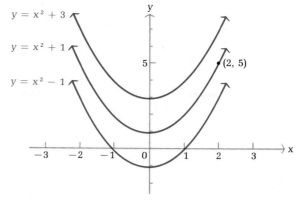

FIGURE 1
$y = x^2 + C$

◆

PROBLEM 4   Find the equation of the curve that passes through (2, 6) if the slope of the curve at any point x is given by $dy/dx = 3x^2$.
◆

In certain situations, it is easier to determine the rate at which something happens than how much of it has happened in a given length of time (for example, population growth rates, business growth rates, rate of healing of a wound, rates of learning or forgetting). If a rate function (derivative) is given and we know the value of the dependent variable for a given value of the independent variable, then — if the rate function is not too complicated — we can often find the original function by integration.

◆ EXAMPLE 5   If the marginal cost of producing x units is given by

Cost Function
$$C'(x) = 0.3x^2 + 2x$$

and the fixed cost is $2,000, find the cost function $C(x)$ and the cost of producing 20 units.

Solution   Recall that marginal cost is the derivative of the cost function and that fixed cost is cost at a 0 production level. Thus, the mathematical problem is to find $C(x)$ given

$$C'(x) = 0.3x^2 + 2x \qquad C(0) = 2,000$$

We now find the indefinite integral of $0.3x^2 + 2x$ and determine the arbitrary integration constant using $C(0) = 2,000$:

$$C'(x) = 0.3x^2 + 2x$$

$$C(x) = \int (0.3x^2 + 2x)\, dx$$

$$= 0.1x^3 + x^2 + K \qquad \text{Since C represents the cost, we use K for the constant of integration.}$$

But

$$C(0) = (0.1)0^3 + 0^2 + K = 2{,}000$$

Thus, $K = 2{,}000$, and the particular cost function is

$$C(x) = 0.1x^3 + x^2 + 2{,}000$$

We now find $C(20)$, the cost of producing 20 units:

$$C(20) = (0.1)20^3 + 20^2 + 2{,}000$$
$$= \$3{,}200 \qquad \blacklozenge$$

PROBLEM 5 Find the revenue function $R(x)$ when the marginal revenue is

$$R'(x) = 400 - 0.4x$$

and no revenue results at a 0 production level. What is the revenue at a production level of 1,000 units? $\qquad \blacklozenge$

◆ EXAMPLE 6

Advertising

An FM radio station is launching an aggressive advertising campaign in order to increase the number of daily listeners. The station currently has 27,000 daily listeners, and management expects the number of daily listeners, $S(t)$, to grow at the rate of

$$S'(t) = 60t^{1/2}$$

listeners per day, where t is the number of days since the campaign began. How long should the campaign last if the station wants the number of daily listeners to grow to 41,000?

Solution We must solve the equation $S(t) = 41{,}000$ for t, given that

$$S'(t) = 60t^{1/2} \qquad \text{and} \qquad S(0) = 27{,}000$$

First, we use integration to find $S(t)$:

$$S(t) = \int 60t^{1/2}\, dt$$

$$= 60\,\frac{t^{3/2}}{\frac{3}{2}} + C$$

$$= 40t^{3/2} + C$$

Since

$$S(0) = 40(0)^{3/2} + C = 27{,}000$$

we have $C = 27{,}000$, and

$$S(t) = 40t^{3/2} + 27{,}000$$

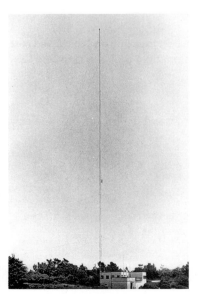

Now we solve the equation $S(t) = 41{,}000$ for t:

$$40t^{3/2} + 27{,}000 = 41{,}000$$
$$40t^{3/2} = 14{,}000$$
$$t^{3/2} = 350$$
$$t = 350^{2/3} \qquad \text{Use a calculator.}$$
$$= 49.664\ 419...$$

Thus, the advertising campaign should last approximately 50 days. ◆

PROBLEM 6

The current monthly circulation of the magazine *Computing News* is 640,000 copies. Due to competition from a new magazine in the same field, the monthly circulation of *Computing News*, $C(t)$, is expected to decrease at the rate of

$$C'(t) = -6{,}000t^{1/3}$$

copies per month, where t is the time in months since the new magazine began publication. How long will it take for the circulation of *Computing News* to decrease to 460,000 copies per month? ◆

Common Errors

1. $\displaystyle \int e^x \, dx \neq \frac{e^{x+1}}{x+1} + C$

   The power rule only applies to power functions of the form x^n where the exponent n is a real constant not equal to -1 and the base x is the variable. The function e^x is an exponential function with variable exponent x and constant base e. The correct form for this problem is

   $$\int e^x \, dx = e^x + C$$

2. $\displaystyle \int x(x^2 + 2) \, dx \neq \frac{x^2}{2} \left(\frac{x^3}{3} + 2x \right) + C$

   The integral of a product is not equal to the product of the integrals. The correct form for this problem is

   $$\int x(x^2 + 2) \, dx = \int (x^3 + 2x) \, dx = \frac{x^4}{4} + x^2 + C$$

Answers to Matched Problems

1. (A) $x + C$ (B) $\frac{3}{5}t^5 + C$ (C) $(x^6/3) - x^3 + x + C$ (D) $\frac{5}{2}w^{8/5} + C$
   (E) $\frac{6}{5}x^{5/3} + x^{-3} + C$
2. (A) $\frac{1}{3}x^3 - 4x^2 + C$ (B) $6x^{4/3} - 12x^{1/2} + C$ 3. $5 \ln|x| - 4e^x + C$
4. $y = x^3 - 2$ 5. $R(x) = 400x - 0.2x^2$; $R(1{,}000) = \$200{,}000$
6. $t = (40)^{3/4} \approx 16$ months

A   *Find each indefinite integral. (Check by differentiating.)*

1. $\int 7\,dx$ 2. $\int \pi\,dx$ 3. $\int x^6\,dx$

4. $\int x^3\,dx$ 5. $\int 8t^3\,dt$ 6. $\int 10t^4\,dt$

7. $\int (2u + 1)\,du$ 8. $\int (1 - 2u)\,du$ 9. $\int (3x^2 + 2x - 5)\,dx$

10. $\int (2 + 4x - 6x^2)\,dx$ 11. $\int (s^4 - 8s^5)\,ds$ 12. $\int (t^5 + 6t^3)\,dt$

13. $\int 3e^t\,dt$ 14. $\int 2e^t\,dt$ 15. $\int 2z^{-1}\,dz$

16. $\int \dfrac{3}{s}\,ds$

Find all the antiderivatives for each derivative.

17. $\dfrac{dy}{dx} = 200x^4$ 18. $\dfrac{dx}{dt} = 42t^5$ 19. $\dfrac{dP}{dx} = 24 - 6x$

20. $\dfrac{dy}{dx} = 3x^2 - 4x^3$ 21. $\dfrac{dy}{du} = 2u^5 - 3u^2 - 1$ 22. $\dfrac{dA}{dt} = 3 - 12t^3 - 9t^5$

23. $\dfrac{dy}{dx} = e^x + 3$ 24. $\dfrac{dy}{dx} = x - e^x$ 25. $\dfrac{dx}{dt} = 5t^{-1} + 1$

26. $\dfrac{du}{dv} = \dfrac{4}{v} + \dfrac{v}{4}$

B   *Find each indefinite integral. (Check by differentiation.)*

27. $\int 6x^{1/2}\,dx$ 28. $\int 8t^{1/3}\,dt$

29. $\int 8x^{-3}\,dx$ 30. $\int 12u^{-4}\,du$

31. $\int \dfrac{du}{\sqrt{u}}$ 32. $\int \dfrac{dt}{\sqrt[3]{t}}$

33. $\int \dfrac{dx}{4x^3}$ 34. $\int \dfrac{6\,dm}{m^2}$

35. $\int \dfrac{du}{2u^5}$ 36. $\int \dfrac{dy}{3y^4}$

37. $\int \left(3x^2 - \dfrac{2}{x^2}\right) dx$ 38. $\int \left(4x^3 + \dfrac{2}{x^3}\right) dx$

39. $\int \left(10x^4 - \dfrac{8}{x^5} - 2\right) dx$

40. $\int \left(\dfrac{6}{x^4} - \dfrac{2}{x^3} + 1\right) dx$

41. $\int \left(3\sqrt{x} + \dfrac{2}{\sqrt{x}}\right) dx$

42. $\int \left(\dfrac{2}{\sqrt[3]{x}} - \sqrt[3]{x^2}\right) dx$

43. $\int \left(\sqrt[3]{x^2} - \dfrac{4}{x^3}\right) dx$

44. $\int \left(\dfrac{12}{x^5} - \dfrac{1}{\sqrt[3]{x^2}}\right) dx$

45. $\int \dfrac{e^x - 3x}{4} dx$

46. $\int \dfrac{e^x - 3x^2}{2} dx$

47. $\int (2z^{-3} + z^{-2} + z^{-1}) \, dz$

48. $\int (3x^{-2} - x^{-1}) \, dx$

In Problems 49–58, find the particular antiderivative of each derivative that satisfies the given condition.

49. $\dfrac{dy}{dx} = 2x - 3; \quad y(0) = 5$

50. $\dfrac{dy}{dx} = 5 - 4x; \quad y(0) = 20$

51. $C'(x) = 6x^2 - 4x; \quad C(0) = 3{,}000$

52. $R'(x) = 600 - 0.6x; \quad R(0) = 0$

53. $\dfrac{dx}{dt} = \dfrac{20}{\sqrt{t}}; \quad x(1) = 40$

54. $\dfrac{dR}{dt} = \dfrac{100}{t^2}; \quad R(1) = 400$

55. $\dfrac{dy}{dx} = 2x^{-2} + 3x^{-1} - 1; \quad y(1) = 0$

56. $\dfrac{dy}{dx} = 3x^{-1} + x^{-2}; \quad y(1) = 1$

57. $\dfrac{dx}{dt} = 4e^t - 2; \quad x(0) = 1$

58. $\dfrac{dy}{dt} = 5e^t - 4; \quad y(0) = -1$

59. Find the equation of the curve that passes through (2, 3) if its slope is given by

$$\dfrac{dy}{dx} = 4x - 3$$

for each x.

60. Find the equation of the curve that passes through (1, 3) if its slope is given by

$$\dfrac{dy}{dx} = 12x^2 - 12x$$

for each x.

C   Find each indefinite integral.

61. $\int \dfrac{2x^4 - x}{x^3} dx$

62. $\int \dfrac{x^{-1} - x^4}{x^2} dx$

63. $\int \dfrac{x^5 - 2x}{x^4} dx$

64. $\int \dfrac{1 - 3x^4}{x^2} dx$

65. $\int \dfrac{x^2 e^x - 2x}{x^2} dx$

66. $\int \dfrac{1 - xe^x}{x} dx$

For each derivative, find an antiderivative that satisfies the given condition.

67. $\dfrac{dM}{dt} = \dfrac{t^2 - 1}{t^2}$; $M(4) = 5$

68. $\dfrac{dR}{dx} = \dfrac{1 - x^4}{x^3}$; $R(1) = 4$

69. $\dfrac{dy}{dx} = \dfrac{5x + 2}{\sqrt[3]{x}}$; $y(1) = 0$

70. $\dfrac{dx}{dt} = \dfrac{\sqrt{t^3} - t}{\sqrt{t^3}}$; $x(9) = 4$

71. $p'(x) = -\dfrac{10}{x^2}$; $p(1) = 20$

72. $p'(x) = \dfrac{10}{x^3}$; $p(1) = 15$

APPLICATIONS

Business & Economics

73. *Profit function.* If the marginal profit for producing x units is given by

$$P'(x) = 50 - 0.04x \qquad P(0) = 0$$

where $P(x)$ is the profit in dollars, find the profit function P and the profit on 100 units of production.

74. *Natural resources.* The world demand for wood is increasing. In 1975, the demand was 12.6 billion cubic feet, and the rate of increase in demand is given approximately by

$$d'(t) = 0.009t$$

where t is time in years after 1975 (data from the U.S. Department of Agriculture and Forest Service). Noting that $d(0) = 12.6$, find $d(t)$. Also find $d(25)$, the demand in the year 2000.

75. *Revenue function.* The marginal revenue from the sale of x digital sports watches is given by

$$R'(x) = 100 - \tfrac{1}{5}x \qquad R(0) = 0$$

where $R(x)$ is the revenue in dollars. Find the revenue function and the price–demand equation. What is the price when the demand is 700 units?

76. *Cost function.* The marginal average cost for producing x digital sports watches is given by

$$\overline{C}'(x) = -\dfrac{1,000}{x^2} \qquad \overline{C}(100) = 25$$

where $\overline{C}(x)$ is the average cost in dollars. Find the average cost function and the cost function. What are the fixed costs?

77. *Sales analysis.* The monthly sales of a particular personal computer are expected to decline at the rate of

$$S'(t) = -25t^{2/3}$$

computers per month, where t is time in months and $S(t)$ is the number of computers sold each month. The company plans to stop manufacturing this computer when the monthly sales reach 800 computers. If the monthly sales now ($t = 0$) are 2,000 computers, find $S(t)$. How long will the company continue to manufacture this computer?

78. *Sales analysis.* The rate of change of the monthly sales of a new home video game cartridge is given by

$$S'(t) = 500t^{1/4} \qquad S(0) = 0$$

where t is the number of months since the game was released and $S(t)$ is the number of cartridges sold each month. Find $S(t)$. When will the monthly sales reach 20,000 cartridges?

79. *Labor costs and learning.* A defense contractor is starting production on a new missile control system. On the basis of data collected while assembling the first 16 control systems, the production manager obtained the following function describing the rate of labor use:

$$g(x) = 2,400x^{-1/2}$$

where $g(x)$ is the number of labor-hours required to assemble the xth unit of the control system. For example, after assembling 16 units, the rate of assembly is 600 labor-hours per unit, and after assembling 25 units, the rate of assembly is 480 labor-hours per unit. The more units assembled, the more efficient the process because of learning. If 19,200 labor-hours are required to assemble the first 16 units, how many labor-hours, $L(x)$, will be required to assemble the first x units? The first 25 units?

80. *Labor costs and learning.* If the rate of labor use in Problem 79 is

$$g(x) = 2,000x^{-1/3}$$

and if the first 8 control units require 12,000 labor-hours, how many labor-hours, $L(x)$, will be required for the first x control units? The first 27 control units?

Life Sciences

81. *Weight–height.* For an average person, the rate of change of weight W (in pounds) with respect to height h (in inches) is given approximately by

$$\frac{dW}{dh} = 0.0015h^2$$

Find $W(h)$ if $W(60) = 108$ pounds. Also find the weight for a person who is 5 feet 10 inches tall.

82. *Wound healing.* If the area A of a healing wound changes at a rate given approximately by

$$\frac{dA}{dt} = -4t^{-3} \qquad 1 \leqslant t \leqslant 10$$

where t is time in days and $A(1) = 2$ square centimeters, what will the area of the wound be in 10 days?

83. *Urban growth.* The rate of growth of the population, $N(t)$, of a newly incorporated city t years after incorporation is estimated to be

$$\frac{dN}{dt} = 400 + 600\sqrt{t} \qquad 0 \leqslant t \leqslant 9$$

If the population was 5,000 at the time of incorporation, find the population 9 years later.

84. *Learning.* A beginning high school language class was chosen for an experiment in learning. Using a list of 50 words, the experiment involved measuring the rate of vocabulary memorization at different times during a continuous 5 hour study session. It was found that the average rate of learning for the whole class was inversely proportional to the time spent studying and was given approximately by

$$V'(t) = \frac{15}{t} \qquad 1 \leqslant t \leqslant 5$$

If the average number of words memorized after 1 hour of study was 15 words, what was the average number of words learned after t hours of study for $1 \leqslant t \leqslant 5$? After 4 hours of study? (Round answer to the nearest whole number.)

S E C T I O N   11-2   # Integration by Substitution

◆   DIFFERENTIALS
◆   INTEGRATION BY SUBSTITUTION
◆   SUBSTITUTION TECHNIQUES
◆   APPLICATION

The properties of the indefinite integral discussed in the last section enable us to find the indefinite integrals of a fairly large set of functions, but there are still many functions with indefinite integrals that cannot be found by the application of these properties. In this section we will substantially increase our capability of finding indefinite integrals by means of a method called *substitution*. Before considering this method, we must first discuss the meaning of the symbol dx in the indefinite integral $\int f(x)\, dx$.

◆ **DIFFERENTIALS**

Recall from Chapter 9 that the derivative of a function $y = f(x)$ is defined as

$$f'(x) = \lim_{h \to 0} \frac{f(x + h) - f(x)}{h}$$

This derivative also can be represented by the symbol dy/dx, which suggests that the derivative is the limit of the ratio of two separate quantities, $f(x + h) - f(x)$ and h. We now define dy and dx as two separate quantities with the property that their ratio is still equal to $f'(x)$.

■ **Differentials**

If $y = f(x)$ defines a differentiable function, then:
1. The **differential dx** of the independent variable x is an arbitrary real number.
2. The **differential dy** of the dependent variable y is defined as the product of $f'(x)$ and dx; that is,

$$dy = f'(x) \, dx$$

The differential dy is actually a function of two independent variables, x and dx; a change in either one or both will affect the value of dy. However, we will be interested in using differential notation, not in evaluating differentials for specific values of x and dx. Since the differential of a dependent variable is defined in terms of its derivative, finding differentials involves the same operations as finding derivatives — only the notation is different.

◆ **EXAMPLE 7**
(A) If $y = f(x) = x^2$, then $dy = f'(x) \, dx = 2x \, dx$.
(B) If $u = g(x) = e^{3x}$, then $du = g'(x) \, dx = 3e^{3x} \, dx$.
(C) If $w = h(t) = \ln(4 + 5t)$, then

$$dw = h'(t) \, dt = \frac{5 \, dt}{4 + 5t}$$

◆

PROBLEM 7
(A) Find dy for $y = f(x) = x^3$.
(B) Find du for $u = h(x) = \ln(2 + x^2)$.
(C) Find dv for $v = g(t) = e^{-5t}$.

◆

◆ **INTEGRATION BY SUBSTITUTION**

In Sections 9-6 and 10-7, we saw that the chain rule was a very important tool for finding derivatives. Now we will see that the chain rule also has applications to integration. To begin, consider the indefinite integral

$$\int \frac{2x}{x^2 + 5} \, dx \tag{1}$$

This integral cannot be evaluated directly by any of the techniques discussed in Section 11-1, but it can be simplified by making a *substitution*. If we let

$$u = x^2 + 5$$

then the differential of u is

$$du = 2x \, dx$$

If we rewrite the integrand in (1) so that $2x$ is grouped with dx, then we can express the integrand entirely in terms of u and du:

$$\int \frac{2x}{x^2+5}\, dx = \int \frac{1}{x^2+5}\, 2x\, dx \qquad \text{Substitute } u = x^2 + 5 \text{ and } du = 2x\, dx.$$

$$= \int \frac{1}{u}\, du$$

This last integral looks just like the integral in formula 6 from Section 11-1, except that u is not an independent variable; it is a function of x. For now, let us assume that formula 6 applies in this situation. Then we have

$$\int \frac{2x}{x^2+5}\, dx = \int \frac{1}{u}\, du \qquad \text{Use formula 6.}$$

$$= \ln|u| + C \qquad \text{Substitute } u = x^2 + 5.$$
$$= \ln(x^2+5) + C \qquad \text{Absolute value signs can be omitted since } x^2 + 5 > 0 \text{ for all } x.$$

We use the chain rule to check this result:

$$D_x \ln(x^2 + 5) = \frac{1}{x^2+5}\, D_x(x^2+5) = \frac{1}{x^2+5}\, 2x = \frac{2x}{x^2+5}$$

which is the integrand in (1). Thus, our assumption concerning formula 6 was valid for this problem. Is this always the case? Yes, it is, as we will now verify. Since the chain rule implies that

$$D_x \ln|f(x)| = \frac{1}{f(x)}\, f'(x)$$

it follows that

$$\int \frac{1}{f(x)}\, f'(x)\, dx = \ln|f(x)| + C$$

or, substituting u for $f(x)$ and du for $f'(x)\, dx$,

$$\int \frac{1}{u}\, du = \ln|u| + C$$

Similarly, all the integration formulas from Section 11-1 are valid if the variable of integration is a function u and du is its differential. We restate these formulas in the box for convenient reference.

For k and C constants:

1. $\displaystyle\int k\ du = ku + C$

2. $\displaystyle\int u^n\ du = \frac{u^{n+1}}{n+1} + C \qquad n \neq -1$

3. $\displaystyle\int kf(u)\ du = k\int f(u)\ du$

4. $\displaystyle\int [f(u) \pm g(u)]\ du = \int f(u)\ du \pm \int g(u)\ du$

5. $\displaystyle\int e^u\ du = e^u + C$

6. $\displaystyle\int \frac{1}{u}\ du = \ln|u| + C$

These formulas are valid if u is an independent variable or if u is a function of another variable and du is its differential with respect to that variable.

The substitution method for evaluating certain indefinite integrals is outlined in the following box:

■ **Integration by Substitution**

Step 1. Select a substitution that appears to simplify the integrand. In particular, try to select u so that du is a factor in the integrand.

Step 2. Express the integrand entirely in terms of u and du, completely eliminating the original variable and its differential.

Step 3. Evaluate the new integral, if possible.

Step 4. Express the antiderivative found in step 3 in terms of the original variable.

◆ E X A M P L E   8   Use a substitution to find the following:

(A) $\displaystyle\int (3x + 4)^6 3\ dx$ (B) $\displaystyle\int e^{t^2} 2t\ dt$

Solutions   (A) If we let $u = 3x + 4$, then $du = 3\,dx$, and

$$\int (3x + 4)^6 3\,dx = \int u^6\,du \qquad \text{Use formula 2.}$$

$$= \frac{u^7}{7} + C$$

$$= \frac{(3x + 4)^7}{7} + C \qquad \text{Since } u = 3x + 4$$

Check: $D_x \dfrac{(3x + 4)^7}{7} = \dfrac{7(3x + 4)^6}{7} D_x(3x + 4) = (3x + 4)^6 3$

(B) If we let $u = t^2$, then $du = 2t\,dt$, and

$$\int e^{t^2} 2t\,dt = \int e^u\,du \qquad \text{Use formula 5.}$$

$$= e^u + C$$

$$= e^{t^2} + C \qquad \text{Since } u = t^2$$

Check: $D_t\, e^{t^2} = e^{t^2} D_t\, t^2 = e^{t^2} 2t$ ◆

PROBLEM   8   Use a substitution to find each indefinite integral.

(A) $\displaystyle \int (2x^3 - 3)^4 6x^2\,dx$ (B) $\displaystyle \int e^{5w} 5\,dw$ ◆

Integration by substitution is an effective procedure for some indefinite integrals, but not all. Substitution is not helpful for $\int e^{x^2}\,dx$ or $\int (\ln x)\,dx$, for example.

◆ SUBSTITUTION TECHNIQUES

In order to use the substitution method, **the integrand must be expressed entirely in terms of u and du.** In some cases, the integrand will have to be modified before making a substitution and using one of the integration formulas. Example 9 illustrates this process.

◆ EXAMPLE   9   Integrate: (A) $\displaystyle \int \frac{1}{4x + 7}\,dx$ (B) $\displaystyle \int te^{-t^2}\,dt$ (C) $\displaystyle \int 4x^2 \sqrt{x^3 + 5}\,dx$

Solutions   (A) If $u = 4x + 7$, then $du = 4\,dx$. We are missing a factor of 4 in the integrand to match formula 6 exactly. Recalling that a constant factor can be moved across an integral sign, we proceed as follows:

$$\int \frac{1}{4x + 7}\,dx = \int \frac{1}{4x + 7} \frac{4}{4}\,dx$$

$$= \frac{1}{4} \int \frac{1}{4x + 7} 4\,dx \qquad \text{Substitute } u = 4x + 7 \text{ and } du = 4\,dx.$$

$$\int \frac{1}{4x+7}\, dx = \frac{1}{4}\int \frac{1}{u}\, du \qquad \text{Use formula 6.}$$

$$= \tfrac{1}{4}\ln|u| + C$$
$$= \tfrac{1}{4}\ln|4x+7| + C \qquad \text{Since } u = 4x + 7$$

Check: $\displaystyle D_x \frac{1}{4}\ln|4x+7| = \frac{1}{4}\frac{1}{4x+7}D_x(4x+7) = \frac{1}{4}\frac{1}{4x+7}\, 4 = \frac{1}{4x+7}$

(B) If $u = -t^2$, then $du = -2t\, dt$. Proceed as in part A:

$$\int te^{-t^2}\, dt = \int e^{-t^2}\frac{-2}{-2}t\, dt$$

$$= -\frac{1}{2}\int e^{-t^2}(-2t)\, dt \qquad \text{Substitute } u = -t^2 \text{ and } du = -2t\, dt.$$

$$= -\frac{1}{2}\int e^{u}\, du \qquad \text{Use formula 5.}$$

$$= -\tfrac{1}{2}e^{u} + C$$
$$= -\tfrac{1}{2}e^{-t^2} + C \qquad \text{Since } u = -t^2$$

Check: $\displaystyle D_t(-\tfrac{1}{2}e^{-t^2}) = -\tfrac{1}{2}e^{-t^2}D_t(-t^2) = -\tfrac{1}{2}e^{-t^2}(-2t) = te^{-t^2}$

(C) $\displaystyle \int 4x^2\sqrt{x^3+5}\, dx = 4\int \sqrt{x^3+5}(x^2)\, dx$

Move the 4 across the integral sign and proceed as before.

$$= 4\int \sqrt{x^3+5}\,\frac{3}{3}(x^2)\, dx$$

$$= \frac{4}{3}\int \sqrt{x^3+5}(3x^2)\, dx \qquad \text{Substitute } u = x^3 + 5 \text{ and } du = 3x^2\, dx.$$

$$= \frac{4}{3}\int \sqrt{u}\, du$$

$$= \frac{4}{3}\int u^{1/2}\, du \qquad \text{Use formula 2.}$$

$$= \frac{4}{3}\frac{u^{3/2}}{\frac{3}{2}} + C$$

$$= \tfrac{8}{9}u^{3/2} + C$$
$$= \tfrac{8}{9}(x^3+5)^{3/2} + C \qquad \text{Since } u = x^3 + 5$$

Check: $D_x[\tfrac{8}{9}(x^3+5)^{3/2}] = \tfrac{4}{3}(x^3+5)^{1/2}D_x(x^3+5)$
$$= \tfrac{4}{3}(x^3+5)^{1/2}3x^2 = 4x^2\sqrt{x^3+5} \qquad \blacklozenge$$

PROBLEM 9   Integrate:

(A) $\displaystyle \int e^{-3x}\, dx$ (B) $\displaystyle \int \frac{x}{x^2-9}\, dx$ (C) $\displaystyle \int 5t^2(t^3+4)^{-2}\, dt$ $\blacklozenge$

Even if it is not possible to find a substitution that makes an integrand match one of the integration formulas exactly, a substitution may sufficiently simplify the integrand so that other techniques can be used.

◆ EXAMPLE 10   Find: $\displaystyle\int \frac{x}{\sqrt{x+2}}\, dx$

Solution   Proceeding as before, if we let $u = x + 2$, then $du = dx$ and

$$\int \frac{x}{\sqrt{x+2}}\, dx = \int \frac{x}{\sqrt{u}}\, du$$

Notice that this substitution is not yet complete, because we have not expressed the integrand entirely in terms of u and du. As we noted earlier, only a constant factor can be moved across an integral sign, so we cannot move x outside the integral sign (as much as we would like to). Instead, we must return to the original substitution, solve for x in terms of u, and use the resulting equation to complete the substitution:

$$u = x + 2 \qquad \text{Solve for } x \text{ in terms of } u.$$
$$u - 2 = x \qquad \text{Substitute this expression for } x.$$

Thus,

$$\int \frac{x}{\sqrt{x+2}}\, dx = \int \frac{u-2}{\sqrt{u}}\, du \qquad\qquad \text{Simplify the integrand.}$$

$$= \int \frac{u-2}{u^{1/2}}\, du$$

$$= \int (u^{1/2} - 2u^{-1/2})\, du$$

$$= \boxed{\int u^{1/2}\, du - 2 \int u^{-1/2}\, du}$$

$$= \frac{u^{3/2}}{\frac{3}{2}} - 2\, \frac{u^{1/2}}{\frac{1}{2}} + C$$

$$= \tfrac{2}{3}(x+2)^{3/2} - 4(x+2)^{1/2} + C \qquad \text{Since } u = x + 2$$

Check: $D_x[\tfrac{2}{3}(x+2)^{3/2} - 4(x+2)^{1/2}] = (x+2)^{1/2} - 2(x+2)^{-1/2}$

$$= \frac{x+2}{(x+2)^{1/2}} - \frac{2}{(x+2)^{1/2}}$$

$$= \frac{x}{(x+2)^{1/2}} \qquad\qquad ◆$$

PROBLEM 10   Find: $\displaystyle\int x\sqrt{x+1}\, dx$ ◆

◆ APPLICATION

◆ E X A M P L E 11

Price–Demand

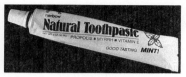

The market research department for a supermarket chain has determined that for one store the marginal price $p'(x)$ at x tubes per week for a certain brand of toothpaste is given by

$$p'(x) = -0.015e^{-0.01x}$$

Find the price–demand equation if the weekly demand is 50 when the price of a tube is \$2.35. Find the weekly demand when the price of a tube is \$1.89.

Solution

$$p(x) = \int -0.015e^{-0.01x}\, dx$$

$$= -0.015 \int e^{-0.01x}\, dx$$

$$= -0.015 \int e^{-0.01x}\, \frac{-0.01}{-0.01}\, dx$$

$$= \frac{-0.015}{-0.01} \int e^{-0.01x}(-0.01)\, dx \qquad \text{Substitute } u = -0.01x \text{ and } du = -0.01\, dx.$$

$$= 1.5 \int e^u\, du$$

$$= 1.5e^u + C$$
$$= 1.5e^{-0.01x} + C \qquad\qquad \text{Since } u = -0.01x$$

We find C by noting that

$$p(50) = 1.5e^{-0.01(50)} + C = \$2.35$$
$$C = \$2.35 - 1.5e^{-0.5} \qquad \text{Use a calculator.}$$
$$C = \$2.35 - 0.91$$
$$C = \$1.44$$

Thus,

$$p(x) = 1.5e^{-0.01x} + 1.44$$

To find the demand when the price is \$1.89, we solve $p(x) = \$1.89$ for x:

$$1.5e^{-0.01x} + 1.44 = 1.89$$
$$1.5e^{-0.01x} = 0.45$$
$$e^{-0.01x} = 0.3$$
$$-0.01x = \ln 0.3$$
$$x = -100 \ln 0.3 \approx 120 \text{ tubes}$$

◆

PROBLEM 11 The marginal price $p'(x)$ at a supply level of x tubes per week for a certain brand of toothpaste is given by

$$p'(x) = 0.001e^{0.01x}$$

Find the price–supply equation if the supplier is willing to supply 100 tubes per week at a price of $1.65 each. How many tubes would the supplier be willing to supply at a price of $1.98 each? ◆

Integrals of the form encountered in Example 11 occur so frequently that it is worthwhile to state a general formula for integrals of this type:

$$\int e^{au}\, du = \frac{1}{a} e^{au} + C \qquad \text{where } a \text{ is a constant, } a \neq 0$$

Common Errors

1. $\displaystyle \int (x^2 + 3)^2\, dx = \int (x^2 + 3)^2\, \frac{2x}{2x}\, dx$

$\displaystyle \qquad \neq \frac{1}{2x} \int (x^2 + 3)^2 2x\, dx$

Remember: **A variable factor cannot be moved across an integral sign!** The correct procedure for this problem is

$$\int (x^2 + 3)^2\, dx = \int (x^4 + 6x^2 + 9)\, dx$$

$$= \frac{x^5}{5} + 2x^3 + 9x + C$$

2. $\displaystyle \int \frac{1}{10x + 3}\, dx = \int \frac{1}{u}\, dx \qquad u = 10x + 3$

$\displaystyle \qquad \neq \ln|u| + C$

Remember: **An integral must be expressed entirely in terms of u and du before applying integration formulas 1–6.** The correct procedure for this problem is

$$\int \frac{1}{10x + 3}\, dx = \frac{1}{10} \int \frac{1}{10x + 3}\, 10\, dx \qquad u = 10x + 3,\ du = 10\, dx$$

$$= \frac{1}{10} \int \frac{1}{u}\, du$$

$$= \tfrac{1}{10} \ln|u| + C$$

$$= \tfrac{1}{10} \ln|10x + 3| + C$$

Answers to Matched Problems

7. (A) $dy = 3x^2\, dx$ (B) $du = \dfrac{2x\, dx}{2 + x^2}$ (C) $dv = -5e^{-5t}\, dt$

8. (A) $\frac{1}{5}(2x^3 - 3)^5 + C$ (B) $e^{5w} + C$

9. (A) $-\frac{1}{3}e^{-3x} + C$ (B) $\frac{1}{2}\ln|x^2 - 9| + C$ (C) $-\frac{5}{3}(t^3 + 4)^{-1} + C$

10. $\frac{2}{5}(x + 1)^{5/2} - \frac{2}{3}(x + 1)^{3/2} + C$ 11. $p(x) = 0.1e^{0.01x} + 1.38$; 179 tubes

A *Find each indefinite integral, and check the result by differentiating.*

1. $\displaystyle\int (x^2 - 4)^5 2x \, dx$ 2. $\displaystyle\int (x^3 + 1)^4 3x^2 \, dx$ 3. $\displaystyle\int e^{4x} 4 \, dx$

4. $\displaystyle\int e^{-3x}(-3) \, dx$ 5. $\displaystyle\int \frac{1}{2t + 3} \, 2 \, dt$ 6. $\displaystyle\int \frac{1}{5t - 7} \, 5 \, dt$

B 7. $\displaystyle\int (3x - 2)^7 \, dx$ 8. $\displaystyle\int (5x + 3)^9 \, dx$ 9. $\displaystyle\int (x^2 + 3)^7 x \, dx$

10. $\displaystyle\int (x^3 - 5)^4 x^2 \, dx$ 11. $\displaystyle\int 10e^{-0.5t} \, dt$ 12. $\displaystyle\int 4e^{0.01t} \, dt$

13. $\displaystyle\int \frac{1}{10x + 7} \, dx$ 14. $\displaystyle\int \frac{1}{100 - 3x} \, dx$ 15. $\displaystyle\int xe^{2x^2} \, dx$

16. $\displaystyle\int x^2 e^{4x^3} \, dx$ 17. $\displaystyle\int \frac{x^2}{x^3 + 4} \, dx$ 18. $\displaystyle\int \frac{x}{x^2 - 2} \, dx$

19. $\displaystyle\int \frac{t}{(3t^2 + 1)^4} \, dt$ 20. $\displaystyle\int \frac{t^2}{(t^3 - 2)^5} \, dt$ 21. $\displaystyle\int \frac{x^2}{(4 - x^3)^2} \, dx$

22. $\displaystyle\int \frac{x}{(5 - 2x^2)^5} \, dx$ 23. $\displaystyle\int x\sqrt{x + 4} \, dx$ 24. $\displaystyle\int x\sqrt{x - 9} \, dx$

25. $\displaystyle\int \frac{x}{\sqrt{x - 3}} \, dx$ 26. $\displaystyle\int \frac{x}{\sqrt{x + 5}} \, dx$ 27. $\displaystyle\int x(x - 4)^9 \, dx$

28. $\displaystyle\int x(x + 6)^8 \, dx$ 29. $\displaystyle\int e^{2x}(1 + e^{2x})^3 \, dx$ 30. $\displaystyle\int e^{-x}(1 - e^{-x})^4 \, dx$

31. $\displaystyle\int \frac{1 + x}{4 + 2x + x^2} \, dx$ 32. $\displaystyle\int \frac{x^2 - 1}{x^3 - 3x + 7} \, dx$ 33. $\displaystyle\int (2x + 1)e^{x^2 + x + 1} \, dx$

34. $\displaystyle\int (x^2 + 2x)e^{x^3 + 3x^2} \, dx$ 35. $\displaystyle\int (e^x - 2x)^3(e^x - 2) \, dx$

36. $\displaystyle\int (x^2 - e^x)^4(2x - e^x) \, dx$ 37. $\displaystyle\int \frac{x^3 + x}{(x^4 + 2x^2 + 1)^4} \, dx$

38. $\displaystyle\int \frac{x^2 - 1}{(x^3 - 3x + 7)^2} \, dx$

C 39. $\displaystyle\int x\sqrt{3x^2 + 7} \, dx$ 40. $\displaystyle\int x^2\sqrt{2x^3 + 1} \, dx$ 41. $\displaystyle\int x(x^3 + 2)^2 \, dx$

42. $\displaystyle\int x(x^2 + 2)^2 \, dx$ 43. $\displaystyle\int x^2(x^3 + 2)^2 \, dx$ 44. $\displaystyle\int (x^2 + 2)^2 \, dx$

45. $\displaystyle\int \frac{x^3}{\sqrt{2x^4 + 3}} \, dx$ 46. $\displaystyle\int \frac{x^2}{\sqrt{4x^3 - 1}} \, dx$ 47. $\displaystyle\int \frac{(\ln x)^3}{x} \, dx$

48. $\displaystyle\int \frac{e^x}{1 + e^x} \, dx$ 49. $\displaystyle\int \frac{1}{x^2} e^{-1/x} \, dx$ 50. $\displaystyle\int \frac{1}{x \ln x} \, dx$

Find the antiderivative of each derivative.

51. $\dfrac{dx}{dt} = 7t^2(t^3 + 5)^6$ **52.** $\dfrac{dm}{dn} = 10n(n^2 - 8)^7$ **53.** $\dfrac{dy}{dt} = \dfrac{3t}{\sqrt{t^2 - 4}}$

54. $\dfrac{dy}{dx} = \dfrac{5x^2}{(x^3 - 7)^4}$ **55.** $\dfrac{dp}{dx} = \dfrac{e^x + e^{-x}}{(e^x - e^{-x})^2}$ **56.** $\dfrac{dm}{dt} = \dfrac{\ln(t - 5)}{t - 5}$

Use substitution techniques to derive the integration formulas in Problems 57 and 58. Then check your work by differentiation.

57. $\displaystyle\int e^{au}\, du = \dfrac{1}{a} e^{au} + C, \quad a \neq 0$

58. $\displaystyle\int \dfrac{1}{au + b}\, du = \dfrac{1}{a} \ln|au + b| + C, \quad a \neq 0$

APPLICATIONS

Business & Economics

59. *Price–demand equation.* The marginal price for a weekly demand of x bottles of baby shampoo in a drug store is given by

$$p'(x) = \dfrac{-6{,}000}{(3x + 50)^2}$$

Find the price–demand equation if the weekly demand is 150 when the price of a bottle of shampoo is \$4. What is the weekly demand when the price is \$2.50?

60. *Price–supply equation.* The marginal price at a supply level of x bottles of baby shampoo per week is given by

$$p'(x) = \dfrac{300}{(3x + 25)^2}$$

Find the price–supply equation if the distributor of the shampoo is willing to supply 75 bottles a week at a price of \$1.60 per bottle. How many bottles would the supplier be willing to supply at a price of \$1.75 per bottle?

61. *Cost function.* The weekly marginal cost of producing x pairs of tennis shoes is given by

$$C'(x) = 12 + \dfrac{500}{x + 1}$$

where C(x) is cost in dollars. If the fixed costs are \$2,000 per week, find the cost function. What is the average cost per pair of shoes if 1,000 pairs of shoes are produced each week?

62. *Revenue function.* The weekly marginal revenue from the sale of x pairs of tennis shoes is given by

$$R'(x) = 40 - 0.02x + \dfrac{200}{x + 1} \qquad R(0) = 0$$

where $R(x)$ is revenue in dollars. Find the revenue function. Find the revenue from the sale of 1,000 pairs of shoes.

63. *Marketing.* An automobile company is ready to introduce a new line of cars with a national sales campaign. After test marketing the line in a carefully selected city, the marketing research department estimates that sales (in millions of dollars) will increase at the monthly rate of

$$S'(t) = 10 - 10e^{-0.1t} \qquad 0 \leqslant t \leqslant 24$$

t months after the national campaign has started. What will be the total sales, $S(t)$, t months after the beginning of the national campaign if we assume 0 sales at the beginning of the campaign? What are the estimated total sales for the first 12 months of the campaign?

64. *Marketing.* Repeat Problem 63 if the monthly rate of increase in sales is found to be approximated by

$$S'(t) = 20 - 20e^{-0.05t} \qquad 0 \leqslant t \leqslant 24$$

65. *Oil production.* Using data from the first 3 years of production as well as geological studies, the management of an oil company estimates that oil will be pumped from a producing field at a rate given by

$$R(t) = \frac{100}{t+1} + 5 \qquad 0 \leqslant t \leqslant 20$$

where $R(t)$ is the rate of production (in thousands of barrels per year) t years after pumping begins. How many barrels of oil, $Q(t)$, will the field produce the first t years if $Q(0) = 0$? How many barrels will be produced the first 9 years?

66. *Oil production.* In Problem 65, if the rate is found to be

$$R(t) = \frac{120t}{t^2 + 1} + 3 \qquad 0 \leqslant t \leqslant 20$$

how many barrels of oil, $Q(t)$, will the field produce the first t years if $Q(0) = 0$? How many barrels will be produced the first 5 years?

Life Sciences

67. *Biology.* A yeast culture is growing at the rate of $W'(t) = 0.2e^{0.1t}$ grams per hour. If the starting culture weighs 2 grams, what will be the weight of the culture, $W(t)$, after t hours? After 8 hours?

68. *Medicine.* The rate of healing for a skin wound (in square centimeters per day) is approximated by $A'(t) = -0.9e^{-0.1t}$. If the initial wound has an area of 9 square centimeters, what will its area, $A(t)$, be after t days? After 5 days?

69. *Pollution.* A contaminated lake is treated with a bactericide. The rate of decrease in harmful bacteria t days after the treatment is given by

$$\frac{dN}{dt} = -\frac{2{,}000t}{1 + t^2} \qquad 0 \leqslant t \leqslant 10$$

where $N(t)$ is the number of bacteria per milliliter of water. If the initial count was 5,000 bacteria per milliliter, find $N(t)$ and then find the bacteria count after 10 days.

70. *Pollution.* An oil tanker aground on a reef is losing oil and producing an oil slick that is radiating outward at a rate given approximately by

$$\frac{dR}{dt} = \frac{60}{\sqrt{t+9}} \qquad t \geq 0$$

where R is the radius (in feet) of the circular slick after t minutes. Find the radius of the slick after 16 minutes if the radius is 0 when $t = 0$.

Social Sciences

71. *Learning.* In a particular business college, it was found that an average student enrolled in an advanced typing class progressed at a rate of $N'(t) = 6e^{-0.1t}$ words per minute per week, t weeks after enrolling in a 15 week course. If at the beginning of the course a student could type 40 words per minute, how many words per minute, $N(t)$, would the student be expected to type t weeks into the course? After completing the course?

72. *Learning.* In the same business college, it was also found that an average student enrolled in a beginning shorthand class progressed at a rate of $N'(t) = 12e^{-0.06t}$ words per minute per week, t weeks after enrolling in a 15 week course. If at the beginning of the course a student could take dictation in shorthand at 0 words per minute, how many words per minute, $N(t)$, would the student be expected to handle t weeks into the course? After completing the course?

73. *College enrollment.* The projected rate of increase in enrollment in a new college is estimated by

$$\frac{dE}{dt} = 5,000(t+1)^{-3/2} \qquad t \geq 0$$

where $E(t)$ is the projected enrollment in t years. If enrollment is 2,000 when $t = 0$, find the projected enrollment 15 years from now.

SECTION 11-3 Definite Integrals

◆ DEFINITE INTEGRALS
◆ PROPERTIES
◆ DEFINITE INTEGRALS AND SUBSTITUTION
◆ APPLICATIONS

◆ DEFINITE INTEGRALS

We start this discussion with a simple example, out of which will evolve a new integral form, called the *definite integral*. Our approach in this section will be intuitive and informal; these concepts will be made more precise in Section 11-5.

Suppose a manufacturing company's marginal cost equation for a particular product is given by

$$C'(x) = 2 - 0.2x \qquad 0 \leqslant x \leqslant 8$$

where the marginal cost is in thousands of dollars and production is x units per day. What is the total change in cost per day going from a production level of 2 units per day to 6 units per day? If $C = C(x)$ is the cost function, then

$$\left(\begin{array}{l}\text{Total net change in cost} \\ \text{between } x = 2 \text{ and } x = 6\end{array}\right) = C(6) - C(2) = C(x)\Big|_2^6 \tag{1}$$

The special symbol $C(x)|_2^6$ is a convenient way of representing the center expression, and it will prove useful to us later.

To evaluate (1), we need to find the antiderivative of $C'(x)$; that is,

$$C(x) = \int (2 - 0.2x)\, dx = 2x - 0.1x^2 + K \tag{2}$$

Thus, we are within a constant of knowing the original cost function. However, we do not need to know the constant K to solve the original problem (1). We compute $C(6) - C(2)$ for $C(x)$ found in (2):

$$\begin{aligned}C(6) - C(2) &= [2(6) - 0.1(6)^2 + K] - [2(2) - 0.1(2)^2 + K] \\ &= 12 - 3.6 + K - 4 + 0.4 - K \\ &= \$4.8 \text{ thousand per day increase in costs for a production} \\ &\qquad \text{increase from 2 to 6 units per day}\end{aligned}$$

The unknown constant K canceled out! Thus, we conclude that any antiderivative of $C'(x) = 2 - 0.2x$ will do, since antiderivatives of a given function can differ by at most a constant (see Section 11-1). Thus, we really do not have to find the constant in the original cost function to solve the problem.

Since $C(x)$ is an antiderivative of $C'(x)$, the above discussion suggests the following notation:

$$C(6) - C(2) = C(x)\Big|_2^6 = \int_2^6 C'(x)\, dx \tag{3}$$

The integral form on the right in (3) is called a *definite integral* — it represents the number found by evaluating an antiderivative of the integrand at 6 and 2 and taking the difference as indicated.

The **definite integral** of a continuous function f over an interval from $x = a$ to $x = b$ is the net change of an antiderivative of f over the interval. Symbolically, if $F(x)$ is an antiderivative of $f(x)$, then

$$\int_a^b f(x)\, dx = F(x)\Big|_a^b = F(b) - F(a) \qquad \text{where} \quad F'(x) = f(x)$$

INTEGRAND: $f(x)$ UPPER LIMIT: b LOWER LIMIT: a

The numbers a and b are referred to as the **limits of integration.** Note that the lower limit a can be larger than the upper limit b, as long as f is continuous on an interval containing both a and b.

In Section 11-5, we will formally define a definite integral as a limit of a special sum. Then the relationship in the box above turns out to be the most important theorem in calculus—the *fundamental theorem of calculus.* Our intent in this and the next section is to give you some intuitive experience with the definite integral concept and its use. You will then be better able to understand a formal definition and to appreciate the significance of the fundamental theorem.

◆ E X A M P L E   12   Evaluate: $\displaystyle\int_{-1}^{2} (3x^2 - 2x)\, dx$

Solution   We choose the simplest antiderivative of the integrand $3x^2 - 2x$, namely, $x^3 - x^2$, since any antiderivative will do:

$$\int_{-1}^{2} (3x^2 - 2x)\, dx = (x^3 - x^2)\Big|_{-1}^{2}$$

$$= (2^3 - 2^2) - [(-1)^3 - (-1)^2] \qquad \text{Be careful of sign errors}$$
$$= 4 - (-2) = 6 \qquad\qquad\qquad\qquad \text{here.} \qquad ◆$$

P R O B L E M   12   Evaluate: $\displaystyle\int_{-2}^{2} (2x - 1)\, dx$ ◆

On Definite and Indefinite Integrals

Do not confuse a definite integral with an indefinite integral. The definite integral $\int_a^b f(x)\, dx$ is a real number; the indefinite integral $\int f(x)\, dx$ is a whole set of functions—all the antiderivatives of $f(x)$.

◆ PROPERTIES

In the next box we state several useful properties of the definite integral. Note that properties 3 and 4 parallel properties 3 and 4 given earlier for the indefinite integral.

Definite Integral Properties

1. $\displaystyle \int_a^a f(x)\, dx = 0$

2. $\displaystyle \int_a^b f(x)\, dx = -\int_b^a f(x)\, dx$

3. $\displaystyle \int_a^b kf(x)\, dx = k\int_a^b f(x)\, dx \qquad k \text{ a constant}$

4. $\displaystyle \int_a^b [f(x) \pm g(x)]\, dx = \int_a^b f(x)\, dx \pm \int_a^b g(x)\, dx$

5. $\displaystyle \int_a^b f(x)\, dx = \int_a^c f(x)\, dx + \int_c^b f(x)\, dx$

These properties are justified as follows: If $F'(x) = f(x)$, then:

1. $\displaystyle \int_a^a f(x)\, dx = F(x)\Big|_a^a = F(a) - F(a) = 0$

2. $\displaystyle \int_a^b f(x)\, dx = F(x)\Big|_a^b = F(b) - F(a) = -[F(a) - F(b)] = -\int_b^a f(x)\, dx$

3. $\displaystyle \int_a^b kf(x)\, dx = kF(x)\Big|_a^b = kF(b) - kF(a) = k[F(b) - F(a)] = k\int_a^b f(x)\, dx$

And so on.

◆ **E X A M P L E   13**   Evaluate: $\displaystyle \int_1^2 \left(2x + 3e^x - \frac{4}{x}\right) dx$

Solution
$$\int_1^2 \left(2x + 3e^x - \frac{4}{x}\right) dx = 2\int_1^2 x\, dx + 3\int_1^2 e^x\, dx - 4\int_1^2 \frac{1}{x}\, dx$$

$$= 2\frac{x^2}{2}\Big|_1^2 + 3e^x\Big|_1^2 - 4\ln|x|\Big|_1^2$$

$$= (2^2 - 1^2) + (3e^2 - 3e^1) - (4\ln 2 - 4\ln 1)$$
$$= 3 + 3e^2 - 3e - 4\ln 2 \approx 14.24$$

◆

P R O B L E M   13   Evaluate: $\displaystyle \int_1^3 \left(4x - 2e^x + \frac{5}{x}\right) dx$

◆

◆ DEFINITE INTEGRALS AND SUBSTITUTION

The evaluation of a definite integral is a two-step process: First, find an antiderivative, and then find the net change in that antiderivative. If substitution techniques are required to find the antiderivative, there are two different ways to proceed. The next example illustrates both methods.

◆ E X A M P L E 14   Evaluate: $\displaystyle\int_0^5 \frac{x}{x^2 + 10}\, dx$

Solution   We will solve this problem using substitution in two different ways:

Method 1.   Use substitution in an indefinite integral to find an antiderivative as a function of x; then evaluate the definite integral:

$$\int \frac{x}{x^2 + 10}\, dx = \frac{1}{2}\int \frac{1}{x^2 + 10}\, 2x\, dx \qquad \text{Substitute } u = x^2 + 10 \text{ and}$$
$$du = 2x\, dx.$$
$$= \frac{1}{2}\int \frac{1}{u}\, du$$
$$= \frac{1}{2}\ln|u| + C$$
$$= \frac{1}{2}\ln(x^2 + 10) + C \qquad \text{Since } u = x^2 + 10 > 0$$

We choose $C = 0$ and use the antiderivative $\frac{1}{2}\ln(x^2 + 10)$ to evaluate the definite integral:

$$\int_0^5 \frac{x}{x^2 + 10}\, dx = \frac{1}{2}\ln(x^2 + 10)\Big|_0^5$$
$$= \frac{1}{2}\ln 35 - \frac{1}{2}\ln 10 \approx 0.626$$

Method 2.   Substitute directly in the definite integral, changing both the variable of integration and the limits of integration: In the definite integral

$$\int_0^5 \frac{x}{x^2 + 10}\, dx$$

the upper limit is $x = 5$ and the lower limit is $x = 0$. When we make the substitution $u = x^2 + 10$ in this definite integral, we must change the limits of integration to the corresponding values of u:

$$x = 5 \quad \text{implies} \quad u = 5^2 + 10 = 35 \qquad \text{New upper limit}$$
$$x = 0 \quad \text{implies} \quad u = 0^2 + 10 = 10 \qquad \text{New lower limit}$$

Thus, we have

$$\int_0^5 \frac{x}{x^2 + 10} \, dx = \tfrac{1}{2} \int_0^5 \frac{1}{x^2 + 10} \, 2x \, dx$$

$$= \tfrac{1}{2} \int_{10}^{35} \frac{1}{u} \, du$$

$$= \frac{1}{2} \left(\ln|u| \Big|_{10}^{35} \right)$$

$$= \tfrac{1}{2}(\ln 35 - \ln 10) \approx 0.626 \qquad \blacklozenge$$

P R O B L E M   14   Use both methods described in Example 14 to evaluate

$$\int_0^1 \frac{1}{2x + 4} \, dx \qquad \blacklozenge$$

$\blacklozenge$ E X A M P L E   15   Use method 2 described in Example 14 to evaluate

$$\int_{-4}^1 \sqrt{5 - t} \, dt$$

Solution   If $u = 5 - t$, then $du = -dt$, and

$t = 1$	implies	$u = 5 - 1 = 4$	New upper limit
$t = -4$	implies	$u = 5 - (-4) = 9$	New lower limit

Notice that the lower limit for u is larger than the upper limit. Be careful not to reverse these two values when substituting in the definite integral.

$$\int_{-4}^1 \sqrt{5 - t} \, dt = -\int_{-4}^1 \sqrt{5 - t} \, (-dt)$$

$$= -\int_9^4 \sqrt{u} \, du$$

$$= -\int_9^4 u^{1/2} \, du$$

$$= -\left(\frac{u^{3/2}}{\frac{3}{2}} \Big|_9^4 \right)$$

$$= -[\tfrac{2}{3}(4)^{3/2} - \tfrac{2}{3}(9)^{3/2}]$$

$$= -[\tfrac{16}{3} - \tfrac{54}{3}] = \tfrac{38}{3} \approx 12.667 \qquad \blacklozenge$$

P R O B L E M   15   Use method 2 described in Example 14 to evaluate

$$\int_2^5 \frac{1}{\sqrt{6 - t}} \, dt \qquad \blacklozenge$$

◆ APPLICATIONS

In Section 11-1, we used the indefinite integral of the rate of change (derivative) of a function to find the original function. Now we can use the definite integral of the derivative of a function to find the total net change in the function. That is, the **total change** in a function $F(x)$ as x changes from a to b is

$$\int_a^b F'(x)\, dx = F(x)\Big|_a^b = F(b) - F(a)$$

◆ E X A M P L E   16

Change in Profit

A company manufactures x television sets per month. The monthly marginal profit (in dollars) is given by

$$P'(x) = 165 - 0.1x \qquad 0 \leqslant x \leqslant 4{,}000$$

The company is currently manufacturing 1,500 sets per month, but is planning to increase production. Find the total change in the monthly profit if monthly production is increased to:

(A) 1,600 sets   (B) 2,000 sets

Solutions

(A) $P(1{,}600) - P(1{,}500) = \displaystyle\int_{1{,}500}^{1{,}600} (165 - 0.1x)\, dx$

$$= (165x - 0.05x^2)\Big|_{1{,}500}^{1{,}600}$$

$$= [165(1{,}600) - 0.05(1{,}600)^2]$$
$$\quad - [165(1{,}500) - 0.05(1{,}500)^2]$$
$$= 136{,}000 - 135{,}000$$
$$= 1{,}000$$

Thus, increasing monthly production from 1,500 units to 1,600 units will increase the monthly profit by only $1,000.

(B) $P(2{,}000) - P(1{,}500) = \displaystyle\int_{1{,}500}^{2{,}000} (165 - 0.1x)\, dx$

$$= (165x - 0.05x^2)\Big|_{1{,}500}^{2{,}000}$$

$$= [165(2{,}000) - 0.05(2{,}000)^2]$$
$$\quad - [165(1{,}500) - 0.05(1{,}500)^2]$$
$$= 130{,}000 - 135{,}000 = -5{,}000$$

Thus, increasing monthly production from 1,500 units to 2,000 units will decrease the monthly profit by $5,000. ◆

PROBLEM   16   Repeat Example 16 if

$$P'(x) = 300 - 0.2x \qquad 0 \leq x \leq 3{,}000$$

and monthly production is increased from 1,400 sets to:

(A) 1,500 sets   (B) 1,700 sets ◆

◆ EXAMPLE   17   An amusement company maintains records for each video game it installs in an arcade. Suppose that $C(t)$ and $R(t)$ represent the total accumulated costs and revenues (in thousands of dollars), respectively, t years after a particular game has been installed and that

Useful Life

$$C'(t) = 2 \qquad R'(t) = 9e^{-0.5t}$$

The value of t for which $C'(t) = R'(t)$ is called the **useful life** of the game.

(A) Find the useful life of the game to the nearest year.
(B) Find the total profit accumulated during the useful life of the game.

Solutions   (A) $R'(t) = C'(t)$

$$9e^{-0.5t} = 2$$
$$e^{-0.5t} = \tfrac{2}{9} \qquad \text{Convert to equivalent logarithmic form.}$$
$$-0.5t = \ln \tfrac{2}{9}$$
$$t = -2 \ln \tfrac{2}{9} \approx 3 \text{ years}$$

Thus, the game has a useful life of 3 years.

(B) The total profit accumulated during the useful life of the game is

$$P(3) - P(0) = \int_0^3 P'(t)\, dt$$

$$= \int_0^3 [R'(t) - C'(t)]\, dt$$

$$= \int_0^3 (9e^{-0.5t} - 2)\, dt$$

$$= \left(\frac{9}{-0.5} e^{-0.5t} - 2t \right)\Big|_0^3 \qquad Recall: \quad \int e^{ax}\, dx = \frac{1}{a} e^{ax} + C$$

$$= (-18e^{-0.5t} - 2t)\Big|_0^3$$

$$= (-18e^{-1.5} - 6) - (-18e^0 - 0)$$
$$= 12 - 18e^{-1.5} \approx 7.984 \quad \text{or} \quad \$7{,}984 \qquad ◆$$

P R O B L E M   17   Repeat Example 17 if $C'(t) = 1$ and $R'(t) = 7.5e^{-0.5t}$.   ◆

Common Errors

1. $\displaystyle\int_0^2 e^x\,dx = e^x\Big|_0^2 \neq e^2$

Do not forget to evaluate the antiderivative at both the upper and lower limits of integration, and do not assume that the antiderivative is 0 just because the lower limit is 0. The correct procedure for this integral is

$$\int_0^2 e^x\,dx = e^x\Big|_0^2 = e^2 - e^0 = e^2 - 1$$

2. $\displaystyle\int_2^5 \frac{1}{2x+3}\,dx \neq \frac{1}{2}\int_2^5 \frac{1}{u}\,du \qquad u = 2x + 3,\ du = 2\,dx$

If a substitution is made in a definite integral, the limits of integration also must be changed. The new limits are determined by the particular substitution used in the integral. The correct procedure for this integral is

$$\int_2^5 \frac{1}{2x+3}\,dx = \frac{1}{2}\int_7^{13}\frac{1}{u}\,du \qquad \begin{array}{l} x = 5 \text{ implies } u = 2(5) + 3 = 13 \\ x = 2 \text{ implies } u = 2(2) + 3 = 7 \end{array}$$

$$= \frac{1}{2}\left(\ln|u|\Big|_7^{13}\right)$$

$$= \tfrac{1}{2}(\ln 13 - \ln 7)$$

Answers to Matched Problems   12. -4 13. $16 + 2e - 2e^3 + 5\ln 3 \approx -13.241$
14. $\frac{1}{2}(\ln 6 - \ln 4) \approx 0.203$ 15.   2
16.   (A) \$1,000   (B) $-$\$3,000
17.   (A) $-2\ln\frac{2}{15} \approx 4$ yr   (B) $11 - 15e^{-2} \approx 8.970$ or \$8,970

Evaluate.

A   1. $\displaystyle\int_2^3 2x\,dx$ 2. $\displaystyle\int_1^2 3x^2\,dx$ 3. $\displaystyle\int_3^4 5\,dx$

4. $\displaystyle\int_{12}^{20} dx$ 5. $\displaystyle\int_1^3 (2x - 3)\,dx$ 6. $\displaystyle\int_1^3 (6x + 5)\,dx$

7. $\int_0^4 (3x^2 - 4)\, dx$ **8.** $\int_0^2 (6x^2 - 2x)\, dx$ **9.** $\int_{-3}^4 (4 - x^2)\, dx$

10. $\int_{-1}^2 (x^2 - 4x)\, dx$ **11.** $\int_0^1 24x^{11}\, dx$ **12.** $\int_0^2 30x^5\, dx$

13. $\int_0^1 e^{2x}\, dx$ **14.** $\int_{-1}^1 e^{5x}\, dx$ **15.** $\int_1^{3.5} 2x^{-1}\, dx$

16. $\int_1^2 \dfrac{dx}{x}$

B   **17.** $\int_1^2 (2x^{-2} - 3)\, dx$ **18.** $\int_1^2 (5 - 16x^{-3})\, dx$ **19.** $\int_1^4 3\sqrt{x}\, dx$

20. $\int_4^{25} \dfrac{2}{\sqrt{x}}\, dx$ **21.** $\int_2^3 12(x^2 - 4)^5 x\, dx$ **22.** $\int_0^1 32(x^2 + 1)^7 x\, dx$

23. $\int_3^9 \dfrac{1}{x - 1}\, dx$ **24.** $\int_2^8 \dfrac{1}{x + 1}\, dx$ **25.** $\int_{-5}^{10} e^{-0.05x}\, dx$

26. $\int_{-10}^{25} e^{-0.01x}\, dx$ **27.** $\int_{-6}^0 \sqrt{4 - 2x}\, dx$ **28.** $\int_{-4}^2 \dfrac{1}{\sqrt{8 - 2x}}\, dx$

29. $\int_{-1}^7 \dfrac{x}{\sqrt{x + 2}}\, dx$ **30.** $\int_0^3 x\sqrt{x + 1}\, dx$

31. $\int_0^1 (e^{2x} - 2x)^2 (e^{2x} - 1)\, dx$ **32.** $\int_0^1 \dfrac{2e^{4x} - 3}{e^{2x}}\, dx$

33. $\int_{-2}^{-1} (x^{-1} + 2x)\, dx$ **34.** $\int_{-3}^{-1} (-3x^{-2} + x^{-1})\, dx$

C   **35.** $\int_2^3 x\sqrt{2x^2 - 3}\, dx$ **36.** $\int_0^1 x\sqrt{3x^2 + 2}\, dx$

37. $\int_0^1 \dfrac{x - 1}{x^2 - 2x + 3}\, dx$ **38.** $\int_1^2 \dfrac{x + 1}{2x^2 + 4x + 4}\, dx$

39. $\int_{-1}^1 \dfrac{e^{-x} - e^x}{(e^{-x} + e^x)^2}\, dx$ **40.** $\int_6^7 \dfrac{\ln(t - 5)}{t - 5}\, dt$

APPLICATIONS

Business & Economics   **41.** *Salvage value.* A new piece of industrial equipment will depreciate in value rapidly at first, then less rapidly as time goes on. Suppose the rate (in dollars per year) at which the book value of a new milling machine changes is given approximately by

$$V'(t) = f(t) = 500(t - 12) \qquad 0 \leqslant t \leqslant 10$$

where $V(t)$ is the value of the machine after t years. What is the total loss in value of the machine in the first 5 years? In the second 5 years? Set up appropriate integrals and solve.

42. *Maintenance costs.* Maintenance costs for an apartment house generally increase as the building gets older. From past records, a managerial service determines that the rate of increase in maintenance costs (in dollars per year) for a particular apartment complex is given approximately by

$$M'(x) = f(x) = 90x^2 + 5,000$$

where x is the age of the apartment complex in years and $M(x)$ is the total (accumulated) cost of maintenance for x years. Write a definite integral that will give the total maintenance costs from 2 to 7 years after the apartment complex was built, and evaluate it.

43. *Useful life.* The total accumulated costs $C(t)$ and revenues $R(t)$ (in thousands of dollars), respectively, for a coin-operated photocopying machine satisfy

$$C'(t) = \tfrac{1}{11}t \quad \text{and} \quad R'(t) = 5te^{-t^2}$$

where t is time in years. Find the useful life of the machine to the nearest year. What is the total profit accumulated during the useful life of the machine?

44. *Useful life.* The total accumulated costs $C(t)$ and revenues $R(t)$ (in thousands of dollars), respectively, for a coal mine satisfy

$$C'(t) = 3 \quad \text{and} \quad R'(t) = 15e^{-0.1t}$$

where t is the number of years the mine has been in operation. Find the useful life of the mine to the nearest year. What is the total profit accumulated during the useful life of the mine?

45. *Labor costs and learning.* A defense contractor is starting production on a new missile control system. On the basis of data collected while assembling the first 16 control systems, the production manager obtained the following function for rate of labor use:

$$g(x) = 2,400x^{-1/2}$$

where $g(x)$ is the number of labor-hours required to assemble the xth unit of a control system. Approximately how many labor-hours will be required to assemble the 17th through the 25th control units? [*Hint:* Let $a = 16$ and $b = 25$.]

46. *Labor costs and learning.* If the rate of labor use in Problem 45 is

$$g(x) = 2,000x^{-1/3}$$

approximately how many labor-hours will be required to assemble the 9th through the 27th control units? [*Hint:* Let $a = 8$ and $b = 27$.]

47. *Oil production.* Using data from the first 3 years of production as well as geological studies, the management of an oil company estimates that oil

will be pumped from a producing field at a rate given by

$$R(t) = \frac{100}{t+1} + 5 \qquad 0 \le t \le 20$$

where $R(t)$ is the rate of production (in thousands of barrels per year) t years after pumping begins. Approximately how many barrels of oil will the field produce during the first 10 years of production? From the end of the 10th year to the end of the 20th year of production?

48. *Oil production.* In Problem 47, if the rate is found to be

$$R(t) = \frac{120t}{t^2 + 1} + 3 \qquad 0 \le t \le 20$$

approximately how many barrels of oil will the field produce during the first 5 years of production? The second 5 years of production?

49. *Marketing.* An automobile company is ready to introduce a new line of cars with a national sales campaign. After test marketing the line in a carefully selected city, the marketing research department estimates that sales (in millions of dollars) will increase at the monthly rate of

$$S'(t) = 10 - 10e^{-0.1t} \qquad 0 \le t \le 24$$

t months after the national campaign is started. What will be the approximate total sales during the first 12 months of the campaign? The second 12 months of the campaign?

50. *Marketing.* Repeat Problem 49 if the monthly rate of increase in sales is found to be approximated by

$$S'(t) = 20 - 20e^{-0.05t} \qquad 0 \le t \le 24$$

Life Sciences

51. *Natural resource depletion.* The instantaneous rate of change of demand for wood in the United States since 1970 ($t = 0$) in billions of cubic feet per year is estimated to be given by

$$Q'(t) = 12 + 0.006t^2 \qquad 0 \le t \le 50$$

where $Q(t)$ is the total amount of wood consumed t years after 1970. How many billions of cubic feet of wood will have been consumed from 1980 to 1990?

52. *Natural resource depletion.* Repeat Problem 51 for the time interval from 1990 to 2000.

53. *Biology.* A yeast culture weighing 2 grams is removed from a refrigerator unit and is expected to grow at the rate of $W'(t) = 0.2e^{0.1t}$ grams per hour at a higher controlled temperature. How much will the weight of the culture increase during the first 8 hours of growth? How much will the weight of

the culture increase from the end of the 8th hour to the end of the 16th hour of growth?

54. *Medicine.* The rate of healing for a skin wound (in square centimeters per day) is approximated by $A'(t) = -0.9e^{-0.1t}$. The initial wound has an area of 9 square centimeters. How much will the area change during the first 5 days? The second 5 days?

Social Sciences

55. *Learning.* In a particular business college, it was found that an average student enrolled in an advanced typing class progressed at a rate of $N'(t) = 6e^{-0.1t}$ words per minute per week, t weeks after enrolling in a 15 week course. At the beginning of the course, an average student could type 40 words per minute. How much improvement would be expected during the first 5 weeks of the course? The second 5 weeks of the course? The last 5 weeks of the course?

56. *Learning.* In the same business college, it was also found that an average student enrolled in a beginning shorthand class progressed at a rate of $N'(t) = 12e^{-0.06t}$ words per minute per week, t weeks after enrolling in a 15 week course. At the beginning of the course, none of the students could take any dictation by shorthand. How much improvement would be expected during the first 5 weeks of the course? The second 5 weeks of the course? The last 5 weeks of the course?

SECTION 11-4 Area and the Definite Integral

◆ AREA UNDER A CURVE

◆ AREA BETWEEN A CURVE AND THE X AXIS

◆ AREA BETWEEN TWO CURVES

◆ APPLICATION: DISTRIBUTION OF INCOME

◆ AREA UNDER A CURVE

Consider the graph of $f(x) = x$ from $x = 0$ to $x = 4$ (Fig. 2). We can easily compute the area of the triangle bounded by $f(x) = x$, the x axis ($y = 0$), and the line $x = 4$, using the formula for the area of a triangle:

$$A = \frac{bh}{2} = \frac{4 \cdot 4}{2} = 8$$

Now, let us integrate $f(x) = x$ from $x = 0$ to $x = 4$:

$$\int_0^4 x \, dx = \frac{x^2}{2}\Big|_0^4 = \frac{(4)^2}{2} - \frac{(0)^2}{2} = 8$$

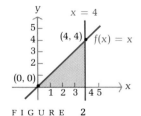

FIGURE 2

We get the same result! It turns out that this is not a coincidence. In general, we can prove the following:

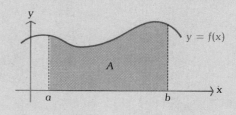

Area under a Curve

If f is continuous and $f(x) \geq 0$ over the interval $[a, b]$, then the area bounded by $y = f(x)$, the x axis $(y = 0)$, and the vertical lines $x = a$ and $x = b$ is given exactly by

$$A = \int_a^b f(x) \, dx$$

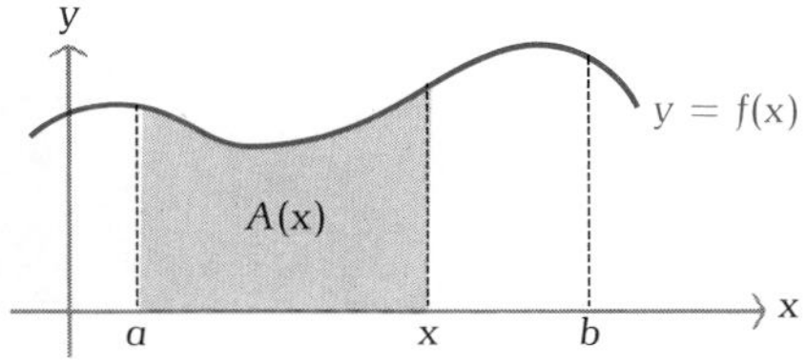

Let us see why the definite integral gives us the area exactly. Let $A(x)$ be the area under the graph of $y = f(x)$ from a to x, as indicated in Figure 3. If we can show that $A(x)$ is an antiderivative of $f(x)$, then we can write

$$\int_a^b f(x) \, dx = A(x) \Big|_a^b = A(b) - A(a)$$

$$= \left(\begin{array}{c} \text{Area from} \\ x = a \text{ to } x = b \end{array} \right) - \left(\begin{array}{c} \text{Area from} \\ x = a \text{ to } x = a \end{array} \right)$$

$$= A - 0 = A$$

To show that $A(x)$ is an antiderivative of $f(x)$—that is, $A'(x) = f(x)$—we use the definition of a derivative (Section 9-3) and write

$$A'(x) = \lim_{h \to 0} \frac{A(x + h) - A(x)}{h}$$

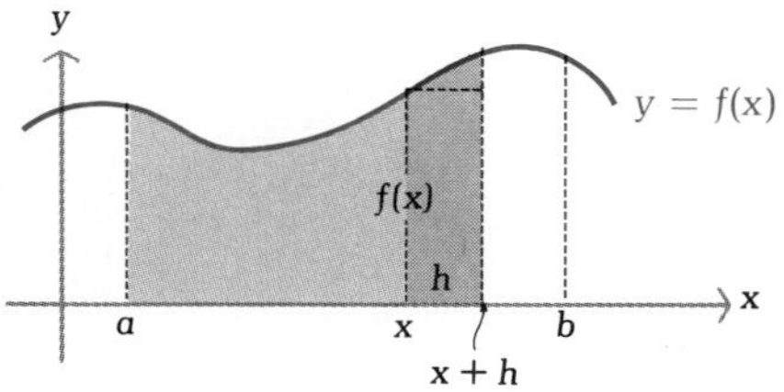

Geometrically, $A(x + h) - A(x)$ is the area from x to $x + h$ (see Fig. 4). This area is given approximately by the area of the rectangle $h \cdot f(x)$, and the smaller h is, the better the approximation. Using

$$A(x + h) - A(x) \approx h \cdot f(x)$$

and dividing both sides by h, we obtain

$$\frac{A(x + h) - A(x)}{h} \approx f(x)$$

Now, if we let $h \to 0$, then the left side has $A'(x)$ as a limit, which is equal to the right side. Hence,

$$A'(x) = f(x)$$

FIGURE 3
$A(x) =$ Area from a to x

FIGURE 4

that is, $A(x)$ is an antiderivative of $f(x)$. Thus,

$$\int_a^b f(x)\,dx = A(x)\Big|_a^b = A(b) - A(a) = A - 0 = A$$

This is a remarkable result: The area under the graph of $y = f(x)$, $f(x) \geq 0$, can be obtained simply by evaluating the antiderivative of $f(x)$ at the end points of the interval $[a, b]$. We have now solved, at least in part, the third basic problem of calculus stated at the beginning of Chapter 9.

◆ E X A M P L E 18

Find the area bounded by $f(x) = 6x - x^2$ and $y = 0$ for $1 \leq x \leq 4$.

Solution

We sketch a graph of the region first. (The solution of every area problem should begin with a sketch.) Then,

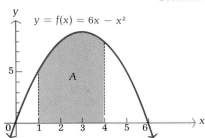

$$A = \int_1^4 (6x - x^2)\,dx = \left(3x^2 - \frac{x^3}{3}\right)\Big|_1^4$$

$$= \left[3(4)^2 - \frac{(4)^3}{3}\right] - \left[3(1)^2 - \frac{(1)^3}{3}\right]$$

$$= 48 - \frac{64}{3} - 3 + \frac{1}{3}$$

$$= 48 - 21 - 3 = 24 \qquad ◆$$

P R O B L E M 18

Find the area bounded by $f(x) = x^2 + 1$ and $y = 0$ for $-1 \leq x \leq 3$. ◆

◆ AREA BETWEEN A CURVE AND THE x AXIS

The condition $f(x) \geq 0$ is essential to the relationship between an area under a graph and the definite integral. How can we find the area between the graph of f and the x axis if $f(x) \leq 0$ on $[a, b]$ or if $f(x)$ is both positive and negative on $[a, b]$? To begin, suppose $f(x) \leq 0$ and A is the area between the graph of f and the x axis for $a \leq x \leq b$, as illustrated in Figure 5A.

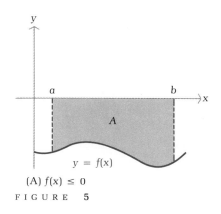

(A) $f(x) \leq 0$ (B) $g(x) \geq 0$

F I G U R E 5

If we let $g(x) = -f(x)$, then A is also the area between the graph of g and the x axis for $a \leqslant x \leqslant b$ (see Fig. 5B). Since $g(x) \geqslant 0$ for $a \leqslant x \leqslant b$, we can use the definite integral of g to find A:

$$A = \int_a^b g(x)\, dx = \int_a^b [-f(x)]\, dx$$

Thus, **the area between the graph of a negative function and the x axis is equal to the definite integral of the negative of the function.** Finally, if $f(x)$ is positive for some values of x and negative for others, the area between the graph of f and the x axis can be obtained by dividing $[a, b]$ into intervals over which f is always positive or always negative, finding the area over each interval, and then summing these areas.

♦ E X A M P L E   19   Find the area between the graph of $f(x) = x^2 - 2x$ and the x axis over the indicated int█████████████

(A) $[1, 2]$ (B) $[-1, 1]$

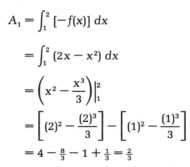

Solutions   We begin by sketching the graph of f, as shown in the margin.

(A) From the graph, we see that $f(x) \leqslant 0$ for $1 \leqslant x \leqslant 2$, so we integrate $-f(x)$:

$$A_1 = \int_1^2 [-f(x)]\, dx$$

$$= \int_1^2 (2x - x^2)\, dx$$

$$= \left(x^2 - \frac{x^3}{3} \right)\Big|_1^2$$

$$= \left[(2)^2 - \frac{(2)^3}{3} \right] - \left[(1)^2 - \frac{(1)^3}{3} \right]$$

$$= 4 - \tfrac{8}{3} - 1 + \tfrac{1}{3} = \tfrac{2}{3}$$

(B) Since the graph shows that $f(x) \geqslant 0$ on $[-1, 0]$ and $f(x) \leqslant 0$ on $[0, 1]$, the computation of this area will require two integrals:

$$A = A_2 + A_3$$

$$= \int_{-1}^0 f(x)\, dx + \int_0^1 [-f(x)]\, dx$$

$$= \int_{-1}^0 (x^2 - 2x)\, dx + \int_0^1 (2x - x^2)\, dx$$

$$= \left(\frac{x^3}{3} - x^2 \right)\Big|_{-1}^0 + \left(x^2 - \frac{x^3}{3} \right)\Big|_0^1$$

$$= \tfrac{4}{3} + \tfrac{2}{3} = 2$$

♦

PROBLEM   19   Find the area between the graph of $f(x) = x^2 - 9$ and the x axis over the indicated intervals:

(A) [0, 2]   (B) [2, 4]   ◆

◆ AREA BETWEEN TWO CURVES

Consider the area bounded by $y = f(x)$ and $y = g(x)$, where $f(x) \geq g(x) \geq 0$, for
$a \leq x \leq b$, as indicated in Figure 6.

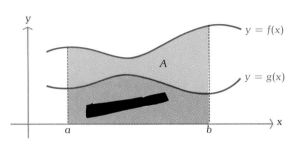

FIGURE   6

$$\begin{pmatrix} \text{Area A between} \\ f(x) \text{ and } g(x) \end{pmatrix} = \begin{pmatrix} \text{Area} \\ \text{under } f(x) \end{pmatrix} - \begin{pmatrix} \text{Area} \\ \text{under } g(x) \end{pmatrix}$$ Areas are from $x = a$
to $x = b$ above the x axis.

$$= \int_a^b f(x)\,dx - \int_a^b g(x)\,dx$$ Use definite integral
property 4 (Section 11-3).

$$= \int_a^b [f(x) - g(x)]\,dx$$

It can be shown that the above result does not require $f(x)$ or $g(x)$ to remain
positive over the interval $[a, b]$. A more general result is stated in the box:

■ **Area between Two Curves**

If f and g are continuous and $f(x) \geq g(x)$ over the interval $[a, b]$, then the
area bounded by $y = f(x)$ and $y = g(x)$ for $a \leq x \leq b$ is given exactly by

$$A = \int_a^b [f(x) - g(x)]\,dx$$

◆ E X A M P L E   20   Find the area bounded by $f(x) = \frac{1}{2}x + 3$, $g(x) = -x^2 + 1$, $x = -2$, and $x = 1$.

Solution   We first sketch the area, then set up and evaluate an appropriate definite integral:

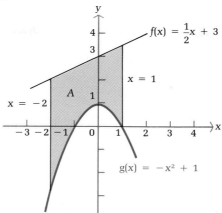

We observe from the graph that $f(x) \geqslant g(x)$ for $-2 \leqslant x \leqslant 1$, so

$$A = \int_{-2}^{1} [f(x) - g(x)]\, dx = \int_{-2}^{1} \left[\left(\frac{x}{2} + 3 \right) - (-x^2 + 1) \right] dx$$

$$= \int_{-2}^{1} \left(x^2 + \frac{x}{2} + 2 \right) dx$$

$$= \left(\frac{x^3}{3} + \frac{x^2}{4} + 2x \right)\Big|_{-2}^{1} = \left(\frac{1}{3} + \frac{1}{4} + 2 \right) - \left(\frac{-8}{3} + \frac{4}{4} - 4 \right) = \frac{33}{4}$$ ◆

P R O B L E M   20   Find the area bounded by $f(x) = x^2 - 1$, $g(x) = -\frac{1}{2}x - 3$, $x = -1$, and $x = 2$.   ◆

◆ E X A M P L E   21   Find the area bounded by $f(x) = 5 - x^2$ and $g(x) = 2 - 2x$.

Solution   First, graph f and g on the same coordinate system, as shown in the margin. Since the statement of the problem does not include any limits on the values of x, we must determine the appropriate values from the graph. The graph of f is a parabola and the graph of g is a line, as shown in the figure. The area bounded by these two graphs extends from the intersection point on the left to the intersection point on the right. To find these intersection points, we solve the equation $f(x) = g(x)$ for x:

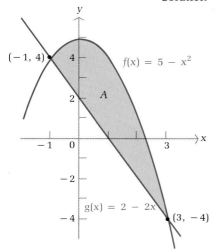

$$f(x) = g(x)$$
$$5 - x^2 = 2 - 2x$$
$$x^2 - 2x - 3 = 0$$
$$x = -1, 3$$

(Note that the area between the graphs for $x < -1$ is unbounded on the left, and the area between the graphs for $x > 3$ is unbounded on the right.) The figure shows that $f(x) \geq g(x)$ over the interval $[-1, 3]$, so we have

$$A = \int_{-1}^{3} [f(x) - g(x)] \, dx = \int_{-1}^{3} [5 - x^2 - (2 - 2x)] \, dx$$

$$= \int_{-1}^{3} (3 + 2x - x^2) \, dx$$

$$= \left(3x + x^2 - \frac{x^3}{3} \right) \Big|_{-1}^{3}$$

$$= \left[3(3) + (3)^2 - \frac{(3)^3}{3} \right] - \left[3(-1) + (-1)^2 - \frac{(-1)^3}{3} \right]$$

$$= 9 + 9 - 9 + 3 - 1 - \tfrac{1}{3} = \tfrac{32}{3} \qquad \blacklozenge$$

PROBLEM   21   Find the area bounded by $f(x) = 6 - x^2$ and $g(x) = x$. $\qquad \blacklozenge$

$\blacklozenge$ EXAMPLE   22   Find the area bounded by $f(x) = x^2 - x$ and $g(x) = 2x$ for $-2 \leq x \leq 3$.

Solution   The graphs of f and g are shown in the figure. Examining the graph, we see that $f(x) \geq g(x)$ on the interval $[-2, 0]$, but $g(x) \geq f(x)$ on the interval $[0, 3]$. Thus, two integrals are required to compute this area:

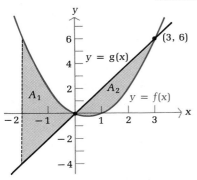

$$A_1 = \int_{-2}^{0} [f(x) - g(x)] \, dx \qquad f(x) \geq g(x) \text{ on } [-2, 0]$$

$$= \int_{-2}^{0} [x^2 - x - 2x] \, dx$$

$$= \int_{-2}^{0} (x^2 - 3x) \, dx$$

$$= \left(\frac{x^3}{3} - \frac{3}{2} x^2 \right) \Big|_{-2}^{0}$$

$$= (0) - \left[\frac{(-2)^3}{3} - \frac{3}{2} (-2)^2 \right]$$

$$= \tfrac{8}{3} + 6 = \tfrac{26}{3}$$

$$A_2 = \int_{0}^{3} [g(x) - f(x)] \, dx \qquad g(x) \geq f(x) \text{ on } [0, 3]$$

$$= \int_{0}^{3} [2x - (x^2 - x)] \, dx$$

$$= \int_{0}^{3} (3x - x^2) \, dx$$

$$= \left(\frac{3}{2} x^2 - \frac{x^3}{3} \right) \Big|_{0}^{3}$$

$$= \left[\frac{3}{2} (3)^2 - \frac{(3)^3}{3} \right] - (0)$$

$$= \tfrac{27}{2} - 9 = \tfrac{9}{2}$$

The total area between the two graphs is

$$A = A_1 + A_2 = \tfrac{26}{3} + \tfrac{9}{2} = \tfrac{79}{6}$$ ◆

PROBLEM 22   Find the area bounded by $f(x) = 2x^2$ and $g(x) = 4 - 2x$ for $-2 \leqslant x \leqslant 2$.   ◆

◆ APPLICATION: DISTRIBUTION OF INCOME

Economists often use a graph called a **Lorenz curve** to provide a graphical description of the distribution of income among various groups of people. For example, the distribution of personal income in the United States given in Table 1 can be represented by the Lorenz curve shown in Figure 7. In both Table 1 and Figure 7, **x represents the cumulative percentage of families at or below a given income level** and **y represents the cumulative percentage of total personal income received by all these families.** For example, the point $(0.51, 0.24)$ on the Lorenz curve in Figure 7 indicates that the bottom 51% of families (those with incomes under $25,000) received 24% of the total income, the point $(0.71, 0.44)$ indicates that the bottom 71% of families received 44% of the total income, and so on.

Absolute equality of income would occur if every family received the same income. That is, 20% of the families receive 20% of the income, 40% of the families receive 40% of the income, and so on. This distribution is represented by the graph of $y = x$ in Figure 7. The area between the Lorenz curve and the line $y = x$ can be used to measure how much the distribution of income differs from absolute equality.

More precisely, the **coefficient of inequality** of income distribution is defined to be the ratio of the area between the line $y = x$ and the Lorenz curve to the area between the line $y = x$ and the x axis. If we are given a function f whose graph is a Lorenz curve, then the area between the line $y = x$ and the Lorenz curve is $\int_0^1 [x - f(x)]\, dx$. Since the area between the graph of the line $y = x$ and the x axis from $x = 0$ to $x = 1$ is $\tfrac{1}{2}$, it follows that the coefficient of inequality is given by

TABLE 1

Income Distribution in the United States (1983)

INCOME LEVEL	x	y
Under $10,000	0.16	0.03
Under $15,000	0.28	0.08
Under $25,000	0.51	0.24
Under $35,000	0.71	0.44
Under $50,000	0.88	0.69
Under $75,000	0.97	0.88

Source: U.S. Bureau of the Census

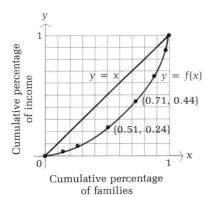

FIGURE 7
Lorenz curve, $y = f(x)$

$$\frac{\text{Area between } y = x \text{ and } y = f(x)}{\text{Area between } y = x \text{ and } x \text{ axis}} = \frac{\int_0^1 [x - f(x)]\, dx}{\tfrac{1}{2}} = 2 \int_0^1 [x - f(x)]\, dx$$

Coefficient of Inequality

If $y = f(x)$ is the equation of a Lorenz curve, then

$$\text{Coefficient of inequality} = 2 \int_0^1 [x - f(x)]\, dx$$

The coefficient of inequality is always a number between 0 and 1. If the coefficient is near 0, then the income distribution is close to absolute equality. The closer the coefficient is to 1, the greater the inequality (or disparity) of the distribution. The coefficient of inequality can be used to compare income distributions at various points in time, between different groups of people, before and after taxes are paid, between different countries, and so on.

◆ E X A M P L E   23

Distribution of Income

The Lorenz curve for the distribution of income in a certain country in 1990 is given by $f(x) = x^{2.6}$. Economists predict that the Lorenz curve for the country in the year 2010 will be given by $g(x) = x^{1.8}$. Find the coefficient of inequality for each curve, and interpret the results.

Solution

The Lorenz curves are shown in the figure below:

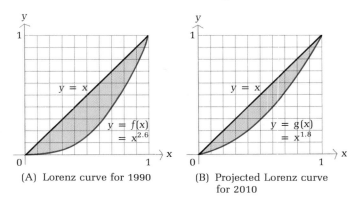

(A) Lorenz curve for 1990   (B) Projected Lorenz curve for 2010

The coefficient of inequality in 1990 is (see figure A)

$$2 \int_0^1 [x - f(x)]\, dx = 2 \int_0^1 [x - x^{2.6}]\, dx$$

$$= 2 \left(\frac{1}{2} x^2 - \frac{1}{3.6} x^{3.6} \right) \Big|_0^1$$

$$= 2 \left(\frac{1}{2} - \frac{1}{3.6} \right) \approx 0.444$$

The projected coefficient of inequality in 2010 is (see figure B)

$$2 \int_0^1 [x - g(x)]\, dx = 2 \int_0^1 [x - x^{1.8}]\, dx$$

$$= 2 \left(\frac{1}{2} x^2 - \frac{1}{2.8} x^{2.8} \right) \Big|_0^1$$

$$= 2 \left(\frac{1}{2} - \frac{1}{2.8} \right) \approx 0.286$$

If this projection is correct, the coefficient of inequality will decrease, and income will be more equally distributed in the year 2010 than in 1990. ◆

PROBLEM 23   Repeat Example 23 if the projected Lorenz curve in the year 2010 is given by $g(x) = x^{3.8}$.   ◆

Answers to Matched Problems

18. $A = \int_{-1}^{3} (x^2 + 1)\, dx = \frac{40}{3}$

19. (A) $A = \int_{0}^{2} (9 - x^2)\, dx = \frac{46}{3}$
     (B) $A = \int_{2}^{3} (9 - x^2)\, dx + \int_{3}^{4} (x^2 - 9)\, dx = 6$

20. $A = \int_{-1}^{2} \left[(x^2 - 1) - \left(-\frac{x}{2} - 3 \right) \right] dx = \frac{39}{4}$

21. $A = \int_{-3}^{2} [(6 - x^2) - x]\, dx = \frac{125}{6}$

22. $A = \int_{-2}^{1} [(4 - 2x) - 2x^2]\, dx + \int_{1}^{2} [2x^2 - (4 - 2x)]\, dx = \frac{38}{3}$

23. Coefficient of inequality ≈ 0.583; income will be less equally distributed in 2010.

Find the area bounded by the graphs of the indicated equations.

A
1. $y = 2x + 4$, $y = 0$, $1 \leq x \leq 3$
2. $y = -2x + 6$, $y = 0$, $0 \leq x \leq 2$
3. $y = 3x^2$ $y = 0$, $1 \leq x \leq 2$
4. $y = 4x^3$, $y = 0$, $1 \leq x \leq 2$
5. $y = x^2 + 2$, $y = 0$, $-1 \leq x \leq 0$
6. $y = 3x^2 + 1$, $y = 0$, $-2 \leq x \leq 0$
7. $y = 4 - x^2$, $y = 0$, $-1 \leq x \leq 2$
8. $y = 12 - 3x^2$, $y = 0$, $-2 \leq x \leq 1$
9. $y = e^x$, $y = 0$, $-1 \leq x \leq 2$
10. $y = e^{-x}$, $y = 0$, $-2 \leq x \leq 1$
11. $y = 1/t$, $y = 0$, $0.5 \leq t \leq 1$
12. $y = 1/t$, $y = 0$, $0.1 \leq t \leq 1$

B
13. $y = 12$, $y = -2x + 8$, $-1 \leq x \leq 2$
14. $y = 3$, $y = 2x + 6$, $-1 \leq x \leq 2$
15. $y = 3x^2$, $y = 12$
16. $y = x^2$, $y = 9$
17. $y = 4 - x^2$, $y = -5$
18. $y = x^2 - 1$, $y = 3$
19. $y = x^2 + 1$, $y = 2x - 2$, $-1 \leq x \leq 2$
20. $y = x^2 - 1$, $y = x - 2$, $-2 \leq x \leq 1$
21. $y = -x$, $y = 0$, $-2 \leq x \leq 1$
22. $y = -x + 1$, $y = 0$, $-1 \leq x \leq 2$
23. $y = e^{0.5x}$, $y = -1/x$, $1 \leq x \leq 2$
24. $y = 1/x$, $y = -e^x$, $0.5 \leq x \leq 1$

C

25. $y = x^2 - 4$, $y = 0$, $0 \le x \le 3$
26. $y = 4\sqrt[3]{x}$, $y = 0$, $-1 \le x \le 8$
27. $y = 10 - 2x$, $y = 4 + 2x$, $0 \le x \le 4$
28. $y = 3x$ $y = x + 5$, $0 \le x \le 5$
29. $y = 5x - x^2$, $y = x + 3$, $0 \le x \le 5$
30. $y = x^2 - 6x + 9$, $y = 5 - x$, $0 \le x \le 5$
31. $y = x^2 + 2x + 3$, $y = 2x + 4$
32. $y = 8 + 4x - x^2$, $y = x^2 - 2x$
33. $y = x^2 - 4x - 10$, $y = 14 - 2x - x^2$
34. $y = 6 + 6x - x^2$, $y = 13 - 2x$
35. $y = x^3$, $y = 4x$
36. $y = x^3 + 1$, $y = x + 1$
37. $y = x^3 - 3x^2 - 9x + 12$, $y = x + 12$
38. $y = x^3 - 6x^2 + 9x$, $y = x$

APPLICATIONS

Business & Economics

Photo courtesy of Shell Oil Company

39. *Oil production.* Using data from the first 3 years of production as well as geological studies, the management of an oil company estimates that oil will be pumped from a producing field at a rate given by

$$R(t) = \frac{100}{t + 10} + 10 \qquad 0 \le t \le 15$$

where $R(t)$ is the rate of production (in thousands of barrels per year) t years after pumping begins. Find the area between the graph of R and the t axis over the interval [5, 10] and interpret the results.

40. *Oil production.* In Problem 39, if the rate is found to be

$$R(t) = \frac{100t}{t^2 + 25} + 4 \qquad 0 \le t \le 25$$

find the area between the graph of R and the t axis over the interval [5, 15] and interpret the results.

41. *Useful life.* An amusement company maintains records for each video game it installs in an arcade. Suppose that $C(t)$ and $R(t)$ represent the total accumulated costs and revenues (in thousands of dollars), respectively, t years after a particular game has been installed. If

$$C'(t) = 2 \qquad \text{and} \qquad R'(t) = 9e^{-0.3t}$$

find the area between the graphs of C' and R' over the interval on the t axis from 0 to the useful life of the game and interpret the results.

42. *Useful life.* Repeat Problem 41 if

$$C'(t) = 2t \qquad \text{and} \qquad R'(t) = 5te^{-0.1t^2}$$

43. *Income distribution.* As part of a study of the effects of World War II on the economy of the United States, an economist used data from the U.S. Bureau of the Census to produce the following Lorenz curves for distribution of income in the United States in 1935 and in 1947:

$$f(x) = x^{2.4} \qquad \text{Lorenz curve for 1935}$$
$$g(x) = x^{1.6} \qquad \text{Lorenz curve for 1947}$$

Find the coefficient of inequality for each Lorenz curve and interpret the results.

44. *Income distribution.* Using data from the U.S. Bureau of the Census, an economist produced the following Lorenz curves for distribution of income in the United States in 1962 and in 1972:

$$f(x) = \tfrac{3}{10}x + \tfrac{7}{10}x^2 \qquad \text{Lorenz curve for 1962}$$
$$g(x) = \tfrac{1}{2}x + \tfrac{1}{2}x^2 \qquad \text{Lorenz curve for 1972}$$

Find the coefficient of inequality for each Lorenz curve and interpret the results.

45. *Distribution of wealth.* Lorenz curves also can be used to provide a relative measure of the distribution of the total assets of a country. Using data in a report by the U.S. Congressional Joint Economic Committee, an economist produced the following Lorenz curves for the distribution of total assets in the United States in 1963 and in 1983:

$$f(x) = x^{10} \qquad \text{Lorenz curve for 1963}$$
$$g(x) = x^{12} \qquad \text{Lorenz curve for 1983}$$

Find the coefficient of inequality for each Lorenz curve and interpret the results.

46. *Income distribution.* The government of a small country is planning sweeping changes in the tax structure in order to provide a more equitable distribution of income. The Lorenz curves for the current income distribution and for the projected income distribution after enactment of the tax changes are given below. Find the coefficient of inequality for each Lorenz curve. Will the proposed changes provide a more equitable income distribution?

$$f(x) = x^{2.3} \qquad\qquad \text{Current Lorenz curve}$$
$$g(x) = 0.4x + 0.6x^2 \qquad \text{Projected Lorenz curve after changes in tax laws}$$

Life Sciences 47. *Biology.* A yeast culture is growing at a rate of $W'(t) = 0.3e^{0.1t}$ grams per hour. Find the area between the graph of W' and the t axis over the interval $[0, 10]$ and interpret the results.

48. *Natural resource depletion.* The instantaneous rate of change of the demand for lumber in the United States since 1970 ($t = 0$) in billions of cubic feet per year is estimated to be given by

$$Q'(t) = 12 + 0.006t^2 \qquad 0 \leqslant t \leqslant 50$$

Find the area between the graph of Q' and the t axis over the interval [15, 20] and interpret the results.

Social Sciences   **49.** *Learning.* A beginning high school language class was chosen for an experiment on learning. Using a list of 50 words, the experiment involved measuring the rate of vocabulary memorization at different times during a continuous 5 hour study session. It was found that the average rate of learning for the whole class was inversely proportional to the time spent studying and was given approximately by

$$V'(t) = \frac{15}{t} \qquad 1 \leqslant t \leqslant 5$$

Find the area between the graph of V' and the t axis over the interval [2, 4] and interpret the results.

SECTION   11-5   **Definite Integral as a Limit of a Sum**

◆ RECTANGLE RULE FOR APPROXIMATING DEFINITE INTEGRALS
◆ DEFINITE INTEGRAL AS A LIMIT OF A SUM
◆ RECOGNIZING A DEFINITE INTEGRAL
◆ AVERAGE VALUE OF A CONTINUOUS FUNCTION

Up to this point, in order to evaluate a definite integral

$$\int_a^b f(x)\, dx$$

we need to find an antiderivative of the function f so that we can write

$$\int_a^b f(x)\, dx = F(x)\, \Big|_a^b = F(b) - F(a) \qquad F'(x) = f(x)$$

But suppose we cannot find an antiderivative of f (it may not even exist in a convenient or closed form). For example, how would you evaluate the following?

$$\int_2^8 \sqrt{x^3 + 1}\, dx \qquad \int_1^5 \left(\frac{x}{x+1}\right)^3 dx \qquad \int_0^5 e^{-x^2}\, dx$$

We now introduce the *rectangle rule* for approximating definite integrals, and out of this discussion will evolve a new way of looking at definite integrals.

◆ RECTANGLE RULE FOR APPROXIMATING DEFINITE INTEGRALS

In Section 11-4, we saw that any definite integral of a positive continuous function f over an interval $[a, b]$ always can be interpreted as the area bounded by $y = f(x)$, $y = 0$, $x = a$, and $x = b$ (see Fig. 8). What we need is a way of approximating such areas, given $y = f(x)$ and an interval $[a, b]$.

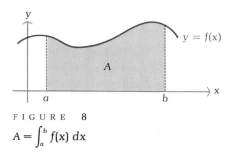

FIGURE 8

$$A = \int_a^b f(x)\, dx$$

Let us start with a concrete example and generalize from the experience. We begin with a simple definite integral we can evaluate exactly:

$$\int_1^5 (x^2 + 3)\, dx = \left(\frac{x^3}{3} + 3x\right)\Big|_1^5$$

$$= \left[\frac{(5)^3}{3} + 3(5)\right] - \left[\frac{(1)^3}{3} + 3(1)\right]$$

$$= \left(\frac{125}{3} + 15\right) - \left(\frac{1}{3} + 3\right)$$

$$= \frac{160}{3} = 53\frac{1}{3}$$

This integral represents the area bounded by $y = x^2 + 3$ and $y = 0$, $1 \le x \le 5$, as indicated in Figure 9.

Since areas of rectangles are easy to compute, we cover the area in Figure 9 with rectangles (see Fig. 10) so that the top of each rectangle has a point in common with the graph of $y = f(x)$. As our first approximation, we divide the interval $[1, 5]$ into two equal subintervals, each with length $(b - a)/2 = (5 - 1)/2 = 2$, and use the midpoint of each subinterval to compute the altitude of the rectangle sitting on top of that subinterval (see Fig. 10):

$$\int_1^5 (x^2 + 3)\, dx \approx A_1 + A_2$$

$$= f(2) \cdot 2 + f(4) \cdot 2$$
$$= 2[f(2) + f(4)]$$
$$= 2(7 + 19) = 52$$

This approximation is less than 3% off the exact area we found above ($53\frac{1}{3}$).

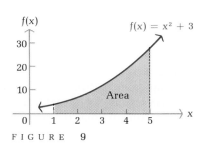

FIGURE 9

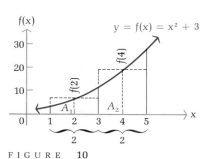

FIGURE 10

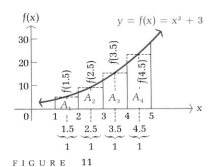

$f(x)$

$y = f(x) = x^2 + 3$

FIGURE 11

Now let us divide the interval [1, 5] into four equal subintervals, each of length $(b - a)/4 = (5 - 1)/4 = 1$, and use the midpoint* of each subinterval to compute the altitude of the rectangle corresponding to that subinterval (see Fig. 11):

$$\int_1^5 (x^3 + 3)\, dx \approx A_1 + A_2 + A_3 + A_4$$

$$= f(1.5) \cdot 1 + f(2.5) \cdot 1 + f(3.5) \cdot 1 + f(4.5) \cdot 1$$
$$= f(1.5) + f(2.5) + f(3.5) + f(4.5)$$
$$= 5.25 + 9.25 + 15.25 + 23.25 = 53$$

Now we are less than 1% off the exact area $(53\frac{1}{3})$.

We would expect the approximations to continue to improve as we use more and more rectangles with smaller and smaller bases. We now state the **rectangle rule** for approximating definite integrals of a continuous function f over the interval from $x = a$ to $x = b$.

■ **Rectangle Rule**

Divide the interval from $x = a$ to $x = b$ into n equal subintervals of length $\Delta x = (b - a)/n$. Let c_k be any point in the kth subinterval. Then

$$\int_a^b f(x)\, dx \approx f(c_1)\Delta x + f(c_2)\Delta x + \cdots + f(c_n)\Delta x$$

$$= \Delta x[f(c_1) + f(c_2) + \cdots + f(c_n)]$$

The rectangle rule is valid for any continuous function f. If f is positive on $[a, b]$, then each term $f(c_k)\Delta x$ in the approximating sum can be interpreted as the area of a rectangle, and the sum of these terms approximates the area under the graph of f. If f is not positive on $[a, b]$, then $f(c_k)\Delta x$ no longer represents the area of a rectangle (area is always positive), but it is still true that the sum of these terms approximates the definite integral of f from a to b.

◆ E X A M P L E 24

Use the rectangle rule to approximate

$$\int_0^2 \frac{12x - 12}{x + 4}\, dx$$

with $n = 4$ and c_k the midpoint of each subinterval. Compute the approximation to three decimal places.

Solution Step 1. Find Δx, the length of each subinterval:

$$\Delta x = \frac{b - a}{n} = \frac{2 - 0}{4} = 0.5$$

---

* We actually do not need to choose the midpoint of each subinterval; any point from each subinterval will do. The midpoint is often a convenient point to choose, because then the rectangle tops are usually above part of the graph and below part of the graph. This tends to cancel some of the error that occurs.

Step 2. Use the midpoint of each interval for c_k:

MIDPOINTS: $c_1 = 0.25$ $c_2 = 0.75$ $c_3 = 1.25$ $c_4 = 1.75$

SUBINTERVALS: $[0, 0.5]$ $[0.5, 1]$ $[1, 1.5]$ $[1.5, 2]$

Step 3. Use the rectangle rule with $n = 4$ (see the figure below):

$$\int_0^2 \frac{12x - 12}{x + 4}\, dx \approx f(c_1)\Delta x + f(c_2)\Delta x + f(c_3)\Delta x + f(c_4)\Delta x$$
$$= [f(c_1) + f(c_2) + f(c_3) + f(c_4)]\Delta x$$
$$= [f(0.25) + f(0.75) + f(1.25) + f(1.75)](0.5)$$
$$= [(-2.117\ 647) + (-0.631\ 579) + (0.571\ 429) + (1.565\ 217)](0.5)$$
$$= (-0.612\ 58)(0.5) = -0.306 \qquad \text{To three decimal places}$$

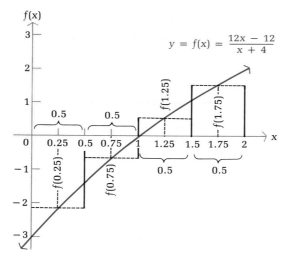

◆

P R O B L E M   24   Use the rectangle rule to approximate

$$\int_{-2}^2 \frac{4x}{x + 6}\, dx$$

with $n = 4$ and c_k the midpoint of each subinterval. Compute the approximation to three decimal places. ◆

One important application of the rectangle rule is the approximation of definite integrals involving **tabular functions** — that is, functions defined by tables rather than by formulas. The next example illustrates this approach.

◆ E X A M P L E   25   A developer is interested in estimating the area of the irregularly shaped prop-

Real Estate   erty shown in Figure 12A. A surveyor used the straight horizontal road at the bottom of the property as the x axis and measured the vertical distance across

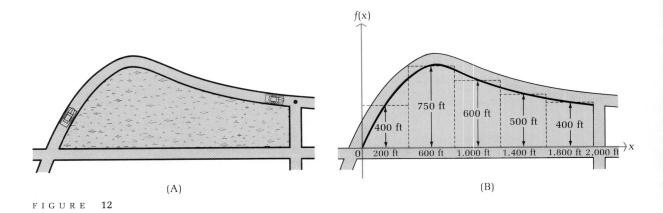

(A)

(B)

FIGURE 12

the property at 400 foot intervals, starting at 200 (see Fig. 12B). These distances can be viewed as the values of the continuous function f whose graph forms the top of the property. Use the rectangle rule to approximate the area of the property.

Solution   List the values of the function f in a table:

x	200	600	1,000	1,400	1,800
f(x)	400	750	600	500	400

The area of the property is given by

$$A = \int_0^{2,000} f(x)\, dx$$

Using the rectangle rule with $n = 5$ and $\Delta x = 400$, and the values of f in the table, we have

$$A \approx \Delta x[f(200) + f(600) + f(1,000) + f(1,400) + f(1,800)]$$
$$= 400[400 + 750 + 600 + 500 + 400]$$
$$= 1,060,000 \text{ square feet} \qquad \blacklozenge$$

PROBLEM   25   To obtain a more accurate approximation of the area of the property shown in Figure 12A, the surveyor measured the vertical distances at 200 foot intervals, starting at 100. The results are listed in the table. Use these values and the rectangle rule with $n = 10$ and $\Delta x = 200$ to approximate the area of the property.

x	100	300	500	700	900	1,100	1,300	1,500	1,700	1,900
f(x)	225	500	700	725	650	575	525	475	425	375

$\blacklozenge$

◆ DEFINITE INTEGRAL AS A LIMIT OF A SUM

In using the rectangle rule to approximate a definite integral, one might expect

$$\lim_{\Delta x \to 0} [f(c_1)\Delta x + f(c_2)\Delta x + \cdots + f(c_n)\Delta x] = \int_a^b f(x)\, dx$$

This idea motivates the formal definition of a definite integral that we referred to in Section 11-3.

Definition of a Definite Integral

Let f be a continuous function defined on the closed interval $[a, b]$, and let

1. $a = x_0 < x_1 < \cdots < x_{n-1} < x_n = b$
2. $\Delta x_k = x_k - x_{k-1}$ for $k = 1, 2, \ldots, n$
3. $\Delta x_k \to 0$ as $n \to \infty$
4. $x_{k-1} \le c_k \le x_k$ for $k = 1, 2, \ldots, n$

Then

$$\int_a^b f(x)\, dx = \lim_{n \to \infty} [f(c_1)\Delta x_1 + f(c_2)\Delta x_2 + \cdots + f(c_n)\Delta x_n]$$

is called a **definite integral.**

In the definition of a definite integral, we divide the closed interval $[a, b]$ into n subintervals of arbitrary lengths in such a way that the length of each subinterval $\Delta x_k = x_k - x_{k-1}$ tends to 0 as n increases without bound. From each of the n subintervals we then select a point c_k and form the sum

$$f(c_1)\Delta x_1 + f(c_2)\Delta x_2 + \cdots + f(c_n)\Delta x_n$$

which is called a **Riemann sum** [named after the celebrated German mathematician Georg Riemann (1826–1866)].

Under the conditions stated in the definition, it can be shown that the limit of the Riemann sum always exists, and it is a real number. The limit is independent of the nature of the subdivisions of $[a, b]$ as long as condition 3 holds, and it is independent of the choice of c_k as long as condition 4 holds. In particular, when using Riemann sums to approximate definite integrals, we can always choose subintervals of equal length and c_k the midpoint of each subinterval, as we did in the previous examples in this section.

In a more formal treatment of the subject, we would then prove the remarkable **fundamental theorem of calculus,** stated below. This theorem shows that the limit in the definition of a definite integral can be determined exactly by evaluating an antiderivative of $f(x)$, if it exists, at the end points of the interval $[a, b]$ and taking the difference.

Under the conditions stated in the definition of a definite integral:

DEFINITION:

$$\int_a^b f(x)\, dx = \lim_{n\to\infty} [f(c_1)\Delta x_1 + f(c_2)\Delta x_2 + \cdots + f(c_n)\Delta x_n]$$

THEOREM:

$$= F(b) - F(a) \qquad \text{where } F'(x) = f(x)$$

Now we are free to evaluate a definite integral by using the fundamental theorem if an antiderivative of $f(x)$ can be found; otherwise, we can approximate it using the formal definition in the form of the rectangle rule.

◆ RECOGNIZING A DEFINITE INTEGRAL

Recall that the derivative of a function f was defined in Section 9-3 by

$$f'(x) = \lim_{h\to 0} \frac{f(x+h) - f(x)}{h}$$

This form is generally not easy to compute directly, but is easy to recognize in certain practical problems (slope, instantaneous velocity, rates of change, and so on). Once it is recognized that we are dealing with a derivative, we then proceed to try to compute it using derivative formulas and rules.

Similarly, evaluating a definite integral using the definition

$$\int_a^b f(x)\, dx = \lim_{n\to\infty}[f(c_1)\Delta x_1 + f(c_2)\Delta x_2 + \cdots + f(c_n)\Delta x_n] \tag{1}$$

is generally not easy; but the form on the right occurs naturally in many practical problems. We can use the fundamental theorem to evaluate the integral (once it is recognized) if an antiderivative can be found; otherwise, we will approximate it using the rectangle rule. We will now illustrate these points by finding the *average value* of a continuous function.

◆ AVERAGE VALUE OF A CONTINUOUS FUNCTION

Suppose the temperature F (in degrees Fahrenheit) in the middle of a small shallow lake from 8 AM ($t = 0$) to 6 PM ($t = 10$) during the month of May is given approximately as shown in Figure 13.

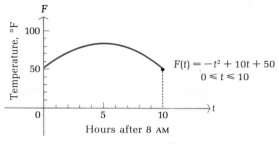

$$F(t) = -t^2 + 10t + 50$$
$$0 \leqslant t \leqslant 10$$

FIGURE 13

How can we compute the average temperature from 8 AM to 6 PM? We know that the average of a finite number of values

$$a_1, a_2, \ldots, a_n$$

is given by

$$\text{Average} = \frac{a_1 + a_2 + \cdots + a_n}{n}$$

But how can we handle a continuous function with infinitely many values? It would seem reasonable to divide the time interval $[0, 10]$ into n equal subintervals, compute the temperature at a point in each subinterval, and then use the average of these values as an approximation of the average value of the continuous function $F = F(t)$ over $[0, 10]$. We would expect the approximations to improve as n increases. In fact, we would be inclined to define the limit of the average of n values as $n \to \infty$ as the *average value* of F over $[0, 10]$, if the limit exists. This is exactly what we will do:

$$\left(\begin{array}{c} \text{Average temperature} \\ \text{for } n \text{ values} \end{array} \right) = \frac{1}{n} [F(t_1) + F(t_2) + \cdots + F(t_n)] \qquad (2)$$

where t_k is a point in the kth subinterval. We will call the limit of (2) as $n \to \infty$ the *average temperature over the time interval* $[0, 10]$.

Form (2) looks sort of like form (1), but we are missing the Δt_k. We take care of this by multiplying (2) by $(b-a)/(b-a)$, which will change the form of (2) without changing its value:

$$\frac{b-a}{b-a} \cdot \frac{1}{n} [F(t_1) + F(t_2) + \cdots + F(t_n)] = \frac{1}{b-a} \cdot \frac{b-a}{n} [F(t_1) + F(t_2) + \cdots + F(t_n)]$$

$$= \frac{1}{b-a} \left[F(t_1) \frac{b-a}{n} + F(t_2) \frac{b-a}{n} + \cdots + F(t_n) \frac{b-a}{n} \right]$$

$$= \frac{1}{b-a} [F(t_1)\Delta t + F(t_2)\Delta t + \cdots + F(t_n)\Delta t]$$

Thus,

$$\begin{pmatrix} \text{Average temperature} \\ \text{over } [a, b] = [0, 10] \end{pmatrix} = \lim_{n \to \infty} \left\{ \frac{1}{b-a} \left[F(t_1)\Delta t + F(t_2)\Delta t + \cdots + F(t_n)\Delta t \right] \right\}$$

$$= \frac{1}{b-a} \{ \lim_{n \to \infty} [F(t_1)\Delta t + F(t_2)\Delta t + \cdots + F(t_n)\Delta t] \}$$

Now the limit inside the braces is of form (1)—that is, a definite integral. Thus,

$$\begin{pmatrix} \text{Average temperature} \\ \text{over } [a, b] = [0, 10] \end{pmatrix} = \frac{1}{b-a} \int_a^b F(t) \, dt$$

$$= \frac{1}{10-0} \int_0^{10} (-t^2 + 10t + 50) \, dt \qquad \text{We now evaluate the definite}$$

$$= \frac{1}{10} \left(-\frac{t^3}{3} + 5t^2 + 50t \right) \Big|_0^{10} \qquad \begin{array}{l} \text{integral using the fundamental} \\ \text{theorem.} \end{array}$$

$$= \frac{200}{3} \approx 67°\text{F}$$

In general, proceeding as above for an arbitrary continuous function f over an interval $[a, b]$, we obtain the following general formula:

Average Value of a Continuous Function f over $[a, b]$

$$\frac{1}{b-a} \int_a^b f(x) \, dx$$

◆ E X A M P L E   26   Find the average value of $f(x) = x - 3x^2$ over the interval $[-1, 2]$.

Solution
$$\frac{1}{b-a} \int_a^b f(x) \, dx = \frac{1}{2-(-1)} \int_{-1}^2 (x - 3x^2) \, dx$$

$$= \frac{1}{3} \left(\frac{x^2}{2} - x^3 \right) \Big|_{-1}^2 = -\frac{5}{2} \qquad ◆$$

P R O B L E M   26   Find the average value of $g(t) = 6t^2 - 2t$ over the interval $[-2, 3]$.   ◆

◆ E X A M P L E   27   Given the demand function

Average Price
$$p = D(x) = 100e^{-0.05x}$$

find the average price (in dollars) over the demand interval $[40, 60]$.

Solution

$$\text{Average price} = \frac{1}{b-a}\int_a^b D(x)\,dx$$

$$= \frac{1}{60-40}\int_{40}^{60} 100e^{-0.05x}\,dx$$

$$= \frac{100}{20}\int_{40}^{60} e^{-0.05x}\,dx \qquad \text{Use }\int e^{ax}\,dx = \frac{1}{a}e^{ax},\ a \neq 0.$$

$$= -\frac{5}{0.05}e^{-0.05x}\Big|_{40}^{60}$$

$$= 100(e^{-2} - e^{-3}) \approx \$8.55 \qquad \blacklozenge$$

PROBLEM 27 Given the supply equation

$$p = S(x) = 10e^{0.05x}$$

find the average price (in dollars) over the supply interval [20, 30]. ◆

◆ **EXAMPLE 28**

Advertising

A metropolitan newspaper currently has a daily circulation of 50,000 papers (weekdays and Sunday). The management of the paper decides to initiate an aggressive advertising campaign to increase circulation. Suppose that the daily circulation (in thousands of papers) t days after the beginning of the campaign is given by

$$S(t) = 100 - 50e^{-0.01t}$$

What is the average daily circulation during the first 30 days of the campaign? During the second 30 days of the campaign?

Solution

$$\left(\begin{array}{c}\text{Average daily circulation}\\ \text{over } [a, b] = [0, 30]\end{array}\right) = \frac{1}{b-a}\int_a^b S(t)\,dt$$

$$= \frac{1}{30}\int_0^{30}(100 - 50e^{-0.01t})\,dt$$

$$= \frac{1}{30}(100t + 5{,}000e^{-0.01t})\Big|_0^{30}$$

$$= \frac{1}{30}(3{,}000 + 5{,}000e^{-0.3} - 5{,}000)$$

$$\approx 56.8 \quad \text{or} \quad 56{,}800 \text{ papers}$$

$$\left(\begin{array}{c}\text{Average daily circulation}\\ \text{over } [a, b] = [30, 60]\end{array}\right) = \frac{1}{60-30}\int_{30}^{60}(100 - 50e^{-0.01t})\,dt$$

$$= \frac{1}{30}(100t + 5{,}000e^{-0.01t})\Big|_{30}^{60}$$

$$= \frac{1}{30}(6{,}000 + 5{,}000e^{-0.6} - 3{,}000 - 5{,}000e^{-0.3})$$

$$\approx 68.0 \quad \text{or} \quad 68{,}000 \text{ papers} \qquad \blacklozenge$$

Refer to Example 28. Satisfied with the increase in circulation, management decides to terminate the advertising campaign. Suppose that the daily circulation (in thousands of papers) t days after the end of the advertising campaign is given by

$$S(t) = 65 + 8e^{-0.02t}$$

What is the average daily circulation during the first 30 days after the end of the campaign? During the second 30 days after the end of the campaign?   ◆

Answers to Matched Problems   24. -0.589 25. $1,035,000 \text{ ft}^2$ 26.   13   27. $35.27

28. $\dfrac{2,350 - 400e^{-0.6}}{30} \approx 71.0$ or 71,000 papers;

$\dfrac{1,950 - 400e^{-1.2} + 400e^{-0.6}}{30} \approx 68.3$ or 68,300 papers

---

EXERCISE   11-5

For Problems 1–8:

(A) *Use the rectangle rule to approximate each definite integral for the indicated number of subintervals n. Choose c_k as the midpoint of each subinterval.*
(B) *Evaluate each integral exactly using an antiderivative.*

A

1. $\displaystyle\int_1^5 3x^2 \, dx; \quad n = 2$

2. $\displaystyle\int_2^6 x^2 \, dx; \quad n = 2$

3. $\displaystyle\int_1^5 3x^2 \, dx; \quad n = 4$

4. $\displaystyle\int_2^6 x^2 \, dx; \quad n = 4$

B

5. $\displaystyle\int_0^4 (4 - x^2) \, dx; \quad n = 2$

6. $\displaystyle\int_0^4 (3x^2 - 12) \, dx; \quad n = 2$

7. $\displaystyle\int_0^4 (4 - x^2) \, dx; \quad n = 4$

8. $\displaystyle\int_0^4 (3x^2 - 12) \, dx; \quad n = 4$

In Problems 9–12, use the rectangle rule with $n = 4$ and the values given for f in each table to approximate the indicated definite integral. Choose c_k as the midpoint of each interval.

9. $\displaystyle\int_0^8 f(x) \, dx$

x	1	3	5	7
$f(x)$	4.5	3.2	2.4	1.6

10. $\displaystyle\int_1^9 f(x) \, dx$

x	2	4	6	8
$f(x)$	3.2	4.5	7.9	9.4

11. $\displaystyle\int_1^5 f(x) \, dx$

x	1.5	2.5	3.5	4.5
$f(x)$	12.5	16.7	15.4	10.7

12. $\displaystyle\int_0^4 f(x) \, dx$

x	0.5	1.5	2.5	3.5
$f(x)$	9.4	14.7	11.5	6.4

Find the average value of each function over the indicated interval.

13. $f(x) = 500 - 50x$; [0, 10]
14. $g(x) = 2x + 7$; [0, 5]
15. $f(t) = 3t^2 - 2t$; [−1, 2]
16. $g(t) = 4t - 3t^2$; [−2, 2]
17. $f(x) = \sqrt[3]{x}$; [1, 8]
18. $g(x) = \sqrt{x + 1}$; [3, 8]
19. $f(x) = 4e^{-0.2x}$; [0, 10]
20. $f(x) = 64e^{0.08x}$; [0, 10]

Use the rectangle rule to approximate (to three decimal places) each quantity in Problems 21–24. Use n = 4 and choose c_k as the midpoint of each subinterval.

21. The average value of $f(x) = (x + 1)/(x^2 + 1)$ for [−1, 1]
22. The average value of $f(x) = x/(x + 1)$ for [0, 4]
23. The area under the graph of $f(x) = \ln(1 + x^3)$ from $x = 0$ to $x = 2$
24. The area under the graph of $f(x) = 1/(2 + x^3)$ from $x = -1$ to $x = 1$

C *In Problems 25–28, use the rectangle rule to approximate (to three decimal places) each definite integral for the indicated number of subintervals n. Choose c_k as the midpoint of each subinterval.*

25. $\int_0^1 e^{-x^2}\, dx$; $n = 5$
26. $\int_0^1 e^{x^2}\, dx$; $n = 5$

27. $\int_0^1 e^{-x^2}\, dx$; $n = 10$
28. $\int_0^1 e^{x^2}\, dx$; $n = 10$

29. Find the average value of $f'(x)$ over the interval [a, b] for any differentiable function f.

30. Show that the average value of $f(x) = Ax + B$ over the interval [a, b] is

$$f\left(\frac{a + b}{2}\right)$$

APPLICATIONS

Business & Economics

31. *Inventory.* A store orders 600 units of a product every 3 months. If the product is steadily depleted to 0 by the end of each 3 months, the inventory on hand, I, at any time t during the year is illustrated as follows:

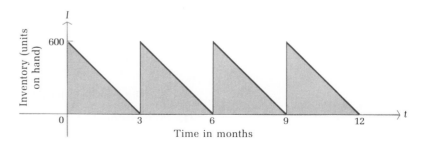

(A) Write an inventory function (assume it is continuous) for the first 3 months. [The graph is a straight line joining (0, 600) and (3, 0).]

(B) What is the average number of units on hand for a 3 month period?

32. Repeat Problem 31 with an order of 1,200 units every 4 months.

33. *Cash reserves.* Suppose cash reserves (in thousands of dollars) are approximated by

$$C(x) = 1 + 12x - x^2 \qquad 0 \leqslant x \leqslant 12$$

where x is the number of months after the first of the year. What is the average cash reserve for the first quarter?

34. Repeat Problem 33 for the second quarter.

35. *Average cost.* The total cost (in dollars) of manufacturing x auto body frames is

$$C(x) = 60,000 + 300x$$

(A) Find the average cost per unit if 500 frames are produced. [*Hint:* Recall that $\overline{C}(x)$ is the average cost per unit.]

(B) Find the average value of the cost function over the interval [0, 500].

36. *Average cost.* The total cost (in dollars) of printing x dictionaries is $C(x) = 20,000 + 10x$.

(A) Find the average cost per unit if 1,000 dictionaries are produced.

(B) Find the average value of the cost function over the interval [0, 1,000].

37. *Continuous compound interest.* If $100 is deposited in an account that earns interest at an annual nominal rate of 8% compounded continuously, find the amount in the account after 5 years and the average amount in the account during this 5 year period. [See Section 10-5 for a discussion of continuous compound interest.]

38. *Continuous compound interest.* If $500 is deposited in an account that earns interest at an annual nominal rate of 12% compounded continuously, find the amount in the account after 10 years and the average amount in the account during this 10 year period.

39. *Supply function.* Given the supply function

$$p = S(x) = 10(e^{0.02x} - 1)$$

find the average price (in dollars) over the supply interval [20, 30].

40. *Demand function.* Given the demand function

$$p = D(x) = \frac{1,000}{x}$$

find the average price (in dollars) over the demand interval [400, 600].

41. *Advertising.* The number of hamburgers (in thousands) sold each day by a chain of restaurants t days after the beginning of an advertising campaign is

given by

$$S(t) = 20 - 10e^{-0.1t}$$

What is the average number of hamburgers sold each day during the first week of the advertising campaign? During the second week of the campaign?

42. *Advertising.* The number of hamburgers (in thousands) sold each day by a chain of restaurants t days after the end of an advertising campaign is given by

$$S(t) = 10 + 8e^{-0.2t}$$

What is the average number of hamburgers sold each day during the first week after the end of the advertising campaign? During the second week after the end of the campaign?

43. *Profit.* Let $R(t)$ and $C(t)$ represent the total accumulated revenues and costs (in dollars), respectively, for a producing oil well, where t is time in years. The graphs of the derivatives of R and C over a 5 year period are shown in the figure in the margin. Use the rectangle rule with $n = 5$ and c_k as the midpoint of each subinterval to approximate the total accumulated profits from the well over this 5 year period. Estimate necessary function values from the graphs.

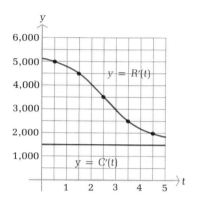

44. *Revenue.* Use the figure in Problem 43 and the rectangle rule with $n = 5$ and c_k as the midpoint of each subinterval to approximate the average annual revenue from the oil well.

45. *Real estate.* A surveyor produced the table below by measuring the vertical distance (in feet) across a piece of real estate at 600 foot intervals, starting at 300 (see the figure in the margin). Use these values and the rectangle rule to estimate the area of the property.

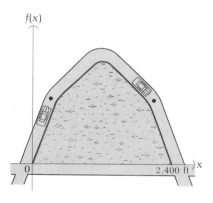

x	300	900	1,500	2,100
$f(x)$	900	1,700	1,700	900

46. *Real estate.* Repeat Problem 45 for the following table of measurements:

x	200	600	1,000	1,400	1,800	2,200
$f(x)$	600	1,400	1,800	1,800	1,400	600

Life Sciences

47. *Temperature.* If the temperature $C(t)$ in an aquarium is made to change according to

$$C(t) = t^3 - 2t + 10 \qquad 0 \leqslant t \leqslant 2$$

(in degrees Celsius) over a 2 hour period, what is the average temperature over this period?

48. *Medicine.* A drug is injected into the bloodstream of a patient through her right arm. The concentration of the drug in the bloodstream of the left arm *t* hours after the injection is given by

$$C(t) = \frac{0.14t}{t^2 + 1}$$

What is the average concentration of the drug in the bloodstream of the left arm during the first hour after the injection? During the first 2 hours after the injection?

49. *Medicine — respiration.* Physiologists use a machine called a pneumotachograph to produce a graph of the rate of flow *R(t)* of air into the lungs (inspiration) and out of the lungs (expiration). The figure in the margin gives the graph of the inspiration phase of the breathing cycle for an individual at rest. The area under this graph represents the total volume of air inhaled during the inspiration phase. Use the rectangle rule with *n* = 3 and c_k the midpoint of each subinterval to approximate the area under the graph. Estimate the necessary function values from the graph.

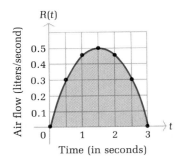

50. *Medicine — respiration.* Use the result obtained in Problem 49 to approximate the average volume of air in the lungs during the inspiration phase.

Social Sciences

51. *Politics.* Public awareness of a Congressional candidate before and after a successful campaign was approximated by

$$P(t) = \frac{8.4t}{t^2 + 49} + 0.1 \qquad 0 \leqslant t \leqslant 24$$

where *t* is time in months after the campaign started and *P(t)* is the fraction of people in the Congressional district who could recall the candidate's name. What is the average fraction of people who could recall the candidate's name during the first 7 months after the campaign began? During the first 2 years after the campaign began?

52. *Population composition.* Because of various factors (such as birth rate expansion, then contraction; family flights from urban areas; and so on), the number of children in a large city was found to increase and then decrease rather drastically. If the number of children over a 6 year period was found to be given approximately by

$$N(t) = -\tfrac{1}{4}t^2 + t + 4 \qquad 0 \leqslant t \leqslant 6$$

what was the average number of children in the city over the 6 year time period? [Assume *N* = *N(t)* is continuous.]

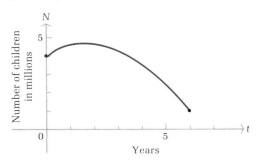

Consumers' and Producers' Surplus

- ◆ CONSUMERS' SURPLUS
- ◆ PRODUCERS' SURPLUS
- ◆ MARKET EQUILIBRIUM

◆ CONSUMERS' SURPLUS

Let $p = D(x)$ be the price–demand equation for a product, where x is the number of units of the product that consumers will purchase at a price of \$$p$ per unit. Notice that we are expressing price $p = D(x)$ as a function of demand x (see Section 9-7). Suppose $\bar{p}$ is the current price and $\bar{x}$ is the number of units that can be sold at that price. The price–demand curve in Figure 14 shows that if the price is higher than $\bar{p}$, then the demand x is less than $\bar{x}$, but some consumers are still willing to pay the higher price. The consumers who were willing to pay more than $\bar{p}$ have saved money. We want to determine the total amount saved by all the consumers who were willing to pay a higher price for this product.

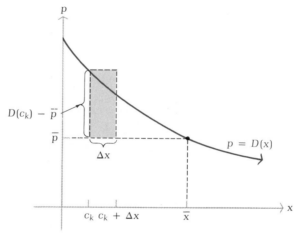

FIGURE   14

To do this, consider the interval $[c_k, c_k + \Delta x]$, where $c_k + \Delta x < \bar{x}$. If the price remained constant over this interval, then the savings on each unit would be the difference between $D(c_k)$, the price consumers are willing to pay, and $\bar{p}$, the price they actually pay. Since Δx represents the number of units purchased by consumers over the interval, the total savings to consumers over this interval is approximately equal to

$$[D(c_k) - \bar{p}]\Delta x \qquad \text{(Savings per unit)} \times \text{(Number of units)}$$

which is the area of the shaded rectangle shown in Figure 14. If we divide the interval $[0, \bar{x}]$ into n equal subintervals, then the total savings to consumers is approximately equal to

$$[D(c_1) - \bar{p}]\Delta x + [D(c_2) - \bar{p}]\Delta x + \cdots + [D(c_n) - \bar{p}]\Delta x$$

which we recognize as a Riemann sum for the integral

$$\int_0^{\bar{x}} [D(x) - \bar{p}] \, dx$$

Thus, we define the *consumers' surplus* to be this integral.

■ Consumers' Surplus

If $(\bar{x}, \bar{p})$ is a point on the graph of the price–demand equation $p = D(x)$ for a particular product, then the **consumers' surplus, CS,** at a price level of $\bar{p}$ is

$$CS = \int_0^{\bar{x}} [D(x) - \bar{p}] \, dx$$

which is the area between $p = \bar{p}$ and $p = D(x)$ from $x = 0$ to $x = \bar{x}$.

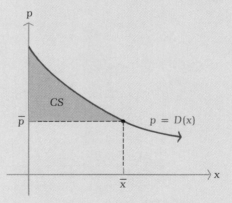

The consumers' surplus represents the total savings to consumers who are willing to pay a higher price for the product.

◆ E X A M P L E   29   Find the consumers' surplus at a price level of $8 for the price–demand equation

$$p = D(x) = 20 - \tfrac{1}{20}x$$

Solution   *Step 1.*   Find $\overline{x}$, the demand when the price is $\overline{p} = 8$:

$$\overline{p} = 20 - \tfrac{1}{20}\overline{x}$$
$$8 = 20 - \tfrac{1}{20}\overline{x}$$
$$\tfrac{1}{20}\overline{x} = 12$$
$$\overline{x} = 240$$

Step 2.   Sketch a graph:

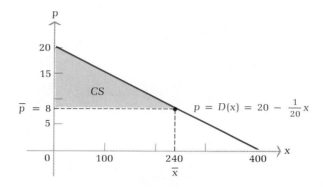

Step 3.   Find the consumers' surplus (the shaded area in the graph):

$$CS = \int_0^{\overline{x}} [D(x) - \overline{p}]\, dx$$

$$= \int_0^{240} (20 - \tfrac{1}{20}x - 8)\, dx$$

$$= \int_0^{240} (12 - \tfrac{1}{20}x)\, dx$$

$$= (12x - \tfrac{1}{40}x^2)\big|_0^{240}$$
$$= 2{,}880 - 1{,}440 = \$1{,}440$$

Thus, the total savings to consumers who are willing to pay a higher price for the product is $1,440.   ◆

P R O B L E M   29   Repeat Example 29 for a price level of $4.   ◆

◆ PRODUCERS' SURPLUS

If $p = S(x)$ is the price–supply equation for a product, $\overline{p}$ is the current price, and $\overline{x}$ is the current supply, then some suppliers are still willing to supply some units at a lower price. The additional money that these suppliers gain from the higher price is called the *producers' surplus* and can be expressed in terms of a definite integral (proceeding as we did for the consumers' surplus).

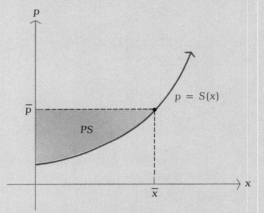

Producers' Surplus

If $(\bar{x}, \bar{p})$ is a point on the graph of the price–supply equation $p = S(x)$, then the **producers' surplus, *PS***, at a price level of $\bar{p}$ is

$$PS = \int_0^{\bar{x}} [\bar{p} - S(x)]\, dx$$

which is the area between $p = \bar{p}$ and $p = S(x)$ from $x = 0$ to $x = \bar{x}$.

The producers' surplus represents the total gain to producers who are willing to supply units at a lower price.

◆ E X A M P L E   30   Find the producers' surplus at a price level of $20 for the price–supply equation

$$p = S(x) = 2 + \tfrac{1}{5,000}x^2$$

Solution   *Step 1.*   Find $\bar{x}$, the supply when the price is $\bar{p} = 20$:

$$\bar{p} = 2 + \tfrac{1}{5,000}\bar{x}^2$$
$$20 = 2 + \tfrac{1}{5,000}\bar{x}^2$$
$$\tfrac{1}{5,000}\bar{x}^2 = 18$$
$$\bar{x}^2 = 90,000$$
$$\bar{x} = 300 \quad \text{There is only one solution since } \bar{x} \geqslant 0.$$

Step 2. Sketch a graph:

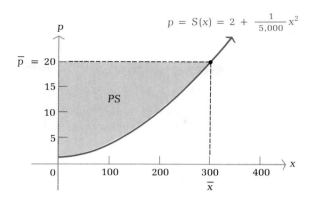

$$p = S(x) = 2 + \frac{1}{5,000}x^2$$

Step 3. Find the producers' surplus (the shaded area in the graph):

$$PS = \int_0^{\bar{x}} [\bar{p} - S(x)]\, dx$$

$$= \int_0^{300} [20 - (2 + \tfrac{1}{5,000}x^2)]\, dx$$

$$= \int_0^{300} (18 - \tfrac{1}{5,000}x^2)\, dx$$

$$= (18x - \tfrac{1}{15,000}x^3)\big|_0^{300}$$
$$= 5,400 - 1,800 = \$3,600$$

Thus, the total gain to producers who are willing to supply units at a lower price is $3,600. ◆

PROBLEM 30 Repeat Example 30 for a price level of $4. ◆

◆ MARKET EQUILIBRIUM

In a free competitive market, the price of a product is determined by the relationship between supply and demand. If $p = D(x)$ and $p = S(x)$ are the price–demand and price–supply equations, respectively, for a product and if $(\bar{x}, \bar{p})$ is the point of intersection of these equations, then $\bar{p}$ is called the **equilibrium price** and $\bar{x}$ is called the **equilibrium quantity.** If the price stabilizes at the equilibrium price $\bar{p}$, then this is the price level that will determine both the consumers' surplus and the producers' surplus.

◆ EXAMPLE 31 Find the equilibrium price and then find the consumers' surplus and producers' surplus at the equilibrium price level if

$$p = D(x) = 20 - \tfrac{1}{20}x \quad \text{and} \quad p = S(x) = 2 + \tfrac{1}{5,000}x^2$$

Solution   *Step 1.*   Find the equilibrium point. Set $D(x)$ equal to $S(x)$ and solve:

$$D(x) = S(x)$$
$$20 - \tfrac{1}{20}x = 2 + \tfrac{1}{5,000}x^2$$
$$\tfrac{1}{5,000}x^2 + \tfrac{1}{20}x - 18 = 0$$
$$x^2 + 250x - 90,000 = 0$$
$$x = 200, -450$$

Since x cannot be negative, the only solution is $x = 200$. The equilibrium price can be determined by using $D(x)$ or $S(x)$. We will use both to check our work:

$$\bar{p} = D(200) \qquad\qquad\qquad \bar{p} = S(200)$$
$$= 20 - \tfrac{1}{20}(200) = 10 \qquad = 2 + \tfrac{1}{5,000}(200)^2 = 10$$

Thus, the equilibrium price is $\bar{p} = 10$, and the equilibrium quantity is $\bar{x} = 200$.

Step 2.   Sketch a graph:

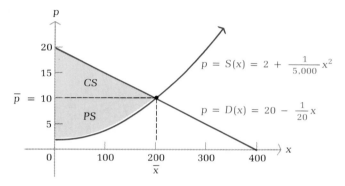

Step 3.   Find the consumers' surplus:

$$CS = \int_0^{\bar{x}} [D(x) - \bar{p}] \, dx$$

$$= \int_0^{200} (20 - \tfrac{1}{20}x - 10) \, dx$$

$$= \int_0^{200} (10 - \tfrac{1}{20}x) \, dx$$

$$= (10x - \tfrac{1}{40}x^2)|_0^{200}$$
$$= 2,000 - 1,000 = \$1,000$$

Step 4.   Find the producers' surplus:

$$PS = \int_0^{\bar{x}} [\bar{p} - S(x)] \, dx$$

$$= \int_0^{200} [10 - (2 + \tfrac{1}{5,000}x^2)] \, dx$$

$$PS = \int_0^{200} \left(8 - \tfrac{1}{5,000}x^2\right) dx$$

$$= \left(8x - \tfrac{1}{15,000}x^3\right)\Big|_0^{200}$$

$$= 1,600 - \tfrac{1,600}{3} \approx \$1,067 \qquad \text{Rounded to the nearest dollar} \qquad \blacklozenge$$

PROBLEM 31　　Repeat Example 31 for

$$p = D(x) = 25 - \tfrac{1}{1,000}x^2 \qquad \text{and} \qquad p = S(x) = 5 + \tfrac{1}{10}x \qquad \blacklozenge$$

Answers to Matched Problems　　29.　\$2,560　　30.　\$133　　31.　$\bar{p} = 15$; $CS = \$667$; $PS = \$500$

EXERCISE 11-6

APPLICATIONS

Business & Economics

1. Find the consumers' surplus at a price level of $\bar{p} = \$150$ for the price-demand equation

   $$p = D(x) = 400 - \tfrac{1}{20}x$$

2. Find the consumers' surplus at a price level of $\bar{p} = \$120$ for the price-demand equation

   $$p = D(x) = 200 - \tfrac{1}{50}x$$

3. Find the producers' surplus at a price level of $\bar{p} = \$65$ for the price-supply equation

   $$p = S(x) = 10 + \tfrac{1}{10}x + \tfrac{1}{3,600}x^2$$

4. Find the producers' surplus at a price level of $\bar{p} = \$55$ for the price-supply equation

   $$p = S(x) = 15 + \tfrac{1}{10}x + \tfrac{3}{1,000}x^2$$

5. Find the consumers' surplus and the producers' surplus at the equilibrium price level for

   $$p = D(x) = 50 - \tfrac{1}{10}x \qquad \text{and} \qquad p = S(x) = 11 + \tfrac{1}{20}x$$

6. Find the consumers' surplus and the producers' surplus at the equilibrium price level for

   $$p = D(x) = 20 - \tfrac{1}{600}x^2 \qquad \text{and} \qquad p = S(x) = 2 + \tfrac{1}{300}x^2$$

7. Find the consumers' surplus and the producers' surplus at the equilibrium price level (rounded to the nearest dollar) for

   $$p = D(x) = 80e^{-0.001x} \qquad \text{and} \qquad p = S(x) = 30e^{0.001x}$$

8. Find the consumers' surplus and the producers' surplus at the equilibrium price level (rounded to the nearest dollar) for

$$p = D(x) = 185e^{-0.005x} \qquad \text{and} \qquad p = S(x) = 25e^{0.005x}$$

9. Suppose that in a monopolistic market, the production level for a certain product is always chosen so that the profit is maximized. The price-demand equation for this product is

$$p = 100 - \tfrac{1}{10}x$$

and the cost equation is

$$C(x) = 5{,}000 + 60x$$

Find the consumers' surplus at the production level that maximizes profit.

10. Suppose that in a certain market, the production level for a product is always chosen so that the revenue is maximized. The price-demand equation for this product is

$$p = 100 - \tfrac{1}{10}x$$

Find the consumers' surplus at the production level that maximizes revenue.

11-5 *Definite Integral as a Limit of a Sum.* Rectangle rule; tabular function; definite integral (as a limit of a Riemann sum); fundamental theorem of calculus; average value of a continuous function

11-6 *Consumers' and Producers' Surplus.* Consumers' surplus; producers' surplus; equilibrium price; equilibrium quantity

Integration Formulas

$$\int k \, dx = kx + C$$

$$\int kf(x) \, dx = k \int f(x) \, dx$$

$$\int [f(x) \pm g(x)] \, dx = \int f(x) \, dx \pm \int g(x) \, dx$$

$$\int u^n \, du = \frac{u^{n+1}}{n+1} + C \qquad n \neq -1$$

$$\int e^u \, du = e^u + C \qquad \text{and} \qquad \int e^{au} \, du = \frac{1}{a} e^{au} + C \qquad a \neq 0$$

$$\int \frac{1}{u} \, du = \ln|u| + C \qquad u \neq 0$$

Chapter Review

Work through all the problems in this chapter review and check your answers in the back of the book. (Answers to all review problems are there.) Where weaknesses show up, review appropriate sections in the text.

A   *Find each integral in Problems 1–6.*

1. $\int (3t^2 - 2t) \, dt$

2. $\int_2^5 (2x - 3) \, dx$

3. $\int (3t^{-2} - 3) \, dt$

4. $\int_1^4 x \, dx$

5. $\int e^{-0.5x} \, dx$

6. $\int_1^5 \frac{2}{u} \, du$

7. Find a function $y = f(x)$ that satisfies both conditions:

$$\frac{dy}{dx} = 3x^2 - 2 \qquad f(0) = 4$$

8. Find the area bounded by the graphs of $y = 3x^2 + 1$ and $y = 0, -1 \leqslant x \leqslant 2$.

9. Approximate $\int_1^5 (x^2 + 1) \, dx$ using the rectangle rule with $n = 2$ and c_k the midpoint of each subinterval.

x	3	7	11	15
$f(x)$	1.2	3.4	2.6	0.5

10. Use the table of values in the margin and the rectangle rule with $n = 4$ and c_k the midpoint of each subinterval to approximate $\int_1^{17} f(x) \, dx$.

11. Find the average value of $f(x) = 6x^2 + 2x$ over the interval $[-1, 2]$.

B *Find each integral in Problems 12–17.*

12. $\int \sqrt[3]{6x - 5} \, dx$

13. $\int_0^1 10(2x - 1)^4 \, dx$

14. $\int \left(\frac{2}{x^2} - 2xe^{x^2} \right) dx$

15. $\int_0^4 x\sqrt{x^2 + 4} \, dx$

16. $\int (e^{-2x} + x^{-1}) \, dx$

17. $\int_0^{10} 10e^{-0.02x} \, dx$

18. Find a function $y = f(x)$ that satisfies both conditions:

$$\frac{dy}{dx} = 3x^{-1} - x^{-2} \qquad f(1) = 5$$

19. Find the equation of the curve that passes through $(2, 10)$ if its slope is given by

$$\frac{dy}{dx} = 6x + 1$$

for each x.

20. Approximate $\int_0^1 e^{2x^2} \, dx$ to three decimal places using $n = 5$ ansd c_k the midpoint of each subinterval.

21. Find the average value of $f(x) = 3x^{1/2}$ over the interval $[1, 9]$.

C
22. Find the area bounded by the graphs of $y = x^2 - 4$ and $y = 0$, $-2 \leqslant x \leqslant 4$.

Find each integral in Problems 23–32.

23. $\int_0^3 \frac{x}{1 + x^2} \, dx$

24. $\int_0^3 \frac{x}{(1 + x^2)^2} \, dx$

25. $\int x^3(2x^4 + 5)^5 \, dx$

26. $\int \frac{e^{-x}}{e^{-x} + 3} \, dx$

27. $\int \frac{e^x}{(e^x + 2)^2} \, dx$

28. $\int \frac{(\ln x)^2}{x} \, dx$

29. $\int x(x^3 - 1)^2 \, dx$

30. $\int \frac{x}{\sqrt{6 - x}} \, dx$

31. $\int_0^7 x\sqrt{16 - x} \, dx$

32. $\int_{-1}^1 x(x + 1)^4 \, dx$

33. Find a function $y = f(x)$ that satisfies both conditions:

$$\frac{dy}{dx} = 9x^2 e^{x^3} \qquad f(0) = 2$$

34. Find the area bounded by the graphs of $y = 6 - x^2$ and $y = x^2 - 2$, $0 \leqslant x \leqslant 3$.

Business & Economics

35. *Profit function.* If the marginal profit for producing x units per day is given by

$$P'(x) = 100 - 0.02x \qquad P(0) = 0$$

where $P(x)$ is the profit in dollars, find the profit function P and the profit on 10 units of production per day.

36. *Resource depletion.* An oil well starts out producing oil at the rate of 60,000 barrels of oil per year, but the production rate is expected to decrease by 4,000 barrels per year. Thus, if $P(t)$ is the total production (in thousands of barrels) in t years, then

$$P'(t) = f(t) = 60 - 4t \qquad 0 \le t \le 15$$

Write a definite integral that will give the total production after 15 years of operation, and evaluate it.

37. *Profit and production.* The weekly marginal profit for an output of x units is given approximately by

$$P'(x) = 150 - \frac{x}{10} \qquad 0 \le x \le 40$$

What is the total change in profit for a production change from 10 units per week to 40 units? Set up a definite integral and evaluate it.

38. *Useful life.* The total accumulated costs $C(t)$ and revenues $R(t)$ in thousands of dollars), respectively, for a coal mine satisfy

$$C'(t) = 3 \qquad \text{and} \qquad R'(t) = 20e^{-0.1t}$$

where t is the number of years the mine has been in operation. Find the useful life of the mine to the nearest year. What is the total profit accumulated during the useful life of the mine?

39. *Marketing.* The market research department for an automobile company estimates that the sales (in millions of dollars) of a new automobile will increase at the monthly rate of

$$S'(t) = 4e^{-0.08t} \qquad 0 \le t \le 24$$

t months after the introduction of the automobile. What will be the total sales $S(t)$, t months after the automobile is introduced if we assume that there were 0 sales at the time the automobile entered the marketplace? What are the estimated total sales during the first 12 months after the introduction of the automobile? How long will it take for the total sales to reach $40 million?

40. *Income distribution.* An economist produced the following Lorenz curves for the current income distribution and the projected income distribution 10 years from now in a certain country:

$$f(x) = \tfrac{1}{10}x + \tfrac{9}{10}x^2 \qquad \text{Current Lorenz curve}$$
$$g(x) = x^{1.5} \qquad\qquad \text{Projected Lorenz curve}$$

Find the coefficient of inequality for each Lorenz curve and interpret the results.

41. *Inventory.* Suppose the inventory of a certain item t months after the first of the year is given approximately by

$$I(t) = 10 + 36t - 3t^2 \qquad 0 \leqslant t \leqslant 12$$

What is the average inventory for the second quarter of the year?

42. *Supply function.* Given the supply function

$$p = S(x) = 8(e^{0.05x} - 1)$$

find the average price (in dollars) over the supply interval [40, 50].

43. *Consumers' and producers' surplus.* Given the price–demand and price–supply equations

$$p = D(x) = 70 - \tfrac{1}{5}x \qquad \text{and} \qquad p = S(x) = 13 + \tfrac{3}{2,500}x^2$$

(A) Find the consumers' surplus at a price level of $\bar{p} = \$50$.
(B) Find the producers' surplus at a price level of $\bar{p} = \$25$.
(C) Find the equilibrium price, and then find the consumers' surplus and producers' surplus at the equilibrium price level.

Life Sciences

44. *Wound healing.* The area of a small, healing surface wound changes at a rate given approximately by

$$\frac{dA}{dt} = -5t^{-2} \qquad 1 \leqslant t \leqslant 5$$

where t is time in days and $A(1) = 5$ square centimeters. What will the area of the wound be in 5 days?

45. *Pollution.* In an industrial area, the concentration $C(t)$ of particulate matter (in parts per million) during a 12 hour period is given in the figure. Use the rectangle rule with $n = 6$ and c_k the midpoint of each subinterval to approximate the average concentration during this 12 hour period. Estimate the necessary function values from the graph.

C(t)

Particulate matter (ppm)

Time (hours)

Social Sciences   46. *Learning.* In a particular business college, it was found that an average student enrolled in a typing class progressed at a rate of $N'(t) = 7e^{-0.1t}$ words per minute t weeks after enrolling in a 15 week course. If at the beginning of the course a student could type 25 words per minute, how many words per minute, $N(t)$, would the student be expected to type t weeks into the course? After completing the course?

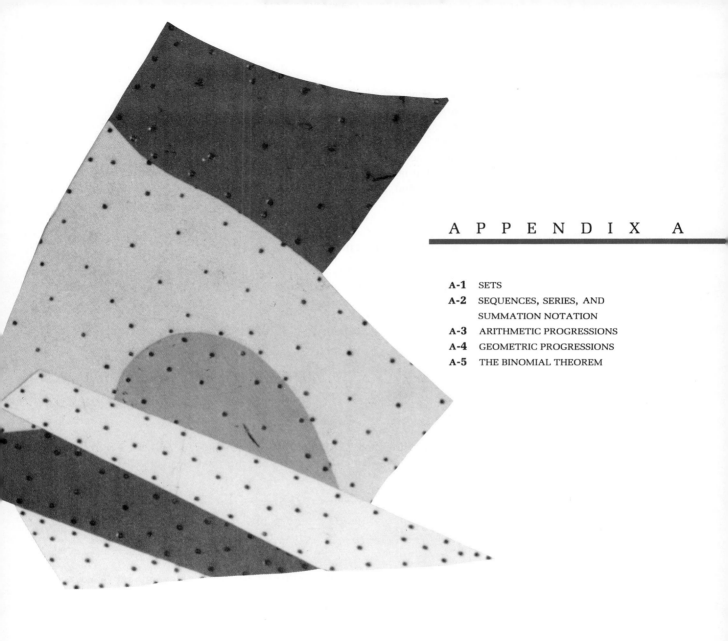

Special Topics

Sets

◆ SET PROPERTIES AND SET NOTATION
◆ SET OPERATIONS
◆ APPLICATION

In this section we will review a few key ideas from set theory. Set concepts and notation not only help us talk about certain mathematical ideas with greater clarity and precision, but are indispensable to a clear understanding of probability.

◆ SET PROPERTIES AND SET NOTATION

We can think of a **set** as any collection of objects specified in such a way that we can tell whether any given object is or is not in the collection. Capital letters, such as A, B, and C, are often used to designate particular sets. Each object in a set is called a **member** or **element** of the set. Symbolically:

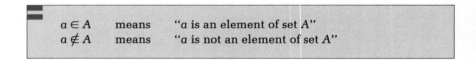

$a \in A$ means   "a is an element of set A"
$a \notin A$ means   "a is not an element of set A"

A set without any elements is called the **empty,** or **null, set.** For example, the set of all people over 10 feet tall is an empty set. Symbolically:

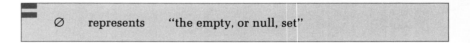

$\varnothing$ represents   "the empty, or null, set"

A set is usually described either by listing all its elements between braces { } or by enclosing a rule within braces that determines the elements of the set. Thus, if $P(x)$ is a statement about x, then

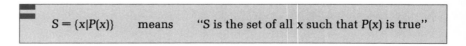

$S = \{x | P(x)\}$ means   "S is the set of all x such that $P(x)$ is true"

Recall that the vertical bar within the braces is read "such that." The following example illustrates the rule and listing methods of representing sets.

Handwritten annotation at top:

How many Subsets $= 2^{\text{no. of elements}}$

ie. $A\ \{1,2,3\}$

Possible Subsets $= 2^3 = 8$

(null set is always one of the subsets)

◆ E X A M P L E 1

$$\{x \mid x \text{ is a weekend day}\} = \{\text{Saturday, Sunday}\}$$
$$\{x \mid x^2 = 4\} = \{-2, 2\}$$
$$\{x \mid x \text{ is an odd positive counting number}\} = \{1, 3, 5, \ldots\} \qquad ◆$$

The three dots (. . .) in the last set given in Example 1 indicate that the pattern established by the first three entries continues indefinitely. The first two sets in Example 1 are **finite sets** (we intuitively know that the elements can be counted, and there is an end); the last set is an **infinite set** (we intuitively know that there is no end in counting the elements). When listing the elements in a set, we do not list an element more than once.

P R O B L E M 1   Let G be the set of all numbers such that $x^2 = 9$.

(A) Denote G by the rule method.   (B) Denote G by the listing method.

(C) Indicate whether the following are true or false: $3 \in G$; $9 \notin G$.   ◆

If each element of a set A is also an element of set B, we say that A is a **subset** of B. For example, the set of all women students in a class is a subset of the whole class. Note that the definition allows a set to be a subset of itself. If set A and set B have exactly the same elements, then the two sets are said to be **equal.** Symbolically:

$A \subset B$	means	"A is a subset of B"
$A = B$	means	"A and B have exactly the same elements"
$A \not\subset B$	means	"A is not a subset of B"
$A \neq B$	means	"A and B do not have exactly the same elements"

It can be proved that

$\varnothing$ **is a subset of every set**

◆ E X A M P L E 2   If $A = \{-3, -1, 1, 3\}$, $B = \{3, -3, 1, -1\}$, and $C = \{-3, -2, -1, 0, 1, 2, 3\}$, then each of the following statements is true:

$A = B$	$A \subset C$	$A \subset B$
$C \neq A$	$C \not\subset A$	$B \subset A$
$\varnothing \subset A$	$\varnothing \subset C$	$\varnothing \not\subset A$

   ◆

P R O B L E M 2   Given $A = \{0, 2, 4, 6\}$, $B = \{0, 1, 2, 3, 4, 5, 6\}$, and $C = \{2, 6, 0, 4\}$, indicate whether the following relationships are true (T) or false (F):

(A) $A \subset B$ (B) $A \subset C$ (C) $A = C$

(D) $C \subset B$ (E) $B \not\subset A$ (F) $\varnothing \subset B$ ◆

◆ E X A M P L E   3   List all the subsets of the set $\{a, b, c\}$.

Solution $\{a, b, c\}, \{a, b\}, \{a, c\}, \{b, c\}, \{a\}, \{b\}, \{c\}, \varnothing$ ◆

P R O B L E M   3   List all the subsets of the set $\{1, 2\}$.   ◆

◆ SET OPERATIONS

The **union** of sets A and B, denoted by $A \cup B$, is the set of all elements formed by combining all the elements of A and all the elements of B into one set. Symbolically:

Union

$$A \cup B = \{x | x \in A \textbf{ or } x \in B\}$$

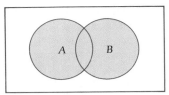

FIGURE   1
$A \cup B$ is the shaded region.

Here we use the word **or** in the way it is always used in mathematics; that is, x may be an element of set A or set B or both.

Venn diagrams are useful in visualizing set relationships. The union of two sets can be illustrated as shown in Figure 1. Note that

$$A \subset A \cup B \quad \text{and} \quad B \subset A \cup B$$

The **intersection** of sets A and B, denoted by $A \cap B$, is the set of elements in set A that are also in set B. Symbolically:

Intersection

$$A \cap B = \{x | x \in A \textbf{ and } x \in B\}$$

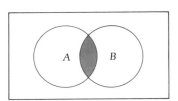

FIGURE   2
$A \cap B$ is the shaded region.

This relationship is easily visualized in the Venn diagram shown in Figure 2. Note that

$$A \cap B \subset A \quad \text{and} \quad A \cap B \subset B$$

If $A \cap B = \varnothing$, then the sets A and B are said to be **disjoint;** this is illustrated in Figure 3.

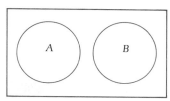

FIGURE 3
$A \cap B = \emptyset$; A and B are disjoint.

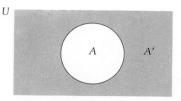

FIGURE 4
The complement of A is A'.

The set of all elements under consideration is called the **universal set** U. Once the universal set is determined for a particular case, all other sets under discussion must be subsets of U.

We now define one more operation on sets, called the *complement*. The **complement** of A (relative to U), denoted by A', is the set of elements in U that are not in A (see Fig. 4). Symbolically:

Complement

$$A' = \{x \in U \,|\, x \notin A\}$$

[handwritten: i.e.]

[handwritten:
$U = \{1, 2, 3, 4, 5\}$
$A = \{1, 2\}$
$A' = \{3, 4, 5\}$ *]*

◆ EXAMPLE 4   If $A = \{3, 6, 9\}$, $B = \{3, 4, 5, 6, 7\}$, $C = \{4, 5, 7\}$, and $U = \{1, 2, 3, 4, 5, 6, 7, 8, 9\}$, then

$A \cup B = \{3, 4, 5, 6, 7, 9\}$
$A \cap B = \{3, 6\}$
$A \cap C = \emptyset$ A and C are disjoint
$\quad B' = \{1, 2, 8, 9\}$ ◆

PROBLEM 4   If $R = \{1, 2, 3, 4\}$, $S = \{1, 3, 5, 7\}$, $T = \{2, 4\}$, and $U = \{1, 2, 3, 4, 5, 6, 7, 8, 9\}$, find:

(A) $R \cup S$ (B) $R \cap S$ (C) $S \cap T$ (D) S' ◆

◆ APPLICATION

◆ EXAMPLE 5

Marketing Survey

From a survey of 100 college students, a marketing research company found that 75 students owned stereos, 45 owned cars, and 35 owned cars and stereos.

(A) How many students owned either a car or a stereo?
(B) How many students did not own either a car or a stereo?

Solutions   Venn diagrams are very useful for this type of problem. If we let

U = Set of students in sample (100)
S = Set of students who own stereos (75)
C = Set of students who own cars (45)
$S \cap C$ = Set of students who own cars and stereos (35)

then:

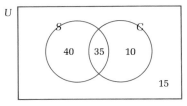

Place the number in the intersection first; then work outward:

$$40 = 75 - 35$$
$$10 = 45 - 35$$
$$15 = 100 - (40 + 35 + 10)$$

(A) The number of students who own either a car or a stereo is the number of students in the set $S \cup C$. You might be tempted to say that this is just the number of students in S plus the number of students in C, $75 + 45 = 120$, but this sum is larger than the sample we started with! What is wrong? We have actually counted the number in the intersection (35) twice. The correct answer, as seen in the Venn diagram, is

$$40 + 35 + 10 = 85$$

(B) The number of students who do not own either a car or a stereo is the number of students in the set $(S \cup C)'$; that is, 15.   ◆

PROBLEM 5   Referring to Example 5:

(A) How many students owned a car but not a stereo?
(B) How many students did not own both a car and a stereo?   ◆

Note in Example 5 and Problem 5 that the word **and** is associated with intersection and the word **or** is associated with union.

Answers to Matched Problems
1. (A) $\{x | x^2 = 9\}$ (B) $\{-3, 3\}$ (C) True; True
2. All are true   3. $\{1, 2\}, \{1\}, \{2\}, \varnothing$
4. (A) $\{1, 2, 3, 4, 5, 7\}$ (B) $\{1, 3\}$ (C) $\varnothing$ (D) $\{2, 4, 6, 8, 9\}$
5. (A) 10, the number in $S' \cap C$ (B) 65, the number in $(S \cap C)'$

EXERCISE   A-1

A   *Indicate true (T) or false (F).*

1. $4 \in \{2, 3, 4\}$
2. $6 \notin \{2, 3, 4\}$
3. $\{2, 3\} \subset \{2, 3, 4\}$
4. $\{3, 2, 4\} = \{2, 3, 4\}$

5. $\{3, 2, 4\} \subset \{2, 3, 4\}$ 6. $\{3, 2, 4\} \in \{2, 3, 4\}$
7. $\varnothing \subset \{2, 3, 4\}$ 8. $\varnothing = \{0\}$

In Problems 9–14, write the resulting set using the listing method.

9. $\{1, 3, 5\} \cup \{2, 3, 4\}$ 10. $\{3, 4, 6, 7\} \cup \{3, 4, 5\}$
11. $\{1, 3, 4\} \cap \{2, 3, 4\}$ 12. $\{3, 4, 6, 7\} \cap \{3, 4, 5\}$
13. $\{1, 5, 9\} \cap \{3, 4, 6, 8\}$ 14. $\{6, 8, 9, 11\} \cap \{3, 4, 5, 7\}$

B   In Problems 15–20, write the resulting set using the listing method.

15. $\{x|x - 2 = 0\}$ 16. $\{x|x + 7 = 0\}$ 17. $\{x|x^2 = 49\}$ 18. $\{x|x^2 = 100\}$
19. $\{x|x$ is an odd number between 1 and 9, inclusive$\}$
20. $\{x|x$ is a month starting with M$\}$

21. For $U = \{1, 2, 3, 4, 5\}$ and $A = \{2, 3, 4\}$, find A'.
22. For $U = \{7, 8, 9, 10, 11\}$ and $A = \{7, 11\}$, find A'.

Problems 23–34 refer to the Venn diagram in the margin. How many elements are in each of the indicated sets?

23. A 24. U 25. A' 26. B'
27. $A \cup B$ 28. $A \cap B$ 29. $A' \cap B$ 30. $A \cap B'$
31. $(A \cap B)'$ 32. $(A \cup B)'$ 33. $A' \cap B'$ 34. U' $\varnothing$

35. If $R = \{1, 2, 3, 4\}$ and $T = \{2, 4, 6\}$, find:
    (A) $\{x|x \in R$ **or** $x \in T\}$ (B) $R \cup T$
36. If $R = \{1, 3, 4\}$ and $T = \{2, 4, 6\}$, find:
    (A) $\{x|x \in R$ **and** $x \in T\}$ ⁴ (B) $R \cap T$ ⁴
37. For $P = \{1, 2, 3, 4\}$, $Q = \{2, 4, 6\}$, and $R = \{3, 4, 5, 6\}$, find $P \cup (Q \cap R)$.
38. For P, Q, and R in Problem 37, find $P \cap (Q \cup R)$.

C   Venn diagrams may be of help in Problems 39–44.

39. If $A \cup B = B$, can we always conclude that $A \subset B$?
40. If $A \cap B = B$, can we always conclude that $B \subset A$?
41. If A and B are arbitrary sets, can we always conclude that $A \cap B \subset B$?
42. If $A \cap B = \varnothing$, can we always conclude that $B = \varnothing$?
43. If $A \subset B$ and $x \in A$, can we always conclude that $x \in B$?
44. If $A \subset B$ and $x \in B$, can we always conclude that $x \in A$?

45. How many subsets does each of the following sets have? Also, try to discover a formula in terms of n for a set with n elements.
    (A) $\{a\}$ (B) $\{a, b\}$ (C) $\{a, b, c\}$

46. How do the sets $\varnothing$, $\{\varnothing\}$, and $\{0\}$ differ from each other?

Business & Economics

Marketing survey. Problems 47–58 refer to the following survey: A marketing survey of 1,000 car commuters found that 600 listen to the news, 500 listen to music, and 300 listen to both. Let

N = Set of commuters in the sample who listen to news
M = Set of commuters in the sample who listen to music

Following the procedures in Example 5, find the number of commuters in each set described below.

47. $N \cup M$ **48.** $N \cap M$ **49.** $(N \cup M)'$
50. $(N \cap M)'$ **51.** $N' \cap M$ **52.** $N \cap M'$
53. Set of commuters who listen to either news or music
54. Set of commuters who listen to both news and music
55. Set of commuters who do not listen to either news or music
56. Set of commuters who do not listen to both news and music
57. Set of commuters who listen to music but not news
58. Set of commuters who listen to news but not music

59. *Committee selection.* The management of a company, a president and three vice presidents, denoted by the set $\{P, V_1, V_2, V_3\}$, wish to select a committee of 2 people from among themselves. How many ways can this committee be formed? That is, how many 2-person subsets can be formed from a set of 4 people?

60. *Voting coalition.* The management of the company in Problem 59 decides for or against certain measures as follows: The president has 2 votes and each vice president has 1 vote. Three favorable votes are needed to pass a measure. List all minimal winning coalitions; that is, list all subsets of $\{P, V_1, V_2, V_3\}$ that represent exactly 3 votes.

Life Sciences

Blood types. When receiving a blood transfusion, a recipient must have all the antigens of the donor. A person may have one or more of the three antigens A, B, and Rh, or none at all. Eight blood types are possible, as indicated in the following Venn diagram, where U is the set of all people under consideration:

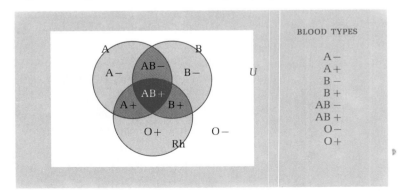

An A— person has A antigens but no B or Rh; an O+ person has Rh but neither A nor B; an AB— person has A and B antigens but no Rh; and so on.

Using the Venn diagram, indicate which of the eight blood types are included in each set.

61. $A \cap Rh$ **62.** $A \cap B$ **63.** $A \cup Rh$ **64.** $A \cup B$

65. $(A \cup B)'$ **66.** $(A \cup B \cup Rh)'$ **67.** $A' \cap B$ **68.** $Rh' \cap A$

Social Sciences

Group structures. R. D. Luce and A. D. Perry, in a study on group structure (Psychometrika, 1949, 14:95–116), used the idea of sets to formally define the notion of a clique within a group. Let G be the set of all persons in the group and let $C \subset G$. Then C is a clique provided that:

1. *C contains at least 3 elements.*
2. *For every $a, b \in C$, a **R** b and b **R** a.*
3. *For every $a \notin C$, there is at least one $b \in C$ such that a **R̸** b or b **R̸** or both.*

*[Note: Interpret "a **R** b" to mean "a relates to b," "a likes b," "a is as wealthy as b," and so on. Of course, "a **R̸** b" means "a does not relate to b," and so on.]*

69. Translate statement 2 into ordinary English.

70. Translate statement 3 into ordinary English.

Sequences, Series, and Summation Notation

◆ SEQUENCES

◆ SERIES AND SUMMATION NOTATION

If someone asked you to list all natural numbers that are perfect squares, you might begin by writing

1, 4, 9, 16, 25, 36

But you would soon realize that it is impossible to actually list all the perfect squares, since there are an infinite number of them. However, you could represent this collection of numbers in several different ways. One common method is to write

$1, 4, 9, \ldots, n^2, \ldots$ $n \in N$

where N is the set of natural numbers. A list of numbers such as this is generally called a *sequence*. Sequences and related topics form the subject matter of this section.

◆ SEQUENCES

Consider the function f given by

$$f(n) = 2n + 1 \tag{1}$$

where the domain of f is the set of natural numbers N. Note that

$$f(1) = 3, \quad f(2) = 5, \quad f(3) = 7, \quad \ldots$$

The function f is an example of a sequence. In general, a **sequence** is a function with domain a set of successive integers. Instead of the standard function notation used in equation (1), sequences are usually defined in terms of a special notation.

The range value $f(n)$ is usually symbolized more compactly with a symbol such as a_n. Thus, in place of equation (1), we write

$$a_n = 2n + 1$$

and the domain is understood to be the set of natural numbers unless something is said to the contrary or the context indicates otherwise. The elements in the range are called **terms of the sequence;** a_1 is the first term, a_2 is the second term, and a_n is the **nth term,** or **general term.**

$$a_1 = 2(1) + 1 = 3 \qquad \text{First term}$$
$$a_2 = 2(2) + 1 = 5 \qquad \text{Second term}$$
$$a_3 = 2(3) + 1 = 7 \qquad \text{Third term}$$
$$\cdot$$
$$\cdot$$
$$\cdot$$
$$a_n = 2n + 1 \qquad \text{General term}$$

When the terms in a sequence are written in their natural order with respect to domain values,

$$a_1, a_2, a_3, \ldots, a_n, \ldots$$

or

$$3, 5, 7, \ldots, 2n + 1, \ldots$$

the ordered list of elements is often informally referred to as a sequence. A sequence also may be represented in the abbreviated form $\{a_n\}$, where a symbol for the nth term is written within braces. For example, we could refer to the sequence $3, 5, 7, \ldots, 2n + 1, \ldots$ as the sequence $\{2n + 1\}$.

If the domain of a sequence is a finite set of successive integers, then the sequence is called a **finite sequence.** If the domain is an infinite set of successive integers, then the sequence is called an **infinite sequence.** The sequence $\{2n + 1\}$ discussed above is an infinite sequence.

◆ E X A M P L E   6   Write the first four terms of each sequence:

(A) $a_n = 3n - 2$ (B) $\left\{\dfrac{(-1)^n}{n}\right\}$

Solutions   (A) 1, 4, 7, 10   (B) $-1, \dfrac{1}{2}, \dfrac{-1}{3}, \dfrac{1}{4}$ ◆

PROBLEM 6   Write the first four terms of each sequence:

(A) $a_n = -n + 3$ (B) $\left\{\dfrac{(-1)^n}{2^n}\right\}$ ◆

Now that we have seen how to use the general term to find the first few terms in a sequence, we consider the reverse problem. That is, can a sequence be defined just by listing the first three or four terms of the sequence? And can we then use these initial terms to find a formula for the nth term? In general, without other information, the answer to the first question is no. Many different sequences may start off with the same terms. For example, each of the following sequences starts off with the same three terms:

$2, 4, 8, \ldots, 2^n, \ldots$

$2, 4, 8, \ldots, n^2 - n + 2, \ldots$

$2, 4, 8, \ldots, 5n + \dfrac{6}{n} - 9, \ldots$

However, these are certainly different sequences. (You should verify that these sequences agree for the first three terms and differ in the fourth term by evaluating the general term for each sequence at $n = 1, 2, 3,$ and 4.) Thus, simply listing the first three terms (or any other finite number of terms) does not specify a particular sequence. In fact, it can be shown that given any list of m numbers, there are an infinite number of sequences whose first m terms agree with these given numbers.

What about the second question? That is, given a few terms, can we find the general formula for at least one sequence whose first few terms agree with the given terms? The answer to this question is a qualified yes. If we can observe a simple pattern in the given terms, then we usually can construct a general term that will produce that pattern. The next example illustrates this approach.

◆ EXAMPLE 7   Find the general term of a sequence whose first four terms are:

(A) $3, 4, 5, 6, \ldots$ (B) $5, -25, 125, -625, \ldots$

Solutions   (A) Since these terms are consecutive integers, one solution is $a_n = n, n \geq 3$. If we want the domain of the sequence to be all natural numbers, then another solution is $b_n = n + 2$.

(B) Each of these terms can be written as the product of a power of 5 and a power of -1:

$$
\begin{aligned}
5 &= (-1)^0 5^1 = a_1 \\
-25 &= (-1)^1 5^2 = a_2 \\
125 &= (-1)^2 5^3 = a_3 \\
-625 &= (-1)^3 5^4 = a_4
\end{aligned}
$$

If we choose the domain to be all natural numbers, then a solution is

$a_n = (-1)^{n-1} 5^n$ ◆

Find the general term of a sequence whose first four terms are:

(A) 3, 6, 9, 12, . . .   (B) 1, −2, 4, −8, . . .   ◆

In general, there is usually more than one way of representing the nth term of a given sequence (see the solution of part A in Example 7). However, unless something is stated to the contrary, we assume the domain of the sequence is the set of natural numbers N.

◆ SERIES AND SUMMATION NOTATION

The sum of the terms of a sequence is called a **series.** If the sequence is finite, the corresponding series is a **finite series.** If the sequence is infinite, the corresponding series is an **infinite series.** We will consider only finite series in this section. For example,

1, 3, 5, 7, 9   Finite sequence
1 + 3 + 5 + 7 + 9   Finite series

Notice that we can easily evaluate this series by adding the five terms:

1 + 3 + 5 + 7 + 9 = 25

Series are often represented in a compact form called **summation notation.** Consider the following examples:

$$\sum_{k=3}^{6} k^2 = 3^2 + 4^2 + 5^2 + 6^2$$

$$= 9 + 16 + 25 + 36 = 86$$

$$\sum_{k=0}^{2} (4k + 1) = (4 \cdot 0 + 1) + (4 \cdot 1 + 1) + (4 \cdot 2 + 1)$$

$$= 1 + 5 + 9 = 15$$

In each case, the terms of the series on the right are obtained from the expression on the left by successively replacing the **summing index k** with integers, starting with the number indicated below the **summation sign Σ** and ending with the number that appears above Σ. The summing index may be represented by letters other than k and may start at any integer and end at any integer greater than or equal to the starting integer. Thus, if we are given the finite sequence

$$\frac{1}{2}, \frac{1}{4}, \frac{1}{8}, \cdots, \frac{1}{2^n}$$

the corresponding series is

$$\frac{1}{2} + \frac{1}{4} + \frac{1}{8} + \cdots + \frac{1}{2^n} = \sum_{j=1}^{n} \frac{1}{2^j}$$

where we have used j for the summing index.

◆ E X A M P L E   8   Write

$$\sum_{k=1}^{5} \frac{k}{k^2 + 1}$$

without summation notation. Do not evaluate the sum.

Solution $$\sum_{k=1}^{5} \frac{k}{k^2 + 1} = \frac{1}{1^2 + 1} + \frac{2}{2^2 + 1} + \frac{3}{3^2 + 1} + \frac{4}{4^2 + 1} + \frac{5}{5^2 + 1}$$

$$= \frac{1}{2} + \frac{2}{5} + \frac{3}{10} + \frac{4}{17} + \frac{5}{26}$$ ◆

P R O B L E M   8   Write

$$\sum_{k=1}^{5} \frac{k + 1}{k}$$

without summation notation. Do not evaluate the sum.   ◆

If the terms of a series are alternately positive and negative, we call the series an **alternating series.** The next example deals with the representation of such a series.

◆ E X A M P L E   9   Write the alternating series

$$\frac{1}{2} - \frac{1}{4} + \frac{1}{6} - \frac{1}{8} + \frac{1}{10} - \frac{1}{12}$$

using summation notation with:

(A) The summing index k starting at 1
(B) The summing index j starting at 0

Solutions   (A) $(-1)^{k+1}$ provides the alternation of sign, and $1/(2k)$ provides the other part of each term. Thus, we can write

$$\frac{1}{2} - \frac{1}{4} + \frac{1}{6} - \frac{1}{8} + \frac{1}{10} - \frac{1}{12} = \sum_{k=1}^{6} \frac{(-1)^{k+1}}{2k}$$

(B) $(-1)^j$ provides the alternation of sign, and $1/[2(j + 1)]$ provides the other part of each term. Thus, we can write

$$\frac{1}{2} - \frac{1}{4} + \frac{1}{6} - \frac{1}{8} + \frac{1}{10} - \frac{1}{12} = \sum_{j=0}^{5} \frac{(-1)^j}{2(j + 1)}$$ ◆

PROBLEM 9 Write the alternating series

$$1 - \frac{1}{3} + \frac{1}{9} - \frac{1}{27} + \frac{1}{81}$$

using summation notation with:

(A) The summing index k starting at 1
(B) The summing index j starting at 0 ◆

Summation notation provides a compact notation for the sum of any list of numbers, even if the numbers are not generated by a formula. For example, suppose the results of an examination taken by a class of 10 students are given in the following list:

87, 77, 95, 83, 86, 73, 95, 68, 75, 86

If we let $a_1, a_2, a_3, \ldots, a_{10}$ represent these 10 scores, then the average test score is given by

$$\frac{1}{10} \sum_{k=1}^{10} a_k = \frac{1}{10} (87 + 77 + 95 + 83 + 86 + 73 + 95 + 68 + 75 + 86)$$

$$= \frac{1}{10} (825) = 82.5$$

More generally, in statistics, the **arithmetic mean $\bar{a}$** of a list of n numbers $a_1, a_2, \ldots, a_n$ is defined as

$$\bar{a} = \frac{1}{n} \sum_{k=1}^{n} a_k$$

◆ EXAMPLE 10 Find the arithmetic mean of 3, 5, 4, 7, 4, 2, 3, and 6.

Solution $\bar{a} = \frac{1}{8} \sum_{k=1}^{8} a_k = \frac{1}{8} (3 + 5 + 4 + 7 + 4 + 2 + 3 + 6) = \frac{1}{8} (34) = 4.25$ ◆

PROBLEM 10 Find the arithmetic mean of 9, 3, 8, 4, 3, and 6. ◆

Answers to Matched Problems

6. (A) 2, 1, 0, -1 (B) $\frac{-1}{2}, \frac{1}{4}, \frac{-1}{8}, \frac{1}{16}$
7. (A) $a_n = 3n$ (B) $a_n = (-2)^{n-1}$ 8. $2 + \frac{3}{2} + \frac{4}{3} + \frac{5}{4} + \frac{6}{5}$

9. (A) $\sum_{k=1}^{5} \frac{(-1)^{k-1}}{3^{k-1}}$ (B) $\sum_{j=0}^{4} \frac{(-1)^j}{3^j}$ 10. 5.5

EXERCISE A-2

A Write the first four terms for each sequence.

1. $a_n = 2n + 3$ 2. $a_n = 4n - 3$ 3. $a_n = \dfrac{n+2}{n+1}$

4. $a_n = \dfrac{2n+1}{2n}$ 5. $a_n = (-3)^{n+1}$ 6. $a_n = (-\frac{1}{4})^{n-1}$

7. Write the 10th term of the sequence in Problem 1.
8. Write the 15th term of the sequence in Problem 2.
9. Write the 99th term of the sequence in Problem 3.
10. Write the 200th term of the sequence in Problem 4.

Write each series in expanded form without summation notation, and evaluate.

11. $\displaystyle\sum_{k=1}^{6} k$

12. $\displaystyle\sum_{k=1}^{5} k^2$

13. $\displaystyle\sum_{k=4}^{7} (2k-3)$

14. $\displaystyle\sum_{k=0}^{4} (-2)^k$

15. $\displaystyle\sum_{k=0}^{3} \frac{1}{10^k}$

16. $\displaystyle\sum_{k=1}^{4} \frac{1}{2^k}$

Find the arithmetic mean of each list of numbers.

17. 5, 4, 2, 1, and 6

18. 7, 9, 9, 2, and 4

19. 96, 65, 82, 74, 91, 88, 87, 91, 77, and 74

20. 100, 62, 95, 91, 82, 87, 70, 75, 87, and 82

B *Write the first five terms of each sequence.*

21. $a_n = \dfrac{(-1)^{n+1}}{2^n}$

22. $a_n = (-1)^n(n-1)^2$

23. $a_n = n[1 + (-1)^n]$

24. $a_n = \dfrac{1 - (-1)^n}{n}$

25. $a_n = \left(-\dfrac{3}{2}\right)^{n-1}$

26. $a_n = \left(-\dfrac{1}{2}\right)^{n+1}$

In Problems 27–42, find the general term of a sequence whose first four terms agree with the given terms.

27. $-2, -1, 0, 1, \ldots$

28. $4, 5, 6, 7, \ldots$

29. $4, 8, 12, 16, \ldots$

30. $-3, -6, -9, -12, \ldots$

31. $\frac{1}{2}, \frac{3}{4}, \frac{5}{6}, \frac{7}{8}, \ldots$

32. $\frac{1}{2}, \frac{2}{3}, \frac{3}{4}, \frac{4}{5}, \ldots$

33. $1, -2, 3, -4, \ldots$

34. $-2, 4, -8, 16, \ldots$

35. $1, -3, 5, -7, \ldots$

36. $3, -6, 9, -12, \ldots$

37. $1, \frac{2}{5}, \frac{4}{25}, \frac{8}{125}, \ldots$

38. $\frac{4}{3}, \frac{16}{9}, \frac{64}{27}, \frac{256}{81}, \ldots$

39. $x, x^2, x^3, x^4, \ldots$

40. $1, 2x, 3x^2, 4x^3, \ldots$

41. $x, -x^3, x^5, -x^7, \ldots$

42. $x, \dfrac{x^2}{2}, \dfrac{x^3}{3}, \dfrac{x^4}{4}, \ldots$

Write each series in expanded form without summation notation. Do not evaluate.

43. $\displaystyle\sum_{k=1}^{5} (-1)^{k+1}(2k-1)^2$

44. $\displaystyle\sum_{k=1}^{4} \frac{(-2)^{k+1}}{2k+1}$

45. $\displaystyle\sum_{k=2}^{5} \frac{2^k}{2k+3}$

46. $\displaystyle\sum_{k=3}^{7} \frac{(-1)^k}{k^2 - k}$

47. $\displaystyle\sum_{k=1}^{5} x^{k-1}$

48. $\displaystyle\sum_{k=1}^{3} \frac{1}{k} x^{k+1}$

49. $\displaystyle\sum_{k=0}^{4} \frac{(-1)^k x^{2k+1}}{2k+1}$

50. $\displaystyle\sum_{k=0}^{4} \frac{(-1)^k x^{2k}}{2k+2}$

Write each series using summation notation with:

(A) The summing index k starting at $k = 1$
(B) The summing index j starting at $j = 0$

51. $2 + 3 + 4 + 5 + 6$

52. $1^2 + 2^2 + 3^2 + 4^2$

53. $1 - \frac{1}{2} + \frac{1}{3} - \frac{1}{4}$

54. $1 - \frac{1}{3} + \frac{1}{5} - \frac{1}{7} + \frac{1}{9}$

Write each series using summation notation with the summing index k starting at $k = 1$.

55. $2 + \dfrac{3}{2} + \dfrac{4}{3} + \cdots + \dfrac{n+1}{n}$

56. $1 + \dfrac{1}{2^2} + \dfrac{1}{3^2} + \cdots + \dfrac{1}{n^2}$

57. $\dfrac{1}{2} - \dfrac{1}{4} + \dfrac{1}{8} - \cdots + \dfrac{(-1)^{n+1}}{2^n}$

58. $1 - 4 + 9 - \cdots + (-1)^{n+1}n^2$

C Some sequences are defined by a ***recursion formula*** — that is, a formula that defines each term of the sequence in terms of one or more of the preceding terms. For example, if $\{a_n\}$ is defined by

$$a_1 = 1 \qquad \text{and} \qquad a_n = 2a_{n-1} + 1 \qquad \text{for } n \geqslant 2$$

then

$$a_2 = 2a_1 + 1 = 2 \cdot 1 + 1 = 3$$
$$a_3 = 2a_2 + 1 = 2 \cdot 3 + 1 = 7$$
$$a_4 = 2a_3 + 1 = 2 \cdot 7 + 1 = 15$$

and so on. In Problems 59–62, write the first five terms of each sequence.

59. $a_1 = 2$ and $a_n = 3a_{n-1} + 2$ for $n \geqslant 2$
60. $a_1 = 3$ and $a_n = 2a_{n-1} - 2$ for $n \geqslant 2$
61. $a_1 = 1$ and $a_n = 2a_{n-1}$ for $n \geqslant 2$
62. $a_1 = 1$ and $a_n = -\frac{1}{3}a_{n-1}$ for $n \geqslant 2$

If A is a positive real number, then the terms of the sequence defined by

$$a_1 = \frac{A}{2} \qquad \text{and} \qquad a_n = \frac{1}{2}\left(a_{n-1} + \frac{A}{a_{n-1}}\right) \qquad \text{for } n \geqslant 2$$

can be used to approximate $\sqrt{A}$ to any decimal place accuracy desired. In Problems 63 and 64, compute the first four terms of this sequence for the indicated value of A, and compare the fourth term with the value of $\sqrt{A}$ obtained from a calculator.

63. $A = 2$

64. $A = 6$

Arithmetic Progressions

- ARITHMETIC PROGRESSIONS — DEFINITIONS
- SPECIAL FORMULAS
- APPLICATION

◆ ARITHMETIC PROGRESSIONS — DEFINITIONS

Consider the sequence of numbers

$$1, 4, 7, 10, 13, \ldots, 1 + 3(n - 1), \ldots$$

where each number after the first is obtained from the preceding one by adding 3 to it. This is an example of an *arithmetic progression*. In general:

Arithmetic Progression

A sequence of numbers

$$a_1, a_2, a_3, \ldots, a_n, \ldots$$

is called an **arithmetic progression** if there is a constant d, called the **common difference,** such that

$$a_n - a_{n-1} = d$$

That is,

$$a_n = a_{n-1} + d \qquad \text{for every } n > 1 \tag{1}$$

◆ EXAMPLE 11   Which of the following can be the first four terms of an arithmetic progression, and what is its common difference?

(A) 2, 4, 8, 10, . . .   (B) 3, 8, 13, 18, . . .

Solution   The terms in B, with $d = 5$ ◆

PROBLEM 11   Which of the following can be the first four terms of an arithmetic progression, and what is its common difference?

(A) 15, 13, 11, 9, . . .   (B) 3, 9, 27, 81, . . .   ◆

◆ SPECIAL FORMULAS

Arithmetic progressions have a number of convenient properties. For example, it is easy to derive formulas for the nth term and the sum of any number of consecutive terms. To obtain a formula for the nth term of an arithmetic progression, we note that if a_1 is the first term and d is the common difference, then

$$a_2 = a_1 + d$$
$$a_3 = a_2 + d = (a_1 + d) + d = a_1 + 2d$$
$$a_4 = a_3 + d = (a_1 + 2d) + d = a_1 + 3d$$

This suggests that:

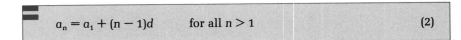

$$a_n = a_1 + (n-1)d \qquad \text{for all } n > 1 \tag{2}$$

◆ E X A M P L E   12   Find the 21st term in the arithmetic progression: 3, 8, 13, 18, . . .

Solution   Find the common difference d and use formula (2):

$$d = 5 \qquad n = 21 \qquad a_1 = 3$$

Thus

$$a_{21} = 3 + (21 - 1)5$$
$$= 103$$

◆

P R O B L E M   12   Find the 51st term in the arithmetic progression: 15, 13, 11, 9, . . .   ◆

We now derive two simple and very useful formulas for the sum of n consecutive terms of an arithmetic progression. Let

$$S_n = a_1 + a_2 + \cdots + a_{n-1} + a_n$$

be the sum of n terms of an arithmetic progression with common difference d. Then,

$$S_n = a_1 + (a_1 + d) + \cdots + [a_1 + (n-2)d] + [a_1 + (n-1)d]$$

Reversing the order of the sum, we obtain

$$S_n = [a_1 + (n-1)d] + [a_1 + (n-2)d] + \cdots + (a_1 + d) + a_1$$

Something interesting happens if we combine these last two equations by addition (adding corresponding terms on the right sides):

$$2S_n = [2a_1 + (n-1)d] + [2a_1 + (n-1)d] + \cdots$$
$$+ [2a_1 + (n-1)d] + [2a_1 + (n-1)d]$$

All the terms on the right side are the same, and there are n of them. Thus,

$$2S_n = n[2a_1 + (n-1)d]$$

and we have the following general formula:

$$S_n = \frac{n}{2}[2a_1 + (n-1)d] \tag{3}$$

Replacing

$$[a_1 + (n-1)d] \qquad \text{in} \qquad \frac{n}{2}[a_1 + a_1 + (n-1)d]$$

by a_n from equation (2), we obtain a second useful formula for the sum:

$$S_n = \frac{n}{2}(a_1 + a_n) \tag{4}$$

◆ E X A M P L E   13   Find the sum of the first 30 terms in the arithmetic progression:
3, 8, 13, 18, . . .

    Solution   Use formula (3) with $n = 30$, $a_1 = 3$, and $d = 5$:

$$S_{30} = \frac{30}{2}[2 \cdot 3 + (30-1)5] = 2{,}265 \qquad \blacklozenge$$

P R O B L E M   13   Find the sum of the first 40 terms in the arithmetic progression:
15, 13, 11, 9, . . .   ◆

◆ E X A M P L E   14   Find the sum of all the even numbers between 31 and 87.

    Solution   First, find n using equation (2):

$$a_n = a_1 + (n-1)d$$
$$86 = 32 + (n-1)2$$
$$n = 28$$

Now find s_{28} using formula (4):

$$s_n = \frac{n}{2}(a_1 + a_n)$$

$$S_{28} = \frac{28}{2}(32 + 86) = 1{,}652 \qquad \blacklozenge$$

P R O B L E M   14   Find the sum of all the odd numbers between 24 and 208.   ◆

◆ APPLICATION

◆ E X A M P L E   15

Loan Repayment

A person borrows $3,600 and agrees to repay the loan in monthly installments over a period of 3 years. The agreement is to pay 1% of the unpaid balance each month for using the money and $100 each month to reduce the loan. What is the total cost of the loan over the 3 years?

Solution   Let us look at the problem relative to a time line:

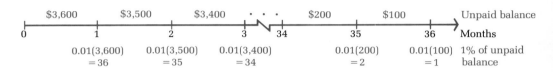

The total cost of the loan is

$$1 + 2 + \cdots + 34 + 35 + 36$$

The terms form an arithmetic progression with $n = 36$, $a_1 = 1$, and $a_{36} = 36$, so we can use formula (4):

$$S_n = \frac{n}{2}(a_1 + a_n)$$

$$S_{36} = \frac{36}{2}(1 + 36) = \$666$$

And we conclude that the total cost of the loan over the period of 3 years is $666.

◆

P R O B L E M   15   Repeat Example 15 with a loan of $6,000 over a period of 5 years.   ◆

Answers to Matched Problems   11.   The terms in A, with $d = -2$ 12. -85 13. -960
14.   10,672   15. $1,830

E X E R C I S E   A-3

A   1.   Determine which of the following can be the first three terms of an arithmetic progression. Find the common difference d and the next two terms for those that are.

(A) 5, 8, 11, . . .   (B) 4, 8, 16, . . .
(C) $-2, -4, -8$, . . .   (D) 8, $-2, -12$, . . .

2.   Repeat Problem 1 for:

(A) 11, 16, 21, . . .   (B) 16, 8, 4, . . .
(C) 2, $-3, -8$, . . .   (D) $-1, -2, -4$, . . .

Let $a_1, a_2, a_3, \ldots, a_n, \ldots$ be an arithmetic progression and S_n be the sum of the first n terms. In Problems 3–8, find the indicated quantities.

3. $a_1 = 7$; $d = 4$; $a_2 = ?$; $a_3 = ?$
4. $a_1 = -2$; $d = -3$; $a_2 = ?$; $a_3 = ?$

B

5. $a_1 = 2$; $d = 4$; $a_{21} = ?$; $S_{31} = ?$
6. $a_1 = 8$; $d = -10$; $a_{15} = ?$; $S_{23} = ?$
7. $a_1 = 18$; $a_{20} = 75$; $S_{20} = ?$
8. $a_1 = 203$; $a_{30} = 261$; $S_{30} = ?$

9. Find $f(1) + f(2) + f(3) + \cdots + f(50)$ if $f(x) = 2x - 3$.
10. Find $g(1) + g(2) + g(3) + \cdots + g(100)$ if $g(t) = 18 - 3t$.
11. Find the sum of all the odd integers between 12 and 68.
12. Find the sum of all the even integers between 23 and 97.

C

13. Show that the sum of the first n odd positive integers is n^2, using appropriate formulas from this section.
14. Show that the sum of the first n positive even integers is $n + n^2$, using formulas in this section.

APPLICATIONS

Business & Economics

15. *Salary analysis.* You are confronted with two job offers. Firm A will start you at $24,000 per year and guarantees you a $900 raise each year for 10 years. Firm B will start you at only $22,000 per year, but guarantees you a $1,300 raise each year for 10 years. Over the period of 10 years, what is the total amount each firm will pay you?

16. *Salary analysis.* In Problem 15, what would be your annual salary in each firm for the 10th year?

17. *Loan repayment.* If you borrow $4,800 and repay the loan by paying $200 per month to reduce the loan and 1% of the unpaid balance each month for the use of the money, what is the total cost of the loan over 24 months?

18. *Loan repayment.* Repeat Problem 17 replacing 1% with 1.5%.

SECTION A-4

Geometric Progressions

◆ GEOMETRIC PROGRESSIONS — DEFINITION
◆ SPECIAL FORMULAS
◆ INFINITE GEOMETRIC PROGRESSIONS

◆ GEOMETRIC PROGRESSIONS — DEFINITION

Consider the sequence of numbers

$$2, 6, 18, 54, \ldots, 2(3)^{n-1}, \ldots$$

where each number after the first is obtained from the preceding one by multiplying it by 3. This is an example of a *geometric progression*. In general:

■ Geometric Progression

A sequence of numbers

$$a_1, a_2, a_3, \ldots, a_n, \ldots$$

is called a **geometric progression** if there exists a nonzero constant r, called a **common ratio**, such that

$$\frac{a_n}{a_{n-1}} = r$$

That is,

$$a_n = r a_{n-1} \qquad \text{for every } n \geq 1 \tag{1}$$

◆ E X A M P L E   16   Which of the following can be the first four terms of a geometric progression, and what is its common ratio?

(A) $5, 3, 1, -1, \ldots$ (B) $1, 2, 4, 8, \ldots$

Solution   The terms in B, with $r = 2$ ◆

P R O B L E M   16   Which of the following can be the first four terms of a geometric progression, and what is its common ratio?

(A) $4, -2, 1, -\frac{1}{2}, \ldots$ (B) $2, 4, 6, 8, \ldots$ ◆

◆ SPECIAL FORMULAS

Like arithmetic progressions, geometric progressions have several useful properties. It is easy to derive formulas for the nth term in terms of n and for the sum of any number of consecutive terms. To obtain a formula for the nth term of a geometric progression, we note that if a_1 is the first term and r is the common ratio, then

$$a_2 = r a_1$$
$$a_3 = r a_2 = r(r a_1) = r^2 a_1 = a_1 r^2$$
$$a_4 = r a_3 = r(r^2 a_1) = r^3 a_1 = a_1 r^3$$

This suggests that:

$$a_n = a_1 r^{n-1} \qquad \text{for all } n > 1 \tag{2}$$

◆ E X A M P L E   17   Find the 8th term in the geometric progression: $\frac{1}{2}, \frac{1}{4}, \frac{1}{8}, \ldots$

Solution   Find the common ratio r and use formula (2):

$$r = \tfrac{1}{2} \qquad n = 8 \qquad a_1 = \tfrac{1}{2}$$

Thus,

$$a_8 = (\tfrac{1}{2})(\tfrac{1}{2})^{8-1} = \tfrac{1}{256}$$

◆

P R O B L E M   17   Find the 7th term in the geometric progression: $\frac{1}{32}, -\frac{1}{16}, \frac{1}{8}, \ldots$ ◆

◆ E X A M P L E   18   If the 1st and 10th terms of a geometric progression are 2 and 4, respectively, find the common ratio r.

Solution
$$a_n = a_1 r^{n-1}$$
$$4 = 2 \cdot r^{10-1}$$
$$2 = r^9$$
$$r = 2^{1/9} \approx 1.08 \qquad \text{Use a calculator.}$$
◆

P R O B L E M   18   If the 1st and 8th terms of a geometric progression are 1,000 and 2,000, respectively, find the common ratio r.   ◆

We now derive two very useful formulas for the sum of n consecutive terms of a geometric progression. Let

$$a_1, a_1 r, a_1 r^2, \ldots, a_1 r^{n-2}, a_1 r^{n-1}$$

be n terms of a geometric progression. Their sum is

$$S_n = a_1 + a_1 r + a_1 r^2 + \cdots + a_1 r^{n-2} + a_1 r^{n-1}$$

If we multiply both sides by r, we obtain

$$r S_n = a_1 r + a_1 r^2 + a_1 r^3 + \cdots + a_1 r^{n-1} + a_1 r^n$$

Now combine these last two equations by subtraction to obtain

$$r S_n - S_n = (a_1 r + a_1 r^2 + a_1 r^3 + \cdots + a_1 r^{n-1} + a_1 r^n)$$
$$\qquad\qquad - (a_1 + a_1 r + a_1 r^2 + \cdots + a_1 r^{n-2} + a_1 r^{n-1})$$
$$(r - 1)S_n = a_1 r^n - a_1$$

Notice how many terms drop out on the right side. Solving for S_n, we have:

$$S_n = \frac{a_1(r^n - 1)}{r - 1} \qquad r \neq 1 \tag{3}$$

Since $a_n = a_1 r^{n-1}$, or $ra_n = a_1 r^n$, formula (3) also can be written in the form:

$$S_n = \frac{ra_n - a_1}{r - 1} \qquad r \neq 1 \qquad\qquad (4)$$

◆ E X A M P L E   19   Find the sum of the first ten terms of the geometric progression:
1, 1.05, 1.05^2, . . .

Solution   Use formula (3) with $a_1 = 1$, $r = 1.05$, and $n = 10$:

$$S_n = \frac{a_1(r^n - 1)}{r - 1}$$

$$S_{10} = \frac{1(1.05^{10} - 1)}{1.05 - 1}$$

$$\approx \frac{0.6289}{0.05} \approx 12.58$$

◆

P R O B L E M   19   Find the sum of the first eight terms of the geometric progression:
100, 100(1.08), $100(1.08)^2$, . . .

◆

◆ INFINITE GEOMETRIC PROGRESSIONS

Given a geometric progression, what happens to the sum S_n of the first n terms as n increases without stopping? To answer this question, let us write formula (3) in the form

$$S_n = \frac{a_1 r^n}{r - 1} - \frac{a_1}{r - 1}$$

It is possible to show that if $|r| < 1$ (that is, $-1 < r < 1$), then r^n will approach 0 as n increases. (See what happens, for example, if you let $r = \frac{1}{2}$ and then increase n.) Thus, the first term above will approach 0 and S_n can be made as close as we please to the second term, $-a_1/(r - 1)$ [which can be written as $a_1/(1 - r)$], by taking n sufficiently large. Thus, if the common ratio r is between -1 and 1, we define the sum of an infinite geometric progression to be:

$$S_\infty = \frac{a_1}{1 - r} \qquad |r| < 1 \qquad\qquad (5)$$

If $r \leqslant -1$ or $r \geqslant 1$, then an infinite geometric progression has no sum.

◆ E X A M P L E   20

Economy Stimulation

The government has decided on a tax rebate program to stimulate the economy. Suppose you receive $600 and you spend 80% of this, and each of the people who receive what you spend also spend 80% of what they receive, and this process continues without end. According to the **multiplier doctrine** in economics, the effect of your $600 tax rebate on the economy is multiplied many times. What is the total amount spent if the process continues as indicated?

Solution

We need to find the sum of an infinite geometric progression with the first amount spent being $a_1 = (0.8)(\$600) = \$480$ and $r = 0.8$. Using formula (5), we obtain

$$S_\infty = \frac{a_1}{1 - r}$$

$$= \frac{\$480}{1 - 0.8} = \$2,400$$

Thus, assuming the process continues as indicated, we would expect the $600 tax rebate to result in about $2,400 of spending. ◆

P R O B L E M   20

Repeat Example 20 with a tax rebate of $1,000. ◆

Answers to Matched Problems

16. The terms in A, with $r = -\frac{1}{2}$ 17. 2   18. Approx. 1.104
19. 1,063.66   20. $4,000

A

1. Determine which of the following can be the first three terms of a geometric progression. Find the common ratio r and the next two terms for those that are.

   (A) $1, -2, 4, \ldots$ (B) $7, 6, 5, \ldots$ (C) $2, 1, \frac{1}{2}, \ldots$
   (D) $2, -4, 6, \ldots$

2. Repeat Problem 1 for:

   (A) $4, -1, -6, \ldots$ (B) $15, 5, \frac{5}{3}, \ldots$ (C) $\frac{1}{4}, -\frac{1}{2}, 1, \ldots$
   (D) $\frac{1}{2}, \frac{2}{3}, \frac{3}{4}, \ldots$

Let $a_1, a_2, a_3, \ldots, a_n, \ldots$ be a geometric progression and S_n be the sum of the first n terms. In Problems 3–12, find the indicated quantities.

3. $a_1 = 3$; $r = -2$; $a_2 = ?$; $a_3 = ?$; $a_4 = ?$
4. $a_1 = 32$; $r = -\frac{1}{2}$; $a_2 = ?$; $a_3 = ?$; $a_4 = ?$
5. $a_1 = 1$; $a_7 = 729$; $r = -3$; $S_7 = ?$
6. $a_1 = 3$; $a_7 = 2,187$; $r = 3$; $S_7 = ?$

B

7. $a_1 = 100$; $r = 1.08$; $a_{10} = ?$
8. $a_1 = 240$; $r = 1.06$; $a_{12} = ?$
9. $a_1 = 100$; $a_9 = 200$; $r = ?$
10. $a_1 = 100$; $a_{10} = 300$; $r = ?$
11. $a_1 = 500$; $r = 0.6$; $S_{10} = ?$; $S_\infty = ?$
12. $a_1 = 8{,}000$; $r = 0.4$; $S_{10} = ?$; $S_\infty = ?$

13. Find the sum of each infinite geometric progression (if it exists).

    (A) 2, 4, 8, . . .   (B) $2, -\frac{1}{2}, \frac{1}{8}, \ldots$

14. Repeat Problem 13 for:

    (A) 16, 4, 1, . . .   (B) $1, -3, 9, \ldots$

C

15. Find $f(1) + f(2) + \cdots + f(10)$ if $f(x) = (\frac{1}{2})^x$.
16. Find $g(1) + g(2) + \cdots + g(10)$ if $g(x) = 2^x$.

APPLICATIONS

Business & Economics

17. *Economy stimulation.* The government, through a subsidy program, distributes $5,000,000. If we assume each individual or agency spends 70% of what is received, and 70% of this is spent, and so on, how much total increase in spending results from this government action? (Let $a_1 = $ $3,500,000.)

18. *Economy stimulation.* Repeat Problem 17 using $10,000,000 as the amount distributed and 80%.

19. *Cost-of-living adjustment.* If the cost-of-living index increased 5% for each of the past 10 years and you had a salary agreement that increased your salary by the same percentage each year, what would your present salary be if you had a salary of $20,000 per year 10 years ago? What would be your total earnings in the past 10 years? [*Hint:* $r = 1.05$]

20. *Depreciation.* In **straight-line depreciation,** an asset less its salvage value at the end of its useful life is depreciated (for tax purposes) in equal annual amounts over its useful life. Thus, a $100,000 company airplane with a salvage value of $20,000 at the end of 10 years would be depreciated at $8,000 per year for each of the 10 years.

    Since certain assets, such as airplanes, cars, and so on, depreciate more rapidly during the early years of their useful life, several methods of depreciation that take this into consideration are available to the taxpayer. One such method is called the **method of declining-balance.** The rate used cannot exceed double that used for straight-line depreciation (ignoring salvage value) and is applied to the remaining value of an asset after the previous year's depreciation has been deducted. In our airplane example, the annual rate of straight-line depreciation over the period of 10 years is 10%. Let us assume we can double this rate for the method of declining-balance. At some point before the salvage value is reached (taxpayer's choice), we must switch over to the straight-line method to depreciate the final amount of the asset.

The table illustrates the two methods of depreciation for the company airplane.

| YEAR END | STRAIGHT-LINE METHOD | | DECLINING-BALANCE METHOD | | |
	Amount Depreciated	Asset Value	Amount Depreciated	Asset Value	
0	$ 0	$100,000	$ 0	$100,000	
1	0.1(80,000) = 8,000	92,000	0.2(100,000) = 20,000	80,000	
2	0.1(80,000) = 8,000	84,000	0.2(80,000) = 16,000	64,000	
3	0.1(80,000) = 8,000	76,000	0.2(64,000) = 12,800	51,200	
.	.	.	.	.	
.	.	.	.	.	
.	.	.	.	.	
7	0.1(80,000) = 8,000	44,000	0.2(26,214) = 5,243	20,972	Shift to straight-line; otherwise next entry will drop below salvage value.
8	0.1(80,000) = 8,000	36,000	$\frac{972}{3}$ = 324	20,648	
9	0.1(80,000) = 8,000	28,000	$\frac{972}{3}$ = 324	20,324	
10	0.1(80,000) = 8,000	20,000	$\frac{972}{3}$ = 324	20,000	

Arithmetic progression · · · · · · · · · · · · · · Geometric progression above dashed line

(A) For the declining-balance method, find the sum of the depreciation amounts above the dashed line using formula (4), and then add the entries below the line to this result.
(B) Repeat part A using formula (3).
(C) Find the asset value under declining-balance at the end of the 5th year using formula (2).
(D) Find the asset value under straight-line at the end of the 5th year using formula (2) in the preceding section.

SECTION A-5 The Binomial Theorem

◆ FACTORIAL
◆ BINOMIAL THEOREM — DEVELOPMENT

The binomial form

$(a + b)^n$

where n is a natural number, appears more frequently than you might expect. The coefficients in the expansion play an important role in probability studies. The *binomial formula*, which we will derive informally, enables us to expand $(a + b)^n$ directly for n any natural number. Since the formula involves *factorials*, we digress for a moment here to introduce this important concept.

◆ FACTORIAL

For n a natural number, **n factorial,** denoted by **$n!$**, is the product of the first n natural numbers. **Zero factorial** is defined to be 1. That is:

$$n! = n \cdot (n-1) \cdot \cdots \cdot 2 \cdot 1$$
$$1! = 1$$
$$0! = 1$$

It is also useful to note that:

$$n! = n \cdot (n-1)! \qquad n \geq 1$$

◆ E X A M P L E   21   Evaluate each:

(A) $5! = 5 \cdot 4 \cdot 3 \cdot 2 \cdot 1 = 120$ (B) $\dfrac{8!}{7!} = \dfrac{8 \cdot 7!}{7!} = 8$

(C) $\dfrac{10!}{7!} = \dfrac{10 \cdot 9 \cdot 8 \cdot 7!}{7!} = 720$ ◆

P R O B L E M   21   Evaluate each: (A) $4!$ (B) $\dfrac{7!}{6!}$ (C) $\dfrac{8!}{5!}$ ◆

The following important formula involving factorials has applications in many areas of mathematics and statistics. We will use this formula to provide a more concise form for the expressions encountered later in this discussion.

For n and r integers satisfying $0 \leq r \leq n$,

$$C_{n,r} = \frac{n!}{r!(n-r)!}$$

◆ E X A M P L E   22   (A) $C_{9,2} = \dfrac{9!}{2!(9-2)!} = \dfrac{9!}{2!7!} = \dfrac{9 \cdot 8 \cdot 7!}{2 \cdot 7!} = 36$

(B) $C_{5,5} = \dfrac{5!}{5!(5-5)!} = \dfrac{5!}{5!0!} = \dfrac{5!}{5!} = 1$ ◆

P R O B L E M   22   Find: (A) $C_{5,2}$ (B) $C_{6,0}$ ◆

◆ BINOMIAL THEOREM — DEVELOPMENT

Let us expand $(a + b)^n$ for several values of n to see if we can observe a pattern that leads to a general formula for the expansion for any natural number n:

$$(a + b)^1 = a + b$$
$$(a + b)^2 = a^2 + 2ab + b^2$$
$$(a + b)^3 = a^3 + 3a^2b + 3ab^2 + b^3$$
$$(a + b)^4 = a^4 + 4a^3b + 6a^2b^2 + 4ab^3 + b^4$$
$$(a + b)^5 = a^5 + 5a^4b + 10a^3b^2 + 10a^2b^3 + 5ab^4 + b^5$$

Observations

1. The expansion of $(a + b)^n$ has $(n + 1)$ terms.
2. The power of a decreases by 1 for each term as we move from left to right.
3. The power of b increases by 1 for each term as we move from left to right.
4. In each term, the sum of the powers of a and b always equals n.
5. Starting with a given term, we can get the coefficient of the next term by multiplying the coefficient of the given term by the exponent of a and dividing by the number that represents the position of the term in the series of terms. For example, in the expansion of $(a + b)^4$ above, the coefficient of the third term is found from the second term by multiplying 4 and 3, and then dividing by 2 [that is, the coefficient of the third term $= (4 \cdot 3)/2 = 6$].

We now postulate these same properties for the general case:

$$(a + b)^n = a^n + \frac{n}{1} a^{n-1}b + \frac{n(n-1)}{1 \cdot 2} a^{n-2}b^2 + \frac{n(n-1)(n-2)}{1 \cdot 2 \cdot 3} a^{n-3}b^3 + \cdots + b^n$$

$$= \frac{n!}{0!(n-0)!} a^n + \frac{n!}{1!(n-1)!} a^{n-1}b + \frac{n!}{2!(n-2)!} a^{n-2}b^2 + \frac{n!}{3!(n-3)!} a^{n-3}b^3 + \cdots + \frac{n!}{n!(n-n)!} b^n$$

$$= C_{n,0}a^n + C_{n,1}a^{n-1}b + C_{n,2}a^{n-2}b^2 + C_{n,3}a^{n-3}b^3 + \cdots + C_{n,n}b^n$$

And we are led to the formula in the binomial theorem (a formal proof requires mathematical induction, which is beyond the scope of this book):

▮ **Binomial Theorem**

For all natural numbers n,

$$(a + b)^n = C_{n,0}a^n + C_{n,1}a^{n-1}b + C_{n,2}a^{n-2}b^2 + C_{n,3}a^{n-3}b^3 + \cdots + C_{n,n}b^n$$

◆ E X A M P L E   23   Use the binomial formula to expand $(u + v)^6$.

Solution

$$(u + v)^6 = C_{6,0}u^6 + C_{6,1}u^5v + C_{6,2}u^4v^2 + C_{6,3}u^3v^3 + C_{6,4}u^2v^4 + C_{6,5}uv^5 + C_{6,6}v^6$$
$$= u^6 + 6u^5v + 15u^4v^2 + 20u^3v^3 + 15u^2v^4 + 6uv^5 + v^6 \qquad ◆$$

PROBLEM   23   Use the binomial formula to expand $(x + 2)^5$.   ◆

◆ EXAMPLE   24   Use the binomial formula to find the sixth term in the expansion of $(x - 1)^{18}$.

Solution   Sixth term $= C_{18,5} x^{13}(-1)^5 = \dfrac{18!}{5!(18 - 5)!} x^{13}(-1)$

$$= -8{,}568x^{13}$$ ◆

PROBLEM   24   Use the binomial formula to find the fourth term in the expansion of $(x - 2)^{20}$.

◆

Answers to Matched Problems

21.   (A) 24   (B) 7   (C) 336   22. (A) 10   (B) 1
23. $x^5 + 5x^4 \cdot 2 + 10x^3 \cdot 2^2 + 10x^2 \cdot 2^3 + 5x \cdot 2^4 + 2^5$
$$= x^5 + 10x^4 + 40x^3 + 80x^2 + 80x + 32$$
24. $-9{,}120x^{17}$

EXERCISE   A-5

A   *Evaluate.*

1. 6!

2. 7!

3. $\dfrac{10!}{9!}$

4. $\dfrac{20!}{19!}$

5. $\dfrac{12!}{9!}$

6. $\dfrac{10!}{6!}$

7. $\dfrac{5!}{2!3!}$

8. $\dfrac{7!}{3!4!}$

9. $\dfrac{6!}{5!((6 - 5)!}$

10. $\dfrac{7!}{4!(7 - 4)!}$

11. $\dfrac{20!}{3!17!}$

12. $\dfrac{52!}{50!2!}$

B   *Evaluate*

13. $C_{5,3}$

14. $C_{7,3}$

15. $C_{6,5}$

16. $C_{7,4}$

17. $C_{5,0}$

18. $C_{5,5}$

19. $C_{18,15}$

20. $C_{18,3}$

Expand each expression using the binomial formula.

21. $(a + b)^4$

22. $(m + n)^5$

23. $(x - 1)^6$

24. $(u - 2)^5$

25. $(2a - b)^5$

26. $(x - 2y)^5$

Find the indicated term in each expansion.

27. $(x - 1)^{18}$;   fifth term

28. $(x - 3)^{20}$;   third term

29. $(p + q)^{15}$;   seventh term

30. $(p + q)^{15}$;   thirteenth term

31. $(2x + y)^{12}$;   eleventh term

32. $(2x + y)^{12}$;   third term

C 33. Show that $C_{n,0} = C_{n,n}$.

34. Show that $C_{n,r} = C_{n,n-r}$.

35. The triangle below is called **Pascal's triangle.** Can you guess what the next two rows at the bottom are? Compare these numbers with the coefficients of binomial expansions.

$$
\begin{array}{ccccccccc}
 & & & & 1 & & & & \\
 & & & 1 & & 1 & & & \\
 & & 1 & & 2 & & 1 & & \\
 & 1 & & 3 & & 3 & & 1 & \\
1 & & 4 & & 6 & & 4 & & 1 \\
\end{array}
$$

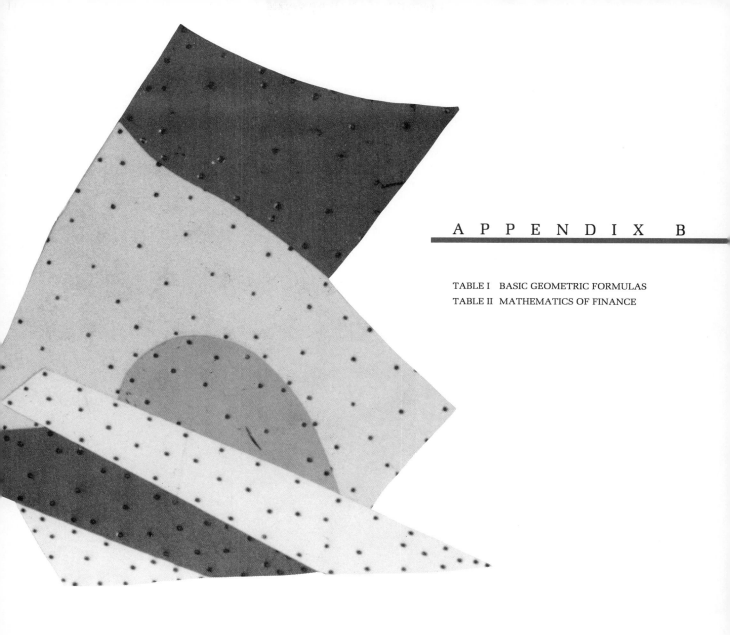

Tables

TABLE I

Basic Geometric Formulas

◆ 1. SIMILAR TRIANGLES

(A) Two triangles are similar if two angles of one triangle have the same measure as two angles of the other.

(B) If two triangles are similar, their corresponding sides are proportional:

$$\frac{a}{a'} = \frac{b}{b'} = \frac{c}{c'}$$

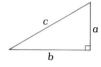

◆ 2. PYTHAGOREAN THEOREM

$$c^2 = a^2 + b^2$$

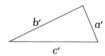

◆ 3. RECTANGLE

$A = ab$ Area

$P = 2a + 2b$ Perimeter

◆ 4. PARALLELOGRAM

$h =$ Height

$A = ah = ab \sin \theta$ Area

$P = 2a + 2b$ Perimeter

◆ 5. TRIANGLE

$h =$ Height

$A = \frac{1}{2}hc$ Area

$P = a + b + c$ Perimeter

$s = \frac{1}{2}(a + b + c)$ Semiperimeter

$A = \sqrt{s(s - a)(s - b)(s - c)}$ Area — Heron's formula

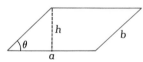

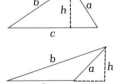

◆ 6. TRAPEZOID

Base a is parallel to base b.

$h =$ Height

$A = \frac{1}{2}(a + b)h$ Area

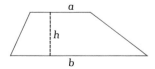

◆ 7. CIRCLE

R = Radius

D = Diameter

$D = 2R$

$A = \pi R^2 = \frac{1}{4}\pi D^2$ Area

$C = 2\pi R = \pi D$ Circumference

$\dfrac{C}{D} = \pi$ For all circles

$\pi \approx 3.141\ 59$

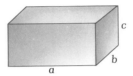

◆ 8. RECTANGULAR SOLID

$V = abc$ Volume

$T = 2ab + 2ac + 2bc$ Total surface area

◆ 9. RIGHT CIRCULAR CYLINDER

R = Radius of base

h = Height

$V = \pi R^2 h$ Volume

$S = 2\pi Rh$ Lateral surface area

$T = 2\pi R(R + h)$ Total surface area

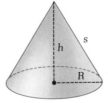

◆ 10. RIGHT CIRCULAR CONE

R = Radius of base

h = Height

s = Slant height

$V = \frac{1}{3}\pi R^2 h$ Volume

$S = \pi Rs = \pi R\sqrt{R^2 + h^2}$ Lateral surface area

$T = \pi R(R + s) = \pi R(R + \sqrt{R^2 + h^2})$ Total surface area

◆ 11. SPHERE

R = Radius

D = Diameter

$D = 2R$

$V = \frac{4}{3}\pi R^3 = \frac{1}{6}\pi D^3$ Volume

$S = 4\pi R^2 = \pi D^2$ Surface area

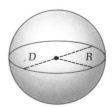

TABLE II

Mathematics of Finance

		$i = 0.0025$ ($\frac{1}{4}\%$)					
n	$(1 + i)^n$	$s_{\overline{n}|i}$	$a_{\overline{n}|i}$	n	$(1 + i)^n$	$s_{\overline{n}|i}$	$a_{\overline{n}|i}$
1	1.002 500	1.000 000	0.997 506	51	1.135 804	54.321 654	47.826 604
2	1.005 006	2.002 500	1.992 525	52	1.138 644	55.457 459	48.704 842
3	1.007 519	3.007 506	2.985 062	53	1.141 490	56.596 102	49.580 890
4	1.010 038	4.015 025	3.975 124	54	1.144 344	57.737 593	50.454 753
5	1.012 563	5.025 063	4.962 718	55	1.147 205	58.881 936	51.326 437
6	1.015 094	6.037 625	5.947 848	56	1.150 073	60.029 141	52.195 947
7	1.017 632	7.052 719	6.930 522	57	1.152 948	61.179 214	53.063 288
8	1.020 176	8.070 351	7.910 745	58	1.155 830	62.332 162	53.928 467
9	1.022 726	9.090 527	8.888 524	59	1.158 720	63.487 993	54.791 489
10	1.025 283	10.113 253	9.863 864	60	1.161 617	64.646 713	55.652 358
11	1.027 846	11.138 536	10.836 772	61	1.164 521	65.808 329	56.511 080
12	1.030 416	12.166 383	11.807 254	62	1.167 432	66.972 850	57.367 661
13	1.032 992	13.196 799	12.775 316	63	1.170 351	68.140 282	58.222 106
14	1.035 574	14.229 791	13.740 963	64	1.173 277	69.310 633	59.074 420
15	1.038 163	15.265 365	14.704 203	65	1.176 210	70.483 910	59.924 608
16	1.040 759	16.303 529	15.665 040	66	1.179 150	71.660 119	60.772 676
17	1.043 361	17.344 287	16.623 481	67	1.182 098	72.839 270	61.618 630
18	1.045 969	18.387 648	17.579 533	68	1.185 053	74.021 368	62.462 474
19	1.048 584	19.433 617	18.533 200	69	1.188 016	75.206 421	63.304 213
20	1.051 205	20.482 201	19.484 488	70	1.190 986	76.394 437	64.143 853
21	1.053 834	21.533 407	20.433 405	71	1.193 964	77.585 423	64.981 400
22	1.056 468	22.587 240	21.379 955	72	1.196 948	78.779 387	65.816 858
23	1.059 109	23.643 708	22.324 145	73	1.199 941	79.976 335	66.650 232
24	1.061 757	24.702 818	23.265 980	74	1.202 941	81.176 276	67.481 528
25	1.064 411	25.764 575	24.205 466	75	1.205 948	82.379 217	68.310 751
26	1.067 072	26.828 986	25.142 609	76	1.208 963	83.585 165	69.137 907
27	1.069 740	27.896 059	26.077 416	77	1.211 985	84.794 128	69.962 999
28	1.072 414	28.965 799	27.009 891	78	1.215 015	86.006 113	70.786 034
29	1.075 096	30.038 213	27.940 041	79	1.218 053	87.221 129	71.607 017
30	1.077 783	31.113 309	28.867 871	80	1.221 098	88.439 181	72.425 952
31	1.080 478	32.191 092	29.793 388	81	1.224 151	89.660 279	73.242 845
32	1.083 179	33.271 570	30.716 596	82	1.227 211	90.884 430	74.057 700
33	1.085 887	34.354 749	31.637 503	83	1.230 279	92.111 641	74.870 524
34	1.088 602	35.440 636	32.556 112	84	1.233 355	93.341 920	75.681 321
35	1.091 323	36.529 237	33.472 431	85	1.236 438	94.575 275	76.490 095
36	1.094 051	37.620 560	34.386 465	86	1.239 529	95.811 713	77.296 853
37	1.096 787	38.714 612	35.298 220	87	1.242 628	97.051 242	78.101 599
38	1.099 528	39.811 398	36.207 700	88	1.245 735	98.293 871	78.904 339
39	1.102 277	40.910 927	37.114 913	89	1.248 849	99.539 605	79.705 076
40	1.105 033	42.013 204	38.019 863	90	1.251 971	100.788 454	80.503 816
41	1.107 796	43.118 237	38.922 557	91	1.255 101	102.040 425	81.300 565
42	1.110 565	44.226 033	39.822 999	92	1.258 239	103.295 526	82.095 327
43	1.113 341	45.336 598	40.721 196	93	1.261 384	104.553 765	82.888 106
44	1.116 125	46.449 939	41.617 154	94	1.264 538	105.815 150	83.678 909
45	1.118 915	47.566 064	42.510 876	95	1.267 699	107.079 688	84.467 740
46	1.121 712	48.684 979	43.402 370	96	1.270 868	108.347 387	85.254 603
47	1.124 517	49.806 692	44.291 641	97	1.274 046	109.618 255	86.039 504
48	1.127 328	50.931 208	45.178 695	98	1.277 231	110.892 301	86.822 448
49	1.130 146	52.058 536	46.063 536	99	1.280 424	112.169 532	87.603 440
50	1.132 972	53.188 683	46.946 170	100	1.283 625	113.449 956	88.382 483

Continued

		$i = 0.005$ $(\frac{1}{2}\%)$					
n	$(1 + i)^n$	$s_{\overline{n}\,i}$	$a_{\overline{n}\,i}$	n	$(1 + i)^n$	$s_{\overline{n}\,i}$	$a_{\overline{n}\,i}$
1	1.005 000	1.000 000	0.995 025	51	1.289 642	57.928 389	44.918 195
2	1.010 025	2.005 000	1.985 099	52	1.296 090	59.218 031	45.689 747
3	1.015 075	3.015 025	2.970 248	53	1.302 571	60.514 121	46.457 459
4	1.020 151	4.030 100	3.950 496	54	1.309 083	61.816 692	47.221 353
5	1.025 251	5.050 250	4.925 866	55	1.315 629	63.125 775	47.981 445
6	1.030 378	6.075 502	5.896 384	56	1.322 207	64.441 404	48.737 757
7	1.035 529	7.105 879	6.862 074	57	1.328 818	65.763 611	49.490 305
8	1.040 707	8.141 409	7.822 959	58	1.335 462	67.092 429	50.239 110
9	1.045 911	9.182 116	8.779 064	59	1.342 139	68.427 891	50.984 189
10	1.051 140	10.228 026	9.730 412	60	1.348 850	69.770 031	51.725 561
11	1.056 396	11.279 167	10.677 027	61	1.355 594	71.118 881	52.463 245
12	1.061 678	12.335 562	11.618 932	62	1.362 372	72.474 475	53.197 258
13	1.066 986	13.397 240	12.556 151	63	1.369 184	73.836 847	53.927 620
14	1.072 321	14.464 226	13.488 708	64	1.376 030	75.206 032	54.654 348
15	1.077 683	15.536 548	14.416 625	65	1.382 910	76.582 062	55.377 461
16	1.083 071	16.614 230	15.339 925	66	1.389 825	77.964 972	56.096 976
17	1.088 487	17.697 301	16.258 632	67	1.396 774	79.354 797	56.812 912
18	1.093 929	18.785 788	17.172 768	68	1.403 758	80.751 571	57.525 285
19	1.099 399	19.879 717	18.082 356	69	1.410 777	82.155 329	58.234 115
20	1.104 896	20.979 115	18.987 419	70	1.417 831	83.566 105	58.939 418
21	1.110 420	22.084 011	19.887 979	71	1.424 920	84.983 936	59.641 212
22	1.115 972	23.194 431	20.784 059	72	1.432 044	86.408 856	60.339 514
23	1.121 552	24.310 403	21.675 681	73	1.439 204	87.840 900	61.034 342
24	1.127 160	25.431 955	22.562 866	74	1.446 401	89.280 104	61.725 714
25	1.132 796	26.559 115	23.445 638	75	1.453 633	90.726 505	62.413 645
26	1.138 460	27.691 911	24.324 018	76	1.460 901	92.180 138	63.098 155
27	1.144 152	28.830 370	25.198 028	77	1.468 205	93.641 038	63.779 258
28	1.149 873	29.974 522	26.067 689	78	1.475 546	95.109 243	64.456 974
29	1.155 622	31.124 395	26.933 024	79	1.482 924	96.584 790	65.131 317
30	1.161 400	32.280 017	27.794 054	80	1.490 339	98.067 714	65.802 305
31	1.167 207	33.441 417	28.650 800	81	1.497 790	99.558 052	66.469 956
32	1.173 043	34.608 624	29.503 284	82	1.505 279	101.055 842	67.134 284
33	1.178 908	35.781 667	30.351 526	83	1.512 806	102.561 122	67.795 308
34	1.184 803	36.960 575	31.195 548	84	1.520 370	104.073 927	68.453 042
35	1.190 727	38.145 378	32.035 371	85	1.527 971	105.594 297	69.107 505
36	1.196 681	39.336 105	32.871 016	86	1.535 611	107.122 268	69.758 711
37	1.202 664	40.532 785	33.702 504	87	1.543 289	108.657 880	70.406 678
38	1.208 677	41.735 449	34.529 854	88	1.551 006	110.201 169	71.051 421
39	1.214 721	42.944 127	35.353 089	89	1.558 761	111.752 175	71.692 956
40	1.220 794	44.158 847	36.172 228	90	1.566 555	113.310 936	72.331 300
41	1.226 898	45.379 642	36.987 291	91	1.574 387	114.877 490	72.966 467
42	1.233 033	46.606 540	37.798 300	92	1.582 259	116.451 878	73.598 475
43	1.239 198	47.839 572	38.605 274	93	1.590 171	118.034 137	74.227 338
44	1.245 394	49.078 770	39.408 232	94	1.598 121	119.624 308	74.853 073
45	1.251 621	50.324 164	40.207 196	95	1.606 112	121.222 430	75.475 694
46	1.257 879	51.575 785	41.002 185	96	1.614 143	122.828 542	76.095 218
47	1.264 168	52.833 664	41.793 219	97	1.622 213	124.442 684	76.711 660
48	1.270 489	54.097 832	42.580 318	98	1.630 324	126.064 898	77.325 035
49	1.276 842	55.368 321	43.363 500	99	1.638 476	127.695 222	77.935 358
50	1.283 226	56.645 163	44.142 786	100	1.646 668	129.333 698	78.542 645

				$i = 0.0075$ (¾%)			
n	$(1 + i)^n$	$s_{\overline{n}i}$	$a_{\overline{n}i}$	n	$(1 + i)^n$	$s_{\overline{n}i}$	$a_{\overline{n}i}$
1	1.007 500	1.000 000	0.992 556	51	1.463 854	61.847 214	42.249 575
2	1.015 056	2.007 500	1.977 723	52	1.474 833	63.311 068	42.927 618
3	1.022 669	3.022 556	2.955 556	53	1.485 894	64.785 901	43.600 614
4	1.030 339	4.045 225	3.926 110	54	1.497 038	66.271 796	44.268 599
5	1.038 067	5.075 565	4.889 440	55	1.508 266	67.768 834	44.931 612
6	1.045 852	6.113 631	5.845 598	56	1.519 578	69.277 100	45.589 689
7	1.053 696	7.159 484	6.794 638	57	1.530 975	70.796 679	46.242 868
8	1.061 599	8.213 180	7.736 613	58	1.542 457	72.327 659	46.891 184
9	1.069 561	9.274 779	8.671 576	59	1.554 026	73.870 111	47.534 674
10	1.077 583	10.344 339	9.599 580	60	1.565 681	75.424 137	48.173 374
11	1.085 664	11.421 922	10.520 675	61	1.577 424	76.989 818	48.807 319
12	1.093 807	12.507 586	11.434 913	62	1.589 254	78.567 242	49.436 545
13	1.102 010	13.601 393	12.342 345	63	1.601 174	80.156 496	50.061 086
14	1.110 276	14.703 404	13.243 022	64	1.613 183	81.757 670	50.680 979
15	1.118 603	15.813 679	14.136 995	65	1.625 281	83.370 852	51.296 257
16	1.126 992	16.932 282	15.024 313	66	1.637 471	84.996 134	51.906 955
17	1.135 445	18.059 274	15.905 025	67	1.649 752	86.633 605	52.513 107
18	1.143 960	19.194 718	16.779 181	68	1.662 125	88.283 356	53.114 746
19	1.152 540	20.338 679	17.646 830	69	1.674 591	89.945 482	53.711 907
20	1.161 184	21.491 219	18.508 020	70	1.687 151	91.620 073	54.304 622
21	1.169 893	22.652 403	19.362 799	71	1.699 804	93.307 223	54.892 925
22	1.178 667	23.822 296	20.211 215	72	1.712 553	95.007 028	55.476 849
23	1.187 507	25.000 963	21.053 315	73	1.725 397	96.719 580	56.056 426
24	1.196 414	26.188 471	21.889 146	74	1.738 337	98.444 977	56.631 688
25	1.205 387	27.384 884	22.718 755	75	1.751 375	100.183 314	57.202 668
26	1.214 427	28.590 271	23.542 189	76	1.764 510	101.934 689	57.769 397
27	1.223 535	29.804 698	24.359 493	77	1.777 744	103.699 199	58.331 908
28	1.232 712	31.028 233	25.170 713	78	1.791 077	105.476 943	58.890 231
29	1.241 957	32.260 945	25.975 893	79	1.804 510	107.268 021	59.444 398
30	1.251 272	33.502 902	26.775 080	80	1.818 044	109.072 531	59.994 440
31	1.260 656	34.754 174	27.568 318	81	1.831 679	110.890 575	60.540 387
32	1.270 111	36.014 830	28.355 650	82	1.845 417	112.722 254	61.082 270
33	1.279 637	37.284 941	29.137 122	83	1.859 258	114.567 671	61.620 119
34	1.289 234	38.564 578	29.912 776	84	1.873 202	116.426 928	62.153 965
35	1.298 904	39.853 813	30.682 656	85	1.887 251	118.300 130	62.683 836
36	1.308 645	41.152 716	31.446 805	86	1.901 405	120.187 381	63.209 763
37	1.318 460	42.461 361	32.205 266	87	1.915 666	122.088 787	63.731 774
38	1.328 349	43.779 822	32.958 080	88	1.930 033	124.004 453	64.249 900
39	1.338 311	45.108 170	33.705 290	89	1.944 509	125.934 486	64.764 169
40	1.348 349	46.446 482	34.446 938	90	1.959 092	127.878 995	65.274 609
41	1.358 461	47.794 830	35.183 065	91	1.973 786	129.838 087	65.781 250
42	1.368 650	49.153 291	35.913 713	92	1.988 589	131.811 873	66.284 119
43	1.378 915	50.521 941	36.638 921	93	2.003 503	133.800 462	66.783 245
44	1.389 256	51.900 856	37.358 730	94	2.018 530	135.803 965	67.278 655
45	1.399 676	53.290 112	38.073 181	95	2.033 669	137.822 495	67.770 377
46	1.410 173	54.689 788	38.782 314	96	2.048 921	139.856 164	68.258 439
47	1.420 750	56.099 961	39.486 168	97	2.064 288	141.905 085	68.742 867
48	1.431 405	57.520 711	40.184 782	98	2.079 770	143.969 373	69.223 689
49	1.442 141	58.952 116	40.878 195	99	2.095 369	146.049 143	69.700 932
50	1.452 957	60.394 257	41.566 447	100	2.111 084	148.144 512	70.174 623

A39

Continued

		$i = 0.01\,(1\%)$					
n	$(1+i)^n$	$s_{\overline{n}|i}$	$a_{\overline{n}|i}$	n	$(1+i)^n$	$s_{\overline{n}|i}$	$a_{\overline{n}|i}$
1	1.010 000	1.000 000	0.990 099	51	1.661 078	66.107 814	39.798 136
2	1.020 100	2.010 000	1.970 395	52	1.677 689	67.768 892	40.394 194
3	1.030 301	3.030 100	2.940 985	53	1.694 466	69.446 581	40.984 351
4	1.040 604	4.060 401	3.901 966	54	1.711 410	71.141 047	41.568 664
5	1.051 010	5.101 005	4.853 431	55	1.728 525	72.852 457	42.147 192
6	1.061 520	6.152 015	5.795 476	56	1.745 810	74.580 982	42.719 992
7	1.072 135	7.213 535	6.728 195	57	1.763 268	76.326 792	43.287 121
8	1.082 857	8.285 671	7.651 678	58	1.780 901	78.090 060	43.848 635
9	1.093 685	9.368 527	8.566 018	59	1.798 710	79.870 960	44.404 589
10	1.104 622	10.462 213	9.471 305	60	1.816 697	81.669 670	44.955 038
11	1.115 668	11.566 835	10.367 628	61	1.834 864	83.486 367	45.500 038
12	1.126 825	12.682 503	11.255 077	62	1.853 212	85.321 230	46.039 642
13	1.138 093	13.809 328	12.133 740	63	1.871 744	87.174 443	46.573 903
14	1.149 474	14.947 421	13.003 703	64	1.890 462	89.046 187	47.102 874
15	1.160 969	16.096 896	13.865 053	65	1.909 366	90.936 649	47.626 608
16	1.172 579	17.257 864	14.717 874	66	1.928 460	92.846 015	48.145 156
17	1.184 304	18.430 443	15.562 251	67	1.947 745	94.774 475	48.658 570
18	1.196 147	19.614 748	16.398 269	68	1.967 222	96.722 220	49.166 901
19	1.208 109	20.810 895	17.226 008	69	1.986 894	98.689 442	49.670 199
20	1.220 190	22.019 004	18.045 553	70	2.006 763	100.676 337	50.168 514
21	1.232 392	23.239 194	18.856 983	71	2.026 831	102.683 100	50.661 895
22	1.244 716	24.471 586	19.660 379	72	2.047 099	104.709 931	51.150 391
23	1.257 163	25.716 302	20.455 821	73	2.067 570	106.757 031	51.634 051
24	1.269 735	26.973 465	21.243 387	74	2.088 246	108.824 601	52.112 922
25	1.282 432	28.243 200	22.023 156	75	2.109 128	110.912 847	52.587 051
26	1.295 256	29.525 632	22.795 204	76	2.130 220	113.021 975	53.056 486
27	1.308 209	30.820 888	23.559 608	77	2.151 522	115.152 195	53.521 274
28	1.321 291	32.129 097	24.316 443	78	2.173 037	117.303 717	53.981 459
29	1.334 504	33.450 388	25.065 785	79	2.194 768	119.476 754	54.437 088
30	1.347 849	34.784 892	25.807 708	80	2.216 715	121.671 522	54.888 206
31	1.361 327	36.132 740	26.542 285	81	2.238 882	123.888 237	55.334 858
32	1.374 941	37.494 068	27.269 589	82	2.261 271	126.127 119	55.777 087
33	1.388 690	38.869 009	27.989 693	83	2.283 884	128.388 391	56.214 937
34	1.402 577	40.257 699	28.702 666	84	2.306 723	130.672 274	56.648 453
35	1.416 603	41.660 276	29.408 580	85	2.329 790	132.978 997	57.077 676
36	1.430 769	43.076 878	30.107 505	86	2.353 088	135.308 787	57.502 650
37	1.445 076	44.507 647	30.799 510	87	2.376 619	137.661 875	57.923 415
38	1.459 527	45.952 724	31.484 663	88	2.400 385	140.038 494	58.340 015
39	1.474 123	47.412 251	32.163 033	89	2.424 389	142.438 879	58.752 490
40	1.488 864	48.886 373	32.834 686	90	2.448 633	144.863 267	59.160 881
41	1.503 752	50.375 237	33.499 689	91	2.473 119	147.311 900	59.565 229
42	1.518 790	51.878 989	34.158 108	92	2.497 850	149.785 019	59.965 573
43	1.533 978	53.397 779	34.810 008	93	2.522 829	152.282 869	60.361 954
44	1.549 318	54.931 757	35.455 454	94	2.548 057	154.805 698	60.754 410
45	1.564 811	56.481 075	36.094 508	95	2.573 538	157.353 755	61.142 980
46	1.580 459	58.045 885	36.727 236	96	2.599 273	159.927 293	61.527 703
47	1.596 263	59.626 344	37.353 699	97	2.625 266	162.526 565	61.908 617
48	1.612 226	61.222 608	37.973 959	98	2.651 518	165.151 831	62.285 759
49	1.628 348	62.834 834	38.588 079	99	2.678 033	167.803 349	62.659 168
50	1.644 632	64.463 182	39.196 118	100	2.704 814	170.481 383	63.028 879

TABLE II

Continued

			$i = 0.0125\ (1\frac{1}{4}\%)$				
n	$(1 + i)^n$	$s_{\overline{n}i}$	$a_{\overline{n}i}$	n	$(1 + i)^n$	$s_{\overline{n}i}$	$a_{\overline{n}i}$
1	1.012 500	1.000 000	0.987 654	51	1.884 285	70.742 812	37.543 581
2	1.025 156	2.012 500	1.963 115	52	1.907 839	72.627 097	38.067 734
3	1.037 971	3.037 656	2.926 534	53	1.931 687	74.534 936	38.585 417
4	1.050 945	4.075 627	3.878 058	54	1.955 833	76.466 623	39.096 708
5	1.064 082	5.126 572	4.817 835	55	1.980 281	78.422 456	39.601 687
6	1.077 383	6.190 654	5.746 010	56	2.005 034	80.402 737	40.100 431
7	1.090 850	7.268 038	6.662 726	57	2.030 097	82.407 771	40.593 019
8	1.104 486	8.358 888	7.568 124	58	2.055 473	84.437 868	41.079 524
9	1.118 292	9.463 374	8.462 345	59	2.081 167	86.493 341	41.560 024
10	1.132 271	10.581 666	9.345 526	60	2.107 181	88.574 508	42.034 592
11	1.146 424	11.713 937	10.217 803	61	2.133 521	90.681 689	42.503 300
12	1.160 755	12.860 361	11.079 312	62	2.160 190	92.815 210	42.966 223
13	1.175 264	14.021 116	11.930 185	63	2.187 193	94.975 400	43.423 430
14	1.189 955	15.196 380	12.770 553	64	2.214 532	97.162 593	43.874 992
15	1.204 829	16.386 335	13.600 546	65	2.242 214	99.377 125	44.320 980
16	1.219 890	17.591 164	14.420 292	66	2.270 242	101.619 339	44.761 462
17	1.235 138	18.811 053	15.229 918	67	2.298 620	103.889 581	45.196 506
18	1.250 577	20.046 192	16.029 549	68	2.327 353	106.188 201	45.626 178
19	1.266 210	21.296 769	16.819 308	69	2.356 444	108.515 553	46.050 547
20	1.282 037	22.562 979	17.599 316	70	2.385 900	110.871 998	46.469 676
21	1.298 063	23.845 016	18.369 695	71	2.415 724	113.257 898	46.883 630
22	1.314 288	25.143 078	19.130 563	72	2.445 920	115.673 621	47.292 474
23	1.330 717	26.457 367	19.882 037	73	2.476 494	118.119 542	47.696 271
24	1.347 351	27.788 084	20.624 235	74	2.507 450	120.596 036	48.095 082
25	1.364 193	29.135 435	21.357 269	75	2.538 794	123.103 486	48.488 970
26	1.381 245	30.499 628	22.081 253	76	2.570 528	125.642 280	48.877 995
27	1.398 511	31.880 873	22.796 299	77	2.602 660	128.212 809	49.262 218
28	1.415 992	33.279 384	23.502 518	78	2.635 193	130.815 469	49.641 696
29	1.433 692	34.695 377	24.200 018	79	2.668 133	133.450 662	50.016 490
30	1.451 613	36.129 069	24.888 906	80	2.701 485	136.118 795	50.386 657
31	1.469 759	37.580 682	25.569 290	81	2.735 254	138.820 280	50.752 254
32	1.488 131	39.050 441	26.241 274	82	2.769 444	141.555 534	51.113 337
33	1.506 732	40.538 571	26.904 962	83	2.804 062	144.324 978	51.469 963
34	1.525 566	42.045 303	27.560 456	84	2.839 113	147.129 040	51.822 185
35	1.544 636	43.570 870	28.207 858	85	2.874 602	149.968 153	52.170 060
36	1.563 944	45.115 506	28.847 267	86	2.910 534	152.842 755	52.513 639
37	1.583 493	46.679 449	29.478 783	87	2.946 916	155.753 289	52.852 977
38	1.603 287	48.292 642	30.102 501	88	2.983 753	158.700 206	53.188 125
39	1.623 328	49.886 229	30.718 520	89	3.021 049	161.683 958	53.519 136
40	1.643 619	51.489 557	31.326 933	90	3.058 813	164.705 008	53.846 060
41	1.664 165	53.133 177	31.927 835	91	3.097 048	167.763 820	54.168 948
42	1.684 967	54.797 341	32.521 319	92	3.135 761	170.860 868	54.487 850
43	1.706 029	56.482 308	33.107 475	93	3.174 958	173.996 629	54.802 815
44	1.727 354	58.188 337	33.686 395	94	3.214 645	177.171 587	55.113 892
45	1.748 946	59.915 691	34.258 168	95	3.254 828	180.386 232	55.421 127
46	1.770 808	61.664 637	34.822 882	96	3.295 513	183.641 059	55.724 570
47	1.792 943	63.435 445	35.380 624	97	3.336 707	186.936 573	56.024 267
48	1.815 355	65.228 388	35.931 481	98	3.378 416	190.273 280	56.320 264
49	1.838 047	67.043 743	36.475 537	99	3.420 646	193.651 696	56.612 606
50	1.861 022	68.881 790	37.012 876	100	3.463 404	197.072 342	56.901 339

Continued

			$i = 0.015\ (1\frac{1}{2}\%)$			
n	$(1 + i)^n$	$s_{\overline{n}i}$	$a_{\overline{n}i}$	n	$(1 + i)^n$	$s_{\overline{n}i}$	$a_{\overline{n}i}$

n	$(1 + i)^n$	$s_{\overline{n}i}$	$a_{\overline{n}i}$	n	$(1 + i)^n$	$s_{\overline{n}i}$	$a_{\overline{n}i}$
1	1.015 000	1.000 000	0.985 222	51	2.136 821	75.788 070	35.467 673
2	1.030 225	2.015 000	1.955 883	52	2.168 873	77.924 892	35.928 742
3	1.045 678	3.045 225	2.912 200	53	2.201 406	80.093 765	36.382 997
4	1.061 364	4.090 903	3.854 385	54	2.234 428	82.295 171	36.830 539
5	1.077 284	5.152 267	4.782 645	55	2.267 944	84.529 599	37.271 467
6	1.093 443	6.229 551	5.697 187	56	2.301 963	86.797 543	37.705 879
7	1.109 845	7.322 994	6.598 214	57	2.336 493	89.099 506	38.133 871
8	1.126 493	8.432 839	7.485 925	58	2.371 540	91.435 999	38.555 538
9	1.143 390	9.559 332	8.360 517	59	2.407 113	93.807 539	38.970 973
10	1.160 541	10.702 722	9.222 185	60	2.443 220	96.214 652	39.380 269
11	1.177 949	11.863 262	10.071 118	61	2.479 868	98.657 871	39.783 516
12	1.195 618	13.041 211	10.907 505	62	2.517 066	101.137 740	40.180 804
13	1.213 552	14.236 830	11.731 532	63	2.554 822	103.654 806	40.572 221
14	1.231 756	15.450 382	12.543 382	64	2.593 144	106.209 628	40.957 853
15	1.250 232	16.682 138	13.343 233	65	2.632 042	108.802 772	41.337 786
16	1.268 986	17.932 370	14.131 264	66	2.671 522	111.434 814	41.712 105
17	1.288 020	19.201 355	14.907 649	67	2.711 595	114.106 336	42.080 891
18	1.307 341	20.489 376	15.672 561	68	2.752 269	116.817 931	42.444 228
19	1.326 951	21.796 716	16.426 168	69	2.793 553	119.570 200	42.802 195
20	1.346 855	23.123 667	17.168 639	70	2.835 456	122.363 753	43.154 872
21	1.367 058	24.470 522	17.900 137	71	2.877 988	125.199 209	43.502 337
22	1.387 564	25.837 580	18.620 824	72	2.921 158	128.077 197	43.844 667
23	1.408 377	27.225 144	19.330 861	73	2.964 975	130.998 355	44.181 938
24	1.429 503	28.633 521	20.030 405	74	3.009 450	133.963 331	44.514 224
25	1.450 945	30.063 024	20.719 611	75	3.054 592	136.972 781	44.841 600
26	1.472 710	31.513 969	21.398 632	76	3.100 411	140.027 372	45.164 138
27	1.494 800	32.986 678	22.067 617	77	3.146 917	143.127 783	45.481 910
28	1.517 222	34.481 479	22.726 717	78	3.194 120	146.274 700	45.794 985
29	1.539 981	35.998 701	23.376 076	79	3.242 032	149.468 820	46.103 433
30	1.563 080	37.538 681	24.015 838	80	3.290 663	152.710 852	46.407 323
31	1.586 526	39.101 762	24.646 146	81	3.340 023	156.001 515	46.706 723
32	1.610 324	40.688 288	25.267 139	82	3.390 123	159.341 536	47.001 697
33	1.634 479	42.298 612	25.878 954	83	3.440 975	162.731 661	47.292 313
34	1.658 996	43.933 092	26.481 728	84	3.492 590	166.172 636	47.578 633
35	1.683 881	45.592 088	27.075 595	85	3.544 978	169.665 226	47.860 722
36	1.709 140	47.275 969	27.660 684	86	3.598 153	173.210 204	48.138 643
37	1.734 777	48.985 109	28.237 127	87	3.652 125	176.808 357	48.412 456
38	1.760 798	50.719 885	28.805 052	88	3.706 907	180.460 482	48.682 222
39	1.787 210	52.480 684	29.364 583	89	3.762 511	184.167 390	48.948 002
40	1.814 018	54.267 894	29.915 845	90	3.818 949	187.929 900	49.209 855
41	1.841 229	56.081 912	30.458 961	91	3.876 233	191.748 849	49.467 837
42	1.868 847	57.923 141	30.994 050	92	3.934 376	195.625 082	49.722 007
43	1.896 880	59.791 988	31.521 232	93	3.993 392	199.559 458	49.972 421
44	1.925 333	61.688 868	32.040 622	94	4.053 293	203.552 850	50.219 134
45	1.954 213	63.614 201	32.552 337	95	4.114 092	207.606 142	50.462 201
46	1.983 526	65.568 414	33.056 490	96	4.175 804	211.720 235	50.701 675
47	2.013 279	67.551 940	33.553 192	97	4.238 441	215.896 038	50.937 611
48	2.043 478	69.565 219	34.042 554	98	4.302 017	220.134 479	51.170 060
49	2.074 130	71.608 698	34.524 683	99	4.366 547	224.436 496	51.399 074
50	2.105 242	73.682 828	34.999 688	100	4.432 046	228.803 043	51.624 704

Continued

		$i = 0.0175$ $(1\tfrac{3}{4}\%)$					
n	$(1+i)^n$	$s_{\overline{n}\rvert i}$	$a_{\overline{n}\rvert i}$	n	$(1+i)^n$	$s_{\overline{n}\rvert i}$	$a_{\overline{n}\rvert i}$
1	1.017 500	1.000 000	0.982 801	51	2.422 453	81.283 014	33.554 014
2	1.035 306	2.017 500	1.948 699	52	2.464 846	83.705 466	33.959 719
3	1.053 424	3.052 806	2.897 984	53	2.507 980	86.170 312	34.358 446
4	1.071 859	4.106 230	3.830 943	54	2.551 870	88.678 292	34.750 316
5	1.090 617	5.178 089	4.747 855	55	2.596 528	91.230 163	35.135 446
6	1.109 702	6.268 706	5.648 998	56	2.641 967	93.826 690	35.513 951
7	1.129 122	7.378 408	6.534 641	57	2.688 202	96.468 658	35.885 947
8	1.148 882	8.507 530	7.405 053	58	2.735 245	99.156 859	36.251 545
9	1.168 987	9.656 412	8.260 494	59	2.783 112	101.892 104	36.610 855
10	1.189 444	10.825 399	9.101 223	60	2.831 816	104.675 216	36.963 986
11	1.210 260	12.014 844	9.927 492	61	2.881 373	107.507 032	37.311 042
12	1.231 439	13.225 104	10.739 550	62	2.931 797	110.388 405	37.652 130
13	1.252 990	14.456 543	11.537 641	63	2.983 104	113.320 202	37.987 351
14	1.274 917	15.709 533	12.322 006	64	3.034 308	116.303 306	38.316 807
15	1.297 228	16.984 449	13.092 880	65	3.088 426	119.338 614	38.640 597
16	1.319 929	18.281 677	13.850 497	66	3.142 473	122.427 039	38.958 817
17	1.343 028	19.601 607	14.595 083	67	3.197 466	125.569 513	39.271 565
18	1.366 531	20.944 635	15.326 863	68	3.253 422	128.766 979	39.578 934
19	1.390 445	22.311 166	16.046 057	69	3.310 357	132.020 401	39.881 016
20	1.414 778	23.701 611	16.752 881	70	3.368 288	135.330 758	40.177 903
21	1.439 537	25.116 389	17.447 549	71	3.427 233	138.699 047	40.469 683
22	1.464 729	26.555 926	18.130 269	72	3.487 210	142.126 280	40.756 445
23	1.490 361	28.020 655	18.801 248	73	3.548 236	145.613 490	41.038 276
24	1.516 443	29.511 016	19.460 686	74	3.610 330	149.161 726	41.315 259
25	1.542 981	31.027 459	20.108 782	75	3.673 511	152.772 056	41.587 478
26	1.569 983	32.570 440	20.745 732	76	3.737 797	156.445 567	41.855 015
27	1.597 457	34.140 422	21.371 726	77	3.803 209	160.183 364	42.117 951
28	1.625 413	35.737 880	21.986 955	78	3.869 765	163.986 573	42.376 364
29	1.653 858	37.363 293	22.591 602	79	3.937 486	167.856 338	42.630 334
30	1.682 800	39.017 150	23.185 849	80	4.006 392	171.793 824	42.879 935
31	1.712 249	40.699 950	23.769 876	81	4.076 504	175.800 216	43.125 243
32	1.742 213	42.412 200	24.343 859	82	4.147 843	179.876 720	43.366 332
33	1.772 702	44.154 413	24.907 970	83	4.220 430	184.024 563	43.603 275
34	1.803 725	45.927 115	25.462 378	84	4.294 287	188.244 992	43.836 142
35	1.835 290	47.730 840	26.007 251	85	4.369 437	192.539 280	44.065 005
36	1.867 407	49.566 129	26.542 753	86	4.445 903	196.908 717	44.289 931
37	1.900 087	51.433 537	27.069 045	87	4.523 706	201.354 620	44.510 989
38	1.933 338	53.333 624	27.586 285	88	4.602 871	205.878 326	44.728 244
39	1.967 172	55.266 962	28.094 629	89	4.683 421	210.481 196	44.941 764
40	2.001 597	57.234 134	28.594 230	90	4.765 381	215.164 617	45.151 610
41	2.036 625	59.235 731	29.085 238	91	4.848 775	219.929 998	44.357 848
42	2.072 266	61.272 357	29.567 801	92	4.933 629	224.778 773	45.560 539
43	2.108 531	63.344 623	30.042 065	93	5.019 967	229.712 401	45.759 743
44	2.145 430	65.453 154	30.508 172	94	5.107 816	234.732 368	45.955 521
45	2.182 975	67.598 584	30.966 263	95	5.197 203	239.840 185	46.147 933
46	2.221 177	69.781 559	31.416 474	96	5.288 154	245.037 388	46.337 035
47	2.260 048	72.002 736	31.858 943	97	5.380 697	250.325 542	46.522 884
48	2.299 599	74.262 784	32.293 801	98	5.474 859	255.706 239	46.705 537
49	2.339 842	76.562 383	32.721 181	99	5.570 669	261.181 099	46.885 049
50	2.380 789	78.902 225	33.141 209	100	5.668 156	266.751 768	47.061 473

Continued

				$i = 0.02\ (2\%)$			
n	$(1 + i)^n$	$s_{\overline{n}i}$	$a_{\overline{n}i}$	n	$(1 + i)^n$	$s_{\overline{n}i}$	$a_{\overline{n}i}$
1	1.020 000	1.000 000	0.980 392	51	2.745 420	87.270 989	31.787 849
2	1.040 400	2.020 000	1.941 561	52	2.800 328	90.016 409	32.144 950
3	1.061 208	3.060 400	2.883 883	53	2.856 335	92.816 737	32.495 049
4	1.082 432	4.121 608	3.807 729	54	2.913 461	95.673 072	32.838 283
5	1.104 081	5.204 040	4.713 460	55	2.971 731	98.586 534	33.174 788
6	1.126 162	6.308 121	5.601 431	56	3.031 165	101.558 264	33.504 694
7	1.148 686	7.434 283	6.471 991	57	3.091 789	104.589 430	33.828 131
8	1.171 659	8.582 969	7.325 481	58	3.153 624	107.681 218	34.145 226
9	1.195 093	9.754 628	8.162 237	59	3.216 697	110.834 843	34.456 104
10	1.218 994	10.949 721	8.982 585	60	3.281 031	114.051 539	34.760 887
11	1.243 374	12.168 715	9.786 848	61	3.346 651	117.332 570	35.059 693
12	1.268 242	13.412 090	10.575 341	62	3.413 584	120.679 222	35.352 640
13	1.293 607	14.680 331	11.348 374	63	3.481 856	124.092 806	35.639 843
14	1.319 479	15.973 938	12.106 249	64	3.551 493	127.574 662	35.921 415
15	1.345 868	17.293 417	12.849 264	65	3.622 523	131.126 155	36.197 466
16	1.372 786	18.639 285	13.577 709	66	3.694 974	134.748 679	36.468 103
17	1.400 241	20.012 071	14.291 872	67	3.768 873	138.443 652	36.733 435
18	1.428 246	21.412 312	14.992 031	68	3.844 251	142.212 525	36.993 564
19	1.456 811	22.840 559	15.678 462	69	3.921 136	146.056 776	37.248 592
20	1.485 947	24.297 370	16.351 433	70	3.999 558	149.977 911	37.498 619
21	1.515 666	25.783 317	17.011 209	71	4.079 549	153.977 469	37.743 744
22	1.545 980	27.298 984	17.658 048	72	4.161 140	158.057 019	37.984 063
23	1.576 899	28.844 963	18.292 204	73	4.244 363	162.218 159	38.219 670
24	1.608 437	30.421 862	18.913 926	74	4.329 250	166.462 522	38.450 657
25	1.640 606	32.030 300	19.523 456	75	4.415 835	170.791 773	38.677 114
26	1.673 418	33.670 906	20.121 036	76	4.504 152	175.207 608	38.899 132
27	1.706 886	35.344 324	20.706 898	77	4.594 235	179.711 760	39.116 796
28	1.741 024	37.051 210	21.281 272	78	4.686 120	184.305 996	39.330 192
29	1.775 845	38.792 235	21.844 385	79	4.779 842	188.992 115	39.539 404
30	1.811 362	40.568 079	22.396 456	80	4.875 439	193.771 958	39.744 514
31	1.847 589	42.379 441	22.937 702	81	4.972 948	198.647 397	39.945 602
32	1.884 541	44.227 030	23.468 335	82	5.072 407	203.620 345	40.142 747
33	1.922 231	46.111 570	23.988 564	83	5.173 855	208.692 752	40.336 026
34	1.960 676	48.033 802	24.498 592	84	5.277 332	213.866 607	40.525 516
35	1.999 890	49.994 478	24.998 619	85	5.382 879	219.143 939	40.711 290
36	2.039 887	51.994 367	25.488 842	86	5.490 536	224.526 818	40.893 422
37	2.080 685	54.034 255	25.969 453	87	5.600 347	230.017 354	41.071 982
38	2.122 299	56.114 940	26.440 641	88	5.712 354	235.617 701	41.247 041
39	2.164 745	58.237 238	26.902 589	89	5.826 601	241.330 055	41.418 668
40	2.208 040	60.401 983	27.355 479	90	5.943 133	247.156 656	41.586 929
41	2.252 200	62.610 023	27.799 489	91	6.061 996	253.099 789	41.751 891
42	2.297 244	64.862 223	28.234 794	92	6.183 236	259.161 785	41.913 619
43	2.343 189	67.159 468	28.661 562	93	6.306 900	265.345 021	42.072 175
44	2.390 053	69.502 657	29.079 963	94	6.433 038	271.651 921	42.227 623
45	2.437 854	71.892 710	29.490 159	95	6.561 699	278.084 960	42.380 023
46	2.486 611	74.330 564	29.892 314	96	6.692 933	284.646 659	42.529 434
47	2.536 344	76.817 176	30.286 582	97	6.826 792	291.339 592	42.675 916
48	2.587 070	79.353 519	30.673 120	98	6.963 328	298.166 384	42.819 525
49	2.638 812	81.940 590	31.052 078	99	7.102 594	305.129 712	42.960 319
50	2.691 588	84.579 401	31.423 606	100	7.244 646	312.232 306	43.098 352

Continued

			$i = 0.0225\ (2\tfrac{1}{4}\%)$				
n	$(1 + i)^n$	$s_{\overline{n}|i}$	$a_{\overline{n}|i}$	n	$(1 + i)^n$	$s_{\overline{n}|i}$	$a_{\overline{n}|i}$
1	1.022 500	1.000 000	0.977 995	51	3.110 492	93.799 664	30.155 889
2	1.045 506	2.022 500	1.934 470	52	3.180 479	96.910 157	30.470 307
3	1.069 030	3.068 006	2.869 897	53	3.252 039	100.090 635	30.777 806
4	1.093 083	4.137 036	3.784 740	54	3.325 210	103.342 674	31.078 539
5	1.117 678	5.230 120	4.679 453	55	3.400 027	106.667 885	31.372 654
6	1.142 825	6.347 797	5.554 477	56	3.476 528	110.067 912	31.660 298
7	1.168 539	7.490 623	6.410 246	57	3.554 750	113.544 440	31.941 611
8	1.194 831	8.659 162	7.247 185	58	3.634 732	117.099 190	32.216 735
9	1.221 715	9.853 993	8.065 706	59	3.716 513	120.733 922	32.485 804
10	1.249 203	11.075 708	8.866 216	60	3.800 135	124.450 435	32.748 953
11	1.277 311	12.324 911	9.649 111	61	3.885 638	128.250 570	33.006 311
12	1.306 050	13.602 222	10.414 779	62	3.973 065	132.136 208	33.258 006
13	1.335 436	14.908 272	11.163 598	63	4.062 459	136.109 272	33.504 162
14	1.365 483	16.243 708	11.895 939	64	4.153 864	140.171 731	33.744 902
15	1.396 207	17.609 191	12.612 166	65	4.247 326	144.325 595	33.980 344
16	1.427 621	19.005 398	13.312 631	66	4.342 891	148.572 920	34.210 605
17	1.459 743	20.433 020	13.997 683	67	4.440 606	152.915 811	34.435 800
18	1.492 587	21.892 763	14.667 661	68	4.540 519	157.356 417	34.656 039
19	1.526 170	23.385 350	15.322 896	69	4.642 681	161.896 937	34.871 432
20	1.560 509	24.911 520	15.963 712	70	4.747 141	166.539 618	35.082 085
21	1.595 621	26.472 029	16.590 428	71	4.853 952	171.286 759	35.288 103
22	1.631 522	28.067 650	17.203 352	72	4.963 166	176.140 711	35.489 587
23	1.668 231	29.699 172	17.802 790	73	5.074 837	181.103 877	35.686 638
24	1.705 767	31.367 403	18.389 036	74	5.189 021	186.178 714	35.879 352
25	1.744 146	33.073 170	18.962 383	75	5.305 774	191.367 735	36.067 826
26	1.783 390	34.817 316	19.523 113	76	5.425 154	196.673 509	36.252 153
27	1.823 516	36.600 706	20.071 504	77	5.547 220	202.098 663	36.432 423
28	1.864 545	38.424 222	20.607 828	78	5.672 032	207.645 883	36.608 727
29	1.906 497	40.288 767	21.132 350	79	5.799 653	213.317 916	36.781 151
30	1.949 393	42.195 264	21.645 330	80	5.930 145	219.117 569	36.949 781
31	1.993 255	44.144 657	22.147 022	81	6.063 574	225.047 714	37.114 700
32	2.038 103	46.137 912	22.637 674	82	6.200 004	231.111 288	37.275 990
33	2.083 960	48.176 015	23.117 530	83	6.339 504	237.311 292	37.433 731
34	2.130 849	50.259 976	23.586 826	84	6.482 143	243.650 796	37.588 001
35	2.178 794	52.390 825	24.045 796	85	6.627 991	250.132 939	37.738 877
36	2.227 816	54.569 619	24.494 666	86	6.777 121	256.760 930	37.886 432
37	2.277 942	56.797 435	24.933 658	87	6.929 606	263.538 051	38.030 740
38	2.329 196	59.075 377	25.362 991	88	7.085 522	270.467 657	38.171 873
39	2.381 603	61.404 573	25.782 876	89	7.244 947	277.553 179	38.309 900
40	2.435 189	63.786 176	26.193 522	90	7.407 958	284.798 126	38.444 890
41	2.489 981	66.221 365	26.595 132	91	7.574 637	292.206 083	38.576 910
42	2.546 005	68.711 346	26.987 904	92	7.745 066	299.780 720	38.706 024
43	2.603 290	71.257 351	27.372 033	93	7.919 330	307.525 786	38.832 298
44	2.661 864	73.860 642	27.747 710	94	8.097 515	315.445 117	38.955 792
45	2.721 756	76.522 506	28.115 120	95	8.279 709	323.542 632	39.076 569
46	2.782 996	79.244 262	28.474 444	96	8.466 003	331.822 341	39.194 689
47	2.845 613	82.027 258	28.825 863	97	8.656 488	340.288 344	39.310 209
48	2.909 640	84.872 872	29.169 548	98	8.851 259	348.944 831	39.423 187
49	2.975 107	87.782 511	29.505 670	99	9.050 412	357.796 090	39.533 680
50	3.042 046	90.757 618	29.834 396	100	9.254 046	366.846 502	39.641 741

TABLE II

Continued

			$i = 0.025\ (2\tfrac{1}{2}\%)$				
n	$(1 + i)^n$	$s_{\overline{n}i}$	$a_{\overline{n}i}$	n	$(1 + i)^n$	$s_{\overline{n}i}$	$a_{\overline{n}i}$
1	1.025 000	1.000 000	0.975 610	51	3.523 036	100.921 458	28.646 158
2	1.050 625	2.025 000	1.927 424	52	3.611 112	104.444 494	28.923 081
3	1.076 891	3.075 625	2.856 024	53	3.701 390	108.055 606	29.193 249
4	1.103 813	4.152 516	3.761 974	54	3.793 925	111.756 996	29.456 829
5	1.131 408	5.256 329	4.645 828	55	3.888 773	115.550 921	29.713 979
6	1.159 693	6.387 737	5.508 125	56	3.985 992	119.439 694	29.964 858
7	1.188 686	7.547 430	6.349 391	57	4.085 642	123.425 687	30.209 617
8	1.218 403	8.736 116	7.170 137	58	4.187 783	127.511 329	30.448 407
9	1.248 863	9.954 519	7.970 866	59	4.292 478	131.699 112	30.681 373
10	1.280 085	11.203 382	8.752 064	60	4.399 790	135.991 590	30.908 656
11	1.312 087	12.483 466	9.514 209	61	4.509 784	140.391 380	31.130 397
12	1.344 889	13.795 553	10.257 765	62	4.622 529	144.901 164	31.346 728
13	1.378 511	15.140 442	10.983 185	63	4.738 092	149.523 693	31.557 784
14	1.412 974	16.518 953	11.690 912	64	4.856 545	154.261 786	31.763 691
15	1.448 298	17.931 927	12.381 378	65	4.977 958	159.118 330	31.964 577
16	1.484 506	19.380 225	13.055 003	66	5.102 407	164.096 289	32.160 563
17	1.521 618	20.864 730	13.712 198	67	5.229 967	169.198 696	32.351 769
18	1.559 659	22.386 349	14.353 364	68	5.360 717	174.428 663	32.538 311
19	1.598 650	23.946 007	14.978 891	69	5.494 734	179.789 380	32.720 303
20	1.638 616	25.544 658	15.589 162	70	5.632 103	185.284 114	32.897 857
21	1.679 582	27.183 274	16.184 549	71	5.772 905	190.916 217	33.071 080
22	1.721 571	28.862 856	16.765 413	72	5.917 228	196.689 122	33.240 078
23	1.764 611	30.584 427	17.332 110	73	6.065 159	202.606 351	33.404 954
24	1.808 726	32.349 038	17.884 986	74	6.216 788	208.671 509	33.565 809
25	1.853 944	34.157 764	18.424 376	75	6.372 207	214.888 297	33.722 740
26	1.900 293	36.011 708	18.950 611	76	6.531 513	221.260 504	33.875 844
27	1.947 800	37.912 001	19.464 011	77	6.694 800	227.792 017	34.025 214
28	1.996 495	39.859 801	19.964 889	78	6.862 170	234.486 818	34.170 940
29	2.046 407	41.856 296	20.453 550	79	7.033 725	241.348 988	34.313 113
30	2.097 568	43.902 703	20.930 293	80	7.209 568	248.382 713	34.451 817
31	2.150 007	46.000 271	21.395 407	81	7.389 807	255.592 280	34.587 139
32	2.203 757	48.150 278	21.849 178	82	7.574 552	262.982 087	34.719 160
33	2.258 851	50.354 034	22.291 881	83	7.763 916	270.556 640	34.847 961
34	2.315 322	52.612 885	22.723 786	84	7.958 014	278.320 556	34.973 620
35	2.373 205	54.928 207	23.145 157	85	8.156 964	286.278 569	35.096 215
36	2.432 535	57.301 413	23.556 251	86	8.360 888	294.435 534	35.215 819
37	2.493 349	59.733 948	23.957 318	87	8.569 911	302.796 422	35.332 507
38	2.555 682	62.227 297	24.348 603	88	8.784 158	311.366 333	35.446 348
39	2.619 574	64.782 979	24.730 344	89	9.003 762	320.150 491	35.557 413
40	2.685 064	67.402 554	25.102 775	90	9.228 856	329.154 253	35.665 768
41	2.752 190	70.087 617	25.466 122	91	9.459 578	338.383 110	35.771 481
42	2.820 995	72.839 808	25.820 607	92	9.696 067	347.842 687	35.874 616
43	2.891 520	75.660 803	26.166 446	93	9.938 469	357.538 755	35.975 235
44	2.963 808	78.552 323	26.503 849	94	10.186 931	367.477 223	36.073 400
45	3.037 903	81.516 131	26.833 024	95	10.441 604	377.664 154	36.169 171
46	3.113 851	84.554 034	27.154 170	96	10.702 644	388.105 758	36.262 606
47	3.191 697	87.667 885	27.467 483	97	10.970 210	398.808 402	36.353 762
48	3.271 490	90.859 582	27.773 154	98	11.244 465	409.778 612	36.442 694
49	3.353 277	94.131 072	28.071 369	99	11.525 577	421.023 077	36.529 458
50	3.437 109	97.484 349	28.362 312	100	11.813 716	432.548 654	36.614 105

Continued

<table>
<tr><td colspan="8" align="center">$i = 0.03$ (3%)</td></tr>
<tr><td>n</td><td>$(1 + i)^n$</td><td>$s_{\overline{n}|i}$</td><td>$a_{\overline{n}|i}$</td><td>n</td><td>$(1 + i)^n$</td><td>$s_{\overline{n}|i}$</td><td>$a_{\overline{n}|i}$</td></tr>
<tr><td>1</td><td>1.030 000</td><td>1.000 000</td><td>0.970 874</td><td>51</td><td>4.515 423</td><td>117.180 773</td><td>25.951 227</td></tr>
<tr><td>2</td><td>1.060 900</td><td>2.030 000</td><td>1.913 470</td><td>52</td><td>4.650 886</td><td>121.696 197</td><td>26.166 240</td></tr>
<tr><td>3</td><td>1.092 727</td><td>3.090 900</td><td>2.828 611</td><td>53</td><td>4.790 412</td><td>126.347 082</td><td>26.374 990</td></tr>
<tr><td>4</td><td>1.125 509</td><td>4.183 627</td><td>3.717 098</td><td>54</td><td>4.934 125</td><td>131.137 495</td><td>26.577 660</td></tr>
<tr><td>5</td><td>1.159 274</td><td>5.309 136</td><td>4.579 707</td><td>55</td><td>5.082 149</td><td>136.071 620</td><td>26.774 428</td></tr>
<tr><td>6</td><td>1.194 052</td><td>6.468 410</td><td>5.417 191</td><td>56</td><td>5.234 613</td><td>141.153 768</td><td>26.965 464</td></tr>
<tr><td>7</td><td>1.229 874</td><td>7.662 462</td><td>6.230 283</td><td>57</td><td>5.391 651</td><td>146.388 381</td><td>27.150 936</td></tr>
<tr><td>8</td><td>1.266 770</td><td>8.892 336</td><td>7.019 692</td><td>58</td><td>5.553 401</td><td>151.780 033</td><td>27.331 005</td></tr>
<tr><td>9</td><td>1.304 773</td><td>10.159 106</td><td>7.786 109</td><td>59</td><td>5.720 003</td><td>157.333 434</td><td>27.505 831</td></tr>
<tr><td>10</td><td>1.343 916</td><td>11.463 879</td><td>8.530 203</td><td>60</td><td>5.891 603</td><td>163.053 437</td><td>27.675 564</td></tr>
<tr><td>11</td><td>1.384 234</td><td>12.807 796</td><td>9.252 624</td><td>61</td><td>6.068 351</td><td>168.945 040</td><td>27.840 353</td></tr>
<tr><td>12</td><td>1.425 761</td><td>14.192 030</td><td>9.954 004</td><td>62</td><td>6.250 402</td><td>175.013 391</td><td>28.000 343</td></tr>
<tr><td>13</td><td>1.468 534</td><td>15.617 790</td><td>10.634 955</td><td>63</td><td>6.437 914</td><td>181.263 793</td><td>28.155 673</td></tr>
<tr><td>14</td><td>1.512 590</td><td>17.086 324</td><td>11.296 073</td><td>64</td><td>6.631 051</td><td>187.701 707</td><td>28.306 478</td></tr>
<tr><td>15</td><td>1.557 967</td><td>18.598 914</td><td>11.937 935</td><td>65</td><td>6.829 983</td><td>194.332 758</td><td>28.452 892</td></tr>
<tr><td>16</td><td>1.604 706</td><td>20.156 881</td><td>12.561 102</td><td>66</td><td>7.034 882</td><td>201.162 741</td><td>28.595 040</td></tr>
<tr><td>17</td><td>1.652 848</td><td>21.761 588</td><td>13.166 118</td><td>67</td><td>7.245 929</td><td>208.197 623</td><td>28.733 049</td></tr>
<tr><td>18</td><td>1.702 433</td><td>23.414 435</td><td>13.753 513</td><td>68</td><td>7.463 307</td><td>215.443 551</td><td>28.867 038</td></tr>
<tr><td>19</td><td>1.753 506</td><td>25.116 868</td><td>14.323 799</td><td>69</td><td>7.687 206</td><td>222.906 858</td><td>28.997 124</td></tr>
<tr><td>20</td><td>1.806 111</td><td>26.870 374</td><td>14.877 475</td><td>70</td><td>7.917 822</td><td>230.594 064</td><td>29.123 421</td></tr>
<tr><td>21</td><td>1.860 295</td><td>28.676 486</td><td>15.415 024</td><td>71</td><td>8.155 357</td><td>238.511 886</td><td>29.246 040</td></tr>
<tr><td>22</td><td>1.916 103</td><td>30.536 780</td><td>15.936 917</td><td>72</td><td>8.400 017</td><td>246.667 242</td><td>29.365 088</td></tr>
<tr><td>23</td><td>1.973 587</td><td>32.452 884</td><td>16.443 608</td><td>73</td><td>8.652 016</td><td>255.067 259</td><td>29.480 668</td></tr>
<tr><td>24</td><td>2.032 794</td><td>34.426 470</td><td>16.935 542</td><td>74</td><td>8.911 578</td><td>263.719 277</td><td>29.592 881</td></tr>
<tr><td>25</td><td>2.093 778</td><td>36.459 264</td><td>17.413 148</td><td>75</td><td>9.178 926</td><td>272.630 856</td><td>29.701 826</td></tr>
<tr><td>26</td><td>2.156 591</td><td>38.553 042</td><td>17.876 842</td><td>76</td><td>9.454 293</td><td>281.809 781</td><td>29.807 598</td></tr>
<tr><td>27</td><td>2.221 289</td><td>40.709 634</td><td>18.327 031</td><td>77</td><td>9.737 922</td><td>291.264 075</td><td>29.910 290</td></tr>
<tr><td>28</td><td>2.287 928</td><td>42.930 923</td><td>18.764 108</td><td>78</td><td>10.030 060</td><td>301.001 997</td><td>30.009 990</td></tr>
<tr><td>29</td><td>2.356 566</td><td>45.218 850</td><td>19.188 455</td><td>79</td><td>10.330 962</td><td>311.032 057</td><td>30.106 786</td></tr>
<tr><td>30</td><td>2.427 262</td><td>47.575 416</td><td>19.600 441</td><td>80</td><td>10.640 891</td><td>321.363 019</td><td>30.200 763</td></tr>
<tr><td>31</td><td>2.500 080</td><td>50.002 678</td><td>20.000 428</td><td>81</td><td>10.960 117</td><td>332.003 909</td><td>30.292 003</td></tr>
<tr><td>32</td><td>2.575 083</td><td>52.502 759</td><td>20.388 766</td><td>82</td><td>11.288 921</td><td>342.964 026</td><td>30.380 586</td></tr>
<tr><td>33</td><td>2.652 335</td><td>55.077 841</td><td>20.765 792</td><td>83</td><td>11.627 588</td><td>354.252 947</td><td>30.466 588</td></tr>
<tr><td>34</td><td>2.731 905</td><td>57.730 177</td><td>21.131 837</td><td>84</td><td>11.976 416</td><td>365.880 536</td><td>30.550 086</td></tr>
<tr><td>35</td><td>2.813 862</td><td>60.462 082</td><td>21.487 220</td><td>85</td><td>12.335 709</td><td>377.856 952</td><td>30.631 151</td></tr>
<tr><td>36</td><td>2.898 278</td><td>63.275 944</td><td>21.832 252</td><td>86</td><td>12.705 780</td><td>390.192 660</td><td>30.709 855</td></tr>
<tr><td>37</td><td>2.985 227</td><td>66.174 223</td><td>22.167 235</td><td>87</td><td>13.086 953</td><td>402.898 440</td><td>30.786 267</td></tr>
<tr><td>38</td><td>3.074 783</td><td>69.159 449</td><td>22.492 462</td><td>88</td><td>13.479 562</td><td>415.985 393</td><td>30.860 454</td></tr>
<tr><td>39</td><td>3.167 027</td><td>72.234 233</td><td>22.808 215</td><td>89</td><td>13.883 949</td><td>429.464 955</td><td>30.932 479</td></tr>
<tr><td>40</td><td>3.262 038</td><td>75.401 260</td><td>23.114 772</td><td>90</td><td>14.300 467</td><td>443.348 904</td><td>31.002 407</td></tr>
<tr><td>41</td><td>3.359 899</td><td>78.663 298</td><td>23.412 400</td><td>91</td><td>14.729 481</td><td>457.649 371</td><td>31.070 298</td></tr>
<tr><td>42</td><td>3.460 696</td><td>82.023 196</td><td>23.701 359</td><td>92</td><td>15.171 366</td><td>472.378 852</td><td>31.136 212</td></tr>
<tr><td>43</td><td>3.564 517</td><td>85.483 892</td><td>23.981 902</td><td>93</td><td>15.626 507</td><td>487.550 217</td><td>31.200 206</td></tr>
<tr><td>44</td><td>3.671 452</td><td>89.048 409</td><td>24.254 274</td><td>94</td><td>16.095 302</td><td>503.176 724</td><td>31.262 336</td></tr>
<tr><td>45</td><td>3.781 596</td><td>92.719 861</td><td>24.518 713</td><td>95</td><td>16.578 161</td><td>519.272 026</td><td>31.322 656</td></tr>
<tr><td>46</td><td>3.895 044</td><td>96.501 457</td><td>24.775 449</td><td>96</td><td>17.075 506</td><td>535.850 186</td><td>31.381 219</td></tr>
<tr><td>47</td><td>4.011 895</td><td>100.396 501</td><td>25.024 708</td><td>97</td><td>17.587 771</td><td>552.925 692</td><td>31.438 077</td></tr>
<tr><td>48</td><td>4.132 252</td><td>104.408 396</td><td>25.266 707</td><td>98</td><td>18.115 404</td><td>570.513 463</td><td>31.493 279</td></tr>
<tr><td>49</td><td>4.256 219</td><td>108.540 648</td><td>25.501 657</td><td>99</td><td>18.658 866</td><td>588.628 867</td><td>31.546 872</td></tr>
<tr><td>50</td><td>4.383 906</td><td>112.796 867</td><td>25.729 764</td><td>100</td><td>19.218 632</td><td>607.287 733</td><td>31.598 905</td></tr>
</table>

TABLE II

Continued

		$i = 0.035\ (3\frac{1}{2}\%)$					
n	$(1 + i)^n$	$s_{\overline{n}|i}$	$a_{\overline{n}|i}$	n	$(1 + i)^n$	$s_{\overline{n}|i}$	$a_{\overline{n}|i}$
1	1.035 000	1.000 000	0.966 184	51	5.780 399	136.582 837	23.628 616
2	1.071 225	2.035 000	1.899 694	52	5.982 713	142.363 236	23.795 765
3	1.108 718	3.106 225	2.801 637	53	6.192 108	148.345 950	23.957 260
4	1.147 523	4.214 943	3.673 079	54	6.408 832	154.538 058	24.113 295
5	1.187 686	5.362 466	4.515 052	55	6.633 141	160.946 890	24.264 053
6	1.229 255	6.550 152	5.328 553	56	6.865 301	167.580 031	24.409 713
7	1.272 279	7.779 408	6.114 544	57	7.105 587	174.445 332	24.550 448
8	1.316 809	9.051 687	6.873 956	58	7.354 282	181.550 919	24.686 423
9	1.362 897	10.368 496	7.607 687	59	7.611 682	188.905 201	24.817 800
10	1.410 599	11.731 393	8.316 605	60	7.878 091	196.516 883	24.944 734
11	1.459 970	13.141 992	9.001 551	61	8.153 824	204.394 974	25.067 376
12	1.511 069	14.601 962	9.663 334	62	8.439 208	212.548 798	25.185 870
13	1.563 956	16.113 030	10.302 738	63	8.734 580	220.988 006	25.300 358
14	1.618 695	17.676 986	10.920 520	64	9.040 291	229.722 586	25.410 974
15	1.675 349	19.295 681	11.517 411	65	9.356 701	238.762 876	25.517 849
16	1.733 986	20.971 030	12.094 117	66	9.684 185	248.119 577	25.621 110
17	1.794 676	22.705 016	12.651 321	67	10.023 132	257.803 762	25.720 880
18	1.857 489	24.499 691	13.189 682	68	10.373 941	267.826 894	25.817 275
19	1.922 501	26.357 180	13.709 837	69	10.737 029	278.200 835	25.910 411
20	1.989 789	28.279 682	14.212 403	70	11.112 825	288.937 865	26.000 397
21	2.059 431	30.269 471	14.697 974	71	11.501 774	300.050 690	26.087 340
22	2.131 512	32.328 902	15.167 125	72	11.904 336	311.552 464	26.171 343
23	2.206 114	34.460 414	15.620 410	73	12.320 988	323.456 800	26.252 505
24	2.283 328	36.666 528	16.058 368	74	12.752 223	335.777 788	26.330 923
25	2.363 245	38.949 857	16.481 515	75	13.198 550	348.530 011	26.406 689
26	2.445 959	41.313 102	16.890 352	76	13.660 500	361.728 561	26.479 892
27	2.531 567	43.759 060	17.285 365	77	14.138 617	375.389 061	26.550 621
28	2.620 172	46.290 627	17.667 019	78	14.633 469	389.527 678	26.618 957
29	2.711 878	48.910 799	18.035 767	79	15.145 640	404.161 147	26.684 983
30	2.806 794	51.622 677	18.392 045	80	15.675 738	419.306 787	26.748 776
31	2.905 031	54.429 471	18.736 276	81	16.224 388	434.982 524	26.810 411
32	3.006 708	57.334 502	19.068 865	82	16.792 242	451.206 913	26.869 963
33	3.111 942	60.341 210	19.390 208	83	17.379 970	467.999 155	26.927 500
34	3.220 860	63.453 152	19.700 684	84	17.988 269	485.379 125	26.983 092
35	3.333 590	66.674 013	20.000 661	85	18.617 859	503.367 394	27.036 804
36	3.450 266	70.007 603	20.290 494	86	19.269 484	521.985 253	27.088 699
37	3.571 025	73.457 869	20.570 525	87	19.943 916	541.254 737	27.138 840
38	3.696 011	77.028 895	20.841 087	88	20.641 953	561.198 653	27.187 285
39	3.825 372	80.724 906	21.102 500	89	21.364 421	581.840 606	27.234 092
40	3.959 260	84.550 278	21.355 072	90	22.112 176	603.205 027	27.279 316
41	4.097 834	88.509 537	21.599 104	91	22.886 102	625.317 203	27.323 010
42	4.241 258	92.607 371	21.834 883	92	23.687 116	648.203 305	27.365 227
43	4.389 702	96.848 629	22.062 689	93	24.516 165	671.890 421	27.406 017
44	4.543 342	101.238 331	22.282 791	94	25.374 230	696.406 585	27.445 427
45	4.702 359	105.781 673	22.495 450	95	26.262 329	721.780 816	27.483 504
46	4.866 941	110.484 031	22.700 918	96	27.181 510	748.043 145	27.520 294
47	5.037 284	115.350 973	22.899 438	97	28.132 863	775.224 655	27.555 839
48	5.213 589	120.388 257	23.091 244	98	29.117 513	803.357 517	27.590 183
49	5.396 065	125.601 846	23.276 564	99	30.136 626	832.475 031	27.623 366
50	5.584 927	130.997 910	23.455 618	100	31.191 408	662.611 657	27.653 425

Continued

			$i = 0.04\ (4\%)$				
n	$(1+i)^n$	$s_{\overline{n}|i}$	$a_{\overline{n}|i}$	n	$(1+i)^n$	$s_{\overline{n}|i}$	$a_{\overline{n}|i}$
1	1.040 000	1.000 000	0.961 538	51	7.390 951	159.773 767	21.617 485
2	1.081 600	2.040 000	1.886 095	52	7.686 589	167.164 718	21.747 582
3	1.124 864	3.121 600	2.775 091	53	7.994 052	174.851 306	21.872 675
4	1.169 859	4.246 464	3.629 895	54	8.313 814	182.845 359	21.992 957
5	1.216 653	5.416 323	4.451 822	55	8.646 367	191.159 173	22.108 612
6	1.265 319	6.632 975	5.242 137	56	8.992 222	109.805 540	22.219 819
7	1.315 932	7.898 294	6.002 055	57	9.351 910	208.797 762	22.326 749
8	1.368 569	9.214 226	6.732 745	58	9.725 987	218.149 672	22.429 567
9	1.423 312	10.582 795	7.435 332	59	10.115 026	227.875 659	22.528 430
10	1.480 244	12.006 107	8.110 896	60	10.519 627	237.990 685	22.623 490
11	1.539 454	13.486 351	8.760 477	61	10.940 413	248.510 312	22.714 894
12	1.601 032	15.025 805	9.385 074	62	11.378 029	259.450 725	22.802 783
13	1.665 074	16.626 838	9.985 648	63	11.833 150	270.828 754	22.887 291
14	1.731 676	18.291 911	10.563 123	64	12.306 476	282.661 904	22.968 549
15	1.800 944	20.023 588	11.118 387	65	12.798 735	294.968 380	23.046 682
16	1.872 981	21.824 531	11.632 296	66	13.310 685	307.767 116	23.121 810
17	1.947 900	23.697 512	12.165 669	67	13.843 112	321.077 800	23.194 048
18	2.025 817	25.645 413	12.659 297	68	14.396 836	334.920 912	23.263 507
19	2.106 849	27.671 229	13.133 939	69	14.972 710	349.317 749	23.330 296
20	2.191 123	29.778 079	13.590 326	70	15.571 618	364.290 459	23.394 515
21	2.278 768	31.969 202	14.029 160	71	16.194 483	379.862 077	23.456 264
22	2.369 919	34.247 970	14.451 115	72	16.842 262	396.056 560	23.515 639
23	2.464 716	36.617 889	14.856 842	73	17.515 953	412.898 823	23.572 730
24	2.563 304	39.082 604	15.246 963	74	18.216 591	430.414 776	23.627 625
25	2.665 836	41.645 908	15.622 080	75	18.945 255	448.631 367	23.680 408
26	2.772 470	44.311 745	15.982 769	76	19.703 065	467.576 621	23.731 162
27	2.883 369	47.084 214	16.329 586	77	20.491 187	487.279 686	23.779 963
28	2.998 703	49.967 583	16.663 063	78	21.310 835	507.770 873	23.826 688
29	3.118 651	52.966 286	16.983 715	79	22.163 268	529.081 708	23.872 008
30	3.243 398	56.084 938	17.292 033	80	23.049 799	551.244 977	23.915 392
31	3.373 133	59.328 335	17.588 494	81	23.971 791	574.294 776	23.957 108
32	3.508 059	62.701 469	17.873 552	82	24.930 663	598.266 567	23.997 219
33	3.648 381	66.209 527	18.147 646	83	25.927 889	623.197 230	24.035 787
34	3.794 316	69.857 909	18.411 198	84	26.965 005	649.125 119	24.072 872
35	3.946 089	73.652 225	18.664 613	85	28.043 605	676.090 123	24.108 531
36	4.103 933	77.598 314	18.908 282	86	29.165 349	704.133 728	24.142 818
37	4.268 090	81.702 246	19.142 579	87	30.331 963	733.299 078	24.175 787
38	4.438 813	85.970 336	19.367 864	88	31.545 242	763.631 041	24.207 487
39	4.616 366	90.409 150	19.584 485	89	32.807 051	795.176 282	24.237 969
40	4.801 021	95.025 516	19.792 774	90	34.119 333	827.983 334	24.267 276
41	4.993 061	99.826 536	19.993 052	91	35.484 107	862.102 667	24.295 459
42	5.192 784	104.819 598	20.185 627	92	36.903 471	897.586 774	24.322 557
43	5.400 495	110.012 382	20.370 795	93	38.379 610	934.490 244	24.348 612
44	5.616 515	115.412 877	20.548 841	94	39.914 794	972.869 854	24.373 666
45	5.841 176	121.029 392	20.720 040	95	41.511 386	1012.784 648	24.397 756
46	6.074 823	126.870 568	20.884 654	96	43.171 841	1054.296 034	24.420 919
47	6.317 816	132.945 390	21.042 936	97	44.898 715	1097.467 876	24.443 191
48	6.570 528	139.263 206	21.195 131	98	46.694 664	1142.366 591	24.464 607
49	6.833 349	145.833 734	21.341 472	99	48.562 450	1189.061 254	24.485 199
50	7.106 683	152.667 084	21.482 185	100	50.504 948	1237.623 705	24.504 999

Continued

			$i = 0.045\ (4\tfrac{1}{2}\%)$			
n	$(1+i)^n$	$s_{\overline{n}|i}$	$a_{\overline{n}|i}$	n	$(1+i)^n$	$s_{\overline{n}|i}$	$a_{\overline{n}|i}$
1	1.045 000	1.000 000	0.956 938	51	9.439 105	187.535 665	19.867 950
2	1.092 025	2.045 000	1.872 668	52	9.863 865	196.974 770	19.969 330
3	1.141 166	3.137 025	2.748 964	53	10.307 739	206.838 634	20.066 345
4	1.192 519	4.278 191	3.587 526	54	10.771 587	217.146 373	20.159 181
5	1.246 182	5.470 710	4.389 977	55	11.256 308	227.917 959	20.248 021
6	1.302 260	6.716 892	5.157 872	56	11.762 842	239.174 268	20.333 034
7	1.360 862	8.019 152	5.892 701	57	12.292 170	250.937 110	20.414 387
8	1.422 101	9.380 014	6.595 886	58	12.845 318	263.229 280	20.492 236
9	1.486 095	10.802 114	7.268 790	59	13.423 357	276.074 597	20.566 733
10	1.552 969	12.288 209	7.912 718	60	14.027 408	289.497 954	20.638 022
11	1.622 853	13.841 179	8.528 917	61	14.658 641	303.525 362	20.706 241
12	1.695 881	15.464 032	9.118 581	62	15.318 280	318.184 031	20.771 523
13	1.772 196	17.159 913	9.682 852	63	16.007 603	333.502 283	20.833 993
14	1.851 945	18.932 109	10.222 825	64	16.727 945	349.509 868	20.893 773
15	1.935 282	20.784 054	10.739 546	65	17.480 702	366.237 831	20.950 979
16	2.022 370	22.719 337	11.234 015	66	18.267 334	383.718 533	21.005 722
17	2.113 377	24.741 707	11.707 191	67	19.089 364	401.985 867	21.058 107
18	2.208 479	26.855 084	12.159 992	68	19.948 385	421.075 231	21.108 236
19	2.307 860	29.063 562	12.593 294	69	20.846 063	441.023 617	21.156 207
20	2.411 714	31.371 423	13.007 936	70	21.784 136	461.869 680	21.202 112
21	2.520 241	33.783 137	13.404 724	71	22.764 422	483.653 815	21.246 040
22	2.633 652	36.303 378	13.784 425	72	23.788 821	506.418 237	21.288 077
23	2.752 166	38.937 030	14.147 775	73	24.859 318	530.207 057	21.328 303
24	2.876 014	41.689 196	14.495 478	74	25.977 987	555.066 375	21.366 797
25	3.005 434	44.565 210	14.828 209	75	27.146 996	581.044 362	21.403 634
26	3.140 679	47.570 645	15.146 611	76	28.368 611	608.191 358	21.438 884
27	3.282 010	50.711 324	15.451 303	77	29.645 199	636.559 969	21.472 616
28	3.429 700	53.993 333	15.742 874	78	30.979 233	666.205 168	21.504 896
29	3.584 036	57.423 033	16.021 889	79	32.373 298	697.184 401	21.535 785
30	3.745 318	61.007 070	16.288 889	80	33.830 096	729.557 699	21.565 345
31	3.913 857	64.752 388	16.544 391	81	35.352 451	763.387 795	21.593 632
32	4.089 981	68.666 245	16.788 891	82	36.943 311	798.740 246	21.620 700
33	4.274 030	72.756 226	17.022 862	83	38.605 760	835.683 557	21.646 603
34	4.466 362	77.030 256	17.246 758	84	40.343 019	874.289 317	21.671 390
35	4.667 348	81.496 618	17.461 012	85	42.158 455	914.632 336	21.695 110
36	4.877 378	86.163 966	17.666 041	86	44.055 586	956.790 791	21.717 809
37	5.096 860	91.041 344	17.862 240	87	46.038 087	1000.846 377	21.739 530
38	5.326 219	96.138 205	18.049 990	88	48.109 801	1046.884 464	21.760 316
39	5.565 899	101.464 424	18.229 656	89	50.274 742	1094.994 265	21.780 207
40	5.816 365	107.030 323	18.401 584	90	52.537 105	1145.269 007	21.799 241
41	6.078 101	112.846 688	18.566 109	91	54.901 275	1197.806 112	21.817 455
42	6.351 615	118.924 789	18.723 550	92	57.371 832	1252.707 387	21.834 885
43	6.637 438	125.276 404	18.874 210	93	59.953 565	1310.079 219	21.851 565
44	6.936 123	131.913 842	19.018 383	94	62.651 475	1370.032 784	21.867 526
45	7.248 248	138.849 965	19.156 347	95	65.470 792	1432.684 259	21.882 800
46	7.574 420	146.098 214	19.288 371	96	68.416 977	1498.155 051	21.897 417
47	7.915 268	153.672 633	19.414 709	97	71.495 741	1566.572 028	21.911 403
48	8.271 456	161.587 902	19.535 607	98	74.713 050	1638.067 770	21.924 788
49	8.643 671	169.859 357	19.651 298	99	78.075 137	1712.780 819	21.937 596
50	9.032 636	178.503 028	19.762 008	100	81.588 518	1790.855 956	21.949 853

TABLE II

Continued

	$i = 0.05$ (5%)				$i = 0.06$ (6%)		
n	$(1 + i)^n$	$s_{\overline{n}i}$	$a_{\overline{n}i}$	n	$(1 + i)^n$	$s_{\overline{n}i}$	$a_{\overline{n}i}$
1	1.050 000	1.000 000	0.952 381	1	1.060 000	1.000 000	0.943 396
2	1.102 500	2.050 000	1.859 410	2	1.123 600	2.060 000	1.833 393
3	1.157 625	3.152 500	2.723 248	3	1.191 016	3.183 600	2.673 012
4	1.215 506	4.310 125	3.545 951	4	1.262 477	4.374 616	3.465 106
5	1.276 282	5.525 631	4.329 477	5	1.338 226	5.637 093	4.212 364
6	1.340 096	6.801 913	5.075 692	6	1.418 519	6.975 319	4.917 324
7	1.407 100	8.142 008	5.786 373	7	1.503 630	8.393 838	5.582 381
8	1.477 455	9.549 109	6.463 213	8	1.593 848	9.897 468	6.209 794
9	1.551 328	11.026 564	7.107 822	9	1.689 479	11.491 316	6.801 692
10	1.628 895	12.577 893	7.721 735	10	1.790 848	13.180 795	7.360 087
11	1.710 339	14.206 787	8.306 414	11	1.898 299	14.971 643	7.886 875
12	1.795 856	15.917 127	8.863 252	12	2.012 196	16.869 941	8.383 844
13	1.885 649	17.712 983	9.393 573	13	2.132 928	18.882 138	8.852 683
14	1.979 932	19.598 632	9.898 641	14	2.260 904	21.015 066	9.294 984
15	2.078 928	21.578 564	10.379 658	15	2.396 558	23.275 970	9.712 249
16	2.182 875	23.657 492	10.837 770	16	2.540 352	25.672 528	10.105 895
17	2.292 018	25.040 366	11.274 066	17	2.692·773	28.212 880	10.477 260
18	2.406 619	28.132 385	11.689 587	18	2.854 339	30.905 653	10.827 603
19	2.526 950	30.539 004	12.085 321	19	3.025 600	33.759 992	11.158 116
20	2.653 298	33.065 954	12.462 210	20	3.207 135	36.785 591	11.469 921
21	2.785 963	35.719 252	12.821 153	21	3.399 564	39.992 727	11.764 077
22	2.925 261	38.505 214	13.163 003	22	3.603 537	43.392 290	12.041 582
23	3.071 524	41.430 475	13.488 574	23	3.819 750	46.995 828	12.303 379
24	3.225 100	44.501 999	13.798 642	24	4.048 935	50.815 577	12.550 358
25	3.386 355	47.727 099	14.093 945	25	4.291 871	54.864 512	12.783 356
26	3.555 673	51.113 454	14.375 185	26	4.549 383	59.156 383	13.003 166
27	3.733 456	54.669 126	14.643 034	27	4.822 346	63.705 766	13.210 534
28	3.920 129	58.402 583	14.898 127	28	5.111 687	68.528 112	13.406 164
29	4.116 136	62.322 712	15.141 074	29	5.418 388	73.639 798	13.590 721
30	4.321 942	66.438 848	15.372 451	30	5.743 491	79.058 186	13.764 831
31	4.538 039	70.760 790	15.592 810	31	6.088 101	84.801 677	13.929 086
32	4.764 941	75.298 829	15.802 677	32	6.453 387	90.889 778	14.084 043
33	5.003 189	80.063 771	16.002 549	33	6.840 590	97.343 165	14.230 230
34	5.253 348	85.066 959	16.192 904	34	7.251 025	104.183 755	14.368 141
35	5.516 015	90.320 307	16.374 194	35	7.686 087	111.434 780	14.498 246
36	5.791 816	95.836 323	16.546 852	36	8.147 252	119.120 867	14.620 987
37	6.081 407	101.628 139	16.711 287	37	8.636 087	127.268 119	14.736 780
38	6.385 477	107.709 546	16.867 893	38	9.154 252	135.904 206	14.846 019
39	6.704 751	114.095 023	17.017 041	39	9.703 507	145.058 458	14.949 075
40	7.039 989	120.799 774	17.159 086	40	10.285 718	154.761 966	15.046 297
41	7.391 988	127.839 763	17.294 368	41	10.902 861	165.047 684	15.138 016
42	7.761 588	135.231 751	17.423 208	42	11.557 033	175.950 545	15.224 543
43	8.149 667	142.993 339	17.545 912	43	12.250 455	187.507 577	15.306 173
44	8.557 150	151.143 006	17.662 773	44	12.985 482	199.758 032	15.383 182
45	8.985 008	159.700 156	17.774 070	45	13.764 611	212.743 514	15.455 832
46	9.434 258	168.685 164	17.880 066	46	14.590 487	226.508 125	15.524 370
47	9.905 971	178.119 422	17.981 016	47	15.465 917	241.098 612	15.589 028
48	10.401 270	188.025 393	18.077 158	48	16.393 872	256.564 529	15.650 027
49	10.921 333	198.426 663	18.168 722	49	17.377 504	272.958 401	15.707 572
50	11.467 400	209.347 996	18.255 925	50	18.420 154	290.335 905	15.761 861

TABLE II

Continued

	$i = 0.07$ (7%)				$i = 0.08$ (8%)		
n	$(1 + i)^n$	$s_{\overline{n}|i}$	$a_{\overline{n}|i}$	n	$(1 + i)^n$	$s_{\overline{n}|i}$	$a_{\overline{n}|i}$
1	1.070 000	1.000 000	0.934 579	1	1.080 000	1.000 000	0.925 925
2	1.144 900	2.070 000	1.808 018	2	1.166 400	2.080 000	1.783 265
3	1.225 043	3.214 900	2.624 316	3	1.259 712	3.246 400	2.577 097
4	1.310 796	4.439 943	3.387 211	4	1.360 489	4.506 112	3.312 127
5	1.402 552	5.750 739	4.100 197	5	1.469 328	5.866 601	3.992 710
6	1.500 730	7.153 291	4.766 540	6	1.586 874	7.335 929	4.622 880
7	1.605 781	8.654 021	5.389 289	7	1.713 824	8.922 803	5.206 370
8	1.718 186	10.259 803	5.971 299	8	1.850 930	10.636 628	5.746 639
9	1.838 459	11.977 989	6.515 232	9	1.999 005	12.487 558	6.246 888
10	1.967 151	13.816 448	7.023 582	10	2.158 925	14.486 562	6.710 081
11	2.104 852	15.783 599	7.498 674	11	2.331 639	16.645 487	7.138 964
12	2.252 192	17.888 451	7.942 686	12	2.518 170	18.977 126	7.536 078
13	2.409 845	20.140 643	8.357 651	13	2.719 624	21.495 297	7.903 776
14	2.578 534	22.550 488	8.745 468	14	2.937 194	24.214 920	8.244 237
15	2.759 032	25.129 022	9.107 914	15	3.172 169	27.152 114	8.559 479
16	2.952 164	27.888 054	9.446 649	16	3.425 943	30.324 283	8.851 369
17	3.158 815	30.840 217	9.763 223	17	3.700 018	33.750 226	9.121 638
18	3.379 932	33.999 033	10.059 087	18	3.996 019	37.450 244	9.371 887
19	3.616 528	37.378 965	10.335 595	19	4.315 701	41.446 263	9.603 599
20	3.869 684	40.995 492	10.594 014	20	4.660 957	45.761 964	9.818 147
21	4.140 562	44.865 177	10.835 527	21	5.033 834	50.422 921	10.016 803
22	4.430 402	49.005 739	11.061 240	22	5.436 540	55.456 755	10.200 744
23	4.740 530	53.436 141	11.272 187	23	5.871 464	60.893 296	10.371 059
24	5.072 367	58.176 671	11.469 334	24	6.341 181	66.764 759	10.528 758
25	5.427 433	63.249 038	11.653 583	25	6.848 475	73.105 940	10.674 776
26	5.807 353	68.676 470	11.825 779	26	7.396 353	79.954 415	10.809 978
27	6.213 868	74.483 823	11.986 709	27	7.988 061	87.350 768	10.935 165
28	6.648 838	80.697 691	12.137 111	28	8.627 106	95.338 830	11.051 078
29	7.114 257	87.346 529	12.277 674	29	9.317 275	103.965 936	11.158 406
30	7.612 255	94.460 786	12.409 041	30	10.062 657	113.283 211	11.257 783
31	8.145 113	102.073 041	12.531 814	31	10.867 669	123.345 868	11.349 799
32	8.715 271	110.218 154	12.646 555	32	11.737 083	134.213 537	11.434 999
33	9.325 340	118.933 425	12.753 790	33	12.676 050	145.950 620	11.513 888
34	9.978 114	128.258 765	12.854 009	34	13.690 134	158.626 670	11.586 934
35	10.676 581	138.236 878	12.947 672	35	14.785 344	172.316 804	11.654 568
36	11.423 942	148.913 460	13.035 208	36	15.968 172	187.102 148	11.717 193
37	12.223 618	160.337 402	13.117 017	37	17.245 626	203.070 320	11.775 179
38	13.079 271	172.561 020	13.193 473	38	18.625 276	220.315 945	11.828 869
39	13.994 820	185.640 292	13.264 928	39	20.115 298	238.941 221	11.878 582
40	14.974 458	199.635 112	13.331 709	40	21.724 522	259.056 519	11.924 613
41	16.022 670	214.609 570	13.394 120	41	23.462 483	280.781 040	11.967 235
42	17.144 257	230.632 240	13.452 449	42	25.339 482	304.243 523	12.006 699
43	18.344 355	247.776 496	13.506 962	43	27.366 640	329.583 005	12.043 240
44	19.628 460	266.120 851	13.557 908	44	29.555 972	356.949 646	12.077 074
45	21.002 452	285.749 311	13.605 522	45	31.920 449	386.505 617	12.108 402
46	22.472 623	306.751 763	13.650 020	46	34.474 085	418.426 067	12.137 409
47	24.045 707	329.224 386	13.691 608	47	37.232 012	452.900 152	12.164 267
48	25.728 907	353.270 093	13.730 474	48	40.210 573	490.132 164	12.189 136
49	27.529 930	378.999 000	13.766 799	49	43.427 419	530.342 737	12.212 163
50	29.457 025	406.528 929	13.800 746	50	46.901 613	573.770 156	12.233 485

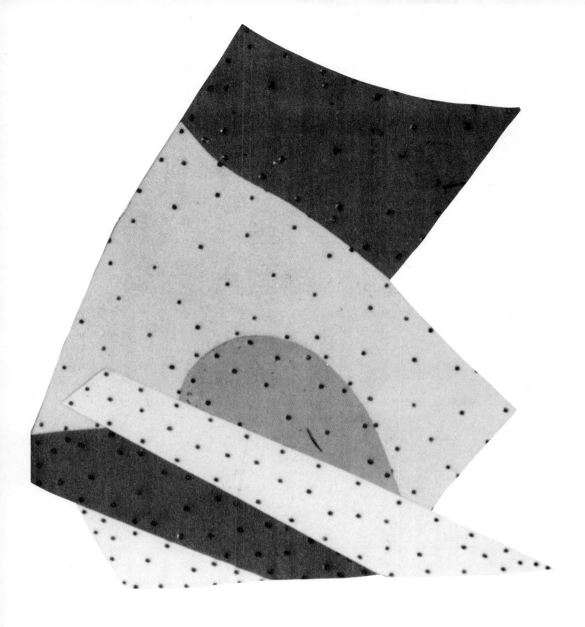

Answers

A53

CHAPTER 1

◆ EXERCISE 1-1

1. vu **3.** $(3 + 7) + y$ **5.** $u + v$ **7.** Associative $(\cdot)$ **9.** Negatives **11.** Subtraction **13.** Division **15.** Distributive **17.** Zero
19. Associative $(\cdot)$ **21.** Distributive **23.** Negatives **25.** Zero **27.** No **29.** (A) F (B) T (C) T
31. $\sqrt{2}$ and π are two examples of infinitely many. **33.** (A) N, Z, Q, R (B) R (C) Q, R (D) Q, R
35. (A) T (B) F, since, for example, $(8 - 4) - 2 \neq 8 - (4 - 2)$. (C) T (D) F, since, for example, $(8 \div 4) \div 2 \neq 8 \div (4 \div 2)$.
37. $\frac{1}{11}$ **39.** 1. Commutative 2. Associative 3. Inverse 4. Identity
41. (A) 2.166 666 666... (B) 4.582 575 69... (C) 0.437 500 000... (D) 0.261 261 261...

◆ EXERCISE 1-2

1. 3 **3.** $x^3 + 4x^2 - 2x + 5$ **5.** $x^3 + 1$ **7.** $2x^5 + 3x^4 - 2x^3 + 11x^2 - 5x + 6$ **9.** $-5u + 2$ **11.** $6a^2 + 6a$ **13.** $a^2 - b^2$ **15.** $6x^2 - 7x - 5$
17. $2x^2 + xy - 6y^2$ **19.** $9y^2 - 4$ **21.** $6m^2 - mn - 35n^2$ **23.** $16m^2 - 9n^2$ **25.** $9u^2 + 24uv + 16v^2$ **27.** $a^3 - b^3$
29. $16x^2 + 24xy + 9y^2$ **31.** 1 **33.** $x^4 - 2x^2y^2 + y^4$ **35.** $5a^2 + 12ab - 10b^2$ **37.** $x^3 - 6x^2y + 12xy^2 - 8y^3$ **39.** $2x^2 - 2xy + 3y^2$
41. $8x^3 - 20x^2 + 20x + 1$ **43.** $4x^3 - 14x^2 + 8x - 6$ **45.** $m + n$ **47.** $0.09x + 0.12(10{,}000 - x) = 1{,}200 - 0.03x$
49. $10x + 30(3x) + 50(4{,}000 - x - 3x) = 200{,}000 - 100x$ **51.** $0.02x + 0.06(10 - x) = 0.6 - 0.04x$

◆ EXERCISE 1-3

1. $3m^2(2m^2 - 3m - 1)$ **3.** $2uv(4u^2 - 3uv + 2v^2)$ **5.** $(7m + 5)(2m - 3)$ **7.** $(a - 4b)(3c + d)$ **9.** $(3y + 2)(y - 1)$ **11.** $(u - 5v)(u + 3v)$
13. Prime **15.** $(wx - y)(wx + y)$ **17.** $(3m - n)^2$ **19.** Prime **21.** $4(z - 3)(z - 4)$ **23.** $2x^2(x - 2)(x - 10)$ **25.** $x(2y - 3)^2$
27. $(2m - 3n)(3m + 4n)$ **29.** $uv(2u - v)(2u + v)$ **31.** $2x(x^2 - x + 4)$ **33.** $(r - t)(r^2 + rt + t^2)$ **35.** $(a + 1)(a^2 - a + 1)$
37. $[(x + 2) - 3y][(x + 2) + 3y]$ **39.** Prime **41.** $(6x - 6y - 1)(x - y + 4)$ **43.** $(y - 2)(y + 2)(y^2 + 1)$ **45.** $a^2(3 + ab)(9 - 3ab + a^2b^2)$

◆ EXERCISE 1-4

1. $8d^6$ **3.** $\dfrac{15x^2 + 10x - 6}{180}$ **5.** $\dfrac{15m^2 + 14m - 6}{36m^3}$ **7.** $\dfrac{1}{x(x - 4)}$ **9.** $\dfrac{x}{x + 5}$ **11.** $\dfrac{x - 5}{(x - 1)^2(x + 1)}$ **13.** $\dfrac{2}{x - 1}$ **15.** $\dfrac{5}{a - 1}$

17. $\dfrac{2}{x + y}$ **19.** $\dfrac{7x^2 - 2x - 3}{6(x + 1)^2}$ **21.** $-\dfrac{1}{x}$ **23.** $\dfrac{-17c + 16}{15(c - 1)}$ **25.** 1 **27.** $\dfrac{2x + 5}{(x + 4)(x + 1)}$ **29.** $\dfrac{1}{x - 3}$ **31.** $\dfrac{1 - m}{m}$ **33.** $-cd$ **35.** $\dfrac{x - y}{x + y}$ **37.** 1

39. x

◆ EXERCISE 1-5

1. $2/x^9$ **3.** $3w^7/2$ **5.** $2/x^3$ **7.** $1/w^5$ **9.** 5 **11.** $1/a^6$ **13.** y^6/x^{12} **15.** x **17.** $a^2\sqrt{a}$ **19.** $3x^2\sqrt{2}$ **21.** $\sqrt{m}/m$ **23.** $\sqrt{6}/3$ **25.** $\sqrt{2x}/x$
27. 8.23×10^{10} **29.** 7.83×10^{-1} **31.** 3.4×10^{-5} **33.** 1 **35.** 10^{14} **37.** $y^6/25x^4$ **39.** 4×10^2 **41.** $4y^3/3x^5$ **43.** $y^{12}/8x^6$ **45.** x
47. uv **49.** $\frac{7}{4} - \frac{1}{4}x^{-3}$ **51.** $\frac{3}{4}x - x^{-1} - \frac{1}{4}x^{-3}$ **53.** $3x^4y^2z\sqrt{2y}$ **55.** $4\sqrt{3x}/x$ **57.** $\sqrt{42xy}/7y$ **59.** $2a\sqrt{3ab}/3b$ **61.** $6m^3n^3$
63. $2a\sqrt{3ab}/3b$ **65.** $\dfrac{15\sqrt{x} + 10x}{9 - 4x}$ **67.** $\dfrac{5\sqrt{6} - 6}{19}$ **69.** 2.4×10^{10}; 24,000,000,000 **71.** 3.125×10^4; 31,250 **73.** t^2/x^2y^{10} **75.** 4

77. $\dfrac{1}{2(x - 3)^2}$ **79.** $\dfrac{bc(c + b)}{c^2 + bc + b^2}$ **81.** $\dfrac{\sqrt{2}}{2}$ **83.** $\dfrac{2\sqrt{x - 2}}{x - 2}$ **85.** $\dfrac{1}{\sqrt{t} + \sqrt{x}}$ **87.** $\dfrac{1}{\sqrt{x + h} + \sqrt{x}}$

1. $6\sqrt[5]{x^3}$ **3.** $\sqrt[5]{(4xy^3)^2}$ **5.** $\sqrt{x^2+y^2}$ (not $x+y$) **7.** $5x^{3/4}$ **9.** $(2x^2y)^{3/5}$ **11.** $x^{1/3}+y^{1/3}$ **13.** 5 **15.** 64 **17.** -6
19. Not a rational number (not even a real number) **21.** $\frac{8}{125}$ **23.** $\frac{1}{27}$ **25.** $x^{2/5}$ **27.** m **29.** $2x/y^2$ **31.** $xy^2/2$ **33.** $2/3x^{7/12}$
35. $2y^2/3x^2$ **37.** $12x-6x^{35/4}$ **39.** $3u-13u^{1/2}v^{1/2}+4v$ **41.** $25m-n$ **43.** $9x-6x^{1/2}y^{1/2}+y$ **45.** $\frac{1}{2}x^{1/3}+x^{-1/3}$
47. $\frac{2}{3}x^{-1/4}+x^{-2/3}$ **49.** $\frac{1}{2}x^{-1/6}-\frac{1}{4}$ **51.** $2mn^2\sqrt[3]{2m}$ **53.** $2m^2n\sqrt[4]{2mn^3}$ **55.** $\sqrt[3]{x^2}$ **57.** $2a^2b\sqrt[3]{4a^2b}$ **59.** $\sqrt[4]{12x^3}/2$ **61.** $\sqrt[4]{(x-3)^3}$
63. $x\sqrt[6]{x}$ **65.** $\sqrt[6]{x^5}/x$ **67.** $\dfrac{x-2}{2(x-1)^{3/2}}$ **69.** $\dfrac{x+6}{3(x+2)^{5/3}}$ **71.** 103.2 **73.** 0.0805 **75.** 4,588

1. (A) $(y+z)x$ (B) $(2+x)+y$ (C) $2x+3x$ **2.** x^3+3x^2+5x-2 **3.** $x^3-3x^2-3x+22$ **4.** $3x^5+x^4-8x^3+24x^2+8x-64$
5. 3 **6.** 1 **7.** $14x^2-30x$ **8.** $9m^2-25n^2$ **9.** $6x^2-5xy-4y^2$ **10.** $4a^2-12ab+9b^2$, **11.** $(3x-2)^2$ **12.** Prime
13. $3n(2n-5)(n+1)$ **14.** $\dfrac{12a^3b-40b^2-5a}{30a^3b^2}$ **15.** $\dfrac{7x-4}{6x(x-4)}$ **16.** $\dfrac{y+2}{y(y-2)}$ **17.** u **18.** $6x^5y^{15}$ **19.** $3u^4/v^2$ **20.** 6×10^2 **21.** x^6/y^4
22. $u^{7/3}$ **23.** $3a^2/b$ **24.** $3\sqrt[5]{x^2}$ **25.** $-3(xy)^{2/3}$ **26.** $3x^2y\sqrt[3]{x^2y}$ **27.** $6x^2y^3\sqrt{xy}$ **28.** $2b\sqrt{3a}$ **29.** $\dfrac{3\sqrt{5}+5}{4}$ **30.** $\sqrt[4]{y^3}$ **31.** Subtraction
32. Commutative (+) **33.** Distributive **34.** Associative (·) **35.** Negatives **36.** Identity (+) **37.** (A) T (B) F
38. 0 and -3 are two examples of infinitely many. **39.** (A) Expressions (a) and (d) (B) None **40.** $4xy-2y^2$ **41.** $m^4-6m^2n^2+n^4$
42. $2x^3-4x^2+12x$ **43.** $x^3-6x^2y+12xy^2-8y^3$ **44.** $(x-y)(7x-y)$ **45.** Prime **46.** $3xy(2x^2+4xy-5y^2)$
47. $3(x+2y)(x^2-2xy+4y^2)$ **48.** $\dfrac{2m}{(m+2)(m-2)^2}$ **49.** $\dfrac{y^2}{x}$ **50.** $\dfrac{-ab}{a^2+ab+b^2}$ **51.** $\frac{1}{4}$ **52.** $\frac{5}{9}$ **53.** $3x^2/2y^2$ **54.** $27a^{1/6}/b^{1/2}$
55. $x+2x^{1/2}y^{1/2}+y$ **56.** $6x+7x^{1/2}y^{1/2}-3y$ **57.** 2×10^{-7} **58.** $-6x^2y^2\sqrt[3]{3x^2y}$ **59.** $x\sqrt[3]{2x^2}$ **60.** $\sqrt[5]{12x^3y^2}/2x$ **61.** $y\sqrt[3]{2x^2y}$
62. $2x-3\sqrt{xy}-5y$ **63.** $\dfrac{6x+3\sqrt{xy}}{4x-y}$ **64.** $\dfrac{4u-12\sqrt{uv}+9v}{4u-9v}$ **65.** $\dfrac{1}{\sqrt{t}+\sqrt{5}}$ **66.** $2-\frac{3}{2}x^{-1/2}$ **67.** 0 **68.** x^3+8x^2-6x+1
69. 3.213×10^6 **70.** 4.434×10^{-5} **71.** -4.541×10^{-6} **72.** 128,800 **73.** 0.3664 **74.** 1.640

C H A P T E R   2

1. $m=5$ **3.** $x<-9$ **5.** $x\le4$ **7.** $x<-3$ or $(-\infty,-3)$ ⟵———→x **9.** $-1\le x\le2$ or $[-1,2]$ ⟵[———]→x **11.** $y=8$
13. $x>-6$ **15.** $y=8$ **17.** $x=10$ **19.** $y\ge3$ **21.** $x=36$ **23.** $m<3$ **25.** $x=10$ **27.** $3\le x<7$ or $[3,7)$ ⟵[———)→x
29. $-20\le C\le20$ or $[-20,20]$ ⟵[———]→C **31.** $y=\frac{3}{4}x-3$ **33.** $y=-(A/B)x+(C/B)=(-Ax+C)/B$ **35.** $C=\frac{5}{9}(F-32)$
37. $B=A/(m-n)$ **39.** $-2<x\le1$ or $(-2,1]$ ⟵(———]→x **41.** 3,000 \$10 tickets; 5,000 \$6 tickets
43. \$7,200 at 10%; \$4,800 at 15% **45.** \$7,800 **47.** 5,000 **49.** 12.6 yr

1. ±2 **3.** $\pm\sqrt{11}$ **5.** $-2,6$ **7.** $0,2$ **9.** $3\pm2\sqrt{3}$ **11.** $-2\pm\sqrt{2}$ **13.** $0,2$ **15.** $\pm\frac{3}{2}$ **17.** $\frac{1}{2},-3$ **19.** $(-1\pm\sqrt{5})/2$ **21.** $(3\pm\sqrt{3})/2$
23. No real solutions **25.** $-4\pm\sqrt{11}$ **27.** $r=\sqrt{A/P}-1$ **29.** \$2 **31.** 8 ft/sec; $4\sqrt{2}$ or 5.66 ft/sec

1.

3.

5. Slope = 2; y intercept = -3

7. Slope = $-\frac{2}{3}$; y intercept = 2

9. $y = -2x + 4$

11. $y = -\frac{3}{5}x + 3$

13.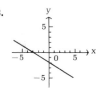

15.

17.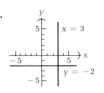

19. $y = -3x + 5$; $m = -3$

21. $y = -\frac{2}{3}x + 4$; $m = -\frac{2}{3}$

23. $y + 1 = -3(x - 4)$; $y = -3x + 11$

25. $y + 5 = \frac{2}{3}(x + 6)$; $y = \frac{2}{3}x - 1$

27. $\frac{1}{3}$

29. $-\frac{1}{5}$

31. $(y - 3) = \frac{1}{3}(x - 1)$; $x - 3y = -8$

33. $(y + 2) = -\frac{1}{5}(x + 5)$; $x + 5y = -15$

35. $x = 3$; $y = -5$

37. $x = -1$; $y = -3$

39. $y = -\frac{1}{2}x + 4$

41. (A) $y = -\frac{1}{2}x + 1$ (B) $y = 2x + 6$

43. (A) $y = \frac{1}{2}x$
(B) $y = -2x - 5$

45.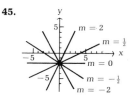

47. $x = 2$

49. $y = 3$

51. (A) $130; $220

(B)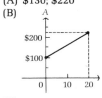

(C) 6

53. (A)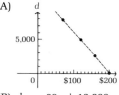

(B) $d = -60p + 12{,}000$

55. $0.2x + 0.1y = 20$

57. (A) 64 g; 35 g

(B)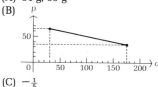

(C) $-\frac{1}{5}$

1. Function **3.** Not a function **5.** Function **7.** Function; Domain = {1, 2, 4, 5}; Range = {3, 5, 7, 9} **9.** Not a function
11. Function; Domain = {1, 2, 3, 4, 5, 6}; Range = {3, 4} **13.** Function **15.** Not a function **17.** Function **19.** 4 **21.** -5 **23.** -6
25. -2 **27.** -12 **29.** -1 **31.** -6 **33.** 12 **35.** $\frac{3}{4}$ **37.** All nonnegative real numbers **39.** All real numbers except $x = -3, 5$
41. $x \geqslant -5$ or $[-5, \infty)$ **43.** All real numbers except $x = \pm 1$ **45.** All real numbers except $x = -4, 1$ **47.** All real numbers
49. A function with domain R **51.** A function with domain R **53.** Not a function; for example, when $x = 1$, $y = \pm 3$
55. A function with domain all real numbers except $x = 4$ **57.** Not a function; for example, when $x = 4$, $y = \pm 3$ **59.** 4 **61.** $h + 2$

63. $h - 1$ **65.** 4 **67.** $8a + 4h - 7$ **69.** $3a^2 + 3ah + h^2$ **71.** $\dfrac{1}{\sqrt{a + h} + \sqrt{a}}$ **73.** $P(w) = 2w + \dfrac{50}{w}$, $w > 0$

75. $A(\ell) = \ell(50 - \ell)$, $0 < \ell < 50$ **77.** $C(x) = 96{,}000 + 80x$; $136{,}000 **79.** $R(x) = x(200 - \frac{1}{40}x)$, $0 \leqslant x \leqslant 8{,}000$

81. (A) $V(x) = x(8 - 2x)(12 - 2x)$ (B) $0 < x < 4$ (C)

x	V(x)
1	60
2	64
3	36

83. $P(x) = 10x - 15 + \dfrac{270}{x}$;

x	P(x)
3	105
4	92.5
5	89
6	90

85. $v = \dfrac{75 - w}{15 + w}$; 1.9032 cm/sec

◆ EXERCISE 2-5

1. Slope: 2
y intercept: -4
x intercept: 2

3. Slope: -2
y intercept: 4
x intercept: 2

5. Slope: $-\frac{2}{3}$
y intercept: 4
x intercept: 6

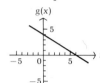

7. $f(x) = -2x + 6$
9. $f(x) = -\frac{1}{2}x + \frac{9}{2}$

11. Axis: $x = 3$
Vertex: $(3, -1)$
Min: $f(3) = -1$
y intercept: 8
x intercepts: 2, 4
Range: $[-1, \infty)$

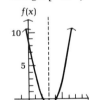

13. Axis: $x = -1$
Vertex: $(-1, 9)$
Max: $h(-1) = 9$
y intercept: 8
x intercepts: -4, 2
Range: $(-\infty, 9]$

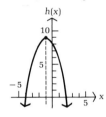

15. Axis: $x = -4$
Vertex: $(-4, 0)$
Min: $f(-4) = 0$
y intercept: 16
x intercept: -4
Range: $[0, \infty)$

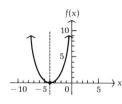

17. Axis: $u = 1$
Vertex: $(1, 3)$
Min: $f(1) = 3$
y intercept: 4
No u intercepts
Range: $[3, \infty)$

19. Axis: $x = 2$
Vertex: $(2, 6)$
Max: $h(2) = 6$
y intercept: 2
x intercepts: $2 \pm \sqrt{6}$
Range: $(-\infty, 6]$

21. Axis: $x = 3$
Vertex: $(3, 9)$
Max: $f(3) = 9$
y intercept: 0
x intercepts: 0, 6
Range: $(-\infty, 9]$

23. Axis: $s = 0$
Vertex: $(0, -4)$
Min: $F(0) = -4$
y intercept: -4
s intercepts: -2, 2
Range: $[-4, \infty)$

25. Axis: $x = 0$
Vertex: $(0, 4)$
Max: $F(0) = 4$
y intercept: 4
x intercepts: -2, 2
Range: $(-\infty, 4]$

27. Domain: [0,5]
Range: {1, 3, 5}

29. Domain: [−2, 2]
Range: [−2, 1]

31. Domain: (−∞, 0) ∪ (0, ∞)
Range: (−∞, −2) ∪ (2, ∞)

33. Domain: All real numbers
Range: [−3, 3]

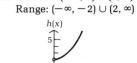

35. Axis: x = 3.5
Vertex: (3.5, −2.25)
Min: f(3.5) = −2.25
y intercept: 10
x intercepts: 2, 5
Range: [−2.25, ∞)

37. Axis: x = −2.5
Vertex: (−2.5, 8.25)
Max: h(−2.5) = 8.25
y intercept: 2
x intercepts: −$\frac{5}{2}$ ± $\frac{1}{2}$√33
Range: (−∞, 8.25]

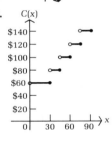

39.

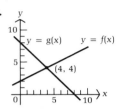

41.

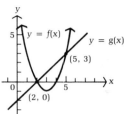

43.

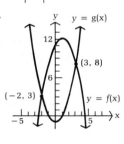

45.

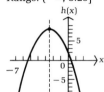

47. (A) C = 360,000 − 900p
(B) R = xp = (9,000 − 30p)p = 9,000p − 30p²
(C)

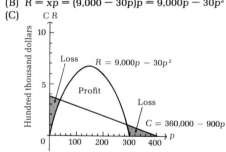

(D) $42, $288
(E) $150

49. (A) C(x) = 240,000 + 20x
(B) R(x) = 35x
(C)

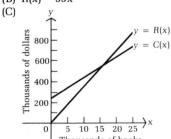

(D) 16,000 books

51.

53. (A) 1 lb; 3 lb
(B)

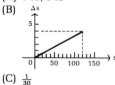

(C) $\frac{1}{30}$

1. $u = 36$ **2.** $x < 4$ or $(-\infty, 4)$ 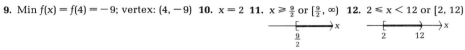 **3.** $x = 0, 5$ **4.** **5.** $y = \frac{1}{2}x + 1$ **6.** **7.** -2

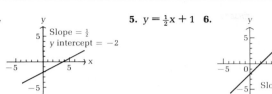

8. 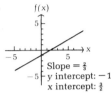   **9.** Min $f(x) = f(4) = -9$; vertex: $(4, -9)$ **10.** $x = 2$ **11.** $x \geq \frac{9}{2}$ or $[\frac{9}{2}, \infty)$ **12.** $2 \leq x < 12$ or $[2, 12)$

13. $y = \frac{2}{3}x - 2$ **14.** $y = 3/(x - 1)$ **15.** $x = \pm\sqrt{7}$ **16.** $x = -4, 5$ **17.** $x = (3 \pm \sqrt{17})/4$ **18.**

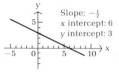

19. $x + 2y = 4$; slope $= -\frac{1}{2}$ **20.** **21.** $y = -\frac{3}{2}x + 2$ **22.** (A) 22   (B) -36 (C) -91 (D) $-\frac{3}{4}$

23. Domain f: R; Domain g: All R except 2   **24.** 2
25. (A) A function with domain all real numbers   (B) Not a function; for example, when $x = 2$, $y = \pm3$
(C) A function with domain all real numbers except $x = -3$

26. Axis: $x = 2$ **27.** Domain: $[-1, 1]$; Range: $[0, 2]$ **28.** $x = \dfrac{-j \pm \sqrt{j^2 - 4k}}{2}$
Vertex: $(2, 8)$
Max: $g(2) = 8$ **29.** $x = 4$
y intercept: 0   **30.** (A) $x - 2y = 8$ (B) $2x + y = 1$
x intercepts: 0, 4   **31.** (A) All real numbers except $x = 3$ (B) $[1, \infty)$
Range: $(-\infty, 8]$ **32.** $2a + 7 + h$
33.

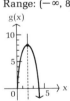

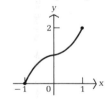

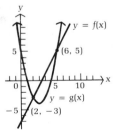

34. \$20,000 at 8%; \$40,000 at 14%

35. $\dfrac{x}{800} = \dfrac{247}{89}$; $x = \$2,220$

36. 10%   **37.** (A) $V(t) = -1,250t + 12,000$ (B) \$5,750

38. (A) $R = \frac{8}{5}C$ (B) \$168

39. \$4.50

40. 2,400 tapes

41. (A) $R(p) = 500p - 10p^2$

(B) $C(p) = 8,000 - 100p$

(C)

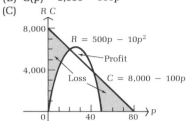

(D) \$20; \$40   (E) \$30

42. (A) $A(x) = 90x - \frac{3}{2}x^2$ (B) $0 < x < 60$ (C) $x = 30, y = 22.5$

C H A P T E R 3

◆ EXERCISE 3-1

1.
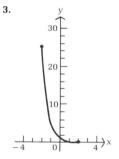

3.
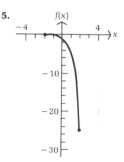

5.
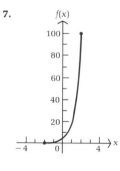

7.
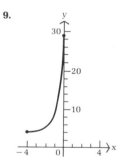

9.
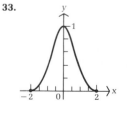

11. 4^{6xy} **13.** 5   **15.** $2^{xz}3^{yz}$ **17.** $x = 1$ **19.** $x = -1, 6$ **21.** $x = 3$ **23.** $x = 2$

25.
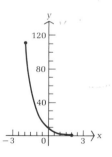

27.
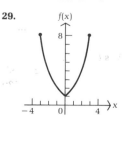

29.
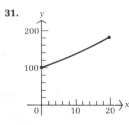

31.

33.

35. $3^{2x} - 3^{-2x}$ **37.** $2(3^{2x}) + 2(3^{-2x})$

39.

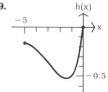

41.

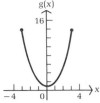

43. 0.2115
45. 0.1377
47. 15.7563
49. (A) $2,633.56   (B) $7,079.54
51. $9,217
53. (A) 33,000,000   (B) 69,000,000
55. (A) 8.49 mg   (B) 0.75 mg

◆ EXERCISE 3-2

1.

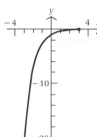

3.

5.

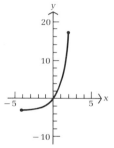

7.

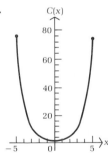

9.

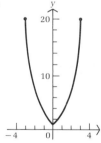

11.

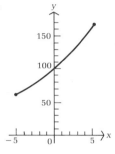

13. $e^x - e^{-x}$
15. $2/e^{2x}$
17. $x = 3$
19. $x = -3, 0$

21.

23. (A) $11,871.65   (B) $20,427.93
25. (A) $10,850.88   (B) $10,838.29
27. $28,847.49

29. N approaches 2 as t increases without bound.

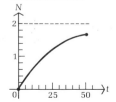

31. (A) 10%   (B) 1%
33. (A) 47,699   (B) 475,757
35. 120 million

◆ EXERCISE 3-3

1. $27 = 3^3$ **3.** $10^0 = 1$ **5.** $8 = 4^{3/2}$ **7.** $\log_7 49 = 2$ **9.** $\log_4 8 = \frac{3}{2}$ **11.** $\log_b A = u$ **13.** 3 **15.** -3 **17.** 3 **19.** $\log_b P - \log_b Q$
21. $5 \log_b L$ **23.** $\log_b p - \log_b q - \log_b r - \log_b s$ **25.** $x = 9$ **27.** $y = 2$ **29.** $b = 10$ **31.** $x = 2$ **33.** $y = -2$ **35.** $b = 100$
37. $5 \log_b x - 3 \log_b y$ **39.** $\frac{1}{3} \log_b N$ **41.** $2 \log_b x + \frac{1}{3} \log_b y$ **43.** $\log_b 50 - 0.2t \log_b 2$ **45.** $\log_b P + t \log_b(1 + r)$ **47.** $\log_e 100 - 0.01t$
49. $x = 2$ **51.** $x = 8$ **53.** $x = 7$ **55.** No solution

57.
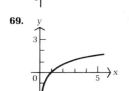

59. (A) 3.547 43 (B) -2.160 32 (C) 5.626 29 (D) -3.197 04
61. (A) 13.4431 (B) 0.0089 (C) 16.0595 (D) 0.1514
63. 4.959 **65.** 7.861 **67.** 3.301

69.
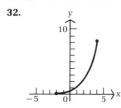

71.
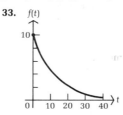

73.
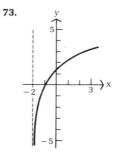

75.
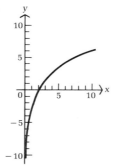

77. $\log_b 1 = 0, b > 0, b \neq 1$ **79.** $y = c \cdot 10^{0.8x}$ **81.** 12 yr **83.** $t = \dfrac{\ln 3}{\ln(1 + r)}$ **87.** Approx. 538 yr

◆ EXERCISE 3-4 CHAPTER REVIEW

1. $v = \ln u$ **2.** $y = \log x$ **3.** $M = e^N$ **4.** $u = 10^v$ **5.** 5^{2x} **6.** e^{2u^2} **7.** $x = 9$ **8.** $x = 6$ **9.** $x = 4$ **10.** $x = 2.157$ **11.** $x = 13.128$
12. $x = 1{,}273.503$ **13.** $x = 0.318$ **14.** $x = 8$ **15.** $x = 3$ **16.** $x = 3$ **17.** $x = -1, 3$ **18.** $x = 0, \frac{3}{2}$ **19.** $x = -2$ **20.** $x = \frac{1}{2}$ **21.** $x = 27$
22. $x = 13.3113$ **23.** $x = 158.7552$ **24.** $x = 0.0097$ **25.** $x = 1.4359$ **26.** $x = 1.4650$ **27.** $x = 92.1034$ **28.** $x = 9.0065$
29. $x = 2.1081$ **30.** $1 + 2e^x - e^{-x}$ **31.** $2e^{-2x} - 2$

32.
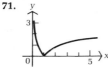

33.
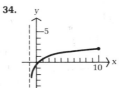

34.

35. $x = 2$
36. $x = 2$
37. $x = 1$
38. $x = 300$
39. $y = ce^{-5t}$

40. If $\log_1 x = y$, then $1^y = x$; that is, $1 = x$ for all positive real numbers x, which is not possible. **41.** \$10,263.65 **42.** \$10,272.17
43. 8 yr **44.** 6.93 yr **45.** (A) $N = 2^{2t}$ or $N = 4^t$ (B) 15 days **46.** $k = 0.009$ 42; 489 ft **47.** 23.4 yr **48.** 23.1 yr

CHAPTER 4

◆ EXERCISE 4-1

1. $I = \$20$ **3.** $r = 0.08$ or 8% **5.** $A = \$112$ **7.** $P = \$888.89$ **9.** $r = I/Pt$ **11.** $P = A/(1 + rt)$ **13.** $\$140$ **15.** $\$9.23$ **17.** $\$7,685.00$
19. 10.125% **21.** 18% **23.** $\$1,680; 36\%$ **25.** 9.126% **27.** $\$9,693.91$ **29.** 18% **31.** 21.335%

◆ EXERCISE 4-2

1. $A = \$112.68$ **3.** $A = \$3,433.50$ **5.** $P = \$2,419.99$ **7.** $P = \$7,351.04$ **9.** 0.75% per month **11.** 1.75% per quarter
13. 9.6% compounded monthly **15.** 9% compounded semiannually
17. (A) $\$126.25; \$26.25$ (B) $\$126.90; \$26.90$ (C) $\$127.05; \$27.05$ **19.** (A) $\$7,147.51$ (B) $\$10,217.39$
21. (A) $\$6,755.64$ (B) $\$4,563.87$ **23.** (A) 10.38% (B) 12.68% **25.** $5\frac{1}{2}$ yr **27.** $n \approx 12$ **29.** (A) $7\frac{1}{4}$ yr (B) 6 yr **31.** $\$22,702.60$
33. $\$196,993.25$ **35.** $\$14.26/\text{ft}^2/\text{mo}$ **37.** 18 yr
39. 9% compounded monthly, since its effective rate is 9.38%; the effective rate of 9.3% compounded annually is 9.3%
41. 2 yr, 10 mo **43.** $\$328,791.70$ **45.** 4.952 yr; 4.959 yr **47.** $\$5,935.34$ **49.** 9.08% **51.** (A) 8.60% (B) 8.60% (C) 8.57%
53. 14.48%

◆ EXERCISE 4-3

1. $FV = \$13,435.19$ **3.** $FV = \$60,401.98$ **5.** $PMT = \$123.47$ **7.** $PMT = \$310.62$ **9.** $n = 17$
11. Value: $\$30,200.99$; interest: $\$10,200.99$ **13.** $\$20,931.01$ **15.** $\$331.46$ **17.** $\$625.28$
19. First year: $\$50.76$; second year: $\$168.09$; third year: $\$296.42$ **21.** (A) $\$413,092$ (B) $\$393,965$ **23.** $\$177.46; \$1,481.92$
25. 3 yr, 5 mo

◆ EXERCISE 4-4

1. $PV = \$3,458.41$ **3.** $PV = \$4,606.09$ **5.** $PMT = \$199.29$ **7.** $PMT = \$586.01$ **9.** $n = 29$ **11.** $\$109,421.92$
13. $\$14,064.67; \$16,800.00$ **15.** (A) $\$36.59$ per month; $\$58.62$ interest (B) $\$38.28$ per month; $\$89.04$ interest
17. $\$273.69$ per month; $\$7,705.68$ interest
19. Amortization schedule:

PAYMENT NUMBER	PAYMENT	INTEREST	UNPAID BALANCE REDUCTION	UNPAID BALANCE
0				$5,000.00
1	$ 758.05	$ 225.00	$ 533.05	4,466.95
2	758.05	201.01	557.04	3,909.91
3	758.05	175.95	582.10	3,327.81
4	758.05	149.75	608.30	2,719.51
5	758.05	122.38	635.67	2,083.84
6	758.05	93.77	664.28	1,419.56
7	758.05	63.88	694.17	725.39
8	758.03	32.64	725.39	0.00
Totals	$6,064.38	$1,064.38	$5,000.00	

21. First year interest $= \$625.07$;
second year interest $= \$400.91$;
third year interest $= \$148.46$
23. $\$85,846.38; \$128,153.62$
25. $\$143.85$ per month; $\$904.80$
27. Monthly payment $= \$841.39$
(A) $\$70,952.33$ (B) $\$55,909.02$ (C) $\$36,813.32$
29. (A) Monthly payment $= \$395.04$;
total interest $= \$64,809.60$
(B) 114 mo, or 9.5 yr;
interest saved $= \$38,375.04$
31. $\$29,799$

1. $A = \$104.50$ **2.** $P = \$800$ **3.** $t = 0.75$ yr, or 9 mo   **4.** $r = 6\%$ **5.** $A = \$1,393.68$ **6.** $P = \$3,193.50$ **7.** $FV = \$69,770.03$
8. $PMT = \$115.00$ **9.** $PV = \$33,944.27$ **10.** $PMT = \$166.07$ **11.** $n \approx 16$ **12.** $n \approx 41$ **13.** $\$3,350.00; \$350.00$ **14.** $\$11.64$ **15.** 15%
16. 20%   **17.** 28.8%   **18.** 8.24%   **19.** $\$4,744.73$ **20.** $\$27,551.32$ **21.** $\$9,422.24$ **22.** $\$10,210.25$ **23.** $\$6,268.21$ **24.** 2 yr, 3 mo
25. 5 yr, 10 mo; 3 yr, 11 mo   **26.** 9.38%
27. 9% compounded quarterly, since its effective rate is 9.31%, while the effective rate of 9.25% compounded annually is 9.25%
28. $\$27,971.23; \$8,771.23$ **29.** $\$526.28$ per month   **30.** 43   **31.** $\$10,988.22; \$12,000$ **32.** $\$99.85$ per month; $\$396.36$
33. Amortization schedule:

34. $\$6,697.11$
35. (A) $\$1,435.63$ (B) $\$74,397.48$ (C) $\$11,625.04$
36. (A) $\$115,573.86$ (B) $\$359.64$ (C) $\$171,228.80$
37. $\$164,402$
38. 6.93 yr; 7.27 yr
39. West Lake S & L: 9.800%;
Security S & L: 9.794%
40. $\$3,176.14$
41. 8.37%

PAYMENT NUMBER	PAYMENT	INTEREST	UNPAID BALANCE REDUCTION	UNPAID BALANCE
0				$1,000.00
1	$ 265.82	$25.00	$ 240.82	759.18
2	265.82	18.98	246.84	512.34
3	265.82	12.81	253.01	259.33
4	265.81	6.48	259.33	0.00
Totals	$1,063.27	$63.27	$1,000.00	

42. 10.45% compounded annually   **43.** (A) $\$571,499$ (B) $\$1,973,277$ **44.** $\$55,347.48; \$185,830.24$ **45.** 1 yr, 1 mo
46. $\$10,318.91; \$2,281.09$ **47.** $\$175.28; \$2,516.80$ **48.** (A) $\$398,807$ (B) $\$374,204$
49. (A) $\$746.79$ per month; $\$896.76$ per month   (B) $\$73,558.78; \$41,482.19$ **50.** $\$19,239$

C H A P T E R   5

1. $x = 3, y = 2$ **3.** $x = 2, y = 4$ **5.** No solution (parallel lines)   **7.** $x = 4, y = 5$ **9.** $x = 1, y = 4$ **11.** $u = 2, v = -3$
13. $m = 8, n = 6$ **15.** $x = 1, y = -5$ **17.** No solution (inconsistent)   **19.** $x = -\frac{4}{3}, y = 1$ **21.** Infinitely many solutions (dependent)
23. $x = 4,000, y = 280$ **25.** $x = 1.1, y = 0.3$ **27.** $x = 0, y = -2, z = 5$ **29.** $x = 2, y = 0, z = -1$ **31.** $a = -1, b = 2, c = 0$
33. $x = 0, y = 2, z = -3$ **35.** No solution (inconsistent)
37. (A) Equilibrium price = $\$6.50$; Equilibrium quantity = 500   **39.** (A) For $x = 120$ units, $C = \$216,000 = R$
(B)

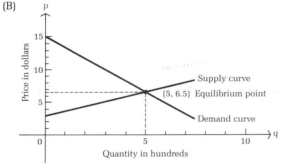

(B)

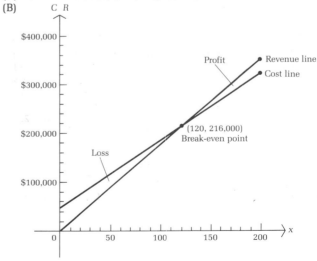

41. 50 one-person boats, 200 two-person boats, 100 four-person boats   **43.** Mix A: 80 g; mix B: 60 g

45. (A)   (B) $d = 141$ cm (approx.)   (C) Vacillate

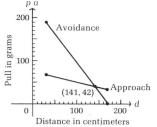

◆ EXERCISE 5-2

1. $\begin{bmatrix} 4 & -6 & | & -8 \\ 1 & -3 & | & 2 \end{bmatrix}$ **3.** $\begin{bmatrix} -4 & 12 & | & -8 \\ 4 & -6 & | & -8 \end{bmatrix}$ **5.** $\begin{bmatrix} 1 & -3 & | & 2 \\ 8 & -12 & | & -16 \end{bmatrix}$ **7.** $\begin{bmatrix} 1 & -3 & | & 2 \\ 0 & 6 & | & -16 \end{bmatrix}$ **9.** $\begin{bmatrix} 1 & -3 & | & 2 \\ 2 & 0 & | & -12 \end{bmatrix}$

11. $\begin{bmatrix} 1 & -3 & | & 2 \\ 3 & -3 & | & -10 \end{bmatrix}$ **13.** $x_1 = 3, x_2 = 2$ **15.** $x_1 = 3, x_2 = 1$ **17.** $x_1 = 2, x_2 = 1$ **19.** $x_1 = 2, x_2 = 4$ **21.** No solution

23. $x_1 = 1, x_2 = 4$ **25.** Infinitely many solutions: $x_2 = s, x_1 = 2s - 3$ for any real number s

27. Infinitely many solutions: $x_2 = s, x_1 = \frac{1}{2}s + \frac{1}{2}$ for any real number s **29.** $x_1 = -1, x_2 = 3$ **31.** No solution

33. Infinitely many solutions: $x_2 = t, x_1 = \frac{3}{2}t + 2$ for any real number t **35.** $x_1 = 2, x_2 = -1$ **37.** $x_1 = 2, x_2 = -1$

39. $x_1 = 1.1, x_2 = 0.3$

◆ EXERCISE 5-3

1. Yes   **3.** No   **5.** No   **7.** Yes   **9.** $x_1 = -2, x_2 = 3, x_3 = 0$ **11.** $x_1 = 2t + 3, x_2 = -t - 5, x_3 = t$ for t any real number   **13.** No solution

15. $x_1 = 2s + 3t - 5, x_2 = s, x_3 = -3t + 2, x_4 = t$ for s and t any real numbers

17. $\begin{bmatrix} 1 & 0 & | & -7 \\ 0 & 1 & | & 3 \end{bmatrix}$ **19.** $\begin{bmatrix} 1 & 0 & 0 & | & -5 \\ 0 & 1 & 0 & | & 4 \\ 0 & 0 & 1 & | & -2 \end{bmatrix}$ **21.** $\begin{bmatrix} 1 & 0 & 2 & | & -\frac{5}{3} \\ 0 & 1 & -2 & | & \frac{1}{3} \\ 0 & 0 & 0 & | & 0 \end{bmatrix}$

23. $x_1 = -2, x_2 = 3, x_3 = 1$ **25.** $x_1 = 0, x_2 = -2, x_3 = 2$ **27.** $x_1 = 2t + 3, x_2 = t - 2, x_3 = t$ for t any real number

29. $x_1 = (-4t - 4)/7, x_2 = (5t + 5)/7, x_3 = t$ for t any real number   **31.** $x_1 = -1, x_2 = 2$ **33.** No solution   **35.** No solution

37. $x_1 = t - 1, x_2 = 2t + 2, x_3 = t$ for t any real number   **39.** $x_1 = -2s + t + 1, x_2 = s, x_3 = t$ for s and t any real numbers

41. $x_1 = 0, x_2 = 2, x_3 = -3$ **43.** $x_1 = 1, x_2 = -2, x_3 = 1$

45. $x_1 = 2s - 3t + 3, x_2 = s + 2t + 2, x_3 = s, x_4 = t$ for s and t any real numbers

47. 20 one-person boats, 220 two-person boats, 100 four-person boats

49. $(t - 80)$ one-person boats, $(-2t + 420)$ two-person boats, t four-person boats, $80 \leq t \leq 210$, t an integer

51. No solution; no production schedule will use all the labor-hours in all departments

53. Federal: \$370,500; state: \$123,500; local: \$58,500   **55.** 8 oz food A, 2 oz food B, 4 oz food C **57.** No solution

59. 8 oz food A, $(-2t + 10)$ oz food B, t oz food C, $0 \leq t \leq 5$

61. $(10 - t)$ barrels of mix A, $(t - 5)$ barrels of mix B, $(25 - 2t)$ barrels of mix C, and t barrels of mix D, where t is an integer, $5 \leq t \leq 10$

63. Company A: 10 hr; company B: 15 hr

◆ EXERCISE 5-4

1. 2×2; 1×4 **3.** 2   **5.** $\begin{bmatrix} 0 & 0 \\ 0 & 0 \end{bmatrix}$ **7.** C, D **9.** A, B **11.** $\begin{bmatrix} -1 & 0 \\ 5 & -3 \end{bmatrix}$ **13.** Not defined

15. $\begin{bmatrix} -2 \\ 3 \\ 0 \end{bmatrix}$ **17.** $\begin{bmatrix} -1 \\ 6 \\ 5 \end{bmatrix}$ **19.** $\begin{bmatrix} -15 & 5 \\ 10 & -15 \end{bmatrix}$ **21.** $\begin{bmatrix} 1 & 3 & -1 & 1 \\ -1 & -5 & 7 & 2 \\ 4 & 8 & 0 & -2 \end{bmatrix}$ **23.** $\begin{bmatrix} 5.4 & 0.7 & -1.8 \\ 7.6 & -4.0 & 7.9 \end{bmatrix}$ **25.** $\begin{bmatrix} 250 & 360 \\ 40 & 350 \end{bmatrix}$

27. $\begin{bmatrix} 2,280 & 3,460 \\ 1,380 & 2,310 \end{bmatrix}$ **29.** $a = -1, b = 1, c = 3, d = -5$ **31.** $x = 2, y = -3$ **33.** No **35.** No

37.
	Guitar	Banjo	
	$33	$26	Materials
	$57	$77	Labor

39.

	Basic car	Air	AM/FM radio	Cruise control (Markup)
Model A	$1,180	$ 82	$ 54	$30
Model B	$1,075	$ 98	$ 81	$30
Model C	$1,715	$106	$108	$36

41.
	Round	Wrinkled	
Yellow	689	225	$= M + N$;
Green	218	68	

	Round	Wrinkled
Yellow	57%	19%
Green	18%	6%

◆ EXERCISE 5-5

1. 10 **3.** -1 **5.** $[12 \quad 13]$ **7.** $\begin{bmatrix} 5 \\ -3 \end{bmatrix}$ **9.** $\begin{bmatrix} 2 & 4 \\ 1 & -5 \end{bmatrix}$ **11.** $\begin{bmatrix} 1 & -5 \\ -2 & -4 \end{bmatrix}$ **13.** $[-7]$ **15.** $\begin{bmatrix} -15 & 6 \\ -20 & 8 \end{bmatrix}$ **17.** 6 **19.** 15

21. $\begin{bmatrix} 0 & 9 \\ 5 & -4 \end{bmatrix}$ **23.** $\begin{bmatrix} 5 & 8 & -5 \\ -1 & -3 & 2 \\ -2 & 8 & -6 \end{bmatrix}$ **25.** $[11]$ **27.** $\begin{bmatrix} 3 & -2 & -4 \\ 6 & -4 & -8 \\ -9 & 6 & 12 \end{bmatrix}$ **29.** Not defined

31. $\begin{bmatrix} 7 \\ 35 \\ 23 \end{bmatrix}$ **33.** $\begin{bmatrix} 0 & 0 & -5 \\ -6 & 15 & 13 \\ 5 & -6 & -14 \end{bmatrix}$ **35.** 0 **37.** $\begin{bmatrix} \frac{1}{3} & \frac{1}{3} \\ \frac{2}{3} & \frac{2}{3} \end{bmatrix}$

39. $AB = \begin{bmatrix} 5 & 7 \\ 2 & 3 \end{bmatrix}$; $BA = \begin{bmatrix} 1 & 3 \\ 2 & 7 \end{bmatrix}$ **41.** Both sides equal $\begin{bmatrix} 0 & 12 \\ 1 & 5 \end{bmatrix}$ **43.** $A^2 - B^2 = \begin{bmatrix} -2 & 0 \\ -8 & -10 \end{bmatrix}$; $(A - B)(A + B) = \begin{bmatrix} 2 & 4 \\ -8 & -14 \end{bmatrix}$

45. (A) $11.80 per boat (B) $[1.5 \quad 1.2 \quad 0.4] \cdot \begin{bmatrix} 9 \\ 12 \\ 6 \end{bmatrix} = \$30.30$ (C) 3×2 (D)

	Plant I	Plant II	
	$11.80	$13.80	One-person boat
	$18.50	$21.60	Two-person boat
	$26.00	$30.30	Four-person boat

Labor costs per boat at each plant

47. (A) $\begin{matrix} A & B & C & D & E \\ [16 & 9 & 11 & 11 & 10] \end{matrix}$, which is the combined inventory in all three stores

(B) $\begin{matrix} W & R \\ [\$108,300 & \$141,340] \end{matrix}$, which is the total wholesale and retail values of the total inventory in all three stores

49. (A) $2,025 (B) $[2,000 \quad 800 \quad 8,000] \cdot \begin{bmatrix} \$0.40 \\ \$0.75 \\ \$0.25 \end{bmatrix} = \$3,400$ (C) $\begin{bmatrix} \$2,025 \\ \$3,400 \end{bmatrix}$ Berkeley Oakland Cost per town (D)

	Telephone call	House call	Letter
[3,000	1,300	13,000]

Number of each type of contact made

1. $\begin{bmatrix} 2 & -3 \\ 4 & 5 \end{bmatrix}$ **3.** $\begin{bmatrix} -2 & 1 & 3 \\ 2 & 4 & -2 \\ 5 & 1 & 0 \end{bmatrix}$ **9.** $x_1 = -8, x_2 = 2$ **11.** $x_1 = 0, x_2 = 4$ **13.** $\begin{bmatrix} 3 & -2 \\ -1 & 1 \end{bmatrix}$ **15.** $\begin{bmatrix} 7 & -3 \\ -2 & 1 \end{bmatrix}$ **17.** $\begin{bmatrix} 7 & 6 & -3 \\ 2 & 2 & -1 \\ -6 & -5 & 3 \end{bmatrix}$

19. $\begin{bmatrix} \frac{3}{2} & -\frac{1}{2} & -\frac{1}{2} \\ -\frac{1}{2} & \frac{1}{2} & \frac{1}{2} \\ -\frac{3}{2} & \frac{1}{2} & \frac{3}{2} \end{bmatrix}$ **21.** (A) $x_1 = -3, x_2 = 2$ (B) $x_1 = -1, x_2 = 2$ (C) $x_1 = -8, x_2 = 3$

23. (A) $x_1 = 17, x_2 = -5$ (B) $x_1 = 7, x_2 = -2$ (C) $x_1 = 24, x_2 = -7$

25. (A) $x_1 = 1, x_2 = 0, x_3 = 0$ (B) $x_1 = -1, x_2 = 0, x_3 = 1$ (C) $x_1 = -1, x_2 = -1, x_3 = 1$

27. (A) $x_1 = 1, x_2 = 1, x_3 = 3$ (B) $x_1 = -1, x_2 = 1, x_3 = -1$ (C) $x_1 = 5, x_2 = -1, x_3 = -5$ **29.** The inverse does not exist.

31. $\begin{bmatrix} 1 & -\frac{1}{2} \\ -2 & \frac{3}{2} \end{bmatrix}$ **33.** $\begin{bmatrix} 1 & 2 & 2 \\ -2 & -3 & -4 \\ -1 & -2 & -1 \end{bmatrix}$ **35.** The inverse does not exist. **37.** $\begin{bmatrix} 1 & 1 & -\frac{1}{2} \\ -2 & -\frac{3}{2} & 1 \\ -1 & -1 & \frac{3}{4} \end{bmatrix}$

41. Concert 1: 6,000 $4 tickets and 4,000 $8 tickets; Concert 2: 5,000 $4 tickets and 5,000 $8 tickets; Concert 3: 3,000 $4 tickets and 7,000 $8 tickets

43. Diet 1: 60 oz mix A and 80 oz mix B; Diet 2: 20 oz mix A and 60 oz mix B; Diet 3: 0 oz mix A and 100 oz mix B

1. $x = 4, y = 4$ **2.** $x = 4, y = 4$ **3.** $\begin{bmatrix} 3 & 3 \\ 4 & 2 \end{bmatrix}$ **4.** Not defined **5.** $\begin{bmatrix} -3 & 0 \\ 1 & -1 \end{bmatrix}$ **6.** $\begin{bmatrix} 4 & 3 \\ 7 & 4 \end{bmatrix}$ **7.** Not defined **8.** $\begin{bmatrix} 5 \\ 5 \end{bmatrix}$

9. $\begin{bmatrix} 2 & 3 \\ 4 & 6 \end{bmatrix}$ **10.** 8 (a real number) **11.** Not defined **12.** $\begin{bmatrix} 3 & -2 \\ -4 & 3 \end{bmatrix}$ **13.** $x_1 = -1, x_2 = 3$ **14.** $x_1 = -1, x_2 = 3$

15. $x_1 = -1, x_2 = 3; x_1 = 1, x_2 = 2; x_1 = 8, x_2 = -10$ **16.** Not defined **17.** $\begin{bmatrix} 10 & -8 \\ 4 & 6 \end{bmatrix}$ **18.** $\begin{bmatrix} -2 & 8 \\ 8 & 6 \end{bmatrix}$

19. 9 (a real number) **20.** [9] (a matrix) **21.** $\begin{bmatrix} 10 & -5 & 1 \\ -1 & -4 & -5 \\ 1 & -7 & -2 \end{bmatrix}$ **22.** $\begin{bmatrix} -\frac{5}{2} & 2 & -\frac{1}{2} \\ 1 & -1 & 1 \\ \frac{1}{2} & 0 & -\frac{1}{2} \end{bmatrix}$

23. (A) $x_1 = 2, x_2 = 1, x_3 = -1$ (B) $x_1 = -5t - 12, x_2 = 3t + 7, x_3 = t$ for t any real number

24. $x_1 = 2, x_2 = 1, x_3 = -1; x_1 = 1, x_2 = -2, x_3 = 1; x_1 = -1, x_2 = 2, x_3 = -2$

25. $\begin{bmatrix} -\frac{11}{12} & -\frac{1}{12} & 5 \\ \frac{10}{12} & \frac{2}{12} & -4 \\ \frac{1}{12} & -\frac{1}{12} & 0 \end{bmatrix}$ **26.** $x_1 = 1,000, x_2 = 4,000, x_3 = 2,000$ **27.** $x_1 = 1,000, x_2 = 4,000, x_3 = 2,000$

28. $0.01x_1 + 0.02x_2 = 4.5$
$0.02x_1 + 0.05x_2 = 10$
$x_1 = 250$ tons of ore A, $x_2 = 100$ tons of ore B

29. (A) $\begin{matrix} X \\ \begin{bmatrix} x_1 \\ x_2 \end{bmatrix} \end{matrix} = \begin{matrix} A^{-1} \\ \begin{bmatrix} 500 & -200 \\ -200 & 100 \end{bmatrix} \end{matrix} \begin{bmatrix} 4.5 \\ 10 \end{bmatrix} = \begin{bmatrix} 250 \\ 100 \end{bmatrix}$
$x_1 = 250$ tons of ore A, $x_2 = 100$ tons of ore B

(B) $\begin{matrix} X \\ \begin{bmatrix} x_1 \\ x_2 \end{bmatrix} \end{matrix} = \begin{matrix} A^{-1} \\ \begin{bmatrix} 500 & -200 \\ -200 & 100 \end{bmatrix} \end{matrix} \begin{bmatrix} 2.3 \\ 5 \end{bmatrix} = \begin{bmatrix} 150 \\ 40 \end{bmatrix}$
$x_1 = 150$ tons of ore A, $x_2 = 40$ tons of ore B

30. (A) $MN = \begin{bmatrix} \$7,620 & \$7,530 \\ \$13,880 & \$13,930 \end{bmatrix} \begin{matrix} \text{Alloy 1} \\ \text{Alloy 2} \end{matrix}$

      Supplier A Supplier B

Cost of each alloy from each supplier

     Supplier A Supplier B

(B) [\$21,500   \$21,460]

Total cost for both alloys from each supplier

31. (A) $[0.25 \quad 0.20 \quad 0.05] \cdot \begin{bmatrix} 15 \\ 12 \\ 4 \end{bmatrix} = \$6.35$

     Calif.   Texas

(B) $\begin{bmatrix} \$3.65 & \$3.00 \\ \$6.35 & \$5.20 \end{bmatrix} \begin{matrix} \text{Model } A \\ \text{Model } B \end{matrix}$

Total labor costs for each model at each plant

32. \$2,000 at 5%; \$3,000 at 10% **33.** \$2,000 at 5%; \$3,000 at 10%

C H A P T E R 6

◆ EXERCISE 6-1

1.
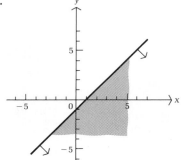

3.
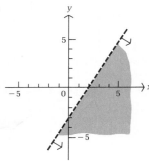

5.
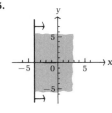

7.
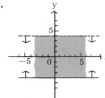

9.
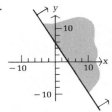

11.
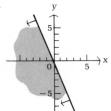

13. IV **15.** I **17.**
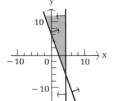

19.
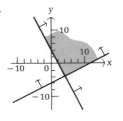

21. IV; (8, 0), (18, 0), (6, 4)   **23.** I; (0, 16), (6, 4), (18, 0)   **25.** Bounded

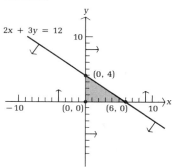

27. Bounded

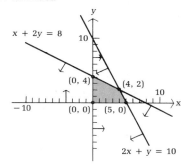

29. Unbounded

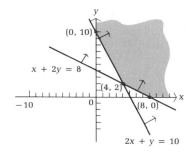

31. Bounded

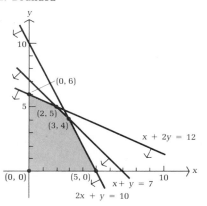

33. Unbounded

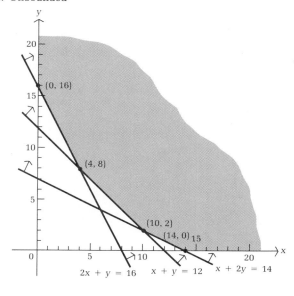

35. Bounded

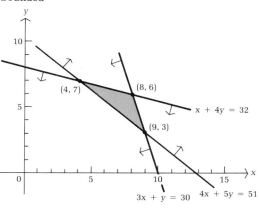

37. Empty

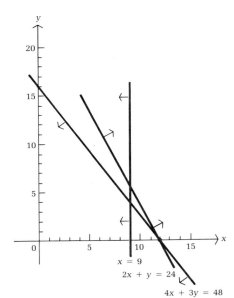

x = 9
2x + y = 24
4x + 3y = 48

39. Unbounded

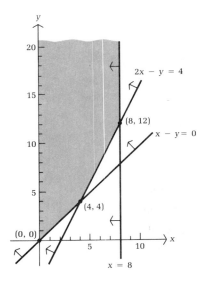

2x − y = 4
(8, 12)
x − y = 0
(4, 4)
(0, 0)
x = 8

41. Bounded

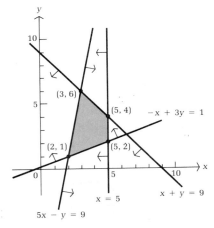

(3, 6)
(5, 4)
−x + 3y = 1
(2, 1)
(5, 2)
x = 5
x + y = 9
5x − y = 9

43. $6x + 4y \leqslant 108$
$x + y \leqslant 24$
$x \geqslant 0$
$y \geqslant 0$

(6, 18)

45. $20x + 10y \geqslant 460$
$30x + 30y \geqslant 960$
$5x + 10y \geqslant 220$
$x \geqslant 0$
$y \geqslant 0$

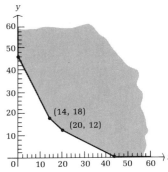

(14, 18)
(20, 12)

47. $10x + 20y \leqslant 800$
$20x + 10y \leqslant 640$
$x \geqslant 0$
$y \geqslant 0$

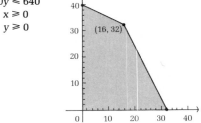

(16, 32)

◆ EXERCISE 6-2

1. 16 **3.** 84 **5.** 32 **7.** 36 **9.** Max $P = 30$ at $x_1 = 4$ and $x_2 = 2$ **11.** Min $z = 14$ at $x_1 = 4$ and $x_2 = 2$; no max

13. Max $P = 260$ at $x_1 = 2$ and $x_2 = 5$ **15.** Min $z = 140$ at $x_1 = 14$ and $x_2 = 0$; no max

17. Min $P = 20$ at $x_1 = 0$ and $x_2 = 2$; Max $P = 150$ at $x_1 = 5$ and $x_2 = 0$ **19.** Feasible region empty; no optimal solutions

21. Min $P = 140$ at $x_1 = 3$ and $x_2 = 8$; Max $P = 260$ at $x_1 = 8$ and $x_2 = 10$, at $x_1 = 12$ and $x_2 = 2$, or at any point on the line segment from (8, 10) to (12, 2)

23. Max $P = 26{,}000$ at $x_1 = 400$ and $x_2 = 600$ **25.** (A) $2a < b$ (B) $\frac{1}{3}a < b < 2a$ (C) $b < \frac{1}{3}a$ (D) $b = 2a$ (E) $b = \frac{1}{3}a$

27. 6 trick, 18 slalom; \$780

29. (A) Plant A: 5 days; Plant B: 4 days; min cost \$8,600 (B) Plant A: 10 days; Plant B: 0 days; min cost \$6,000 (C) Plant A: 0 days; Plant B: 10 days; min cost \$8,000

31. 7 buses, 15 vans; min cost \$9,900 **33.** 1,500 gal by new process, none by old process; max profit \$300

35. (A) 150 bags brand A, 100 bags brand B; max nitrogen 1,500 lb (B) 0 bags brand A, 250 bags brand B; min nitrogen 750 lb

37. 20 cubic yards of A, 12 cubic yards of B; \$1,020 **39.** 48; 16 mice, 32 rats

◆ EXERCISE 6-3

1. (A) 5 (B) 4 (C) 5 basic and 4 nonbasic variables (D) 5 linear equations with 5 variables

3.

	NONBASIC	BASIC	FEASIBLE?
(A)	x_1, x_2	s_1, s_2	Yes
(B)	x_1, s_1	x_2, s_2	Yes
(C)	x_1, s_2	x_2, s_1	No
(D)	x_2, s_1	x_1, s_2	No
(E)	x_2, s_2	x_1, s_1	Yes
(F)	s_1, s_2	x_1, x_2	Yes

5.

	x_1	x_2	s_1	s_2	FEASIBLE?
(A)	0	0	50	40	Yes
(B)	0	50	0	-60	No
(C)	0	20	30	0	Yes
(D)	25	0	0	15	Yes
(E)	40	0	-30	0	No
(F)	20	10	0	0	Yes

7.

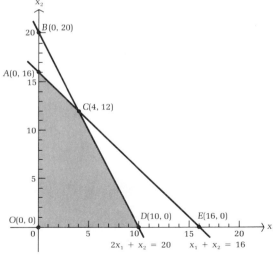

$$x_1 + x_2 + s_1 \qquad = 16$$
$$2x_1 + x_2 \qquad + s_2 = 20$$

x_1	x_2	s_1	s_2	INTERSECTION POINT	FEASIBLE?
0	0	16	20	O	Yes
0	16	0	4	A	Yes
0	20	-4	0	B	No
16	0	0	-12	E	No
10	0	6	0	D	Yes
4	12	0	0	C	Yes

9.

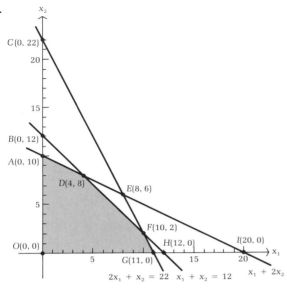

$$2x_1 + x_2 + s_1 \qquad\qquad = 22$$
$$x_1 + x_2 \qquad + s_2 \qquad = 12$$
$$x_1 + 2x_2 \qquad\qquad + s_3 = 20$$

x_1	x_2	s_1	s_2	s_3	INTERSECTION POINT	FEASIBLE?
0	0	22	12	20	O	Yes
0	22	0	-10	-24	C	No
0	12	10	0	-4	B	No
0	10	12	2	0	A	Yes
11	0	0	1	9	G	Yes
12	0	-2	0	8	H	No
20	0	-18	-8	0	I	No
10	2	0	0	6	F	Yes
8	6	0	-2	0	E	No
4	8	6	0	0	D	Yes

◆ EXERCISE **6-4**

---

1. (A) Basic: x_2, s_1, P; nonbasic: x_1, s_2 (B) $x_1 = 0$, $x_2 = 12$, $s_1 = 15$, $s_2 = 0$, and $P = 20$ (C) Additional pivot required

3. (A) Basic: x_2, x_3, s_3, P; nonbasic: x_1, s_1, s_2 (B) $x_1 = 0$, $x_2 = 15$, $x_3 = 5$, $s_1 = 0$, $s_2 = 0$, $s_3 = 12$, and $P = 45$ (C) No solution

5. Exit → s_1
$$
\begin{array}{c}
\quad\;\; \overset{\text{Enter}}{\downarrow} \\
\begin{array}{ccccc}
x_1 & x_2 & s_1 & s_2 & P
\end{array} \\
\begin{bmatrix}
\textcircled{1} & 4 & 1 & 0 & 0 & \big| & 4 \\
3 & 5 & 0 & 1 & 0 & \big| & 24 \\
\hline
-8 & -5 & 0 & 0 & 1 & \big| & 0
\end{bmatrix}
\end{array}
$$
s_2
P

$$
\sim
\begin{array}{c}
x_1 \\ s_2 \\ P
\end{array}
\begin{bmatrix}
1 & 4 & 1 & 0 & 0 & \big| & 4 \\
0 & -7 & -3 & 1 & 0 & \big| & 12 \\
\hline
0 & 27 & 8 & 0 & 1 & \big| & 32
\end{bmatrix}
$$

7. Exit → x_2
$$
\begin{array}{c}
\quad\;\; \overset{\text{Enter}}{\downarrow} \\
\begin{array}{cccccc}
x_1 & x_2 & s_1 & s_2 & s_3 & P
\end{array} \\
\begin{bmatrix}
\textcircled{2} & 1 & 1 & 0 & 0 & 0 & \big| & 4 \\
3 & 0 & 1 & 1 & 0 & 0 & \big| & 8 \\
0 & 0 & 2 & 0 & 1 & 0 & \big| & 2 \\
\hline
-4 & 0 & -3 & 0 & 0 & 1 & \big| & 5
\end{bmatrix}
\end{array}
$$
s_2
s_3
P

$$
\sim
\begin{array}{c}
x_1 \\ s_2 \\ s_3 \\ P
\end{array}
\begin{bmatrix}
1 & \frac{1}{2} & \frac{1}{2} & 0 & 0 & 0 & \big| & 2 \\
0 & -\frac{3}{2} & -\frac{1}{2} & 1 & 0 & 0 & \big| & 2 \\
0 & 0 & 2 & 0 & 1 & 0 & \big| & 2 \\
\hline
0 & 2 & -1 & 0 & 0 & 1 & \big| & 13
\end{bmatrix}
$$

9. (A)
$$2x_1 + x_2 + s_1 \qquad\qquad = 10$$
$$x_1 + 2x_2 \qquad + s_2 \qquad = 8$$
$$-15x_1 - 10x_2 \qquad\qquad + P = 0$$

(B) Exit → s_1
$$
\begin{array}{c}
\quad\;\; \overset{\text{Enter}}{\downarrow} \\
\begin{array}{ccccc}
x_1 & x_2 & s_1 & s_2 & P
\end{array} \\
\begin{bmatrix}
\textcircled{2} & 1 & 1 & 0 & 0 & \big| & 10 \\
1 & 2 & 0 & 1 & 0 & \big| & 8 \\
\hline
-15 & -10 & 0 & 0 & 1 & \big| & 0
\end{bmatrix}
\end{array}
$$
s_2
P

(C) Max $P = 80$ at $x_1 = 4$ and $x_2 = 2$

11. (A)
$$2x_1 + x_2 + s_1 \qquad = 10$$
$$x_1 + 2x_2 \qquad + s_2 \qquad = 8$$
$$-30x_1 - x_2 \qquad + P = 0$$

(B)

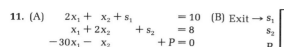

$$\text{Exit} \to \begin{matrix} s_1 \\ s_2 \\ P \end{matrix} \begin{bmatrix} \overset{x_1}{\textcircled{2}} & \overset{x_2}{1} & \overset{s_1}{1} & \overset{s_2}{0} & \overset{P}{0} & 10 \\ 1 & 2 & 0 & 1 & 0 & 8 \\ \hdashline -30 & -1 & 0 & 0 & 1 & 0 \end{bmatrix}$$

(C) Max $P = 150$ at $x_1 = 5$ and $x_2 = 0$

13. Max $P = 260$ at $x_1 = 2$ and $x_2 = 5$

15. No optimal solution exists.

17. Max $P = 7$ at $x_1 = 3$ and $x_2 = 5$

19. Max $P = \frac{190}{3}$ at $x_1 = \frac{40}{3}$, $x_2 = 0$, and $x_3 = \frac{10}{3}$

21. Max $P = 17$ at $x_1 = 4$, $x_2 = 3$, and $x_3 = 0$

23. Max $P = 22$ at $x_1 = 1$, $x_2 = 6$, and $x_3 = 0$

25. Max $P = 26{,}000$ at $x_1 = 400$ and $x_2 = 600$

27. Max $P = 450$ at $x_1 = 0$, $x_2 = 180$, and $x_3 = 30$

29. Max $P = 88$ at $x_1 = 24$ and $x_2 = 8$

31. Let $x_1 = $ Number of A components
$x_2 = $ Number of B components
$x_3 = $ Number of C components

Maximize $P = 7x_1 + 9x_2 + 10x_3$
Subject to $2x_1 + x_2 + 2x_3 \le 1{,}000$
$x_1 + 2x_2 + 2x_3 \le 800$
$x_1, x_2, x_3 \ge 0$

400 A components, 200 B components, and 0 C components; max profit \$4,600

33. Let $x_1 = $ Amount invested in government bonds
$x_2 = $ Amount invested in mutual funds
$x_3 = $ Amount invested in money market funds

Maximize $P = 0.08x_1 + 0.13x_2 + 0.15x_3$
Subject to $x_1 + x_2 + x_3 \le 100{,}000$
$-x_1 + x_2 + x_3 \le 0$
$x_1, x_2, x_3 \ge 0$

\$50,000 in government bonds, \$0 in mutual funds, and \$50,000 in money market funds; max return \$11,500

35. Let $x_1 = $ Number of ads placed in daytime shows
$x_2 = $ Number of ads placed in prime-time shows
$x_3 = $ Number of ads placed in late-night shows

Maximize $P = 14{,}000x_1 + 24{,}000x_2 + 18{,}000x_3$
Subject to $x_1 + x_2 + x_3 \le 15$
$1{,}000x_1 + 2{,}000x_2 + 1{,}500x_3 \le 20{,}000$
$x_1, x_2, x_3 \ge 0$

10 daytime ads, 5 prime-time ads, and 0 late-night ads; max number of potential customers 260,000

37. Let $x_1 = $ Number of colonial houses
$x_2 = $ Number of split-level houses
$x_3 = $ Number of ranch houses

Maximize $P = 20{,}000x_1 + 18{,}000x_2 + 24{,}000x_3$
Subject to $\frac{1}{2}x_1 + \frac{1}{2}x_2 + x_3 \le 30$
$60{,}000x_1 + 60{,}000x_2 + 80{,}000x_3 \le 3{,}200{,}000$
$4{,}000x_1 + 3{,}000x_2 + 4{,}000x_3 \le 180{,}000$
$x_1, x_2, x_3 \ge 0$

20 colonial, 20 split-level, and 10 ranch houses; max profit \$1,000,000

39. Let $x_1 = $ Number of boxes of assortment I
$x_2 = $ Number of boxes of assortment II
$x_3 = $ Number of boxes of assortment III

Maximize $P = 4x_1 + 3x_2 + 5x_3$
Subject to $4x_1 + 12x_2 + 8x_3 \le 4{,}800$
$4x_1 + 4x_2 + 8x_3 \le 4{,}000$
$12x_1 + 4x_2 + 8x_3 \le 5{,}600$
$x_1, x_2, x_3 \ge 0$

200 boxes of assortment I, 100 boxes of assortment II, and 350 boxes of assortment III; max profit \$2,850

41. Let $x_1 = $ Number of grams of food A
$x_2 = $ Number of grams of food B
$x_3 = $ Number of grams of food C

Maximize $P = 3x_1 + 3x_2 + 5x_3$
Subject to $x_1 + 3x_2 + 2x_3 \le 30$
$2x_1 + x_2 + x_3 \le 24$
$x_1, x_2, x_3 \ge 0$

6 g food A, 0 g food B, and 12 g food C; max protein 78 units

43. Let $x_1 =$ Number of undergraduate students
$x_2 =$ Number of graduate students
$x_3 =$ Number of faculty members

Maximize $P = 18x_1 + 25x_2 + 30x_3$
Subject to $x_1 + x_2 + x_3 \leq 20$
$60x_1 + 90x_2 + 120x_3 \leq 1,620$
$x_1, x_2, x_3 \geq 0$

6 undergraduate and 14 graduate students, 0 faculty members; max number of interviews 458

◆ EXERCISE 6-5 CHAPTER REVIEW

1. Bounded

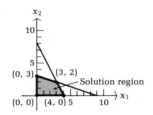

2. Unbounded

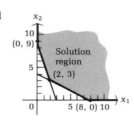

3. Max $P = 24$ at $x_1 = 4$ and $x_2 = 0$
4. $2x_1 + x_2 + s_1 = 8$
$x_1 + 2x_2 + s_2 = 10$
5. 2 basic and 2 nonbasic variables

6.

x_1	x_2	s_1	s_2	FEASIBLE?
0	0	8	10	Yes
0	8	0	−6	No
0	5	3	0	Yes
4	0	0	6	Yes
10	0	−12	0	No
2	4	0	0	Yes

7.

8. Max $P = 24$ at $x_1 = 4$ and $x_2 = 0$

9. Basic variables: x_2, s_2, s_3, P; nonbasic variables: x_1, x_3, s_1

Enter
↓

$$\begin{array}{c} x_2 \\ s_2 \\ \text{Exit} \to s_3 \\ P \end{array} \left[\begin{array}{ccccccc|c} x_1 & x_2 & x_3 & s_1 & s_2 & s_3 & P & \\ 2 & 1 & 3 & -1 & 0 & 0 & 0 & 20 \\ 3 & 0 & 4 & 1 & 1 & 0 & 0 & 30 \\ \boxed{2} & 0 & 5 & 2 & 0 & 1 & 0 & 10 \\ \hline -8 & 0 & -5 & 3 & 0 & 0 & 1 & 50 \end{array} \right] \sim \begin{array}{c} x_2 \\ s_2 \\ x_1 \\ P \end{array} \left[\begin{array}{ccccccc|c} x_1 & x_2 & x_3 & s_1 & s_2 & s_3 & P & \\ 0 & 1 & -2 & -3 & 0 & -1 & 0 & 10 \\ 0 & 0 & -\frac{7}{2} & -2 & 1 & -\frac{3}{2} & 0 & 15 \\ 1 & 0 & \frac{5}{2} & 1 & 0 & \frac{1}{2} & 0 & 5 \\ \hline 0 & 0 & 15 & 11 & 0 & 4 & 1 & 90 \end{array} \right]$$

10. (A) $x_1 = 0, x_2 = 2, s_1 = 0, s_2 = 5, P = 12$; additional pivoting required **11.** Min $C = 40$ at $x_1 = 0$ and $x_2 = 20$
(B) $x_1 = 0, x_2 = 0, s_1 = 0, s_2 = 7, P = 22$; no optimal solution exists
(C) $x_1 = 6, x_2 = 0, s_1 = 15, s_2 = 0, P = 10$; optimal solution
12. Max $P = 26$ at $x_1 = 2$ and $x_2 = 5$ **13.** Max $P = 26$ at $x_1 = 2$ and $x_2 = 5$ **14.** Min $C = 51$ at $x_1 = 9$ and $x_2 = 3$
15. No optimal solution exists. **16.** Max $P = 23$ at $x_1 = 4, x_2 = 1$, and $x_3 = 0$
17. Max $P = 36$ at $x_1 = 6$ and $x_2 = 8$

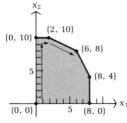

18. Let x_1 = Number of regular sails   Maximize $P = 100x_1 + 200x_2$
        x_2 = Number of competition sails Subject to $2x_1 + 3x_2 \leqslant 150$
                                           $4x_1 + 9x_2 \leqslant 360$
                                              $x_1, x_2 \geqslant 0$

19. Let x_1 = Amount invested in oil stock   Maximize $P = 0.12x_1 + 0.09x_2 + 0.05x_3$
        x_2 = Amount invested in steel stock   Subject to $x_1 + x_2 + x_3 \leqslant 150{,}000$
        x_3 = Amount invested in government bonds $x_1 + x_2 - x_3 \leqslant 0$
                                            $2x_1 - x_2 \leqslant 0$
                                             $x_1, x_2, x_3 \geqslant 0$

20. Let x_1 = Number of grams of mix A Minimize $C = 0.02x_1 + 0.04x_2$
        x_2 = Number of grams of mix B Subject to $3x_1 + 4x_2 \geqslant 300$
                                          $2x_1 + 5x_2 \geqslant 200$
                                          $6x_1 + 10x_2 \geqslant 900$
                                             $x_1, x_2 \geqslant 0$

C H A P T E R 7

◆ EXERCISE 7-1

1. 24 **3.** 9 **5.** 990 **7.** 10 **9.** 35 **11.** 1 **13.** 60 **15.** 6,497,400 **17.** 10 **19.** 270,725 **21.** $5 \cdot 3 \cdot 4 \cdot 2 = 120$
23. $P_{10,3} = 10 \cdot 9 \cdot 8 = 720$ **25.** $C_{7,3} = 35$; $P_{7,3} = 210$ **27.** $C_{10,2} = 45$ **29.** $6 \cdot 5 \cdot 4 \cdot 3 = 360$; $6 \cdot 6 \cdot 6 \cdot 6 = 1{,}296$
31. $P_{10,5} = 30{,}240$; $10^5 = 100{,}000$ **33.** $C_{13,5} = 1{,}287$ **35.** $26 \cdot 26 \cdot 26 \cdot 10 \cdot 10 \cdot 10 = 17{,}576{,}000$; $26 \cdot 25 \cdot 24 \cdot 10 \cdot 9 \cdot 8 = 11{,}232{,}000$
37. $C_{13,5}C_{13,2} = 100{,}386$ **39.** $C_{8,3}C_{10,4}C_{7,2} = 246{,}960$ **41.** $12 \cdot 11 = 132$ **43.** (A) $C_{8,2} = 28$ (B) $C_{8,3} = 56$ (C) $C_{8,4} = 70$
45. $P_{5,2} = 20$; $P_{5,3} = 60$; $P_{5,4} = 120$; $P_{5,5} = 120$ **47.** (A) $P_{8,5} = 6{,}720$ (B) $C_{8,5} = 56$ (C) $2 \cdot C_{6,4} = 30$
49. (A) 6 combined outcomes: **51.** 12   **55.** (A) 12 classifications:

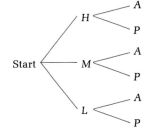

53. (A) $C_{6,3}C_{5,2} = 200$
     (B) $C_{6,4}C_{5,1} = 75$
     (C) $C_{6,5} = 6$
     (D) $C_{11,5} = 462$
     (E) $C_{6,4}C_{5,1} + C_{6,5} = 81$

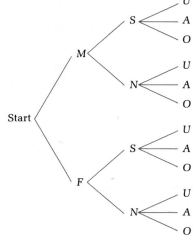

(B) $3 \cdot 2 = 6$

57. 336; 512 **59.** $P_{4,2} = 12$

(B) $2 \cdot 2 \cdot 3 = 12$

◆ EXERCISE 7-2

1. Occurrence of E is certain. **3.** $\frac{1}{2}$
5. (A) Reject; no probability can be negative (B) Reject; $P(R) + P(G) + P(Y) + P(B) \neq 1$ (C) Acceptable **7.** $P(R) + P(Y) = .56$ **9.** $\frac{1}{8}$
11. $1/P_{10,3} \approx .0014$ **13.** $C_{26,5}/C_{52,5} \approx .025$ **15.** $C_{12,5}/C_{52,5} \approx .000\ 305$ **17.** $(2 \cdot 5 \cdot 5 \cdot 1)/(2 \cdot 5 \cdot 5 \cdot 5) = .2$
19. $1/P_{5,5} = 1/5! = .008\ 33$ **21.** $\frac{1}{36}$ **23.** $\frac{5}{36}$ **25.** $\frac{1}{6}$ **27.** $\frac{7}{9}$ **29.** 0 **31.** $\frac{1}{3}$ **33.** $\frac{2}{9}$ **35.** $\frac{2}{3}$ **37.** $\frac{1}{4}$ **39.** $\frac{1}{4}$ **41.** $\frac{3}{4}$ **43.** $\frac{1}{9}$ **45.** $\frac{1}{3}$ **47.** $\frac{1}{9}$
49. $\frac{4}{9}$ **51.** $C_{16,5}/C_{52,5} \approx .001\ 68$ **53.** $48/C_{52,5} \approx .000\ 018\ 5$ **55.** $4/C_{52,5} \approx .000\ 001\ 5$ **57.** $C_{4,2}C_{4,3}/C_{52,5} \approx .000\ 009$
59. (A) $1/P_{12,4} \approx .000\ 084$ (B) $1/12^4 \approx .000\ 048$
61. (A) $C_{6,3}C_{5,2}/C_{11,5} \approx .433$ (B) $C_{6,4}C_{5,1}/C_{11,5} \approx .162$ (C) $C_{6,5}/C_{11,5} \approx .013$ (D) $[C_{6,4}C_{5,1} + C_{6,5}]/C_{11,5} \approx .175$
63. (A) $1/P_{8,3} \approx .0030$ (B) $1/8^3 \approx .0020$ **65.** (A) $P_{6,2}/P_{11,2} \approx .273$ (B) $[C_{5,3} + C_{6,1}C_{5,2}]/C_{11,3} \approx .424$

◆ EXERCISE 7-3

1. .1 **3.** .45 **5.** $P(\text{Point down}) = .389$, $P(\text{Point up}) = .611$; no
7. (A) $P(2 \text{ girls}) \approx .2351$, $P(1 \text{ girl}) \approx .5435$, $P(0 \text{ girls}) \approx .2214$ (B) $P(2 \text{ girls}) = .25$, $P(1 \text{ girl}) = .50$ (C) $P(0 \text{ girls}) = .25$
9. (A) $P(3 \text{ heads}) \approx .132$, $P(2 \text{ heads}) \approx .368$, $P(1 \text{ head}) \approx .38$, $P(0 \text{ heads}) \approx .12$
  (B) $P(3 \text{ heads}) \approx .125$, $P(2 \text{ heads}) \approx .375$, $P(1 \text{ head}) \approx .375$, $P(0 \text{ heads}) \approx .125$
  (C) 3 heads, 125; 2 heads, 375; 1 head, 375; 0 heads, 125
11. 4 heads, 5; 3 heads, 20; 2 heads, 30; 1 head, 20; 0 heads, 5 **13.** (A) .015 (B) .222 (C) .169 (D) .958
15. (A) $P(\text{Red}) = .3$, $P(\text{Pink}) = .44$, $P(\text{White}) = .26$ (B) 250 red, 500 pink, 250 white

◆ EXERCISE 7-4

1. $E(X) = -.1$ **3.** Probability distribution: **5.** Payoff table:

x_i	0	1	2
p_i	$\frac{1}{4}$	$\frac{1}{2}$	$\frac{1}{4}$

$E(X) = 1$

x_i	\$1	$-\$1$
p_i	$\frac{1}{2}$	$\frac{1}{2}$

$E(X) = 0$; game is fair

7. Payoff table:

x_i	$-\$3$	$-\$2$	$-\$1$	\$0	\$1	\$2
p_i	$\frac{1}{6}$	$\frac{1}{6}$	$\frac{1}{6}$	$\frac{1}{6}$	$\frac{1}{6}$	$\frac{1}{6}$

$E(X) = -50¢$; game is not fair

9. $-\$0.50$
11. $-\$0.036$; $\$0.036$
13. Win \$1
15. A_2; \$210

17. Payoff table:

x_i	\$35	$-\$1$
p_i	$\frac{1}{38}$	$\frac{37}{38}$

$E(X) = -5.26¢$

19. Payoff table:

x_i	\$499	\$99	\$19	\$4	$-\$1$
p_i	.0002	.0006	.001	.004	.9942

$E(X) = -80¢$

21. (A)

x_i	0	1	2
p_i	$\frac{7}{15}$	$\frac{7}{15}$	$\frac{1}{15}$

(B) .60

23. (A)

x_i	$-\$5$	$\$195$	$\$395$	$\$595$
p_i	.985	.0149	.000 059 9	.000 000 06

(B) $E(X) \approx -\$2$

25. Payoff table:

x_i	$\$4,850$	$-\$150$
p_i	.01	.99

$E(X) = -\$100$

27. Site A, with $E(X) = \$3.6$ million **29.** 1.54 **31.** For A_1, $E(X) = \$4$, and for A_2, $E(X) = \$4.80$; A_2 is better

---

◆ EXERCISE **7-5**   CHAPTER REVIEW

1. (A) 12 combined (B) $6 \cdot 2 = 12$
outcomes:

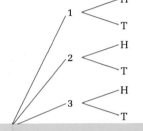

2. 15; 30
3. $6 \cdot 5 \cdot 4 \cdot 3 \cdot 2 \cdot 1 = 720$
4. $P_{6,6} = 6! = 720$
5. $C_{13,5}/C_{52,5} \approx .0005$
6. $1/P_{15,2} \approx .0048$
7. $1/P_{10,3} \approx .0014$; $1/C_{10,3} \approx .0083$
8. .05
9. Payoff table:

x_i	$-\$2$	$-\$1$	$\$0$	$\$1$	$\$2$
p_i	$\frac{1}{5}$	$\frac{1}{5}$	$\frac{1}{5}$	$\frac{1}{5}$	$\frac{1}{5}$

is fair
y of an event cannot be negative;
babilities of simple events must be 1;
of an event cannot be greater than 1

(B) $C_{5,2} = 10$
= .21, $P(1 \text{ head}) = .48$, $P(0 \text{ heads}) = .31$
= .25, $P(1 \text{ head}) = .50$, $P(0 \text{ heads}) = .25$
50; 1 head, 500; 0 heads, 250

$= \frac{2}{15}$ **17.** Payoff table:

x_i	$\$5$	$-\$4$	$\$2$
p_i	.25	.5	.25

$E(X) = -25¢$; game is not fair

18. (A) $\frac{1}{3}$ (B) $\frac{2}{9}$

	9	10	11	12
	$\frac{4}{36}$	$\frac{3}{36}$	$\frac{2}{36}$	$\frac{1}{36}$

20. $2^5 = 32$; 6 **21.** $1 - C_{7,3}/C_{10,3} = \frac{17}{24}$

)) $(C_{6,3} + C_{6,2} \cdot C_{4,1})/C_{10,3} = \frac{2}{3}$ **23.** $P_{4,4}/P_{2,2} = 12$
10

26. (A)

x_i	2	3	4	5	6
p_i	$\frac{9}{36}$	$\frac{12}{36}$	$\frac{10}{36}$	$\frac{4}{36}$	$\frac{1}{36}$

27. $E(X) \approx -\$0.167$; no   **28.** $P_{5,5} = 120$ **29.** (A) .04 (B) .16 (C) .54

(B) $E(X) = \frac{10}{3}$

30. A: $E(X) = \$7.6$ million; B: $E(X) = \$7.8$ million; plan B

31. Payoff table:

x_i	\$270	$-\$30$
p_i	.08	.92

$E(X) = -\$6$

32. $1 - (C_{10,4}/C_{12,4}) \approx .576$ **33.** (A)

x_i	0	1	2
p_i	$\frac{12}{22}$	$\frac{9}{22}$	$\frac{1}{22}$

(B) $E(X) = \frac{1}{2}$

CHAPTER 8

◆ EXERCISE 8-1

1. .997 **3.** (1), $\frac{1}{2}$ **5.** (2), $\frac{7}{10}$ **7.** .4 **9.** .25 **11.** .05 **13.** .2 **15.** .6 **17.** .65 **19.** $\frac{1}{4}$ **21.** $\frac{11}{36}$ **23.** (A) $\frac{3}{5}$; $\frac{5}{3}$ (B) $\frac{1}{3}$; $\frac{3}{1}$ (C) $\frac{2}{3}$; $\frac{3}{2}$ (D) $\frac{11}{9}$; $\frac{9}{11}$
25. (A) $\frac{3}{11}$ (B) $\frac{11}{18}$ (C) $\frac{4}{5}$ or .8 (D) .49 **27.** 1 to 1 **29.** 7 to 1 **31.** 2 to 1 **33.** 1 to 2 **35.** (A) $\frac{1}{8}$ (B) \$8
37. (A) .31; $\frac{31}{69}$ (B) .6; $\frac{3}{2}$ **39.** $\frac{11}{26}$; $\frac{11}{15}$ **41.** $\frac{7}{13}$; $\frac{7}{6}$ **43.** .78 **45.** $\frac{250}{1,000} = .25$ **47.** $P(E) \approx 1 - .13 = .87$

49. $\frac{9}{19} \approx .4737$; $E(X) \approx -\$0.0526$ (The house edge is approximately 5.3%.)   **51.** $P(E) = 1 - \dfrac{12!}{(12 - n)!\,12^n}$

55. (A) $P(C \cup S) = P(C) + P(S) - P(C \cap S) = .45 + .75 - .35 = .85$ (B) $P(C' \cap S') = .15$
57. (A) $P(M_1 \cup A) = P(M_1) + P(A) - P(M_1 \cap A) = .2 + .3 - .05 = .45$
  (B) $P[(M_2 \cap A') \cup (M_3 \cap A')] = P(M_2 \cap A') + P(M_3 \cap A') = .2 + .35 = .55$
59. $P(K' \cap D') = .9$ **61.** .83 **63.** $P(A \cap S) = \frac{50}{1,000} = .05$ **65.** (A) $P(U \cup N) = .22$; $\frac{11}{39}$ (B) $P[(D \cap A) \cup (R \cap A)] = .3$; $\frac{7}{3}$
67. $1 - C_{15,3}/C_{20,3} \approx .6$

◆ EXERCISE 8-2

1. .50 **3.** .20 **5.** .10 **7.** .06 **9.** .50 **11.** .30 **13.** Independent **15.** Dependent **17.** (A) $\frac{1}{2}$ (B) $2(\frac{1}{2})^8 \approx .007\ 81$
19. (A) $\frac{1}{4}$ (B) Dependent **21.** (A) .18 (B) .26 **23.** (A) Yes (B) No **25.** $(\frac{1}{2})(\frac{1}{2}) = \frac{1}{4}$; $\frac{1}{2} + \frac{1}{2} - \frac{1}{4} = \frac{3}{4}$
27. (A) $(\frac{1}{4})(\frac{13}{51}) \approx .0637$ (B) $(\frac{1}{4})(\frac{1}{4}) = .0625$ **29.** (A) $\frac{3}{13}$ (B) Independent **31.** (A) Dependent (B) Independent

33. (A) Start

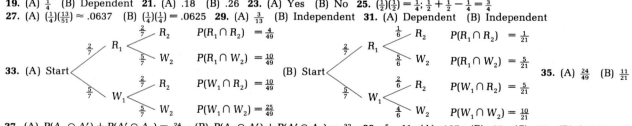

35. (A) $\frac{24}{49}$ (B) $\frac{11}{21}$

37. (A) $P(A_1 \cap A_2') + P(A_1' \cap A_2) = \frac{24}{169}$ (B) $P(A_1 \cap A_2') + P(A_1' \cap A_2) = \frac{32}{221}$ **39.** $\frac{5}{18}$ **41.** (A) .167 (B) .25 (C) .25 (D) \$13.50
43. $P(A|A) = P(A \cap A)/P(A) = P(A)/P(A) = 1$ **45.** $P(A)P(B) \ne 0 = P(A \cap B)$

47. (A)

	H	S	B	Totals
Y	.400	.180	.020	.600
N	.150	.120	.130	.400
Totals	.550	.300	.150	1.000

(B) $P(Y|H) = \dfrac{.400}{.550} \approx .727$ (C) $P(Y|B) = \dfrac{.020}{.150} \approx .133$

(D) $P(S) = .300$; $P(S|Y) = .300$ (E) $P(H) = .550$; $P(H|Y) \approx .667$

(F) $P(B \cap N) = .130$ (G) Yes (H) No (I) No

49. (A) .167 (B) .25 (C) .25 (D) $25,500

51. (A) $P(S|A) = P(S \cap A)/P(A) = \frac{5}{11}$ (B) $P(A|S) = P(A \cap S)/P(S) = \frac{5}{14}$ (C) $P(S|A') = P(S \cap A')/P(A') = \frac{9}{89}$

(D) $P(A|S') = P(A \cap S')/P(S') = \frac{3}{43}$

53. (A)

	A	B	C	Totals
F	.130	.286	.104	.520
F'	.120	.264	.096	.480
Totals	.250	.550	.200	1.000

(B) $P(A|F) = \dfrac{.130}{.520} = .250$; $P(A|F') = \dfrac{.120}{.480} = .250$

(C) $P(C|F) = \dfrac{.104}{.520} = .200$; $P(C|F') = \dfrac{.096}{.480} = .200$

(D) $P(A) = .250$ (E) $P(B) = .550$; $P(B|F') = .550$ (F) $P(F \cap C) = .104$

(G) No; A, B, and C are independent of F and F'.

◆ EXERCISE 8-3

1. $(.6)(.8) = .48$ **3.** $(.6)(.8) + (.4)(.3) = .60$ **5.** .80 **7.** .417 **9.** .375 **11.** .222 **13.** .50 **15.** .278 **17.** .125 **19.** .50 **21.** .375

23.

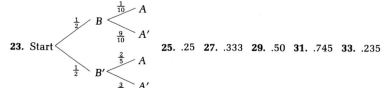

25. .25 **27.** .333 **29.** .50 **31.** .745 **33.** .235

35. $\dfrac{P(U_1 \cap R)}{P(R)} + \dfrac{P(U_1' \cap R)}{P(R)} = \dfrac{P(U_1 \cap R) + P(U_1' \cap R)}{P(R)} = \dfrac{P(R)}{P(R)} = 1$ **37.** .913; .226

39. .091, .545, .364 **41.** .667; .000 412 **43.** .231; .036 **45.** .941; .0588

◆ EXERCISE 8-4   CHAPTER REVIEW

1. (A) .7 (B) .6 **2.** $P(R \cup G) = .8$; odds for $R \cup G$ are 8 to 2 **3.** $\frac{5}{11} \approx .455$ **4.** .27 **5.** .20 **6.** .02 **7.** .03 **8.** .15 **9.** .1304 **10.** .10

11. No, since $P(T|Z) \neq P(T)$ **12.** Yes, since $P(S \cap X) = P(S)P(X)$ **13.** .4 **14.** .2 **15.** .3 **16.** .08 **17.** .18 **18.** .26 **19.** .31 **20.** .43

21. (A) $\frac{2}{13}$; 2 to 11 (B) $\frac{4}{13}$; 4 to 9 (C) $\frac{12}{13}$; 12 to 1 **22.** (A) 1 to 8 (B) $8

23. $A = \{(1, 3), (2, 2), (3, 1), (2, 6), (3, 5), (4, 4), (5, 3), (6, 2), (6, 6)\}$;

$B = \{(1, 5), (2, 4), (3, 3), (4, 2), (5, 1), (6, 6)\}$; $P(A) = \frac{1}{4}$; $P(B) = \frac{1}{6}$; $P(A \cap B) = \frac{1}{36}$; $P(A \cup B) = \frac{7}{18}$

24. (A) .6 (B) $\frac{5}{6}$ **25.** (A) $\frac{1}{13}$ (B) Independent **26.** (A) $\frac{6}{25}$ (B) $\frac{3}{10}$ **27.** Part B **28.** (A) 1.2 (B) 1.2

29. (A) $\frac{3}{5}$ (B) $\frac{1}{3}$ (C) $\frac{7}{15}$ (D) $\frac{9}{14}$ (E) $\frac{5}{8}$ (F) $\frac{3}{10}$ **30.** No **31.** $\frac{93}{200} = .465$ **32.** .564 **33.** $\frac{12}{51} \approx .235$ **34.** $\frac{12}{51} \approx .235$

35. (A) $\frac{1}{4}$; 1 to 3 (B) $3 **36.** $1 - 10!/(5!10^5) \approx .70$ **37.** (A) .9 (B) .3 **38.** (A) .8 (B) .2 (C) .5 **39.** .891

40. $P(A \cap P) = P(A)P(P|A) = .34$ **41.** .955 **42.** $\frac{6}{7} \approx .857$

CHAPTER 9

◆ EXERCISE 9-1

1. (A) 2 (B) 2 (C) Does not exist (D) 4 **3.** (A) 2 (B) 2 (C) 2 (D) Not defined **5.** $c = 0, 1$ **7.** 2, 4 **9.** All x
11. All x, except $x = 5$ **13.** All x, except $x = -2$ and $x = 3$ **15.** (A) 1 (B) 1 (C) 1 **17.** (A) 2 (B) 1 (C) Does not exist
19. (A) 1 (B) 1 (C) 1 **21.** (A) 1 (B) 1 (C) Yes **23.** (A) Does not exist (B) 1 (C) No **25.** (A) 1 (B) 3 (C) No **27.** -2
29. (A)

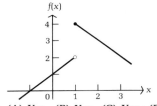

(B) 1
(C) 2
(D) No
(E) All integers

31. 5
33. 7
35. -6
37. 1
39. $(-\infty, \infty)$
41. $[5, \infty)$
43. $(-\infty, \infty)$
45. $(-\infty, 1), (1, 2), (2, \infty)$

47.

x	0.9	0.99	0.999 → 1 ← 1.001	1.01	1.1
f(x)	-1	-1	-1 → ? ← 1	1	1

(A) -1 (B) 1 (C) Does not exist

49.

x	0.9	0.99	0.999 → 1 ← 1.001	1.01	1.1
f(x)	2.71	2.97	2.997 → ? ← 3.003	3.03	3.31

(A) 3 (B) 3 (C) 3

51. Discontinuous at $x = 1$ **53.** Continuous for all x   **55.** Discontinuous at $x = 0$

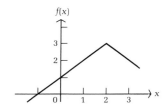

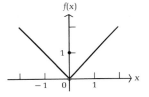

57. (A) Yes (B) Yes (C) Yes (D) Yes **59.** (A) Yes (B) No (C) Yes (D) No (E) Yes **61.** No; no
63. (A)

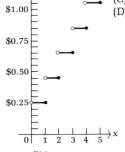

(B) $\lim_{x \to 4.5} P(x) = \$1.05$; $P(4.5) = \$1.05$
(C) $\lim_{x \to 4} P(x)$ does not exist; $P(4) = \$0.85$
(D) Continuous at $x = 4.5$; not continuous at $x = 4$

65. (A)

(B) $\lim_{s \to 10,000} E(s) = \$1,000$; $E(10,000) = \$1,000$
(C) $\lim_{s \to 20,000} E(s)$ does not exist; $E(20,000) = \$2,000$
(D) Yes; no

67. (A) t_2, t_3, t_4, t_6, t_7 (B) $\lim_{t \to t_5} N(t) = 7$; $N(t_5) = 7$ (C) $\lim_{t \to t_3} N(t)$ does not exist; $N(t_3) = 4$

1. −4 **3.** 36 **5.** $\frac{5}{9}$ **7.** $\sqrt{5}$ **9.** $\frac{7}{5}$ **11.** 47 **13.** −4 **15.** $\frac{5}{3}$ **17.** 243 **19.** −3 **21.** 0 **23.** 5 **25.** 2 **27.** 0 **29.** 1 **31.** $\frac{1}{2}$ **33.** 0 **35.** −1
37. −5 **39.** ∞ **41.** 2 **43.** $\frac{3}{5}$ **45.** 4 **47.** $\frac{2}{3}$ **49.** −∞ **51.** ∞ **53.** Does not exist **55.** (A) 0 (B) 0 (C) $y = 0$
57. (A) ∞ (B) ∞ (C) ∞ (D) $x = −1$ **59.** 3 **61.** 4 **63.** 0 **65.** $1/(2\sqrt{2})$ **67.** Does not exist **69.** $\sqrt{3}$ **71.** $\sqrt[3]{4}$ **73.** ∞ **75.** 0 **77.** 0
79. $\frac{1}{4}$ **81.** 12 **83.** (A) −∞ (B) ∞ (C) Does not exist (D) Yes **85.** $2a$ **87.** $1/(2\sqrt{a})$ **89.** (A) \$23 (B) \$3.20 (C) \$5 (D) \$3

91. (A)

COMPOUNDED	n	$A(n)$
Annually	1	\$108.00
Semiannually	2	\$108.16
Quarterly	4	\$108.24
Monthly	12	\$108.30
Weekly	52	\$108.32
Daily	365	\$108.33
Hourly	8,760	\$108.33

(B) \$108.33 **93.** (A) 0.056 (B) 0.07 (C) 0.07 (D) 0
95. (A) 30 (B) 44 (C) 44 (D) 60

1. (A) 3 (B) 3 **3.** (A) $8 + 2h$ (B) 8 **5.** (A) 1 (B) $2 + h$ (C) 2 **7.** $f'(x) = 2; f'(1) = 2, f'(2) = 2, f'(3) = 2$
9. $f'(x) = −2x; f'(1) = −2, f'(2) = −4, f'(3) = −6$ **11.** (A) 5 (B) $3 + h$ (C) 3 (D) $y = 3x − 1$
13. (A) 5 m/sec (B) $3 + h$ m/sec (C) 3 m/sec **15.** $f'(x) = 6 − 2x; f'(1) = 4, f'(2) = 2, f'(3) = 0$
17. $f'(x) = 1/(2\sqrt{x}); f'(1) = \frac{1}{2}, f'(2) = 1/(2\sqrt{2}), f'(3) = 1/(2\sqrt{3})$ **19.** $f'(x) = 1/x^2; f'(1) = 1, f'(2) = \frac{1}{4}, f'(3) = \frac{1}{9}$ **21.** Yes **23.** No **25.** No
27. Yes **29.** (A) $f'(x) = 2x − 4$ (B) −4, 0, 4 (C)

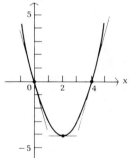

31. $v = f'(x) = 8x − 2$; 6 ft/sec, 22 ft/sec, 38 ft/sec
33. f is nondifferentiable at $x = 1$

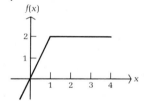

35. No
37. No
39. (A) $2 − 2x$ (B) 2 (C) −2

41. (A) 3 (\$300/board) (B) $4 − h$ (C) $C'(3) = 4$ (\$400/board) (D) $C'(x) = 10 − 2x$
(E) $C'(1) = 8$ (the rate of total cost increase at the level of production of 1 board per day is \$800/board); $C'(2) = 6$ (the rate of total cost increase at the level of production of 2 boards per day is \$600/board); $C'(3) = 4$ (the rate of total cost increase at the level of production of 3 boards per day is \$400/board); $C'(4) = 2$ (the rate of total cost increase at the level of production of 4 boards per day is \$200/board)

1. 0 **3.** 0 **5.** $12x^{11}$ **7.** 1 **9.** $−7x^{-8}$ **11.** $\frac{5}{2}x^{3/2}$ **13.** $−5x^{-6}$ **15.** $8x^3$ **17.** $2x^5$ **19.** $x^4/3$ **21.** $−10x^{-6}$ **23.** $−16x^{-5}$ **25.** x^{-3}
27. $−x^{-2/3}$ **29.** $4x − 3$ **31.** $15x^4 − 6x^2$ **33.** $−12x^{-5} − 4x^{-3}$ **35.** $−\frac{1}{2}x^{-2} + 2x^{-4}$ **37.** $2x^{-1/3} − \frac{5}{3}x^{-2/3}$ **39.** $−\frac{9}{5}x^{-8/5} + 3x^{-3/2}$
41. $−\frac{1}{3}x^{-4/3}$ **43.** $−6x^{-3/2} + 6x^{-3} + 1$
45. (A) $f'(x) = 6 − 2x$ (B) $f'(2) = 2; f'(4) = −2$ (C) $y = 2x + 4; y = −2x + 16$ (D) $x = 3$
47. (A) $f'(x) = 12x^3 − 12x$ (B) $f'(2) = 72; f'(4) = 720$ (C) $y = 72x − 127; y = 720x − 2,215$ (D) $x = −1, 0, 1$
49. (A) $v = f'(x) = 176 − 32x$ (B) $f'(0) = 176$ ft/sec; $f'(3) = 80$ ft/sec (C) $x = 5.5$ sec
51. (A) $v = f'(x) = 3x^2 − 18x + 15$ (B) $f'(0) = 15$ ft/sec; $f'(3) = −12$ ft/sec (C) $x = 1, 5$ sec **53.** $−20x^{-2}$ **55.** $2x − 3 − 10x^{-3}$

59. (A) $C'(x) = 60 - (x/2)$

(B) $C'(60) = \$30/$racket (at a production level of 60 rackets, the rate of change of total cost relative to production is $30 per racket; thus, the cost of producing 1 more racket at this level of production is approx. $30).

(C) $29.75 (the marginal cost of $30 per racket found in part B is a close approximation to this value)

(D) $C'(80) = \$20$ per racket (at a production level of 80 rackets, the rate of change of total cost relative to production is $20 per racket; thus, the cost of producing 1 more racket at this level of production is approx. $20)

61. (A) $N'(x) = 60 - 2x$

(B) $N'(10) = 40$ (at the $10,000 level of advertising, there would be an approximate increase of 40 units of sales per $1,000 increase in advertising); $N'(20) = 20$ (at the $20,000 level of advertising, there would be an approximate increase of only 20 units of sales per $1,000 increase in advertising); the effect of advertising levels off as the amount spent increases

63. (A) -1.37 beats/min   (B) -0.58 beat/min   **65.** (A) 25 items/hr   (B) 8.33 items/hr

---

◆ EXERCISE 9-5

1. $2x^3(2x) + (x^2 - 2)(6x^2) = 10x^4 - 12x^2$ **3.** $(x - 3)(2) + (2x - 1)(1) = 4x - 7$ **5.** $\dfrac{(x - 3)(1) - x(1)}{(x - 3)^2} = \dfrac{-3}{(x - 3)^2}$

7. $\dfrac{(x - 2)(2) - (2x + 3)(1)}{(x - 2)^2} = \dfrac{-7}{(x - 2)^2}$ **9.** $(x^2 + 1)(2) + (2x - 3)(2x) = 6x^2 - 6x + 2$

11. $\dfrac{(2x - 3)(2x) - (x^2 + 1)(2)}{(2x - 3)^2} = \dfrac{2x^2 - 6x - 2}{(2x - 3)^2}$ **13.** $(x^2 + 2)2x + (x^2 - 3)2x = 4x^3 - 2x$

15. $\dfrac{(x^2 - 3)2x - (x^2 + 2)2x}{(x^2 - 3)^2} = \dfrac{-10x}{(x^2 - 3)^2}$ **17.** $(2x + 1)(2x - 3) + (x^2 - 3x)(2) = 6x^2 - 10x - 3$

19. $(2x - x^2)(5) + (5x + 2)(2 - 2x) = -15x^2 + 16x + 4$ **21.** $\dfrac{(x^2 + 2x)(5) - (5x - 3)(2x + 2)}{(x^2 + 2x)^2} = \dfrac{-5x^2 + 6x + 6}{(x^2 + 2x)^2}$

23. $\dfrac{(x^2 - 1)(2x - 3) - (x^2 - 3x + 1)(2x)}{(x^2 - 1)^2} = \dfrac{3x^2 - 4x + 3}{(x^2 - 1)^2}$ **25.** $f'(x) = (1 + 3x)(-2) + (5 - 2x)(3); \; y = -11x + 29$

27. $f'(x) = \dfrac{(3x - 4)(1) - (x - 8)(3)}{(3x - 4)^2}; \; y = 5x - 13$ **29.** $f'(x) = (2x - 15)(2x) + (x^2 + 18)(2) = 6(x - 2)(x - 3); \; x = 2, \, x = 3$

31. $f'(x) = \dfrac{(x^2 + 1)(1) - x(2x)}{(x^2 + 1)^2} = \dfrac{1 - x^2}{(x^2 + 1)^2}; \; x = -1, \, x = 1$ **33.** $7x^6 - 3x^2$ **35.** $-27x^{-4}$

37. $(2x^4 - 3x^3 + x)(2x - 1) + (x^2 - x + 5)(8x^3 - 9x^2 + 1)$ **39.** $\dfrac{(4x^2 + 5x - 1)(6x - 2) - (3x^2 - 2x + 3)(8x + 5)}{(4x^2 + 5x - 1)^2}$

41. $9x^{1/3}(3x^2) + (x^3 + 5)(3x^{-2/3})$ **43.** $\dfrac{(x^2 - 3)(2x^{-2/3}) - 6x^{1/3}(2x)}{(x^2 - 3)^2}$ **45.** $x^{-2/3}(3x^2 - 4x) + (x^3 - 2x^2)(-\frac{2}{3}x^{-5/3})$

47. $\dfrac{(x^2 + 1)[(2x^2 - 1)(2x) + (x^2 + 3)(4x)] - (2x^2 - 1)(x^2 + 3)(2x)}{(x^2 + 1)^2}$

49. (A) $S'(t) = \dfrac{7{,}200 - 200t^2}{(t^2 + 36)^2}$

(B) $S(2) = 10; \; S'(2) = 4;$ at $t = 2$ months, monthly sales are 10,000 and increasing at 4,000 albums per month

(C) $S(8) = 16; \; S'(8) = -0.56;$ at $t = 8$ months, monthly sales are 16,000 and decreasing at 560 albums per month

51. (A) $d'(x) = \dfrac{-50{,}000(2x + 10)}{(x^2 + 10x + 25)^2} = \dfrac{-100{,}000}{(x + 5)^3}$

(B) $d'(5) = -100$ radios per $1 increase in price; $d'(10) = -30$ radios per $1 increase in price

53. (A) $C'(t) = \dfrac{0.14 - 0.14t^2}{(t^2 + 1)^2}$

(B) $C'(0.5) = 0.0672$ (concentration is increasing at 0.0672 unit/hr); $C'(3) = -0.0112$ (concentration is decreasing at 0.0112 unit/hr)

55. (A) $N'(x) = \dfrac{(x+32)(100)-(100x+200)}{(x+32)^2} = \dfrac{3{,}000}{(x+32)^2}$ (B) $N'(4) = 2.31$; $N'(68) = 0.30$

◆ EXERCISE 9-6

1. $6(2x+5)^2$ **3.** $-8(5-2x)^3$ **5.** $30x(3x^2+5)^4$ **7.** $8(x^3-2x^2+2)^7(3x^2-4x)$ **9.** $(2x-5)^{-1/2}$ **11.** $-8x^3(x^4+1)^{-3}$
13. $f'(x) = 6(2x-1)^2$; $y = 6x-5$; $x = \frac{1}{2}$ **15.** $f'(x) = 2(4x-3)^{-1/2}$; $y = \frac{2}{3}x+1$; none **17.** $24x(x^2-2)^3$ **19.** $-6(x^2+3x)^{-4}(2x+3)$
21. $x(x^2+8)^{-1/2}$ **23.** $(3x+4)^{-2/3}$ **25.** $\frac{1}{2}(x^2-4x+2)^{-1/2}(2x-4) = (x-2)/(x^2-4x+2)^{1/2}$ **27.** $(-1)(2x+4)^{-2}(2) = -2/(2x+4)^2$

29. $-15x^2(x^3+4)^{-6}$ **31.** $(-1)(4x^2-4x+1)^{-2}(8x-4) = -4/(2x-1)^3$ **33.** $-2(x^2-3x)^{-3/2}(2x-3) = \dfrac{-2(2x-3)}{(x^2-3x)^{3/2}}$

35. $f'(x) = (4-x)^3 - 3x(4-x)^2$; $y = -16x+48$ **37.** $f'(x) = \dfrac{(2x-5)^3 - 6x(2x-5)^2}{(2x-5)^6}$; $y = -17x+54$

39. $f'(x) = (2x+2)^{1/2} + x(2x+2)^{-1/2}$; $y = \frac{5}{2}x - \frac{1}{2}$
41. $f'(x) = 2x(x-5)^3 + 3x^2(x-5)^2 = 5x(x-5)^2(x-2)$; $x = 0$, $x = 2$, $x = 5$

43. $f'(x) = \dfrac{(2x+5)^2 - 4x(2x+5)}{(2x+5)^4} = \dfrac{5-2x}{(2x+5)^3}$; $x = \frac{5}{2}$ **45.** $f'(x) = \dfrac{x-4}{\sqrt{x^2-8x+20}}$; $x = 4$

47. $18x^2(x^2+1)^2 + 3(x^2+1)^3 = 3(x^2+1)^2(7x^2+1)$ **49.** $\dfrac{2x^3 4(x^3-7)^3 3x^2 - (x^3-7)^4 6x^2}{4x^6} = \dfrac{3(x^3-7)^3(3x^3+7)}{2x^4}$

51. $(2x-3)^2[3(2x^2+1)^2(4x)] + (2x^2+1)^3[2(2x-3)(2)] = 4(2x^2+1)^2(2x-3)(8x^2-9x+1)$

53. $4x^2[\frac{1}{2}(x^2-1)^{-1/2}(2x)] + (x^2-1)^{1/2}(8x) = \dfrac{12x^3-8x}{\sqrt{x^2-1}}$ **55.** $\dfrac{(x-3)^{1/2}(2) - 2x[\frac{1}{2}(x-3)^{-1/2}]}{x-3} = \dfrac{x-6}{(x-3)^{3/2}}$

57. $(2x-1)^{1/2}(x^2+3)(11x^2-4x+9)$
59. (A) $C'(x) = (2x+16)^{-1/2} = 1/\sqrt{2x+16}$
(B) $C'(24) = \frac{1}{8}$ or \$12.50 per calculator (the rate of change of total cost relative to production at a production level of 24 calculators is \$12.50 per calculator; the cost of producing 1 more calculator at this level of production is approx. \$12.50); $C'(42) = \frac{1}{10}$ or \$10.00 per calculator (the rate of change of total cost relative to production at a production level of 42 calculators is \$10.00 per calculator; the cost of producing 1 more calculator at this level of production is approx. \$10.00)

61. $4{,}000(1 + \frac{1}{12}i)^{47}$ **63.** $\dfrac{(4 \times 10^6)x}{(x^2-1)^{5/3}}$

65. (A) $f'(n) = n(n-2)^{-1/2} + 2(n-2)^{1/2} = \dfrac{3n-4}{(n-2)^{1/2}}$
(B) $f'(11) = \frac{29}{3}$ (rate of learning is $\frac{29}{3}$ units/min at the $n = 11$ level); $f'(27) = \frac{77}{5}$ (rate of learning is $\frac{77}{5}$ units/min at the $n = 27$ level)

◆ EXERCISE 9-7

1. (A) \$29.50 (B) \$30
3. (A) \$420
(B) $\overline{C}'(500) = -0.24$; at a production level of 500 units, a unit increase in production will decrease average cost by approx. 24¢
5. (A) $R'(1{,}600) = 20$; at a production level of 1,600 units, a unit increase in production will increase revenue by approx. \$20
(B) $R'(2{,}500) = -25$; at a production level of 2,500 units, a unit increase in production will decrease revenue by approx. \$25
7. (A) \$4.50 (B) \$5
9. (A) $P'(450) = 0.5$; at a production level of 450 units, a unit increase in production will increase profit by approx. 50¢
(B) $P'(750) = -2.5$; at a production level of 750 units, a unit increase in production will decrease profit by approx. \$2.50
11. (A) \$1.25
(B) $\overline{P}'(150) = 0.015$; at a production level of 150 units, a unit increase in production will increase average profit by approx. 1.5¢
13. (A) $C'(x) = 60$ (B) $R(x) = 200x - (x^2/30)$ (C) $R'(x) = 200 - (x/15)$
(D) $R'(1{,}500) = 100$ (at a production level of 1,500 units, a unit increase in production will increase revenue by approx. \$100); $R'(4{,}500) = -100$ (at a production level of 4,500 units, a unit increase in production will decrease revenue by approx. \$100)

13. (E) Break-even points: (600, 108,000) and (3,600, 288,000)

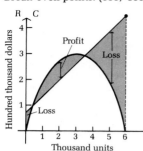

(F) $P(x) = -(x^2/30) + 140x - 72,000$
(G) $P'(x) = -(x/15) + 140$
(H) $P'(1,500) = 40$ (at a production level of 1,500 units, a unit increase in production will increase profit by approx. $40); $P'(3,000) = -60$ (at a production level of 3,000 units, a unit increase in production will decrease profit by approx. $60)

15. (A) $p = 20 - (x/50)$ (B) $R(x) = 20x - (x^2/50)$ (C) $C(x) = 4x + 1,400$
(D) Break-even points: (100, 1,800) and (700, 4,200)

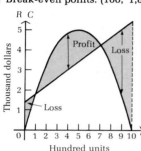

(E) $P(x) = 16x - (x^2/50) - 1,400$
(F) $P'(250) = 6$ (at a production level of 250 units, a unit increase in production will increase profit by approx. $6); $P'(475) = -3$ (at a production level of 475 units, a unit increase in production will decrease profit by approx. $3

◆ EXERCISE 9-8 CHAPTER REVIEW

1. $12x^3 - 4x$ **2.** $x^{-1/2} - 3 = \dfrac{1}{x^{1/2}} - 3$ **3.** 0 **4.** $-x^{-3} + x$ **5.** $(2x - 1)(3) + (3x + 2)(2) = 12x + 1$

6. $(x^2 - 1)(3x^2) + (x^3 - 3)(2x) = 5x^4 - 3x^2 - 6x$ **7.** $\dfrac{(x^2 + 2)2 - 2x(2x)}{(x^2 + 2)^2} = \dfrac{4 - 2x^2}{(x^2 + 2)^2}$ **8.** $(-1)(3x + 2)^{-2}(3) = \dfrac{-3}{(3x + 2)^2}$

9. $3(2x - 3)^2(2) = 6(2x - 3)^2$ **10.** $-2(x^2 + 2)^{-3}(2x) = \dfrac{-4x}{(x^2 + 2)^3}$ **11.** (A) Does not exist (B) 3 (C) No

12. (A) 2 (B) Not defined (C) No **13.** (A) 1 (B) 1 (C) Yes **14.** $12x^3 + 6x^{-4}$

15. $(2x^2 - 3x + 2)(2x + 2) + (x^2 + 2x - 1)(4x - 3) = 8x^3 + 3x^2 - 12x + 7$ **16.** $\dfrac{(x - 1)^2(2) - (2x - 3)(2)(x - 1)}{(x - 1)^4} = \dfrac{4 - 2x}{(x - 1)^3}$

17. $x^{-1/2} - 2x^{-3/2} = \dfrac{1}{\sqrt{x}} - \dfrac{2}{\sqrt{x^3}}$ **18.** $(x^2 - 1)[2(2x + 1)(2)] + (2x + 1)^2(2x) = 2(2x + 1)(4x^2 + x - 2)$

19. $\dfrac{1}{3}(x^3 - 5)^{-2/3}(3x^2) = \dfrac{x^2}{\sqrt[3]{(x^3 - 5)^2}}$ **20.** $-8x^{-3}$ **21.** $\dfrac{(2x - 3)(4)(x^2 + 2)^3(2x) - (x^2 + 2)^4(2)}{(2x - 3)^2} = \dfrac{2(x^2 + 2)^3(7x^2 - 12x - 2)}{(2x - 3)^2}$

22. (A) $m = f'(1) = 2$ (B) $y = 2x + 3$ **23.** (A) $m = f'(1) = 16$ (B) $y = 16x - 12$ **24.** $x = 5$ **25.** $x = -5, x = 3$
26. $x = -2, x = 2$ **27.** $x = 0, x = 3, x = \frac{15}{2}$ **28.** (A) $v = f'(x) = 32x - 4$ (B) $f'(3) = 92$ ft/sec
29. (A) $v = f'(x) = 96 - 32x$ (B) $x = 3$ sec **30.** (A) 4 (B) 6 (C) Does not exist (D) 6 (E) No
31. (A) 3 (B) 3 (C) 3 (D) 3 (E) Yes **32.** (A) ∞ (B) ∞ (C) ∞ **33.** (A) $-\infty$ (B) ∞ (C) Does not exist **34.** $(-\infty, \infty)$

35. $(-\infty, -5), (-5, \infty)$ **36.** $(-\infty, -2), (-2, 3), (3, \infty)$ **37.** $[3, \infty)$ **38.** $(-\infty, \infty)$ **39.** $\dfrac{2(3) - 3}{3 + 5} = \dfrac{3}{8}$ **40.** $2(3^2) - 3 + 1 = 16$ **41.** -1 **42.** 4

43. 4 **44.** $\frac{1}{6}$ **45.** Does not exist **46.** $1/(2\sqrt{7})$ **47.** $\sqrt{2}$ **48.** 3 **49.** ∞ **50.** $\frac{2}{3}$ **51.** 0 **52.** ∞ **53.** $2x - 1$ **54.** $1/(2\sqrt{x})$ **55.** No **56.** No
57. No **58.** Yes

59. Discontinuous at $x = 0$ **60.** Continuous for all x **61.** $7x(x - 4)^3(x + 3)^2$

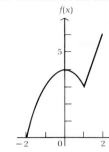

62. $\dfrac{x^4(2x + 5)}{(2x + 1)^5}$

63. $\dfrac{1}{x^2\sqrt{x^2 - 1}}$

64. $\dfrac{4}{(x^2 + 4)^{3/2}}$

65. (A) 1 (B) 1 (C) 1 (D) Yes
66. (A) Yes (B) Yes (C) Yes (D) Yes
67. (A) 1 (B) -1 (C) Does not exist (D) No
68. (A) \$179.90 (B) \$180

69. (A) $C'(x) = 2$; $\overline{C}(x) = 2 + 56x^{-1}$; $\overline{C}'(x) = -56x^{-2}$
(B) $R(x) = xp = 20x - x^2$; $R'(x) = 20 - 2x$; $\overline{R}(x) = 20 - x$; $\overline{R}'(x) = -1$
(C) $P(x) = R(x) - C(x) = 18x - x^2 - 56$; $P'(x) = 18 - 2x$; $\overline{P}(x) = 18 - x - 56x^{-1}$; $\overline{P}'(x) = -1 + 56x^{-2}$
(D) Solving $R(x) = C(x)$, we find break-even points at $x = 4, 14$.
(E) $P'(7) = 4$ (increasing production increases profit); $\overline{P}'(9) = 0$ (stable); $P'(11) = -4$ (increasing production decreases profit)
(F)

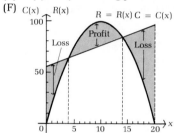

70. (A) 2 components/day (B) 3.2 components/day (C) 40 components/day
71. $C'(9) = -1$ ppm/m; $C'(99) = -0.001$ ppm/m **72.** (A) 10 items/hr (B) 5 items/hr

C H A P T E R   10

◆ EXERCISE 10-1

1. (a, b); (d, f); (g, h); **3.** c, d, f **5.** b, f **7.** $-3 < x < 4$; $(-3, 4)$ **9.** $x < 3$ or $x > 7$; $(-\infty, 3) \cup (7, \infty)$
11. $-5 < x < 0$ or $x > 3$; $(-5, 0) \cup (3, \infty)$ **13.** Decreasing on $(-\infty, 8)$; increasing on $(8, \infty)$; local minimum at $x = 8$
15. Increasing on $(-\infty, 5)$; decreasing on $(5, \infty)$; local maximum at $x = 5$ **17.** Increasing for all x; no local extrema
19. Decreasing for all x; no local extrema
21. Increasing on $(-\infty, -2)$ and $(2, \infty)$; decreasing on $(-2, 2)$; local maximum at $x = -2$; local minimum at $x = 2$
23. Increasing on $(-\infty, -2)$ and $(4, \infty)$; decreasing on $(-2, 4)$; local maximum at $x = -2$; local minimum at $x = 4$
25. Increasing on $(-\infty, -1)$ and $(0, 1)$; decreasing on $(-1, 0)$ and $(1, \infty)$; local maxima at $x = -1$ and $x = 1$; local minimum at $x = 0$

27. Increasing on $(-\infty, 4)$
Decreasing on $(4, \infty)$
Horizontal tangent at $x = 4$

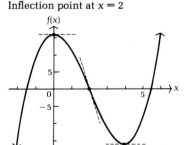

29. Increasing on $(-\infty, -1)$, $(1, \infty)$
Decreasing on $(-1, 1)$
Horizontal tangents at $x = -1, 1$

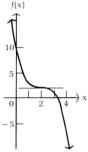

31. Decreasing for all x
Horizontal tangent at $x = 2$

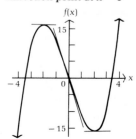

33. No critical values; increasing on $(-\infty, -2)$ and $(-2, \infty)$ no local extrema
35. Critical values: $x = -2$, $x = 2$; increasing on $(-\infty, -2)$ and $(2, \infty)$; decreasing on $(-2, 0)$ and $(0, 2)$; local maximum at $x = -2$; local minimum at $x = 2$
37. Critical value: $x = -2$; increasing on $(-2, 0)$; decreasing on $(-\infty, -2)$ and $(0, \infty)$; local minimum at $x = -2$
39. Critical values: $x = 0$, $x = 4$; increasing on $(-\infty, 0)$ and $(4, \infty)$; decreasing on $(0, 2)$ and $(2, 4)$; local maximum at $x = 0$; local minimum at $x = 4$
41. Critical values: $x = 0$, $x = 4$, $x = 6$; increasing on $(0, 4)$ and $(6, \infty)$; decreasing on $(-\infty, 0)$ and $(4, 6)$; local maximum at $x = 4$; local minima at $x = 0$ and $x = 6$
43. Critical value: $x = 2$; increasing on $(2, \infty)$; decreasing on $(-\infty, 2)$; local minimum at $x = 2$
45. Critical value: $x = 1$; increasing on $(0, 1)$; decreasing on $(1, \infty)$; local maximum at $x = 1$
47. Critical value: $x = 80$; decreasing for $0 < x < 80$; increasing for $80 < x < 150$; local minimum at $x = 80$
49. $P(x)$ is increasing over (a, b) if $P'(x) = R'(x) - C'(x) > 0$ over (a, b); that is, if $R'(x) > C'(x)$ over (a, b).
51. Critical value: $t = 1$; increasing for $0 < t < 1$; decreasing for $1 < t < 24$; local maximum at $t = 1$
53. Critical value: $t = 7$; increasing for $0 < t < 7$; decreasing for $7 < t < 24$; local maximum at $t = 7$

◆ EXERCISE 10-2

1. (a, c), (c, d), (e, g) **3.** d, e, g **5.** $6x - 4$ **7.** $40x^3$ **9.** $6x$ **11.** $24x^2(x^2 - 1) + 6(x^2 - 1)^2 = 6(x^2 - 1)(5x^2 - 1)$ **13.** $6x^{-3} + 12x^{-4}$
15. $f(2) = -2$ is a local minimum **17.** $f(-1) = 2$ is a local maximum; $f(2) = -25$ is a local minimum **19.** No local extrema
21. $f(-2) = -6$ is a local minimum; $f(0) = 10$ is a local maximum; $f(2) = -6$ is a local minimum **23.** $f(0) = 2$ is a local minimum
25. $f(-4) = -8$ is a local maximum; $f(4) = 8$ is a local minimum **27.** Concave upward for all x; no inflection points
29. Concave upward on $(6, \infty)$; concave downward on $(-\infty, 6)$; inflection point at $x = 6$
31. Concave upward on $(-\infty, -2)$ and $(2, \infty)$; concave downward on $(-2, 2)$; inflection points at $x = -2$ and $x = 2$
33. Concave upward on $(0, 2)$; concave downward on $(-\infty, 0)$ and $(2, \infty)$; inflection points at $x = 0$ and $x = 2$
35. Local maximum at $x = 0$
Local minimum at $x = 4$
Inflection point at $x = 2$

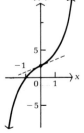

37. Inflection point at $x = 0$

39. Inflection point at $x = 2$

41. Local maximum at $x = -2$
Local minimum at $x = 2$
Inflection point at $x = 0$

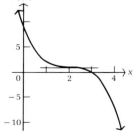

43. Inflection points at $x = -2$ and $x = 2$ **45.** Inflection points at $x = -6$, $x = 0$, and $x = 6$

47. (A) Local maximum at $x = 60$ (B) Concave downward on the whole interval $(0, 80)$

49. (A) Increasing on $(10, 25)$; decreasing on $(25, 40)$

     (B) Inflection point at $x = 25$

     (C)

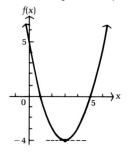

     (D) Max $N'(x) = N'(25) = 2{,}025$

51. (A) Increasing on $(0, 10)$; decreasing on $(10, 20)$

     (B) Inflection point at $t = 10$

     (C)

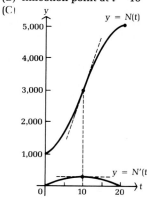

     (D) $N'(10) = 300$

53. (A) Increasing on $(5, \infty)$; decreasing on $(0, 5)$

     (B) Inflection point at $n = 5$

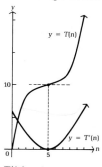

     (C) $T'(5) = 0$

◆ EXERCISE 10-3

1. (b, d), $(d, 0)$, (g, ∞) **3.** $x = 0$ **5.** (a, d), (e, h) **7.** $x = a$, $x = h$ **9.** $x = d$, $x = e$

11. Horizontal asymptote: $y = 2$; vertical asymptote: $x = -2$

13. Horizontal asymptote: $y = 1$; vertical asymptotes: $x = -1$ and $x = 1$ **15.** No horizontal or vertical asymptotes

17. Horizontal asymptote: $y = 0$; no vertical asymptotes   **19.** No horizontal asymptote; vertical asymptote: $x = 3$

21. Domain: All real numbers

     y intercept: 5

     x intercepts: 1, 5

     Decreasing on $(-\infty, 3)$

     Increasing on $(3, \infty)$

     Local minimum at $x = 3$

     Concave upward on $(-\infty, \infty)$

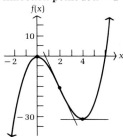

23. Domain: All real numbers

     y intercept: 0

     x intercepts: 0, 6

     Increasing on $(-\infty, 0)$ and $(4, \infty)$

     Decreasing on $(0, 4)$

     Local maximum at $x = 0$

     Local minimum at $x = 4$

     Concave upward on $(2, \infty)$

     Concave downward on $(-\infty, 2)$

     Inflection point at $x = 2$

25. Domain: All real numbers

     y intercept: 16

     x intercepts: -4, 2

     Increasing on $(-\infty, -2)$ and $(2, \infty)$

     Decreasing on $(-2, 2)$

     Local maximum at $x = -2$

     Local minimum at $x = 2$

     Concave upward on $(0, \infty)$

     Concave downward on $(-\infty, 0)$

     Inflection point at $x = 0$

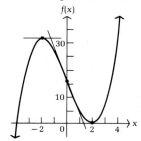

27. Domain: All real numbers
y intercept: 0

29. Domain: All real numbers except 3
y intercept: −1

31. Domain: All real numbers except 2
y intercept: 0

53. (A)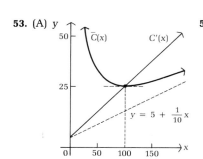

(B) 25 at x = 100

55.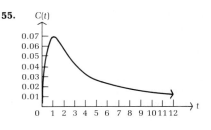

57.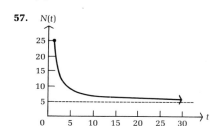

◆ EXERCISE 10-4

1. Min $f(x) = f(2) = 1$; no maximum   **3.** Max $f(x) = f(4) = 26$; no minimum   **5.** No absolute extrema exist.   **7.** Max $f(x) = f(2) = 16$
9. Min $f(x) = f(2) = 14$
11. (A) Max $f(x) = f(5) = 14$; Min $f(x) = f(-1) = -22$ (B) Max $f(x) = f(1) = -2$; Min $f(x) = f(-1) = -22$
(C) Max $f(x) = f(5) = 14$; Min $f(x) = f(3) = -6$
13. (A) Max $f(x) = f(0) = 126$; Min $f(x) = f(2) = -26$ (B) Max $f(x) = f(7) = 49$; Min $f(x) = f(2) = -26$
(C) Max $f(x) = f(6) = 6$; Min $f(x) = f(3) = -15$
15. Exactly in half   **17.** 15 and −15   **19.** A square of side 25 cm; maximum area = 625 cm²   **21.** 3,000 pairs; $3.00/pair
23. $35; $6,125   **25.** 40 trees; 1,600 lb   **27.** $(10 - 2\sqrt{7})/3 = 1.57$ in. squares
29. 20 ft by 40 ft (with the expensive side being one of the short sides)   **31.** 10,000 books in 5 printings
33. (A) x = 5.1 mi   (B) x = 10 mi   **35.** 4 days; 20 bacteria/cm³   **37.** 50 mice per order   **39.** 1 month; 2 ft   **41.** 4 yr from now

◆ EXERCISE 10-5

1. $1,221.40; $1,648.72; $2,225.54   **3.** 11.55   **5.** 10.99   **7.** 0.14

9.

n	$[1 + (1/n)]^n$
10	2.593 74
100	2.704 81
1,000	2.716 92
10,000	2.718 15
100,000	2.718 27
1,000,000	2.718 28
10,000,000	2.718 28
↓	↓
∞	e = 2.718 281 828 459 . . .

11. $55,463.90   **13.** $9,931.71   **15.** $r = \frac{1}{4} \ln 1.5 \approx 0.1014$ or 10.14%

17. (A)

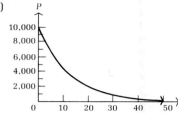

(B) $\lim_{t \to \infty} 10{,}000e^{-0.08t} = 0$

19. 2.77 yr **21.** 13.86% **23.** $A = Pe^{rt}$; $2P = Pe^{rt}$; $e^{rt} = 2$; $\ln e^{rt} = \ln 2$; $rt = \ln 2$; $t = (\ln 2)/r$ **25.** 34.66 yr **27.** 3.47%
29. $t = -(\ln 0.5)/0.000\ 433\ 2 \approx 1{,}600$ yr **31.** $r = (\ln 0.5)/30 \approx -0.0231$ **33.** Approx. 521 yr

◆ EXERCISE 10-6

1. $6e^x - \dfrac{7}{x}$ **3.** $2exe^{-1} + 3e^x$ **5.** $\dfrac{5}{x}$ **7.** $\dfrac{2 \ln x}{x}$ **9.** $x^3 + 4x^3 \ln x = x^3(1 + 4 \ln x)$ **11.** $x^3e^x + 3x^2e^x = x^2e^x(x + 3)$

13. $\dfrac{(x^2 + 9)e^x - 2xe^x}{(x^2 + 9)^2} = \dfrac{e^x(x^2 - 2x + 9)}{(x^2 + 9)^2}$ **15.** $\dfrac{x^3 - 4x^3 \ln x}{x^8} = \dfrac{1 - 4 \ln x}{x^5}$ **17.** $3(x + 2)^2 \ln x + \dfrac{(x + 2)^3}{x} = (x + 2)^2 \left(3 \ln x + \dfrac{x + 2}{x} \right)$

19. $(x + 1)^3e^x + 3(x + 1)^2e^x = (x + 1)^2e^x(x + 4)$ **21.** $\dfrac{2xe^x - (x^2 + 1)e^x}{(e^x)^2} = \dfrac{2x - x^2 - 1}{e^x}$ **23.** $(\ln x)^3 + 3(\ln x)^2 = (\ln x)^2(\ln x + 3)$

25. $-15e^x(4 - 5e^x)^2$ **27.** $\dfrac{1}{2x\sqrt{1 + \ln x}}$ **29.** xe^x **31.** $4x \ln x$ **33.** $y = ex$ **35.** $y = \dfrac{1}{e}\,x$

37. Max $f(x) = f(e^3) = e^3 \approx 20.086$ **39.** Min $f(x) = f(1) = e \approx 2.718$ **41.** Max $f(x) = f(e^{1/2}) = 2e^{-1/2} \approx 1.213$

43. Domain: All real numbers
y intercept: 0
x intercept: 0
Horizontal asymptote: $y = 1$
Decreasing on $(-\infty, \infty)$
Concave downward on $(-\infty, \infty)$

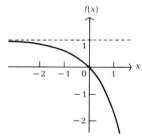

45. Domain: $(0, \infty)$
Vertical asymptote: $x = 0$
Increasing on $(1, \infty)$
Decreasing on $(0, 1)$
Local minimum at $x = 1$
Concave upward on $(0, \infty)$

47. Domain: All real numbers
y intercept: 3
x intercept: 3
Horizontal asymptote: $y = 0$
Increasing on $(-\infty, 2)$
Decreasing on $(2, \infty)$
Local maximum at $x = 2$
Concave upward on $(-\infty, 1)$
Concave downward on $(1, \infty)$
Inflection point at $x = 1$

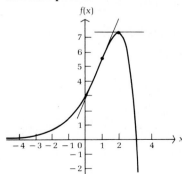

49. Domain: $(0, \infty)$
x intercept: 1
Increasing on $(e^{-1/2}, \infty)$
Decreasing on $(0, e^{-1/2})$
Local minimum at $x = e^{-1/2}$
Concave upward on $(e^{-3/2}, \infty)$
Concave downward on $(0, e^{-3/2})$
Inflection point at $x = e^{-3/2}$

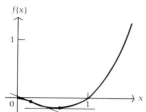

51. $p = \$2$ **53.** Min $\overline{C}(x) = \overline{C}(e^7) \approx \$99.91$ **55.** (A) At \$3.68 each, the maximum revenue will be \$3,680/wk (in the test city).

(B)

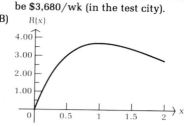

57. At the 40 lb weight level, blood pressure would increase at the rate of 0.44 mm of mercury/lb of weight gain. At the 90 lb weight level, blood pressure would increase at the rate of 0.19 mm of mercury/lb of weight gain.

59. (A) After 1 hr, the concentration is decreasing at the rate of 1.60 mg/ml/hr; after 4 hr, the concentration is decreasing at the rate of 0.08 mg/ml/hr.

(B)

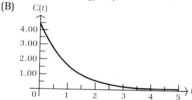

61. $dR/dS = k/S$

◆ EXERCISE 10-7

1. $y = u^3$; $u = 2x + 5$ **3.** $y = \ln u$; $u = 2x^2 + 7$ **5.** $y = e^u$; $u = x^2 - 2$ **7.** $y = (2 + e^x)^2$; $dy/dx = 2e^x(2 + e^x)$

9. $y = e^{2-x^4}$; $dy/dx = -4x^3 e^{2-x^4}$ **11.** $y = \ln(4x^5 - 7)$; $\dfrac{dy}{dx} = \dfrac{20x^4}{4x^5 - 7}$ **13.** $\dfrac{1}{x-3}$ **15.** $\dfrac{-2}{3-2t}$ **17.** $6e^{2x}$ **19.** $-8e^{-4t}$ **21.** $-3e^{-0.03x}$

23. $\dfrac{4}{x+1}$ **25.** $4e^{2x} - 3e^x$ **27.** $(6x - 2)e^{3x^2 - 2x}$ **29.** $\dfrac{2t+3}{t^2 + 3t}$ **31.** $\dfrac{x}{x^2 + 1}$

33. $\dfrac{4[\ln(t^2 + 1)]^3(2t)}{t^2 + 1} = \dfrac{8t[\ln(t^2 + 1)]^3}{t^2 + 1}$ **35.** $4(e^{2x} - 1)^3(2e^{2x}) = 8e^{2x}(e^{2x} - 1)^3$ **37.** $\dfrac{(x^2 + 1)(2e^{2x}) - e^{2x}(2x)}{(x^2 + 1)^2} = \dfrac{2e^{2x}(x^2 - x + 1)}{(x^2 + 1)^2}$

39. $(x^2 + 1)(-e^{-x}) + e^{-x}(2x) = e^{-x}(2x - x^2 - 1)$ **41.** $\dfrac{e^{-x}}{x} - e^{-x} \ln x$ **43.** $\dfrac{-2x}{(1 + x^2)[\ln(1 + x^2)]^2}$ **45.** $\dfrac{-2x}{3(1 - x^2)[\ln(1 - x^2)]^{2/3}}$

47. Domain: $(-\infty, \infty)$
   y intercept: 0
   x intercept: 0
   Horizontal asymptote: $y = 1$
   Increasing on $(-\infty, \infty)$
   Concave downward on $(-\infty, \infty)$

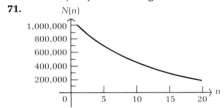

49. Domain: $(-\infty, 1)$
   y intercept: 0
   x intercept: 0
   Vertical asymptote: $x = 1$
   Decreasing on $(-\infty, 1)$
   Concave downward on $(-\infty, 1)$

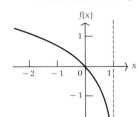

51. Domain: $(-\infty, \infty)$
   y intercept: 1
   Horizontal asymptote: $y = 0$
   Increasing on $(-\infty, 0)$
   Decreasing on $(0, \infty)$
   Local maximum at $x = 0$
   Concave upward on $(-\infty, -1)$ and $(1, \infty)$
   Concave downward on $(-1, 1)$
   Inflection points at $x = -1$ and $x = 1$

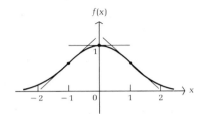

53. $y = 1 + [\ln(2 + e^x)]^2$; $\dfrac{dy}{dx} = \dfrac{2e^x \ln(2 + e^x)}{2 + e^x}$ **55.** $\dfrac{1}{\ln 2}\left(\dfrac{6x}{3x^2 - 1}\right)$ **57.** $(2x + 1)(10^{x^2 + x})(\ln 10)$

61. A maximum revenue of \$735.80 is realized at a production level of 20 units at \$36.79 each.

63. $-\$27{,}145/\text{yr}$; $-\$18{,}196/\text{yr}$; $-\$11{,}036/\text{yr}$ **65.** (A) 23 days; \$26,685; about 50%

(B)

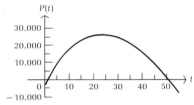

67. 2.27 mm of mercury/yr; 0.81 mm of mercury/yr; 0.41 mm of mercury/yr

69. $A'(t) = 2(\ln 2)5{,}000e^{2t\ln 2} = 10{,}000(\ln 2)2^{2t}$; $A'(1) = 27{,}726$ bacteria/hr (rate of change at the end of the first hour); $A'(5) = 7{,}097{,}827$ bacteria/hr (rate of change at the end of the fifth hour)

71.

◆ EXERCISE 10-8   CHAPTER REVIEW

1. (a, c_1), (c_3, c_6) **2.** (c_1, c_3), (c_6, b) **3.** (a, c_2), (c_4, c_5), (c_7, b) **4.** c_3 **5.** c_6 **6.** c_1, c_3, c_5 **7.** c_6 **8.** c_2, c_4, c_5, c_7

9. $f''(x) = 12x^2 + 30x$ **10.** $y'' = 8/x^3$ **11.** \$3,136.62; \$4,919.21; \$12,099.29 **12.** $\dfrac{2}{x} + 3e^x$ **13.** $2e^{2x-3}$ **14.** $\dfrac{2}{2x + 7}$

15. (A) $y = \ln(3 + e^x)$ (B) $\dfrac{dy}{dx} = \dfrac{e^x}{3 + e^x}$ **16.** $-3 < x < 4$; $(-3, 4)$ **17.** $-3 < x < 0$ or $x > 5$; $(-3, 0) \cup (5, \infty)$

18. (A) All real numbers (B) y intercept: 0; x intercepts: 0, 9 (C) No horizontal or vertical asymptotes

19. (A) 3, 9 (B) 3, 9 (C) Increasing on $(-\infty, 3)$ and $(9, \infty)$; decreasing on $(3, 9)$
(D) Local maximum at $x = 3$; local minimum at $x = 9$

20. (A) Concave downward on $(-\infty, 6)$; concave upward on $(6, \infty)$ (B) Inflection point at $x = 6$

21.

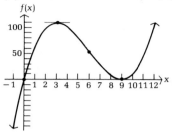

22. (A) All real numbers except -2 (B) y intercept: 0; x intercept: 0 (C) Horizontal asymptote: $y = 3$; vertical asymptote: $x = -2$

23. (A) None (B) -2 (C) Increasing on $(-\infty, -2)$ and $(-2, \infty)$ (D) None

24. (A) Concave upward on $(-\infty, -2)$; concave downward on $(-2, \infty)$ (B) No inflection points

25.

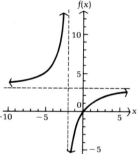

26. Min $f(x) = f(2) = -4$; Max $f(x) = f(5) = 77$ **27.** Min $f(x) = f(2) = 8$

28. Horizontal asymptote: $y = 0$; no vertical asymptotes

29. No horizontal asymptotes; vertical asymptotes: $x = -3$ and $x = 3$

30. Domain: All real numbers
y intercept: 100
Horizontal asymptote: $y = 0$
Decreasing on $(-\infty, \infty)$
Concave upward on $(-\infty, \infty)$

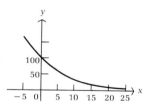

31. $\dfrac{7[(\ln z)^6 + 1]}{z}$ **32.** $x^5(1 + 6 \ln x)$

33. $\dfrac{e^x(x - 6)}{x^7}$ **34.** $\dfrac{6x^2 - 3}{2x^3 - 3x}$

35. $(3x^2 - 2x)e^{x^3 - x^2}$ **36.** $\dfrac{1 - 2x \ln 5x}{xe^{2x}}$

37. Max $f'(x) = f'(2) = 12$

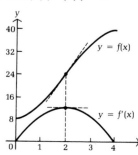

38. Each number is 20; minimum sum is 40

39. Domain: All real numbers
y intercept: -3
x intercepts: -3, 1
No vertical or horizontal asymptotes
Increasing on $(-2, \infty)$
Decreasing on $(-\infty, -2)$
Local minimum at $x = -2$
Concave upward on $(-\infty, -1)$ and $(1, \infty)$
Concave downward on $(-1, 1)$
Inflection points at $x = -1$ and $x = 1$

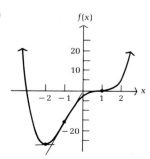

40. Max $f(x) = f(e^{4.5}) = 2e^{4.5} \approx 180.03$ **41.** Max $f(x) = f(0.5) = 5e^{-1} \approx 1.84$

42. Domain: All real numbers
y intercept: 0
x intercept: 0
Horizontal asymptote: $y = 5$
Increasing on $(-\infty, \infty)$
Concave downward on $(-\infty, \infty)$

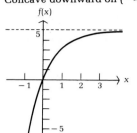

43. Domain: $(0, \infty)$
x intercept: 1
Increasing on $(e^{-1/3}, \infty)$
Decreasing on $(0, e^{-1/3})$
Local minimum at $x = e^{-1/3}$
Concave upward on $(e^{-5/6}, \infty)$
Concave downward on $(0, e^{-5/6})$
Inflection point at $x = e^{-5/6}$

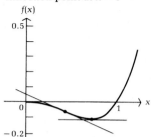

44. (A) $y = [\ln(4 - e^x)]^3$ (B) $\dfrac{dy}{dx} = \dfrac{-3e^x[\ln(4 - e^x)]^2}{4 - e^x}$

45. $2x(5^{x^2-1})(\ln 5)$

46. $\left(\dfrac{1}{\ln 5}\right) \dfrac{2x - 1}{x^2 - x}$

47. $\dfrac{2x + 1}{2(x^2 + x)\sqrt{\ln(x^2 + x)}}$

48. Max $P(x) = P(3,000) = \$175,000$

49. Min $\overline{C}(x) = \overline{C}(200) = 50$

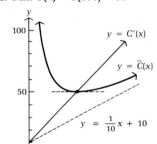

50. \$49; \$6,724
51. 12 orders/yr
52. (A) 15 yr   (B) 13.9 yr
53. $A'(t) = 10e^{0.1t}$; $A'(1) = \$11.05/\text{yr}$; $A'(10) = \$27.18/\text{yr}$
54. $R'(x) = (1,000 - 20x)e^{-0.02x}$

55. A maximum revenue of \$18,394 is realized at a production level of 50 units at \$367.88 each.

56.

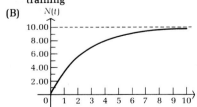

57. Min $\overline{C}(x) = \overline{C}(e^5) \approx \$49.66$
58. 3 days
59. -1.111 mg/ml/hr; -0.335 mg/ml/hr
60. 2 yr from now

61. (A) Increasing at the rate of 2.68 units/day at the end of 1 day of training; increasing at the rate of 0.54 unit/day after 5 days of training

(B)

CHAPTER 11

◆ EXERCISE 11-1

1. $7x + C$ **3.** $(x^7/7) + C$ **5.** $2t^4 + C$ **7.** $u^2 + u + C$ **9.** $x^3 + x^2 - 5x + C$ **11.** $(s^5/5) - \frac{4}{3}s^6 + C$ **13.** $3e^t + C$ **15.** $2\ln|z| + C$
17. $y = 40x^5 + C$ **19.** $P = 24x - 3x^2 + C$ **21.** $y = \frac{1}{3}u^6 - u^3 - u + C$ **23.** $y = e^x + 3x + C$ **25.** $x = 5\ln|t| + t + C$ **27.** $4x^{3/2} + C$
29. $-4x^{-2} + C$ **31.** $2\sqrt{u} + C$ **33.** $-(x^{-2}/8) + C$ **35.** $-(u^{-4}/8) + C$ **37.** $x^3 + 2x^{-1} + C$ **39.** $2x^5 + 2x^{-4} - 2x + C$
41. $2x^{3/2} + 4x^{1/2} + C$ **43.** $\frac{3}{5}x^{5/3} + 2x^{-2} + C$ **45.** $(e^x/4) - (3x^2/8) + C$ **47.** $-z^{-2} - z^{-1} + \ln|z| + C$ **49.** $y = x^2 - 3x + 5$
51. $C(x) = 2x^3 - 2x^2 + 3{,}000$ **53.** $x = 40\sqrt{t}$ **55.** $y = -2x^{-1} + 3\ln|x| - x + 3$ **57.** $x = 4e^t - 2t - 3$ **59.** $y = 2x^2 - 3x + 1$
61. $x^2 + x^{-1} + C$ **63.** $\frac{1}{2}x^2 + x^{-2} + C$ **65.** $e^x - 2\ln|x| + C$ **67.** $M = t + t^{-1} + \frac{3}{4}$ **69.** $y = 3x^{5/3} + 3x^{2/3} - 6$ **71.** $p(x) = 10x^{-1} + 10$
73. $P(x) = 50x - 0.02x^2$; $P(100) = \$4{,}800$ **75.** $R(x) = 100x - (x^2/10)$; $p = 100 - (x/10)$; $p = \$30$
77. $S(t) = 2{,}000 - 15t^{5/3}$; $80^{3/5} \approx 14$ months **79.** $L(x) = 4{,}800x^{1/2}$; $L(25) = 24{,}000$ labor-hours **81.** $W(h) = 0.0005h^3$; $W(70) = 171.5$ lb
83. $19{,}400$

◆ EXERCISE 11-2

1. $\frac{1}{6}(x^2 - 4)^6 + C$ **3.** $e^{4x} + C$ **5.** $\ln|2t + 3| + C$ **7.** $\frac{1}{24}(3x - 2)^8 + C$ **9.** $\frac{1}{16}(x^2 + 3)^8 + C$ **11.** $-20e^{-0.5t} + C$ **13.** $\frac{1}{10}\ln|10x + 7| + C$
15. $\frac{1}{4}e^{2x^2} + C$ **17.** $\frac{1}{3}\ln|x^3 + 4| + C$ **19.** $-\frac{1}{18}(3t^2 + 1)^{-3} + C$ **21.** $\frac{1}{3}(4 - x^3)^{-1} + C$ **23.** $\frac{2}{5}(x + 4)^{5/2} - \frac{8}{3}(x + 4)^{3/2} + C$
25. $\frac{2}{3}(x - 3)^{3/2} + 6(x - 3)^{1/2} + C$ **27.** $\frac{1}{11}(x - 4)^{11} + \frac{2}{5}(x - 4)^{10} + C$ **29.** $\frac{1}{8}(1 + e^{2x})^4 + C$ **31.** $\frac{1}{2}\ln|4 + 2x + x^2| + C$ **33.** $e^{x^2+x+1} + C$
35. $\frac{1}{4}(e^x - 2x)^4 + C$ **37.** $-\frac{1}{12}(x^4 + 2x^2 + 1)^{-3} + C$ **39.** $\frac{1}{9}(3x^2 + 7)^{3/2} + C$ **41.** $\frac{1}{8}x^8 + \frac{4}{5}x^5 + 2x^2 + C$ **43.** $\frac{1}{9}(x^3 + 2)^3 + C$
45. $\frac{1}{4}(2x^4 + 3)^{1/2} + C$ **47.** $\frac{1}{4}(\ln x)^4 + C$ **49.** $e^{-1/x} + C$ **51.** $\frac{1}{3}(t^3 + 5)^7 + C$ **53.** $y = 3(t^2 - 4)^{1/2} + C$ **55.** $p = -(e^x - e^{-x})^{-1} + C$
59. $p(x) = 2{,}000/(3x + 50)$; 250 bottles **61.** $C(x) = 12x + 500\ln(x + 1) + 2{,}000$; $\overline{C}(1{,}000) = \$17.45$
63. $S(t) = 10t + 100e^{-0.1t} - 100$, $0 \le t \le 24$; $S(12) \approx \$50$ million **65.** $Q(t) = 100\ln(t + 1) + 5t$, $0 \le t \le 20$; $Q(9) \approx 275$ thousand barrels
67. $W(t) = 2e^{0.1t}$; $W(8) \approx 4.45$ g **69.** $N(t) = 5{,}000 - 1{,}000\ln(1 + t^2)$; $N(10) \approx 385$ bacteria/ml
71. $N(t) = 100 - 60e^{-0.1t}$, $0 \le t \le 15$; $N(15) \approx 87$ words/min **73.** $E(t) = 12{,}000 - 10{,}000(t + 1)^{-1/2}$; $E(15) = 9{,}500$ students

◆ EXERCISE 11-3

1. 5 **3.** 5 **5.** 2 **7.** 48 **9.** $-\frac{7}{3} \approx -2.333$ **11.** 2 **13.** $\frac{1}{2}(e^2 - 1)$ **15.** $2\ln 3.5$ **17.** -2 **19.** 14 **21.** $5^6 = 15{,}625$ **23.** $\ln 4$
25. $20(e^{0.25} - e^{-0.5}) \approx 13.55$ **27.** $\frac{56}{3} \approx 18.667$ **29.** $\frac{28}{3} \approx 9.333$ **31.** $\frac{1}{6}[(e^2 - 2)^3 - 1]$ **33.** $-3 - \ln 2$ **35.** $\frac{1}{6}(15^{3/2} - 5^{3/2})$
37. $\frac{1}{2}(\ln 2 - \ln 3)$ **39.** 0 **41.** $\int_0^5 500(t - 12)\, dt = -\$23{,}750$; $\int_5^{10} 500(t - 12)\, dt = -\$11{,}250$
43. Useful life $= \sqrt{\ln 55} \approx 2$ yr; Total profit $= \frac{51}{22} - \frac{5}{2}e^{-4} \approx 2.272$ or $\$2{,}272$ **45.** $4{,}800$ labor-hours
47. $100\ln 11 + 50 \approx 290$ thousand barrels; $100\ln 21 - 100\ln 11 + 50 \approx 115$ thousand barrels
49. $20 + 100e^{-1.2} \approx \$50$ million; $120 + 100e^{-2.4} - 100e^{-1.2} \approx \$99$ million **51.** 134 billion ft^3
53. $2e^{0.8} - 2 \approx 2.45$ g; $2e^{1.6} - 2e^{0.8} \approx 5.45$ g
55. An increase of $60 - 60e^{-0.5} \approx 24$ words/min; an increase of $60e^{-0.5} - 60e^{-1} \approx 14$ words/min; an increase of $60e^{-1} - 60e^{-1.5} \approx 9$
words/min

◆ EXERCISE 11-4

1. 16 **3.** 7 **5.** $\frac{7}{2}$ **7.** 9 **9.** $e^2 - e^{-1}$ **11.** $-\ln 0.5$ **13.** 15 **15.** 32 **17.** 36 **19.** 9 **21.** $\frac{5}{2}$ **23.** $2e + \ln 2 - 2e^{0.5}$ **25.** $\frac{23}{3}$ **27.** 17 **29.** $\frac{28}{3}$
31. $\frac{4}{3}$ **33.** $\frac{343}{3}$ **35.** 8 **37.** $\frac{407}{4}$
39. Total production from the end of the 5th year to the end of the 10th year is $50 + 100\ln 20 - 100\ln 15 \approx 79$ thousand barrels.

41. Total profit over the 5 year useful life of the game is $20 - 30e^{-1.5} \approx 13.306$ or \$13,306.

43. 1935: 0.412; 1947: 0.231; income is more equally distributed in 1947.

45. 1963: 0.818; 1983: 0.846; total assets are less equally distributed in 1983.

47. Total weight gain during the first 10 hr is $3e - 3 \approx 5.15$ g.

49. Average number of words learned during the second 2 hr is $15 \ln 4 - 15 \ln 2 \approx 10$.

◆ EXERCISE 11-5

1. (A) 120 (B) 124 **3.** (A) 123 (B) 124 **5.** (A) -4 (B) -5.33 **7.** (A) -5 (B) -5.33 **9.** 23.4 **11.** 55.3 **13.** 250 **15.** 2

17. $\frac{45}{28} \approx 1.61$ **19.** $2(1 - e^{-2}) \approx 1.73$ **21.** 0.791 **23.** 1.650 **25.** 0.748 **27.** 0.747 **29.** $[f(b) - f(a)]/(b - a)$

31. (A) $I = -200t + 600$ (B) $\frac{1}{3}\int_0^3(-200t + 600)\,dt = 300$ **33.** \$16,000 **35.** (A) \$420 (B) \$135,000 **37.** \$149.18; \$122.96

39. $50e^{0.6} - 50e^{0.4} - 10 \approx \$6.51$

41. $(40 + 100e^{-0.7})/7 \approx 12.8$ or 12,800 hamburgers; $(140 + 100e^{-1.4} - 100e^{-0.7})/7 \approx 16.4$ or 16,400 hamburgers **43.** \$10,000

45. 3,120,000 ft² **47.** 10°C **49.** 1.1 liters **51.** $0.6 \ln 2 + 0.1 \approx 0.516$; $(4.2 \ln 625 + 2.4 - 4.2 \ln 49)/24 \approx 0.546$

◆ EXERCISE 11-6

1. \$625,000 **3.** \$9,500 **5.** $\bar{p} = 24$; CS = \$3,380; PS = \$1,690

7. $\bar{p} \approx 49$; CS $= 55,990 - 80,000e^{-0.49} \approx \$6,980$; PS $= 54,010 - 30,000e^{0.49} \approx \$5,041$ **9.** \$2,000

◆ EXERCISE 11-7 CHAPTER REVIEW

1. $t^3 - t^2 + C$ **2.** 12 **3.** $-3t^{-1} - 3t + C$ **4.** $\frac{15}{2}$ **5.** $-2e^{-0.5x} + C$ **6.** $2 \ln 5$ **7.** $y = f(x) = x^3 - 2x + 4$ **8.** 12 **9.** 44 **10.** 30.8 **11.** 7

12. $\frac{1}{8}(6x - 5)^{4/3} + C$ **13.** 2 **14.** $-2x^{-1} - e^{x^2} + C$ **15.** $(20^{3/2} - 8)/3$ **16.** $-\frac{1}{2}e^{-2x} + \ln|x| + C$ **17.** $-500(e^{-0.2} - 1) \approx 90.63$

18. $y = f(x) = 3\ln|x| + x^{-1} + 4$ **19.** $y = 3x^2 + x - 4$ **20.** 2.317 **21.** $\frac{13}{2}$ **22.** $\frac{64}{3}$ **23.** $\frac{1}{2}\ln 10$ **24.** 0.45 **25.** $\frac{1}{48}(2x^4 + 5)^6 + C$

26. $-\ln(e^{-x} + 3) + C$ **27.** $-(e^x + 2)^{-1} + C$ **28.** $\frac{1}{3}(\ln x)^3 + C$ **29.** $\frac{1}{8}x^8 - \frac{2}{5}x^5 + \frac{1}{2}x^2 + C$ **30.** $\frac{2}{3}(6 - x)^{3/2} - 12(6 - x)^{1/2} + C$

31. $\frac{1,234}{15} \approx 82.267$ **32.** $\frac{64}{15}$ **33.** $y = 3e^{x^3} - 1$ **34.** $\frac{46}{3}$

35. $P(x) = 100x - 0.01x^2$; $P(10) = \$999$ **36.** $\int_0^{15}(60 - 4t)\,dt = 450$ thousand barrels **37.** $\int_{10}^{40}\left(150 - \frac{x}{10}\right)dx = \$4,425$

38. Useful life $= 10 \ln \frac{20}{3} \approx 19$ yr; total profit $= 143 - 200e^{-1.9} \approx 113.086$ or \$113,086

39. $S(t) = 50 - 50e^{-0.08t}$; $50 - 50e^{-0.96} \approx \$31$ million; $-(\ln 0.2)/0.8 \approx 20$ months

40. Current: 0.3; projected: 0.2; income will be more equally distributed 10 yr from now. **41.** 109 items

42. $16e^{2.5} - 16e^2 - 8 \approx \$68.70$ **43.** (A) \$1,000 (B) \$800 (C) $\bar{p} = 40$; CS = \$2,250; PS = \$2,700 **44.** 1 cm² **45.** 6.5 ppm

46. $N(t) = 95 - 70e^{-0.1t}$; $N(15) \approx 79$ words/min

APPENDIX A

◆ EXERCISE A-1

1. T **3.** T **5.** T **7.** T **9.** $\{1, 2, 3, 4, 5\}$ **11.** $\{3, 4\}$ **13.** $\varnothing$ **15.** $\{2\}$ **17.** $\{-7, 7\}$ **19.** $\{1, 3, 5, 7, 9\}$ **21.** $A' = \{1, 5\}$ **23.** 40 **25.** 60
27. 60 **29.** 20 **31.** 95 **33.** 40 **35.** (A) $\{1, 2, 3, 4, 6\}$ (B) $\{1, 2, 3, 4, 6\}$ **37.** $\{1, 2, 3, 4, 6\}$ **39.** Yes **41.** Yes **43.** Yes
45. (A) 2 (B) 4 (C) 8; 2^n **47.** 800 **49.** 200 **51.** 200 **53.** 800 **55.** 200 **57.** 200 **59.** 6 **61.** A+, AB+
63. A−, A+, B+, AB−, AB+, O+ **65.** O+, O− **67.** B−, B+ **69.** Everybody in the clique relates to each other.

◆ EXERCISE A-2

1. 5, 7, 9, 11 **3.** $\frac{3}{2}, \frac{4}{3}, \frac{5}{4}, \frac{6}{5}$ **5.** 9, -27, 81, -243 **7.** 23 **9.** $\frac{101}{100}$ **11.** $1 + 2 + 3 + 4 + 5 + 6 = 21$ **13.** $5 + 7 + 9 + 11 = 32$
15. $1 + \frac{1}{10} + \frac{1}{100} + \frac{1}{1,000} = \frac{1,111}{1,000}$ **17.** 3.6 **19.** 82.5 **21.** $\frac{1}{2}, -\frac{1}{4}, \frac{1}{8}, -\frac{1}{16}, \frac{1}{32}$ **23.** 0, 4, 0, 8, 0 **25.** $1, -\frac{3}{2}, \frac{9}{4}, -\frac{27}{8}, \frac{81}{16}$ **27.** $a_n = n - 3$
29. $a_n = 4n$ **31.** $a_n = (2n - 1)/2n$ **33.** $a_n = (-1)^{n+1}n$ **35.** $a_n = (-1)^{n+1}(2n - 1)$ **37.** $a_n = (\frac{2}{5})^{n-1}$ **39.** $a_n = x^n$ **41.** $a_n = (-1)^{n+1}x^{2n-1}$
43. $1 - 9 + 25 - 49 + 81$ **45.** $\frac{4}{7} + \frac{8}{9} + \frac{16}{11} + \frac{32}{13}$ **47.** $1 + x + x^2 + x^3 + x^4$ **49.** $x - \dfrac{x^3}{3} + \dfrac{x^5}{5} - \dfrac{x^7}{7} + \dfrac{x^9}{9}$

51. (A) $\displaystyle\sum_{k=1}^{5} (k + 1)$ (B) $\displaystyle\sum_{j=0}^{4} (j + 2)$ **53.** (A) $\displaystyle\sum_{k=1}^{4} \dfrac{(-1)^{k+1}}{k}$ (B) $\displaystyle\sum_{j=0}^{3} \dfrac{(-1)^j}{j + 1}$ **55.** $\displaystyle\sum_{k=1}^{n} \dfrac{k + 1}{k}$ **57.** $\displaystyle\sum_{k=1}^{n} \dfrac{(-1)^{k+1}}{2^k}$ **59.** 2, 8, 26, 80, 242
61. 1, 2, 4, 8, 16 **63.** $1, \frac{3}{2}, \frac{17}{12}, \frac{577}{408}$; $a_4 = \frac{577}{408} \approx 1.414\ 216$, $\sqrt{2} \approx 1.414\ 214$

◆ EXERCISE A-3

1. (A) $d = 3$; 14, 17 (B) Not an arithmetic progression (C) Not an arithmetic progression (D) $d = -10$; -22, -32
3. $a_2 = 11$; $a_3 = 15$ **5.** $a_{21} = 82$; $S_{31} = 1,922$ **7.** $S_{20} = 930$ **9.** 2,400 **11.** 1,120
13. Use $a_1 = 1$ and $d = 2$ in $S_n = (n/2)[2a_1 + (n - 1)d]$. **15.** Firm A: $280,500; firm B: $278,500
17. $48 + $46 + \cdots + $4 + $2 = $600

◆ EXERCISE A-4

1. (A) $r = -2$; $a_4 = -8$, $a_5 = 16$ (B) Not a geometric progression (C) $r = \frac{1}{2}$; $a_4 = \frac{1}{4}$, $a_5 = \frac{1}{8}$ (D) Not a geometric progression
3. $a_2 = -6$; $a_3 = 12$; $a_4 = -24$ **5.** $S_7 = 547$ **7.** $a_{10} = 199.90$ **9.** $r = 1.09$ **11.** $S_{10} = 1,242$; $S_\infty = 1,250$
13. (A) Does not exist (B) $S_\infty = \frac{8}{5} = 1.6$ **15.** 0.999 **17.** About $11,670,000 **19.** $31,027; $251,558

◆ EXERCISE A-5

1. 720 **3.** 10 **5.** 1,320 **7.** 10 **9.** 6 **11.** 1,140 **13.** 10 **15.** 6 **17.** 1 **19.** 816
21. $C_{4,0}a^4 + C_{4,1}a^3b + C_{4,2}a^2b^2 + C_{4,3}ab^3 + C_{4,4}b^4 = a^4 + 4a^3b + 6a^2b^2 + 4ab^3 + b^4$ **23.** $x^6 - 6x^5 + 15x^4 - 20x^3 + 15x^2 - 6x + 1$

25. $32a^5 - 80a^4b + 80a^3b^2 - 40a^2b^3 + 10ab^4 - b^5$ **27.** $3,060x^{14}$ **29.** $5,005p^9q^6$ **31.** $264x^2y^{10}$ **33.** $C_{n,0} = \dfrac{n!}{0!n!} = 1$; $C_{n,n} = \dfrac{n!}{n!0!} = 1$

35. 1 5 10 10 5 1; 1 6 15 20 15 6 1

Index